Geodynamic Processes in the Andes of Central Chile and Argentina

The Geological Society of London
Books Editorial Committee

Chief Editor
RICK LAW (USA)

Society Books Editors
JIM GRIFFITHS (UK)
DAVE HODGSON (UK)
HOWARD JOHNSON (UK)
PHIL LEAT (UK)
NICK RICHARDSON (UK)
DANIELA SCHMIDT (UK)
RANDELL STEPHENSON (UK)
ROB STRACHAN (UK)
MARK WHITEMAN (UK)

Society Books Advisors
GHULAM BHAT (India)
MARIE-FRANÇOISE BRUNET (France)
MAARTEN DE WIT (South Africa)
JAMES GOFF (Australia)
MARIO PARISE (Italy)
SATISH-KUMAR (Japan)
MARCO VECOLI (Saudi Arabia)
GONZALO VEIGA (Argentina)

Geological Society books refereeing procedures

The Society makes every effort to ensure that the scientific and production quality of its books matches that of its journals. Since 1997, all book proposals have been refereed by specialist reviewers as well as by the Society's Books Editorial Committee. If the referees identify weaknesses in the proposal, these must be addressed before the proposal is accepted.

Once the book is accepted, the Society Book Editors ensure that the volume editors follow strict guidelines on refereeing and quality control. We insist that individual papers can only be accepted after satisfactory review by two independent referees. The questions on the review forms are similar to those for *Journal of the Geological Society*. The referees' forms and comments must be available to the Society's Book Editors on request.

Although many of the books result from meetings, the editors are expected to commission papers that were not presented at the meeting to ensure that the book provides a balanced coverage of the subject. Being accepted for presentation at the meeting does not guarantee inclusion in the book.

More information about submitting a proposal and producing a book for the Society can be found on its website: www.geolsoc.org.uk.

It is recommended that reference to all or part of this book should be made in one of the following ways:

SEPÚLVEDA, S. A., GIAMBIAGI, L. B., MOREIRAS, S. M., PINTO, L., TUNIK, M., HOKE, G. D. & FARÍAS, M. (eds) 2015. *Geodynamic Processes in the Andes of Central Chile and Argentina*. Geological Society, London, Special Publications, **399**.

SURIANO, J., LIMARINO, C. O., TEDESCO, A. M. & ALONSO, M. S. 2015. Sedimentation model of piggyback basins: Cenozoic examples of San Juan Precordillera, Argentina. *In*: SEPÚLVEDA, S. A., GIAMBIAGI, L. B., MOREIRAS, S. M., PINTO, L., TUNIK, M., HOKE, G. D. & FARÍAS, M. (eds) *Geodynamic Processes in the Andes of Central Chile and Argentina*. Geological Society, London, Special Publications, **399**, 221–244. First published online February 28, 2014, http://dx.doi.org/10.1144/SP399.17

GEOLOGICAL SOCIETY SPECIAL PUBLICATION NO. 399

Geodynamic Processes in the Andes of Central Chile and Argentina

EDITED BY

S. A. SEPÚLVEDA
Universidad de Chile, Chile

L. B. GIAMBIAGI
IANIGLA-CCT Mendoza-CONICET, Argentina

S. M. MOREIRAS
IANIGLA-CCT Mendoza-CONICET, Argentina

L. PINTO
Universidad de Chile, Chile

M. TUNIK
Universidad Nacional de Rio Negro, Argentina

G. D. HOKE
Syracuse University, USA

and

M. FARÍAS
Universidad de Chile, Chile

2015
Published by
The Geological Society
London

THE GEOLOGICAL SOCIETY

The Geological Society of London (GSL) was founded in 1807. It is the oldest national geological society in the world and the largest in Europe. It was incorporated under Royal Charter in 1825 and is Registered Charity 210161.

The Society is the UK national learned and professional society for geology with a worldwide Fellowship (FGS) of over 10 000. The Society has the power to confer Chartered status on suitably qualified Fellows, and about 2000 of the Fellowship carry the title (CGeol). Chartered Geologists may also obtain the equivalent European title, European Geologist (EurGeol). One fifth of the Society's fellowship resides outside the UK. To find out more about the Society, log on to www.geolsoc.org.uk.

The Geological Society Publishing House (Bath, UK) produces the Society's international journals and books, and acts as European distributor for selected publications of the American Association of Petroleum Geologists (AAPG), the Indonesian Petroleum Association (IPA), the Geological Society of America (GSA), the Society for Sedimentary Geology (SEPM) and the Geologists' Association (GA). Joint marketing agreements ensure that GSL Fellows may purchase these societies' publications at a discount. The Society's online bookshop (accessible from www.geolsoc.org.uk) offers secure book purchasing with your credit or debit card.

To find out about joining the Society and benefiting from substantial discounts on publications of GSL and other societies worldwide, consult www.geolsoc.org.uk, or contact the Fellowship Department at: The Geological Society, Burlington House, Piccadilly, London W1J 0BG: Tel. +44 (0)20 7434 9944; Fax +44 (0)20 7439 8975; E-mail: enquiries@geolsoc.org.uk.

For information about the Society's meetings, consult *Events* on www.geolsoc.org.uk. To find out more about the Society's Corporate Affiliates Scheme, write to enquiries@geolsoc.org.uk.

Published by The Geological Society from:
The Geological Society Publishing House, Unit 7, Brassmill Enterprise Centre, Brassmill Lane, Bath BA1 3JN, UK

The Lyell Collection: www.lyellcollection.org
Online bookshop: www.geolsoc.org.uk/bookshop
Orders: Tel. +44 (0)1225 445046, Fax +44 (0)1225 442836

The publishers make no representation, express or implied, with regard to the accuracy of the information contained in this book and cannot accept any legal responsibility for any errors or omissions that may be made.

© The Geological Society of London 2015. No reproduction, copy or transmission of all or part of this publication may be made without the prior written permission of the publisher. In the UK, users may clear copying permissions and make payment to The Copyright Licensing Agency Ltd, Saffron House, 6–10 Kirby Street, London EC1N 8TS UK, and in the USA to the Copyright Clearance Center, 222 Rosewood Drive, Danvers, MA 01923, USA. Other countries may have a local reproduction rights agency for such payments. Full information on the Society's permissions policy can be found at: www.geolsoc.org.uk/permissions

British Library Cataloguing in Publication Data

A catalogue record for this book is available from the British Library.
ISBN 978-1-86239-653-1
ISSN 0305-8719

Distributors

For details of international agents and distributors see:
www.geolsoc.org.uk/agentsdistributors

Typeset by Techset Composition India (P) Ltd, Bangalore and Chennai, India.
Printed by Berforts Information Press Ltd, Oxford, UK

Contents

SEPÚLVEDA, S. A., GIAMBIAGI, L. B., MOREIRAS, S. M., PINTO, L., TUNIK, M., HOKE, G. D. & FARÍAS, M. Geodynamic processes in the Andes of Central Chile and Argentina: an introduction   1

CHARRIER, R., RAMOS, V. A., TAPIA, F. & SAGRIPANTI, L. Tectono-stratigraphic evolution of the Andean Orogen between 31 and 37°S (Chile and Western Argentina)   13

GIAMBIAGI, L., TASSARA, A., MESCUA, J., TUNIK, M., ALVAREZ, P. P., GODOY, E., HOKE, G., PINTO, L., SPAGNOTTO, S., PORRAS, H., TAPIA, F., JARA, P., BECHIS, F., GARCÍA, V. H., SURIANO, J., MOREIRAS, S. M. & PAGANO, S. D. Evolution of shallow and deep structures along the Maipo–Tunuyán transect (33°40′S): from the Pacific coast to the Andean foreland   63

JARA, P., LIKERMAN, J., WINOCUR, D., GHIGLIONE, M. C., CRISTALLINI, E. O., PINTO, L. & CHARRIER, R. Role of basin width variation in tectonic inversion: insight from analogue modelling and implications for the tectonic inversion of the Abanico Basin, 32°–34°S, Central Andes   83

WINOCUR, D. A., LITVAK, V. D. & RAMOS, V. A. Magmatic and tectonic evolution of the Oligocene Valle del Cura basin, main Andes of Argentina and Chile: evidence for generalized extension   109

NAIPAUER, M., TUNIK, M., MARQUES, J. C., ROJAS VERA, E. A., VUJOVICH, G. I., PIMENTEL, M. M. & RAMOS, V. A. U–Pb detrital zircon ages of Upper Jurassic continental successions: implications for the provenance and absolute age of the Jurassic–Cretaceous boundary in the Neuquén Basin   131

GODOY, E. The north-western margin of the Neuquén Basin in the headwater region of the Maipo drainage, Chile   155

SÁNCHEZ, M., LINCE KLINGER, F., MARTINEZ, M. P., ALVAREZ, O., RUIZ, F., WEIDMANN, C. & FOLGUERA, A. Geophysical characterization of the upper crust in the transitional zone between the Pampean flat slab and the normal subduction segment to the south (32–34°S): Andes of the Frontal Cordillera to the Sierras Pampeanas   167

ALVAREZ, O., GIMENEZ, M. E., MARTINEZ, M. P., LINCEKLINGER, F. & BRAITENBERG, C. New insights into the Andean crustal structure between 32° and 34°S from GOCE satellite gravity data and EGM2008 model   183

SAGRIPANTI, L., AGUIRRE-URRETA, B., FOLGUERA, A. & RAMOS, V. A. The Neocomian of Chachahuén (Mendoza, Argentina): evidence of a broken foreland associated with the Payenia flat-slab   203

SURIANO, J., LIMARINO, C. O., TEDESCO, A. M. & ALONSO, M. S. Sedimentation model of piggyback basins: Cenozoic examples of San Juan Precordillera, Argentina   221

COSTA, C. H., AHUMADA, E. A., GARDINI, C. E., VÁZQUEZ, F. R. & DIEDERIX, H. Quaternary shortening at the orogenic front of the Central Andes of Argentina: the Las Peñas Thrust System   245

CORTÉS, J. M., TERRIZZANO, C. M., PASINI, M. M., YAMIN, M. G. & CASA, A. L. Quaternary tectonics along oblique deformation zones in the Central Andean retro-wedge between 31°30′S and 35°S   267

ZÁRATE, M. A., MEHL, A. & PERUCCA, L. Quaternary evolution of the Cordillera Frontal piedmont between c. 33° and 34°S Mendoza, Argentina   293

GARCÍA, V. H. & CASA, A. L. Quaternary tectonics and seismic potential of the Andean retrowedge at 33–34°S   311

MOREIRAS, S. M. & SEPÚLVEDA, S. A. Megalandslides in the Andes of central Chile and Argentina (32°–34°S) and potential hazards 329

HERMANNS, R. L., FAUQUÉ, L. & WILSON, C. G. J. ^{36}Cl terrestrial cosmogenic nuclide dating suggests Late Pleistocene to Early Holocene mass movements on the south face of Aconcagua mountain and in the Las Cuevas–Horcones valleys, Central Andes, Argentina 345

MOREIRAS, S. M. & PÁEZ, M. S. Historical damage and earthquake environmental effects related to shallow intraplate seismicity of central western Argentina 369

HOKE, G. D., GRABER, N. R., MESCUA, J. F., GIAMBIAGI, L. B., FITZGERALD, P. G. & METCALF, J. R. Near pure surface uplift of the Argentine Frontal Cordillera: insights from (U–Th)/He thermochronometry and geomorphic analysis 383

CARRETIER, S., TOLORZA, V., RODRÍGUEZ, M. P., PEPIN, E., AGUILAR, G., REGARD, V., MARTINOD, J., RIQUELME, R., BONNET, S., BRICHAU, S., HÉRAIL, G., PINTO, L., FARÍAS, M., CHARRIER, R. & GUYOT, J. L. Erosion in the Chilean Andes between 27°S and 39°S: tectonic, climatic and geomorphic control 401

RODRÍGUEZ, M. P., AGUILAR, G., URRESTY, C. & CHARRIER, R. Neogene landscape evolution in the Andes of north-central Chile between 28 and 32°S: interplay between tectonic and erosional processes 419

Index 447

Geodynamic processes in the Andes of Central Chile and Argentina: an introduction

SERGIO A. SEPÚLVEDA[1]*, LAURA B. GIAMBIAGI[2], STELLA M. MOREIRAS[2], LUISA PINTO[1], MAISA TUNIK[3], GREGORY D. HOKE[4] & MARCELO FARÍAS[1]

[1]*Departamento de Geología, Universidad de Chile, Plaza Ercilla 803, Santiago, Chile*

[2]*IANIGLA-CCT Mendoza-CONICET, Av. Ruiz Leal s/n, Parque Gral San Martin (5500) Mendoza, Argentina*

[3]*Universidad Nacional de Río Negro, Av. Roca 1242, (8332) Roca, Río Negro, Argentina*

[4]*Department of Earth Sciences, Syracuse University, Syracuse, NY, USA*

*Corresponding author (e-mail: sesepulv@ing.uchile.cl)

Abstract: The Andes, the world's largest non-collisional orogen, is considered the paradigm for geodynamic processes associated with the subduction of an oceanic plate below a continental plate margin. In the framework of UNESCO-sponsored IGCP 586-Y project, this Special Publication includes state-of-the-art reviews and original articles from a range of Earth Science disciplines that investigate the complex interactions of tectonics and surface processes in the subduction-related orogen of the Andes of central Chile and Argentina (c. 27–39°S). This introduction provides the geological context of the transition from flat slab to normal subduction angles, where this volume is focused, along with a brief description of the individual contributions ranging from internal geodynamics and tectonics, Quaternary tectonics and related geohazards, to landscape evolution of this particular segment of the Andes.

Convergent continental margins are one of the first-order expressions of the movement of Earth's tectonic plates atop a convecting mantle. The topography of convergent margins is due to the interactions of rock uplift, climate and surface processes that are dominantly driven by upper crustal and internal lithospheric deformation as well as erosional processes (e.g. Molnar & Lyon-Caen 1988; Beaumont et al. 2004; Whipple & Meade 2006). The Andes of central Chile and Argentina (c. 32–36°S, Figs 1 & 2) straddle a potentially important transition in geodynamic boundary conditions imposed on the upper plate: a drastic change in the geometry of the subducting Nazca plate from 'flat' to normal subduction angles, from c. 5–10° to 30° (Cahill & Isacks 1992; Fig. 1). This segment of the Andes is therefore a particularly suitable setting to evaluate the interplay between constructive deep mechanisms, resulting in rock uplift and lateral expansion of the Andes, and the mechanisms of exhumation and erosion that shape the landscape in an active, convergent margin.

The Central Andes are composed of different morphostructural units (from west to east; Figs 1 & 2): the Chilean Coastal Cordillera, the Principal Cordillera (spanning Chile and Argentina), the Frontal Cordillera, the Argentine Precordillera and the Pampean Ranges (Jordan et al. 1983). Although previous studies attempt to constrain the magnitudes of orogenic shortening for each morphostructural unit, limited amounts of geochronology complicate attempts to constrain more tightly the rates of shortening and topographic uplift. There are several outstanding questions regarding the interplay between deep and surface processes in the development of the Andean Orogen over different timescales, as follows.

- What is the nature of the interaction between tectonic and surface processes in this sector of the Andes? In particular, are surface and tectonic processes part of a strongly coupled system?
- What control does the geometry of the subducting slab exert over the timing and style of deformation in the South American plate?
- How do deformation and denudation evolve in space and time? Are there differences between the flat slab (32–33°S), the transitional (33–34°S) and normal (34–36°S) segments in this respect?
- What role do pre-existing crustal and lithospheric heterogeneities play in the development of the orogen? In particular, is there a link between heterogeneities in the overriding plate and the geometry of the subducting plate (flat v. normal subduction)?

From: SEPÚLVEDA, S. A., GIAMBIAGI, L. B., MOREIRAS, S. M., PINTO, L., TUNIK, M., HOKE, G. D. & FARÍAS, M. (eds) 2015. *Geodynamic Processes in the Andes of Central Chile and Argentina.* Geological Society, London, Special Publications, **399**, 1–12. First published online December 15, 2014, http://dx.doi.org/10.1144/SP399.21
© 2015 The Geological Society of London. For permissions: http://www.geolsoc.org.uk/permissions.
Publishing disclaimer: www.geolsoc.org.uk/pub_ethics

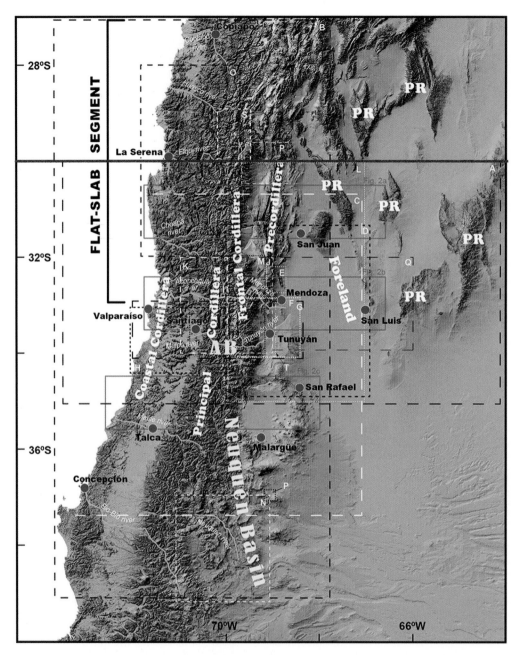

Fig. 1. Digital elevation model of the region showing morphologic zones, principal localities and rivers mentioned in this Special Publication. Study areas are indicated by boxes in segmented lines: A, Alvarez *et al.*; B, Carretier *et al.*; C, Charrier *et al.*; D, Cortés *et al.*; E, Costa *et al.*: F, García *et al.*; G, Giambiagi *et al.*; H, Godoy; I, Hermanns *et al.*; J, Hoke *et al.*; K, Jara *et al.*; L, Moreiras & Páez; M, Moreiras & Sepúlveda; N, Nainpauer *et al.*; O, Rodríguez *et al.*; P, Sagripanti *et al.*; Q, Sánchez *et al.*; R, Suriano *et al.*; S, Winocur *et al.*; T, Zárate *et al*. The position of the flat-slab segment (Cahill & Isacks 1992) is represented. Red boxes mark the position of the 1°-wide swath topographic profiles shown in Figure 2. Abbreviations: AB, Abanico Basin; PR, Pampean Range.

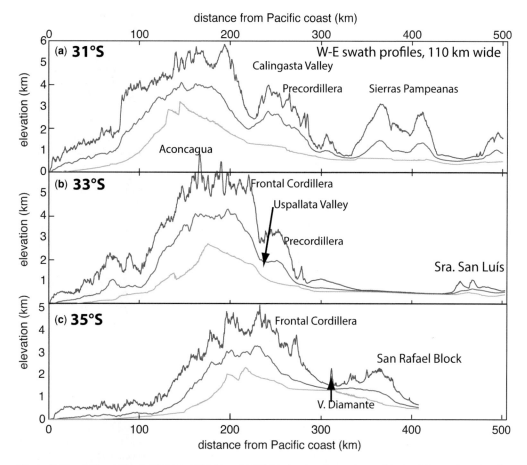

Fig. 2. Topographic profiles at 31°S (**a**), 33°S (**b**) and 35°S (**c**) showing the maximum (blue), minimum (green) and mean (red) elevations constructed using a 250 m resolution resampling of the SRTM 90 m digital elevations model available from http://www.cgiar-csi.org/. Major topographic features are labelled on the swath profiles for reference. Location of swath profiles is indicated in Fig. 1.

- What role do mega-landslides play in shaping the landscape and erosion rates, and how do they relate to active tectonics, climate and river incision?
- How do Andean sedimentary basins evolve in relation to spatial and temporal variations in deformation and topographic uplift during the Miocene to present?

In the framework of UNESCO-sponsored IGCP 586-Y project 'The tectonics and geomorphology of the Andes (32–34°S): Interplay between short-term and long-term processes', this collection of state-of-the-art reviews and original articles investigates the complex interactions between tectonics and surface processes in the non-collisional Andean Orogen of central Chile and Argentina (c. 27–39°S) with a particular emphasis on the area between 32°S and 34°S.

Internal geodynamics and tectonic evolution

The Andes are considered the geodynamic archetype of a convergent, non-collisional mountain range and are generated by subduction of the oceanic lithosphere of the Nazca Plate beneath the continental lithosphere of the South American Plate. Consequently, the present-day architecture of the Andes Mountains is largely the result of convergence between the Pacific–Nazca and South American plates. However, these mountains preserve evidence of previous periods of contractional, extensional and strike-slip deformation along the proto-Pacific margin of southern Gondwana. The tectonic style of the different morphostructural units comprising the orogen is strongly influenced by pre-Andean structures, especially those developed during the Late Devonian–early Carboniferous Chanic Orogeny and the late Carboniferous–early Permian San Rafael Orogeny of the Gondwanan orogenic cycle (Ramos 1988; Mpodozis & Ramos 1989; Giambiagi et al. 2011, 2014a; Heredia et al. 2012). The first chapter by **Charrier et al. (2014)** presents a review of the major geological features and tectonic events that occurred in the Andean region between 31°S and 37°S since early Phanerozoic time and prior to the Neogene Andean deformation. The authors synthesize current knowledge of pre-Andean orogenic cycles, as well as the influence of these cycles over Andean Orogeny since Neogene time, while emphasizing the importance of the relationship between magmatism, metamorphism, sedimentation and deformation.

Although the Andes have been described as a consequence of crustal shortening that leads to crustal thickening and surface uplift (Isacks 1988; Sheffels 1990; Allmendinger et al. 1997), the mechanisms by which this crustal shortening is achieved remain controversial. **Giambiagi et al. (2014b)** propose an integrated kinematic-mechanical model of the Maipo–Tunuyán transect (33°40′S) across the Andes, describing the relation between horizontal shortening, surface uplift, crustal thickening and activity of the magmatic arc, while accounting for the main deep crustal processes that have shaped the Andes since early Miocene time.

During Eocene–early Miocene time, a protracted extensional event took place with deformation concentrated in the western sector of the Principal Cordillera, with the opening of the Abanico extensional basin (Charrier et al. 2002; Muñoz-Sáez et al. 2014). This event has been linked to a segmented rollback-subduction event (Mpodozis & Cornejo 2012). **Jara et al. (2014)** study the role of width variation in the Eocene–lower Miocene Abanico Basin (Fig. 1) on the development of inverted structures in the Andes between 32 and 33°S. They use analogue modelling to propose that the basin aperture increased southwards and was subsequently inverted by homogeneous shortening.

During this extensional regime, Oligocene–lower Miocene retro-arc volcanism developed in the El Indio Belt. **Winocur et al. (2014)** shed light on the age and evolution of this belt where major mining districts are located, based on extensive fieldwork as well as geochronological and geochemical data. After many years of walking and studying the Andes, **Godoy (2014)** remarks on the along-strike lithological variations of the Neuquén Basin (Fig. 1) during Kimmeridgian–Hauterivian time. He recognizes thick Hauterivian andesitic flows in the high Cordillera and proposes that a volcanic event could be responsible for the isolation of the Neuquén Basin during Aptian–Albian time, despite a worldwide sea-level highstand.

South America, and especially the Andean region, are generally characterized by a lack of gravity data necessary to describe the crust's internal structure. **Sánchez et al. (2014)** geophysically characterize the upper crust of the Andean back-arc between 32 and 34°S. Their gravimetric studies allow them to highlight and delineate two NW-trending features in the upper crust: (1) the San Pedro Ridge and the Tunuyán lineament; and (2) a lateral density contrast between the basement of the Western Sierras Pampeanas and that of the Precordillera, which they interpret as the geophysical expression of the contact between the Cuyania and Pampia terrains. **Alvarez et al. (2014)** present a new interpretation of Andean crustal structure between 32 and 34°S from GOCE satellite gravity data. In addition, they discuss the use of global gravity models to study the crustal structure at a regional scale.

Late Cenozoic compressional deformation led to development of an extensive retro-arc thrust-belt/foreland basin system (Jordan et al. 1983; Ramos et al. 1996). Knowing the temporal–spatial pattern of Neogene orogenic exhumation and its relation to structural deformation, topographic evolution, basin development and sediment dispersal patterns in the southern Central Andes is crucial to reconstruct the geodynamic processes that controlled the building and erosion of the orogen. **Suriano et al. (2014)** analyse two Quaternary and Oligocene–Miocene piggyback basins along the Precordillera at 30°S; the first is controlled by climatic changes, and the second is related to tectonic activity in the Precordillera fold–thrust belt (Fig. 1). The authors postulate that understanding the dynamics of sediment accumulation in piggyback basins and the controlling allocyclic changes is essential to understand the geodynamics of an evolving orogen.

Sagripanti et al. (2014) explain the presence of Lower Cretaceous marine sedimentary deposits in the foreland of the Neuquén Basin. These deformed deposits are 70 km from the Andean orogenic front and 2 km above sea level due to local uplift on high-angle basement reverse faults that reactivated a previous early Mesozoic rift system. The authors postulate that the increase in compression could be related to the decrease in the subduction angle and the expansion of the magmatic arc during the late Miocene Payenia flat-slab period.

Provenance studies of coastal facies sediments associated with the uplift of the Andes (Rodríguez et al. 2012) and the Neuquén Basin (Tunik & Lazo 2008; Tunik et al. 2010; Di Giulio et al. 2012) shed light on the sediment sources necessary for palaeogeographic reconstructions. Within that framework, **Naipauer et al. (2014)** present an important contribution to the understanding of the tectonic and sedimentary history of a main Andean basin through a combination of field data, petrography and provenance characteristics of detrital zircons from a Kimmeridgian lowstand wedge within the Tordillo Formation from the Neuquén Basin (Fig. 1), west-central Argentina. They present new U–Pb detrital zircon ages indicating that the most important source region of sediment supply to the basin was the Jurassic Andean magmatic arc, which has significant implications for the provenance and absolute age of the Jurassic–Cretaceous boundary in the Neuquén Basin.

Quaternary tectonics and geohazards

The Andes of central Chile and Argentina is one of the most hazardous seismic zones of South America (Holtkamp et al. 2011). The strongest earthquake ever recorded occurred just south of the study area in 1960 (Valdivia, M_w 9.5), and was associated with the plate boundary to seismicity. The February 2010 Maule earthquake (M_w 8.8) occurred in this region, and generated a tsunami as well as shallow landslides, liquefaction and ground rupture (DIC-UChile 2012). While high-magnitude inter-plate earthquakes predominate in the Pacific coast, intra-plate seismicity related to active Quaternary faults is characteristic of the Andean Main Range and foreland areas (e.g. Barrientos et al. 2004; Alvarado et al. 2005, 2007; Ahumada & Costa 2009; Farías et al. 2010). For instance, new findings reveal that the San Ramon Fault, which crosses the eastern urban area of Santiago, Chile is capable of generating earthquakes of up to M_w c. 7.5 (Vargas et al. 2014). Additional sources of seismicity are medium-depth intra-plate earthquakes that mainly affect the Chilean central valley area, the Chilean coastal area and Andes Principal Cordillera (Leyton et al. 2010).

The present active orogenic front is located in the piedmont of central-western Argentina (Cortés et al. 1999; Costa et al. 2000a). This area shows intense neotectonic activity evidenced by Quaternary faulting and seismic activity (Costa et al. 2000a, b; Siame et al. 2006), corresponding to shallow intra-plate earthquakes at depths of <30 km. This active orogenic front is composed of a series of east- and west-verging reverse faults with geomorphological evidence of Quaternary displacements (Moreiras et al. 2014). According to geometry of these active faults, the probable maximum earthquake magnitudes range from M_w 5.4 to 6.8, depending on the fault (Moreiras et al. 2014).

The contribution by **Cortés et al. (2014)** presents a synoptic overview of Quaternary structures across the Precordillera and the low foothills south of the Precordillera. They document the growing lateral expansion of deformation and the subsequent creation of topographic highs and depressions, with a particular emphasis on the Uspallata–Calingasta Valley. They attribute this deformation to pre-existing structural elements of the Triassic Cuyo Basin.

At the other end of the Barreal–Las Peñas deformation zone, **Costa et al. (2014)** analysed the Quaternary shortening of the orogenic front of the Central Andes of Argentina (32°15′–32°40′S). The 40 km NNW-trending Las Peñas Thrust System, one of the key structures along the Andean orogenic front, is described in detail. They suggest the Las Peñas Thrust System represents the latest stage of the eastwards migration of an imbricated fan structure, responsible for the neotectonic uplift of the Las Peñas–Las Higueras range in the Southern Precordillera.

Zárate et al. (2014) recognized fluvio-aeolian deposits fractured and folded by faulting together with some landforms of volcanic origin between 33 and 34°S. Alluvial fans, related to several Quaternary aggradational cycles, are the most remarkable geomorphological units described in this study. Several tectonic features are present and give rise to conspicuous morphological features. Some streams are structurally controlled by faults, while several drainage anomalies indicate active tectonic processes. Although precise data are still needed, several episodes of Pleistocene–present-day erosion are identified.

García & Casa (2014) combine field and subsurface data in an analysis of neotectonic structures in the Andean foothills between 33 and 34°S latitude. They demonstrate a decrease in neotectonic deformation and seismicity south of the flat-slab transition and estimate Quaternary rock uplift rates. Based on the estimated uplift rates and fault lengths, the authors infer that earthquakes with M_w of 6.4–7.25 are responsible for generating the larger faults present with their study area.

There is a necessity for a comprehensive regional catalogue of historical earthquakes and their environmental ground effects, such as liquefaction, ground cracking or landslides, in central Chile and Argentina. Specific palaeo-seismological studies are uncommon, and historic observations in remote mountain areas are extremely rare. Nevertheless, a catalogue of environmental effects is fundamental to understanding and quantifying the degree of hazard and vulnerability of this region. **Moreiras & Páez (2014)** provide a comprehensive review of historic destructive shallow crustal earthquakes in central-western Argentina and their related hazards associated to both shaking and environment effects such as liquefaction and landslides. They discuss the shortcomings of seismic hazard assessments, particularly for potential earthquakes related to blind faults, in a region of increasing population.

Landslides inventory maps have been compiled at different scales. In Argentina, the Mendoza river valley (Moreiras 2004, 2005, 2009a) has been studied extensively, with many deposits which were initially interpreted as glacial in origin reinterpreted as large rock avalanches (Moreiras *et al.* 2008). Quaternary faults have been proposed to trigger many of the rock avalanches (Moreiras 2006, 2009b). In Chile, Antinao & Gosse (2009) prepared a large landslide inventory and suggested a relationship with regional faults and seismicity, yet only one historic earthquake in 1958 (M_w 6.3, Alvarado *et al.* 2009) triggered landslides and these were only of medium to large size (volume of a few millions of cubic metres; Sepúlveda *et al.* 2008). In this context, a major question is whether huge palaeo-landslide deposits observed throughout the Andean region are linked to seismic activity or climatic events, that is, whether these deposits can be used as unambiguous indicators of tectonic or climatic forcing.

Mega-landslides ($>10^6$ m^3) are a common geomorphic feature of the Central Andes in Chile and Argentina (32–34°S). Different types and volumes of landslides are usually present in the uplands, located mainly in glacial valleys (Fig. 3). **Moreiras & Sepúlveda (2014)** revise and integrate an inventory of mega palaeo-landslides in this portion of Central Andes and discuss traditional hypotheses used to explaining landslide occurrences in the Central Andes. Whereas earthquakes have been widely viewed as the main triggering mechanism in Chilean collapses, palaeo-climatic conditions are considered as the main cause of mega-landslides in Argentina. The authors identify the necessity of future multidisciplinary, specific studies focused on resolving the controversy between seismic- or climate-landslide-triggering mechanism in the Central Andes.

Hermanns *et al.* (2014) determine the origin and ages of mass movements in the Aconcagua mountain region, Argentina. Deposits previously interpreted as glacial drifts (Penitentes, Horcones and Almacenes) along the Las Cuevas–Horcones valleys are reinterpreted as giant landslides based on morphological, sedimentological and mineralogical studies. A combination of ^{36}Cl terrestrial cosmogenic nuclide (TCN) and U-series dating indicates late Pleistocene–early Holocene ages for these deposits. These new findings suggest that hillslope collapse could be more important than glaciation in the landscape evolution of the central Andes, which is still poorly understood.

External geodynamics and landscape evolution

The study of the magnitude and timing of surface uplift and its erosional response and relief growth, in conjunction with structural geology and shallow seismicity studies, has been carried out in the Chilean flank of the study region (e.g. Farías *et al.* 2008). Such studies have determined that the main stage of uplift occurred during late Miocene time and that the river incision in response to uplift was very slow, taking more than 5 million years to arrive at the present-day Pacific–Atlantic watershed. This implies that exhumation is not co-eval to uplift, but it is delayed because of lithological resistance to erosion and climatic settings. The restricted amount of exhumation allows the preservation of geomorphical markers of uplift and erosion, which permit the quantification of this kind of processes. The 2 km magnitude of surface uplift deduced by Farías *et al.* (2008) at *c*. 34°S on the western flanks of the Andes has been confirmed by a recent stable isotope palaeo-altimetry study on the eastern flanks of the range (Hoke *et al.* 2014a). The linkage between climate and tectonics in the Andes remains unresolved. Some research suggests that the development of relief on the western flank of the Andes may be linked with the amount of shortening and precipitation rates (e.g. Farías *et al.* 2012), while other studies (Carretier *et al.* 2013, 2014) demonstrate that precipitation is negatively correlated with millennial-scale erosion rates.

Hoke *et al.* (2014b) present new apatite (U–Th)/He thermochronology data and a geomorphic analysis of two large rivers from the Frontal Cordillera of Argentina that bear on the timing and origin of mountain uplift. Their data demonstrate that the Frontal Cordillera was never buried by a thick package of foreland basin sediments, as is commonly assumed; rather, there was always a positive or near-surface topographic feature since at least Miocene time. These findings constitute a new view on the evolution of the proximal parts of the Andean foreland, demonstrating that sediment deposition was restricted to relatively small, deep intermontane basins between the rising ranges and not a continuous foreland. Moreover, this contribution provides evidence that uplift of the Frontal Cordillera occurred during late Miocene time, when exhumation rates increased. These results are important for investigators working in a variety of different fields in the geosciences, specifically those studying

Fig. 3. (**a**) Cortaderas mega-landslide (A: Holocene) in the Yeso valley, Chile, reactivated during the 1958 Las Melosas shallow crustal earthquake (B: scarp, C: deposit). (**b**) the Blanca rock avalanche in Mendoza river basin, Argentina. The deposit running along the Blanca gully reached the valley damming the Cuevas River during late Pleistocene time.

Andean orogenesis and the role of tectonics and climate.

Carretier et al. (2014) analyse the relationship between erosion and tectonic, climatic and geomorphical controls (c. 27°S and 39°S). The catchment-mean decadal and millennial erosion rates for central Chile and two sites of Argentina are studied with respect to the long-term evolution of the topography over millions of years. In this paper, they discuss why decadal and millennial erosion rates are negatively correlated with mean runoff south of 32°S (Pepin et al. 2010; Carretier et al. 2013) and explore the correlation of these parameters with vegetation, runoff, lithology and topography. In addition, the north–south pattern of erosion rates with cumulative erosion estimates is contrasted with structural and thermochronological data. They consider new data from the Bio-Bío catchment and previously published catchment-mean decadal and millennial erosion rates (Pepin et al. 2010, 2013; Aguilar et al. 2013; Carretier et al. 2013).

Interplay between tectonic and erosional processes is also brought to the forefront by **Rodríguez et al. (2014)**. They combine geomorphological analysis of palaeo-surfaces and U–Pb zircon geochronology of overlying tuffs to reconstruct the Neogene landscape evolution of north-central Chile (28–32°S). At this latitude, the authors demonstrate the presence of three major uplift stages: (1) eastern Coastal Cordillera and the Frontal Cordillera during early–middle? Miocene time; (2) the Coastal Cordillera during late Miocene time; and (3) uplift of a Pleistocene pediplain post-500 ka. These stages correlate with episodes of increased deformation throughout the Central Andes, starting after a late Oligocene–early Miocene episode of increased plate convergence.

Concluding remarks

The landscape of convergent margins is the result of the complex interplay between internal tectonic processes, surface deformation and erosive processes driven by the interactions of tectonics and climate. The Andes of central Chile and Argentina, where a drastic change in the subduction angles and precipitation gradients occur, is a key locality for the study of such interactions between internal and external geodynamic processes.

This Special Publication aims to improve our understanding of tectonic and landscape evolution of the Andean range at different timescales, as well as the mutual relationship between internal and external mechanisms in Cenozoic deformation, mountain building, topographic evolution, basin development and mega-landslides across the flat slab to normal subduction segments. The geodynamic processes of the Andes of central Chile and Argentina are analysed within a number of subdisciplines of the Earth Sciences including tectonics, petrology, geophysics, geochemistry, structural geology, geomorphology, engineering geology, stratigraphy and sedimentology.

This book is the result of collaborative work supported by Unesco IGCP project 586-Y. In addition, the editors received support from projects FONDECYT 11085022, 1120272 (M. Farías), 1090165 (L. Pinto), NSF award OISE-0601957, ACS-PRF Grant 52480-DNI8 (G. Hoke) and PIP 638 y PICT 1079 (L. Giambiagi).

References

AGUILAR, G., RIQUELME, R., MARTINOD, J. & DARROZES, J. 2013. Rol del clima y la tectónica en la evolución geomorfológica de los Andes Semiáridos chilenos entre los 27–32°S. *Andean Geology*, **40**, 79–101.

AHUMADA, E. A. & COSTA, C. H. 2009. Deformación cuaternaria en la culminación norte del corrimiento las Peñas, frente orogénico andino, Precordillera Argentina. *XII Congreso Geológico Chileno*, Santiago, Chile.

ALLMENDINGER, R. W., ISACKS, B. L., JORDAN, T. E. & KAY, S. M. 1997. The evolution of the Altiplano-Puna plateau of the Central Andes. *Annual Reviews of Earth Science*, **25**, 139–174.

ALVARADO, P., BECK, S., ZANDT, G., ARAUJO, M. & TRIEP, E. 2005. Crustal deformation in the south-central Andes backarc terranes as viewed from regional broadband seismic waveform modelling. *Geophysical Journal International*, **163**, 580–598, http://dx.doi.org/10.1111/j.1365-246X.2005.02759.x

ALVARADO, P., BECK, S. & ZANDT, G. 2007. Crustal structure of the south-central Andes Cordillera and backarc region from regional waveform modelling. *Geophysical Journal International*, **170**, 858–875, http://dx.doi.org/10.1111/j.1365-246X.2007.03452.x

ALVARADO, P., BARRIENTOS, S., SAEZ, M., ASTROZA, M. & BECK, S. 2009. Source study and tectonic implications of the historic 1958 Las Melosas crustal earthquake, Chile, compared to earthquake damage. *Physics of the Earth and Planetary Interiors*, **175**, 26–36.

ALVAREZ, O., GIMENEZ, M. E., MARTINEZ, M. P., LINCEKLINGER, F. & BRAITENBERG, C. 2014. New insights into the Andean crustal structure between 32° and 34°S from GOCE satellite gravity data and EGM2008 model. *In*: SEPÚLVEDA, S. A., GIAMBIAGI, L. B., MOREIRAS, S. M., PINTO, L., TUNIK, M., HOKE, G. D. & FARÍAS, M. (eds) *Geodynamic Processes in the Andes of Central Chile and Argentina*. Geological Society, London, Special Publications, **399**. First published online February 6, 2014, http://dx.doi.org/10.1144/SP399.3

ANTINAO, J. L. & GOSSE, J. 2009. Large rockslides in the Southern Central Andes of Chile (32–34.5°S): tectonic control and significance for Quaternary landscape evolution. *Geomorphology*, **104**, 117–133.

BARRIENTOS, S., VERA, E., ALVARADO, P. & MONFRET, T. 2004. Crustal seismicity in central Chile. *Journal of South American Earth Sciences*, **16**, 759–768.

BEAUMONT, C., JAMIESON, R. A., NGUYEN, M. H. & MEDVEDEV, S. 2004. Crustal channel flows: 1. numerical models with applications to the tectonics of the Himalayan-Tibetan orogeny. *Journal of Geophysical Research*, **109**, B06406, http://dx.doi.org/10.1029/2003JB002809

CAHILL, T. & ISACKS, B. L. 1992. Seismicity and shape of the subducted Nazca Plate. *Journal of Geophysical Research: Solid Earth*, **97**, 17503–17529, http://dx.doi.org/10.1029/92jb00493

CARRETIER, S., REGARD, V. ET AL. 2013. Slope and climate variability control of erosion in the Andes of central Chile. *Geology*, **41**, 195–198, http://dx.doi.org/10.1130/g33735.1

CARRETIER, S., TOLORZA, V. ET AL. 2014. Erosion in the Chilean Andes between 27°S and 39°S: tectonic, climatic and geomorphic control. *In*: SEPÚLVEDA, S. A., GIAMBIAGI, L. B., MOREIRAS, S. M., PINTO, L., TUNIK, M., HOKE, G. D. & FARÍAS, M. (eds) *Geodynamic Processes in the Andes of Central Chile and Argentina*. Geological Society, London, Special Publications, **399**. First published online April 9, 2014, http://dx.doi.org/10.1144/SP399.16

CHARRIER, R., BAEZA, O. ET AL. 2002. Evidence for Cenozoic extensional basin development and tectonic inversion south of the flat-slab segment, southern Central Andes, Chile (33°–36°S.L.). *Journal of South American Earth Sciences*, **15**, 117–139.

CHARRIER, R., RAMOS, V. A., TAPIA, F. & SAGRIPANTI, L. 2014. Tectono-stratigraphic evolution of the Andean Orogen between 31 and 37°S (Chile and Western Argentina). *In*: SEPÚLVEDA, S. A., GIAMBIAGI, L. B., MOREIRAS, S. M., PINTO, L., TRNIK, M., HOKE, G. D. & FARÍAS, M. (eds) *Geodynamic Processes in the Andes of Central Chile and Argentina*. Geological Society, London, Special Publications, **399**. First published online August 27, 2014, http://dx.doi.org/10.1144/SP399.20

CORTÉS, J. M., VINCIGUERRA, P., YAMÍN, M. & PASINI, M. M. 1999. Tectónica cuaternaria de la Región Andina del Nuevo Cuyo (28°–38° LS). *In*: *Geología Argentina, Servicio Geológico Minero Argentino, Subsecretaría de Minería*, Buenos Aires.

CORTÉS, J. M., TERRIZZANO, C. M., PASINI, M. M., YAMIN, M. G. & CASA, A. L. 2014. Quaternary tectonics along oblique deformation zones in the Central Andean retrowedge between 31°30′S and 35°S. *In*: SEPÚLVEDA, S. A., GIAMBIAGI, L. B., MOREIRAS, S. M., PINTO, L., TUNIK, M., HOKE, G. D. & FARÍAS, M. (eds) *Geodynamic Processes in the Andes of Central Chile and Argentina*. Geological Society, London, Special Publications, **399**. First published online February 5, 2014, http://dx.doi.org/10.1144/SP399.10

COSTA, C. H., MACHETTE, M. N. ET AL. 2000a. Map and Database of Quaternary Faults and Folds in Argentina. USGS Open-File Report 00-0108.

COSTA, C. H., GARDINI, C. E., DIEDERIX, H. & CORTÉS, J. M. 2000b. The Andean orogenic front at Sierra de Las Peñas-Las Higueras, Mendoza, Argentina. *Journal of South American Earth Sciences*, **13**, 287–292.

COSTA, C. H., AHUMADA, E. A., GARDINI, C. E., VÁZQUEZ, F. R. & DIEDERIX, H. 2014. Quaternary shortening at the orogenic front of the Central Andes of Argentina: the Las Peñas Thrust System. *In*: SEPÚLVEDA, S. A., GIAMBIAGI, L. B., MOREIRAS, S. M., PINTO, L., TUNIK, M., HOKE, G. D. & FARÍAS, M. (eds) *Geodynamic Processes in the Andes of Central Chile and Argentina*. Geological Society, London, Special Publications, **399**. First published online February 6, 2014, http://dx.doi.org/10.1144/SP399.5

DI GIULIO, A., RONCHI, A., SANFILIPPO, A., TIEPOLO, M., PIMENTEL, M. & RAMOS, V. A. 2012. Detrital zircon provenance from the Neuquén Basin (south-central Andes): Cretaceous geodynamic evolution and sedimentary response in a retroarc-foreland basin. *Geology*, **40**, 559–562.

DIC-UCHILE 2012. Mw = 8.8 Terremoto en Chile, 27 de febrero 2010. MORONI, M. O., CASARES, A. (eds) Departamento de Ingeniería Civil, Facultad de Ciencias Físicas y Matemáticas Universidad de Chile.

FARÍAS, M., CHARRIER, R. ET AL. 2008. Late Miocene high and rapid surface uplift and its erosional response in the Andes of central Chile (33 degrees-35 degrees S). *Tectonics*, **27**, http://dx.doi.org/10.1029/2006tc 002046

FARÍAS, M., COMTE, D. ET AL. 2010. Crustal-scale structural architecture in central Chile based on seismicity and surface geology: Implications for Andean mountain building. *Tectonics*, **29**, http://dx.doi.org/10.1029/2009tc002480

FARÍAS, M., CHARRIER, R. ET AL. 2012. Contribución de largo-plazo de la segmentación climática en Chile central a la construcción Andina. *XIII Congreso Geológico Chileno*, Antofagasta, 194–196.

GARCÍA, V. H. & CASA, A. L. 2014. Quaternary tectonics and seismic potential of the Andean retrowedge at 33–34°S. *In*: SEPÚLVEDA, S. A., GIAMBIAGI, L. B., MOREIRAS, S. M., PINTO, L., TUNIK, M., HOKE, G. D. & FARÍAS, M. (eds) *Geodynamic Processes in the Andes of Central Chile and Argentina*. Geological Society, London, Special Publications, **399**. First published online February 27, 2014, http://dx.doi.org/10.1144/SP399.11

GIAMBIAGI, L., MESCUA, J., BECHIS, F., MARTINEZ, A. & FOLGUERA, A. 2011. Pre-Andean deformation of the Precordillera southern sector, southern Central Andes. *Geosphere*, **7**, 219–239, http://dx.doi.org/10.1130/ges00572.1

GIAMBIAGI, L., MESCUA, J. ET AL. 2014a. Reactivation of Paleozoic structures during Cenozoic deformation in the Cordón del Plata and Southern Precordillera ranges (Mendoza, Argentina). *Journal of Iberian Geology*, **40**, 309–320.

GIAMBIAGI, L., TASSARA, A. ET AL. 2014b. Evolution of shallow and deep structures along the Maipo–Tunuyán transect (33°40′S): from the Pacific coast to the Andean foreland. *In*: SEPÚLVEDA, S. A., GIAMBIAGI, L. B., MOREIRAS, S. M., PINTO, L., TUNIK, M., HOKE, G. D. & FARÍAS, M. (eds) *Geodynamic Processes in the Andes of Central Chile and Argentina*. Geological Society, London, Special Publications, **399**. First published online February 27, 2014, http://dx.doi.org/10.1144/SP399.14

GODOY, E. 2014. The north-western margin of the Neuquén Basin in the headwater region of the Maipo drainage, Chile. *In*: SEPÚLVEDA, S. A., GIAMBIAGI, L. B., MOREIRAS, S. M., PINTO, L., TUNIK, M., HOKE, G. D. & FARÍAS, M. (eds) *Geodynamic Processes in the Andes of Central Chile and Argentina*.

Geological Society, London, Special Publications, **399**. First published online February 27, 2014, http://dx.doi.org/10.1144/SP399.13

Heredia, N., Farias, P., Garcia-Sansegundo, J. & Giambiagi, L. 2012. The basement of the Andean Frontal Cordillera in the Cordon del Plata (Mendoza, Argentina): geodynamic evolution. *Andean Geology*, **39**, 242–257, http://dx.doi.org/10.5027/andgeoV39n2-a03

Hermanns, R. L., Fauqué, L. & Wilson, C. G. J. 2014. ^{36}Cl terrestrial cosmogenic nuclide dating suggests late Pleistocene to Early Holocene mass movements on the south face of Aconcagua mountain and in the Las Cuevas–Horcones valleys, Central Andes, Argentina. *In*: Sepúlveda, S. A., Giambiagi, L. B., Moreiras, S. M., Pinto, L., Tunik, M., Hoke, G. D. & Farías, M. (eds) *Geodynamic Processes in the Andes of Central Chile and Argentina*. Geological Society, London, Special Publications, **399**. First published online May 23, 2014, updated June 13, 2014, http://dx.doi.org/10.1144/SP399.19

Hoke, G. D., Giambiagi, L. B., Garzione, C. N., Mahoney, J. B. & Strecker, M. R. 2014a. Neogene Paleoelevation of intermontane basins in a narrow, compressional mountain range. *Earth and Planetary Science Letters*, **406**, 153–164, http://dx.doi.org/10.1016/j.epsl.2014.08.032

Hoke, G. D., Graber, N. R., Mescua, J. F., Giambiagi, L. B., Fitzgerald, P. G. & Metcalf, J. R. 2014b. Near pure surface uplift of the Argentine Frontal Cordillera: insights from (U–Th)/He thermochronometry and geomorphic analysis. *In*: Sepúlveda, S. A., Giambiagi, L. B., Moreiras, S. M., Pinto, L., Tunik, M., Hoke, G. D. & Farías, M. (eds) *Geodynamic Processes in the Andes of Central Chile and Argentina*. Geological Society, London, Special Publications, **399**. First published online February 5, 2014, http://dx.doi.org/10.1144/SP399.4

Holtkamp, S. G., Pritchard, M. E. & Lohman, R. B. 2011. Earthquake swarms in South America. *Geophysical Journal International*, **187**, 128–146.

Isacks, B. 1988. Uplift of the Central Andean plateau and bending of the Bolivian Orocline. *Journal of Geophysical Research*, **93**, 3211–3231.

Jara, P., Likerman, J., Winocur, D., Ghiglione, M. C., Cristallini, E. O., Pinto, L. & Charrier, R. 2014. Role of basin width variation in tectonic inversion: insight from analogue modelling and implications for the tectonic inversion of the Abanico Basin, 32°–34°S, Central Andes. *In*: Sepúlveda, S. A., Giambiagi, L. B., Moreiras, S. M., Pinto, L., Tunik, M., Hoke, G. D. & Farías, M. (eds) *Geodynamic Processes in the Andes of Central Chile and Argentina*. Geological Society, London, Special Publications, **399**. First published online February 27, 2014, http://dx.doi.org/10.1144/SP399.7

Jordan, T. E., Isacks, B. L., Allmendinger, R. W., Brewer, J. A., Ramos, V. A. & Ando, C. J. 1983. Andean tectonics related to geometry of subducted Nazca plate. *Geological Society of America Bulletin*, **94**, 341–361.

Leyton, F., Ruiz, S. & Sepúlveda, S. A. 2010. Reevaluación del peligro sísmico probabilístico en Chile central. *Andean Geology*, **37**, 455–472.

Molnar, P. & Lyon-Caen, H. 1988. Some simple physical aspects of the support, structure and evolution of mountain belts. *In*: Clark, S. P., Jr., Burchfiel, B. C. & Suppe, J. (eds) *Processes in Continental Lithospheric Deformation*. GSA Special Paper, Boulder, CO, **281**, 179–208, http://dx.doi.org/10.1130/SPE218-p179

Moreiras, S. M. 2004. *Zonificación de peligrosidad y de riesgo de procesos de remoción en masa en el valle del río Mendoza*. Provincia de Mendoza. PhD thesis, Universidad Nacional de San juan.

Moreiras, S. M. 2005. Climatic effect of ENSO associated with landslide occurrence in the Central Andes, Mendoza province, Argentina. *Landslides*, **2**, 53–59.

Moreiras, S. M. 2006. Frequency of debris flows and rockfall along the Mendoza river valley (Central Andes), Argentina. Special Issue Holocene Environmental Catastrophes in South America. *Quaternary International*, **158**, 110–121.

Moreiras, S. M. 2009a. Análisis estadístico probabilístico de las variables que condicionan la inestabilidad de las laderas en los valles de los ríos Las Cuevas y Mendoza. *Revista de la Asociación Geológica*, **65**, 321–327.

Moreiras, S. M. 2009b. Clustering of Pleistocene rock avalanches in the Central Andes, Argentina. *In*: Salfity, J. A. & Marquillas, R. A. (eds) *Cenozoic Geology of Central Andes of Argentina*. SCS Publisher, Salta, 265–282.

Moreiras, S. M. & Páez, M. S. 2014. Historical damage and earthquake environmental effects related to shallow intraplate seismicity of central western Argentina. *In*: Sepúlveda, S. A., Giambiagi, L. B., Moreiras, S. M., Pinto, L., Tunik, M., Hoke, G. D. & Farías, M. (eds) *Geodynamic Processes in the Andes of Central Chile and Argentina*. Geological Society, London, Special Publications, **399**. First published online February 19, 2014, http://dx.doi.org/10.1144/SP399.6

Moreiras, S. M. & Sepúlveda, S. A. 2014. Megalandslides in the Andes of central Chile and Argentina (32°–34°S) and potential hazards. *In*: Sepúlveda, S. A., Giambiagi, L. B., Moreiras, S. M., Pinto, L., Tunik, M., Hoke, G. D. & Farías, M. (eds) *Geodynamic Processes in the Andes of Central Chile and Argentina*. Geological Society, London, Special Publications, **399**. First published online May 13, 2014, http://dx.doi.org/10.1144/SP399.18

Moreiras, S. M., Lenzano, M. G. & Riveros, N. 2008. Inventario de procesos de remoción en masa en el Parque provincial Aconcagua, provincia de Mendoza – Argentina. *Multiequina. Latin American Journal of Natural Resources*, **17**, 129–146.

Moreiras, S. M., Giambiagi, L. B., Spagnotto, S., Nacif, S., Mescua, J. F. & Toural, R. 2014. Caracterización de fuentes sismogénicas en el frente orogénico activo de los Andes Centrales a la latitud de la ciudad de Mendoza (32°50′-33°S). *Andean Geology*, **41**, 345–361.

Mpodozis, C. & Cornejo, P. 2012. Cenozoic tectonics and porphyry copper systems of the chilean andes. *Society of Economic Geologists, Special Publication*, **16**, 329–360.

MPODOZIS, C. & RAMOS, V. A. 1989. The Andes of Chile and Argentina. *In*: ERICKSEN, G. E., CAÑAS, M. T. & REINEMUD, J. A. (eds) *Geology of the Andes and its relation to Hydrocarbon and Mineral Resources*. Circum-Pacific Council for Energy and Mineral Resources, Houston, Texas, 59–90.

MUÑOZ-SÁEZ, C., PINTO, L., CHARRIER, R. & NALPAS, T. 2014. Influence of depositional load on the development of a shortcut fault system during the inversion of an extensional basin: the Eocene-Oligocene Abanico Basin case, central Chile Andes (33°–35°S). *Andean Geology*, **41**, 1–28.

NAIPAUER, M., TUNIK, M., MARQUES, J. C., ROJAS VERA, E. A., VUJOVICH, G. I., PIMENTEL, M. M. & RAMOS, V. A. 2014. U–Pb detrital zircon ages of Upper Jurassic continental successions: implications for the provenance and absolute age of the Jurassic-Cretaceous boundary in the Neuquén Basin. *In*: SEPÚLVEDA, S. A., GIAMBIAGI, L. B., MOREIRAS, S. M., PINTO, L., TUNIK, M., HOKE, G. D. & FARÍAS, M. (eds) *Geodynamic Processes in the Andes of Central Chile and Argentina*. Geological Society, London, Special Publications, **399**. First published online March 4, 2014, http://dx.doi.org/10.1144/SP399.1

PEPIN, E., CARRETIER, S., GUYOT, J. & ESCOBAR, F. 2010. Specific suspended sediment yields of the Andean rivers of Chile and their relationship to climate, slope and vegetation. *Hydrological Sciences Journal*, **55**, 1190–1205.

PEPIN, E., CARRETIER, S. *ET AL*. 2013. Pleistocene landscape entrenchment: a geomorphological mountain to foreland field case, the Las Tunas system, Argentina. *Basin Research*, **25**, 613–637, http://dx.doi.org/10.1111/bre.12019

RAMOS, V. 1988. The tectonics of the Central Andes: 30°–33°S latitude. *In*: CLARK, S. & BURCHFIEL, D. (eds) *Processes in Continental Lithospheric Deformation*. Geological Society of America, Boulder, Special Paper, **218**, 31–54, http://dx.doi.org/10.1130/SPE218-p31

RAMOS, V. A., CEGARRA, M. & CRISTALLINI, E. 1996. Cenozoic tectonics of the High Andes of west-central Argentina (30–36°S latitude). *Tectonophysics*, **259**, 185–200.

RODRÍGUEZ, M. P., PINTO, L. & ENCINAS, A. 2012. Cenozoic erosion in the Andean forearc in Central Chile (33°–34°S): sediment provenance inferred by heavy mineral studies. *In*: RASBURY, T. E., HEMMING, S. R. & RIGGS, N. R. (eds) *Mineralogical and Geochemical Approaches to Provenance*. Geological Society of America, Special Papers, **487**, 141–162, http://dx.doi.org/10.1130/2012.2487(09)

RODRÍGUEZ, M. P., AGUILAR, G., URRESTY, C. & CHARRIER, R. 2014. Neogene landscape evolution in the Andes of north-central Chile between 28 and 32°S: interplay between tectonic and erosional processes. *In*: SEPÚLVEDA, S. A., GIAMBIAGI, L. B., MOREIRAS, S. M., PINTO, L., TUNIK, M., HOKE, G. D. & FARÍAS, M. (eds) *Geodynamic Processes in the Andes of Central Chile and Argentina*. Geological Society, London, Special Publications, **399**. First published online April 4, 2014, http://dx.doi.org/10.1144/SP399.15

SAGRIPANTI, L., AGUIRRE-URRETA, B., FOLGUERA, A. & RAMOS, V. A. 2014. The Neocomian of Chachahuén (Mendoza, Argentina): evidence of a broken foreland associated with the Payenia flat-slab. *In*: SEPÚLVEDA, S. A., GIAMBIAGI, L. B., MOREIRAS, S. M., PINTO, L., TUNIK, M., HOKE, G. D. & FARÍAS, M. (eds) *Geodynamic Processes in the Andes of Central Chile and Argentina*. Geological Society, London, Special Publications, **399**. First published online February 5, 2014, http://dx.doi.org/10.1144/SP399.9

SÁNCHEZ, M., LINCE KLINGER, F., MARTINEZ, M. P., ALVAREZ, O., RUIZ, F., WEIDMANN, C. & FOLGUERA, A. 2014. Geophysical characterization of the upper crust in the transitional zone between the Pampean flat slab and the normal subduction segment to the south (32–34°S): Andes of the Frontal Cordillera to the Sierras Pampeanas. *In*: SEPÚLVEDA, S. A., GIAMBIAGI, L. B., MOREIRAS, S. M., PINTO, L., TUNIK, M., HOKE, G. D. & FARÍAS, M. (eds) *Geodynamic Processes in the Andes of Central Chile and Argentina*. Geological Society, London, Special Publications, **399**. First published online February 6, 2014, http://dx.doi.org/10.1144/SP399.12

SEPÚLVEDA, S. A., ASTROZA, M., KAUSEL, E., CAMPOS, J., CASAS, E. A., REBOLLEDO, S. & VERDUGO, R. 2008. New findings on the 1958 Las Melosas earthquake sequence, Central Chile: implications for Seismic Hazard related to shallow crustal earthquakes in subduction zones. *Journal of Earthquake Engineering*, **12**, 432–455, http://dx.doi.org/10.1080/13632460701512951

SHEFFELS, B. 1990. Lower bound on the amount of crustal shortening in the Central Bolivian Andes. *Geology*, **18**, 812–815.

SIAME, L., BELLIER, O. & SEBRIER, M. 2006. Active tectonics in the Argentine Precordillera and western Sierras Pampeanas. *Revista de la Asociación Geológica Argentina*, **61**, 604–619.

SURIANO, J., LIMARINO, C. O., TEDESCO, A. M. & ALONSO, M. S. 2014. Sedimentation model of piggyback basins: Cenozoic examples of San Juan Precordillera, Argentina. *In*: SEPÚLVEDA, S. A., GIAMBIAGI, L. B., MOREIRAS, S. M., PINTO, L., TUNIK, M., HOKE, G. D. & FARÍAS, M. (eds) *Geodynamic Processes in the Andes of Central Chile and Argentina*. Geological Society, London, Special Publications, **399**. First published online February 28, 2014, http://dx.doi.org/10.1144/SP399.17

TUNIK, M. & LAZO, D. 2008. Microfacies, fósiles y paleoambientes de la Formación Agrio en el sector centro-norte de Neuquén. Actas XII Reunión Argentina de Sedimentología, Buenos Aires, **177**.

TUNIK, M. A., FOLGUERA, A., NAIPAUER, M., PIMENTEL, M. & RAMOS, V. A. 2010. Early uplift and orogenic deformation in the Neuquén basin: constraints on the Andean uplift from U–Pb and Hf analyses of detrital zircons. *Tectonophysics*, **489**, 258–273.

VARGAS, G., KLINGER, Y., ROCKWELL, T. K., FORMAN, S. L., REBOLLEDO, S., BAIZE, S., LACASSIN, R. & ARMIJO, R. 2014. Probing large intraplate earthquakes at the west flank of the Andes. *Geology*, http://dx.doi.org/10.1130/G35741.1

WHIPPLE, K. & MEADE, B. 2006. Orogen response to changes in climatic and tectonic forcing. *Earth and Planetary Science Letters*, **243**, 218–228, http://dx.doi.org/10.1016/j.epsl.2005.12.022

WINOCUR, D. A., LITVAK, V. D. & RAMOS, V. A. 2014. Magmatic and tectonic evolution of the Oligocene Valle del Cura basin, main Andes of Argentina and Chile: evidence for generalized extension. *In*: SEPÚLVEDA, S. A., GIAMBIAGI, L. B., MOREIRAS, S. M., PINTO, L., TUNIK, M., HOKE, G. D. & FARÍAS, M. (eds) *Geodynamic Processes in the Andes of Central Chile and Argentina*. Geological Society, London, Special Publications, **399**. First published online February 17, 2014, http://dx.doi.org/10.1144/ SP399.2

ZÁRATE, M. A., MEHL, A. & PERUCCA, L. 2014. Quaternary evolution of the Cordillera Frontal piedmont between *c.* 33° and 34°S Mendoza, Argentina. *In*: SEPÚLVEDA, S. A., GIAMBIAGI, L. B., MOREIRAS, S. M., PINTO, L., TUNIK, M., HOKE, G. D. & FARÍAS, M. (eds) *Geodynamic Processes in the Andes of Central Chile and Argentina*. Geological Society, London, Special Publications, **399**. First published online February 5, 2014, http://dx.doi.org/10.1144/ SP399.8

Tectono-stratigraphic evolution of the Andean Orogen between 31 and 37°S (Chile and Western Argentina)

REYNALDO CHARRIER[1,2]*, VICTOR A. RAMOS[3], FELIPE TAPIA[1] & LUCÍA SAGRIPANTI[3]

[1]*Departamento de Geología, Universidad de Chile, Plaza Ercilla 803, Santiago, Chile*

[2]*Escuela de Ciencias de la Tierra, Universidad Andres Bello, Campus República, Salvador Sanfuentes 2357, Santiago, Chile*

[3]*Laboratorio de Tectónica Andina, Instituto de Estudios Andinos Don Pablo Groeber, Universidad de Buenos Aires-CONICET, Intendente Güirales 2160, Ciudad Universitaria, C1428EGA, Capital Federal, Buenos Aires, Argentina*

*Corresponding author (e-mail: rcharrie@ing.uchile.cl)

Abstract: In this classic segment, many tectonic processes, like flat-subduction, terrane accretion and steepening of the subduction, among others, provide a robust framework for their understanding. Five orogenic cycles, with variations in location and type of magmatism, tectonic regimes and development of different accretionary prisms, show a complex evolution. Accretion of a continental terrane in the Pampean cycle exhumed lower to middle crust in Early Cambrian. The Ordovician magmatic arc, associated metamorphism and foreland basin formation characterized the Famatinian cycle. In Late Devonian, the collision of Chilenia and associated high-pressure/low-temperature metamorphism contrasts with the late Palaeozoic accretionary prisms. Contractional deformation in Early to Middle Permian was followed by extension and rhyolitic (Choiyoi) magmatism. Triassic to earliest Jurassic rifting was followed by subduction and extension, dominated by Pacific marine ingressions, during Jurassic and Early Cretaceous. The Late Cretaceous was characterized by uplift and exhumation of the Andean Cordillera. An Atlantic ingression occurred in latest Cretaceous. Cenozoic contraction and uplift pulses alternate with Oligocene extension. Late Cenozoic subduction was characterized by the Pampean flat-subduction, the clockwise block tectonic rotations in the normal subduction segments and the magmatism in Payenia. These processes provide evidence that the Andean tectonic model is far from a straightforward geological evolution.

This chapter will focus on the general framework of the Andes along this segment of the orogen, as an introduction to the different specific contributions of the following chapters. This part of the southern Central Andes of Argentina and Chile has received the attention of numerous overviews and syntheses in recent years, such as the publications of Mpodozis & Ramos (1989), Ramos *et al.* (1996*b*), Ramos (1988*b*, 1999) and Charrier *et al.* (2007, 2009). However, several articles have been published in recent years that complement and modify the tectonic and palaeogeographic reconstructions and interpretations presented in previous syntheses. Therefore, the objective of this introductory chapter will be to present a brief updated summary, taking into consideration the most recent advances in the geological knowledge of the region.

Tectonic cycles

The continental margin of southern South America was an active plate margin during most of its history. The Neoproterozoic to late Palaeozoic evolution is punctuated by a succession of tectonic regimes in which extension and compression alternate through time. As a result, terrane accretion and westward arc migration alternate with periods of rifting and extensional basin formation. Although accretion of some terranes has been documented until Jurassic time, as in Patagonia (Madre de Dios terrane; Thompson & Hervé 2002), the post-Triassic history is characterized by the eastward retreat of the continental margin and eastward arc migration, attributed to a combination of shallowing of the subducting plate and subduction erosion. The period between the latest Permian and earliest Jurassic

From: SEPÚLVEDA, S. A., GIAMBIAGI, L. B., MOREIRAS, S. M., PINTO, L., TUNIK, M., HOKE, G. D. & FARÍAS, M. (eds) 2015. *Geodynamic Processes in the Andes of Central Chile and Argentina*. Geological Society, London, Special Publications, **399**, 13–61. First published online August 27, 2014, http://dx.doi.org/10.1144/SP399.20
© 2015 The Geological Society of London. For permissions: http://www.geolsoc.org.uk/permissions.
Publishing disclaimer: www.geolsoc.org.uk/pub_ethics

corresponds to an episode of arrested continental drift, which, however, does not mean that subduction along the continental margin ceased. Different palaeogeographic organizations were developed at that time, and a widely distributed magmatism with essentially different affinities occurred. It is therefore possible to differentiate major stages in the tectonostratigraphic evolution of the Chilean–Argentine Andes, which can be related to the following episodes of supercontinent evolution: (a) breakup of Rodinia; (b) Gondwanaland assembly; and (c) post-Pangaea breakup (Fig. 1).

These stages can in turn be subdivided into shorter tectonic cycles separated from each other

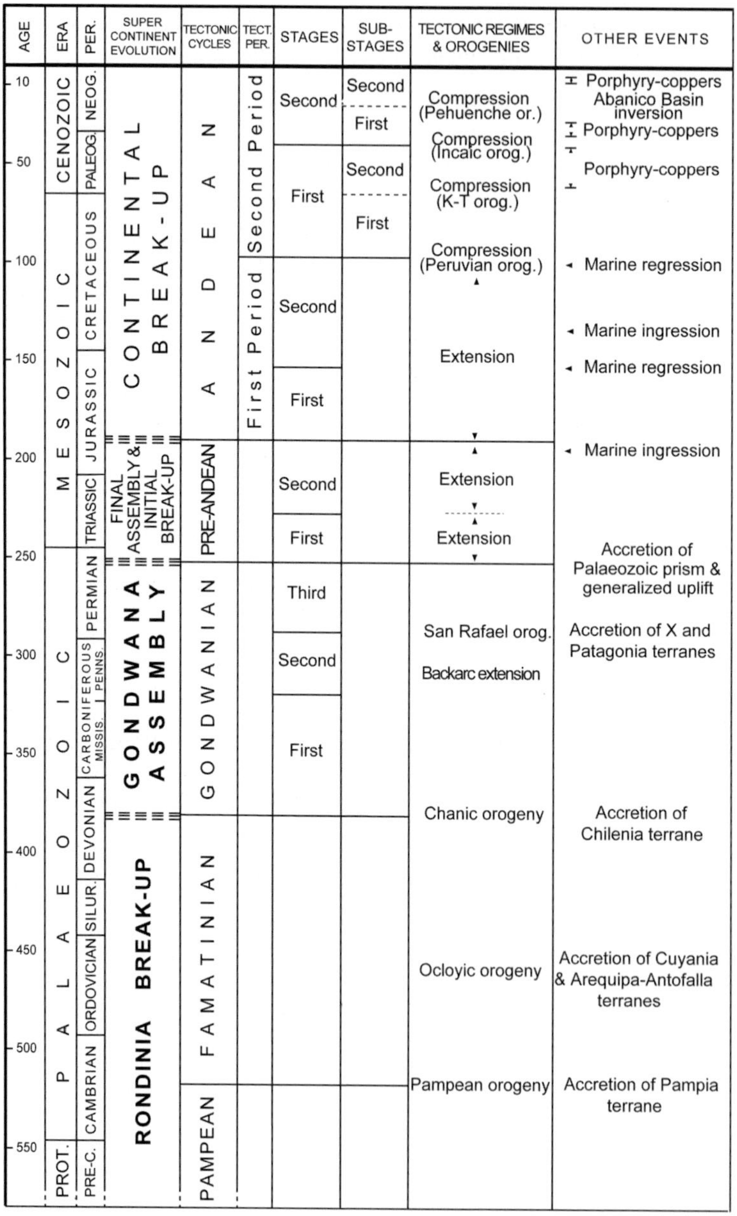

Fig. 1. Tectonic cycles, orogenies and events in the evolution of the continental margin of southern South America compared with the supercontinent evolution.

by regional unconformities or by significant palaeogeographic changes that indicate the occurrence of drastic tectonic events in the continental margin. These tectonic events have been related to modifications in the dynamics of the lithospheric plates (see James 1971; Rutland 1971; Charrier 1973; Aguirre et al. 1974; Frutos 1981; Jordan et al. 1983a, 1997; Malumián & Ramos 1984; Ramos et al. 1986; Isacks 1988; Ramos 1988a; Mpodozis & Ramos 1989). As a result of that, tectonic cycles have been identified along the western southern South America, which according to Mpodozis & Ramos (1989) and Charrier et al. (2007) are (Fig. 1): *Pampean* (Neoproterozoic–Early Cambrian), *Famatinian* (latest Cambrian–Early Devonian), *Gondwanian* (Late Devonian–early Late Permian), *pre-Andean* (latest Permian–earliest Jurassic) and *Andean* (Early Jurassic–Present).

Morphostructural units

The northern part of the analysed segment (31–33°S) is located in the flat-slab subduction segment developed between c. 27 and c. 33°S. There the passive Juan Fernández Ridge has been subducting the continental margin since c. 12 Ma with an eastward dip of c. 20° to the east beneath the forearc to almost horizontal beneath the retroarc (Jordan et al. 1983a; Yáñez et al. 2001; Pardo et al. 2002). This Andean segment, known as the Pampean flat-slab segment, is characterized by the absence of recent volcanic activity and of a Central Depression, existing further south in the normal subduction segment (south of 33°S). Therefore, in this segment no differentiation can be made easily between a Coastal and a Principal Cordillera, and the Coastal Cordillera has been extended further east (Rodríguez 2013; Rodríguez et al. 2013, 2014). East of this extended Coastal Cordillera, other morphostructural units are developed, which gradually disappear southwards or are not developed further south in the normal subduction segment. These are, from west to east: the Frontal Cordillera, the Precordillera and the Sierras Pampeanas (Fig. 2). The alignment of the crustal earthquake epicentres parallel to the projection of the subducted Juan Fernández Ridge shows the strong control that this subduction exerted in the more recent morphologic, magmatic and tectonic features of the Andean cordillera in the flat-slab segment (Alvarado et al. 2009).

South of the flat-slab subduction segment (south of c. 33°S), the Wadati–Benioff zone dips c. 30°E (Cahill & Isacks 1992). As indicated, in this segment, a Central Depression is well developed separating the Coastal Cordillera, to the west, from the Principal Cordillera, to the east, with the edifices of the volcanic arc (Fig. 2). The western flank of the Principal Cordillera is located in Chile, while the eastern flank is located in Argentina.

Pampean tectonic cycle (Neoproterozoic–Early Cambrian)

The Pampean tectonic cycle as proposed by Aceñolaza & Tosselli (1976) includes sedimentation, magmatism and important deformation that took place in Early to Middle Cambrian in northwestern Argentina (Coira et al. 1982). The orogenic deformation at that time, in the central segment analysed here (31–37°S), was located along the Eastern Pampean Ranges, and its extension to the south (Fig. 2) (Ramos 1988a; Rapela et al. 1998; Chernicoff et al. 2012).

The Eastern Sierras Pampeanas comprises an orogenic belt characterized by metamorphic rocks of middle-to-high amphibolite facies, low-grade metapelites and granulite facies meta-basic rocks (Gordillo 1984; Kraemer et al. 1995; Rapela et al. 1998). Medium-grade para- and ortho-gneisses and schists constitute the dominant lithology; large massifs composed of garnet-cordierite pelitic migmatites are also characteristic (Rapela et al. 1998). Geochemical and isotopic studies show a typical calc-alkaline magmatic arc related to subduction (Lira et al. 1996; Rapela et al. 1998). Ultrabasic rocks dominated by harzburgites, chromitites and serpentinites have been interpreted as a disrupted ophiolitic sequence (Escayola et al. 1996; Ramos et al. 2000).

This orogenic belt has been interpreted as (a) the result of a collision between a Pampean block against the Río de La Plata Craton (Ramos 1988a; Ramos & Vujovich 1993; Rapela et al. 1998); (b) in the context of a subduction of a mid-ocean ridge beneath the palaeo-Pacific Gondwana margin (e.g. Gromet & Simpson 2000; Simpson et al. 2003; Piñán-Llamas & Simpson 2006; Schwartz et al. 2008); or (c) the result of complex strike-slip tectonics between Kalahari and Río de la Plata craton (Rapela et al. 2007; Casquet et al. 2012; Spagnuolo et al. 2012a).

In recent years more precise studies of the pre- and post-collisional suites seem to indicate that the main episode of deformation is bracketed between 537 and 530 Ma (Iannizzotto et al. 2013), and the tectonic evolution comprises a complex geological history to fit all of the observations (Escayola et al. 2007; Ramos et al. 2010). This model implies a primitive island arc that under western subduction collided against the Río de la Plata Craton by the end of the Ediacarian (Escayola et al. 2007). This oceanic terrane of the Eastern Sierras Pampeanas known as the Córdoba terrane

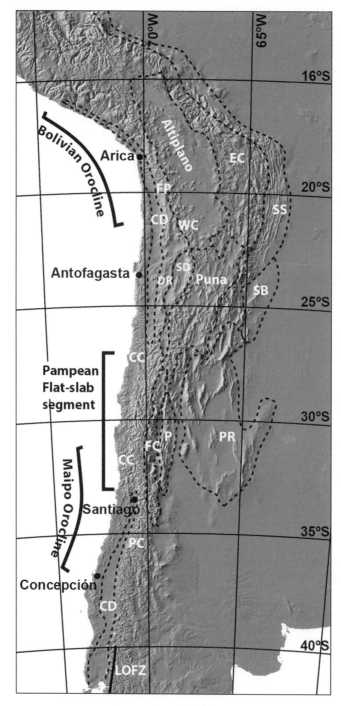

Fig. 2. Digital elevation model of the Andes between 16 and 40°S with indication of the main geographical, tectonic and morphostructural features. Abbreviations: CC, Coastal Cordillera; CD, Central Depression; DR, Domeyko Range; EC, Eastern Cordillera; FC, Frontal Cordillera; FP, Forearc Precordillera (western flank of the Altiplano); LOFZ, Liquiñe–Ofqui Fault Zone; P, Precordillera in Argentina; PC, Principal Cordillera; PR, Pampean Ranges; SB, Santa Barbara System; SD, Salar Depressions; SS, Subandean System; WC, Western Cordillera. Rectangle: Andean region considered in this chapter.

is composed of several belts of ophiolites (Mutti 1997; Ramos *et al.* 2000, 2010). This collision was followed by the final collision of Pampia through an east-dipping subduction against the Río de la Plata craton as proposed by Kraemer *et al.* (1995), Escayola *et al.* (2007) and Ramos *et al.* (2010). Among the post-collisional effects, the emplacement of mafic bodies of Ocean-Island Basalts (OIB) signature emplaced at about 520 Ma could be interpreted as evidence of slab breakoff (Tibaldi *et al.* 2008), associated with general anatexis and crustal delamination as indicated by extensive rhyolitic plateaux preserved in the northern sector of Eastern Sierras Pampeanas, such as the Oncán Rhyolites and Los Burros Rhyodacites of 532–512 Ma (Leal *et al.* 2004). These regions have been affected by ductile shear deformation along some weakness zones during the early Palaeozoic (Martino 2003).

Famatinian tectonic cycle (Cambrian to Late Devonian)

The Famatinian orogenic cycle as proposed by Aceñolaza & Tosselli (1976) comprises the evolution of two important sedimentary sequences, separated by an important angular unconformity. The basal sequence comprises Early Cambrian to Middle Ordovician carbonatic and clastic platform deposits of the Cuyo Precordillera deformed during the Ocloyic diastrophism at about 460 Ma (Astini *et al.* 1996; Ramos 2004, and references therein). Both sequences were deformed during the Middle to Late Devonian, developing the Chanic unconformity that separates these deposits from the late Palaeozoic sequences. The different sedimentary, magmatic and metamorphic rocks of these two sequences will be described from west to east.

Frontal Cordillera

In Cordón del Carrizalito region, in the southern Frontal Cordillera, north of the Río Diamante valley, is exposed a sequence of turbidites of the Las Lagunitas Formation (Fig. 3) (Volkheimer 1978). Graptolites of Ordovician age have been found in these turbidites, previously interpreted as Carboniferous deposits (Tickyj *et al.* 2009*a*). These rocks are characterized by low-grade metamorphism, and are intruded by pre-tectonic to syn-tectonic granitoids such as the Carrizalito Tonalite and Pampa de Los Avestruces Granite deformed during the Chanic orogeny (Tickyj *et al.* 2009*b*; Tickyj 2011). The studies of García-Sansegundo *et al.* (2014*a*) described the Chanic structures as west-vergent, in contrast with the east-vergent late Palaeozoic structures.

North of 34°S at Cordón del Plata region, in northern Frontal Cordillera (Fig. 3), Heredia *et al.* (2002, 2012) described the Vallecitos beds, low-grade metamorphic rocks correlated with Devonian turbidites of western Precordillera. These rocks as well as the Las Lagunitas Formation were deformed during the Middle to Late Devonian Chanic deformation.

Exposures of metamorphic basement rocks that include ultrabasic rocks occur within the Frontal Cordillera as a medium-grade unit further south (Polanski 1964, 1972; Bjerg *et al.* 1990; López & Gregori 2004), known as the Guarguaraz Complex (Willner *et al.* 2011). This complex consists of garnet–micaschist and quartzitic schist, metasediments with intercalated lenses of garnet bearing amphibolite with a N- or E-MORB (normal or enriched mid-ocean ridge basalt) geochemical signature, serpentinite with tremolite-/talc-bearing wall rocks and marbles and calc-silicates. Willner *et al.* (2011) interpreted this sequence as related to a high-pressure–low-temperature metamorphism, with a metamorphic peak dated in 385 Ma, followed by decompression and retrograde metamorphism between 348 and 337 Ma, associated with the new stage of subduction along the Pacific margin. These high pressure–low temperature conditions (12–14 kbar–550 °C) are interpreted as a collisional metamorphism, distinctive from the subduction complexes of the western series (basal accretion) younger than 307 Ma and the eastern series (frontal accretion) younger than 330–345 Ma (Hervé *et al.* 2013).

Cuyo Precordillera

This region comprises a series of platform sedimentary rocks, which are unconformably deposited on scarce syn-rift deposits, mainly preserved in the northern Precordillera. The Cerro de la Totora Formation includes some evaporites and red-beds of Early Cambrian age (515 Ma), which fill the half-graben system developed on the Grenvillian age basement (Rapalini & Astini 1998; Rapalini 2012).

The platform deposits comprise mainly an Early Cambrian to Early Ordovician carbonate sequence bearing typical *Olenellus* trilobites of Laurentia derivation (Bordonaro 1980, 1992). Various studies have characterized the complex third-order sequences of these carbonates (Cañas 1999; Keller 1999), the palaeogeographic connections of their fauna (Benedetto *et al.* 1999) and their tectonosequences (Astini *et al.* 1995; Thomas & Astini 1996).

These platform deposits are bounded to the west by slope deposits (Alonso *et al.* 2008; Voldman *et al.* 2010) and the Famatinian ophiolites (Haller & Ramos 1984, 1993). This sequence of mafic, ultramafic and granulitic rocks of Middle to Late

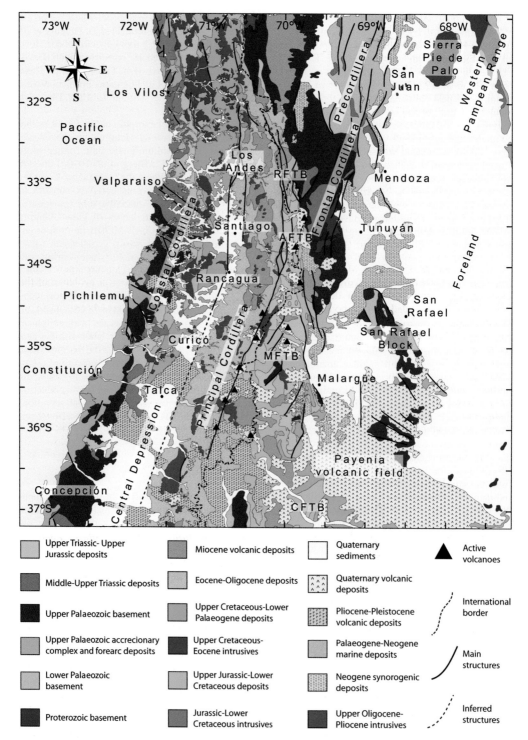

Fig. 3. Geological map of the Andean region between 31 and 37°S. *Abbreviations*: RFTB, Ramada fold-and-thrust belt; AFTB, Aconcagua fold-and-thust belt; MFTB, Malargüe fold-and-thust belt; CFTB, Chos-Malal fold-and-thrust belt.

Ordovician age extends for more than 1000 km along the eastern slope of the Andes near the boundary between the Precordillera and the Frontal Cordillera (Fig. 3) (Ramos et al. 2000). These rocks show a dominant E-MORB composition, as well as the Silurian and Devonian mafic rocks and pillow lavas further to the west (Cortés & Kay 1994). The time of emplacement of these ophiolites is constrained by the Middle to Late Devonian age of the low-grade metamorphism of these rocks (Buggish et al. 1994; Davis et al. 1999; Robinson et al. 2005).

The subsequent clastic deposits above the platform successions were characterized by two syn-orogenic sequences; the older comprises the sandstones, conglomerates, glacial deposits and turbidites with olistostromes and olistoliths of Late Ordovician to Silurian age, represented by numerous formations (Astini et al. 1996). A second syn-orogenic sequence is represented by the turbidites and clastic deposits of Early to Middle Devonian age. Both sequences are separated by an angular unconformity seen in the Cuesta del Tambolar region between the eroded tilted carbonates and the Silurian deposits, interpreted there as the expression of a peripheral bulge associated with the deformation of the Sierra de Pie de Palo (Fig. 3) (Astini et al. 1995). Both sequences are affected by the Chanic unconformity, associated with an important period of uplift, stacking and deformation of the Frontal Cordillera and western Precordillera with a characteristic west-vergence (Von Gosen 1992; García-Sansegundo et al. 2014a).

Western Sierras Pampeanas

The western Sierras Pampeanas as defined by Caminos (1979) encompasses the Sierra de Pie de Palo and the belt of westernmost Sierras Pampeanas, such as the Valle Fértil, La Huerta and other basement uplifts further to the east. This region is the locus of the Famatinian orogeny, which has been preserved in metamorphic facies that, based on their characteristics, can be divided into several belts.

The western slope of Sierra de Pie de Palo preserved a belt of limestones and quartzites, known as the Caucete Group (Vujovich & Ramos 1994), heavily deformed and overridden by the Pie de Palo Complex, an ophiolite sequence of Grenville age (Vujovich & Kay 1998). Based on detrital zircons and geochemistry, the Caucete Group has been correlated with the sedimentary Cambrian and Ordovician sequences of Precordillera (Naipauer et al. 2010a, b). The metamorphic conditions of the Caucete Group indicate high pressures and modest temperatures in amphibolite facies. These conditions occurred during subduction of a cold sedimentary slab in an A-subduction zone setting (Van Staal et al. 2011) around 460 Ma (Ramos 2004). The data indicate a clockwise pressure–temperature ($P-T$) trajectory reaching a peak pressure of roughly 13 kbar at 450 °C, and then heating as pressure declined, reaching a maximum temperature of roughly 500–560 °C at pressures of 8–10 kbar. This fact has been explained by these authors as continuous subduction of a cold sedimentary slab (Caucete Group) after amalgamation with the Pie de Palo Complex in the subduction channel, followed by underthrusting of progressively more buoyant Precordillera crust.

Further to the east the remaining western Sierras Pampeanas have been extensively studied. Two dominant rock types characterized this region, a western magmatic belt that consists of latest Cambrian, Early to Middle Ordovician calc-alkaline granitoids and metamorphic rocks produced in the same interval (Ramos 1988a, 2004). After the pioneering work of Pankhurst & Rapela (1998) and Quenardelle & Ramos (1999), several new studies have characterized the geological evolution of the Ordovician magmatic arc. The petrological conditions of these granitoids have been studied by Otamendi et al. (2008, 2009a, b, 2010a, b, and references therein) complemented by the analyses of Verdecchia et al. (2007) and Casquet et al. (2012, and references therein). The wall rocks of these granitoids were characterized by Verdecchia et al. (2007), who found fossil shelly faunas preserved in the metamorphic rocks of Ordovician age. Further north, the Ordovician rocks are preserved in sedimentary facies and sedimentological studies show that sedimentation took place in an extensional environment (Mángano & Buatois 1996). Similar conclusions were obtained for the Ordovician age of the metamorphism (Collo et al. 2008), and the extensional regime of the magmatic rocks (Collo et al. 2009).

Tectonic evolution of the Famatinian cycle

Although several tectonic models have been proposed to explain the early Palaeozoic history of central Chile and western Argentina, sometimes with contrasting interpretations, such as accretions of allochthonous terranes (Ramos 1988a; Astini et al. 1995) or by continent–continent collisions involving Laurentia and Gondwana (Dalla Salda et al. 1992a, b), in recent years some consensus has been obtained for the evolution of that region (see Ramos & Dalla Salda 2011).

Most of the authors agree that the main deformation that affected the early Palaeozoic the Precordillera platform and slope deposits occurred at about 460 Ma almost in the upper part of the Darriwillian (Thomas & Astini 2003; Ramos 2004). This deformation explains the stacking of different thrust sheets of the eastern Precordillera and the development of olistoliths in the foreland basin during Late Ordovician as part of the Ocloyic deformation (Thomas & Astini 2007). These authors explain the deformation by a collision of the Cuyania (or a larger Precordillera) terrane, an allochthonous terrane derived from the Ouachita embayment of Laurentia that collided against the proto-margin of Gondwana (Astini et al. 1995; Thomas & Astini 1996). The first palaeomagnetic data obtained from Cerro Totora syn-rift deposits confirmed this origin for Cuyania (Rapalini & Astini 1998), and, although the new data from the polar apparent curve of Gondwana poses some uncertainties in the location of Cuyania in the Early Cambrian, the recent palaeomagnetic analysis of Rapalini (2012) shows that the origin in the Ouachita embayment is still the best alternative to explain most of the existing data.

The western margin of Sierras Pampeanas shows that the Cambrian–Ordovician calc-alkaline magmatic arc lasted between the Furongian (497–485 Ma) and the Darriwilian (457–458 Ma), reaching the maximum activity in the Early Ordovician (Pankhurst & Rapela 1998; Quenardelle & Ramos 1999; Rapela et al. 2010; Dahlquist et al. 2013, and references therein). The metamorphic peak was around 460 Ma (Ramos 2004; Van Staal et al. 2011; and references therein). This peak was associated with the collision of the Cuyania terrane (Von Gosen et al. 2002; Chernicoff & Ramos 2003) by the end of Middle Ordovician.

The consensus that has been obtained for the Ocloyic deformation is greater than that for the various alternatives still discussed for the Silurian and Devonian evolution, which ends with the Chanic deformation at Middle to Late Devonian times. There is agreement to relate that deformation with the collision of the Chilenia terrane, but is not clear where the magmatic arc was located. Most authors have agreed with the proposal of Cucchi (1972), who dated by K–Ar the deformation and low-grade metamorphism of western Precordillera as Late Devonian. New Ar–Ar ages and metamorphic studies confirm the age and the low-grade metamorphism related to the collision (Buggish et al. 1994; Robinson et al. 2005; Voldman et al. 2009).

The main problem that has persisted since the early proposal of Ramos et al. (1984, 1986) is the polarity of subduction. These authors proposed an eastern subduction, a criterion that was followed by subsequent studies (see Willner et al. 2011, and references therein). On the other hand, the hypothesis of a western subduction beneath the Chilenia terrane proposed by Astini et al. (1995) was followed by Davis et al. (1999), Gerbi et al. (2002), Heredia et al. (2012) and González-Menéndez et al. (2013), among others. The uncertainty about the location and age of the magmatic arc make these two alternatives difficult to reconcile. Heredia et al. (2012) interpreted a Devonian arc developed in the Frontal Cordillera based on the clasts found in the Devonian (?) Vallecitos beds, but the ages of this arc and these beds are not well constrained. Since the peak of high-pressure metamorphism occurred at 385–390 Ma (Willner et al. 2011) in Middle Devonian times, and is related to the collision of Chilenia and Cuyania, the arc inferred from the Vallecitos clasts could be Late Devonian and associated with east-dipping subduction from the Pacific side. The other criterion to establish the polarity of subduction is the dominant Chanic vergence of deformation. García-Sansegundo et al. (2014a) describe a dominant west-vergence for the early Palaeozoic rocks of the Frontal Cordillera in the Cordón del Carrizalito region. Until the age and location of the magmatic arc are established, it will not be possible to address the polarity of the subduction. Moreover, the propagation of the uplift and deformation of the Frontal Cordillera produced the syn-orogenic deposits of the Angualasto Group in the Precordillera at the Early Carboniferous (Limarino et al. 2006).

Gondwanian tectonic cycle (Mississippian–Lopingian)

The Gondwanan units north of c. 33°S differ considerably from those exposed south of this latitude and will be described in two different segments.

Northern segment

North of c. 33°S the following Gondwanan units were recognized from west to east on both sides of the Andean Cordillera.

Accretionary prism rocks. Polyphase deformed metamorphic rocks of the Choapa Metamorphic Complex are exposed along or next to the coast line, northwards of Los Vilos (Fig. 3). This complex consists mostly of grey-coloured phyllites, schists, fine-grained gneisses and metabasites (Muñoz-Cristi 1942; Thiele & Hervé 1984; Hervé 1988; Hervé et al. 1988; Irwin et al. 1988; Godoy & Charrier 1991; Rivano & Sepúlveda 1991; Rebolledo & Charrier 1994; Charrier et al. 2007; Hervé et al. 2007; Richter et al. 2007; Willner et al. 2008, 2012; García-Sansegundo et al. 2014b), with protoliths of fine- to coarse-grained sedimentary deposits for the phyllites, quartz-mica schists and gneisses, and basic to ultrabasic volcanic rocks, with occasional pillow structures, for the green-coloured, amphibole schists. The Choapa Metamorphic Complex resulted from basal and frontal accretion in a subduction complex and has been subjected to different $P-T$ metamorphic conditions. The presence of phengite-rich muscovite and garnet relics indicates the first stage of very high pressures, probably in the subduction channel, followed by retrogression to greenschist facies conditions. This occurred between 308 and 274 Ma (from Late Pennsylvanian to the end of Cisuralian, in the Early Permian) with a high-pressure–low-temperature peak at c. 279 Ma (Willner et al. 2008, 2012). Exhumation and deformation, like broken formation-type breccias (mélange-type 1 of Cowan 1985), of the metamorphic complex continued during Mesozoic times and resetting events on minerals have ages that correspond to known extensional or compressional Mesozoic events in this region of the Andes (Willner et al. 2012). Recent U–Pb Sensitive High Resolution Ion Micro Probe (SHRIMP) age determinations on detrital zircons from the Choapa Complex immediately north of the region considered here indicate that sedimentation occurred until at least Early Triassic and, thus, that metamorphic processes affected the accretionary prism until at least early Mesozoic times (Emparan & Calderón 2014).

Forearc basin deposits. Two sedimentary stages have been recognized in the forearc region. In the first one, the meta-sedimentary unit Agua Dulce Metaturbidites, and the not metamorphic, strongly folded turbiditic Arrayán Formation, were deposited in a forearc basin (Rivano & Sepúlveda 1991; Rebolledo & Charrier 1994). These units have maximum depositional ages of 337 and 343 Ma, respectively (Willner et al. 2008). Supply of the Arrayán Formation is from the NW and the deposits accumulated on the western side of the basin in Carboniferous time (Rebolledo & Charrier 1994; Willner et al. 2008, 2012). Platformal deposits of the forearc Arrayán basin are exposed on the eastern side of the basin in the present day western Frontal Cordillera, intruded by granitoids of Cisuralian age (Elqui plutonic complex). These slightly contact-metamorphosed deposits consist of a rhythmic alternation of slates and sandstones of at least 1500 m thick, which have been included in the Hurtado Formation (Mpodozis & Cornejo 1988). These deposits are considered to be the prolongation of a series of similar outcrops representing a transgressive–regressive event, some of which further north contain fossil remains indicating a Middle Devonian to Mississippian age, and a provenance of sediments from a volcanic source located to the east and SE (Charrier et al. 2007 and references therein). The continuous Devonian to Mississippian sedimentation in the retrowedge Arrayán basin (Charrier et al. 2007; García-Sansegundo et al. 2014b) demonstrates that the Late Devonian Chanic deformation did not affect the western margin of the Chilenia terrane.

The second sedimentary stage occurred in third stage of the Gondwanian cycle (Charrier et al. 2007) and is characterized by coarse- to fine-grained, fossiliferous marine deposits with turbiditic and calcareous intercalations exposed close to the coast, north of Los Vilos. The Quebrada Mal Paso Beds and Huentelauquén Formation (Muñoz-Cristi 1973; Charrier 1977; Mundaca et al. 1979; Rivano & Sepúlveda 1983, 1985, 1991; Irwin et al. 1988; Méndez-Bedia et al. 2009) unconformably overlie the Arrayán Formation, and mark a major palaeogeographic change at the moment of deposition. A maximum depositional age of 303 Ma was obtained by Willner et al. (2008). Although Rivano & Sepúlveda (1983, 1985, 1991) favour a Late Pennsylvanian to Cisuralian (Early Permian) age, other authors consider that its fossil content indicates a Permian (Fuenzalida 1940; Muñoz-Cristi 1942, 1968; Minato & Tazawa 1977; Thiele & Hervé 1984; Mundaca et al. 1979) or a mid-Permian age (Díaz-Martínez et al. 2000). A Guadalupian age would be consistent with its maximum depositional age of 303 Ma (Willner et al. 2008) and its unconformable superposition on the Arrayán Formation.

Main magmatic arc rocks. The magmatic arc is not exposed in the study region, although it is well represented further north in the Frontal Cordillera, between 29 and 31°S, where it forms the Elqui–Limarí and Chollay batholiths, and still further north the Montosa–El Potro batholith (Nasi et al. 1985; Mpodozis & Kay 1990, 1992). The Elqui plutonic complex includes series of plutons that range in age from the Mississipian to the Late Triassic (Hervé et al. 2014), indicating a protracted magmatic history in this region of the Frontal Cordillera. New U–Pb ages obtained by Hervé et al. (2014) plus those obtained by previous authors (Pankhurst et al. 1996; Pineda & Calderón 2008; Coloma et al. 2012) form four groups, falling into the Late Mississippian, Late Pennsylvanian to Cisuralian, Late Lopingian to Middle Triassic, and Late Triassic, respectively. The second group with ages between 301 and 284 Ma predates the San Rafael orogeny, coincides in time with the evolution of the Choapa metamorphic complex and corresponds to the next Gondwanan unit to the east. Geochemical features indicate development in a magmatic arc associated with a subduction zone on a gradually thickening crust.

The rhyolitic welded ash-tuffs and flows of the Guanaco Sonso Formation (lower portion of the Pastos Blancos Group) exposed in the Elqui drainage basin (c. 30°S) are related to this plutonic activity. They yielded K–Ar biotite ages of 281 ± 6, 262 ± 6 and 260 ± 6 Ma, and a U–Pb zircon age of 265.8 ± 5.6 Ma, which except for the first one correspond to the Guadalupian (Martin *et al.* 1999). Based on these ages, the Guanaco Sonso Formation should instead be included in the post-tectonic Guadalupian to early Lopingian third stage of Gondwanian evolution of Charrier *et al.* (2007) and thus be related to the Colangüil plutonic activity, which yielded ages between *c.* 279 and *c.* 252 Ma (Sato *et al.* 1990; Sato & Llambías 1993, 2014), rather than to the pre-tectonic second group of Hervé *et al.* (2014).

In the Elqui valley, immediately north of the region considered here, K–Ar hornblende and biotite age determinations in this batholith yielded Permian ages of 297 ± 9 and 258 ± 4 Ma, respectively (Nasi *et al.* 1985, 1990), which coincide within errors with the age of the second group of Hervé *et al.* (2014). These ages also coincide fairly well with a recent U–Pb age determination of 282.7 ± 5.8 Ma for a rhyolitic volcanic sequence that still further north overlies the western El Tránsito Metamorphic Complex (Salazar *et al.* 2009). The felsic character of these units makes it difficult to differentiate them from each other and the wide age range covered by them suggests that there existed a long lasting felsic volcanic activity from at least Pennsylvanian to Triassic times. Late Pennsylvanian to Cisuralian activity corresponds to the second group of Hervé *et al.* (2014) and is pre-tectonic relative to the San Rafael orogeny, whereas the Guadalupian to early Lopingian activity is post-tectonic. The latter, according to its age, would correspond to a volcanic activity coeval with intrusion of the Colangüil plutonic activity, deposition of the Huentelauquén Formation and the subduction-related lower portion of the Choiyoi Group (Kay *et al.* 1989; Kleiman & Japas 2009). A third stage of activity would include the felsic upper portion of the Choiyoi Group in Argentina and the Matahuaico Formation, exposed further west in Chile in the Elqui river drainage, which has been assigned to the early stage of the next tectonic Pre-Andean cycle (Charrier *et al.* 2007).

Retroarc magmatic rocks. The Colangüil batholith and associated volcanic rocks are exposed north of 31°S to east of the previous arc rocks in the Cordillera Frontal along the Argentine slope in the province of San Juan. The batholith is composed of several granitoids varying from granodiorites to granites as the Los Puentes, Los Lavaderos, Las Opeñas, Agua Blanca and Chita plutons (Sato *et al.* 1990), with K–Ar ages varying between 272 and 247 Ma (Sato & Llambías 1993). The coarse-grained granodiorites have an Early to Middle Permian age, and they are typically calc-alkaline with magmatic arc affinities, associated with the final stage of subduction in a retroarc setting, and as early post-orogenic products respect to the orogenic San Rafael deformation (Sato & Llambías 1993). The granites are commonly fine grained and characterized by granophyric textures consistent with intrusions at shallow levels, and have a transitional signature from calc-alkaline to alkaline (A-type). They were interpreted as reflecting an evolution to an extensional post-orogenic setting during Late Permian times. These granitoids are associated with lava flows, ignimbrites and tuffs of andesitic and rhyolitic composition, which follow the same trend as the plutonic rocks. Based on the geochemical characteristics, the calc-alkaline series was assigned to a pre-Choiyoi field, typical of an arc setting, while the younger transitional series was assigned to a Choiyoi field of within-plate affinities in an extensional regime (Kay *et al.* 1989).

Retroarc basin deposits. Several authors have described the sedimentation in the retroarc region of this segment of the Andes as part of the Calingasta and Uspallata basins developed in the Frontal Cordillera and the Precordillera, as well as an intraplate basin known as the Paganzo basin developed between the Precordillera and the Sierras Pampeanas (Fig. 3) (Ramos *et al.* 1986; López Gamundi *et al.* 1994, among others). These authors recognized two different stages separated by the San Rafael orogenic deformation in the Middle Permian. The first stage is unconformably overlying the Devonian deposits and comprises mainly Late Carboniferous to Early Permian sequences, where a complete record of different glacial stages has been recognized. The Early Carboniferous is only preserved in the eastern Precordillera (Fig. 3) where the Angualasto Group of mainly Visean age recorded the first infills of the retroarc marine basin with an early glacial stage (Fernández-Seveso & Tankard 1995; Limarino *et al.* 2013). The upper part of the first stage comprises widely developed marine deposits in the Precordillera, and recorded the glacial diamictites, overlain by transgressive postglacial shales (Pazos 2002; Limarino & Spalletti 2006). These transgressive facies were succeeded by deltaic and fluvial sequences bearing coal beds where recent U–Pb ages indicate a late Bashkirian age (Gulbranson *et al.* 2010; Spalletti *et al.* 2012; Limarino *et al.* 2013).

The sequences above the San Rafael unconformity of Early–Middle Permian age are mainly volcaniclastic, pyroclastic and volcanic rocks associated with the lower and upper parts of the Choiyoi Group (Kay *et al.* 1989; Kleiman & Japas 2009). The lower part of the Choiyoi volcanic rocks is subduction related, while the upper part corresponds to an extensional regime that controlled the extensive rhyolitic plateaux in the foreland area (Mpodozis & Ramos 1989; Mpodozis & Kay 1992; Llambías 1999).

There is no agreement on the dominant tectonic regime after the Late Devonian Chanic compressive deformation. Fernández-Seveso & Tankard (1995) interpreted that the described sequences reflect changes from transtensional to extensional regimes. A similar extensional regime was proposed by Astini (1996), Astini et al. (2011), and Martina et al. (2011) for the Mississippian (348–342 Ma) based on the occurrence of rhyolites at about 28°S in the southern Puna, coeval with the well-known occurrence of A-type granites in the Sierras Pampeanas (Grosse et al. 2009). On the other hand, after the Chanic deformation during Middle to Late Devonian times (Ramos et al. 1986) compressional deformation and flexural loading produced the foreland basin where the Mississippian deposits of the Angualasto Group and El Ratón Formation have accumulated (Heredia et al. 2012). As a result of this deformation, the Protoprecordillera was uplifted and remained as a positive area until the end of the Carboniferous, when it collapsed by extensional faulting (Limarino et al. 2013).

Southern segment

South of 33°S and further south of the considered region, four Gondwanan units are exposed continuously paralleling the coast from west to east (see Fig. 3): a metamorphic complex; a north–south elongated Coastal Batholith that intrudes the former; an extensive batholith emplaced in the Frontal Cordillera; and an extensive volcanic episode in the Frontal Cordillera and further east in the San Rafael Block (Fig. 3) represented by the Choiyoi Group.

Metamorphic complex. In this segment the metamorphic complex includes the eastern and a western series that form a paired metamorphic belt (González-Bonorino 1970, 1971; González-Bonorino & Aguirre 1970; Aguirre et al. 1972; Hervé et al. 1974, 1984, 2003, 2013; Hervé 1977; Kato & Godoy 1995; Willner 2005; Willner et al. 2004, 2005, 2008; Glodny et al. 2005, 2006, 2008), interpreted recently as the result of frontal and basal accretion in a subduction system, respectively (Richter et al. 2007; Willner et al. 2008).

The western series, which was deposited shortly after the eastern series (Hervé et al. 2013), consists of polyphase deformed and metamorphosed sandstones and pelites, metacherts, metabasites, occasionally with pillow structures and scarce serpentine bodies, formed by basal accretion under a higher $P-T$ metamorphic gradient, while the eastern series consists mainly of polyphase deformed metaturbidites, with recognizable primary structures and lenses of calc-silicate rocks, deposited in a retrowedge or forearc basin, metamorphosed by a low $P-T$ gradient (Glodny et al. 2006; Richter et al. 2007; Hervé et al. 2013). All SHRIMP U–Pb age determinations on igneous detrital zircons from the accretionary complex yielded peaks older than Mesozoic. The youngest peak obtained from the eastern series was dated at 330–345 Ma, while the youngest peak in the western series was dated at 307 Ma (Hervé et al. 2013). All these ages are older or coeval with the Coastal Batholith, and coincide with ages determined for the Agua Dulce metaturbidites and the non-metamorphic Arrayán Formation north of c. 33°S, respectively. Moreover, ages of detrital zircons in both series indicate a major input from the Famatinian orogenic belt and subordinately from Pampean and Grenvillian sources (Hervé et al. 2013).

Ar–Ar dating of white mica in the metamorphic series indicates for the western series a peak of high-$P-T$ metamorphism between 320 and 288 Ma and for the eastern series a peak of high-temperature metamorphism between 302 and 294 Ma (Willner et al. 2005). According to the age and the contact metamorphism affecting the eastern series, the sedimentation, at least in the western series, began before emplacement of the Coastal Batholith (see below).

South of the Lanalhue lineament (c. 38°S) conditions seem to have been different from those to north of this latitude. Here the metamorphic complex bends to the east and consists mainly of the western series, which is here much younger than further north (Hervé et al. 2013).

Deposits in the Coastal Cordillera close to Concepción, in the southern part of the considered region, form the newly proposed Patagual–El Venado unit (Mardonez et al. 2012). These deposits overlie unconformably the eastern series and are unconformably covered in the Bio Bío valley by the marine, Carnian Santa Juana Formation (Nielsen 2005). This unit consists of a tightly folded and slightly metamorphosed (epizone close to the anchizone) alternation of pelitic and thick psamitic layers, which differentiates them from the eastern series and the overlying early Late Triassic deposits. Its loosely constrained age, between the Early Permian (age of the thermal metamorphism in the eastern series of the metamorphic complex, south of 33°S) and the early Late Triassic Santa Juana Formation makes its age assignment difficult. Considering that this unit overlies the eastern series, which was affected by thermal metamorphism between 302 and 294 Ma (Willner et al. 2005), its maximum age is early Sakmarian, in the Early Permian. According to the lithologic description, strong deformation and low-grade metamorphism, which are reminiscent of the Agua Dulce metaturbidites and the Arrayán Formation, we suggest that the tectonic setting for this unit is the late Palaeozoic forearc or retrowedge basin. Another possibility is that these deposits accumulated in a rift basin during the first stage of the Pre-Andean cycle.

Its strong deformation and low metamorphic grade suggest that the Lopingian to Early Triassic deposits close to the continental margin would at that time still have been affected by processes related to subduction activity, as has been shown by the presence of Triassic detritic zircons in rocks of the Choapa Metamorphic Complex (Emparan & Calderón 2014).

Coastal batholith. This batholith consists of a series of plutons of calc-alkaline character, meta- to peraluminous composition, and granitic to quartz-dioritic lithologies exposed along the Coastal Cordillera, between 33 and 38°20′S, intruding to the east the metamorphic complex (Fig. 3). Further south (38°S), the batholith curves to the east and can be followed southwards along the Principal Cordillera. This shift is probably controlled by the NW-orientated Lanalhue lineament (Glodny et al. 2008; Hervé et al. 2013). According to the recent SHRIMP U–Pb age determinations, the Coastal Batholith was emplaced in a c. 19 Ma period, between 319.6 ± 3.3 and 300.8 ± 2.4 Ma, in Pennsylvanian time, according to Deckart et al. (2014). This age differs considerably from the ages recently obtained by Hervé et al. (2014) for the Elqui plutonic complex, which fall into the Early Permian (Cisuralian; 301–284 Ma). Ages for the metamorphic peaks on the western series indicate that metamorphism overlapped with emplacement of plutons of the Coastal Batholith (Hervé et al. 2013).

Frontal Cordillera batholith

Further east, along the Frontal Cordillera, there are several granitoid stocks and batholiths, exposed south of 33°S latitude (Fig. 3). These granitoids outcrop from the Cordón del Plata to the Cordón del Portillo region, and continuous further south until the Río Diamante valley. They have been described by Polanski (1964, 1972) and Caminos (1965) as typical calc-alkaline metaluminous granitoids emplaced in the Carboniferous deposits. Petford & Gregori (1994) and Gregori et al. (1996) compared this belt of Frontal Cordillera granitoids with the Coastal Batholith of Peru and concluded that they share similar La/Yb ratios, Al_2O_3 contents and several petrographic characteristics that show a typical subduction setting. These authors interpreted these granitoids as emplaced in a several thousand metres-thick sedimentary sequence formed in an extensional regime. The first U–Pb ages of the Frontal Cordillera of this segment were presented by Orme & Atherton (1999), ranging in age between 276 and 262 Ma and with εNd between −2.5 and −3.5. These postectonic granitoids were also recognized in the Frontal Cordillera by Gregori & Benedini (2013), who interpreted these Cisuralian and Guadalupian granodiorites, tonalites and monzogranites of I-type as emplaced subsequent to the San Rafael orogeny that closed the Carboniferous basin between 284 and 276 Ma. The new ages seem to discard the presence of Carboniferous granitoids in this sector of Frontal Cordillera assumed by Caminos (1979).

Choiyoi volcanic rocks. These volcanic rocks, recognized in this segment of the Andes by Groeber (1953), are widely represented in the Frontal Cordillera and further east in the San Rafael Block as well as in the foothills of western Sierras Pampeanas. The volcanic rocks of the Choiyoi Group formed after the San Rafael tectonic phase. This group is subdivided into two essentially different units. A lower unit consisting of volcanic rocks of basic to intermediate composition and calc-alkaline signature (Poma & Ramos 1994) developed in late Cisuralian and Guadalupian times in an arc setting in association with subduction of oceanic lithosphere, an upper volcanic unit consisting of silicic volcanic and volcaniclastic deposits and subvolcanic intrusives derived from crustal melting under extensional tectonic conditions deposited in Lopingian to Anisian times (Kay et al. 1989; Mpodozis & Kay 1990, Llambías et al. 1993, 2003; Llambías & Sato 1995; Spalletti 1999; Martínez 2005; Martínez et al. 2006; Giambiagi & Martínez 2008). New U–Pb ages have been presented by Rocha Campos et al. (2011), which confirm a Guadalupian age for this section of the older Choiyoi volcanic rocks. The partly coeval age of this older Choiyoi unit with deposition of the Huentelauquén deposits, its location further east of the retrowedge or foreland basin, and its calc-alkaline signature indicate that it corresponds to a subduction related magmatic arc developed during closure of the retrowedge basin and uplift of the continental margin at the final stage of Gondwanan evolution (third stage of Charrier et al. 2007). The younger Choiyoi unit has, in turn, been assigned to the next pre-Andean tectonic cycle.

Gondwanan tectonic evolution

Based on the marked difference between the regions north and south of 33°S, it is difficult to reconcile the late Palaeozoic tectonic evolution of the two regions in one single coherent model, which is evidence that more information is necessary to solve this problem. However, some general conclusions can be drawn on the basis of the available information. The post-Devonian age of the units described from the western slope of the Andes suggests that these developed on the rear side of Chilenia. The recent chronologic data for the metamorphic complexes and plutonic belts north and south of 33°S indicate that:

(1) Maximum depositional ages for the deposits accreted by basal and frontal accretion in both regions are approximately the same; however, the peaks of high $P-T$ metamorphism in the paired belt, south of 33°S, are older than in the Choapa Metamorphic Complex, to the north of this latitude.
(2) Emplacement of the Coastal Batholith, south of 33°S, is older (Pennsylvanian) than the Elqui plutonic complex (Cisuralian), in the Frontal Cordillera, north of 33°S. This evidence indicates that the Choapa Complex and the paired metamorphic complex south of 33°S probably belong to two different accretionary complexes, and that the plutonic belts correspond to two different subduction-related magmatic arcs.
(3) The late stage of emplacement of the Elqui plutonic complex in late Cisuralian and Guadalupian times is coeval with sedimentation of the Huentelauqén Formation, in the retro-wedge basin, and the lower Choiyoi Group would thus correspond to the extrusive products of the magmatic arc.

Additionally, two features – the narrow width of the outcrops of the western series north of the NW-orientated Lanalhue lineament (c. 38°S) and the considerably wider outcrops of this series to the south of the lineament that interrupts the southward prolongation of the Coastal Batholith and apparently caused its bend towards the Principal Cordillera (Hervé et al. 2013), and the much younger depositional age of the western series to the south of the lineament containing abundant Permian detrital zircons (Hervé et al. 2013) – suggest that: (a) the accretionary complex was considerably wider than the present day outcrops; (b) its age is younger towards the SW; and (c) the probable orientation of the accretionary complex and the Gondwana coast was NNW-SSE.

Although there is no consensus on the tectonic regime during the Carboniferous, there is agreement that the San Rafael orogeny caused in Early to Middle Permian (late Cisuralian to early Guadalupian) generalized uplift of the region (Ramos 1988b; Mpodozis & Kay 1992; López Gamundi et al. 1994; Limarino et al. 2006, 2013; Giambiagi et al. 2011; Willner et al. 2008, 2012; García-Sansegundo et al. 2014a, among others). This tectonic event was responsible for deformation of the Arrayán Formation in the forearc (Charrier et al. 2007) and the cataclastic fabric in the El Volcán plutonic unit of the Elqui plutonic complex (Mpodozis & Kay 1990).

It has been proposed that the late Palaeozoic evolution of this sector of the Andes can be explained by a period of subduction in the Carboniferous times associated with extension after the Chanic compressive orogenic episode in the Middle to Late Devonian, with an active subduction-related magmatic arc. This arc expanded and migrated towards the foreland, reaching the Central Precordillera and the western side of the Sierras Pampeanas during Early Permian times, associated with the San Rafael orogenic deformation. Subsequent extension and formation of extensive rhyolitic plateaux were associated with delamination of the lower crust owing to injection of hot anhydrous asthenosphere. These processes were previously interpreted as produced by orogenic collapse and slab break-off by Mpodozis & Ramos (1989), Kay et al. (1989) and Mpodozis & Kay (1992). However, this evidence together with the new available U–Pb ages favour the interpretation advanced by Ramos & Folguera (2009), where these episodes can easily be explained by a period of slab shallowing followed by steepening of the subduction zone, as proposed by Martínez et al. (2006).

Pre-Andean tectonic cycle (Lopingian–late Early Jurassic)

The term Pre-Andean is used for a period of arrested or very slow subduction during which extensional tectonic conditions prevailed following deformation of the Late Permian deposits, closure of the retro-wedge or forearc basin, exhumation of the continental margin and intense erosion on the Palaeozoic units. Groeber (1922) presented the first description of these deposits along the continental margin. Extensional conditions on the thickened crust of the continental margin determined the reactivation of pre-existent weakness zones like the sutures of the Palaeozoic terranes accreted to western Gondwana (Ramos & Kay 1991; Ramos 1994). The weakness zone defined a different palaeogeographic organization compared with those that prevailed earlier and later, that is, in the Gondwanan and Andean tectonic cycles, consisting of NW-trending rift basins (Charrier 1979; Uliana & Biddle 1988; Suárez & Bell 1992) (Fig. 4). Additionally, extension favoured the development of a widely distributed felsic magmatism that resulted predominantly from intense crustal melting (upper Choiyoi magmatic province; Rapela & Kay 1988; Kay et al. 1989; Llambías 1999; Spalletti 1999). End of this cycle is marked by resumption or more intense subduction activity along the continental margin and development of the Early Jurassic magmatic arc. The pre-Andean cycle reflects the tectonic conditions determined by the assembly of Gondwana, but also the initial processes that later resulted in its breakup.

Generalized extension and the existence of weakness zones (sutures) on the continental margin resulted in the development of the following basins and successions of associated basins (Fig. 4): (a) El Quereo–Los Molles, next to the coast, at c. 32°S; (b) La Ramada, further east in the high Andes, (c) Cuyo, in the Precordillera and Andean foreland, in the San Juan–Mendoza region; (d) Bermejo, SW of La Rioja, in the northeastern San Juan province; and (e) Curepto–Bio Bío–Temuco, further south, which extends south-southeastwards from the coast up to the high Andes, at 40°S. Deposits in these basins are generally marine next

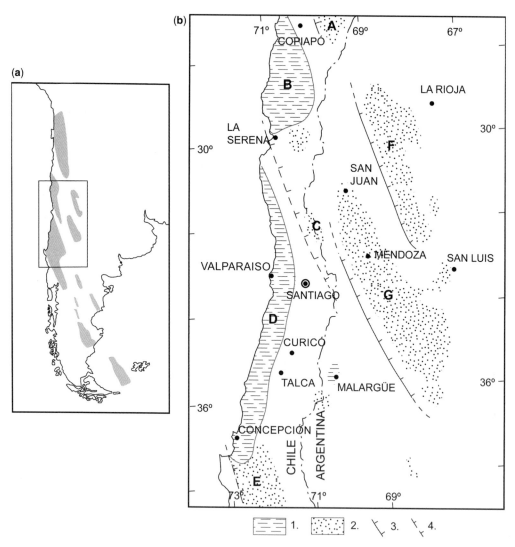

Fig. 4. Generalized palaeogeographic sketch for the Triassic in the Andean region of Argentina and Chile. (**a**) Southern part of South America showing in grey the approximate distribution of the basins, based on Uliana & Biddle (1988). Rectangle corresponds to the area represented in (b). (**b**). Distribution of the marine and continental deposits in the Triassic basins. 1, Marine deposits; 2, continental deposits; 3, observed basin bounding faults; 4, inferred faults. A, Profeta–La Ternera, basin; B, San Félix–Rivadavia basin; C, La Ramada basin; D, El Quereo–Los Molles basin; E, Curepto–Bio Bío–Temuco basin; F, Bermejo basin; and G, Cuyo basin; based on Charrier et al. (2007).

to the coast and continental towards their south-southeastern prolongation. The extensional faults that control the basins in Chile have not been clearly identified and the structural pattern is certainly more complicated than represented here. In Argentina, sedimentary polarity, seismic data and structural analyses indicate that some of the basins consist of hemigrabens (Milana & Alcober 1994; López Gamundi 1994).

Two rift stages have been detected in the evolution of the Pre-Andean cycle, each one consisting of a phase of tectonic subsidence followed by a phase of thermal subsidence (Charrier et al. 2007), a model that coincides with the one developed by Milana & Alcober (1994) and Milana (1998) for the Bermejo basin, in Argentina (Fig. 4). According to Charrier et al. (2007), in Chile, the first stage would have begun in Lopingian times (Late Permian) and ended by Ladinian to Carnian times (Middle Triassic to early Late Triassic), while the second one would have lasted from Late Triassic (post-Carnian) to Pliensbachian, in Early Jurassic times. Each phase of tectonic subsidence would have been accompanied by a pulse of felsic volcanism followed by marine or continental deposits, depending on the distance from the Triassic coast.

Sedimentary deposits of the first stage are known in the Coastal Cordillera north of 32°S in the El Quereo–Los Molles basin. These consist of a series of marine outcrops that can be grouped into the El Quereo Formation (Muñoz-Cristi 1942, 1973; Cecioni & Westermann 1968; Mundaca et al. 1979; Irwin et al. 1988; García 1991; Rivano & Sepúlveda 1991). These deposits reveal an Early? to Middle Triassic transgression–regression sedimentary cycle beginning with breccias and separated from the deposits of the second stage by the thick felsic volcanic and volcaniclastic Pichidangui Formation, associated with initiation of the second tectonic subsidence phase of the pre-Andean cycle.

Second-stage deposits have a latest Triassic to Early Jurassic age and therefore correspond to the late portion of the second stage, probably to the thermal subsidence phase occurring after the felsic volcanic event (Charrier et al. 2007). In the high Andes, at 31°S, Early Jurassic transgressive marine deposits of the Tres Cruces Formation are exposed overlying conformably the Late Triassic, continental, sedimentary and volcanic Las Breas Formation. This formation contains rests of Dicroïdium flora and was recently dated at 219.5 ± 1.7 Ma (Norian) (U–Pb SHRIMP zircon crystallization age) on a dacitic volcanic breccia from the base of the formation (Hervé et al. 2014). The Las Cruces Formation is in turn covered by backarc volcanic and volcaniclastic deposits of the Late Jurassic Algarrobal Formation (Dedios 1967; Letelier 1977; Mpodozis & Cornejo 1988; Pineda & Emparan 2006). At the coast, at 32°S, overlying the Pichidangui Formation, is the transgressive–regressive Los Molles Formation (Cecioni & Westermann 1968; Bell & Suárez 1995). The mostly marine deposits of the second stage, exposed along the Coastal Cordillera, south of 35°S, overlie, between 35 and 36°15'S, silicic volcanic deposits assigned to the bimodal volcanic Pichidangui Formation (Vicente 1974; Vergara et al. 1995) (La Totora–Pichidangui volcanic pulse) related to the upper Choiyoi magmatic province (Charrier et al. 2007). Further south of 36°15'S, they rest on Palaeozoic intrusive and metamorphic rocks (Fig. 3). The presence of a rather continuous series of marine deposits assigned to the second stage along the Coastal Cordillera, between 35 and 37°S (Curepto–Bio Bío–Temuco basin; see Charrier et al. 2007, Fig. 3.11, p. 39), suggests that the region covered by marine deposits during the thermal subsidence phase of the second stage was considerable and that the sea extended beyond the faults that controlled subsidence of the basin. We assign the marine Retian Malargüe rift deposits on the eastern versant of the Principal Cordillera, in Argentina, at c. 36°S (Riccardi & Iglesia Llanos 1999) (Fig. 4) to the earliest rift stages of the Jurassic backarc basin rather than to the second stage of the Pre-Andean cycle as previously proposed by Charrier et al. (2007). These deposits belong to one of the depocentres related to the Neuquén basin (Fig. 5).

Continental Triassic to Early Jurassic deposits were accumulated in the south-southeastward prolongation of the basins or in smaller isolated basins. These contain generally abundant volcanic and volcaniclastic deposits, like the Carnian–Norian Los Tilos sequence in the high Andes, at c. 30°S, somewhat north of the considered region (Martin et al. 1999), probably deposited in the prolongation of the San Félix basin (see Charrier et al. 2007, Fig. 3.11, p. 39). Other deposits apparently filled separated basins, like the La Ramada, in the high Andes, at 32°S (Álvarez 1996; Álvarez & Ramos 1999), the various depocentres of Cuyo basin located in the Frontal Cordillera, Precordillera and Andean foreland, between 31 and 36°S (Legarreta et al. 1992; Kokogián et al. 1993, 1999; Manceda & Figueroa 1995; Barredo et al. 2012), and the depocentres in the Sierras Pampeanas (Milana & Alcober 1994; Milana 1998), between 29 and 33°S (Fig. 4).

The best dated deposits are in the Cuyo basin in the Potrerillos depocentre, and were included in the Uspallata Group. These are separated from Devonian turbiditic deposits (Villavicencio Formation) and volcanic rocks of the Choiyoi Group by a normal fault and unconformably overlain by Miocene foreland deposits of the Mariño Formation.

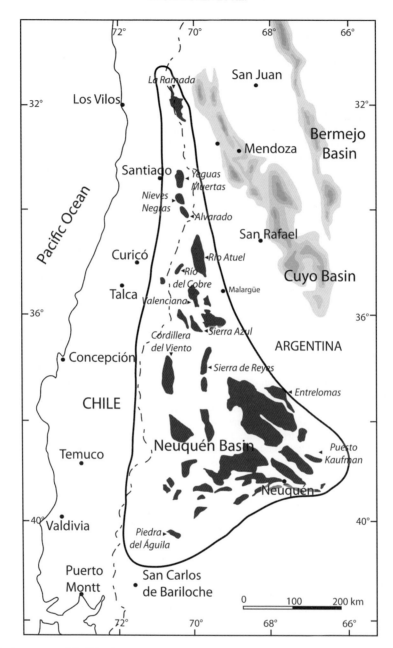

Fig. 5. Main depocentres of the Triassic–Early Jurassic rifts with the location of Alto del Tigre High (after Giambiagi et al. 2003b).

The Uspallata Group forms a 1385 m-thick continental sedimentary succession rich in fossil plants consisting of the following formations, from bottom to top (Spalletti et al. 1999, 2005, 2008): Río Mendoza (314 m; Anisian; 243 ± 4.7 Ma; Ávila et al. 2006), Cerro de las Cabras (190 m), Potrerillos (735 m; late Middle to early Late Triassic; 239.2 ± 4.5, 239.7 ± 2.2 and 230.3 ± 2.3 Ma), Cacheuta (44 m; early Late Triassic) and Río Blanco (102 m). The succession consists at the bottom and top of alluvial conglomerates and in the middle part of fluvial medium- to fine-grained deposits with tuff intercalations. The Río Mendoza, Cerro de las Cabras and Potrerillos formations correspond to the syn-rift phase of the basin, while the two upper units correspond to the sag phase. A rich fossil flora recovered in the Potrerillos and Cacheuta formations was used to propose a biozonation of the whole Triassic (Spalletti et al. 1999, 2005).

However, recent precise U–Pb dating in zircons from the Rincón Blanco, one of the northern depocentres of the Cuyo basin, has presented new ages from the base to the top of the sequence (Barredo et al. 2012). The SHRIMP age obtained for the base is 246.4 ± 1.1 Ma; the middle sequences yielded a SHRIMP age of 239.5 ± 1.9 Ma and Laser Ablation age of 238.0 ± 5.4 Ma; and the top, for both methods yielded 230.3 ± 1.5 and 230.3 ± 3.4 Ma (Barredo et al. 2012). These geochronological data, which are similar to those for the Potrerillos depocentre, indicate that almost the whole sedimentation is circumscribed to the Middle Triassic and the base of the Late Triassic, reaching neither the Norian nor the Rhaetian. The age differences observed between the Pre-Andean evolution in Chile and Argentina possibly indicate that extension began earlier in Chile than in Argentina. However, more precise chronological support is needed to make this decision.

Deposits in the Coastal Cordillera close to Concepción, in the southern part of the considered region, form the newly proposed Patagual–El Venado unit (Mardonez et al. 2012). These deposits overlay unconformably the eastern series and are unconformably covered in the Bio Bío valley by the marine, Carnian Santa Juana Formation (Nielsen 2005). This unit consists of a tightly folded and slightly metamorphosed (epizone close to the anchizone) alternation of pelitic and thick psamitic layers, which differentiates them from the eastern series and the overlying early Late Triassic deposits. Its loosely constrained age, which must be younger than the Early Permian, which is the age of the thermal metamorphism on the eastern series of the metamorphic complex, south of 33°S (between 302 and 294 Ma; Willner et al. 2005), and older than the early Late Triassic Santa Juana Formation, makes its age assignment difficult. According to these considerations, these deposits could correspond to (a) Permian forearc basin accumulations equivalent to the Huentelauquén Formation (third stage of the Gondwanan cycle) or (b) Latest Permian to Early Triassic deposits, which in this case would have accumulated during the first stage of the Pre-Andean cycle and, therefore, represent evidence for rifting in the first stage of the Pre-Andean cycle in this region.

In the Elqui plutonic complex, in the Frontal Cordillera north of 33°S, the plutonic and subvolcanic units included in the Late Lopingian to Middle Triassic age group (264–242 Ma) of Hervé et al. (2014) (Ingaguás superunit of Mpodozis & Kay 1990, 1992) represent the intrusive equivalents of the felsic upper Choiyoi Group (Nasi et al. 1985; Mpodozis & Kay 1990) and possibly of the Mathuaico Formation. These plutons, along with the Laguna gabbro, form an epizonal association of intrusive rocks derived from deep, garnet-bearing levels in a thickened crust, and hypersilicic, calc-alkaline to transitional A-type granites (Nasi et al. 1985, 1990; Mpodozis & Cornejo 1988; Mpodozis & Kay 1990, 1992). These units consist predominantly of biotite–hornblende granodiorites and monzogranites, and syenogranites; graphic granites are known from the youngest El Colorado units. The given age range comprises a 22 Ma time lapse that goes from the Capitanian (late Middle Permian) to the Anisian/Ladinian boundary (Middle Triassic), and therefore coincides with the first stage of the Pre-Andean cycle.

Norian and somewhat younger Triassic ages have been also obtained along the coast: (a) in the northern part of the considered region in the A-type to transitional Altos de Talinay Plutonic Complex (Gana 1991; Emparan & Pineda 2006; Emparan & Calderón 2014), and further south in the Tranquilla and Millahue units (Parada et al. 1988, 1991, 1999, 2007; Rivano & Sepúlveda 1991); (b) in the Norian 'Dioritas Gnéisicas de Cartagena' (Gana & Tosdal 1996), next to San Antonio, at 33°30′S, and (c) in the Norian fayalite, anorogenic A-type Cobquecura pluton (Vásquez & Franz 2008). All these widely distributed plutonic units demonstrate the great regional extension of the Choiyoi Magmatic Province.

Andean tectonic cycle (late Early Jurassic–present)

The beginning of this long-lasting cycle was determined in Chile by the initiation of subduction-related volcanism in the late Early Jurassic (Pliensbachian) (Charrier et al. 2007, and references therein). However, on the eastern versant of the Andes in central Argentina, a post-Choiyoi extensional event occurred in the Late Triassic that formed new depocentres in the region where later the backarc basin developed during Jurassic times. This extensional event and the initiation of sedimentation in the backarc region are considered in Argentina to mark the beginning of the Andean cycle. Thus, the age of the beginning of this cycle remains a question that needs further investigation: (a) do the Late Triassic deposits in Argentina belong to the second stage of the Pre-Andean cycle; or (b) did the new geodynamic conditions at this moment cause extension of the continental margin already in the Late Triassic, whereas subduction magmas reached the surface of the crust somewhat later in Pliensbachian time?

The Pre-Andean cycle, which began with renewal or intensification of subduction underneath the central Argentine–Chilean continental margin, reflects evolution of the continental margin during continental breakup and continental drift (Fig. 1). Subduction created conditions for arc magmatism, active almost uninterruptedly right through to the present day, and extensional tectonic conditions along the continental margin. During early evolution of this cycle (Pliensbachian to late Early or early Late Cretaceous), in northern and central Chile, the arc was located along the present day Coastal Cordillera, parallel to the western margin of Gondwana with a backarc basin on its eastern side. In contrast, the later evolution (Late Cretaceous and Cenozoic) is characterized by gradual eastward shift of the magmatic arc and by the development of retroarc foreland basins on the eastern side of the arc. These two major periods correspond to the Early and Late periods, respectively, described by Coira et al. (1982) for this tectonic cycle. However, each of these periods can be subdivided into shorter stages, which can be differentiated from each other by major palaeogeographic changes (Fig. 1). These changes are a consequence of major modifications of the convergence and subduction pattern in this region.

Early period (Pliensbachian–late Early Cretaceous)

The early period of the Andean tectonic cycle ended in late Early or early Late Cretaceous with the Peruvian tectonic phase (Fig. 1) that caused a considerable crustal thickening, major palaeogeographic reorganization and modification of the tectonic regime in this Andean region (Coira et al. 1982; Mpodozis & Ramos 1989). Evolution of this early period in central Chile and Argentina is characterized by a dominating extensional tectonic regime, a rather thin crust, the development of a magmatic arc slightly oblique to the present day Pacific coast and a backarc basin on its eastern side (Mpodozis & Ramos 1989, 2008; Ramos 2010). This orientation of the arc–backarc basin palaeogeographic pair suggests that the late Proterozoic and Palaeozoic sutures and other major structures along the western margin of Gondwana still exerted a control on the tectonic evolution of this region. The tectonic evolution was further controlled by a rather loose plate coupling (negative trench roll-back) caused by the subduction of a considerably old and dense oceanic plate that had remained practically quiet during the pre-Andean cycle.

The two stages in which this Early period has been subdivided are reflected in the late Early to Late Jurassic and latest Jurassic to late Early Cretaceous by two magmatic episodes along the arc and two transgression–regression cycles in the backarc. This separation is clear for the backarc basin deposits, although it is less evident for the arc deposits and plutonic units. However, the plutonic units of the second stage are slightly shifted to the east of the plutonic units of the first stage (Sernageomin 2002). In the arc region, the beginning of this cycle is defined by the first appearance of subduction related lavas covering and interrupting sedimentation of the westernmost marine deposits of the second stage of the pre-Andean cycle (i.e. Pan de Azúcar, Profeta and Los Molles formations; see Charrier et al. 2007), whereas in the eastern flank of the Andes it seems to have begun before. In fact, structuration of the backarc basins of the Early Period began there in the latest Triassic–earliest Jurassic (Pre-Cuyano sedimentary cycle), suggesting that the tectonic conditions (tectonic subsidence) had changed before the subduction-related magmas could reach the surface. The first stage (late Sinemurian–Pliensbachian to Kimmeridgian) is characterized by intense activity in the arc and development of a transgressive–regressive marine cycle in the backarc basin. At the end of this stage a second phase of tectonic subsidence followed by thermal subsidence began. This second stage (Kimmeridgian to Aptian–Albian) is characterized by apparently less activity in the arc, and by a second transgression–regression marine cycle in the backarc basin.

Palaeogeography during this Early Period was apparently controlled by NNW-orientated structures, somehow like in the Pre-Andean cycle. This is demonstrated by the reduction in the width of the presently remaining arc towards the north, and its almost complete absence in the Arica region, in northernmost Chile. This plus the eastward shift of the magmatic arc in the second stage, allows identification in the second stage and south of c. 30°S (latitude of La Serena) of another depocentre, the Lo Prado basin, located west of the arc (i.e. in the Coastal Cordillera) and therefore in a forearc position (Charrier 1984; Charrier et al. 2007). Additionally, in this region, the backarc basin, which is known as the Mendoza–Neuquén basin, gradually bends southeastwards, and becomes considerably wider than further north. This basin, which extends eastwards into Argentina, represents the southernmost part of the Jurassic–Early Cretaceous backarc basin, which is traceable without interruption along the eastern side of the magmatic arc from at least southern Ecuador to southern Argentina, at c. 40°S, and, possibly, still further south (Vicente 2005).

First stage (Pliensbachian–Kimmeridgian)

Magmatic arc. Along the coast a wide swath consisting mostly of Middle–Late Jurassic and Early Cretaceous plutons and associated volcanic units represents the arc activity (Fig. 3). In the northern part of the study region, Jurassic plutonic units can be continuously followed up to 34°S (Fig. 3). Further south, exposures are patchy, up to 38°S, where they begin to occur further east, in the high Andes, following the bend formed by the Palaeozoic units (south of the region included in Fig. 3; see Sernageomin 2002, for more detail). North of 33°S, they consist of monzodiorites and granodiorites of Late Jurassic age assigned to the Puerto Oscuro and Cavilolén units, with Sr initial ratios of 0.7034 and 0.7035 (Parada *et al.* 1988; Rivano & Sepúlveda 1991). These initial ratios, being lower than those obtained in the Late Triassic Tranquilla and Millahue units (0.7063 and 0.7050), indicate a different magma source directly derived from the upper mantle and associated with the recently renewed subduction activity. South of 33°S, they consist of several units (Laguna Verde, El Sauce, Peñuelas, Limache and Lliu-Lliu) comprising I-type, calc-alkaline diorites, tonalities, granodiorites and granites (Gana & Tosdal 1996). According to these authors, intrusion of these units occurred in only 6 myr, between 162 and 156 Ma (Oxfordian to Kimmeridgian times), implying a very rapid pulse of ascent of large amounts of magma. This magmatic pulse can be related to the event of rift-associated subsidence that facilitated extrusion of magmas in the backarc at this time (see Oliveros *et al.* 2012; Rossel *et al.* 2013) and permitted the marine ingression in the backarc basin.

Volcanic deposits associated with the Jurassic arc are well exposed in the Coastal Cordillera south of 31°S (Fig. 3). These are the Middle–Late Jurassic Ajial Formation (Thomas 1958; Piracés 1977; Vergara *et al.* 1995) and the Late Jurassic Horqueta Formation (Piracés 1977), separated from each other by the Middle Jurassic marine Cerro Calera Formation (Piracés 1976; Nasi & Thiele 1982) that represents a westward advance of the backarc deposits into the arc domain. Further south, at 35°S, the Jurassic arc volcanic activity is represented by the Middle Jurassic Altos de Hualmapu Formation (Morel 1981). The mentioned arc deposits conformably overlie Sinemurian marine deposits assigned to the late stage of the pre-Andean cycle in the Curepto region (Thiele 1965). The Horqueta volcanic activity coincides with the rapid plutonic pulse detected by Gana & Tosdal (1996) in the Coastal Cordillera west of Santiago (see above). The Jurassic arc formed a rather low relief, which probably indicates high rates of subsidence (Oliveros *et al.* 2007; Charrier *et al.* 2007).

Backarc deposits. The first transgression–regression cycle (*first stage*) in the backarc basin in this region is represented on the western side of the cordillera by the following marine deposits, from north to south: (a) lower member of the Lagunilla Formation (Aguirre 1960); (b) Río Colina Formation (Thiele 1980); (c) Nieves Negras Formation (Álvarez *et al.* 1997; Charrier *et al.* 2002) formerly Leñas–Espinoza Formation of Klohn (1960) and Charrier (1982); (d) Nacientes del Teno Formation, at 35°S (Klohn 1960) and 36°S (Muñoz & Niemeyer 1984); (e) Valle Grande Formation (González & Vergara 1962), at 35°30′S; and (f) Nacientes del Bio Bío Formation (De la Cruz & Suárez 1997; Suárez & Emparan 1997), at 38°30′S. The base of these formations is not exposed, except for the Nacientes del Teno, which unconformably overlies rhyolitic rocks of possible Triassic age at 35°S (Davidson 1971; Davidson & Vicente 1973), and of confirmed Triassic age (Cajón de Troncoso Beds), between 36 and 37°S (Muñoz & Niemeyer 1984). These formations consist of thick successions of sandstones (some of them turbiditic), marls and limestones, and represent a transgression–regression cycle that ends with thick Oxfordian evaporitic deposits, generally named '*Yeso Principal*' (Schiller 1912) or more formally Auquilco Formation (Groeber 1946), in Argentina, and the Santa Elena Member of the Nacientes del Teno Formation in Chile (Klohn 1960; Davidson 1971; Davidson & Vicente 1973). This gypsum unit, which is the middle member of the Lagunilla Formation, at 33°S (Aguirre 1960), is overlain by the upper member of the Lagunilla Formation (Aguirre 1960) and its southern equivalent, the Río Damas Formation (Klohn 1960), which consists of breccias and alluvial fan deposits that grade towards the east into the red, finer-grained and thinner fluvial sandstones of the Tordillo Formation (Klohn 1960; Arcos 1987). At its type locality (Río de las Damas, next to Termas del Flaco, at 35°S), the Río Damas Formation consists of a c. 3000 m-thick red continental, detrital succession, with coarse and fine intercalations, which includes at the top a member comprising >1000 m of andesitic lavas culminating in breccias containing enormous angular blocks, some over 4 m in diameter (Arcos 1987). Close to its contact with the Baños del Flaco Formation, dinosaur tracks are well exposed (Casamiquela & Fasola 1968; Moreno & Pino 2002; Moreno & Benton 2005). Thus, the Río Damas Formation and northern equivalents represent the final deposits of the first transgression–regression cycle in the backarc basin. Because of the thick backarc volcanic intercalation at the upper part of this formation (Rossel *et al.* 2014), which is overlain by coarse breccias, it has been considered to also represent the initial deposits of the Second stage of the Early Period, associated with the tectonic subsidence phase (Charrier et al. 2007). These deposits are conformably overlain by Late Jurassic to Early Cretaceous marine sediments that form the bulk of the second transgression–regression cycle of the second stage.

Along the eastern slope of the Principal Cordillera the depocentres of the first transgression are represented by a series of alternate sub-basins. The Ramada depocentre in the northern section (31°30′–32°30′S) has a complete section of continental and volcanic deposits of the Triassic Rancho de Lata Formation, unconformably overlain by the Los Patillos Formation of Early Jurasssic age. This last unit has a rich fauna of ammonites studied by Álvarez (1996) that includes from the Aalenian until the Callovian. The Alto del Tigre High is located in the central segment at the latitude of the Mount Aconcagua (32°30′–33°30′S) (Fig. 5). It is a positive area where there is no deposition of the Triassic–Early Jurassic deposits known since the early work of Groeber (1918).

Further south, the Yeguas Muertas, Nieves Negras, Alvarado, Río del Cobre, Río Atuel–La Valenciana, Palauco, Sierra Azul, Sierra de Reyes and Cordillera del Viento are some of the depocentres of the Nequén basin located south of 33°S (Fig. 5). In particular, the Río Atuel–La Valenciana is an important subsidence zone west of Malargüe (34°–35°30′S) with a thick sequence of Triassic rift deposits that continue with the different units of the Cuyo Group represented by Arroyo Malo, El Freno, Puesto Araya and Tres Esquina Formations (Fig. 6) (Legarreta et al. 1993). The marine sequence dated by ammonites goes from Rhaetian to Toarcian in a complex fluvio-deltaic array (Lanés 2005).

It is interesting to remark that the Cuyo Group is overlain by the Callovian transgression of the La Manga Formation of the Lotena Group (Fig. 6). This unit includes a thin sequence of limestones between 50 and 100 m thick that covers all the depocentres from La Ramada towards the south, including the Alto del Tigre High (Giambiagi et al. 2003a, b). This carbonate platform is followed by a generalized regression that ended with the Auquilco Formation, a thick gypsum deposit up to 200 m thick widely distributed in the Principal Cordillera (Legarreta et al. 1993).

Second stage (Kimmeridgian–Albian)

From 30°S southwards the distribution of the marine deposits accumulated during the second stage forms two clearly separated depositional areas: one in the Coastal Cordillera, and the other in the Principal Cordillera, and mostly on its eastern side (Charrier 1984; Mpodozis & Ramos 1989; Charrier & Muñoz 1994; Charrier et al. 2007). The two basins are separated from each other by a volcanic domain that we propose to name the Lo Prado volcanic arc. Therefore, it is possible to identify three palaeogeographic domains at this moment, from west to east: (a) the Lo Prado forearc basin, bounded to the west by a relief formed on older units; (b) the Lo Prado volcanic arc; and (c) the Mendoza–Neuquén backarc basin, the latter including volcanic activity.

Forearc (Lo Prado forearc basin). Immediately north of the here considered region, in the Elqui river valley transect, at 30°S, on top of the Jurassic arc deposits, the forearc or rather the transitional deposits between the arc and a marine basin to the west are represented by the >4000 m-thick, volcanic, principally basaltic andesites and marine sedimentary Arqueros Formation (Berriasian–Albian) and the >2000 m-thick, continental, mostly red and volcanic Quebrada Marquesa Formation (Hauterivian–early Albian) (Aguirre & Egert 1965, 1970; Thomas 1967; Emparan & Pineda 2006; Rivano & Sepúlveda 1991). The Quebrada Marquesa Formation interfingers with and covers gradually the Arqueros Formation; in its lower portion it contains abundant marine intercalations, some of which are fossiliferous, and consists in its upper portion of a thick succession of pyroxene–olivine bearing basaltic andesites and pyroxene–amphibole bearing andesites (Aguirre & Egert 1965; Emparan & Pineda 1999, 2006; Emparan & Calderón 2014). In this region, the Early Cretaceous backarc deposits (Río Tascadero Formation) are exposed over 60 km to the east in the high cordillera (Mpodozis & Cornejo 1988), where, like the Arqueros Formation, they interfinger with and grade upwards to red continental, volcanic and volcaniclastic facies (Pucalume Formation). In our palaeogeographic reconstruction the Marquesa Formation represents the late Early Cretaceous western facies associated with the magmatic arc, and the Pucalume Formation represents either the easternmost deposits of the magmatic arc or volcanic deposits that resulted from volcanic activity in the backarc.

Southern deposits with similar facies and the same stratigraphic position that the Arqueros Formation are the Lo Prado (Berriasian to Valanginian) and the La Lajuela (Early Cretaceous) formations. Both formations are constrained to the Coastal Cordillera south of c. 34°S and correspond to the oldest forearc basin deposits, which contain abundant lavas derived from the volcanic arc flanking the basin to the east. These units overlie, from north to south, the Late Jurassic arc-related Agua Salada Volcanic Complex (Emparan & Pineda 2006; Emparan & Calderón 2014), the Horqueta (Thomas 1958; Piracés 1977; Nasi & Thiele 1982;

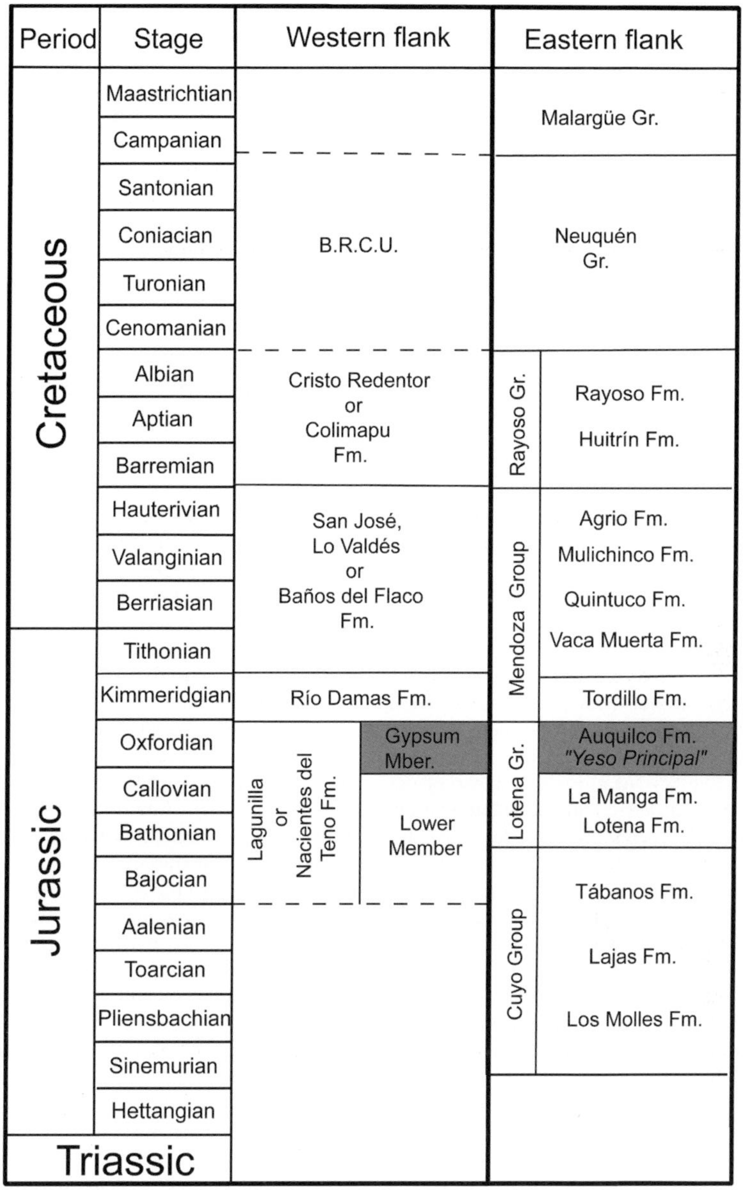

Period	Stage	Western flank	Eastern flank
Cretaceous	Maastrichtian	B.R.C.U.	Malargüe Gr.
Cretaceous	Campanian	B.R.C.U.	Malargüe Gr.
Cretaceous	Santonian	B.R.C.U.	Neuquén Gr.
Cretaceous	Coniacian	B.R.C.U.	Neuquén Gr.
Cretaceous	Turonian	B.R.C.U.	Neuquén Gr.
Cretaceous	Cenomanian	B.R.C.U.	Neuquén Gr.
Cretaceous	Albian	Cristo Redentor or Colimapu Fm.	Rayoso Gr. — Rayoso Fm.
Cretaceous	Aptian	Cristo Redentor or Colimapu Fm.	Rayoso Gr. — Huitrín Fm.
Cretaceous	Barremian	Cristo Redentor or Colimapu Fm.	Rayoso Gr. — Huitrín Fm.
Cretaceous	Hauterivian	San José, Lo Valdés or Baños del Flaco Fm.	Mendoza Group — Agrio Fm.
Cretaceous	Valanginian	San José, Lo Valdés or Baños del Flaco Fm.	Mendoza Group — Mulichinco Fm.
Cretaceous	Berriasian	San José, Lo Valdés or Baños del Flaco Fm.	Mendoza Group — Quintuco Fm.
Jurassic	Tithonian		Mendoza Group — Vaca Muerta Fm.
Jurassic	Kimmeridgian	Río Damas Fm.	Tordillo Fm.
Jurassic	Oxfordian	Lagunilla or Nacientes del Teno Fm. — Gypsum Mber.	Lotena Gr. — Auquilco Fm. "Yeso Principal"
Jurassic	Callovian	Lagunilla or Nacientes del Teno Fm. — Lower Member	Lotena Gr. — La Manga Fm.
Jurassic	Bathonian	Lagunilla or Nacientes del Teno Fm. — Lower Member	Lotena Gr. — Lotena Fm.
Jurassic	Bajocian		Cuyo Group — Tábanos Fm.
Jurassic	Aalenian		Cuyo Group — Lajas Fm.
Jurassic	Toarcian		Cuyo Group — Lajas Fm.
Jurassic	Pliensbachian		Cuyo Group — Los Molles Fm.
Jurassic	Sinemurian		
Jurassic	Hettangian		
Triassic			

Fig. 6. Stratigraphic succession of the Jurassic to Cretaceous deposits in the Principal Cordillera in central Chile and Argentina, between 32 and 37°S. Based on Aguirre (1960), Klohn (1960), González & Vergara (1962), González (1963), Davidson (1971), Davidson & Vicente (1973), Thiele (1980), Charrier (1981a) and Charrier et al. (2002).

Rivano & Sepúlveda 1991; Rivano 1996; Vergara et al. 1995) and Altos de Hualmapu formations (Bravo 2001). The Arqueros, Lo Prado and La Lajuela formations are conformably overlain by thick continental volcanic and volcaniclastic deposits of the Quebrada Marquesa (Rivano & Sepúlveda 1991; Emparan & Calderón 2014), Veta Negra Formation (Piracés 1977; Nasi & Thiele 1982; Vergara et al. 1995) and El Culenar Beds (Bravo 2001), respectively. These extremely thick successions form a narrow band of almost continuous outcrops up to almost 36°S (Fig. 3). The stratigraphic position of the Veta Negra Formation and the age of the oldest granitoids that intrude this unit bracket its age to the Berriasian–Albian (Vergara et al. 1995). Notwithstanding the stratigraphic position of the Veta Negra Formation, radioisotopic age determinations in this unit indicate that it is slightly older (c. 119 Ma; Aguirre et al. 1999; Fuentes et al. 2001, 2005) than the older lavas of the northern equivalent Arqueros Formation (117–115 Ma) (Morata & Aguirre 2003), suggesting a northward progression of volcanism and tectonic extension (Morata & Aguirre 2003; Morata et al. 2008).

The rather deep depositional conditions detected for the lower Lo Prado marine deposits and the several thousand metres-thick pile encompassed by the Lo Prado and Veta Negra formations indicate intense subsidence in the forearc basin (Vergara et al. 1995). Additionally, the rather primitive geochemical composition of the Lo Prado rocks with low MgO and high K content, that fall in the classification of high-K and shoshonitic porphyric basaltic andesites and andesites, and their low initial Sr ratios (c. 0.7036) are indicative of intense crustal extension in the basin during Early Cretaceous time (Morata & Aguirre 2003; Parada et al. 2005). Similarly, the flood-basalt-type flows of the overlying Veta Negra lavas (Åberg et al. 1984), the presence in this formation of approximately north–south-orientated dyke swarms suggestive of fissural eruptions parallel to the strike of the basin (Vergara et al. 1995), and its geochemical features (low La/Yb ratios, low $^{87}Sr/^{86}Sr$ ratio of 0.70374 and more primitive Sr–Nd ratios than those of the Jurassic lavas), indicating an attenuated crust (Vergara et al. 1995; Morata & Aguirre 2003), are additional evidence for extensional tectonic conditions during Early Cretaceous times in the Lo Prado forearc basin. This basin corresponds to the aborted marginal basin of Åberg et al. (1984) and the intra-arc basin of Charrier (1984).

On the eastern flank of the Coastal Cordillera, the 3000 m-thick Las Chilcas Formation overlies with an apparently conformable contact the Veta Negra Formation (Wall et al. 1999). The lower volcanic portion of this formation has been dated (U–Pb) at 109.6 ± 0.1 and 106.5 ± 0.2 Ma (Wall et al. 1999), while lavas from its upper portion yielded K–Ar ages in plagioclase of 95 ± 3 Ma (Gallego 1994). Based on these ages, we suggest that the Las Chilcas Formation formed during the transition from the Early to the Late period of Andean evolution. The lower portion, consisting of basaltic and andesitic lavas and dacitic and rhyolithic pyroclastics, developed in apparent conformity with the underlying Veta Negra Formation, would be the continuation of the volcanic activity developed in the Lo Prado forearc basin, whereas the coarse conglomerates of its upper part would correspond to syn-orogenic deposits associated with the Peruvian orogeny. We discuss this point below.

Lo Prado magmatic arc. Early Cretaceous plutonic rocks form an almost continuous swath of hypabissal dioritic to granitic plutons, which mostly intrude the Early Cretaceous deposits mentioned above (Rivano & Sepúlveda 1991; Gana & Tosdal 1996; Rivano et al. 1993; Emparan & Calderón 2014). K–Ar age determinations on these plutons yielded ages 134–86 Ma (Parada et al. 1988; Bravo 2001). Recent $^{40}Ar/^{39}Ar$ dates in the northward prolongation of this swath, immediately north of the here considered region, yielded ages between 139 and 100 Ma (Valanginian–Albian) (Emparan & Calderón 2014). Low Sr initial ratios obtained on Early Cretaceous granitoids in the northern part of the region (31°–32°S) indicate an upper mantle origin, with virtually no continental crust involvement for these magmas (Parada et al. 1988; Creixell et al. 2011). This conclusion confirms the idea that the crust was thin and the tectonic conditions were extensional during development of the forearc basin, as well as in the Principal Cordillera, where thermal subsidence dominated the retroarc basins.

Arc volcanic rocks are also exposed further east and only in the northern part of the region (north of 32°S). These have been assigned to the Pucalume Formation, unless they correspond to backarc volcanic activity (Charrier et al. 2007). In the Aconcagua area the Late Jurassic–Early Cretaceous successions are interbedded with backarc basalts and pyroclastic deposits (Cristallini & Ramos 1996).

Backarc basin (Mendoza–Neuquén basin). Backarc basin deposits in this region are exposed in Chile in the High Andes close to the international boundary and extend eastwards into western Argentina (Fig. 3). These generally consist mostly of sedimentary marine facies; however, at some localities they interfinger with volcanic deposits that, depending on their location, are interpreted as the easternmost arc deposits or to products of backarc volcanic activity.

In the high cordillera, in northern part of the region, between 31 and 32°S, volcanic deposits of the Late Jurassic Algarrobal Formation are exposed. These deposits interfinger and are covered by the conglomeratic Mostazal Formation (Mpodozis & Cornejo 1988; Oliveros et al. 2012), which is a western lateral facies of the Kimmeridgian, finer-grained, continental Tordillo Formation exposed next to the international border, between 31 and 33°S (Rivano & Sepúlveda 1991; Rivano et al. 1993; Lo Forte et al. 1996; Aguirre-Urreta & Lo Forte 1996). The great distance (>80 km) that separates these volcanic deposits from those of the arc in this region suggests that they correspond to backarc volcanism, as has been observed in some regions along the central Argentina–Chilean Andes (Charrier et al. 2007; Mescua et al. 2008;

Mpodozis & Ramos 2008; Oliveros *et al.* 2012, among others).

Second-stage backarc marine sedimentary deposits in the northern part of the region consist of the Early Crertaceous Río Tascadero Formation (Mpodozis & Cornejo 1988; Rivano & Sepúlveda 1991). This unit forms a NNW-orientated swath that extends south-southeastwards into Argentina. Southwards, on the western flank of the cordillera, at 31°30′S, there is another similarly orientated swath, extending also into Argentina and consisting predominantly of volcanic and volcaniclastic deposits with mostly coarse detritic intercalations (breccias, conglomerates, and sandstones), and minor and finer marine fossiliferous calcareous sandstones, assigned to the Los Pelambres Formation (Rivano & Sepúlveda 1991).

Similar deposits exposed along the international border and next to it on the eastern flank of the cordillera at 33°S have been assigned to the Juncal Formation (Ramos *et al.* 1990; Aguirre-Urreta & Lo Forte 1996; Cristallini & Ramos 1996). Further east, the volcanic intercalations rapidly disappear eastwards, indicating that the source of the lavas and coarse volcaniclastic sediments was located to the west. Their source can either be the volcanic Lo Prado arc or volcanic edifices developed in the backarc basin to the east of the main magmatic arc. Because of the considerable distance separating the volcanic arc in the Coastal Cordillera and the volcanic deposits exposed along the axis of the Principal Cordillera, we favour the existence of volcanic activity in the backarc at this time coexisting with marine sedimentation. If this interpretation is correct it would emphasize the importance of volcanism in the backarc during the early period of the Andean Cycle (Ramos 1999; Charrier *et al.* 2007; Oliveros *et al.* 2012; Rossel *et al.* 2013).

Further south, between 33 and 36°15′S, along the western side of the Principal Cordillera (Fig. 3), but with a more external position than the previously mentioned units, Late Jurassic (Kimmeridgian) to late Early Cretaceous (Aptian–Albian), red continental and marine backarc deposits have been assigned to the upper member of the Lagunillas (Aguirre 1960) and Río Damas (Klohn 1960) formations, and the overlying marine San José (Aguirre 1960), Lo Valdés (González 1963; Hallam *et al.* 1986) and Baños del Flaco (Klohn 1960; González & Vergara 1962; Covacevich *et al.* 1976; Charrier 1981*a*; Arcos 1987) formations, and the red, detritic, continental Colimapu Formation (Klohn 1960) (see Fig. 6). These form a continuous, although considerably thrusted and folded, swath of outcrops that extends further east into Argentina. The marine rocks consist of thick, richly fossiliferous successions of predominantly calcareous neritic to shallow (external platform) sediments. Lavas in the Lo Valdés Formation form a thin intercalation in its lower portion (Biró-Bagóczky 1964), and volcanic intercalations in the Baños del Flaco Formation, a few kilometres to the south of the previous locality (*c.* 34°S), form a 440 m-thick succession of volcanic breccias and silicic lavas between marine fossiliferous sediments of Tithonian–Neocomian age (Charrier 1981*b*). In these formations the volcanic intercalations and volcanic components in the detritic deposits rapidly disappear eastwards, indicating again that the source of the lavas and coarse volcaniclastic sediments was located towards the west. At Termas del Flaco, in the Tinguiririca river valley, at 35°S, only the lowest portion of the Baños del Flaco is exposed. Its upper portion, and probably also the overlying Colimapu Formation, have been eroded in this place (Charrier *et al.* 1996). The final regressive episode led to the deposition of a second, generally thin band of gypsum ('Yeso Secundario' or 'Yeso Barremiano') at the base of the 1500 m-thick, red detrital Colimapu Formation (Klohn 1960; González & Vergara 1962; González 1963; Charrier 1981*b*), which corresponds to the generally fine-grained continental deposits with thin calcareous intercalations containing ostracodes that followed the regression in Aptian–Albian times. In the Maule river valley (36°S), a recent dating with detritic zircons yielded an Aptian maximum age for deposition of these deposits (Astaburuaga 2014). This formation is a lateral equivalent of the Huitrín–Rayoso Formation in western Argentina (Fig. 6).

At the end of the first Andean stage, a major plate reorganization associated with a great increase in generation of oceanic crust in the proto-Pacific (Larson 1991) and rapid westward drift of South America modified the tectonic conditions in the continental margin of South America. This geodynamic event, known as the Peruvian orogeny (Steinmann 1929; see also Groeber 1951; Charrier & Vicente 1972; Vicente *et al.* 1973; Ramos 1988*b*, 2010; Reutter 2001; Tunik *et al.* 2010), caused along western South America uplift of the continental margin, the marine regression referred above and definite emersion in the backarc basin, compressive deformation of the existing units, and crustal thickening. As a result of this phase, the first Andean mountain range was formed.

Late period (early Late Cretaceous: Present)

The Peruvian orogeny separates the Early and Late periods into which Coira *et al.* (1982) subdivided the evolution of the Andean tectonic cycle. After this episode the palaeogeographic organization in this region of the Andes changed completely: the backarc basin was inverted, the magmatic arc shifted considerably eastwards, a new mountain range was developed, a continental retroarc foreland basin was formed to the east of the arc instead of a backarc basin, and a rather wide forearc region west of the arc was produced as a result of eastward arc migration. Oblique subduction also prevailed at this time, although the movement of the oceanic plates towards the continent was mostly northeastward orientated, producing dextral displacement along trench-parallel faults. Moreover, some authors suggest from plan view reconstructions an orthogonal subduction for this moment, thus, dextral displacement could be less important (Arriagada et al. 2008; Martinod et al. 2010). The Late Period has been subdivided into two stages separated from each other by a major orogenic phase that occurred in middle Eocene, the Incaic orogeny (Fig. 1). Each one of these stages can be, in turn, subdivided into two substages by orogenic episodes that occurred at approximately the Cretaceous–Cenozoic boundary ('K–T' orogeny; Cornejo et al. 2003) and the Oligocene–Miocene boundary (Pehuenche orogeny; Fig. 1), respectively.

First stage (early Late Cretaceous–middle Eocene)

During the first stage, in the region between 31 and 37°S, a high sea-level stand in latest Cretaceous–earliest Cenozoic times caused a slight marine ingression along the western border of the present day Coastal Cordillera, and an extended marine incursion of Atlantic origin, on the eastern side of the mountain range that reached the axis of the present-day Principal Cordillera. This marine ingression from the Atlantic side was favoured by the tectonic loading and subsequent subsidence that developed a long foredeep along the eastern foothills of the Andes (Aguirre-Urreta et al. 2011).

For a better understanding of the following description of geological units, we will describe them separately in two different segments: a northern (31°–34°S) and a southern segment (34°–37°S).

Northern segment. The arc is represented by two parallel, close to each other, series of small and medium-sized plutonic outcrops located along the eastern flank of the Coastal Cordillera, immediately to the east of the previous arc representatives (Fig. 3). These outcrops can be followed up to 34°S, where they disappear or have not been identified. They consist of monzodiorites and subordinated granodiorites, gabbros, diorites and hypabissal andesitic and dioritic bodies (Rivano & Sepúlveda 1991; Rivano et al. 1993; Sellés & Gana 2001), and have been included by these authors in the Cogotí superunit, and more recently in the Illapel Plutonic Complex by Morata et al. (2010) and Ferrando et al. (2014).

Stratified deposits corresponding to this stage are represented, between 31 and 33°S, by the following Late Cretaceous volcanic, volcaniclastic and sedimentary stratigraphic units: (a) upper part of the Las Chilcas; (b) the Viñita and its equivalent the Salamanca; and (c) the Lo Valle formations, with ages ranging from 95.3 to 64.6 ± 5 Ma (Drake et al. 1976; Rivano & Sepúlveda 1991; Rivano et al. 1993; Gallego 1994; Mpodozis et al. 2009; Jara & Charrier 2014). The coarse and thick conglomeratic deposits of the upper Las Chilcas Formation, with a marine calcareous intercalation and some basaltic and andesitic–basaltic lavas at the top, would correspond to syn-orogenic deposits accumulated in a retroarc foreland basin developed with the Peruvian orogeny, which was deep enough to be invaded by the sea, and volcanic-arc deposits in its eastern and western border, respectively.

In fact, in this region, apatite fission track ages indicate for the western Coastal Cordillera the existence of a cooling event that began at 106–98 Ma (Gana & Zentilli 2000). This age is complemented by studies in the Caleu pluton on the eastern Coastal Cordillera indicating that crystallization occurred in the interval 94.2–97.3 Ma and that cooling occurred until about 90 Ma (Parada et al. 2005; Ferrando et al. 2014). These data have been confirmed by Willner et al. (2005), who reported for the eastern and western series of the metamorphic complex a cooling event between 113 and 80 Ma. This event is probably related to an exhumation process associated with uplift that can be associated with the Peruvian orogeny. Considering that it coincides with the age of the Las Chilcas Formation, we propose that the Las Chilcas Formation formed during the end of the Early period of Andean evolution and the beginning of the Late period, and represents the transition from an extensional to a compressional tectonic regime. A similar view has been proposed for the Caleu pluton (Parada et al. 2005). The calc-alkaline, silicic pyroclastic deposits, intermediate lavas and continental sediments of the Lo Valle Formation (Thomas 1958; Godoy 1982; Moscoso et al. 1982; Rivano 1996; Gana & Wall 1997) cover unconformably the Las Chilcas Formation. The Lo Valle Formation represents the deposits of the Late Cretaceous volcanic arc. K–Ar and Ar–Ar age determinations from samples collected at 33°S yielded 70.5 ± 2.5, 64.6 ± 5, 72.4 ± 1.4 and 71.4 ± 1.4 Ma (Vergara & Drake 1978; Drake et al. 1976; Gana & Wall 1997). According to these ages the unconformity that separates the Las Chilcas Formation from the overlying Lo Valle Formation represents a 20 Ma hiatus (Gana & Wall 1997).

Southern segment. Further south, between 33 and 37°S, Late Cretaceous to Palaeogene deposits are exposed on the western and eastern flanks of the Coastal Cordillera, and in the Principal Cordillera. On the western flank of the Coastal Cordillera to south of Santiago, deposits of both Late Cretaceous to early Paleocene age and of late Paleocene (?) to Eocene age have been reported. The Late Cretaceous to Paleocene outcrops consist of fossiliferous marine plataformal deposits related to the eustatic high stand developed at this time, and are exposed in the following localities, from north to south: Algarrobo (Levi & Aguirre 1960; Tavera 1980; Wall *et al.* 1996; Yury-Yáñez *et al.* 2012), Topocalma (Charrier 1973; Cecioni 1978; Tavera 1979), Faro Carranza, south of Constitución (Chanco Formation; Cecioni 1983), and in the Arauco region, at the latitude of Concepción (*c.* 37°S) (Quiriquina Formation; Steinmann *et al.* 1895; Wetzel 1930; Muñoz-Cristi 1946, 1956; Biró-Bagóczky 1982; Stinnesbeck 1986; Finger *et al.* 2007; Salazar *et al.* 2010; Buatois & Encinas 2011). The Quiriquina Formation overlies the late Palaeozoic metamorphic complex and is unconformably overlain by the late Paleocene (?) to Eocene Concepción Group.

In the coastal region at the latitude of Concepción, late Paleocene (?)–Eocene deposits of the Concepción Group comprise alternations of continental and marine deposits accumulated in extensional basins formed along the coast in late Paleocene (?) to Eocene times. This outstanding succession containing hydrocarbon and important coal reserves is characterized by an alternation of transgressive and regressive episodes, controlled by eustatic changes, local subsidence and uplift of tectonic blocks (Wenzel *et al.* 1975; Pineda 1983*a*, *b*), and general uplift of the Andean range. It consists of the following formations, some of which interfinger with each other: Pilpilco (early Eocene, littoral marine sequence, partly continental), Curanilahue (early Eocene, a mainly continental sequence, coal-bearing strata), Boca Lebu (early Eocene, marine transgressive sequence), Trihueco (middle Eocene, a mainly continental sequence, coal-bearing strata) and Millongue (middle to late Eocene, marine sequence) (Tavera 1942; Muñoz-Cristi 1946, 1973; Pineda 1983*a*, *b*; Arévalo 1984; Finger *et al.* 2007). Along the eastern flank of the Coastal Cordillera up 35°15′S, the Late Cretaceous Lo Valle Formation is further exposed (Bravo 2001).

In the Principal Cordillera at 35°S, upward fining and thinning red-coloured fluvial deposits that unconformably rest on Jurassic terms of the Baños del Flaco Formation and unconformably underlie early Oligocene mammal bearing levels of the Abanico Formation have been informally named Brownish-red Clastic Unit by Charrier *et al.* (1996). Similar deposits crop out next to the water divide at 36°S overlying Middle to Late Jurassic rocks of the Nacientes del Teno Formation and underlying the Late Cenozoic volcanic deposits assigned to the Campanario Formation (Drake 1976; Hildreth *et al.* 1998). These have been included in the Estero Cristales Beds by Muñoz & Niemeyer (1984). According to their stratigraphic position, these deposits can be assigned a Late Cretaceous age and correlated with the Late Cretaceous Neuquén Group on the eastern side of the cordillera (Charrier *et al.* 1996; Mescua 2011). On the western side of the cordillera, these units can be correlated with the upper Las Chilcas and the Viñita formations, further north. In westernmost Argentina, between 33 and 38°S on the eastern side of the Principal Cordillera, marine deposits of the Saldeño Formation (Tunik 2003) and the Malargüe Group (Bertels 1969, 1970; Aguirre-Urreta *et al.* 2011) testify to the far-reaching nature of the Late Cretaceous to early Cenozoic Atlantic transgression. The absence of these deposits in Chile suggests the existence of a relief that stopped the advance of the sea further west.

Along the western flank of the Andes, in Argentina between 35 and 38°S, the Late Cretaceous, 1500 m-thick red fluvial detritic deposits of the Neuquén Group, and the overlying Bajada del Agrio Group correspond to the retroarc foreland deposits related to the Peruvian orogeny (Ramos 1981; Ramos & Folguera 2005; Tunik *et al.* 2010; Di Giulio *et al.* 2012) (Fig. 1). New U–Pb detrital zircons age determinations indicate an early phase of westward-sourced deposits followed by deposition of sediments originated to the east of the retroarc foreland basin associated with uplift of the peripheral bulge as a consequence of the Late Cretaceous thrust front migration (Di Giulio *et al.* 2012).

In middle Eocene, the Incaic orogeny put an end to this stage. This event coincides with the peak of high convergence rate associated with a considerable reduction of the obliquity of convergence after 45 Ma (Pardo-Casas & Molnar 1987).

Second stage (middle Eocene–Present)

The northern part of the considered region, between 31 and 33°S, is located in the southern part of the flat-slab segment, where the Central Depression and the volcanic arc are not developed. This is the region where the Frontal Cordillera, the Precordillera and the Pampean ranges are developed (Fig. 2). The southern part, instead, between 33 and 37°S, is located in the transition to and in the normal subduction segment, where the Central Depression and the volcanic arc (along the axis of the Principal Cordillera) have developed. In this stage, the Maipo orocline occurred in close relationship with the Pampean flat-slab. Palaeomagnetic rotations are observed within the normal subduction segment (Arriagada et al. 2013). In the considered region, second-stage deposits are located in all morphostructural units. We will describe the deposits from west to east.

Western Coastal Cordillera. No Oligocene deposits exist along the coastal region in Chile, probably because of the eustatic low stand at this time. Between 33°40′ and 34°15′S, exposures of Miocene marine sediments are known as the Navidad Formation *sensu* Darwin (1846) and Tavera (1979). Recent work subdivided these deposits into the Navidad, Licancheo and Rapel formations (Encinas et al. 2006a). The Navidad Formation has been assigned a late Miocene age and the Licancheo and Rapel formations a Pliocene age by Finger et al. (2003) and Encinas et al. (2006a), whereas Gutiérrez et al. (2013) assigned an early to middle Miocene age to the Navidad Formation and a late Miocene age to the Licancheo and Rapel formations. The latter is overlain by the late Miocene to Pliocene, transitional marine to continental, richly tuffaceous La Cueva Formation (Tavera 1979), which interfingers and is overlain to the east by continental deposits of the Potrero Alto Beds of uncertain Miocene–Pliocene to Pleistocene age (Wall et al. 1996). According to Encinas et al. (2006a), the Navidad Formation was deposited on a rapidly subsiding basin, which reached depths of 1500 m; in contrast, Gutiérrez et al. (2013) favour a shallow coastal to outer shelf environment for this formation. The Navidad Formation can be correlated with the Ranquil Formation, to the south at 37°S.

The source of sediments in the Navidad basin based on the analysis of heavy mineral assemblages is the nearshore basement rocks (metamorphic and intrusive units) in the Coastal Cordillera and the central Chilean forearc (Rodríguez et al. 2012). Sediment supply from the latter began with erosion at the present-day eastern Central Depression (western Abanico Formation), later at the eastern Central Depression–western Principal Cordillera border (Miocene plutons), and finally in the western Principal Cordillera (Farellones Formation) (Rodríguez et al. 2012). The eastwards shift of the sediment source indicates the slow and gradual retreat experienced by the nick-points, which, according to Farías et al. (2008), arrived in the western Principal Cordillera between 2 and 6 myr after onset of surface uplift, at c. 7.6 Ma, and >2 myr later in the eastern Principal Cordillera.

Along the coast of central Chile, Plio-Pleitocene events are widely documented by: (a) marine deposits of the upper Coquimbo Formation (at 30°S) and correlatives, like the Pliocene upper La Cueva formations (at 34°S) and further south the Tubul Formation (at 37°S) in the Arauco region; (b) fluvial deposits exposed in coastal–near river drainages, like the Confluencia and Caleta Horcón formations (Rivano 1996), and the Potrero Alto Beds; and (c) shoreline and fluvial geomorphic features that testify to a considerable uplift of the forearc. Five well-preserved marine terraces (wave-cut platforms) (Darwin 1846; Paskoff 1970, 1977; Fuenzalida et al. 1965; Ota et al. 1995; Saillard et al. 2009, 2012) and pedimentary surfaces developed in fluvial drainages connected with the marine terraces (Rodríguez et al. 2013) have been reported. Cosmogenic datings on these features yielded 6, 122, 232, 321 and 690 ka, and allow reconstruction of a non-steady history of uplift to c. 100–150 m during interglacial periods after 400 ± 100 ka (Saillard et al. 2009; Regard et al. 2010; Rodríguez et al. 2013).

Central Depression. Until recently, this morphostructural feature, which is developed south of 33°S, has been considered to be of tectonic origin (i.e. Carter & Aguirre 1965) and related to the subduction of the Juan Fernández Ridge (i.e. Jordan et al. 1983b). However, based on analyses of uplift markers and nick-point progression supported by geochronological and thermochronological dating, it has been recently interpreted as an erosional feature (Farías et al. 2008). However, the Quaternary alluvial and fluvial deposits derived from the Principal Cordillera that build up the c. 400 m thick infill, as well as new tectonic evidence, show that the Central Depression is tectonically controlled (see Giambiagi et al. 2014). Explosive volcanic activity in the volcanic arc produced abundant lahar and volcanic avalanche deposits such as La Cueva Formation in the coastal region (Encinas et al. 2006b). Between 33°30′ and 37°S, tuff and ash-flow deposits covered the Central Depression, that is, the Pudahuel–Machalí Ignimbrite (Stern et al. 1984), Teno (MacPhail & Saa 1967; Marangunic et al. 1979), Tinguiririca (Abele 1982) and Laja (MacPhail 1966) lahars.

The origin of the Pudahuel–Machalí Ignimbrite (Stern et al. 1984) has been associated with the Maipo Caldera, in the high Andes, at 34°S. The ignimbrite flow was channelized along the main river valleys towards the coastal region, and into Argentina, along the Yaucha and Papagayos valleys. The tuff deposits in Pudahuel, next to Santiago, and Machalí, east of Rancagua, yielded apatite fission track ages of 0.44 ± 0.08 and 0.47 ± 0.007 Ma, respectively (Stern et al. 1984).

Principal and Frontal Cordillera. In the northernmost part of the region (31°–32°S), Oligocene plutonic rocks (El Maitén–Junquillar and Bocatoma units) form extensive outcrops that intrude Permo-Triassic volcanics of the Choiyoi Group and younger Mesozoic units (Fig. 3) (Mpodozis & Cornejo 1988; Martin *et al.* 1997; Bissig *et al.* 2001). Further south, these outcrops disappear and two alignments of scattered intrusives of early and middle to late Miocene age (the latter to the east of the former) are exposed along the western flank of the Principal Cordillera. The La Obra and the San Gabriel plutons in the Maipo river valley next to Santiago belong to this group of intrusives. Some of these plutonic bodies are associated with super-giant late Miocene to Pliocene porphyry Cu–Mo ore bodies such as Los Pelambres, Río Blanco–Los Bronces and El Teniente. These ore deposits developed within hydrothermal alteration zones linked to multiphase stocks, breccia pipes and diatreme structures in rocks of Oligocene to Miocene age (Cuadra 1986; Serrano *et al.* 1996; Vivallo *et al.* 1999; Camus 2002, 2003; Skewes *et al.* 2002; Maksaev *et al.* 2004; Charrier *et al.* 2009; Mpodozis & Cornejo 2012).

Some of the stratified units of this stage exposed between 31 and 33°S have been previously considered to represent much older ages. Recent studies have shown that they formed essentially during Oligocene to Miocene times (Mpodozis *et al.* 2009; Jara & Charrier 2014; Jara *et al.* 2014). According to this age and their volcanic and volcaniclastic nature with only subordinated sedimentary intercalations, they can be correlated with similar deposits exposed between 29 and 30°S on the eastern flank of the cordillera in the Valle del Cura region (Winocur *et al.* 2014), and with the Abanico and Farellones formations well exposed 33°S along the western flank of the Principal Cordillera and to the north (Fig. 3).

The dominantly volcanic, middle–late Eocene to Oligocene Abanico Formation, and the Miocene Farellones Formation make up the pre-Pliocene Cenozoic deposits in the Principal Cordillera of Central Chile (31°–36°S) (Aguirre 1960; Klohn 1960; González & Vergara 1962; Charrier 1973, 1981*a*, *b*; Thiele 1980; Charrier *et al.* 2002; Godoy 2011). The Abanico Formation consists of a locally strongly folded, *c.* 3000 m-thick succession of volcanic, pyroclastic volcaniclastic and sedimentary deposits including abundant subvolcanic intrusions of the same age (Vergara *et al.* 2004), with a well-developed paragenesis of low-grade metamorphic minerals (Padilla & Vergara 1985; Levi *et al.* 1989; Aguirre *et al.* 2000; Fuentes *et al.* 2001, 2005; Bevins *et al.* 2003; Fuentes 2004; Muñoz *et al.* 2006, 2010). The outcrops of this formation form two north–south orientated swaths separated by the Farellones Formation (Fig. 3) (Sernageomin 2002). This formation contains abundant mammalian rests (Flynn *et al.* 2007; Charrier *et al.* 2012). At the western side of the Abanico outcrops, 34.3 ± 2.2 Ma old basal Abanico deposits unconformably overlie the 72.4 ± 1.4 and 71.4 ± 1.4 Ma old Lo Valle Formation (Gana & Wall 1997). At the eastern side of the Abanico exposures, the oldest age obtained for the base of the Abanico Formation is 37.67 ± 0.31 Ma (Charrier *et al.* 1996). The Abanico Formation was deposited in an extensional basin formed while the crust was relatively thin, persisting throughout the Oligocene epoch, and underwent subsequent tectonic inversion in late Oligocene to early Miocene time (Pehuenche orogeny, see Fig. 1). This diachronic extensional event has not been recognized in northern Chile (Charrier *et al.* 2009, 2013) and seems to have been concentrated between 28 and 39°S, and probably extended further south, up to 43°S (Godoy 2011).

The younger Farellones Formation is a 2400 m thick, gently folded, almost entirely volcanic unit forming a continuous north–south trending swath between approximately 32 and 35°S (Fig. 3) (Thiele 1980; Charrier 1981*a*, *b*; Vergara *et al.* 1988). The deposits of the Farellones Formation typically cover the Abanico Formation, up to 35°S, where this formation is not exposed any more. The views about the contact are controversial (Charrier *et al.* 2002; 2007). It has generally been reported as unconformable (Aguirre 1960; Klohn 1960; Jaros & Zelman 1967; Charrier 1973, 1981*a*, *b*; Thiele 1980; Quiroga 2013). Growth strata have been observed at several localities between the upper Abanico and the lower Farellones deposits (Fock *et al.* 2006; Quiroga 2013). The contractional event evidenced by the growth strata has been associated with the inversion of pre-existing normal faults that participated in the development of the Abanico basin (Charrier *et al.* 2002; Fock *et al.* 2006; Farías *et al.* 2010). Inversion along the eastern side of the basin linked to El Diablo fault triggered the development of the east-vergent thrust–fold belt systems during middle Miocene on the eastern flank of the Principal Cordillera (Ramos *et al.* 1996*b*; Giambiagi *et al.* 2003*a*; Mescua 2011; Muñoz-Saez *et al.* 2014; Giambiagi *et al.* 2014). Furthermore, inversion along the San Ramón Fault on the western side of the Abanico basin caused west-vergent thrusting of the Abanico Formation deposits over the Central Depression (Rauld 2002; Armijo *et al.* 2010). Recent activity has been detected on the San Ramón Fault (Vargas & Rebolledo 2012).

South of 36°S, the prolongation of the younger part of Abanico Formation is the Cura–Mallín Formation (Niemeyer & Muñoz 1983; Muñoz & Niemeyer 1984; Charrier *et al.* 2002; Radic *et al.* 2002; Flynn *et al.* 2008). This formation accumulated in the southern prolongation of the Abanico Extensional basin (Elgueta 1990; Vergara *et al.* 1997; Jordan *et al.* 2001; Charrier *et al.* 2002; Radic *et al.* 2002; Croft *et al.* 2003; Flynn *et al.* 2008) that in this region was inverted during late Miocene times (Burns & Jordan 1999; Radic *et al.* 2002).

The Cura–Mallín Formation is conformably overlain by the andesitic and conglomeratic Trapa–Trapa Formation, which in turn is unconformably overlain by the late Miocene Campanario Formation, which is a southern equivalent of the upper Farellones Formation and the Pliocene Cola de Zorro Formation (González & Vergara 1962; Muñoz & Niemeyer 1984; Astaburuaga 2014). A nearly coeval stratigraphic series has been recognized in the Andacollo region on the Argentine side of the Andes at 37°S (Jordan *et al.* 2001), where these authors applied the same formational names as used in Chile. These Argentine deposits unconformably overlie the early Cenozoic Serie Andesítica, and are overlain by the late Miocene sedimentary and volcaniclastic Pichi Neuquén Formation.

Scattered Plio-Pleistocene volcanic activity on the western Principal Cordillera has been reported for areas located next to the El Teniente ore deposit at 34°S (Camus 1977; Charrier & Munizaga 1979; Charrier 1981*b*; Cuadra 1986; Godoy & Lara 1994; Gómez 2001) and Sierras de Bellavista at 34°45′S (Klohn 1960; Vergara 1969; Charrier 1973; Malbran 1986; Eyquem 2009). These volcanic centres form a north–south alignment suggesting a tectonic control for this activity. Finally, the volcanoes of the present-day magmatic arc lie east of the eastern outcrops of the Abanico Formation, covering Mesozoic units and forming the northern part of the Southern Volcanic Zone (Stern *et al.* 2007). From north to south the most important of these volcanoes are named Tupungato, San José, Maipo and Maipo caldera, Tinguiririca, Planchón–Peteroa, Descabezado Grande, Cerro Azul, Descabezado Chico, San Pedro, Longaví and Chillán, some of which are aligned with the El Diablo fault (see Fig. 3). Isotopic studies on late Neogene magmatic rocks from the Chilean Principal Cordillera reveal a source contamination probably resulting from the deep westwards underthrusting of the basement beneath the orogen, a process that is coeval with thickening and uplifting events in the Andes (Muñoz *et al.* 2013).

Thermochronometric studies oriented to constrain the tectonic-related exhumation history in central Chile, between 28.5° and 32°S, reveal that uplift of the Coastal Cordillera occurred mostly in Palaeogene times (apatite fission track (AFT) ages between *c.* 60 and 40 Ma and apatite He (AHe) ages around 30 Ma) and that little exhumation occurred during the rest of the Neogene in that region, while exhumation ages from the Frontal Cordillera are younger AFT ages between *c.* 40 and 8 Ma and AHe ages from *c.* 20 to 6 Ma). Thermal modelling of AFT and AHe data allows recognition of three main episodes of accelerated cooling affecting different areas of the Frontal Cordillera: *c.* 30, *c.* 22–17 and *c.* 7 Ma. The first of them coincides with the early stages of development of a late Oligocene extensional intra-arc basin along the eastern Frontal Cordillera, between 29 and 30°S, and is interpreted as a consequence of tectonic exhumation. The early and late Miocene periods of accelerated cooling along the Frontal Cordillera correlate with periods of contractional deformation widely recognized throughout the Central Andes and, thus, interpreted as a consequence of surface-uplift (Rodríguez 2013; Rodríguez *et al.* 2014).

^{10}Be content in bed-load from different rivers and across a major climatic gradient on the western flank of the mountain range in central Chile has been analysed in order to determine erosion rates associated with uplift (Carretier *et al.* 2013). This study confirms the primary role of slope as a control of erosion even under contrasting climates and supports the view that the influence of runoff variability on millennial erosion rates increases with aridity. However, even if current erosion rates are decoupled from precipitation rates, climate still plays a fundamental role by accelerating the erosion response to uplift (Carretier *et al.* 2014).

Foreland. In Neogene times in the Andean foreland some processes related to shallowing and steepening of the subducting slab occurred, which are described next.

Pampean flat-slab subduction. The magmatic arc that was developed during previous stages of the Andean cycle along the western slope of the Andes expanded and shifted during late Miocene to the Quaternary to the Argentine side between 31 and 33°30′S latitude (Fig. 7) associated with a period of shallowing of the subduction zone (Jordan *et al.* 1983*a*, *b*).

Magmatic activity ended in the Principal Cordillera at about 8.6 Ma and in Sierras Pampeanas at 1.9 Ma with the last subduction-related volcanism more than 750 km away from the trench (Ramos *et al.* 2002). The first migration of the volcanic arc at these latitudes is recorded in the Aconcagua volcanic rocks. Huge amounts of andesitic and dacitic rocks were erupted at about 15.8 ± 0.4 and 8.9 ± 0.5 Ma in the Aconcagua massif on the eastern side of the Principal Cordillera, at 33°S (see Fig. 3). This area constitutes the new volcanic front 50 km east of the Farellones arc in the western slope. The retroarc magmatism of Paramillos,

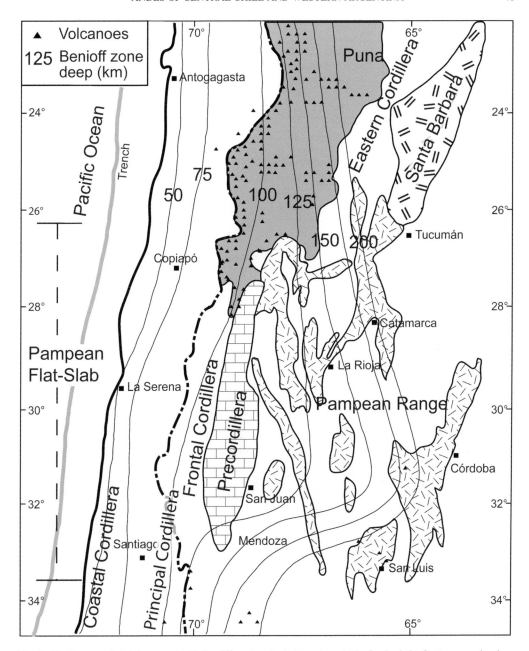

Fig. 7. The Pampean flat-slab segment with the different geological provinces in the foreland, the Quaternary volcanic arc and the isobath to the subducted oceanic slab (after Ramos et al. 2002).

west of the city of Mendoza, was shut off at 15.2 Ma, almost at the same time that the migration took place (Ramos et al. 1996b). Geochemical studies show a typical calc-alkaline signature of these volcanic rocks (Kay & Abbruzzi 1996).

The shifting of the magmatic arc was preceded by an important deformation in the western half of the Aconcagua fold-and-thrust belt (Fig. 3) (Ramos et al. 1996a). At about 8.6 Ma, the thin-skinned Aconcagua fold-and-thrust belt detached in Jurassic evaporites ceased. As a result of that, the orogenic front migrated about 25 km from Las Cuevas to Río de Las Vacas (Ramos et al. 1996b). Syn-orogenic deposits were preserved in isolated exposures between Principal and Frontal Cordilleras, in the Uspallata–Calingasta depression and in the present foreland basin in the foothills around Mendoza (Fig. 3). Magnetostratigraphic studies performed in the foothills show that sedimentation started at 15.7 Ma in the Mariño Formation with distal fluvial and eolian deposits with a low sedimentation rate. This unit was followed

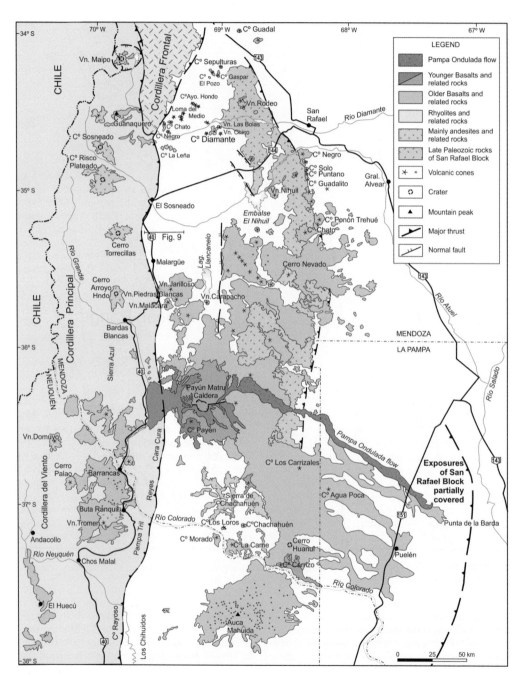

Fig. 8. The Miocene magmatic arc rocks, the Quaternary Payenia basaltic province and the Present thrust front (after Ramos & Folguera 2011).

by the conglomerates and sandstones of La Pilona Formation at 11.7 Ma, which exhibit a marked increase in accumulation rate with time (Irigoyen *et al.* 2000).

There is no magmatic activity in the Principal and Frontal Cordilleras. Magmatic arc rocks are concentrated in the Sierras Pampeanas between 8 and 6 Ma. Volcanic arc rocks are widespread in the Sierra de San Luis (Urbina & Sruoga 2009) and consist of high-K andesites and dacites with typical shoshonites in the Sierra del Morro, which recorded the latest eruption at 1.9 Ma along the flat-slab subduction segment (Ramos *et al.* 1991, 2002; Kay *et al.* 1991).

The magmatic expansion was accompanied by the development of a broken foreland where several late Cenozoic basins were formed related to tectonic loading and dynamic subsidence (Dávila *et al.* 2005).

The neotectonic activity is presently concentrated between Precordillera and Sierras Pampeanas, where large intracrustal earthquakes have occurred (Alvarado *et al.* 2009). Surface fault ruptures and other neotectonic features indicate important Quaternary deformation (Schmidt *et al.* 2011).

Payenia palaeoflat-slab subduction. South of 33°30′S latitude a different geological setting is observed (Fig. 8). The Principal Cordillera is flanked by the Frontal Cordillera up to 34°30′S, where south of this latitude the foothills of the Andes are in contact with an extensive basaltic plateau of Quaternary age between 34 and 37°S (Ramos & Folguera 2011). This basaltic plateau is known as the Payenia volcanic province.

The Principal Cordillera between 35 and 37.5°S latitudes recorded in the Miocene an important expansion of the magmatic arc from the Chilean slope to the foreland eastern foothills (Spagnuolo *et al.* 2012*b*; Folguera & Ramos 2011). After a period of extension during the Oligocene, where the within-plate alkaline basaltic rocks of Palauco Formation erupted, a series of granitic stocks were emplaced in the eastern slope of the Principal Cordillera. Subvolcanic bodies of calc-alkaline andesites with ages ranging from middle to late Miocene are found in the foreland extra-andean plains. The eruption of these rocks is also linked to another phase of contraction and deformation (Spagnuolo *et al.* 2012*b*). The within-plate basalts of the Payenia volcanic province are unconformably overlying previous deposits and extend over an area larger than 40 000 km^2 between 33°30′ and 38°S latitudes. The huge Payún Matru caldera is the main feature related to the Payenia retroarc basalts (Bertotto *et al.* 2009; Llambías *et al.* 2010). The basalts have an estimated volcanic volume of about 8387 km^3 erupted through more than 800 volcanic centres in the last *c.* 2 Ma (Ramos & Folguera 2011; Gudnason *et al.* 2012; Søager *et al.* 2013).

The sedimentary basin evolution at these latitudes during the Cenozoic shows the transit from a single foreland basin to a broken foreland basin associated with the uplift and exhumation of the San Rafael Block (Silvestro & Atencio 2009; Ramos *et al.* 2014). Geophysical studies have demonstrated that the steepening of a subducted slab is the more appropriate process to explain the extension and characteristics of the Payenia basaltic retroarc province (Burd *et al.* 2008).

Concluding remarks

The analysis of the geology of this sector of the southern Central Andes shows continuous subduction along the proto-Pacific margin of Gondwana and the present margin of South America during most of the Cenozoic. However, it is possible to identify an accretionary orogen during the Palaeozoic, with docking of different terranes, and a Meso-Cenozoic subduction with different tectonic regimes. These variations defined tectonic cycles with different processes, where extensional and compressive regimes alternate through time. The accretion of continental basement terranes produced the obduction of slices of oceanic crust, high-pressure–low-temperature metamorphism, and important shifting to the trench of the magmatic arcs. Periods of shallow subduction, partially combined in late Palaeozoic times with the processes of accretionary orogenesis, produced broken forelands under extreme contraction and subsequent extension and widespread rhyolitic volcanism. The Mesozoic was characterized by subduction with generalized extension until late Early Cretaceous, when evidence of contraction and orogeny led to the present Andean tectonic setting with dominant contraction. The development of segments with flat-slab subduction with no arc-magmatism alternates with segments where, after periods of flat subduction, the steepening of the subduction zone produced generalized extension and backarc basaltic magmatism.

All these processes indicate the complexities of the classic Andean tectonic setting, where the simple subduction of oceanic crust under continental crust controls the orogenesis. The analysed segment, one of the classical sectors of the Central Andes, with excellent exposures and where many of these processes have been first recognized, enhanced the importance of understanding the relationship among magmatism, metamorphism, sedimentation and deformation.

The authors thank the participants of the IGCP 586-Y 'Geodynamic Processes in the Andes, 32–34°S' for their invitation to prepare this introductory chapter to this book. They thank two anonymous reviewers for their valuable suggestions that considerably improved the manuscript and several colleagues who provided in-press manuscripts on subjects related to the considered region. The authors acknowledge: FONDECYT Project no. 1120272, 'Extension, inversion and propagation: key tectonic styles on the development of the Andean Cordillera of central Chile-Argentina (32–36°S)'.

References

ABELE, G. 1982. El lahar Tinguiririca: su significado entre los lahares chilenos. *Informaciones Geográficas*, **29**, 21–34.

ÄBERG, G., AGUIRRE, L., LEVI, B. & NYSTRÖM, J. O. 1984. Spreading subsidence and generation of ensialic marginal basin: an example from the Early Cretaceous of Central Chile. *In*: KOKELAAR, B. P. & HOWELLS, M. F. (eds) *Marginal Basin Geology*. Geological Society, London, Special Publications, **16**, 185–193.

ACEÑOLAZA, G. & TOSSELLI, A. C. 1976. Consideraciones estratigráficas y tectónicas sobre el Paleozoico inferior del Noroeste Argentino. *In: II Congreso Latinoamericano de Geología*, Caracas, 1973, **2**, 755–764.

AGUIRRE, L. 1960. Geología de los Andes de Chile Central, provincia de Aconcagua. *Boletín del Instituto de Investigaciones Geológicas*, **9**, 1–70.

AGUIRRE, L. & EGERT, E. 1965. Cuadrángulo Quebrada Marquesa, Provincia de Coquimbo. Instituto de Investigaciones Geológicas, Santiago, Carta Geológica de Chile, escala 1:50.000, Carta 15.

AGUIRRE, L. & EGERT, E. 1970. Cuadrángulo Lambert (La Serena), Provincia de Coquimbo. Instituto de Investigaciones Geológicas, Santiago, Carta Geológica de Chile, escala 1:50.000, Carta 23.

AGUIRRE, L., HERVÉ, F. & GODOY, E. 1972. Distribution of metamorphic facies in Chile, an outline. *Kristallinikum*, **9**, 7–19.

AGUIRRE, L., CHARRIER, R. *ET AL*. 1974. Andean Magmatism: its paleogeographical and structural setting in the central part (30–35°S) of the Southern Andes. *Pacific Geology*, **8**, 1–38.

AGUIRRE, L., FÉRAUD, G., MORATA, D., VERGARA, M. & ROBINSON, D. 1999. Time interval between volcanism and burial metamorphism and rate of basin subsidence in a Cretaceous Andean extensional setting. *Tectonophysics*, **313**, 433–447.

AGUIRRE, L., ROBINSON, D., BEVINS, R. E., MORATA, D., VERGARA, M., FONSECA, E. & CARRASCO, J. 2000. A low-grade metamorphic model for the Miocene volcanic sequences in the Andes of central Chile. *New Zealand Journal of Geology and Geophysics*, **43**, 83–93.

AGUIRRE-URRETA, B., TUNIK, M., NAIPAUER, M., PAZOS, P., OTTONE, E., FANNING, M. & RAMOS, V. A. 2011. Malargüe Group (Maastrichtian–Danian) deposits in the Neuquén Andes, Argentina: implications for the onset of the first Atlantic transgression related to Western Gondwana break-up. *Gondwana Research*, **19**, 482–494.

AGUIRRE-URRETA, M. B. & LO FORTE, G. L. 1996. Los depósitos tithoneocomianos. *In*: RAMOS, V. A. (ed.) *Geología de la región del Aconcagua, provincias de San Juan y Mendoza, República Argentina*. Dirección Nacional del Servicio Geológico, Buenos Aires, Anales, **24**, 179–230.

ALONSO, J. L., GALLASTEGUI, J., GARCÍA-SANSEGUNDO, J., FARIAS, P., RODRÍGUEZ FERNÁNDEZ, L. R. & RAMOS, V. A. 2008. Extensional tectonics and gravitational collapse in an Ordovician passive margin: the Western Argentine Precordillera. *Gondwana Research*, **13**, 204–215.

ALVARADO, P., PARDO, M., GILBERT, H., MIRANDA, F., ANDERSON, M., SAEZ, M. & BECK, S. 2009. Flat-slab subduction and crustal models for the seismically active Sierras Pampeanas región of Argentina. *In*: KAY, S. M., RAMOS, V. A. & DICKINSON, W. R. (eds) *Backbone of the Americas: Shallow Subduction, Plateau Uplift, and Ridge and Terrane Collision*. Geological Society of America, Boulder, CO, Memoirs, **204**, 261–278.

ÁLVAREZ, P. P. 1996. Los depósitos triásicos y jurásicos de la Alta Cordillera de San Juan. *In*: RAMOS, V. (ed.) *Geología de la Región del Aconcagua, provincias de San Juan y Mendoza*. Subsecretaría de Minería de la Nación, Dirección Nacional del Servicio Geológico, Buenos Aires, Anales, **24**, 59–137.

ÁLVAREZ, P. P. & RAMOS, V. A. 1999. The Mercedario Rift System in the Principal Cordillera of Argentina and Chile (32°SL). *Journal of South American Earth Sciences*, **12**, 17–31.

ÁLVAREZ, P. P., AGUIRRE-URRETA, M. B., GODOY, E. & RAMOS, V. A. 1997. Estratigrafía del Jurásico de la Cordillera Principal de Argentina y Chile (33°45′–34°00′LS). *In: VIII Congreso Geológico Chileno Antofagasta*. Antofagasta, **1**, 425–429.

ARÉVALO, A. 1984. *Geología de subsuperficie del área al sur del río Lebu, VIII Región*. Thesis, Departamento de Geología, Universidad de Chile, Santiago.

ARCOS, R. 1987. *Geología del Cuadrángulo Termas del Flaco, provincia de Colchagua, VI Región, Chile*. Thesis, Departamento de Geología, Universidad de Chile.

ARMIJO, R., RAULD, R., THIELE, R., VARGAS, G., CAMPOS, J., LACASSIN, R. & KAUSEL, E. 2010. The West Andean Thrust, the San Ramón Fault, and the seismic hazard for Santiago, Chile. *Tectonics*, **29**, TC2007, http://dx.doi.org/10.1029/2008TC002427

ARRIAGADA, C., ROPERCH, P., MPODOZIS, C. & COBBOLD, P. R. 2008. Paleogene building of the Bolivian Orocline: tectonic restoration of the Central Andes in 2-D map view. *Tectonics*, **27**, TC6014, http://dx.doi.org/10.1029/2008TC002269

ARRIAGADA, C., FERRANDO, R., CÓRDOVA, L., MORATA, D. & ROPERCH, P. 2013. The Maipo Orocline: a first scale structural feature in the Miocene to recent geodynamic evolution in the Central Andes. *Andean Geology*, **40**, 419–137.

ASTABURUAGA, D. 2014. *Geología estructural y configuración del límite Mesozoico-Cenozoico de la Cordillera Principal entre 35°30′y 36°S, Región del Maule, Chile*. Master's thesis, Departamento de Geología, Universidad de Chile.

ASTINI, R. A. 1996. Las fases diastróficas del Paleozoico Medio en la Precordillera del oeste argentino. *In: Evidencias estratigráficas. XIII Congreso Geológico Argentino y III Congreso de Exploración de Hidrocarburos*, Buenos Aires, **5**, 509–526.

ASTINI, R. A., BENEDETTO, J. L. & VACCARI, N. E. 1995. The early Paleozoic evolution of the Argentina Precordillera as a Laurentian rifted, drifted, and collided terrane: a geodynamic model. *Geological Society of America, Bulletin*, **107**, 253–273.

ASTINI, R., RAMOS, V. A., BENEDETTO, J. L., VACCARI, N. E. & CAÑAS, F. L. 1996. La Precordillera: un terreno exótico a Gondwana. *In: XIII Congreso Geológico Argentino y III Congreso Exploración de Hidrocarburos*, Buenos Aires, **5**, 293–324.

ASTINI, R. A., MARTINA, F. & DÁVILA, F. M. 2011. La Formación Los Llantenes en la Precordillera de Jagüé (La Rioja) y la identificación de un episodio de extensión en la evolución temprana de las cuencas del Paleozoico superior en el oeste argentino. *Andean Geology*, **38**, 245–267.

ÁVILA, J. N., CHEMALE, F., JR, MALLMANN, G., KAWASHITA, K. & ARMSTRONG, R. 2006. Combined stratigraphic and isotopic studies of Triassic strata, Cuyo Basin, Argentine Precordillera. *Geological Society of America Bulletin*, **118**, 1088–1098.

BARREDO, S., CHEMALE, F., MARSICANO, C., ÁVILA, J. N., OTTONE, E. G. & RAMOS, V. A. 2012. Tectonosequence stratigraphy and U–Pb zircon ages of the Rincón Blanco depocenter, northern Cuyo Rift, Argentina. *Gondwana Research*, **21**, 624–636.

BELL, M. & SUÁREZ, M. 1995. Slope apron deposits of the Lower Jurassic Los Molles Formation, Central Chile. *Revista Geologica de Chile*, **22**, 103–114.

BENEDETTO, J. L., SÁNCHEZ, T. M., CARRERA, M. G., BRUSSA, E. D. & SALAS, M. J. 1999. Paleontological constraints in successive paleogeographic positions of Precordillera terrane during the Early Paleozoic. *In:* RAMOS, V. A. & KEPPIE, D. (eds) *Laurentia-Gondwana Connections Before Pangea*. Geological Society of America, Boulder, CO, Special Papers, **336**, 21–42.

BERTELS, A. 1969. Estratigrafía del límite Cretácico-Terciario en Patagonia septentrional. *Revista Asociación Geológica Argentina*, **24**, 41–54.

BERTELS, A. 1970. Micropaleontología y estratigrafía del límite Cretácico-Terciario en Huantraico (Provincia del Neuquén, Parte II). *Ameghiniana*, **4**, 253–298.

BERTOTTO, G. W., CINGOLANI, C. & BERG, E. 2009. Geochemical variations in Cenozoic back-arc basalts at the border of La Pampa and Mendoza provinces, Argentina. *Journal of South American Earth Sciences*, **28**, 360–373.

BEVINS, R. E., ROBINSON, D., AGUIRRE, L. & VERGARA, M. 2003. Episodic burial metamorphism in the Andes – a viable model? *Geology*, **31**, 705–708.

BIRÓ-BAGÓCZKY, L. 1964. *Estudio sobre el límite Titoniano y el Neocomiano en la Formación Lo Valdés, Provincia de Santiago, principalmente en base a ammonoideos, Región Metropolitana, Chile*. Thesis, Departamento de Geología, Universidad de Chile.

BIRÓ-BAGÓCZKY, L. 1982. Revisión y redefinición de los "Estratos de Quiriquina", Campaniano-Maastrichtiano, en su localidad tipo en la Isla Quiriquina, 36° 35′ Lat. S, Chile, Sudamérica, con un perfil complementario en Cocholgüe. *In: III Congreso Geológico Chileno*, Concepción, **1**, A 29–64.

BISSIG, T., LEE, J. K. W., CLARK, A. H. & HEATHER, K. B. 2001. The Cenozoic history of magmatic activity and hydrothermal alteration in the central Andean flat-slab region: new $^{40}Ar/^{39}Ar$ constraints from the El Indio–Pascua Au (-Ag, Cu) belt, 29°20′–30°30′S. *International Geology Review*, **43**, 312–340.

BJERG, E. A., GREGORI, D. A., LOSADA CALDERON, A. & LABUDIA, C. H. 1990. Las metamorfitas del faldeo oriental de la Cuchilla de Guarguaraz, Cordillera Frontal, Provincia de Mendoza. *Revista de la Asociación Geológica Argentina*, **45**, 234–245.

BORDONARO, O. 1980. El Cámbrico de la quebrada de Zonda, Provincia de San Juan. *Revista de la Asociación Geológica Argentina*, **35**, 26–40.

BORDONARO, O. 1992. El Cámbrico de Sudamérica. *In:* GUTIÉRREZ MARCO, J. C., SAAVEDRA, J. & RÁBANO, I. (eds) *Paleozoico inferior de Ibero América*. Universidad de Extremadura, Mérida, 69–84.

BRAVO, P. 2001. *Geología del borde oriental de la Corillera de la Costa entre los ríos Mataquito y Maule, VII Región*. Thesis, Departamento de Geología, Universidad de Chile.

BUATOIS, L. A. & ENCINAS, A. 2011. Ichnology, sequence stratigraphy and depositional evolution of an Upper Cretaceous rocky shoreline in central Chile: Bioerosion structures in a transgressed metamorphic basement. *Cretaceous Resource*, **32**, 203–212.

BUGGISH, W., VON GOSEN, W., HENJES-KUNST, F. & KRUMM, S. 1994. The age of Early Paleozoic deformation and metamorphism in the Argentine Precordillera – evidence from K–Ar data. *Zentralblatt fur Geologie und Palaeontologie, Teil*, **1**, 275–286.

BURD, A. I., BOOKER, J. R., POMPOSIELLO, M. C., FAVETTO, A., LARSEN, J., GIORDANENGO, G. & OROZCO BERNAL, L. 2008. Electrical conductivity beneth the Payún Matrú volcanic field in the Andean back-arc of Argentina near 36,5°S: insights into the magma source. *In: VII Internatinal Symposium on Andean Geodynamics*, Nice, Extenden Abstract, 90–93.

BURNS, W. M. & JORDAN, T. E. 1999. Extension in the Southern Andes as evidenced by an Oligo-Mioceneage intra-arc basin. *In: IV International Symposium on Andean Geodynamics (ISAG)*, Göttingen, 115–118.

CAHILL, T. & ISACKS, B. 1992. Seismicity and shape of the subducted Nazaca Plate. *Journal of Geophysical Research*, **97**, 17 503–17 529.

CAMINOS, R. 1965. Geología de la vertiente oriental del Cordón del Plata, Cordillera Frontal de Mendoza. *Revista de la Asociación Geológica Argentina*, **20**, 351–392.

CAMINOS, R. 1979. Cordillera Frontal. *In:* TURNER, J. C. M. (ed.) *Geología Regional Argentina*. Academia Nacional de Ciencias, Córdoba, Argentina, **1**, 397–453.

CAMUS, F. 1977. Geología del área de emplazamiento de los depósitos de cuarzo Olla Blanca, provincia de Cachapoal. *Revista Geológica de Chile*, **4**, 43–54.

CAMUS, F. 2002. The Andean porphyry systems. *In:* COOKE, D. R. & PONGRATZ, J. (eds) *Giant Ore Deposits: Characteristics, Genesis and Exploration*. CODES, Australia, Special Publications, **4**, 5–21.

CAMUS, F. 2003. *Geología de los sistemas porfíricos en los Andes de Chile*. Servicio Nacional de Geología y Minería, Santiago.

CAÑAS, F. L. 1999. Facies and sequences of the Late Cambrian-Early Ordovician carbonates of the Argentine Precordillera: a stratigraphic comparison with Laurentian platforms. *In*: RAMOS, V. A. & KEPPIE, J. D. (eds) *Laurentia–Gondwana Connections before Pangea*. Geological Society of America, Boulder, CO, Special Papers, **336**, 43–62.

CARRETIER, S., REGARD, V. *ET AL.* 2013. Slope and climate variability control of erosion in the Andes of central Chile. *Geology*, **41**, 195–198.

CARRETIER, S., TOLORZA, V. *ET AL.* 2014. Erosion in the Chilean Andes between 27°S and 39°S: tectonics, climatic and geomorphic control. *In*: SEPÚLVEDA, S. A., GIAMBIAGI, L. B., MOREIRAS, S. M., PINTO, L., TUNIK, M., HOKE, G. D. & FARÍAS, M. (eds) *Geodynamic Processes in the Andes of Central Chile and Argentina*. Geological Society, London, Special Publications, **399**. First published online April 9, 2014, http://dx.doi.org/10.1144/SP399.16

CARTER, W. & AGUIRRE, L. 1965. Structural Geology of Aconcagua province and its relationship to the central Valley Graben, Chile. *Geological Society of America Bulletin*, **76**, 651–664.

CASAMIQUELA, R. M. & FASOLA, A. 1968. Sobre pisadas de dinosauriosdel Cretácico Inferior de Colchagua (Chile). *Boletín Departamento de Geología, Universidad de Chile*, **30**, 1–24.

CASQUET, C., RAPELA, C. *ET AL.* 2012. A history of Proterozoic terranes in southern South America: from Rodinia to Gondwana. *Geoscience Frontiers*, **3**, 137–145.

CECIONI, G. 1978. Petroleum possibilities of the Darwin's Navidad Formation near Santiago, Chile. *Publicación Ocasional del Museo Nacional de Historia Natural, Chile*, **25**, 3–28.

CECIONI, G. 1983. Chanco Formation, a potential Cretaceous reservoir, central Chile. *Journal of Petroleum Geology*, **6**, 89–93.

CECIONI, G. & WESTERMANN, G. E. G. 1968. The Triassic–Jurassic marine transition of coastal central Chile. *Pacific Geology*, **1**, 41–75.

CHARRIER, R. 1973. Geología de las Provincias O'Higgins y Colchagua. *Instituto de Investigación de Recursos Naturales, Santiago*, **7**, 11–69.

CHARRIER, R. 1977. Geology of region of Huentelauquén, Coquimbo Province, Chile. *In*: ISHIKAWA, T. & AGUIRRE, L. (eds) *Comparative Studies on the Geology of the Circum-Pacific Orogenic Belt in Japan and Chile*. First Report. Japanese Society for Promotion of Science, Tokyo, 81–94.

CHARRIER, R. 1979. El Triásico de Chile y regiones adyacentes de Argentina: una reconstrucción paleogeográfica y paleoclimática. *Comunicaciones*, **26**, 1–37.

CHARRIER, R. 1981*a*. Mesozoic and Cenozoic stratigraphy of the Central Argentinian Chilean Andes (32° 35°S) and chronology of their tectonic evolution. *Zentralblatt fur Geologie und Palaontologie*, Teil **1**, 344–355.

CHARRIER, R. 1981*b*. *Geologie der chilenischen Hauptkordillere zwischen 34°30p südlicher Breite und ihre tektonische, magmatische und paleogeographische Entwicklung*. Berliner Geowissenschaftliche Abhandlungen (A), Berlin, **36**.

CHARRIER, R. 1982. La Formación Leñas-Espinoza: redefinición, petrografía y ambiente de sedimentación. *Revista Geológica de Chile*, **17**, 71–82.

CHARRIER, R. 1984. Areas subsidentes en el borde occidental de la cuenca de tras arco jurásico cretácica, Cordillera Principal Chilena entre 34° y 34°30'S. *In*: *IX Congreso Geológico Argentino*. Bariloche, Argentina, Actas, **2**, 107–124.

CHARRIER, R. & VICENTE, J. C. 1972. Liminary and Geosyncline Andes: major orogenic phases and synchronical evolution of the Central and Austral sectors of the Southern Andes. *Conferencia sobre Problemas de la Tierra Sólida, Buenos Aires*, **2**, 451–470.

CHARRIER, R. & MUNIZAGA, F. 1979. Edades K–Ar de volcanitas cenozoicas del sector cordillerano del río Cachapoal, Chile (34° 15' de latitud Sur). *Revista Geológica de Chile*, **7**, 41–51.

CHARRIER, R. & MUÑOZ, N. 1994. Jurassic–Cretaceous paleogeographic evolution of the Chilean Andes at 23°–24°S and 34°–35°S latitude: a comparative analysis. *In*: REUTTER, K. J., SCHEUBER, E. & WIGGER, P. (eds) *Tectonics of the Southern Central Andes*. Springer, Heidelberg, 233–242.

CHARRIER, R., WYSS, A. R. *ET AL.* 1996. New evidence for late Mesozoic–Early Cenozoic evolution of the Chilean Andes in the Upper Tinguiririca valley (35°S), Central Chile. *Journal of South American Earth Sciences*, **9**, 393–422.

CHARRIER, R., BAEZA, O. *ET AL.* 2002. Evidence for Cenozoic extensional basin development and tectonic inversion south of the flat-slab segment, southern Central Andes, Chile, (33°–36° S.L.). *Journal of South American Earth Sciences*, **15**, 17–139.

CHARRIER, R., PINTO, L. & RODRÍGUEZ, M. P. 2007. Tectonostratigraphic evolution of the Andean Orogen in Chile. *In*: MORENO, T. & GIBBONS, W. (eds) *The Geology of Chile*. Geological Society, London, 21–114.

CHARRIER, R., FARÍAS, M. & MAKSAEV, V. 2009. Evolución tectónica, paleogeográfica y metalogénica durante el Cenozoico en los Andes de Chile norte y central e implicaciones para las regiones adyacentes de Bolivia y Argentina. *Revista de la Asociación Geológica Argentina*, **65**, 5–35.

CHARRIER, R., CROFT, D. A., FLYNN, J. J., PINTO, L. & WYSS, A. R. 2012. Mamíferos fósiles cenozoicos en Chile: implicancias paleontológicas y tectónicas. Continuación de investigaciones iniciadas por Darwin en América del Sur. *In*: VELOSO, A. & SPOTORNO, A. (eds) *Darwin y la evolución: avances en la Universidad de Chile*. Editorial Universitaria, Santiago de Chile, 281–316.

CHARRIER, R., HÉRAIL, G., PINTO, L., GARCÍA, M., RIQUELME, R., FARÍAS, M. & MUÑOZ, N. 2013. Cenozoic tectonic evolution in the Central Andes in northern Chile and west-central Bolivia. Implications for paleogeographic, magmatic and mountain building evolution. *International Journal of Earth Sciences*, **102**, 235–264, http://dx.doi.org/10.1007/s00531-012-0801-4

CHERNICOFF, J. & RAMOS, V. A. 2003. El basamento de la Sierra de San Luis: nuevas evidencias magnéticas y sus

implicancias tectónicas. *Revista de la Asociación Geológica Argentina*, **58**, 511–524.
CHERNICOFF, C. J., ZAPPETTINI, E. O., SANTOS, J. O. S., GODEAS, M. C., BELOUSOVA, E. & MCNAUGHTON, N. K. 2012. Identification and isotopic studies of early Cambrian magmatism (El Carancho Igneous Complex) at the boundary between Pampia terrane and the Río de la Plata craton, La Pampa province, Argentina. *Gondwana Research*, **21**, 378–393.
COIRA, B. L., DAVIDSON, J. D., MPODOZIS, C. & RAMOS, V. A. 1982. Tectonic and magmatic evolution of the Andes of Northern Argentina and Chile. *Earth Science Reviews*, **18**, 303–332.
COLLO, G., ASTINI, R. A., CARDONA, A., DO CAMPO, M. D. & CORDANI, U. 2008. Edad del metamorfismo de las unidades con bajo grado de la región central del Famatina: La impronta del ciclo orogénico oclóyico. *Revista Geológica de Chile*, **35**, 191–213.
COLLO, G., ASTINI, R. A., CAWOOD, P. A., BUCHAN, C. & PIMENTEL, M. 2009. U–Pb detrital zircon ages and Sm–Nd isotopic features in low-grade metasedimentary rocks of the Famatina belt: implications for late Neoproterozoic-early Palaeozoic evolution of the proto-Andean margin of Gondwana. *Journal of the Geological Society*, **166**, 303–319.
COLOMA, F., SALAZAR, E. & CREIXELL, C. 2012. Nuevos antecedentes acerca de la construcción de los plutones Pérmicos y Permo-Triásicos en el valle del río Tránsito, región de Atacama, Chile. *In*: XIII *Congreso Geológico Chileno*, Antofagasta, Tematic Session 3: Magmatism and Metamorphism, 324–326, digital abstract.
CORNEJO, P., MATTHEWS, S. & PÉREZ, C. 2003. The 'K–T' compressive deformation event in northern Chile (24°–27°S). *In*: *X Congreso Geológico Chileno*, Concepción, Tematic Session 1: Tectonics, 13 p., digital.
CORTÉS, J. M. & KAY, S. M. 1994. Una dorsal oceánica como origen de las lavas almohadilladas del Grupo Ciénaga del Medio (Silúrico-Devónico) de la Precordillera de Mendoza. *In*: *VII Congreso Geológico Chileno*, Concepción, Actas, **I**, 1005–1009.
COVACEVICH, V., VARELA, J. & VERGARA, M. 1976. Estratigrafía y sedimentación de la Formación Baños del Flaco al sur del rio Tinguiririca, Cordillera de los Andes, provincia de Curicó, Chile. *In*: *I Congreso Geológico Chileno*, Santiago, **1**, A191–A211.
COWAN, D. S. 1985. Structural styles in Mesozoic and Cenozoic melanges in the western Cordillera of North-America. *Geological Society of America Bulletin*, **96**, 451–462.
CREIXELL, C., PARADA, M. A., MORATA, D., VÁSQUEZ, P., PÉREZ DE ARCE, C. & ARRIAGADA, C. 2011. Middle–Late Jurassic to Early Cretaceous transtension and transpression during arc building in central Chile: evidence from mafic dike swarms. *Andean Geology*, **38**, 16–42.
CRISTALLINI, E. O. & RAMOS, V. A. 1996. Los depósitos continentales cretácicos y volcanitas asociadas. *In*: RAMOS, V. A. (ed.) *Geología de la Región del Aconcagua, Provincias de San Juan y Mendoza*. Dirección Nacional del Servicio Geológico, Buenos Aires, Argentina, Anales, **24**, 231–274.
CROFT, D. A., RADIC, J. P., ZURITA, E., CHARRIER, R., FLYNN, J. & WYSS, A. R. 2003. A Miocene toxodontid (Mammalia: notoungulata) from the sedimentary series of the Cura-Mallín Formation, Lonquimay, Chile. *Revista Geológica de Chile*, **30**, 285–298.
CUADRA, P. 1986. Geocronología K–Ar del yacimiento El Teniente y áreas adyacentes. *Revista Geológica de Chile*, **27**, 3–26.
CUCCHI, R. J. 1972. Edades radimétricas y correlación de metamorfitas de la Precordillera, San Juan–Mendoza, Rep. Argentina. *Revista de la Asociación Geológica Argentina*, **26**, 503–515.
DALLA SALDA, L. H., CINGOLANI, C. & VARELA, R. 1992a. Early Palaeozoic Orogenic belt of the Andes in southwestern South-America – result of Laurentia–Gondwana collision. *Geology*, **20**, 617–620.
DALLA SALDA, L., DALZIEL, I. W. D., CINGOLANI, C. A. & VARELA, R. 1992b. Did the Taconic Appalachians continue into southern South America? *Geology*, **20**, 1059–1062.
DAHLQUIST, J. A., PANKHURST, R. J. *ET AL*. 2013. Hf and Nd isotopes in Early Ordovician to Early Carboniferous granites as monitors of crustal growth in the Proto-Andean margin of Gondwana. *Gondwana Research*, **23**, 1617–1630.
DARWIN, C. 1846. *Geological Observations on South America; Part III, The Geology of the Voyage of the Beagle*. Smith Elder, London.
DAVIDSON, J. 1971. *Tectónica y paleogeografía de la Cordillera Principal en el área de las Nacientes del Teno, Curicó, Chile*. Thesis, Departamento de Geología, Universidad de Chile.
DAVIDSON, J. & VICENTE, J. C. 1973. Características paleogeográficas y estructurales del área fronteriza de las nacientes del Teno (Chile) y Santa Elena (Argentina), (Cordillera Principal, 35° a 35°15′ de Latitud Sur). *In*: *V Congreso Geológico Argentino*. Carlos Paz, Argentina, Actas, **5**, 11–55.
DÁVILA, F. M., ASTINI, R. A. & JORDAN, T. E. 2005. Cargas subcorticales en el antepaís andino y la planicie pampeana: evidencias estratigráficas, topográficas y geofísicas. *Revista de la Asociación Geológica Argentina*, **60**, 775–786.
DAVIS, J. S., ROESKE, S. M., MCCLELLAND, W. C. & SNEE, L. W. 1999. Closing the ocean between the Precordillera terrane and Chilenia: early Devonian ophiolite emplacement and deformation in the southwest Precordillera. *In*: RAMOS, V. A. & KEPPIE, J. D. (eds) *Laurentia-Gondwana Connections before Pangea*. Geological Society of America, Boulder, CO, Special Papers, **336**, 115–138.
DECKART, K., HERVÉ, F., FANNING, C. M., RAMÌREZ, V., CALDERÓN, M. & GODOY, E. 2014. U–Pb geochronology and Hf–O isotopes of zircons from the Pennsylvanian Coastal Batholith, south-central Chile. *Andean Geology*, **41**, 49–82.
DEDIOS, P. 1967. *Cuadrángulo Vicuña, Provgincia de Coquimbo*. Instituto de Investigaciones Geológicas, Santiago, Carta, **16**.
DE LA CRUZ, R. & SUÁREZ, M. 1997. El Jurásico de la cuenca de Neuquén en Lonquimay, Chile: formación Nacientes del Biobio (38°–39°S). *Revista Geológica de Chile*, **24**, 3–24.
DÍAZ-MARTÍNEZ, E., MAMET, B., ISAACSON, P. E. & GRADER, G. W. 2000. Permian marine sedimentation

in northern Chile: new paleontological evidence from the Juan de Morales Formation, and regional paleogeographic implications. *Journal of South American Earth Sciences*, **13**, 511–525.

DI GIULIO, A., RONCHI, A., SANFILIPPO, A., TIEPOLO, M., PIMENTEL, M. & RAMOS, V. A. 2012. Detrital zircon provenance from the Neuquén Basin (south-central Andes): Cretaceous geodynamic evolution and sedimentary response in a retroarc-foreland basin. *Geology*, **40**, 559–562.

DRAKE, R. E. 1976. Chronology of Cenozoic igneous and tectonic events in the Central Chilean Andes-latitudes 35°30′–36°00′S. *Journal of Volcanology and Geothermal Research*, **1**, 265–284.

DRAKE, R. E., CURTIS, G. & VERGARA, M. 1976. Potassium–argon dating of igneous activity in the central Chilean Andes – latitude 33°S. *Journal of Volcanology and Geothermal Research*, **1**, 285–295.

ELGUETA, S. 1990. Sedimentación marina y paleogeograía del Terciario Superior de la Cuenca de Temuco, Chile. *In*: *II Simposio sobre el Terciario en Chile*. Concepción, Chile, Actas, **1**, 85–96.

EMPARAN, C. & CALDERÓN, G. 2014. *Geología del Área Ovalle-Peña Blanca, Región de Coquimbo*. Servicio Nacional de Geología y Minería, Santiago.

EMPARAN, C. & PINEDA, G. 1999. *Área Condoriaco-Rivadavia, Región de Coquimbo*. Servicio Nacional de Geología y Minería, Santiago, Mapas Geológicos, **12**.

EMPARAN, C. & PINEDA, G. 2006. *Geología del Área Andacollo-Puerto Aldea, Región de Coquimbo*. Servicio Nacional de Geología y Minería, Santiago, Carta Geológica de Chile, Serie Geología Básica, **96**.

ENCINAS, A., LE ROUX, J., BUATOIS, L. A., NIELSEN, S. N., FINGER, K. L., FOURTANIER, E. & LAVENU, A. 2006a. Nuevo esquema estratigráfico para los depósitos marinos mio-pliocenos del área de Navidad (33°00′–34°30′S), Chile central. *Revista Geológica de Chile*, **33**, 221–246.

ENCINAS, A., MAKSAEV, V., PINTO, L., LE ROUX, J., MUNIZAGA, F. & ZENTILLI, M. 2006b. Pliocene lahar deposits in the Coastal Cordillera of central Chile: implications for uplift, avalanche deposits, and porphyry copper systems in the Main Andean Cordillera. *Journal of South American Earth Sciences*, **20**, 369–381.

ESCAYOLA, M. P., RAMÉ, G. A. & KRAEMER, P. E. 1996. Caracterización y significado geotectónico de las fajas ultramáficas de las Sierras Pampeanas de Córdoba. *In*: *XIII Congreso Geológico Argentino y III Congreso de Exploración de Hidrocarburos*. Mendoza, Argentina, **3**, 421–438.

ESCAYOLA, M. P., PIMENTEL, M. & ARMSTRONG, R. 2007. Neoproterozoic backarc basin: sensitive high-resolution ion microprobe U–Pb and Sm–Nd isotopic evidence from the Eastern Pampean Ranges, Argentina. *Geology*, **35**, 495–498.

EYQUEM, D. 2009. *Volcanismo cuaternario de Sierras de Bellavista. Comparación geoquímica con el magmatismo contemporáneo del arco comprendido entre los 34°30′ y los 35°30′S*. Thesis, Departamento de Geología, Universidad de Chile, Santiago.

FARÍAS, M., CHARRIER, R. ET AL. 2008. Late Miocene high and rapid surface uplift and its erosional response in the Andes of central Chile (33°–35°S). *Tectonics*, **27**, TC1005, http://dx.doi.org/10.1029/2006TC002046

FARÍAS, M., COMTE, D. ET AL. 2010. Crustal scale architecture in central Chile based on seismicity and surface geology: implications for Andean mountain building. *Tectonics*, **29**, TC3006, http://dx.doi.org/10.1029/2009TC002480

FERNÁNDEZ-SEVESO, F. & TANKARD, A. J. 1995. Tectonics and stratigraphy of the Late Paleozoic Paganzo basin of western Argentina and its regional implications. *In*: TANKARD, A. J., SUÁREZ SORUCO, R. & WELSINK, H. J. (eds) *Petroleum Basins of South America*. American Association of Petroleum Geologists, Tulsa, OK, Memoirs, **62**, 285–301.

FERRANDO, R., ROPERCH, P., MORATA, D., ARRIAGADA, C., RUFFET, G. & CÓRDOBA, M. L. 2014. A paleomagnetic and magnetic fabric study of the Illapel Plutonic complex, Coastal Range, central Chile: implications for emplacement mechanism and regional tectonic evolution during the mid-Cretaceous. *Journal of South American Earth Science*, **50**, 12–26.

FINGER, K., ENCINAS, A., NIELSEN, S. & PETERSON, D. 2003. Microfaunal indications of late Miocene deep-water basins off the central coast of Chile. *In*: *X Congreso Geológico Chileno*. Concepción, Tematic Session 3, 8 p., digital abstract.

FINGER, K. L., NIELSEN, S. N., DEVRIES, T. J., ENCINAS, A. & PETERSON, D. E. 2007. Paleontologic evidence for sedimentary displacement in Neogene forearc basins of Central Chile. *Palaios*, **22**, 3–16, http://dx.doi.org/10.2110/palo.2005.p05-081r

FLYNN, J. J., WYSS, A. R. & CHARRIER, R. 2007. South America's missing mammals. *Scientific American*, **296**, 68–75.

FLYNN, J. J., CHARRIER, R., CROFT, D. A., GANS, P. B., HERRIOTT, T. M., WERTHEIM, J. A. & WYSS, A. R. 2008. Chronologic implications of new Miocene mammals from the Cura–Mallín and Trapa–Trapa Formations, Laguna del Laja area, south Central Chile. *Journal of South American Earth Sciences*, **26**, 412–423.

FOCK, A., CHARRIER, R., FARÍAS, M. & MUÑOZ, M. 2006. Fallas de vergencia Oeste en la Cordillera Principal de Chile Central: inversión de la cuenca de Abanico (33°–34°S). *In*: HONGN, F., BECCHIO, R. & SEGGIARIO, R. (eds) *XII Reunión sobre microtectónica y geología estructural*. Revista de la Asociación Geológica Argentina, Serie D, Publicación Especial, Buenos Aires, Argentina, **9**, 48–55.

FOLGUERA, A. & RAMOS, V. A. 2011. Repeated eastward shifts of arc magmatism in the Southern Andes: a revision to the long-term patter of Andean uplift and magmatism. *Journal of South American Earth Sciences*, **32**, 531–546, http://dx.doi.org/10.1016/j.jsames.2011.04.003

FRUTOS, J. 1981. Andean tectonic as consequence of sea-floor spreading. *Tectonophysics*, **72**, 21–32.

FUENTES, F. 2004. *Petrología y metamorfismo de muy bajo grado de unidades volcánicas oligo-miocenas en la ladera occidental de los Andes de Chile Central (33°S)*. PhD thesis, Universidad de Chile.

FUENTES, F., FÉRAUD, G., AGUIRRE, L. & MORATA, D. 2001. Convergent strategy to date metamorphic minerals in subgreenschit facies metabasites by the $^{40}Ar/^{39}Ar$ method. *In*: *III South American Symposium on Isotope Geology*. Pucón, Chile, electronic abstracts, 34–36.

FUENTES, F., FÉRAUD, G., AGUIRRE, L. & MORATA, D. 2005. $^{40}Ar/^{39}Ar$ dating of volcanism and subsequent very low-grade metamorphism in a subsiding basin: example of the Cretaceous lava series from central Chile. *Chemical Geology*, **214**, 157–177.

FUENZALIDA, H. 1940. Algunos afloramientos Paleozoicos de la desembocadura del Choapa. *Boletín del Museo Nacional de Historia Natural*, **28**, 37–64.

FUENZALIDA, H., COOKE, R., PASKOFF, R., SEGERSTROM, K. & WEISCHET, W. 1965. High stands of Quaternary sea level along the Chilean coast. *In*: WRIGHT, H. E. & FREY, D. (eds) *International Studies on the Quaternary. Papers Prepared on the Occasion of the VII Congress of the International Association for Quaternary Research Boulder*, Colorado. Geological Society of America, Boulder, CO, Special Papers, **84**, 473–496.

GALLEGO, A. 1994. *Paleoambiente y mecanismo de depositación de la secuencia sedimentaria que aflora en el sector de Polpaico, Región Metropolitana, Chile*. Thesis, Departamento de Geología, Universidad de Chile, Santiago.

GANA, P. 1991. Magmatismo bimodal del Triásico Superior – Jurásico Inferior, en la Cordillera de la Costa, Provincias de Elqui y Limarí, Chile. *Revista Geológica de Chile*, **18**, 55–67.

GANA, P. & TOSDAL, R. M. 1996. Geocronología U–Pb y K–Ar en intrusivos del Paleozoico y Mesozoico de la Cordillera de la Costa, región de Valparaíso, Chile. *Revista Geológica de Chile*, **23**, 151–164.

GANA, P. & WALL, R. 1997. Evidencias geocronológicas 40Ar/39Ar y K–Ar de un hiatus Cretácico Superior-Eoceno en Chile Central (33°–33°30′ S). *Revista Geológica de Chile*, **24**, 145–163.

GANA, P. & ZENTILLI, M. 2000. Historia termal y exhumación de intrusivos de la Cordillera de la Costa de Chile central. *In*: *IX Congreso Geológico Chileno*. Puerto Varas, Actas, **2**, 664–668.

GARCÍA, C. 1991. *Geología del sector de quebrada El Teniente, Región de Coquimbo*. Thesis, Departamento de Geología, Universidad de Chile.

GARCÍA-SANSEGUNDO, J., FARÍAS, P., RUBIO-ORDÓÑEZ, A. & HEREDIA, N. 2014a. The Palaeozoic basement of the Andean Frontal Cordillera at 34° S (Cordón del Carrizalito, Mendoza Province, Argentina): geotectonic implications. *Journal of Iberian Geology*, **40**, 321–330. http://dx.doi.org/10.5209/rev_JIGE.2014.v40.n2.45299.

GARCÍA-SANSEGUNDO, J., FARIAS, P., HEREDIA, N., GALLASTEGUI, G., CHARRIER, R., RUBIO-ORDÓÑEZ, A. & CUESTA, A. 2014b. Structure of the Andean Paleozoic basement in the Chilean coast at 31° 30′ S: geodynamic evolution of a subduction margin. *Journal of Iberian Geology*, **40**, 293–308. http://dx.doi.org/10.5209/rev_JIGE.2014.v40.n2.45300

GERBI, C., ROESKE, S. M. & DAVIS, J. S. 2002. Geology and structural history of the southwest Precordillera margin, northern Mendoza Province, Argentina. *Journal of South American Earth Sciences*, **14**, 821–835.

GIAMBIAGI, L. & MARTÍNEZ, A. N. 2008. Permo-Triassic oblique extension in the Uspallata–Potrerillos area, western Argentina. *Journal of South American Earth Sciences*, **26**, 252–260.

GIAMBIAGI, L., MESCUA, J., BECHIS, F., MARTÍNEZ, A. & FOLGUERA, A. 2011. Pre-Andean deformation of the Precordillera southern sector, Southern Central Andes. *Geosphere*, **7**, 219–239.

GIAMBIAGI, L., TASSARA, A. ET AL. 2014. Evolution of shallow and deep structures along the Maipo – Tunuyán transect (33°40′S): from the Pacific coast to the Andean foreland. *In*: SEPÚLVEDA, S. A., GIAMBIAGI, L. B., MOREIRAS, S. M., PINTO, L., TUNIK, M., HOKE, G. D. & FARÍAS, M. (eds) *Geodynamic Processes in the Andes of Central Chile and Argentina*. Geological Society, London, Special Publications, **399**. First published online February 27, 2014, http://dx.doi.org/10.1144/SP399.14

GIAMBIAGI, L. B., RAMOS, V. A., GODOY, E., ÁLVAREZ, P. P. & ORTS, S. 2003a. Cenozoic deformation and tectonic style of the Andes, between 33° and 34° South Latitude. *Tectonics*, **22**, 1041–1051.

GIAMBIAGI, L. B., ALVAREZ, P. P., GODOY, E. & RAMOS, V. A. 2003b. The control of pre-existing extensional structures on the evolution of the southern sector of the Aconcagua fold and thrust belt, Southern Andes. *Tectonophysics*, **369**, 1–19.

GLODNY, J., LOHRMANN, J., ECHTLER, H., GRÄFE, K., SEIFERT, W., COLLAO, S. & FIGUEROA, O. 2005. Internal dynamics of a paleoaccretionary wedge: insights from combined isotope tectonochronology and sandbox modelling of the south-central Chilean fore-arc. *Earth and Planetary Science Letters*, **231**, 23–39.

GLODNY, J., ECHTLER, H. ET AL. 2006. Long-term geological evolution and mass flow balance of the South-Central Andes. *In*: ONCKEN, O., CHONG, G., FRANZ, G., GIESE, P., GÖTZE, H. J., RAMOS, V., STRECKER, M. & WIGGER, P. (eds) *The Andes – Active Subduction Orogeny*. Frontiers in Earth Sciences, Springer, Berlin, **1**, 401–442.

GLODNY, J., ECHTLER, H., COLLAO, S., ARDILES, M., BURÓN, P. & FIGUEROA, O. 2008. Differential Late Paleozoic active margin evolution in South-Central Chile (37°S–40°S) – the Lanalhue Fault Zone. *Journal of South American Earth Sciences*, **26**, 397–411.

GODOY, E. 1982. Geología del área de Montenegro, Cuesta de Chacabuco, Región Metropolitana. El problema de la Formación Lo Valle. *In*: *II Congreso Geológico Chileno*, Concepción, **1**, A124–A146.

GODOY, E. 2011. Structural setting and diachronism in the Central Andean Eocene to Miocene volcano-tectonic basins. *In*: SALFITY, J. A. & MARQUILLAS, R. A. (eds) *Cenozoic Geology of the Central Andes of Argentina*. SCS, Salta, Argentina, 155–167.

GODOY, E. & CHARRIER, R. 1991. Antecedentes mineralógicos para el origen de las metabasitas y metacherts del Complejo Metamórfico del Choapa (Región de Coquimbo, Chile): un prisma de acreción Paleozoico Inferior. *In*: *VI Congreso Geológico Chileno*, Viña del Mar, Actas, 410–414.

GODOY, E. & LARA, L. 1994. Segmentación estructural andina a los 33°–34°: nuevos datos en la Cordillera Principal. *In*: *VII Congreso Geológico Chileno*, Concepción, **2**, 1344–1348.

GÓMEZ, R. 2001. *Geología de las unidades volcanogénicas cenozoicas del área industrial de la Mina El Teniente, entre Colón y Coya, Cordillera Principal* de Rancagua, VI Región. Thesis, Departamento de Geología, Universidad de Chile.

GONZÁLEZ, O. 1963. Observaciones geológicas en el valle del Río Volcán. *Revista Minerales*, **17**, 20–61.

GONZÁLEZ, O. & VERGARA, M. 1962. *Reconocimiento geológico de la Cordillera de los Andes entre los paralelos 35° y 38° latitud S*. Instituto de Geología, Universidad de Chile, Santiago, Publicación, 24.

GONZÁLEZ-BONORINO, F. 1970. Series metamórficas del basamento cristalino de la Cordillera de la Costa de Chile Central. Departamento de Geología, Universidad de Chile. *Publicaciones*, **37**, 1–68.

GONZÁLEZ-BONORINO, F. 1971. Metamorphism of the crystalline basement of Central Chile. *Journal of Petrology*, **12**, 149–175.

GONZÁLEZ-BONORINO, F. & AGUIRRE, L. 1970. Metamorphic facies series of the crystalline basement of Chile. *Geologische Rundschau*, **59**, 979–993.

GONZÁLEZ-MENÉNDEZ, L., GALLASTEGUI, G., CUESTA, A., HEREDIA, N. & RUBIO-ORDÓÑEZ, A. 2013. Petrogenesis of Early Paleozoic basalts and gabbros in the western Cuyania terrane: constraints on the tectonic setting of the southwestern Gondwana margin (Sierra del Tigre, Andean Argentine Precordillera). *Gondwana Research*, **24**, 359–376.

GORDILLO, C. E. 1984. Migmatitas cordieríticas de la Sierra de Córdoba, condiciones físicas de la migmatización. Academia Nacional de Ciencias. *Miscelánea*, **68**, 1–40.

GREGORI, D. & BENEDINI, L. 2013. The Cordon del Portillo Permian magmatism, Mendoza, Argentina, plutonic and volcanic sequences at the western margin of Gondwana. *Journal of South American Earth Sciences*, **42**, 61–73.

GREGORI, D. A., FERNÁNDEZ-TURIEL, J. L., LÓPEZ-SOLER, A. & PETFORD, N. 1996. Geochemistry of Upper Palaeozoic–Lower Triassic granitoids of the Central Frontal Cordillera (33°10′–33°45′), Argentina. *Journal of South American Earth Sciences*, **9**, 141–151.

GROEBER, P. 1918. Estratigrafía del Dogger en la República Argentina. Estudio sintético comparativo. *Dirección General de Minas, Geología e Hidrogeología, Boletín*, **18**, Serie B (Geología), 1–81.

GROEBER, P. 1922. Pérmico y Triásico en la costa de Chile. *Revista de la Sociedad Argentina de Ciencias Naturales*, **5**, 979–994.

GROEBER, P. 1946. Observaciones geológicas a lo largo del meridiano 70°. 1, Hoja Chos Malal. *Revista de la Asociación Geológica Argentina*, **1**, 117–208. Reprint in Asociación Geológica Argentina, Serie C, Reimpresiones (1980) 1, 1–174.

GROEBER, P. 1951. La Alta Cordillera entre las latitudes 34° y 29°30′. Instituto Investigaciones de las Ciencias Naturales. *Revista del Museo Argentino de Ciencias Naturales Bernardino Rivadavia (Ciencias Geológicas)*, **1**, 1–352.

GROEBER, P. 1953. Mesozoico. *Geografía de la República Argentina, Sociedad Argentina de Estudios Geográficos GAEA, Buenos Aires*, **2**, 9–541.

GROMET, L. P. & SIMPSON, C. 2000. Cambrian orogeny in the Sierras Pampeanas, Argentina: ridge subduction or continental collision? *Geological Society of America Abstracts with Programs*, **32**, A-505.

GROSSE, P., SÖLLNER, F., BÁEZ, M. A., TOSELLI, A. J., ROSSI, J. N. & ROSA, J. D. 2009. Lower Carboniferous post-orogenic granites in central-eastern Sierra de Velasco, Sierras Pampeanas, Argentina: U–Pb monazite geochronology, geochemistry and Sr–Nd isotopes. *International Journal of Earth Sciences*, **98**, 1001–1025.

GUDNASON, J., HOLM, P. M., SØAGER, N. & LLAMBÍAS, E. J. 2012. Geochronology of the late Pliocene to recent volcanic activity in the Payenia back-arc volcanic province, Mendoza, Argentina. *Journal of South American Earth Sciences*, **37**, 191–201.

GULBRANSON, E. L., MONTAÑEZ, I. P., SCHMITZ, M. D., LIMARINO, C. O., ISBELL, J. L., MARENSSI, S. A. & CROWLEY, J. L. 2010. High-precision U–Pb calibration of Carboniferous glaciation and climate history, Paganzo Group, NW Argentina. *Geological Society of America Bulletin*, **122**, 1480–1498.

GUTIÉRREZ, N. M., HINOJOSA, L. F., LE ROUX, J. P. & PEDROZA, V. 2013. Evidence for an Early–Middle Miocene age of the Navidad Formation (central Chile): paleontological, paleoclimatic and tectonic implications. *Andean Geology*, **40**, 66–78.

HALLAM, A., BIRÓ-BAGÓCZKY, L. & PÉREZ, E. 1986. Facies analysis of the Valdés Formation (Tithonian–Hauterivian) of the High Cordillera of Central Chile, and the Paleogeographic evolution of the Andean Basin. *Geological Magazine*, **123**, 425–435.

HALLER, M. J. & RAMOS, V. A. 1984. Las ofiolitas famatinianas (Eopaleozoico) de las provincias de San Juan y Mendoza. *In*: *IX Congreso Geológico Argentino*, Bariloche, 66–83.

HALLER, M. J. & RAMOS, V. A. 1993. Las ofiolitas y otras rocas afines. *In*: RAMOS, V. A. (ed.) *Geología y Recursos Naturales de Mendoza. Relatorio XII Congreso Geológico Argentino y II Congreso de Exploración de Hidrocarburos*. Buenos Aires, 31–40.

HEREDIA, N., FERNÁNDEZ, L. R. R., GALLASTEGUI, G., BUSQUETS, P. & COLOMBO, F. 2002. Geological setting of the Argentine Frontal Cordillera in the flat-slab segment (30° 00′–31° 30′ S latitude). *Journal of South American Earth Sciences*, **15**, 79–99.

HEREDIA, N., FARIAS, P., GARCÍA-SANSEGUNDO, J. & GIAMBIAGI, L. 2012. The Basement of the Andean Frontal Cordillera in the Cordón del Plata (Mendoza, Argentina): geodynamic evolution. *Andean Geology*, **39**, 242–257.

HERVÉ, F. 1977. Petrology of the Crystalline Basement of the Nahuelbuta Mountains, Southcentral Chile. *In*: ISHIKAWA, T. & AGUIRRE, L. (eds) *Comparative Studies on the Geology of the Circum Pacific Orogenic Belt in Japan and Chile*. Japan Society for the Promotion of Science, Tokyo, 1–51.

HERVÉ, F. 1988. Late Palaeozoic subduction and accretion in Southern Chile. *Episodes*, **11**, 183–188.

HERVÉ, F., MUNIZAGA, F., GODOY, E. & AGUIRRE, L. 1974. Late Paleozoic K/Ar ages of blueschists from Pichilemu, Central Chile. *Earth and Planetary Science Letters*, **23**, 261–264.

HERVÉ, F., KAWASHITA, K., MUNIZAGA, F. & BASEI, M. 1984. Rb–Sr isotopic ages from late Paleozoic metamorphic rocks of Central Chile. *Journal of the Geological Society London*, **141**, 877–884.

Hervé, F., Munizaga, F., Parada, M. A., Brook, M., Pankhurst, R. J., Snelling, N. J. & Drake, R. 1988. Granitoids of the Coast Range of central Chile: geochronology and geologic setting. *Journal of South American Earth Sciences*, **1**, 185–194.

Hervé, F., Fanning, C. M. & Pankhurst, R. J. 2003. Detrital zircon age patterns and provenance of the metamorphic complexes of southern Chile. *Journal of South American Earth Sciences*, **16**, 107–123.

Hervé, F., Faúndez, V., Calderón, M., Massonne, H.-J. & Willner, A. P. 2007. Metamorphic and plutonic basement complexes. *In*: Moreno, T. & Gibbons, W. (eds) *The Geology of Chile*. Geological Society, London, 5–20.

Hervé, F., Calderón, M., Fanning, C. M., Pankhurst, R. J. & Godoy, E. 2013. Provenance variations in the Late Paleozoic accretionary complex of central Chile as indicated by detrital zircons. *Gondwana Research*, **23**, 1122–1135.

Hervé, F., Fanning, C. M., Calderón, M. & Mpodozis, C. 2014. Early Permian to Late Triassic batholiths of the Chilean Frontal Cordillera (28°–31°S): SHRIMP U–Pb zircon ages and Lu–Hf and O isotope systematics. *Lithos*, **184–187**, 436–446.

Hildreth, W., Singer, B., Godoy, E. & Munizaga, F. 1998. The age and constitution of Cerro Campanario, a mafic stratovolcano in the Andes of Central Chile. *Revista Geológica de Chile*, **25**, 17–28.

Iannizzotto, N. F., Rapela, C. W., Baldo, E. G. A., Galindo, C., Fanning, C. M. & Pankhurst, R. J. 2013. The Sierra Norte–Ambargasta batholith: Late Ediacaran–Early Cambrian magmatism associated with Pampean transpressional tectonics. *Journal of South American Earth Sciences*, **42**, 127–143.

Irigoyen, M. V., Buchan, K. L. & Brown, R. L. 2000. Magnetostratigraphy of Neogene Andean foreland-basin strata, lat 33°S, Mendoza Province, Argentina. *Geological Society of America Bulletin*, **112**, 803–816.

Irwin, J. J., García, C., Hervé, F. & Brook, M. 1988. Geology of part of a long-lived dynamic plate margin – the Coastal Cordillera of north-central Chile, latitude 30°51′–31°S. *Canadian Journal of Earth Sciences*, **25**, 603–624.

Isacks, B. L. 1988. Uplift of the Central Andean Plateau and bending of the Bolivian Orocline. *Journal of Geophysical Research*, **93**, 3211–3231.

James, D. E. 1971. Plate tectonic model for the evolution of the Central Andes. *Geological Society of America Bulletin*, **82**, 3325–3346.

Jara, P. & Charrier, R. 2014. Nuevos antecedentes geocronológicos y estratigráficos para la Alta Cordillera de Chile central a ∼32°10′S. Implicancias paleogeográficas y estructurales. *Andean Geology*, **41**, 174–209, http://dx.doi.org/10.5027/andgeoV41n1-a07

Jara, P., Likerman, J., Winocur, D., Guiglione, M. C., Cristallini, E. O., Pinto, L. & Charrier, R. 2014. Role of basin width variation in tectonic inversion: insight from analogue modelling and implications for the tectonic inversion of the Abanico Basin, 32°–34°S, Central Andes. *In*: Sepúlveda, S. A., Giambiagi, L. B., Moreiras, S. M., Pinto, L., Tunik, M., Hoke, G. D. & Farías, M. (eds) *Geodynamic Processes in the Andes of Central Chile and Argentina*. Geological Society, London, Special Publications, **399**. First published online February 27, 2014, http://dx.doi.org/10.1144/SP399.7

Jaros, J. & Zelman, J. 1967. La relación estructural entre las formaciones Abanico y Farellones en la Cordillera del Mesón, Provincia de Aconcagua, Chile. Departamento de Geología, Universidad de Chile, Santiago de Chile, **34**.

Jordan, T., Isacks, B., Ramos, V. A. & Allmendinger, R. W. 1983a. Mountain building model: the Central Andes. *Episodes*, **1983**, 20–26.

Jordan, T. E., Isacks, B. L., Allmendinger, R. W., Brewer, J. A., Ramos, V. A. & Ando, C. J. 1983b. Andean tectonics related to geometry of subducted Nazca plate. *Geological Society of America Bulletin*, **94**, 341–361.

Jordan, T. E., Reynolds III, J. H. & Enkson, J. P. 1997. Variability in age of initial shortening and uplift in the central Andes, 16–33°30′S. *In*: Ruddiman, W. (ed.) *Tectonic Uplift and Climate Change*. Plenum, New York, 41–61.

Jordan, T. E., Burns, W. M., Veiga, R., Pangaro, F., Copeland, P., Kelley, S. & Mpodozis, C. 2001. Extension and basin formation in the southern Andes caused by increased convergence rate: a Mid-Cenozoic trigger for the Andes. *Tectonics*, **20**, 308–324.

Kato, T. & Godoy, E. 1995. Petrogenesis and tectonic significance of late Paleozoic coarse-crystalline blueschist and amphibolite boulders in the Coastal Range of Chile. *International Geology Review*, **37**, 992–1006.

Kay, S. M. & Abbruzzi, J. M. 1996. Magmatic evidence for Neogene lithospheric evolution of the Central Andean flat-slab between 30 and 32°S. *Tectonophysics*, **259**, 15–28.

Kay, S. M., Ramos, V., Mpodozis, C. & Sruoga, P. 1989. Late Paleozoic to Jurassic silicic magmatism at the Gondwana margin: analogy to the Middle Proterozoic in North America. *Geology*, **17**, 324–328.

Kay, S. M., Mpodozis, C., Ramos, V. A. & Munizaga, F. 1991. Magma source variations for mid to late Tertiary volcanic rocks erupted over a shallowing subduction zone and through a thickening crust in the Main Andean Cordillera (28–33°S). *In*: Harmon, R. S. & Rapela, C. W. (eds) *Andean Magmatism and its Tectonic Setting*. Geological Society of America, Boulder, CO, Special Papers, **265**, 113–137.

Keller, M. 1999. *Argentine Precordillera, sedimentary and plate tectonic history of a Laurentian crustal fragment in South America*. Geological Society of America, Boulder, CO, Special Papers, **341**, 1–131.

Kleiman, L. E. & Japas, M. S. 2009. The Choiyoi volcanic province at 34°S–36°S (San Rafael, Mendoza, Argentina): implications for the Late Paleozoic evolution of the southwestern margin of Gondwana. *Tectonophysics*, **473**, 283–299.

Klohn, C. 1960. Geología de la Cordillera de los Andes de Chile Central, provincia de Santiago, O'Higgins, Colchagua y Curicó. *Instituto Investigaciones Geológicas Boletín*, **8**, 1–95.

Kokogián, D. A., Fernández-Seveso, F. & Mosquera, A. 1993. Las secuencias sedimentarias triásicas. *In*: *9° Congreso Geológico Argentino*. Mendoza, Relatorio, 65–78.

KOKOGIÁN, D. A., SPALLETTI, L. ET AL. 1999. Depósitos continentales triásicos. *In*: CAMINOS, R. (ed.) *Geología Argentina*. Servicio Geológico Minero Argentino, Buenos Aires, Argentina, Anales, **29**, 377–397.

KRAEMER, P., ESCAYOLA, M. & MARTINO, R. 1995. Hipótesis sobre la evolución tectónica neoproterozoica de las Sierras Pampeanas de Córdoba (30°40′–32°40′S), Argentina. *Revista de la Asociación Geológica Argentina*, **50**, 47–59.

LANÉS, L. 2005. Late Triassic to Early Jurassic sedimentation in northern Neuquén Basin, Argentina: tectono sedimentary evolution of the first trangression. *Geológica Acta*, **3**, 81–106.

LARSON, R. L. 1991. Geological consequences of superplumes. *Geology*, **19**, 963–966.

LEAL, P. R., HARTMANN, L. A., SANTOS, O., MIRÓ, R. & RAMOS, V. A. 2004. Volcanismo postorogénico en el extremo norte de las Sierras Pampeanas Orientales: nuevos datos geocronológicos y sus implicancias tectónicas. *Revista de la Asociación Geológica Argentina*, **58**, 593–607.

LEGARRETA, L., KOKOGIAN, D. A. & DELLAPE, D. A. 1992. Estructuración terciaria de la cuenca Cuyana: ¿Cuánto de inversión tectónica? *Revista Asociación Geología Argentina*, **47**, 83–86.

LEGARRETA, L., GULISANO, C. A. & ULIANA, M. A. 1993. Las secuencias sedimentarias jurásico-cretácicas. *In*: RAMOS, V. A. (ed.) *Relatorio Geología y Recursos Naturales de Mendoza. XII Congreso Geológico Argentino y II Congreso de Exploración de Hidrocarburos*, Buenos Aires, 87–114.

LETELIER, M. 1977. *Petrología y ambiente de depositación y estructura de las Formaciones Matahuaico, Las Breas, Tres Cruces sensu lato e intrusivos permotriásicos en el área de Rivadavia-Alcohuás, vallede Elqui, IV Región, Chile*. Thesis, Departamento de Geología, Universidad de Chile.

LEVI, B. & AGUIRRE, L. 1960. El Conglomerado de Algarrobo y su relación con las formaciones de Cretcacico Superior de Chile central. *I Jornadas Geológicas Argentinas*, **2**, 417–431.

LEVI, B., AGUIRRE, L., NYSTRÖM, J., PADILLA, H. & VERGARA, M. 1989. Low-grade regional metamorphism in the Mesozoic–Cenozoic volcanic sequences of the Central Andes. *Journal of Metamorphic Petrology*, **7**, 487–495.

LIMARINO, C. O. & SPALLETTI, L. A. 2006. Paleogeography of the upper Paleozoic basins of southern South America: an overview. *Journal of South American Earth Sciences*, **22**, 134–155.

LIMARINO, C. O., TRIPALDI, A., MARENSSI, S. & FAUQUÉ, L. 2006. Tectonic, sea-level, and climatic controls on Late Paleozoic sedimentation in the western basins of Argentina. *Journal of South American Earth Sciences*, **22**, 205–226.

LIMARINO, C. O., CÉSARI, S. N., SPALLETTI, L. A., TABOADA, A. C., ISBELL, J. L., GEUNA, S. & GULBRANSON, E. L. 2013. A paleoclimatic review of southern South America during the late Paleozoic: a record from icehouse to extreme greenhouse conditions. *Gondwana Research*, **25**, 1396–1421.

LIRA, R., MILLONE, H. A., KIRSCHBAUM, A. M. & MORENO, R. S. 1996. Calc-alkaline arc granitoid activity in the Sierra Norte-Ambargasta Ranges, central Argentina. *Journal of South American Earth Sciences*, **10**, 157–177.

LLAMBÍAS, E. J. 1999. Las rocas ígneas gondwánicas. El magmatismo gondwánico durante el Paleozoico Superior-Triásico. *In*: CAMINOS, R. (ed.) *Geología Argentina*. Instituto de Geología y Recursos Minerales, Buenos Aires, Argentina, Anales, **29**, 349–363.

LLAMBÍAS, E. J. & SATO, A. M. 1995. El Batolito de Colangüil: transición entre orogénesis y anorogénesis. *Revista de la Asociación Geológica Argentina*, **50**, 111–131.

LLAMBÍAS, E. J., KLEIMAN, L. E. & SALVARREDI, J. A. 1993. El magmatismo gondwánico. *In*: RAMOS, V. A. (ed.) *Geología y Recursos Naturales de Mendoza. Relatorio XII Congreso Geológico Argentino and II Congreso de exploración de Hidrocarburos*. Mendoza, **I**, 53–64.

LLAMBÍAS, E. J., QUENARDELLE, S. & MONTENEGRO, T. 2003. The Choiyoi Group from central Argentina: a subalkaline transitional to alkaline association in the craton adjacent to the active margin of the Gondwana continent. *Journal of South American Earth Sciences*, **16**, 243–257.

LLAMBÍAS, E. J., BERTOTTO, G. W., RISSO, C. & HERNANDO, I. 2010. El volcanismo cuaternario en el retroarco de Payenia: una revisión. *Revista de la Asociación Geológica Argentina*. Zurich, **67**, 278–300.

LO FORTE, G. L., ANSELMI, G. & AGUIRRE-URRETA, M. B. 1996. Tithonian Paleogeography of the Aconcagua Basin, West-Central Andes of Argentina. *In*: RICCARDI, A. C. (ed.) *Advances in Jurassic Research, Geo-Research Forum*, Transtec Publications, Zurich, **1–2**, 369–376.

LÓPEZ, V. & GREGORI, D. A. 2004. Provenance and evolution of the Guarguaraz Complex, Cordillera Frontal, Argentina. *Gondwana Research*, **7**, 1197–1208.

LÓPEZ GAMUNDI, O. 1994. Facies distribution in an asymmetric half graben: the northern Cuyo Basin (Triassic), western Argentina. *In*: *XIV International Sedimentological Congress*, Recife, Abstract, 6–7.

LÓPEZ GAMUNDI, O. R., ESPEJO, I., CONAGHAN, P. J. & POWELL, C. MCA. 1994. Southern South America. *In*: VEEVERS, J. J. & POWELL, C. MCA. (eds) *Permian Triassic Pangean Basins and Foldbelts Along the Panthalassan Margin of Gondwanaland*. Geological Society of America, Boulder, CO, Memoirs, **184**, 281–329.

MACPHAIL, D. D. 1966. El gran lahar del Laja. *Estudios Geográficos*. Departamento de Geografía, Universidad de Chile, Santiago, 133–155.

MACPHAIL, D. D. & SAA, R. 1967. Los Cerrillos de Teno: a laharic landscape of Central Chile. *Annals of the Association of American Geographers*, **59**, 171.

MAKSAEV, V., MUNIZAGA, F., MCWILLIAMS, M., FANNING, M., MATHUR, R., RUIZ, J. & ZENTILLI, M. 2004. New chronology for El Teniente, Chilean Andes, from U/Pb, $^{40}Ar/^{39}Ar$, Re/Os and fission-track dating: implications for the evolution of a supergiant porphyry Cu-Mo deposit. *In*: SILLITOE, R. H., PERELLÓ, J. & VIDAL, C. E. (eds) *Andean Metallogeny: New Discoveries, Concepts, Update*. Society of Economic Geologists, Littleton, CO, Special Publications, **11**, 15–54.

MALBRAN, F. 1986. *Estudio geológico-estructural del área de río Clarillo, con énfasis en la Formación Coya-Machalí, hoya del río Tinguiririca, Chile*. Thesis,

Departamento de Gelogía, Universidad de Chile, Santiago.

MALUMIÁN, N. & RAMOS, V. A. 1984. Magmatic intervals, transgression–regression cycles and oceanic events in the Cetraceous and Tertiary of souther South America. *Earth and Planetary Science Letters*, **67**, 228–237.

MANCEDA, R. & FIGUEROA, D. 1995. Inversion of the Mesozoic Neuquén rift in the Malargüe fold–thrust belt, Mendoza, Argentina. *In*: TANKARD, A. J., SUAREZ, R. & WELSINK, H. J. (eds) *Petroleum Basins of South America*. American Association of Petroleum Geologists, Tulsa, OK, Memoirs, **62**, 369–382.

MÁNGANO, M. G. & BUATOIS, L. A. 1996. Shallow marine event sedimentation in a volcanic arc-related setting: the Ordovician Suri Formation, Famatina Range, northwest Argentina (Famatina System). *Sedimentary Geology*, **105**, 63–90.

MARANGUNIC, C., MORENO, H. & VARELA, J. 1979. Observaciones sobre los depósitos de relleno de la Depresión Longitudinal de Chile entre los ríos Tinguiririca y Maule. *In*: *II Congreso Geológico Chileno*, Arica, **3**, J129–J139.

MARDONEZ, D., VELÁSQUEZ, R., MERINO, R. & BONILLA, R. 2012. Caracterización y condiciones de metamorfismo de una nueva unidad dentro del paleozoico de la Cordillerea de la Costa (Unidad Patagual – El Venado), Región de Biobío, Chile. *In*: *XIII Congreso Geológico Chileno*, Antofagasta, 365–367.

MARTIN, M., CLAVERO, J. & MPODOZIS, C. 1997. Eocene to Late Miocene magmatic development of the El Indio Belt, ~30°S, north central Chile. *In*: *VIII Congreso Geológico Chileno*, Antofagasta, **1**, 149–153.

MARTIN, M. W., CLAVERO, J. & MPODOZIS, C. 1999. Late Paleozoic to Early Jurassic tectonic development of the high Andean Principal Cordillera, El Indio region, Chile (29°–30°S). *Journal of South American Earth Sciences*, **12**, 33–49.

MARTINA, F., VIRAMONTE, J. M., ASTINI, R. A., PIMENTEL, M. M. & DANTAS, E. 2011. Mississippian volcanism in the southern Central Andes: new U–Pb SHRIMP zircon geochronology and whole-rock geochemistry. *Gondwana Research*, **19**, 524–544.

MARTÍNEZ, A. N. 2005. Secuencias volcánicas Permo-Triásicas de los cordones del Portillo y del Plata, Cordillera Frontal, Mendoza: su interpretación tectónica. PhD thesis, Universidad de Buenos Aires.

MARTÍNEZ, A. N., RODRÍGUEZ BLANCO, L. & RAMOS, V. A. 2006. Permo-Troassic magmatism of the Choiyoi Group in the Cordillera Frontal of Mendoza, Argentina: geological variations associated with changes in Paleo-Benioff zone. *In*: *Backbone of the Americas*. Asociación Geológica Argentina and Geological Society of America, Mendoza, Argentina, Abstracts with Programs, 77.

MARTINO, R. D. 2003. Las fajas de deformación dúctil de las Sierras Pampeanas de Córdoba: Una reseña general. *Revista de la Asociación Geológica Argentina*, **58**, 549–571.

MARTINOD, J., HUSSON, L., ROSPERCH, P., GUILLAUME, B. & ESPURT, N. 2010. Horizontal subduction zones, convergence velocity and the building of the Andes. *Earth and Planetary Science Letters*, **299**, 299–309.

MÉNDEZ-BEDIA, I., CHARRIER, R., BUSQUETS, P. & COLOMBO, F. 2009. Barras litorales carbonatadas en el Paleozoico Superior Andino (Formación Huentelauquén, Norte Chico, Chile). *In*: *XII Congreso Geológico Chileno*, Santiago, Actas, 1–4.

MESCUA, J. F. 2011. *Evolución estructural de la Cordillera Principal entre Las Choicas y santa Elenea (35° S), Provinciade Mendoza, Argentina*. PhD thesis, Departamento de Ciencias Geológicas, Universidad de Buenos Aires.

MESCUA, J. F., GIAMBIAGI, L. B. & BECHIS, F. 2008. Evidencias de tectónica extensional en el Jurásico Tardío (Kimeridgiano) del suroeste de la provincia de Mendoza. *Revista de la Asociación Geológica Argentina*, **63**, 512–519.

MILANA, J. P. 1998. Anatomía de parasecuencias en un lago de rift y su relación con la generación de hidrocarburos, cuenca triásica de Ischigualasto, San Juan. *Revista de la Asociación Geológica Argentina*, **53**, 365–387.

MILANA, J. P. & ALCOBER, O. 1994. Modelo tectosedimentario de la cuenca triásica de Ischigualasto (San Juan, Argentina). *Revista de la Asociación Geológica Argentina*, **49**, 217–235.

MINATO, M. & TAZAWA, J. 1977. Fossils of the Huentelauquén Formation at the locality F, Coquimbo Province, Chile. *In*: ISHIKAWA, T. & AGUIRRE, L. (eds) *Comparative Studies of the Circum-Pacific Orogenic Belt in Japan and Chile*. First Report. Japan Society for the Promotion of Science, Tokyo, 95–117.

MORATA, D. & AGUIRRE, L. 2003. Extensional Lower Cretaceous volcanism in the Coastal Range (29°20′–30°S), Chile: geochemistry and petrogenesis. *Journal of South American Earth Sciences*, **16**, 459–476.

MORATA, D., FÉRAUD, G., AGUIRRE, L., ARANCIBIA, G., BELMAR, M., MORALES, S. & CARRILLO, J. 2008. Geochronology of the Lower Cretaceous volcanism from the Coastal Range (29°20′–30°S), Chile. *Revista Geológica de Chile*, **35**, 123–145.

MORATA, D., VARAS, M. J., HIGGINS, M., VALENCIA, V. & VERHOORT, J. 2010. Episodic emplacement of the Illapel Plutonic Complex (Coastal Cordillera, central Chile): Sr and Nd isotopic, and zircon U–Pb geochronological constraints. *In*: *South American Symposium on Isotope Geology*, 7, Brasilia, http://refhub.elsevier.com/S0895-9811(1300169-7/sref26)

MOREL, R. 1981. *Geología del sector Norte de la hoja Gualleco, entre los 35° 00' y 35° 10' lat. Sur, Provincia de Talca, VII Región, Chile*. Thesis, Departamento de Geología, Universidad de Chile.

MORENO, K. & BENTON, M. J. 2005. Occurrence of sauropod dinosaur tracks in the Upper Jurassic of Chile (redescription of *Iguanodonichnus frenki*). *Journal of South American Earth Sciences*, **20**, 253–257.

MORENO, K. & PINO, M. 2002. Huellas de dinosaurios en la Formación Baños del Flaco (Titoniano-Jurásico Superior), VI Región, Chile: paleoetología y paleoambiente. *Revista Geológica de Chile*, **29**, 191–206.

MOSCOSO, R., PADILLA, H. & RIVANO, S. 1982. *Hoja Los Andes, Región de Valparaiso*. Servicio Nacional de Geología y Minería, Santiago, Carta Geológica de Chile, **52**.

MPODOZIS, C. & CORNEJO, P. 1988. *Carta Geológica de Chile, hoja Pisco Elqui, IV Región de Coquimbo a escala 1: 250.000.* Servicio Nacional de Geología y Minería, Santiago de Chile, **68**.

MPODOZIS, C. & CORNEJO, P. 2012. Cenozoic tectonics and porphyry copper systems of the Chilean Andes. *In*: HEDENQUIST, J. W., HARRIS, M. & CAMUS, F. (eds) *Geology and Genesis of Major Copper Deposits and Districts of the World: A tribute to Richard H. Sillitoe*. Society of Economic Geologists, Littleton, CO, Special Publications, **16**, 329–360.

MPODOZIS, C. & KAY, S. M. 1990. Provincias magmáticas ácidas y evolución tectónica de Gondwana: Andes chilenos (28–31°S). *Revista Geológica de Chile*, **17**, 153–180.

MPODOZIS, C. & KAY, S. M. 1992. Late Paleozoic to Triassic evolution of the Gondwana margin: evidence from Chilean Frontal Cordilleran batholiths (28°S to 31°S). *Geological Society of America Bulletin*, **104**, 999–1014.

MPODOZIS, C. & RAMOS, V. A. 1989. The Andes of Chile and Argentina. *In*: ERICKSEN, G. E., CAÑAS PINOCHET, M. T. & REINEMUD, J. A. (eds) *Geology of the Andes and its Relation to Hydrocarbon and Mineral Resources*. Circumpacific Council for Energy and Mineral Resources, Houston, Earth Sciences Series, **11**, 59–90.

MPODOZIS, C. & RAMOS, V. A. 2008. Tectónica jurásica en Argentina y Chile: extensión, subducción oblicua, rifting, deriva y colisiones? *Revista de la Asociación Geológica Argentina*, **63**, 481–497.

MPODOZIS, C., BROCKWAY, H., MARQUARDT, C. & PERELLÓ, J. 2009. Geocronología U/Pb y tectónica de la región Los Pelambres-Cerro Mercedario: implicancias para la evolución cenozoica de los Andes del centro de Chile y Argentina. *In*: *XII Congreso Geológico Chileno*, Santiago, electronic abstract.

MUNDACA, P., PADILLA, H. & CHARRIER, R. 1979. Geología del área comprendida entre Quebrada Angostura, Cerro Talinai y Punta Claditas, Provincia de Choapa. *In*: *II Congreso Geológico Chileno*, Arica, Actas, **1**, A121–A161.

MUÑOZ, J. & NIEMEYER, H. 1984. *Hoja Laguna del Maule, Regiones del Maule y del Bio-Bio*. Servicio Nacional de Geología y Minería, Santiago, Carta, **64**.

MUÑOZ, M., FUENTES, F., VERGARA, M., AGUIRRE, L., NYSTRÖM, J. O., FÉRAUD, G. & DEMANT, A. 2006. Abanico East Formation: petrology and geochemistry of volcanic rocks behind the Cenozoic arc front in the Andean Cordillera, central Chile (33°50′S). *Revista Geológica de Chile*, **33**, 109–140.

MUÑOZ, M., AGUIRRE, L., VERGARA, M., DEMANT, A., FUENTES, F. & FOCK, A. 2010. Prehnite- pumpellyite facies metamorphism in the eastern belt of the Abanico Formation, Andean Cordillera of central Chile (33° 50′S): chemical and scale controls on mineral assemblages, reaction progress and the equilibrium state. *Andean Geology*, **37**, 54–77.

MUÑOZ, M., FARÍAS, M., CHARRIER, R., FANNING, C. M., POLVÉ, M. & DECKART, K. 2013. Isotopic shifts in the Cenozoic Andean arc of central Chile: records of an evolving basement throughout cordilleran arc mountain building. *Geology*, **41**, 931–934.

MUÑOZ-CRISTI, J. 1942. Rasgos generales de la construcción geológica de la Cordillera de la Costa; especialmente en la Provincia de Coquimbo. *In*: *Congreso Panamericano de Ingeniería de Minas y Geología*. Santiago, Chile, Anales, **1**, 285–318.

MUÑOZ-CRISTI, J. 1946. Estado actual del conocimiento sobre la geología de la provincia de Arauco. *Anales Facultad de Ciencias Físicas y Matemáticas, Universidad de Chile*, **3**, 30–63.

MUÑOZ-CRISTI, J. 1956. Chile. *In*: YENKS, W. F. (ed.) *Handbook of South American Geology*. Geological Society of America, Boulder, CO, Memoirs, **65**, 187–214.

MUÑOZ-CRISTI, J. 1968. Evolución geológica del territorio chileno. *Boletín de la Academia de Ciencias de Chile*, **1**, 18–26.

MUÑOZ-CRISTI, J. 1973. *Geología de Chile: Pre-Paleozoico, Paleozoico y Mesozoico*. Editorial Andrés Bello, Santiago.

MUÑOZ-SAEZ, C., PINTO, L., CHARRIER, R. & NALPAS, T. 2014. Importance of volcanic load and shortcut fault development during inversion of the Abanico basin, Central Chile (33°–35°S). *Andean Geology*, **41**, 1–28.

MUTTI, D. I. 1997. La secuencia ofiolítica basal desmembrada de las sierras de Córdoba. *Revista de la Asociación Geológica Argentina*, **52**, 275–285.

NAIPAUER, M., VUJOVICH, G. I., CINGOLANI, C. A. & MCCLELLAND, W. C. 2010a. Detrital zircon analysis from the Neoproterozoic–Cambrian sedimentary cover (Cuyania terrane), Sierra de Pie de Palo, Argentina: evidences of a rift and passive margin system. *Journal South American Earth Sciences*, **29**, 306–326.

NAIPAUER, M., CINGOLANI, C. A., VUJOVICH, G. I. & CHEMALE, F. 2010b. Geochemistry and Nd isotopic signatures of metasedimentary rocks of the Caucete Group, Sierra de Pie de Palo, Argentina: implications for their provenance. *Journal South American Earth Sciences*, **30**, 84–96.

NASI, C. & THIELE, R. 1982. Estratigrafía del Jurásico y Cretácico de la Cordillera de la Costa al sur del rio Maipo, entre Melipilla y Laguna de aculeo (Chile Central). *Revista Geológica de Chile*, **16**, 81–99.

NASI, C., MPODOZIS, C., CORNEJO, P., MOSCOSO, R. & MAKSAEV, V. 1985. El batolito Elqui-Limari (Paleozoico Superior-Triásico): características petrográficas, geoquímicas y significado tectónico. *Revista Geológica de Chile*, **25–26**, 77–111.

NASI, C., MOSCOSO, R. & MAKSAEV, V. 1990. *Hoja Guanta, Regiones de Atacama y Coquimbo*. Servicio Nacional de Geología y Minería, Santiago.

NIELSEN, S. 2005. The Triassic Santa Juana Formation at the lower Biobío River, south central Chile. *Journal of South American Earth Sciences*, **19**, 547–562.

NIEMEYER, H. & MUÑOZ, J. 1983. *Geología de la Hoja Laguna de La Laja*. Servicio Nacional de Geología y Minería, Santiago, Serie Carta Geológica de Chile, **58**.

OLIVEROS, V., MORATA, D., AGUIRRE, L., FÉRAUD, G. & FORNARI, G. 2007. Jurassic to Early Cretaceous subduction-related magmatism in the Coastal Cordillera of northern Chile (18°30′–24°S): geochemistry and petrogenesis. *Revista Geológica de Chile*, **34**, 209–232.

OLIVEROS, V., LABBÉ, M., ROSSEL, P., CHARRIER, R. & ENCINAS, A. 2012. Late Jurassic paleogeographic evolution of the Andean back-arc basin: new constrains from the Lagunillas Formation, northern Chile (27°30′–28°30′S). *Journal of South American Earth Sciences*, **37**, 25–40.

ORME, H. M. & ATHERTON, M. P. 1999. New U–Pb and Sr–Nd Data from the Frontal Cordillera Composite Batholith, Mendoza: implications for Magma Source and Evolution. *In*: *IV International Symposium on Andean Geology*, Göttingen, 555–558.

OTA, Y., MIYAUCHI, T., PASKOFF, R. & KOBA, M. 1995. Plio–Quaternary terraces and their deformation along the Altos de Talinay, North–Central Chile. *Revista Geológica de Chile*, **22**, 89–102.

OTAMENDI, J. E., TIBALDI, A. M., VUJOVICH, G. I. & VIÑAO, G. A. 2008. Metamorphic evolution of migmatites from the deep Famatinian arc crust exposed in Sierras Valle Fértil e La Huerta, San Juan, Argentina. *Journal of South American Earth Sciences*, **25**, 313–335.

OTAMENDI, J. E., VUJOVICH, G. I., DE LA ROSA, J. D., TIBALDI, A. M., CASTRO, A. & MARTINO, R. D. 2009a. Geology and petrology of a deep crustal zone from the Famatinian paleo-arc, Sierras Valle Fértil–la Huerta, San Juan, Argentina. *Journal of South American Earth Sciences*, **27**, 258–279.

OTAMENDI, J. E., DUCEA, M. N., TIBALDI, A. M., BERGANTZ, G., DE LA ROSA, J. D. & VUJOVICH, G. I. 2009b. Generation of tonalitic and dioritic magmas by coupled partial melting of gabbroic and metasedimentary rocks within the deep crust of the Famatinian magmatic arc, Argentina. *Journal of Petrology*, **50**, 841–873.

OTAMENDI, J. E., CRISTOFOLINI, E., TIBALDI, A. M., QUEVEDO, F. & BALIANI, I. 2010a. Petrology of mafic and ultramafic layered rocks from the Jaboncillo Valley, Sierra de Valle Fértil, Argentina: implications for the evolution of magmas in the lower crust of the Famatinian arc. *Journal of South American Earth Sciences*, **29**, 685–704.

OTAMENDI, J. E., PINOTTI, L. P., BASEI, M. A. S. & TIBALDI, A. M. 2010b. Evaluation of petrogenetic models for intermediate and silicic plutonic rocks from the Sierra de Valle Fértil-La Huerta, Argentina: petrologic constraints on the origin of igneous rocks in the Ordovician Famatinian-Puna paleoarc. *Journal of South American Earth Sciences*, **30**, 29–45.

PADILLA, H. & VERGARA, M. 1985. Control estructural y alteración de tipo campo geotérmico en los intrusivos subvolcánicos miocénicos del área de la Cuesta de Chacabuco-Baños del Corazón, Chile central. *Revista Geológica de Chile*, **24**, 3–17.

PANKHURST, R. J. & RAPELA, C. W. 1998. The proto-Andean margin of Gondwana: an introduction. *In*: PANKHURST, R. J. & RAPELA, C. W. (eds) *The Proto-Andean Margin of Gondwana*. Geological Society, London, Special Publications, **142**, 1–9.

PANKHURST, R. J., MILLAR, I. L. & HERVÉ, F. 1996. A Permo-Carboniferous U–Pb age for part of the Guanta Unit of the Elqui–Limari Batholith at Rio del Transito, Northern Chile. *Revista Geologica de Chile*, **23**, 35–42.

PARADA, M. A., RIVANO, S. & SEPÚLVEDA, P. 1988. Mesozoic and Cainozoic plutonic development in the Andes of central Chile. *Journal of South American Earth Sciences*, **1**, 249–260.

PARADA, M. A., LEVI, B. & NYSTRÖM, J. 1991. Geochemistry of the Triassic to Jurassic plutonism of central Chile (30 to 33°S): petrogenetic implications and a tectonic discussion. *In*: HARMON, R. S. & RAPELA, C. W. (eds) *Andean Magmatism and its Tectonic Setting*. Geological Society of America, Boulder, CO, Special Papers, **265**, 99–112.

PARADA, M. A., NYSTROM, J. O. & LEVI, B. 1999. Multiple sources for the Coastal Batholith of central Chile (31–34°S): geochemical and Sr–Nd isotopic evidence and tectonic implications. *Lithos*, **46**, 505–521.

PARADA, M. A., FÉRAUD, G., FUENTES, F., AGUIRRE, L., MORATA, D. & LARRONDO, P. 2005. Ages and cooling history of the Early Cretaceous Caleu pluton: testimony of a switch from a rifted to a compressional continental margin in central Chile. *Journal of the Geological Society of London*, **162**, 273–287.

PARADA, M. A., LÓPEZ-ESCOBAR, L. *ET AL.* 2007. Andean magmatism. *In*: MORENO, T. & GIBBONS, W. (eds) *The Geology of Chile*. Geological Society, London, 115–146.

PARDO, M., COMTE, D. & MONFRET, T. 2002. Seismotectonic and stress distribution in the central Chile subduction zone. *Journal of South American Earth Sciences*, **15**, 11–22.

PARDO-CASAS, F. & MOLNAR, P. 1987. Relative motion of the Nazca (Farallon) and South American plates since Late Cretaceous time. *Tectonics*, **6**, 233–248.

PASKOFF, R. 1970. *Recherches Géomorphologiques Dans le Chili Semi-Aride*. Biscaye Frères, Bordeaux.

PASKOFF, R. 1977. The Quaternary of Chile: the state of research. *Quaternary Research*, **8**, 2–31.

PAZOS, P. J. 2002. The Late Carboniferous glacial to postglacial transition: facies and sequence stratigraphy, Western Paganzo Basin, Argentina. *Gondwana Research*, **5**, 467–487.

PETFORD, N. & GREGORI, D. A. 1994. Geological and geochemical comparison between the coastal batholith (Perú) and the Frontal Cordillera composite batholith (Argentina). *In*: *VII Congreso Geológico Chileno*, Concepción, Actas, **2**, 1428–1432.

PINEDA, V. 1983a. *Evolución paleogeográfica de la península de Arauco durante el Cretácico Superior-Terciario*. Thesis, Departamento de Geología, Universidad de Chile, Santiago.

PINEDA, V. 1983b. Evolución Paleogeográfica de la Cuenca Sedimentaria Cretácico-Terciaria de Arauco. *Geología y Recursos Minerales de Chile, Universidad Concepción*, **1**, 375–390.

PINEDA, G. & CALDERÓN, M. 2008. *Geología del área Monte Patria-El Maqui, Región de Coquimbo*. Servicio Nacional de Geología y Minería, Santiago.

PINEDA, G. & EMPARAN, C. 2006. *Geología del área Vicuña-Pichasca, Región de Coquimbo*. Servicio Nacional de Geología y Minería, Santiago.

PIÑÁN-LLAMAS, A. & SIMPSON, C. 2006. Deformation of Gondwana margin turbidites during the Pampean orogeny, north-central Argentina. *Bulletin Geological Society of America*, **118**, 1270–1279.

PIRACÉS, R. 1976. Estratigrafía de la Cordillera de la Costa entre la cuesta El Melón y Limache, Provincia de Valparaíso, Chile. *In*: *I Congreso Geológico Chileno*, Santiago, Actas, **1**(A), 65–82.

PIRACÉS, R. 1977. *Geología de la Cordillera de la Costa entre Catapilco y Limache, región de Aconcagua*. Thesis, Departamento de Geología, Universidad de Chile.

POLANSKI, J. 1964. Descripción geológica de la Hoja 25a Volcán San José, provincia de Mendoza. *Dirección Nacional de Geología y Minería, Boletín*, **98**, 1–94.

POLANSKI, J. 1972. Descripción geológica de la Hoja 24a-b Cerro Tupungato, provincia de Mendoza. *Dirección Nacional de Geología y Minería, Boletín*, **128**, 1–110.

POMA, S. & RAMOS, V. A. 1994. Las secuencias básicas iniciales del Grupo Choiyoi. Cordón del Portillo, Mendoza: sus implicancias tectónicas. *In: VII Congreso Geológico Chileno*, Concepción, Actas, **2**, 1162–1166.

QUENARDELLE, S. & RAMOS, V. A. 1999. The Ordovician western Sierras Pampeanas magmatic belt: record of Precordillera accretion in Argentina. *In*: RAMOS, V. A. & KEPPIE, D. (eds) *Laurentia Gondwana Connections before Pangea*. Geological Society of America, Boulder, CO, Special Papers, **336**, 63–86.

QUIROGA, R. 2013. *Análisis estructural de los depósitos cenozoicos de la Cordillera Principal entre el cerro Provincia y el cordón El Quempo, Región Metropolitana, Chile (33°18′ a 33°25′S)*. Thesis, Departamento de Geología, Universidad de Chile.

RADIC, J. P., ROJAS, L., CARPINELLI, A. & ZURITA, E. 2002. Evolución tectónica de la Cuenca de Cura-Mallín, región cordillerana chileno argentina (36°30′–39°00′S). *In: XV Congreso Geológico Argentino, El Calafate*, **3**, 233–237.

RAMOS, V. A. 1981. *Descripción geológica de la hoja 33 c. Los Chihuidos Norte, provincia de Neuquén*. Servicio Geológico Nacional, Buenos Aires, **182**.

RAMOS, V. A. 1988a. Tectonics of the Late Proterozoic–Early Paleozoic: a collisional history of southern South America. *Episodes*, **11**, 168–174.

RAMOS, V. A. 1988b. The tectonics of the Central Andes; 30° to 33° S latitude. *In*: CLARK, S. & BURCHFIEL, D. (eds) *Processes in Continental Lithospheric Deformation*. Geological Society of America, Boulder, CO, Special Paper, **218**, 31–54.

RAMOS, V. A. 1994. Terranes of southern Gondwanaland and their control in the Andean structure (30–33°s lat.). *In*: REUTTER, K. J., SCHEUBER, E. & WIGGER, P. J. (eds) *Tectonics of the Southern Central Andes, Structure and Evolution of an Active Continental Margin*. Springer, Berlin, 249–261.

RAMOS, V. A. 1999. Rasgos estructurales del territorio argentino. 1. Evolución tectónica de la Argentina. *In*: CAMINOS, R. (ed.) *Geología Argentina*. Instituto de Geología y Recursos Naturales, Buenos Aires, Argentina, Anales, **29**, 715–784.

RAMOS, V. A. 2004. Cuyania, an exotic block to Gondwana: review of a historical success and the present problems. *Gondwana Research*, **7**, 1009–1026.

RAMOS, V. A. 2010. The tectonic regime along the Andes: present settings as a key for the Mesozoic regimes. *Geological Journal*, **45**, 2–25.

RAMOS, V. A. & DALLA SALDA, L. 2011. Occidentalia: Un terreno acrecionado sobre el margen gondwánico? *In: XVIII Congreso Geológico Argentino*, Neuquén, Actas, 222–223.

RAMOS, V. A. & FOLGUERA, A. 2005. Tectonic evolution of the Andes of Neuquén: constraints derived from the magmatic arc and foreland deformation. *In*: VEIGA, G. D., SPALLETTI, L. A., HOWELL, J. A. & SCHWARZ, E. (eds) *The Neuquén Basin, Argentina: A Case Study in Sequence Stratigraphy and Basin Dynamics*. Geological Society, London, Special Publications, **252**, 15–35.

RAMOS, V. A. & FOLGUERA, A. 2009. Andean flat slab subduction through time. *In*: MURPHY, B. (ed.) *Ancient Orogens and Modern Analogues*. Geological Society, London, Special Publications, **327**, 31–54.

RAMOS, V. A. & FOLGUERA, A. 2011. Payenia volcanic province in Southern Andes: an appraisal of an exceptional Quaternary tectonic setting. *Journal of Volcanology and Geothermal Research*, **201**, 53–64.

RAMOS, V. A. & KAY, S. M. 1991. Triassic rifting and associated basalts in the Cuyo basin, central Argentina. *In*: HARMON, R. S. & RAPELA, C. W. (eds) *Andean Magmatism and its Tectonic Setting*. Geological Society of America, Boulder, CO, Special Papers, **265**, 79–91.

RAMOS, V. A. & VUJOVICH, G. I. 1993. The Pampia Craton within Western Gondwanaland. *In*: ORTEGA-GUTIÉRREZ, F., CONEY, P., CENTENO-GARCÍA, E. & GÓMEZ-CABALLERO, A. (eds) *Proceedings of The First Circum-Pacific and Circum-Atlantic Terrane Conference*, México, 113–116.

RAMOS, V. A., JORDAN, T. E., ALLMENDINGER, R. W., KAY, S. M., CORTÉS, J. M. & PALMA, M. A. 1984. Chilenia: un terreno alóctono en la evolución paleozoica de los Andes Centrales. *In: IX Congreso Geológico Argentino*, Bariloche, Actas, **2**, 84–106.

RAMOS, V. A., JORDAN, T. E., ALLMENDINGER, R. W., MPODOZIS, C., KAY, S., CORTÉS, J. M. & PALMA, M. A. 1986. Paleozoic terranes of the Central Argentine Chilean Andes. *Tectonics*, **5**, 855–880.

RAMOS, V. A., RIVANO, S., AGUIRRE-URRETA, M. B., GODOY, E. & LO FORTE, G. L. 1990. El Mesozoico del cordón del Límite entre Portezuelo Navarro y Monos de Agua (Chile-Argentina). *In: XI Congreso Geológico Argentino*, San Juan, Actas, **2**, 43–46.

RAMOS, V. A., MUNIZAGA, F. & KAY, S. M. 1991. El magmatismo cenozoico a los 33°S de latitud: geocronología y relaciones tectónicas. *In: VI Congreso Geológico Chileno*, Viña del Mar, Actas, **1**, 892–896.

RAMOS, V. A., AGUIRRE-URRETA, M. B. & ÁLVAREZ, P. P. 1996a. *Geología de la Región del Aconcagua*. Dirección Nacional del Servicio Geológico, Buenos Aires, Subsecretaría de Minería de la Nación Anales, **24**.

RAMOS, V. A., CEGARRA, M. & CRISTALLINI, E. 1996b. Cenozoic tectonics of the High Andes of west-central Argentina, (30°–36°S latitude). *Tectonophysics*, **259**, 185–200.

RAMOS, V. A., ESCAYOLA, M., MUTTI, D. & VUJOVICH, G. I. 2000. Proterozoic–early Paleozoic ophiolites in the Andean basement of southern South America. *In*: DILEK, Y., MOORES, E. M., ELTHON, D. & NICOLAS, A. (eds) *Ophiolites and Oceanic Crust: New Insights from Field Studies and Ocean Drilling Program*. Geological Society of America, Boulder, CO, Special Papers, **349**, 331–349.

Ramos, V. A., Cristallini, E. O. & Pérez, D. J. 2002. The Pampean flat-slab of the Central Andes. *Journal of South America Earth Sciences*, **15**, 59–78.

Ramos, V. A., Vujovich, G., Martino, R. & Otamendi, J. 2010. Pampia: a large cratonic block missing in the Rodinia supercontinent. *Journal of Geodynamics*, **50**, 243–255.

Ramos, V. A., Litvak, V., Folguera, A. & Spagnuolo, M. 2014. An Andean tectonic cycle: from crustal thickening to extension in a thin crust (34°–37°SL). *Geoscience Frontiers*, **5**, 351–367. http://dx.doi.org/10.1016/j.gsf.2013.12.009

Rapalini, A. E. 2012. Paleomagnetic evidence for the origin of the Argentine Precordillera, fifteen years later: what is new, what has changed, what is still valid? *Latinmag Letters*, **2**, 1–20.

Rapalini, A. E. & Astini, R. A. 1998. Paleomagnetic confirmation of the Laurentian origin of the Argentine Precordillera. *Earth and Planetary Science Letters*, **155**, 1–14.

Rapela, C. & Kay, S. 1988. Late Paleozoic to Recent magmatic evolution of northern Patagonia. *Episodes*, **11**, 175–182.

Rapela, C. W., Pankhurst, R. J., Casquet, C., Baldo, E., Saaverdra, J., Galindo, C. & Fanning, C. M. 1998. The Pampean Orogeny of the southern Proto-Andean: Cambrian continental collision in the Sierras de Córdoba. In: Pankhurst, R. J. & Rapela, C. W. (eds) *The Proto-Andean Margin of Gondwana*. Geological Society, London, Special Publications, **142**, 181–217.

Rapela, C. W., Pankhurst, R. J. et al. 2007. The Río de la Plata craton and the assembly of SW Gondwana. *Earth-Science Reviews*, **83**, 49–82.

Rapela, C. W., Pankhurst, R. J., Casquet, C., Baldo, E., Galindo, C., Fanning, C. M. & Dahlquist, J. A. 2010. The Western Sierras Pampeanas: protracted Grenville-age history (1330–1030 Ma) of intra-oceanic arcs, subduction–accretion at continental-edge and AMCG intraplate magmatism. *Journal of South American Earth Sciences*, **29**, 105–127.

Rauld, R. A. 2002. *Análisis morfoestructural del frente cordillerano: Santiago oriente entre el río Mapocho y Quebrada de Macul*. Thesis, Departamento de Geología, Universidad de Chile, Santiago.

Rebolledo, S. & Charrier, R. 1994. Evolución del basamento paleozoico en el área de Punta Claditas, Región de Coquimbo, Chile (31–32°S). *Revista Geológica de Chile*, **21**, 55–69.

Regard, V., Saillard, M. et al. 2010. Renewed uplift of the Central Andes Forearc revealed by coastal evolution during the Quaternary. *Earth and Planetary Science Letters*, **297**, 199–210.

Reutter, K. J. 2001. Le Ande centrali: elemento di un'orogenesi di margine continentale attivo. *Acta Naturalia de l'Ateneo Parmense*, **37**, 5–37.

Riccardi, A. C. & Iglesia Llanos, M. P. 1999. Primer hallazgo de amonites en el Triásico de Argentina. *Revista de la Asociación Geológica Argentina*, **54**, 298–300.

Richter, P., Ring, U., Willner, A. P. & Leiss, B. 2007. Structural contacts in subduction complexes and their tectonic significance: the Late Palaeozoic coastal accretionary wedge of central Chile. *Journal of the Geological Society, London*, **164**, 203–214.

Rivano, S. 1996. *Geología de las Hojas Quillota y Portillo*. Servicio Nacional de Geología y Minería. Santiago.

Rivano, S. & Sepúlveda, P. 1983. Hallazgo de foraminíferos del Carbonífero Superior en la Formación Huentelauquén. *Revista Geológica de Chile*, **19–20**, 25–35.

Rivano, S. & Sepúlveda, P. 1985. Las calizas de la Formación Huentelauquén: depósitos de aguas templadas a frías en el Carbonífero Superior–Pérmico Inferior. *Revista Geológica de Chile*, **25–26**, 29–38.

Rivano, S. & Sepúlveda, P. 1991. *Hoja Illapel, Región de Coquimbo. Carta Geológica de Chile*. Servicio Nacional de Geología y Minería, Santiago.

Rivano, S., Sepúlveda, P., Boric, R. & Espiñeira, D. 1993. *Hojas Quillota y Portillo*. Servicio Nacional de Geología y Minería, Santiago, Carta Geológica de Chile, **73**.

Robinson, D., Bevins, R. E. & Rubinstein, N. 2005. Subgreenschist facies metamorphism of metabasites from the Precordillera terrane of western Argentina: constraints on the latter stages of accretion to Gondwana. *European Journal of Mineralogy*, **17**, 441–452.

Rocha Campos, A. C., Basei, M. A. et al. 2011. 30 million years of Permian volcanism recorded in the Choiyoi igneous province (W Argentina) and their source for younger ash fall deposits in the Paraná Basin: SHRIMP U–Pb zircon geochronology evidence. *Gondwana Research*, **19**, 509–523.

Rodríguez, M. P. 2013. *Cenozoic uplift and exhumation above the southern part of the flat slab subduction segment of Chile (28.5–32°S)*. Thesis, Departamento de Geología, Universidad de Chile, Santiago.

Rodríguez, M. P., Pinto, L. & Encinas, A. 2012. Cenozoic erosion in the Andean forearc in Central Chile (33°–34°S): Sediment provenance inferred by heavy mineral studies. In: Rasbury, E. T., Hemming, S. R. & Riggs, N. (eds) *Mineralogical and Geochemical Approaches to Provenance*. Geological Society of America, Boulder, CO, Special Papers, **487**, 141–162.

Rodríguez, M. P., Carretier, S. et al. 2013. Geochronology of pediments and marine terraces in north-central Chile and their implications for Quaternary uplift in the Western Andes. *Geomorphology*, **180–181**, 33–46.

Rodríguez, M. P., Aguilar, G., Urresty, C. & Charrier, R. 2014. Neogene landscape evolution in the Andes of north-central Chile between 28 and 32°S: interplay between tectonic and erosional processes. In: Sepúlveda, S. A., Giambiagi, L. B., Moreiras, S. M., Pinto, L., Tunik, M., Hoke, G. D. & Farías, M. (eds) *Geodynamic Processes in the Andes of Central Chile and Argentina*. Geological Society, London, Special Publications, **399**. First published online April 2, 2014, http://dx.doi.org/10.1144/SP399.15

Rossel, P., Oliveros, V., Ducea, M. N., Charrier, R., Scaillet, S., Retamal, L. & Figueroa, O. 2013. The Early Andean subduction system as an analogue to island arcs: evidence from across-arc geochemical variations in northern Chile. *Lithos*, **179**, 211–230, http://dx.doi.org/10.1016/j.lithos.2013.08.014

Rossel, P., Oliveros, V. et al. 2014. La Formación Rio Damas-Tordillo (33°–35,5°S): antecedentes sobre petrogénesis, proveniencia e implicancias tectónicas. Andean Geology, http://ww.scielo.cl/andgeol.htm

RUTLAND, R. W. R. 1971. Andean Orogeny and ocean floor spreading. *Nature*, **233**, 252–255.

SAILLARD, M., HALL, S. R. ET AL. 2009. Non-steady long-term uplift rates and Pleistocene marine terrace development along the Andean margin of Chile (31°S) inferred from 10Be dating. *Earth and Planetary Science Letters*, **277**, 50–63.

SAILLARD, M., RIOTTE, J., REGARD, V., VIOLETTE, A., HÉRAIL, G., AUDIN, L. & RIQUELME, R. 2012. Beach ridges U–Th dating in Tongoy bay and tectonic implications for a peninsula–bay system, Chile. *Journal of South American Earth Sciences*, **40**, 77–84.

SALAZAR, C., STINNESBECK, W. & QUINZIO-SINN, L. A. 2010. Ammonites from the Maastrichtian (Upper Cretaceous) Quiriquina Formation in central Chile. *Neues Jahrbuch für Geologie und Paläontologie, Abhandlungen*, **257**, 181–236.

SALAZAR, E., ARRIAGADA, C., MPODOZIS, C., MARTÍNEZ, F., PEÑA, M. & ÁLVAREZ, J. 2009. Análisis Estructural del Oroclino de Vallenar: Primeros Resultados. *In*: *XII Congreso Geológico Chileno*, Santiago, electronic actas.

SATO, A. M. & LLAMBÍAS, E. J. 1993. El Grupo Choiyoi, provincia de San Juan: equivalente efusivo del Batolito de Colangüil. *In*: *XII Congreso Geológico Argentino y II Congreso Exploración de Hidrocarburos*. Mendoza, Argentina, Actas, **4**, 156–165.

SATO, A. M., LLAMBÍAS, E. J., SHAW, S. E. & CASTRO, C. E. 1990. El Batolito de Colangüil: modelo del magmatismo neopaleozoico de la provincia de San Juan. *In*: BORDONARO, O. (ed.) *Geología y Recursos Naturales de la provincia de San Juan, Relatorio XI Congreso Geológico Argentino*, San Juan, 100–122.

SATO, A. M., LLAMBÍAS, E. J., BASEI, M. A. S. & CASTRO, C. E. 2014. El magmatismo Choiyoi de la Cordillera Frontal de San Juan. *In*: *XIX Congreso Geológico Argentino*, Córdoba, Session 21: Pre-Andean tectonics, Presentation no. 52, Digital.

SCHWARTZ, J. J., GROMET, L. P. & MIRÓ, R. 2008. Timing and duration of the calc-alkaline arc of the Pampean orogeny: implications for the late Neoproterozoic to Cambrian evolution of western Gondwana. *Journal of Geology*, **116**, 39–61.

SCHILLER, W. 1912. La Alta Cordillera de San Juan y Mendoza y parte de la provincia de San Juan. *Anales del Ministerio de Agricultura, Sección Geología, Mineralogía y Minería*, **7**, 1–68.

SCHMIDT, S., HETZEL, R., MINGORANCE, F. & RAMOS, V. A. 2011. Coseismic displacements and Holocene slip rates for two active thrust faults at the mountain front of the Andean Precordillera (∼33°S). *Tectonics*, **30**, TC5011, http://dx.doi.org/10.1029/2011TC002932

SELLÉS, D. & GANA, P. 2001. *Geología del área Talagante-San Francisco de Mostazal, Regiones metropolitana y del Libertador Bernardo O'Higgins*. Servicio Nacional de Geología y Minería, Santiago, Carta Geológica, Serie Geología Básica, **74**.

SERRANO, L., VARGAS, R. & STAMBUK, V. 1996. The Late Miocene Río Blanco–Los Bronces copper deposit central Chilean Andes. *In*: CAMUS, F., SILLITOE, R. H. & PETERSEN, R. (eds) *Andean Copper Deposits: New Discoveries, Mineralization Styles and Metallogeny*. Society of Economic Geologists, Littleton, CO, Special Publications, **5**, 119–130.

SILVESTRO, J. & ATENCIO, M. 2009. La cuenca cenozoica del Río Grande y Palauco: edad, evolución y control estructural. Faja plegada de Malargüe (36°S). *Revista de la Asociación Geológica Argentina*, **65**, 154–169.

SIMPSON, C., LAW, R. D., GROMET, L. P., MIRO, R. & NORTHRUP, C. J. 2003. Paleozoic deformation in the Sierras de Córdoba and Sierra de La Minas, eastern Sierras Pampeanas, Argentina. *Journal of South American Earth Sciences*, **15**, 749–764.

SERNAGEOMIN. 2002. *Mapa Geológico de Chile, escala 1: 1.000.000*. Servicio Nacional de Geología y Minería, Santiago, **75**.

SKEWES, M. A., ARÉVALO, A., FLOODY, R., ZÚÑIGA, P. & STERN, C. R. 2002. The giant El Teniente, breccia deposit: hypogene copper distribution and emplacement. *In*: GOLDFARB, R. J. & NIELSEN, R. L. (eds) *Integrated Methods for Discovery: Global Exploration in the Twenty-first Century*. Society of Economic Geologists, Littleton, CO, Special Publications, **9**, 299–332.

SØAGER, N., HOLM, P. M. & LLAMBÍAS, E. J. 2013. Payenia volcanic province, southern Mendoza, Argentina: OIB mantle upwelling in a backarc environment. *Chemical Geology*, **349–350**, 36–53.

SPAGNUOLO, C., RAPALINI, A. E. & ASTINI, R. A. 2012a. Assembly of Pampia to the SW Gondwana margin: a case of strike-slip docking? *Gondwana Research*, **21**, 406–421.

SPAGNUOLO, M. G., LITVAK, V. D., FOLGUERA, A., BOTTESI, G. & RAMOS, V. A. 2012b. Neogene magmatic expansion and mountain building processes in the southern Central Andes, 36–37°S, Argentina. *Journal of Geodynamics*, **53**, 81–94.

SPALLETTI, L. 1999. Cuencas triásicas del oeste argentino: origen y evolución. *Acta Geológica Hispánica*, **32**, 29–50.

SPALLETTI, L., ARTABE, A., MOREL, E. & BREA, M. 1999. Paleofloristic biozonation and chronostratigraphy of the Argentine Triassic. *Ameghiniana*, **36**, 419–451.

SPALLETTI, L., MOREL, E., ARTABE, A., ZAVATTIERI, A. & GANUZA, D. 2005. Estratigrafía, facies y paleoflora de la sucesión triásica de Potrerillos, Mendoza, República Argentina. *Revista Geológica de Chile*, **32**, 249–272.

SPALLETTI, L. A., FANNING, C. M. & RAPELA, C. W. 2008. Dating the Triassic continental rift in the southern Andes: the Potrerillos Formation, Cuyo Basin, Argentina. *Geologica Acta*, **6**, 267–283.

SPALLETTI, L. A., LIMARINO, C. O. & COLOMBO PIÑOL, F. 2012. Petrology and geochemistry of Carboniferous siliciclastics from the Argentine Frontal Cordillera: a test of methods for interpreting provenance and tectonic setting. *Journal of South American Earth Sciences*, **36**, 32–54.

STEINMANN, G. 1929. *Geologie von Perú*. Karl Winter, Heidelberg.

STEINMANN, G., DEEKE, W. & MÖRICKE, W. 1895. Das Alter und die Fauna der Quiriquina Schichten in Chile. *Neues Jahrbuch Mineralogie und Palaeontologie*, **10**, 1–118.

STERN, C. R., AMINI, H., CHARRIER, R., GODOY, E., HERVÉ, F. & VARELA, J. 1984. Petrochemistry and age of rhyolitic pyroclastics flows which occur along the drainage valleys of the Río Maipo and Río Cachapoal (Chile) and the Río Chaucha and Río Papagayos (Argentina). *Revista Geológica de Chile*, **23**, 39–52.

STERN, C. R., MORENO, H. *ET AL.* 2007. Chilean volcanoes. *In*: MORENO, T. & GIBBONS, W. (eds) *The Geology of Chile*. Geological Society, London, 147–178.

STINNESBECK, W. 1986. Zu den faunistischen und palökologischen Verhältnissen in der Quriquina Formation (Maastrichtium) Zentral-Chiles. *Palaeontographica*, **194**, 99–237.

SUÁREZ, M. & BELL, M. 1992. Triassic rift-related sedimentary basins in northern Chile (24°–29°S). *Journal of South American Earth Sciences*, **6**, 109–121.

SUÁREZ, M. & EMPARAN, C. 1997. *Hoja Curacautín, Regiones de Araucanía y Bío-Bío*. Carta geológica de Chile, Santiago, Servicio Nacional de Geología y Minería, **71**.

TAVERA, J. 1942. Contribución al estudio de la estratigrafía y paleontología del Terciario de Arauco. *In*: *I Congreso Panamericano de Ingeniería de Minas y Geología*. Santiago, Chile, **2**, 580–632.

TAVERA, J. 1979. Estratigrafía y paleontología de la Formación Navidad, Provincia de Colchagua, Chile (30°50′–34°S). *Boletín del Museo Nacional de Historia Natural, Santiago*, **36**.

TAVERA, J. 1980. *El Cretáceo y Terciario de Alagarrobo*. Imprentas Gráficas, Santiago.

THIELE, R. 1965. *El Triásico-Jurásico del Departamento de Curepto en la Provincia de Talca*. Departamento de Geología, Universidad de Chile, Santiago, Chile, **28**.

THIELE, R. 1980. Hoja Santiago. Región Metropolitana. Carta Geológica de Chile. *Instituto de Investigaciones Geológicas*, **39**, 1–51.

THIELE, R. & HERVÉ, F. 1984. Sedimentación y Tectónica de antearco en los terrenos pre-andinos del Norte Chico. *Revista Geológica de Chile*, **22**, 61–75.

THOMAS, H. 1958. Geología de la Cordillera de la Costa entre el valle de La Ligua y la cuesta de Barriga. *Instituto de Investigaciones Geológicas Boletín*, **2**, 1–80.

THOMAS, H. 1967. *Geología de la Hoja Ovalle, Provincia de Coquimbo*. Instituto de Investigaciones Geológicas, Santiago, Boletín, **23**.

THOMAS, W. A. & ASTINI, R. A. 1996. The Argentine Precordillera: a traveller from the Ouachita embayment of North American Laurentia. *Science*, **273**, 752–757.

THOMAS, W. A. & ASTINI, R. A. 2003. Ordovician accretion of the Argentine Precordillera terrane to Gondwana: a review. *Journal of South American Earth Sciences*, **16**, 67–79.

THOMAS, W. A. & ASTINI, R. A. 2007. Vestiges of an Ordovician west-vergent thin-skinned Ocloyic thrust belt in the Argentine Precordillera, southern Central Andes. *Journal of Structural Geology*, **29**, 1369–1385.

THOMPSON, S. N. & HERVÉ, F. 2002. New time constraints for the age of metamorphism at the ancestral Pacific Gondwana margin of southern Chile. *Revista Geológica de Chile*, **29**, 255–271.

TIBALDI, A. M., OTAMENDI, J. E., GROMET, L. P. & DEMICHELIS, A. H. 2008. Suya Taco and Sol de Mayo mafic complexes from eastern Sierras Pampeanas, Argentina: evidence for the emplacement of primitive OIB-like magmas into deep crustal levels at a late stage of the Pampean orogeny. *Journal of South American Earth Sciences*, **26**, 172–187.

TICKYJ, H. 2011. Granitoides calcoalcalinos tardíofamatinianos en el Cordón del Carrizalito, Cordillera Frontal, Mendoza. *In*: *XVIII Congreso Geológico Argentino*, Neuquén, 1531–1532.

TICKYJ, H., RODRÍGUEZ RAISING, M., CINGOLANI, C. A., ALFARO, M. & URIZ, N. 2009a. Graptolitos ordovícicos en el Sur de la Cordillera frontal de Mendoza. *Revista de la Asociación Geológica Argentina*, **64**, 295–302.

TICKYJ, H., FERNÁNDEZ, M. A., CHEMALE, J. F. & CINGOLANI, C. A. 2009b. Granodiorita Pampa de los Avestruces, Cordillera Frontal, Mendoza: Un intrusivo sintectónico de edad devónica inferior. 14° Reunión de Tectónica y 3° Taller de Campo de Tectónica, p. 27, Córdoba.

TUNIK, M. A. 2003. Interpretación paleoambiental de los depósitos de la Formación Saldeño (Cretácico Superior), en la alta cordillera de Mendoza. *Revista de la Asociación Geológica Argentina*, **58**, 417–433.

TUNIK, M., FOLGUERA, A., NAIPAUER, M., PIMENTEL, M. & RAMOS, V. A. 2010. Early uplift and orogenic deformation in the Neuquén basin: constraints on the Andean uplift from U–Pb and Hf isotopic data of detrital zircons. *Tectonophysics*, **489**, 258–273.

ULIANA, M. A. & BIDDLE, K. T. 1988. Mesozoic–Cenozoic paleogeographic and geodynamic evolution of southern South America. *Revista Brasileira de Geociências*, **18**, 172–190.

URBINA, N. P. & SRUOGA, P. 2009. La faja metalogenética de San Luis, Sierras Pampeanas: mineralización y geocronología en el contexto metalogenético regional. *Revista de la Asociación Geológica Argentina*, **64**, 635–645.

VAN STAAL, C. R., VUJOVICH, G. I. & NAIPAUER, M. 2011. An Alpine-style Ordovician collision complex in the Sierra de Pie de Palo, Argentina: record of subduction of Cuyania beneath the Famatina arc. *Journal of Structural Geology*, **33**, 343–361.

VARGAS, G. & REBOLLEDO, S. 2012. Paleosismología de la Falla San Ramón e implicancias para el peligros sísmico de Santiago. *In*: *XIII Congreso Geológico Chileno*, Antofagasta, electronic abstract.

VÁSQUEZ, P. & FRANZ, G. 2008. The Triassic Cobquecura Pluton (Central Chile): an example of a fayalite-bearing A-type intrusive massif at a continental margin. *Tectonophysics*, **459**, 66–84.

VERDECCHIA, S. O., BALDO, E. G., BENEDETTO, J. L. & BORGHI, P. A. 2007. The first shelly faunas from metamorphic rocks of the Sierras Pampeanas (La Cébila Formation, Sierra de Ambato, Argentina): age and paleogeographic implications. *Ameghiniana*, **44**, 493–498.

VERGARA, M. 1969. *Rocas volcánicas y sedimentario-volcánicas Mesozoicas y Cenozoicas en la latitud 34° 30′S, Chile*. Departamento de Geología, Universidad de Chile, Santiago, Chile, **32**.

VERGARA, M. & DRAKE, R. 1978. *Edades K–Ar y su implicancia en la Geología Regional de Chile central*. Universidad de Chile, Departamento de Geología y Geofísica, Santiago, Comunicaciones, **23**.

VERGARA, M., CHARRIER, R., MUNIZAGA, F., RIVANO, S., SEPÚLVEDA, P., THIELE, R. & DRAKE, R. 1988. Miocene volcanism in the central Chilean Andes (31°30′S–34°35′S). *Journal of South American Earth Sciences*, **1**, 199–209.

VERGARA, M., LEVI, B., NYSTROM, J. & CANCINO, A. 1995. Jurassic and Early Cretaceous island arc volcanism extension and subsidence in the Coastal Range of central Chile. *Geological Society of America Bulletin*, **107**, 1427–1440.

VERGARA, M., LOPEZ ESCOBAR, L. & HICKEY-VARGAS, R. 1997. Geoquímica de las rocas volcanicas miocenas de la cuenca intermontana de Parraly Ñuble. *In*: *VIII Congreso Geológico Chileno*. Antofagasta, Actas, **2**, 1570–1573.

VERGARA, M., LÓPEZ-ESCOBAR, L., PALMA, J. L., HICKEY-VARGAS, R. & ROESCHMANN, C. 2004. Late Tertiary episodes in the area of the city of Santiago de Chile: new geochronological and geochemical data. *Journal of South American Earth Sciences*, **17**, 227–238.

VICENTE, J. C. 1974. Geological cross section of the Andes between Santiago and Mendoza (33° Lat. S.) Guide Book, Excursion D-5, International Association of Volcanology and Chemistry of Earth's Interior. *In*: *Symposium Andean and Antarctic Problems*. Santiago, Chile, 1–10.

VICENTE, J. C. 2005. Dynamic paleogeography of the Jurassic Andean Basin: pattern of transgression and localization of main straits through the magmatic arc. *Revista de la Asociación Geológica Argentina*, **60**, 221–250.

VICENTE, J. C., CHARRIER, R., DAVIDSON, J., MPODOZIS, C. & RIVANO, S. 1973. La orogénesis subhercínica: fase mayor de la evolución paleogeográfica y estructural de los Andes argentino-chilenos centrales. *In*: *V Congreso Geológico Argentino*. Carlos Paz, Argentina, Actas, **5**, 81–98.

VIVALLO, W., ZANETTINI, J. C. M., GARDEWEG, M., MÁRQUEZ, M. J., TASSARA, A. & GONZÁLEZ, R. A. 1999. *Mapa de recursos minerales del área fronteriza argentino-chilena entre los 34° y 56°S*. Servicio Nacional de Geología y Minería, Chile, Publicación Multinacional, **1**.

VOLDMAN, G. G., ALBANESI, G. L. & RAMOS, V. A. 2009. Ordovician metamorphic event in the carbonate platform of the Argentine Precordillera: implications for the geotectonic evolution of the proto-Andean margin of Gondwana. *Geology*, **37**, 311–314.

VOLDMAN, G. G., ALBANESI, G. L. & RAMOS, V. A. 2010. Conodont geothermometry of the lower Paleozoic from the Precordillera (Cuyania terrane), northwestern Argentina. *Journal of South American Earth Sciences*, **29**, 278–288.

VOLKHEIMER, W. 1978. Descripción geológica de la Hoja 27b, Cerro Sosneado, Provincia de Mendoza. *Secretaría de Estado de Minería, Boletín*, **151**, 1–83, Buenos Aires.

VON GOSEN, W. 1992. Structural evolution of the Argentine Precordillera: the Río San Juan Section. *Journal of Structural Geology*, **14**, 643–667.

VON GOSEN, W., LOSKE, W. & PROZZI, C. 2002. New isotopic dating of intrusive rocks in the Sierra de San Luis (Argentina): implications for the geodynamic history of the Eastern Sierras Pampeanas. *Journal of South American Earth Sciences*, **15**, 237–250.

VUJOVICH, G. & KAY, S. M. 1998. A Laurentian? Grenville-age oceanic arc/back-arc terrane in the Sierra de Pie de Palo, Western Sierras Pampeanas, Argentina. *In*: PANKHURST, R. & RAPELA, C. W. (eds) *Protomargin of Gondwana*. Geological Society, London, Special Publications, **142**, 159–180.

VUJOVICH, G. & RAMOS, V. A. 1994. La Faja de Angaco y su relación con las Sierras Pampeanas Occidentales, Argentina. *In*: *VII Congreso Geológico Chileno*, Concepción, Actas, **1**, 215–219.

WALL, R., GANA, P. & GUTIÉRREZ, A. 1996. *Mapa Geológico del área de San Antonio-Melipilla*. Regiones de Valparaiso, Metropolitana y del Libertador General Bernardo O'Higgins, Servicio Nacional de Geología y Minería, Mapa, Santiago, Chile, **2**.

WALL, R., SELLES, D. & GANA, P. 1999. *Mapa área Tiltil–Santiago, región Metropolitana*. Servicio Nacional de Geología y Minería, Chile.

WENZEL, O., WATHELET, J., CHÁVEZ, L. & BONILLA, R. 1975. La sedimentación cíclica Meso-Cenozoica en la región Carbonífera de Arauco–Concepción, Chile. *In*: *II Congreso Americano de Geología Económica*, Buenos Aires, 215–237.

WETZEL, W. 1930. Die Quiriquina Schichten als Sediment und paläontologisches Archiv. *Palaeontographica*, **73**, 49–101.

WILLNER, A. P. 2005. Pressure–temperature evolution of a Late Palaeozoic paired metamorphic belt in north-central Chile (34–35° 30′ S). *Journal of Petrology*, **46**, 1805–1833.

WILLNER, A. P., GLODNY, J., GERYA, T. V., GODOY, E. & MASSONNE, H. 2004. A counterclockwise *PTt* path of high-pressure/low-temperature rocks from the Coastal Cordillera accretionary complex of south-central Chile: constraints for the earliest stage of subduction mass flow. *Lithos*, **75**, 283–310.

WILLNER, A. P., THOMSON, S. N., KRÖNER, A., WARTHO, J.-A., WIJBRANS, J. R. & HERVÉ, F. 2005. Time markers for the evolution and exhumation history of a Late Palaeozoic paired metamorphic belt in North-Central Chile (34°–35°30′S). *Journal of Petrology*, **46**, 1835–1858.

WILLNER, A. P., GERDES, A. & MASSONNE, H. J. 2008. History of crustal growth and recycling at the Pacific convergent margin of South America at latitudes 29–36°S revealed by a U–Pb and Lu–Hf isotope study of detrital zircon from late Paleozoic accretionary systems. *Chemical Geology*, **253**, 114–129.

WILLNER, A. P., GERDES, A., MASSONNE, H.-J., SCHMIDT, A., SUDO, M., THOMSON, S. N. & VUJOVICH, G. 2011. The geodynamics of collision of a microplate (Chilenia) in Devonian times deduced by the pressure–temperature–time evolution within part of a collisional belt (Guarguaraz Complex, W-Argentina). *Contributions to Mineralogy and Petrology*, **162**, 303–327.

WILLNER, A. P., MASSONNE, H.-J., RING, U., SUDO, M. & THOMSON, S. N. 2012. P-T evolution and timing of a late Palaeozoic fore-arc system and its heterogeneous Mesozoic overprint in north-central Chile (latitudes 31–32°S). *Geological Magazine*, **149**, 177–207.

WINOCUR, D. A., LITVAK, V. D. & RAMOS, V. A. 2014. Magmatic and tectonic evolution of the Oligocene Valle del Cura basin, main Andes of Argentina and Chile: evidence for generalized extension. In: SEPÚLVEDA, S. A., GIAMBIAGI, L. B., MOREIRAS, S. M., PINTO, L., TUNIK, M., HOKE, G. D. & FARÍAS, M. (eds) Geodynamic Processes in the Andes of Central Chile and Argentina. Geological Society, London, Special Publications, **399**. First published online February 17, 2014, http://dx.doi.org/10.1144/SP399.2

YÁÑEZ, G., CEMBRANO, S., PARDO, M., RANERO, C. & SELLÉS, D. 2001. The Challenger–Juan Fernández–Maipo major tectonic transition of the Nazca–Andean subduction system at 33°–34°S: geodynamic evidence and implications. *Journal of South American Earth Sciences*, **15**, 23–38.

YURY-YÁÑEZ, R. E., OTERO, R. A., SOTO-ACUÑA, S., SUÁREZ, M. E., RUBILAR-ROGERS, D. & SALLABERRY, M. 2012. First bird remains from the Eocene of Algarrobo, central Chile. *Andean Geology*, **39**, 548–557.

Evolution of shallow and deep structures along the Maipo–Tunuyán transect (33°40'S): from the Pacific coast to the Andean foreland

LAURA GIAMBIAGI[1]*, ANDRÉS TASSARA[2], JOSÉ MESCUA[1], MAISA TUNIK[3], PAMELA P. ALVAREZ[4], ESTANISLAO GODOY[5], GREG HOKE[6], LUISA PINTO[7], SILVANA SPAGNOTTO[8], HERNÁN PORRAS[6], FELIPE TAPIA[6], PAMELA JARA[9], FLORENCIA BECHIS[10], VÍCTOR H. GARCÍA[3], JULIETA SURIANO[11], STELLA MARIS MOREIRAS[1] & SEBASTÍAN D. PAGANO[1]

[1]*IANIGLA, CCT Mendoza, Centro Regional de Investigaciones Científicas y Tecnológicas, Parque San Martín s/n, 5500 Mendoza, Argentina*

[2]*Departamento de Ciencias de la Tierra, Universidad de Concepción, Victor Lamas 1290, Concepción, Chile*

[3]*CONICET, Universidad Nacional de Río Negro, Argentina*

[4]*Tehema S.A., Virginia Subercaseaux 4100, Pirque, Chile*

[5]*Consultant, Virginia Subercaseaux 4100, Pirque, Chile*

[6]*Department of Earth Sciences, Syracuse University, Syracuse, NY, 13244, USA*

[7]*Departamento de Ciencias Geológicas, Universidad de Chile, Chile*

[8]*Departamento de Ciencias Geológicas, Universidad Nacional de San Luis, Argentina*

[9]*Departamento de Ingeniería en Minas, Universidad de Santiago de Chile, Chile*

[10]*IIDyPCA, CONICET, Universidad Nacional de Río Negro, Argentina*

[11]*Departamento de Ciencias Geológicas, Universidad de Buenos Aires, Argentina*

Corresponding author (e-mail: lgiambiagi@mendoza-conicet.gob.ar)

Abstract: We propose an integrated kinematic model with mechanical constrains of the Maipo–Tunuyán transect (33°40'S) across the Andes. The model describes the relation between horizontal shortening, uplift, crustal thickening and activity of the magmatic arc, while accounting for the main deep processes that have shaped the Andes since Early Miocene time. We construct a conceptual model of the mechanical interplay between deep and shallow deformational processes, which considers a locked subduction interface cyclically released during megathrust earthquakes. During the coupling phase, long-term deformation is confined to the thermally and mechanically weakened Andean strip, where plastic deformation is achieved by movement along a main décollement located at the base of the upper brittle crust. The model proposes a passive surface uplift in the Coastal Range as the master décollement decreases its slip eastwards, transferring shortening to a broad area above a theoretical point S where the master detachment touches the Moho horizon. When the crustal root achieves its actual thickness of 50 km between 12 and 10 Ma, it resists further thickening and gravity-driven forces and thrusting shifts eastwards into the lowlands achieving a total Miocene–Holocene shortening of 71 km.

The Andes, the world's largest non-collisional orogen, is considered the paradigm for geodynamic processes associated with the convergence of an oceanic plate (Nazca) below a continental plate margin (South American western margin). Although these mountains have been described as a consequence of crustal shortening that leads to crustal thickening and surface uplift (Isacks 1988; Sheffels 1990; Allmendinger *et al.* 1997), the mechanisms by which this crustal shortening is achieved remain controversial (e.g. Garzione *et al.* 2008; DeCelles *et al.* 2009; Ehlers & Poulsen 2009; Armijo *et al.* 2010; Farías *et al.* 2010).

Tectonic plates near subduction zones are displaced towards the trench by buoyancy forces, such as slab pull and ridge push. These driving forces are resisted by forces associated with mantle viscosity, slab bending around the outer-rise and interplate friction (Forsyth & Uyeda 1975; Heuret & Lallemand 2005; Lamb 2006; Iaffaldano & Bunge

2008; Schellart 2008). Part of the work done by the driving forces is stored as elastic strain energy associated with the deformation of the upper plate above the seismogenic portion of the locked subduction interface. The elastic deformation accumulated during the decades to centuries of the interseismic period is cyclically recovered by the coseismic slip of both plates along the megathrust during earthquakes and their post-seismic relaxation phases (Savage 1983; Wang et al. 2012). If the Earth were purely elastic, then no permanent deformation would accumulate at active continental margins at geological timescales, all the convergence would be absorbed by coseismic interplate slip and no Andean-type mountains would exist. The elastic behaviour is appropriated to describe the rheology of the cold and rigid forearc and foreland regions, whereas the hot and weak arc–back-arc region is dominated by brittle (plastic) deformation of the upper crust and ductile (viscous) deformation of the lower crust and lithospheric mantle (Hyndman et al. 2005). In this context, the long-term structure of the arc–back-arc region is the result of the imbalance between compressive strain accumulated during the interseismic period and extensional strain activated by the co- and post-seismic phases, but summed over thousands or millions of seismic cycles.

Andean-type margins are geodynamically controlled in the long term by a strong coupling between the forearc and the down-going slab (Lamb 2006) and a comparatively rapid advance of the foreland (and the entire upper plate) toward the forearc with respect to the ocean-wards rollback velocity of the subducted slab (Heuret & Lallemand 2005; Schellart 2008). The permanently deforming arc–back-arc region located between the colliding forearc and foreland regions accumulates shortening and thickening to construct the Andes and its crustal roots. However, the mechanisms and structures by which this process actually occurs are not well understood. Our study explores a possible solution. We focus on this problem by studying the Maipo–Tunuyán transect across the Southern Central Andes (33°40′S) at the latitudes of the cities of Santiago and Mendoza, which has been the subject of a number of tectono-structural studies including the pioneering observations of Darwin (Giambiagi et al. 2009). Because of the amount of pre-existing data, this transect constitutes a key area for understanding the role of deep-seated structural and tectonic processes on the constitution of the central Chilean–Argentinean Orogen.

Several conceptual models of deep crustal deformation have been proposed for the Andes between 18° and 38°S. These models can be divided into two types: east-vergent and west-vergent models. Among the first to be proposed are the wedge model (Allmendinger et al. 1990), the duplex model (Schmitz 1994) and the east-vergent décollement model (Allmendinger & Gubbels 1996; Giambiagi et al. 2003b, 2012; Ramos et al. 2004; Farías et al. 2010). The east-vergent models assume a shallow subhorizontal to gently west-dipping detachment located at different depths in the upper crust or at the transition between the upper and lower crust. This crustal-scale detachment has been identified within a shallow brittle–ductile transition (e.g. Tassara 2005; Farías et al. 2010) and it concentrates nearly all the horizontal crustal shortening between the forearc and foreland. The eastwards propagation of deformation into the foreland generates predominantly east-verging upper crustal thrusts and folds. These models propose an underthrusting of the rigid, cold South American craton under the mechanically and thermally weakened Andean sector (Allmendinger & Gubbels 1996). The diametrically opposed west-vergent model proposed by Armijo et al. (2010) argues for the existence of a ramp-flat décollement dipping to the east and the growth of the Andes mountain belt as a bivergent orogen. In this model, it is the forearc (coastal crustal-scale rigid block) that underthrusts beneath the Principal Cordillera.

The aim of this paper is to propose an integrated Miocene to present kinematic model with mechanical constraints for the Maipo–Tunuyán transect (33°40′S) that correlates with the large volume of geological and geophysical studies carried out by numerous authors. For that purpose, we construct a conceptual model of mechanical interplay between deep and shallow deformational processes based on thermomechanical and kinematic modeling.

The Andes between 33° and 34°S latitudes: geological framework

The 33°40′S transect (Fig. 1) is located in a transition zone between a subhorizontal subduction segment north of 33°S and a zone of normal subduction south of 34°S (Stauder 1975; Barazangi & Isacks 1976; Cahill & Isacks 1992; Tassara et al. 2006). The abrupt southwards disappearance of the Precordillera and Sierras Pampeanas has been related to variation in the slab geometry (Jordan et al. 1983; Charrier et al., this volume, in review). At this latitude, the Andes of Argentina and Chile are composed from West to East by the Coastal Range, the Central Depression, the Principal Cordillera, the Frontal Cordillera and the Cerrilladas Pedemontanas range, where the pre-existing extensional structures of the Triassic Cuyo basin are partially inverted (Fig. 2a).

The Coastal Range can be divided into two sectors. The western sector with low topographic

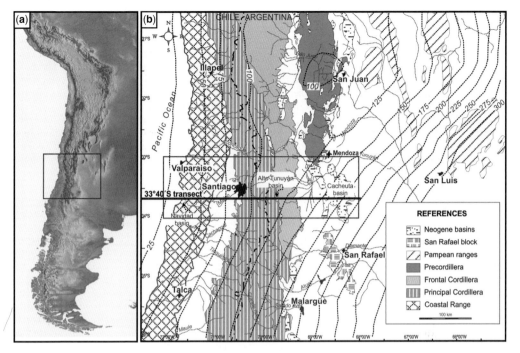

Fig. 1. (a) Shaded relief map of the Andes. Box indicates location of map in Figure 1b. (b) Regional map of the Andes at latitutde 31–36°S highlighting the present-day morphostructural units. The 33°40′S transect crosses the Coastal Range, the Central Depression, the Principal and Frontal cordilleras and the foreland. Dashed black lines represent depth of the subducted Nazca plate (from Tassara & Echaurren 2012). Notice the location of the transect above the transition segment between flat and normal subducted slab. Box indicates location of map in Figure 2.

altitude (<500 m) is composed of a series of Late Pliocene–Pleistocene marine ablation terraces (Wall *et al.* 1996; Rodríguez *et al.* 2012) carved on Late Palaeozoic–Middle Jurassic plutons (Sellés & Gana 2001). The eastern sector with altitudes up to 2000 m and relicts of high-elevated peneplains at different elevations (Brüggen 1950; Farías *et al.* 2008) is made up of Cretaceous plutons within a Late Jurassic–Early Cretaceous sedimentary and volcanic country rock.

The Central Depression at an elevation of 700–500 m.a.s.l separates the Coastal Range from the Principal Cordillera. It consists of a Quaternary sedimentary and ignimbritic cover of up to 500 m thickness beneath the Santiago valley (Araneda *et al.* 2000) and basement rocks cropping out in junction ridges and isolated hills reaching 1600 m a.s.l. (Rodríguez *et al.* 2012).

The Principal Cordillera is also subdivided into western and eastern sectors. The western sector consists of Eocene–Early Miocene volcaniclastic rocks of the Abanico extensional basin (Charrier *et al.* 2002), covered by the Miocene volcanic-arc rocks of the Farellones Formation (21.6–16.6 Ma; Aguirre *et al.* 2000) which young southwards (Godoy 2014). The eastern Principal Cordillera consists of a thick sequence of Mesozoic sedimentary rocks, highly deformed into the Aconcagua fold-and-thrust belt (Giambiagi *et al.* 2003*a*).

The Frontal Cordillera at this latitude corresponds to the Cordón del Portillo range, where Proterozoic metamorphic rocks, Late Palaeozoic marine deposits, Carboniferous–Permian granitoids and Permo-Triassic volcanic rocks crop out (Polanski 1964). This range is uplifted by several east-vergent faults of the Portillo fault system.

The Cerrilladas Pedemontanas range is marked by the inversion of Triassic extensional faults of the Cuyo basin and moderately dipping basement faults (García & Casa 2014). Triassic continental rocks are covered by Neogene to Quaternary synorogenic deposits of the Cacheuta basin (Irigoyen *et al.* 2000). The Quaternary volcanic arc at this latitude is represented by the Marmolejo-Espíritu Santo-San José volcanic centre, located along the crest of the Andes 300 km east of the Chile trench.

Thermomechanical modelling

A necessary first-order constraint for the construction of a crustal-scale balanced cross-section is the

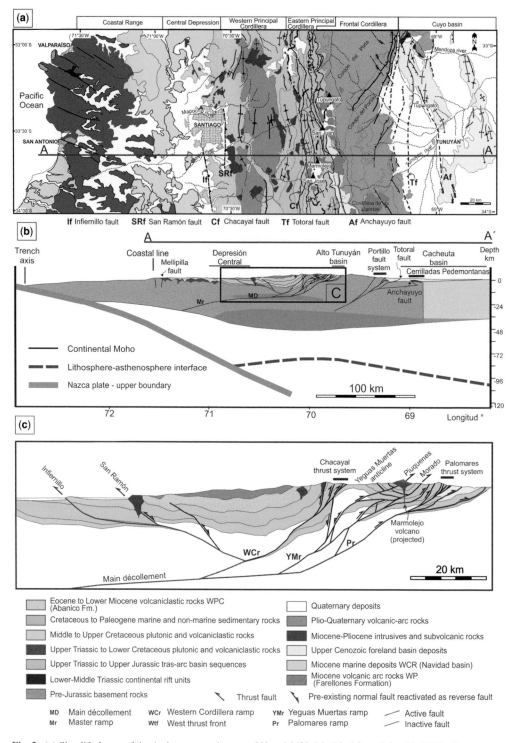

Fig. 2. (a) Simplified map of the Andean orogen between 33° and 34°S. Modified from Polanski (1964), Gana et al. (1996), Wall et al. (1996), Sellés & Gana (2001), Giambiagi & Ramos (2002) and Giambiagi et al. (2003a).
(b) Balanced cross-section from coast to foreland, based on our own data and detailed compilation from Polanski (1964),

expected mechanical structure of the lithosphere, particularly the identification of eventual ductile zones associated with brittle–ductile transitions that can serve as master detachments. The gravity-based and seismically constrained three-dimensional density model of the Andean margin developed by Tassara *et al.* (2006) and recently upgraded by Tassara & Echaurren (2012) contains the geometries for the subducted slab, the lithosphere–asthenosphere boundary of the South American plate (LAB), the continental Moho and the intra-crustal discontinuity, which separates the upper crust (density 2700 kg m^{-3}) from the lower crust (density 3100 kg m^{-3}). Based on these geometries and analytical formulations of the heat transfer equation (Turcotte & Schubert 2002) with appropriate boundary conditions, a 3D thermal model of the Andean margin has been developed (Morales & Tassara 2012; Tassara & Morales 2013). This model accounts for heat conduction from the LAB and subduction megathrust with radiogenic heat generation into the crust and thermal advection by the subducted plate, and predicts the distribution of temperature inside the Andean lithosphere.

The 3D thermal model and the original density model serve as the base to construct a 3D mechanical model based on the concept of the yield strength envelope (Goetze & Evans 1979; Burov & Diament 1995; Karato 2008). This envelope predicts the mechanical behaviour (brittle, elastic or thermally activated ductile creep) of a compositionally layered lithosphere with depth along a given 1D geotherm via the extrapolation of constitutive rheological laws for different Earth materials from experimental conditions to lithospheric space–temporal scales.

We use a preliminary version of the thermomechanical model (Tassara 2012) to extract an east–west cross-section along the Maipo–Tunuyán transect (Fig. 2b). This section shows that most of the upper-middle crust has a brittle-elastic behaviour, particularly for the cold and rigid forearc and foreland regions. However, a ductile behaviour is predicted by the model below the Principal Cordillera within a thin layer (<5 km) at mid-crustal depths (15–20 km) and for the entire lower crust (i.e. deeper than 30 km).

Kinematic model of subduction orogeny

Our kinematic model with thermomechanical constraints (Fig. 3) accounts for the long-term deformation of the Andean convergent margin between the cratonic interior of the South American plate and the forearc. It considers a coupled slab–forearc system with elastic loading and release associated with the megathrust seismic cycle. We kinematically model a velocity gradient ΔV_x between the continental lithospheric plate and the underlying asthenospheric mantle by fixing the slab–forearc interface and applying a westwards movement of the continental plate along an artificial line at the base of the Moho. This artificial line has no geological significance; it has been designed for the purpose of kinematical modelling and represents a broad area with dislocation creeping and a vertical velocity gradient ΔV_x constrained between the Moho and the lithosphere–asthenosphere boundary. Displacement is transmitted along this base using the *trishear* algorithm with p/s (propagation/displacement) ratio between 0 and 0.5 until the singularity point S below the Cordilleran axis (Fig. 3). At the singularity S, which is located above the downdip limit of the coupled seismogenic zone, shortening is transmitted to a ramp-flat master décollement, modelled with the fault parallel flow algorithm as a passive master fault. Above this décollement, crustal block B experiences brittle deformation. Below the decóllement, the crust thickens by ductile deformation. Inside crustal block A, deformation diffuses westwards and upwards in a triangular zone of distributed shear. This deformation represents flexural permanent deformation above the S point.

The geometry of our proposed master décollement coincides with previous conceptual models for deep crustal deformation proposing an eastwards vergence of the orogen (Allmendinger *et al.* 1990; Allmendinger & Gubbels 1996; Ramos *et al.* 2004) and with the active ramp-flat structure dipping 10°W beneath the western Principal Cordillera and 25–30°W beneath the Coastal Range proposed by Farías *et al.* (2010) based on seismological studies. In our model, the décollement ramps up beneath the easternmost sector of the Chilean slope of the Andes. The vertical component of slip in this ramp would be much greater than in the Argentinean slope, where low-angle thrusting develops. The abrupt rising of the international border sector may therefore be a consequence of localized rapid uplift on this segment of the transect.

The rationale follows Isacks' (1988) conceptual model in which brittle crustal horizontal shortening in the back-arc upper crust is compensated by

---

Fig. 2. (*Continued*) Gana & Tosdal (1996), Alvarez *et al.* (1999), Godoy *et al.* (1999), Giambiagi & Ramos (2002), Giambiagi *et al.* (2003a, b), Rauld *et al.* (2006), Godoy (2011). Red lines indicate faults with Quaternary activity. Geophysical data was extracted from the ACHISZS electronic database (www.achiszs.edu.cl/~achiszs/accessdb.html). (**c**) Detail of the cross-section in (b) for the sector of Principal Cordillera. Red lines indicate faults with Quaternary activity.

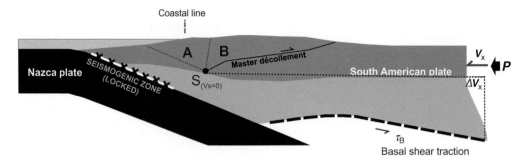

Fig. 3. Conceptual model for the 33°40′S transect. A velocity gradient ΔV_x is applied at the back of the foreland area during the 100% locked interseismic period. We kinematically model deformation by applying movement along an artificial basal line at the Moho, which transmits displacement westwards until the singularity point S. At this point S shortening is transmitted to a ramp-flat master décollement, modelled as a passive master fault. Above this décollement, crustal block B experiences brittle deformation. Below this line, the crust thickens by ductile deformation. Inside crustal block A, deformation diffuses westwards and upwards in a triangular zone of distributed shear.

ductile thickening in the arc and back-arc lower crust, and with the Andean microplate model of Brooks *et al.* (2003) in which a west-dipping décollement below the Andes mountain belt is viscously coupled to the South American craton and allows continuous creeping and stress transmission across the boundary.

We perform a forward model using the 2DMove academic license, taking into account the geological and geophysical constraints and the thermomecanical model. We assume plain strain along the transect, without magmatic additions or subduction erosion. Shortening estimates rely on upper crustal deformation measured all along the transect and the balanced cross-section, which provides a minimum estimate of horizontal shortening. First of all, we reconstruct the Late Cretaceous–Early Paleocene and Eocene–Early Miocene compressional and extensional events using the available geological and geochemical studies. For the Early Miocene–present shortening period, we carry out 36 steps each of 2 km shortening, constrained by the timing of movement along the main structures of previous studies (Godoy *et al.* 1999; Giambiagi & Ramos 2002; Giambiagi *et al.* 2003*a*), isotopic analysis (Ramos *et al.* 1996; Kay *et al.* 2005) and exhumation analysis (Kurtz *et al.* 1997; Maksaev *et al.* 2004, 2009; Hoke *et al.* 2014). With these data we calculate an average shortening rate for each period. Every 2–3 steps, we simulate erosion and sedimentation in accordance with sedimentological, palaeogeographic and provenance studies from the foreland and forearc basins (Irigoyen *et al.* 2000; Giambiagi *et al.* 2001; Rodríguez *et al.* 2012; Porras *et al.*, this volume, in review), and allow flexural-isostatic adjustments of the lithosphere due to local load changes assuming a default value for the Young's modulus $E = 7 \times 10^{10}$ Pa and the effective elastic thickness (T_e) calculated in Tassara *et al.* (2007). Considering that we take into account the effects of erosion and sedimentation, topographic data presented here are purely qualitative.

Structure of the 33°40′S transect

The present-day structure of the 33°40′S transect is represented in Figure 2b, c. The crust below the Andes appears to reach its greatest thickness of 50 km at a longitude near 69°45′W. This longitude closely corresponds to the location of the highest peaks at this latitude that reach an altitude of 6000 m in the Marmolejo volcano. The Coastal Range (Fig. 4a) represents an east-dipping gentle homocline of Upper Palaeozoic–Cretaceous rocks (Wall *et al.* 1999). No major Andean thrust fault has been identified in this range, which is affected mainly by high-angle pre-Andean NNW–NW-trending faults such as the Melipilla fault (Yañez *et al.* 2002), which may have been inherited from the Late Triassic continental-scale rifting.

The western Principal Cordillera comprises the Late Eocene–Miocene volcanic arc and is dominated by the inversion of the Abanico extensional basin (Fig. 4b). Fock *et al.* (2006) proposed that this inversion has a bivergent sense. According to Godoy *et al.* (1999) however, a double vergence for Abanico is developed only south of this latitude; at the transect latitude, the Abanico master faults are only inverted in its western edge.

The Front Range east of Santiago city represents the western thrust front of the Principal Cordillera. It was uplifted by inversion of the east-dipping Abanico master fault system, including the Infiernillo and San Ramón faults (Godoy *et al.* 1999; Fock *et al.* 2006; Rauld *et al.* 2006). The latter of which has been described by Rauld *et al.* (2006)

and Armijo *et al.* (2010) as a feature active during the Holocene. The hanging wall of the San Ramón fault is folded into a tight syncline–anticline pair interpreted to be generated by an underlying basement ramp (Godoy *et al.* 1999). Toward the east, east-dipping faults with small throws are inferred to fold the Abanico and Farellones strata (Armijo *et al.* 2010).

The east-vergent Chacayal thrust system marks the border between western and eastern Principal Cordillera sectors (Giambiagi & Ramos 2002). This thrust system, which uplifts a thick sheet of Upper Jurassic sedimentary rocks (Fig. 4c), runs from the Las Cuevas river (32°50′S) to the Maipo river (34°10′S) for more than 150 km along-strike and corresponds to the most important structure in the western slope of the Principal Cordillera. The Aconcagua fold-and-thrust belt presents an overall geometry of a low-angle eastwards-tapering wedge, characterized by a dense array of east-vergent imbricate low-angle thrusts (Fig. 4d) with subordinate west-vergent out-of-sequence thrusts (Fig. 4d) (Giambiagi & Ramos 2002). The western sector of the belt is dominated by a broad anticline related to the inversion of the Triassic–Jurassic Yeguas Muertas extensional depocentre (Alvarez *et al.* 1999).

The Frontal Cordillera represents a rigid block uplifted by the Portillo fault system, located at its eastern margin (Fig. 4e). This system corresponds to a series of east-vergent deeply seated thrust faults (Fig. 4f), which uplift the pre-Jurassic basement rocks on top of the Middle Miocene–Quaternary sedimentary rocks deposited in the foreland basin (Polanski 1964; Giambiagi *et al.* 2003a). The foreland area comprises Neogene–Quaternary sedimentary deposits gently folded into open anticlines (Fig. 4g). The Triassic Cuyo basin is partially inverted in this sector by reactivation of pre-existing structures, such as the Anchayuyo fault, and generation of new thrusts (Fig. 4h).

Deformational periods

Even although the tectonic evolution of the transect can be regarded as a continuous deformational event from the Early Miocene to the present, with crustal thickening and widening and uplift of the Andean ranges, we can separate this evolution into several periods of deformation during which rock uplift and erosion shape the orogen.

The Late Cretaceous–Paleocene deformational event

The Andean orogeny started in several segments with contractional deformation during Late Cretaceous time when the back-arc basins began to be tectonically inverted (Mpodozis & Ramos 1989). Deformation in the southern Central Andean thrust belts north and south of this transect started during the Upper Cretaceous–Palaeogene shortening event (Mpodozis & Ramos 1989; Tunik *et al.* 2010; Orts *et al.* 2012; Mescua *et al.* 2013). According to Tapia *et al.* (2012), in the studied transect this contractional event was localized in the Coastal Cordillera and the western sector of the Principal Cordillera. However, Godoy (2014) argues that the evidence for this event in the Main Range is dubious and interprets the area to represent a structural knot that resisted K–T (Cretaceous–Tertiary) deformation.

In the eastern Principal Cordillera on the other hand, continental sediments were deposited in the northern Neuquén foreland basin (Tunik *et al.* 2010) which culminated with an Atlantic marine ingression during Maastrichtian time (Tunik 2003; Aguirre-Urreta *et al.* 2011). Further east, geothermocronological analysis indicates that the foreland area east of Frontal Cordillera was stable (not deformed nor uplifted) during the Jurassic–Palaeogene period (Ávila *et al.* 2005). Further structural studies are needed to understand the deformation during this stage, which is beyond the scope of this study.

Given the lack of precise constraints for the deformation during Cretaceous–Palaeogene time in the studied transect, we estimated the shortening based on the assumption of conservation of the crustal area along the section. We used the Tassara & Echaurren (2012) geophysical model to calculate the crustal area at present. Taking into account the shortening estimations based on structural studies (Godoy *et al.* 1999; Giambiagi & Ramos 2002; Giambiagi *et al.* 2003a; Fig. 2b), an initial crustal thickness of $T_0 = 40$ km is needed to accommodate Andean shortening along the transect. However, crustal thickness was not constant across-strike before the Neogene, since a Palaeogene extensional event led to a thin crust (of 30–35 km of thickness) in the western Principal Cordillera (see the following section). In order to preserve crustal area, this region of thin crust should be compensated with a thick region which we associate with the Cretaceous–Palaeogene contractional deformation developed to the west of the extended sector. A block of 42-km-thick crust in the present Coastal Cordillera and Central Depression can account for the missing crustal area. We modelled this block of thicker crust with 10 km of Late Cretaceous–Palaeogene shortening (Fig. 5a).

The Eocene–Early Miocene extensional event

During Eocene–Early Miocene time, a protracted extensional event took place with deformation concentrated in the western sector of the Principal

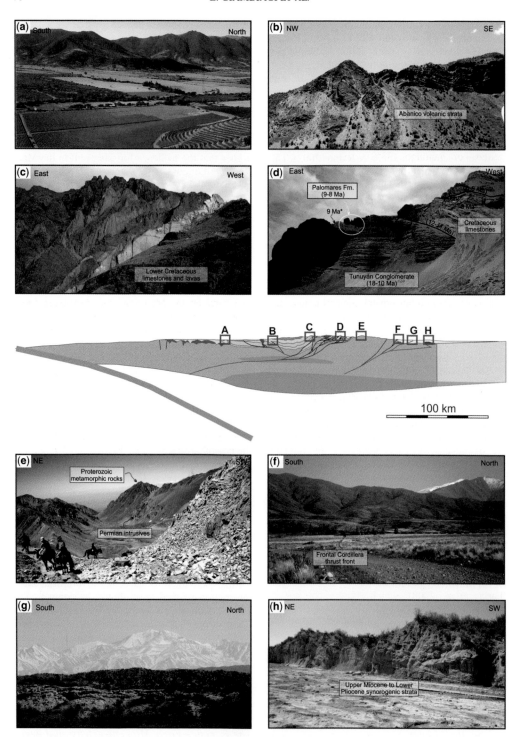

Fig. 4. Photographs of the 33°40′S transect. (**a**) Central Depression and Coastal Range (photograph P. Alvarez). (**b**) Abanico strata deformed into open west-vergent folds in the western Principal Cordillera (photograph J. Suriano). (**c**) Vertical mesozoic beds uplifted by the Chacayal fault and back-tilting by the Aconcagua FTB faults (photograph L. Pinto). (**d**) Palomares fault system in the Aconcagua FTB, uplifting Mesozoic strata over the Neogene synorogenic

Cordillera (Charrier et al. 2002). This event has been linked to a segmented roll back subduction event (Mpodozis & Cornejo 2012) in contraposition to the compressive deformation registered in the Andean margin north of 25°S (Carrapa et al. 2005; Hongn et al. 2007). Normal faulting was associated with crustal thinning and tholeiitic magmatism of the Abanico Formation (Nyström et al. 1993; Kay & Kurtz 1995; Zurita et al. 2000; Muñoz et al. 2006), whose $^{40}Ar/^{39}Ar$ ages at the latitude of Santiago (33°S) range from 35 to 21 Ma (Muñoz et al. 2006). The extensional basin was filled with up to 3000 m of volcaniclastic deposits, and acidic to intermediate lavas with sedimentary intercalations (Charrier et al. 2002). Even although geochemical analyses suggest it was formed upon a c. 30–35-km thick continental crust (Nyström et al. 2003; Kay et al. 2005; Muñoz et al. 2006), there is no evidence of marine sedimentation. Isotopic analysis and comparison between western and eastern outcrops of the Abanico Formation indicate higher crustal contamination, suggesting larger crustal thickness to the east (Fuentes 2004; Muñoz et al. 2006) and an asymmetric geometry of the extensional basin, due to the concentration of extensional deformation in the western master fault system.

During this period, the Coastal Range should have been elevated due to isostatic compensation of the Late Cretaceous–Paleocene crustal thickening event and locally by isostatic rebound close to the Abanico east-dipping master fault. Further east in the Neuquén basin, Paleocene continental distal sediments were deposited tapering towards the east.

The extensional structures of the Abanico basin have been obliterated by the later compressional faults. Inferred master faults have been suggested to be located beneath the Central Depression (Godoy et al. 1999; Fock et al. 2006). Much more speculative is the existence of eastern basin-border faults. Apatite and zircon fission track cooling ages from the eastern Coastal Cordillera constrain an exhumation period between 36 and 42 Ma (Farías et al. 2008), which could be related to uplift of the rift shoulder westwards of the master fault (Fock 2005; Charrier et al. 2007) at the beginning of the extensional period (since Late Eocene).

The Early Miocene (21–18 Ma)

The last major shortening event began during the Early Miocene with the inversion of the Abanico basin (Godoy et al. 1999; Charrier et al. 2002; Fock et al. 2006). After that, chronological data suggest that deformation has accumulated during a single period of shortening between the Early Miocene and the present (Giambiagi et al. 2003a; Porras et al., this volume, in review). The onset of deformation in the western slope is marked by the change from low-K Abanico Formation tholeiites to calc-alkaline dacites of the Teniente Volcanic Complex (Kay et al. 2005, 2006), between 21 and 19 Ma (Charrier et al. 2002, 2005). This inversion occurred coevally with the development of the Farellones volcanic arc of Early–Middle Miocene age (Vergara et al. 1999). A 22.5 Ma U–Pb zircon SHRIMP age of the base of the Farellones Formation along the Ramón–Damas Range anticline marks the beginning of the Farellones arc at this transect (Fock 2005).

In our model, the western Principal Cordillera is related to the generation of the Main decóllement (MD) rooted in the singularly point S at the contact between the Moho and mantle sub-arc lithosphere, c. 40 km depth from the surface, and the Western Cordillera ramp (WCr) inferred to be located below the Western thrust front (WTF). We interpret this event as the generation of an important detachment connected to the Chilean ramp proposed by Farías et al. (2010) and located between 12 and 15 km depth beneath the Principal Cordillera. Eastwards movement of the upper crust relative to the stable and long-lived lower crust MASH (melting, assimilation, storage and homogenization) zone proposed by Muñoz et al. (2012) could explain the width of up to 40 km of the Farellones volcanic arc. The main uplifted sectors for this period are the eastern Coastal Range and the western Principal Cordillera, above the Western Cordillera ramp (the Front Range east of the city of Santiago). The passive uplift of both sectors is in agreement with provenance studies of heavy mineral assemblages for the lower Navidad Formation (Rodríguez et al. 2012), whose radiometric ages indicate deposition during 23–18 Ma (Gutiérrez et al. 2013) and apatite fission track (AFT) age (18.3 ± 2.6 Ma) for an Upper Cretaceous deposit (Fock 2005).

The passage of the pre-existing Abanico master normal faults over the WCr could have reactivated these structures as passive back-thrusts and favoured the intrusion of the La Obra pluton (19.6 Ma; Kurtz et al. 1997). In the foreland, retroarc volcanism of the Contreras Formation (18.3 Ma;

Fig. 4. (*Continued*) units (photograph M. Tunik). (**e**) Frontal Cordillera from the Portillo pass (photograph M. Tunik). (**f**) Frontal Cordillera active thrust front (photograph L.Giambiagi). (**g**) Looking westwards from the Cacheuta basin. Strata on front correspond to uplifted Neogene synorogenic deposits (photograph L. Giambiagi). (**h**) Neogene deposits uplifted and folded at the easternmost sector of the transect (photograph L. Giambiagi).

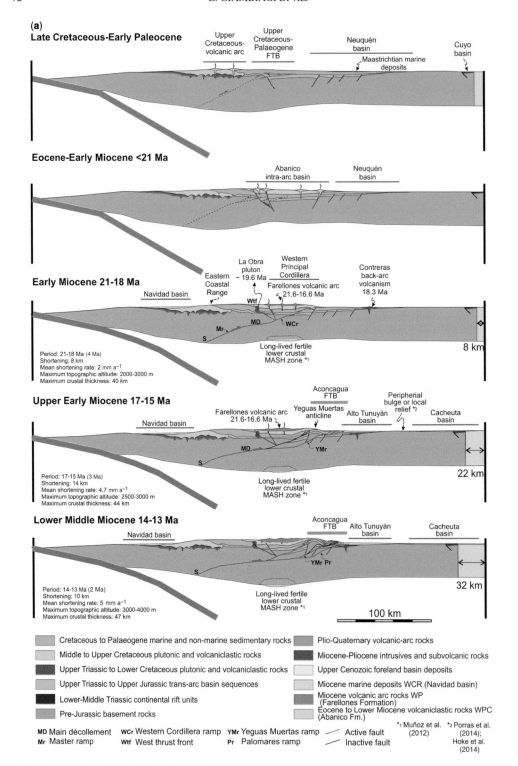

Fig. 5. Model for the evolution of the 33°40′S transect from the Late Cretaceous to the present. (a) Deformational periods from the Late Cretaceous–Early Paleocene compressional event to the early Middle Miocene phase.

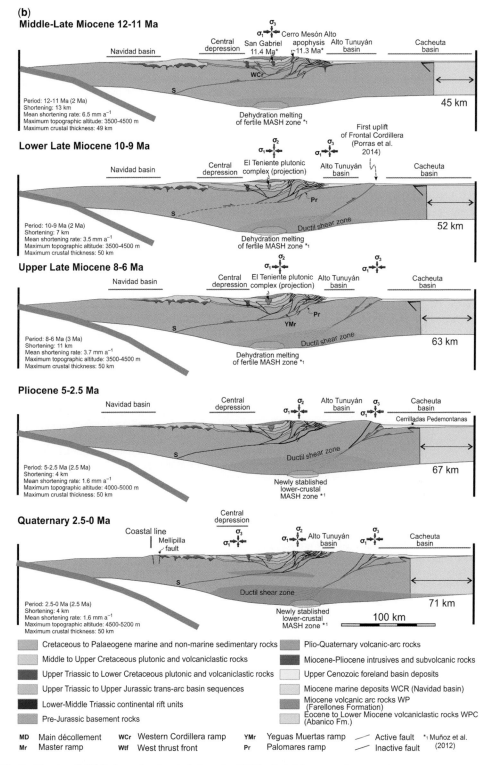

Fig. 5. (*Continued*) (**b**) Deformational periods from the Middle–Late Miocene to the present.

Giambiagi & Ramos 2002) with geochemical signatures of unthickened crust (Ramos et al. 1996) predates the formation of the Alto Tunuyán foreland basin.

We calculated 8 km of shortening concentrated in the western Principal Cordillera (7 km) and the Coastal Range (1 km), with a mean shortening rate of 2 mm a^{-1} for this period. Beneath the western Principal Cordillera, the crust achieved its maximum thickness (40 km) and maximum topographic elevation ranges between 2000 and 3000 m a.s.l.

The Late Early Miocene (17–15 Ma)

This period marks the beginning of deformation along the Aconcagua FTB (fold-and-thrust belt), linked to the prolongation of the Main décollement eastwards and the formation of the Yeguas Muertas ramp beneath the homonymous anticline (Giambiagi et al. 2003a, b). By this time, driving forces cannot supply the energy needed to elevate the western Principal Cordillera by movement along the WCr (supercritical wedge stage), and the orogen begins to grow in width to lower the taper by propagating forwards towards the foreland. According to Muñoz et al. (2012), this shift in the locus of deformation should be related to the rapid ascent of subduction-related mantle-derived magmas that had little interaction with the upper lithosphere, such as those analysed by these authors in the western Principal Cordillera.

The earliest Neogene synorogenic strata appear in the Alto Tunuyán basin at c. 17–16 Ma, with source region restricted to the volcanic rocks of the Principal Cordillera and to the Frontal Cordillera (Porras et al., this volume, in review). For this last source region, Porras et al. (this volume, in review) propose the existence of a peripheral bulge located in what is now the Frontal Cordillera, while Hoke et al. (2014) suggest a significant pre-Middle Miocene local relief. Further studies are required in order to decide which option is the best. During this period, the Cacheuta basin, apparently disconnected from the Alto Tunuyán basin, only received aeolian deposits (Irigoyen et al. 2000).

The uplift of the western Principal Cordillera evidenced in the provenance of the Navidad basin (Rodríguez et al. 2012) is due to overriding of this sector of the range across the Yeguas Muertas ramp. Movement along the Main décollement and this ramp achieves 14 km of shortening in the Aconcagua FTB, with an average shortening rate of 4.7 mm a^{-1}. Maximum topographic elevation is between 2500 and 3000 m in the western Principal Cordillera, and the crust thickens to 44 km below the westernmost Principal Cordillera as suggested by geochemical data (Kay et al. 2005). Flexural–isostatic compensation of this sector of the orogen could explain the exhumation rates of the La Obra pluton (0.5–0.6 mm a^{-1}) during this period (Kurtz et al. 1997).

The coeval uplift of the eastern Coastal Cordillera and the Principal Cordillera leads to the formation of a depocentre filled by up to 3000 m of arc-related volcanics of the Farellones Formation (Vergara et al. 1988; Elgueta et al. 1999; Godoy et al. 1999). Magma migration within the arc may have been enhanced by active movement along the Western Cordillera ramp and the associated back-thrusts.

The Middle Miocene (15–11 Ma)

During this phase, upper crust deformation is localized in the eastern Principal Cordillera with the development of the east-vergent in-sequence Aconcagua FTB faults and associated back-thrusts close to the international border. Cross-cutting relationships between thrust faults suggest the cyclic activation–deactivation of the Western, Yeguas Muertas and Palomares ramps with important shortening (17 km) in the Principal Cordillera. The period between 12 and 11 Ma corresponds to one of quiescence in deformation along the Aconcagua FTB, when the thrust front became inactive and erosion of the previously uplifted area was responsible for the generation of an important unconformity between the Cretaceous strata and the Middle Miocene synorogenic deposits (Giambiagi et al. 2001). Instead, deformation is concentrated in the Chilean slope of the Principal Cordillera with the reactivation of the Western Cordillera ramp and movement along the pre-existing San Ramón and Infiernillo faults (Fock et al. 2006). After this short quiescence period, the reactivation of the Yeguas Muertas and Palomares ramps leads to generation of out-of-sequence thrusts in the Aconcagua FTB.

This is a period of high crustal shortening (23 km) concentrated along the international border and the eastern Principal Cordillera, with the mean shortening rates of 5 mm a^{-1} and 6.5 mm a^{-1} for the 15–13 and 12–11 Ma periods, respectively. The crust almost achieves a thickness of 49 km, with an eastwards shift in the crustal keel. Important space creation due to flexural compensation of the tectonic load, immediately to the east of the Aconcagua FTB, is registered both in the Alto Tunuyán and Cacheuta basins (Irigoyen et al. 2000; Giambiagi et al. 2001).

At the end of this phase, the volcanic activity practically wanes and only very localized activity is recorded during late Middle Miocene–Late Pliocene time, in agreement with a highly compressive stress regime in the arc region. The present volcanic arc develops in the latest Pliocene along the watershed. Instead, barren and mineralization-hosting intrusives, such as La Gloria and San Gabriel plutons and Cerro Mesón Alto porphyry (Deckart et al. 2010), intrude the Miocene Farellones Formation or the Upper Cretaceous sedimentary rocks.

The Late Miocene (10–6 Ma)

At the beginning of this phase around 10–9 Ma, the crust achieves its present maximum thickness of 50 km in accordance with isotopic analysis (Kay et al. 2005). At this point, driving forces can no longer supply the energy needed to thicken the crust and uplift the range and the crustal root is likely to grow laterally in width instead of increasing its depth, with a reduction in shortening rates. This is in accordance with the proposition of Muñoz et al. (2012) for the existence of deep crustal hot zones. According to these authors, the repeated basalt intrusion into the lower crust induces a significant thermal perturbation, with the amphibolitic lower crust reaching temperatures up to 750–870°C and increased melt component. This promotes the ductile behaviour of the lowermost crust and the widening of the crustal root.

Overall, no volcanic activity is registered during this period. Instead, the Teniente Plutonic Complex intrudes the Abanico and Farellones formations between 12 and 7 Ma (Kay & Kurtz 1995; Kurtz et al. 1997; Kay et al. 2005).

During the early Late Miocene (10–9 Ma), reactivation of the Palomares ramp leads to out-of-sequence thrusting in the Aconcagua thrust front and the cannibalization of the Alto Tunuyán foreland basin deposits. Seven kilometres of shortening are achieved during this phase at the thrust front. Provenance analysis of the synorogenic fill of the Alto Tunuyán foreland basin indicates the beginning of uplift of the Frontal Cordillera (Porras et al., this volume, in review), thrusts of which have been inferred to be deeply seated in a ductile shear zone of the lower crust (Giambiagi et al. 2012).

During the latest Late Miocene (8–6 Ma), the sedimentary record indicates important uplift of the Frontal Cordillera between 33°30′ and 34°30′S (Irigoyen et al., 2000; Giambiagi et al. 2003a; Porras et al., this volume, in review). Thermochronological studies in Argentina both bordering and within the study area show that very little exhumation has occurred in this area during Cenozoic time. An apatite (U–Th)/He study in the Frontal Cordillera between 32°50′ and 33°40′ yields pre-Miocene exhumation rates of c. 12 m Ma^{-1}, with an increse to 40 m Ma^{-1} at 25 Ma and an inferred onset of rapid river incision between 10 and 7 Ma, roughly within geological constraints (Hoke et al. 2014).

A couple of million years after the onset of uplift in the Frontal Cordillera, the Principal Cordillera experiences important reactivation of the Yeguas Muertas and Palomares ramps with generation of out-of-sequence thrusts (e.g. Piuquenes and Morado) and the reactivation of the Palomares thrust system (Giambiagi & Ramos 2002).

At this time, movement along the Yeguas Muertas ramp would cause the Western Cordillera ramp to ramp up with a vertical component of slip much greater than the eastern Principal Cordillera and a localized rapid uplift. This is in agreement with fission track data from the Miocene plutons exposed in the western Principal Cordillera, compatible with <3 km of denudation of the El Teniente district since Late Miocene time (Maksaev et al. 2004), and the identification of a main stage of surface uplift in the Chilean slope during Late Miocene–Early Pliocene time with maximum vertical throw 0.7–1.1 km (Farías et al. 2008).

During Early Miocene–Early Pliocene time, the western sector of the Coastal Cordillera is submerged as evidenced by the marine deposits of the Navidad Formation and younger units (Encinas et al. 2008; Gutiérrez et al. 2013), and is subjected to local extensional stress field (Lavenu & Encinas 2005). Flexural elasticity of the rigid cold forearc region causes the western Coastal Range to subside while the eastern sector of the range is uplifted. This is in agreement with geologic evidence for subsidence between 10 and 4 Ma in this sector of the transect (Encinas et al. 2006).

The Pliocene (5–2.5 Ma)

During Pliocene time, the magmatic activity reassumes its current locus along the High Andean drainage divide. Around this time, the Main décollement becomes inactive and there is a lull in deformation in the Principal Cordillera, concomitant with the mineralization of the El Teniente porphyry whose magmatic-hydrothermal centre records 6 Ma of continuous activity during 9–3 Ma (Mpodozis & Cornejo 2012). Compressional deformation occurs only in the eastern Frontal Cordillera with generation of thrusts affecting the Miocene–Early Pliocene synorogenic deposits of the Cacheuta basin, and in the Cerrilladas Pedemontanas with the generation of an angular unconformity in the synorogenic units (Yrigoyen 1993; Irigoyen et al. 2000) and the inversion of the Anchayuyo fault (Giambiagi et al. submitted to Tectonics).

The compressional deformation in the Frontal Cordillera is coeval with movement along NE dextral and WNW sinistral strike-slip faults reported to affect the El Teniente porphyry copper deposits in western Principal Cordillera (Garrido et al. 1994). According to these authors, the strike-slip faults were active before, during and after the formation of the giant porphyry Cu–Mo deposit (6.3–4.3 Ma, Maksaev et al. 2004), indicating a strike-slip regime during Early Pliocene time.

The widening of the crustal root during this stage favours the Coastal Cordillera uplifts by isostatic rebound, evidenced by the emergence of marine deposits between 4.4 and 2.7 Ma (Encinas et al. 2006) and the onset of knickpoint retreat in the western Coastal Range around 4.6 Ma (Farías et al. 2008). This process would have partially blocked the drainage, inducing sedimentation in the Central depression (Farías et al. 2008). The widening of the crustal root could in turn be responsible for the increase in topographic elevation by isostatic compensation. The enhanced erosion related to increased relief during 6–3 Ma suggested by Maksaev et al. (2009) could be attributed to this phenomenon.

The Quaternary (2.5 Ma–present)

The distribution of shallow earthquake epicentres in the Andes between 33° and 34°S gives important clues for the Quaternary deformation along the transect. In the Chilean slope of the Andes, shallow seismic activity is distributed mainly along the western flank of the Principal Cordillera at depths of 12 and 15 km (Barrientos et al. 2004), and beneath the Central depression at depths shallower than 20 km (Farías et al. 2010). This suggests that at least the western portion of the Main décollement is active today. Beneath the Western Principal Cordillera, seismicity is concentrated on the steeper portion of the main décollement. We suspect that slip occurs aseismically on the gently dipping segments of this detachment zone. This is in agreement with minor Quaternary movement along the San Ramón fault, whose scarp has been linked to the vertical offsets (0.7–1.1 km) of the peneplains present in the western Principal Cordillera before 2.3 Ma (Farías et al. 2008).

In the High Andes close to the Chilean–Argentine border, most shallow earthquakes of $M > 5$ show focal mechanisms related to strike-slip kinematics (Barrientos et al. 2004), such as the M_w 6.9 Las Melosas earthquake (Alvarado et al. 2005). Beneath the eastern part of the Frontal Cordillera and the Cuyo basin shallow focal mechanisms are related to compressional kinematics (Alvarado et al. 2007). The neotectonics of the Andean retrowedge at this latitude is characterized by movement along buried faults that fold the Quaternary deposits (García & Casa 2014).

During this phase, the backward tilting of the Principal and Frontal cordilleras by movement along the Portillo thrust system and the isostatic readjustment of the thickened crust could be responsible for the tilting to the west (1–3°) of several flat erosional surfaces located between 2600 and 3200 m a.s.l. in the western Principal Cordillera (Farías et al. 2008). We propose that the increase in gravitational potential energy due to crustal roots formation and mountain uplift prevents the main décollement from propagating eastwards. Instead, reactivation of the WPC back-thrusts should have implied less work against resisting stresses.

Discussion: implications of the model

The kinematic model presented in this work integrates structural, sedimentological, petrological, geochronological and geophysical data for the studied transect. The geological constraints available for the area allowed us to build a complete model that describes the relation between horizontal shortening, uplift, crustal thickening and activity of the magmatic arc, accounting for the main deep processes that have shaped the Andes since Early Miocene times. These geological and geophysical constraints reinforce previous hypotheses of a west-dipping detachment at the transition from brittle-elastic to ductile rheology in the crust below the Andean strip (Allmendinger & Gubbels 1996; Ramos et al. 2004; Farías et al. 2010; Giambiagi et al. 2012), and are consistent with the east-vergent models. Furthermore, the model allows us to discuss some aspects of Andean history in light of our results.

Our model proposes passive surface uplift in the Coastal Range as the master décollement decreases its slip downwards transferring shortening to a broad area above the singularity point S, where the master detachment touches the Moho horizon (Figs 3 & 5). During the 18–5 Ma period, the main phase of deformation is located in the Aconcagua FTB. As the crust thickens from 40 to 50 km and the upper crust shortens to 52 km, the S point slowly migrates westwards (Fig. 5b). This migration generates a slowly propagating wave of surface uplift in the Coastal Range, from its eastern part at c. 18 Ma to its western part at the beginning of Pliocene time, in agreement with provenance studies from the Navidad basin (Rodríguez et al. 2012). Coeval uplift of the Coastal Range and Principal Cordillera creates the Central Depression, inducing thick sedimentary filling as suggested by Farías et al. (2008). This explains why no important coastal-parallel faults have been observed in the Coastal Range, even although it experience significant topographic uplift during Cenozoic time suggested by the exposure of the Miocene marine deposits (Encinas et al. 2008; Gutiérrez et al. 2013). Moreover, the presence of relict continental erosion surfaces with different elevations developed close to sea level and uplifted up to 2.1 km a.s.l. (Farías *et al.* 2008) indicates several pulses of Coastal Range uplift.

The eastward migration of the volcanic arc during Late Miocene–Quaternary time could be related to an eastwards migration of the trench due to tectonic erosion of the continental margin, as proposed by Stern (1989, 2011) and Kay *et al.* (2005); the delamination of an eclogitic root in the lower crust and mantle lithosphere (Kay & Kay 1993; DeCelles *et al.* 2009); and to shortening of the brittle crust above the major decóllement. Migration of the volcanic arc can be explained by our model without invoking the decrease in the angle of subduction of the oceanic slab for this period of time, consistent with the proposal of Godoy (2005). Instead, the overall decrease in the volume of the asthenospheric wedge due to the construction of the crustal root could inhibit the influx of hot asthenosphere into the region and be responsible for the cooling of the subarc mantle and the eastward migration of the arc, as proposed by Stern (1989).

Underthrusting of the mechanically strong South American craton beneath the Andean strip leads to thickening of the crust. Kay *et al.* (1999) and Kay & Mpodozis (2001) argued that the transformation of hydrous lower-crustal amphibolite to garnet-bearing eclogite during crustal thickening can be responsible for the exsolution of fluids. These fluids can substantially decrease the strength of the lower crust. In this way, crustal thickening before Middle Miocene time may have favoured increased deformation during the Middle–Late Miocene (12–10 Ma) period, enhancing horizontal shortening in the Principal Cordillera and widening of the crustal roots. The forces that support the Andes provide an upper limit to the height of the mountain range and also to crustal thickness (Molnar & Lyon-Caen 1988). In our study area, the critical value seems to be of the order 50 km. Once this thickness is achieved, gravitational potential forces created by the buoyant crustal root are higher than unbalanced driving tectonic forces and therefore thrusting shifts eastwards into the lowlands. Even although convergence was steady, faults in the Principal Cordillera become inactive, the main décollement is deactivated and a new décollement is formed in the east to uplift the Frontal Cordillera.

Conclusions

In this paper we propose a kinematic model with thermomechanical constraints for the Miocene–present evolution of the Southern Central Andes at the latitude of the city of Santiago. Our model assumes a main décollement located at the base of the upper crust (15–12 km in depth), which produces the underthrusting of the South American craton beneath the Andean strip. The total amount of horizontal shortening calculated with our master-detachment is 71 km, distributed between the western and eastern slopes of the Principal Cordillera (17 and 35 km, respectively), the Frontal Cordillera and foreland area (16 km). On the other hand, the Coastal Range undergoes passive surface uplift with only 3 km of shortening.

During the 18–5 Ma period, the main phase of deformation is located in the Aconcagua FTB. As the crust thickens from 40 to 50 km and the upper crust shortens by 52 km, the thrust front migrates from the Principal Cordillera to the Frontal Cordillera with a peak of deformation at *c.* 12–10 Ma. After the Andean crust achieves its present thickness of 50 km beneath the western Principal Cordillera, gravitational stresses drive the lateral expansion of the crustal root westwards and eastwards, driving surface uplift of the Central Depression and both slopes of the Principal Cordillera by isostatic response. Afterwards, during Pliocene time (5–2.5 Ma), there was a lull in deformation both in the eastern and western slope of the Andes with a deactivation of the master décollement. Uplift at this time is concentrated in the easternmost sector of the Frontal Cordillera and the Cacheuta basin. During Quaternary time, there is a reactivation of contractional deformation in the actual thrust front and in the Frontal range close to the city of Santiago as evidenced by seismological studies, suggesting the reactivation of the western sector of the main décollement.

This research was supported by grants from CONICET (PIP 638 and PIP 112-201102-00484), the Agencia de Promoción Científica y Tecnológica (PICT-2011-1079) and UNESCO (IGCP-586Y). Bárbara Carrapa and Andrés Folguera are sincerely thanked for their critical and helpful comments and suggestions. We thank Midland Valley Ltd. for the Academic Licence of the Move© Software. We would like to acknowledge the IGCP 586 Y group for fruitful discussions and suggestions.

References

AGUIRRE, L., FERAUD, G., VERGARA, M., CARRASCO, J. & MORATA, D. 2000. 40Ar/39Ar ages of basic flows from the Valle Nevado stratifies sequence (Farellones Formation), Andes of Central Chile. *IX Congreso Geológico Chileno*, **1**, Puerto Varas, Chile, 583–585.

AGUIRRE-URRETA, B., TUNIK, M., NAIPAUER, M., PAZOS, P., OTTONE, E., FANNING, M. & RAMOS, V. A. 2011. Malargüe Group (Maastrichtian–Danian) deposits in the Neuquén Andes, Argentina: implications for the onset of the first Atlantic transgression related to Western Gondwana break-up. *Gondwana Research*, **19**, 482–494.

ALLMENDINGER, R. W. & GUBBELS, T. 1996. Pure and simple shear plateau uplift, Altiplano-Puna, Argentina and Bolivia. *Tectonophysics*, **259**, 1–13.

ALLMENDINGER, R. W., FIGUEROA, D., SNYDER, D., BEER, J., MPODOZIS, C. & ISACKS, B. L. 1990. Foreland shortening and crustal balancing in the Andes at 30°S Latitude. *Tectonics*, **9**, 789–809.

ALLMENDINGER, R. W., ISACKS, B. L., JORDAN, T. E. & KAY, S. M. 1997. The evolution of the Altiplano-Puna plateau of the Central Andes. *Annual Reviews of Earth Science*, **25**, 139–174.

ALVARADO, P., BECK, S., ZANDT, G., ARAUJO, M. & TRIEP, E. 2005. Crustal deformation in the south-central Andes backarc terranes as viewed from regional broadband seismic waveform modeling. *Geophysical Journal International*, **163**, 580–598.

ALVARADO, P., BECK, S. & ZANDT, G. 2007. Crustal structure of the south-central Andes Cordillera and backarc region from regional waveform modeling. *Geophysical Journal International*, **170**, 858–875.

ALVAREZ, P. P., GODOY, E. & GIAMBIAGI, L. 1999. Estratigrafía de la Alta Cordillera de Chile Central a la latitud del paso Piuquenes (33°35′ LS). In: *Proceedings of the XIV Congreso Geológico Argentino*, Salta, Argentina, **1**, 55.

ARANEDA, M., AVENDAÑO, M. S. & MERLO, C. 2000. Modelo gravimétrico de la cuenca de Santiago, etapa II final. In: *Proceedings of the IX Congreso Geológico Chileno*, Puerto Varas, Chile.

ARMIJO, R., RAULD, R., THIELE, G., VARGAS, J., CAMPOS, R., LACASSIN, R. & KAUSEL, E. 2010. The West Andean Thrust, the San Ramón Fault, and the seismogenic hazard for Santiago, Chile. *Tectonics*, **29**, TC2007, http://dx.doi.org/10.1029/2008tc002427

ÁVILA, J. N., CHEMALE, F., MALLMANN, G., BORBA, A. W. & LUFT, F. F. 2005. Thermal evolution of inverted basins: constraints from apatite fission track thermochronology in the Cuyo Basin, Argentine Precordillera. *Radiation Measurements*, **39**, 603–611.

BARAZANGI, B. A. & ISACKS, B. L. 1976. Spatial distribution of earthquakes and subduction of the Nazca Plate beneath South America. *Geology*, **4**, 686–692.

BARRIENTOS, S., VERA, E., ALVARADO, P. & MONFRET, T. 2004. Crustal Seismicity in Central Chile. *Journal of South American Earth Sciences*, **16**, 759–768.

BROOKS, B. A., BEVIS, M. ET AL. 2003. Crustal motion in the Southern Andes (26°–36°S): do the Andes behave like a microplate? *Geochemistry, Geophysics, Geosystems*, **4**, 1085, http://dx.doi.org/10.1029/2003GC000505

BRÜGGEN, J. 1950. *Fundamentos de la Geología de Chile*. Instituto Geográfico Militar, Santiago, Chile.

BUROV, E. & DIAMENT, M. 1995. The effective elastic thickness (Te) of continental lithosphere: what does it really mean? *Journal of Geophysical Research*, **100**, 3905–3927.

CAHILL, T. & ISACKS, B. L. 1992. Seismicity and shape of the subducted Nazca Plate. *Journal of Geophysical Research*, **97**, 17503–17529.

CARRAPA, B., ADELMANN, D., HILLEY, G. E., MORTIMER, E., SOBEL, E. R. & STRECKER, M. R. 2005. Oligocene uplift and development of plateau morphology in the southern central Andes. *Tectonics*, **24**, TC4011, http://dx.doi.org/10.1029/2004TC001762

CHARRIER, R., BAEZA, O. ET AL. 2002. Evidence for Cenozoic extensional basin development and tectonic inversion south of the flat-slab segment, southern Central Andes, Chile (33°–36°S.L.). *Journal of South American Earth Sciences*, **15**, 117–139.

CHARRIER, R., BUSTAMANTE, M. ET AL. 2005. The Abanico extensional basin: regional extension, chronology of tectonic inversion, and relation to shallow seismic activity and Andean uplift. *Neues Jahrbuch für Geologie und Paläontologie*, **236**, 43–77.

CHARRIER, R., PINTO, L. & RODRÍGUEZ, M. P. 2007. Tectonostratigraphic evolution of the Andean Orogen in Chile. In: MORENO, T. & GIBBONS, W. (eds) *The Geology of Chile*. Geological Society, London, 21–113.

CHARRIER, R., RAMOS, V. A., SAGRIPANTI, L. & TAPIA, F. In review. Tectono-stratigraphic evolution of the Andean Orogen between 31° and 37°S (Chile and Western Argentina). In: SEPÚLVEDA, S. A., GIAMBIAGI, L. B., MOREIRAS, S. M., PINTO, L., TUNIK, M., HOKE, G. D. & FARÍAS, M. (eds) *Geodynamic Processes in the Andes of Central Chile and Argentina*. Geological Society, London, Special Publications, **399**.

DECELLES, P. G., DUCEA, M. N., KAPP, P. & ZANDT, G. 2009. Cyclicity in Cordilleran orogenic systems. *Nature Geoscience*, **2**, 251–257, http://dx.doi.org/10.1038/NGEO469.

DECKART, K., GODOY, E., BERTENS, A., JEREZ, D. & SAEED, A. 2010. Barern Miocene granitoids in the Central Andean metallogenic belt, Chile: geochemistry and Nd–Hf and U–Pb isotope systematic. *Andean Geology*, **37**, 1–31.

EHLERS, T. A. & POULSEN, C. J. 2009. Influence of Andean uplift on climate and paleoaltimetry estimates. *Earth and Planetary Science Letters*, **281**, 238–248.

ELGUETA, S., CHARRIER, R., AGUIRRE, R., KIEFFER, G. & VATIN-PERIGNON, N. 1999. Volcanogenic sedimentation model for the Miocene Farellones Formation, Andean Cordillera, central Chile. In: *Proceedings of the IV International Symposium on Andean Geodynamics*, Göttingen.

ENCINAS, A., LE ROUX, J. P., BUATOIS, L. A., NIELSEN, S. N., FINGER, K. L., FOURTANIER, E. & LAVENU, A. 2006. Nuevo esquema estratigráfico para los depósitos marinos mio-pliocenos del area de Navidad (33°00′–34°30′S), Chile central. *Revista Geológica de Chile*, **33**, 221–246.

ENCINAS, A., FINGER, K., NIELSEN, S., LAVENU, A., BUATOIS, L., PETERSON, D. & LE ROUX, J. P. 2008. Rapid and major coastal subsidence during the late Miocene in south-central Chile. *Journal of South American Earth Sciences*, **25**, 157–175.

FARÍAS, M., CHARRIER, R. ET AL. 2008. Late Miocene high and rapid surface uplift and its erosional response in the Andes of central Chile (33°–35°S). *Tectonics*, **27**, TC1005, http://dx.doi.org/10.1029/2006TC002046

FARÍAS, M., COMTE, D. ET AL. 2010. Crustal-scale structural architecture in central Chile based on seismicity and surface geology: implications for Andean mountain building. *Tectonics*, **29**, TC3006, http://dx.doi.org/10.1029/2009TC002480

FOCK, A. 2005. *Cronología y tectónica de la exhumación en el Neógeno de los Andes de Chile central entre los 33° y los 34°S*. MSc thesis, Universidad de Chile, Chile.

Fock, A., Charrier, R., Marsaev, V., Farías, M. & Alvarez, P. 2006. Evolución cenozoica de los Andes de Chile Central (33°–34°S). In: *Proceedings of the IX Congreso Geológico Chileno*, Antofagasta, Chile, **2**, 205–208.

Forsyth, D. & Uyeda, S. 1975. On the relative importance of the driving forces of plate motion. *Geophysical Journal Royal Astronomic Society*, **43**, 163–200.

Fuentes, F. 2004. *Petrología y metamorfismo de muy bajo grado de unidades volcánicas oligoceno-miocenas en la ladera occidental de los Andes de Chile central (33°S)*. PhD thesis, Universidad de Chile, Chile.

Gana, P. & Tosdal, R. 1996. Geocronología U–Pb y k–Ar en intrusivos del Paleozoico y mesozoico de la Cordillera de la Costa, Región de Valparaíso, Chile. *Revista Geológica de Chile*, **23**, 151–164.

Gana, P., Wall, R. & Gutiérrez, A. 1996. Mapa geológico del área Valparaíso-Curacaví, Regiones de Valparaíso y Metropolitana. Servicio Nacional de Geología y Minería, map scale 1:100 000.

García, V. H. & Casa, A. 2014. Quaternary tectonics and seismic potential of the Andean retrowedge at 33°–34°S. In: Sepúlveda, S. A., Giambiagi, L. B., Moreiras, S. M., Pinto, L., Tunik, M., Hoke, G. D. & Farías, M. (eds) *Geodynamic Processes in the Andes of Central Chile and Argentina*. Geological Society, London, Special Publications, **399**. First published online February 27, 2014, http://dx.doi.org/10.1144/SP399.11

Garrido, I., Riveros, M., Cladouhos, T., Espiñeira, D. & Allmendinger, R. 1994. Modelo geológico estructural de El Teniente. In: *Proceedings of the VII Congreso Geológico Chileno, Concepción*, Chile. Actas 2: 1553–1558.

Garzione, C., Hoke, G. *et al.* 2008. The rise of the Andes. *Science*, **320**, 1304–1307.

Giambiagi, L. & Ramos, V. A. 2002. Structural evolution of the Andes between 33°30′ and 33°45′ S, above the transition zone between the flat and normal subduction segment, Argentina and Chile. *Journal of South American Earth Sciences*, **15**, 99–114.

Giambiagi, L., Tunik, M. & Ghiglione, M. 2001. Cenozoic tectonic evolution of the Alto Tunuyán foreland basin above the transition zone between the flat and normal subduction segment (33°30′–34°S), western Argentina. *Journal of South American Earth Sciences*, **14**, 707–724.

Giambiagi, L., Ramos, V. A., Godoy, E., Alvarez, P. P. & Orts, S. 2003a. Cenozoic deformation and tectonic style of the Andes, between 33° and 34° South Latitude. *Tectonics*, **22**, 1041, http://dx.doi.org/10.1029/2001TC001354

Giambiagi, L., Alvarez, P., Godoy, E. & Ramos, V. 2003b. The control of pre-existing extensional structures on the evolution of the southern sector of the Aconcagua fold and thrust belt, southern Andes. *Tectonophysics*, **369**, 1–19.

Giambiagi, L., Tunik, M., Ramos, V. & Godoy, E. 2009. The High Andean Cordillera of central Argentina and Chile along the Piuquenes pass – Cordón del Portillo transect: Darwin's pioneering observations compared with modern geology. *Revista Asociación Geológica Argentina*, **64**, 43–54.

Giambiagi, L., Mescua, J., Bechis, F., Tassara, A. & Hoke, G. 2012. Thrust belts of the Southern Central Andes: along-strike variations in shortening, topography, crustal geometry, and denudation. *Geological Society of America Bulletin*, **124**, 1339–1351.

Godoy, E. 2005. Tectonic erosion in the Central Andes: use and abuse. In: *Proceedings of the VI International Symposium on Andean Geodynamics*, Barcelona, Spain, 330–332.

Godoy, E. 2011. Structural setting and diachronism in the Central Andean Eocene to Miocene volcano-tectonic basins. In: Salfity, J. A. & Marquilla, R. (eds) *Cenozoic Geology of the Central Andes of Argentina*, SCR Publisher, Salta, Argentina, 155–167.

Godoy, E. 2014. The north-western margin of the Neuquén Basin in the headwater regions of the Maipo drainage, Chile. In: Sepúlveda, S. A., Giambiagi, L. B., Moreiras, S. M., Pinto, L., Tunik, M., Hoke, G. D. & Farías, M. (eds) *Geodynamic Processes in the Andes of Central Chile and Argentina*. Geological Society, London, Special Publications, **399**. First published online February 27, 2014, http://dx.doi.org/10.1144/SP399.13

Godoy, E., Yañez, G. & Vera, E. 1999. Inversion of an Oligocene volcano-tectonic basin and uplift of its superimposed Miocene magmatic arc, Chilean Central Andes: first seismic and gravity evidence. *Tectonophysics*, **306**, 217–326.

Goetze, C. & Evans, B. 1979. Stress and temperature in the bending lithosphere as constrained by experimental rock mechanics. *Geophysical Journal of the Royal Astronomical Society*, **59**, 463–478.

Gutiérrez, N., Hinojosa, L., Le Roux, J. P. & Pedroza, V. 2013. Evidence for an Early-Middle Miocene age of the Navidad Formation (central Chile): paleontological, paleoclimatic and tectonic implications. *Andean Geology*, **40**, 66–78.

Heuret, A. & Lallemand, S. 2005. Plate motions, slab dynamics and back-arc deformation. *Physics of the Earth and Planetary Interiors*, **149**, 31–51.

Hoke, G. D., Graber, N. R., Mescua, J. F., Giambiagi, L. B., Fitzgerald, P. G. & Metcalf, J. R. 2014. Near pure surface uplift of the Argentine Frontal Cordillera: insights from (U–Th)/He thermochronology and geomorphic analysis. In: Sepúlveda, S. A., Giambiagi, L. B., Moreiras, S. M., Pinto, L., Tunik, M., Hoke, G. D. & Farías, M. (eds) *Geodynamic Processes in the Andes of Central Chile and Argentina*. Geological Society, London, Special Publications, **399**. First published online February 5, 2014, http://dx.doi.org/10.1144/SP399.4

Hongn, F., Del Papa, C., Powell, J., Petrinovic, I., Mon, R. & Deraco, V. 2007. Middle Eocene deformation and sedimentation in the Puna-Eastern Cordillera transition (23°–26°S): control by preexisting heterogeneities on the pattern of initial Andean shortening. *Geology*, **35**, 271–274.

Hyndman, R. D., Currie, C. A. & Mazzotti, S. P. 2005. Subduction zone backarcs, mobile belts, and orogenic heat. *GSA Today*, **15**, 4–10.

Iaffaldano, G. & Bunge, H. P. 2008. Strong plate coupling along the Nazca-South American convergent margin. *Geology*, **36**, 443–446.

Irigoyen, M. V., Buchan, K. L. & Brown, R. L. 2000. Magnetostratigraphy of Neogene Andean foreland-basin strata, lat 33°S, Mendoza Province, Argentina. *Geological Society of America, Bulletin*, **112**, 803–816.

Isacks, B. 1988. Uplift of the Central Andean plateau and bending of the Bolivian Orocline. *Journal of Geophysical Research*, **93**, 3211–3231.

Jordan, T. E., Isacks, B. L., Allmendinger, R. W., Brewer, J. A., Ramos, V. A. & Ando, C. J. 1983. Andean tectonics related to geometry of subducted Nazca plate. *Geological Society of America, Bulletin*, **94**, 341–361.

Karato, S. 2008. *Deformation of Earth Materials: An Introduction to the Rheology of Solid Earth*. Cambridge University Press, Cambridge.

Kay, R. W. & Kay, S. 1993. Delamination and delamination magmatism. *Tectonophysics*, **219**, 177–189.

Kay, S. & Kurtz, A. 1995. *Magmatic and tectonic characterization of the El Teniente region*. Internal report, Superintendencia de Geología, El Teniente, CODELCO.

Kay, S. M. & Mpodozis, C. 2001. Central Andean ore deposits linked to evolving shallow subduction systems and thickening crust. *GSA Today*, **11**, 4–11.

Kay, S. M., Mpodozis, C. & Coira, B. 1999. Neogene magmatism, tectonism and mineral deposits of the Central Andes (22°–33°S Latitude). *In*: Skinner, B. J. (ed.) *Geology and Ore Deposits of the Central Andes*. Society of Economic Geologists, Lancaster, USA, Special Publications, **7**, 27–59.

Kay, S. M., Godoy, E. & Kurtz, A. 2005. Episodic arc migration, crustal thickening, subduction erosion, and magmatism in the south-central Andes. *Geological Society of America, Bulletin*, **117**, 67–88.

Kay, S. M., Burns, M., Copeland, P. & Mancilla, O. 2006. Upper Cretaceous to Holocene magmatism and evidence for transient Miocene shallowing of the Andean subduction zone under the northern Neuquén Basin. *In*: Kay, S. M. & Ramos, V. A. (eds) *Evolution of an Andean margin: A Tectonic and Magmatic view from the Andes to the Neuquén basin (35°–39°S lat)*. Geological Society of America, Boulder, Colorado, Special Paper, **407**, 19–60.

Kurtz, A., Kay, S. M., Charrier, R. & Farrar, E. 1997. Geochronology of Miocene plutons and exhumation history of the El Teniente region, Central Chile (34°35′S). *Revista Geológica de Chile*, **24**, 73–90.

Lamb, S. 2006. Shear stresses on megathrusts: implications for mountain building behing subduction zones. *Journal of Geophysical Research*, **111**, B07401, http://dx.doi.org/10.1029/2005JB003916

Lavenu, A. & Encinas, A. 2005. Brittle deformation of the Neogene deposits of the Navidad Basin (Coastal Cordillera, 34°S, central Chile). *Revista Geológica de Chile*, **32**, 229–248.

Maksaev, V., Munizag, V. F., Macwilliams, M., Fanning, M., Mathur, R., Ruiz, J. & Zentilli, M. 2004. New chronology for El Teniente, Chilean Andes, from U–Pb, 40Ar/39Ar, Re–Os and fission-track dating. *In*: (eds) *Implications for the Evolution of a Supergiant Porphyry Cu–Mo deposit*. Society of Economic Geologists, Lancaster, USA, Special Publication, **11**, 15–54.

Maksaev, V., Munizaga, F., Zentilli, M. & Charrier, R. 2009. Fission track thermochronology of Neogene plutons in the Principal Andean Cordillera of central Chile (33–35°S): implications for tectonic evolution and porphyry Cu–Mo mineralization. *Andean Geology*, **36**, 153–171.

Mescua, J., Giambiagi, L. & Ramos, V. A. 2013. Late Cretaceous uplift in the Malargüe fold-and-thrust belt (35°S), Southern Central Andes of Argentina and Chile. *Andean Geology*, **40**, 102–116.

Molnar, P. & Lyon-Caen, H. 1988. Some simple physical aspects of the support, structure, and evolution of mountain belts. *Geological Society of America, Special Paper*, **218**, 179–207.

Morales, D. & Tassara, A. 2012. Avances hacia un modelo termal 3D del margen Andino. *In*: *Proceedings of the XII Congreso Geologico Chileno, Antofagasta*, Chile, 498–499.

Mpodozis, C. & Cornejo, P. 2012. Cenozoic tectonics and porphyry copper systems of the Chilean Andes. *In*: Hedenquist, J. W., Harris, M. & Camus, F. (eds) *Geology and Genesis of Major Copper Deposits and Districts of the World: A Tribute to Richard H. Sillitoe*. Society of Economic Geologists, Lancaster, USA, Special Publication, **16**, 329–360.

Mpodozis, C. & Ramos, V. A. 1989. The Andes of Chile and Argentina. *In*: Ericksen, G. E., Cañas, M. T. & Reinemund, J. A. (eds) *Geology of the Andes and its Relation to Hydrocarbon and Mineral Resources*. Circum-Pacific Council for Energy and Mineral Resources, Lancaster, USA, Earth Science Series, **11**, 59–90.

Muñoz, M., Fuentes, F., Vergara, M., Aguirre, L., Nyström, J. O. & Féraud, G. 2006. Abanico East Formation: petrology and geochemistry of volcanic rocks behind the Cenozoic arc front in the Andean Cordillera, central Chile (33°50′S). *Revista Geológica de Chile*, **33**, 109–140.

Muñoz, M., Charrier, R., Fanning, C. M., Maksaev, V. & Deckart, K. 2012. Zircon trace element and O–Hf isotope analyses of mineralized intrusions from El Teniente Ore Deposit, Chilean Andes: constraints on the source and magmatic evolution of porphyry Cu–Mo related magmas. *Journal of Petrology*, **53**, 1091–1122.

Nyström, J., Parada, M. & Vergara, M. 1993. Sr–Nd Isotope Compositions of Cretaceous to Miocene Volcanic Rocks in Central Chile: A Trend Towards a MORB Signature and a Reversal with Time. II International Symposium on Andean Geodynamics (ISAG), Oxford, UK, IRD (eds), 411–414.

Nyström, J. O., Vergara, M., Morata, D. & Levi, B. 2003. Tertiary volcanism in central Chile (33 °15′– 33°45′S): a case of Andean Magmatism. *Geological Society of America, Bulletin*, **115**, 1523–1537.

Orts, D., Folguera, A., Giménez, M. & Ramos, V. A. 2012. Variable structural controls through time in the Southern Central Andes (36°S). *Andean Geology*, **39**, 220–241.

Polanski, J. 1964. Descripción geológica de la Hoja 25 a-b - Volcán de San José, provincia de Mendoza, Dirección Nacional de Geología y Minería, Boletín 98, 1–92, Buenos Aires, Argentina.

PORRAS, H., PINTO, L., TUNIK, M. & GIAMBIAGI, L. In review. Provenance analysis using whole-rock geochemistry and U–Pb dating of the Alto Tunuyán basin: implications for its palaeogeographic evolution. *In*: SEPÚLVEDA, S. A., GIAMBIAGI, L. B., MOREIRAS, S. M., PINTO, L., TUNIK, M., HOKE, G. D. & FARÍAS, M. (eds) *Geodynamic Processes in the Andes of Central Chile and Argentina*. Geological Society, London, Special Publications, **399**.

RAMOS, V. A., GODOY, E., GODOY, V. & PÁNGARO, F. 1996. Evolución tectónica de la Cordillera Principal argentino-chilena a la latitud del Paso de Piuquenes (33°30′S). *In*: *Proceeding of the XIII Congreso Geológico Argentino*, Buenos Aires, Argentina, 337–352.

RAMOS, V. A., ZAPATA, T., CRISTALLINI, E. & INTROCASO, A. 2004. The Andean thrust system: latitudinal variations in structural styles and orogenic shortening. *In*: MCCLAY, K. R. (ed.) *Thrust Tectonics and Hydrocarbon Systems*. American Association of Petroleum Geology, Boulder, USA, Memoir, **82**, 30–50.

RAULD, R., VARGAS, G., ARMIJO, R., ORMEÑO, A., VALDERAS, C. & CAMPOS, J. 2006. Cuantificación de escarpes de falla y deformación reciente en el frente cordillerano de Santiago. *In: Proceedings of the XI Congreso Geológico Chileno*, Antofagasta, Chile, 447–450.

RODRÍGUEZ, M. P., PINTO LINCOÑIR, L. & ENCINAS, A. 2012. Cenozoic erosion in the Andean forearc in Central Chile (33°–34°S): sediment provenance inferred by heavy mineral studies. *In*: RASBURY, E. T., HEMMING, S. R. & RIGGS, N. R. (eds) *Mineralogical and Geochemical Approaches to Provenance*. Geological Society of America, Boulder, Special Paper, **487**, 141–162.

SAVAGE, J. C. 1983. A dislocation model of strain accumulation and release at a subduction zone. *Journal of Geophysical Research*, **88**, 4984–4996.

SCHELLART, W. P. 2008. Overriding plate shortening and extension above subduction zones: a parametric study to explain formation of the Andes Mountains. *Geological Society of America Bulletin*, **120**, 1441–1454.

SCHMITZ, M. 1994. A balanced model of the southern Central Andes. *Tectonics*, **13**, 484–492.

SELLÉS, D. & GANA, P. 2001. *Geología del area Talagante-San Francisco de Mostazal: Regiones Metropolitana y del Libertador Genral Bernardo ÓHiggins. 1:100 000*. SERNAGEOMIN, Carga Geológica de Chile, Seria Geológica Básica.

SHEFFELS, B. 1990. Lower bound on the amount of crustal shortening in the Central Bolivian Andes. *Geology*, **18**, 812–815.

STAUDER, W. 1975. Subduction of the Nazca Plate under Peru as evidenced by focal mechanism and by seismicity. *Journal of Geophysical Research*, **80**, 1053–1064.

STERN, C. R. 1989. Pliocene to present migration of the volcanic front, Andean Southern Volcanic Zone. *Revista Geológica de Chile*, **16**, 145–162.

STERN, C. R. 2011. Subduction erosion: rates, mechanisms, and its role in arc magmatism and the evolution of the continental crust and mantle. *Gondwana Research*, **20**, 284–338.

TAPIA, F., FARÍAS, M. & ASTABURUAGA, D. 2012. Deformación cretácica-paleocena y sus evidencias en la cordillera de los Andes de Chile Central (33.7–36°S). *XI Congreso Geológico Chileno*, Antofagasta, Chile, 232–234.

TASSARA, A. 2005. Interaction between the Nazca and South American plates and formation of the Altiplano–Puna plateau: review of a flexural analysis along the Andean margin (15°–34°S). *Tectonophysics*, **399**, 39–57.

TASSARA, A. 2012. Thermomechanical support of high topography: a review of hypothesis, observations and models on the Altiplano-Puna Plateau. *In: Proceedings of the American Geophysical Union Annual Meeting*, San Francisco.

TASSARA, A. & ECHAURREN, A. 2012. Anatomy of the Andean subduction zone: three-dimensional density model upgraded and compared against global-scale models. *Geophysical Journal International*, **189**, 161–168.

TASSARA, A. & MORALES, D. 2013. 3D temperature model of south-western South America. *Annual Meeting of the European Geosciences Union*, Viena, Austria. *Geophysical Research Abstracts*, V15, EGU2013–945.

TASSARA, A., GÖTZE, H.-J., SCHMIDT, S. & HACKNEY, R. 2006. Three-dimensional density model of the Nazca plate and the Andean continental margin. *Journal of Geophysical Research*, **111**, B09404, http://dx.doi.org/10.1029/2005JB003976

TASSARA, A., SWAIN, C., HACKNEY, R. & KIRBY, J. 2007. Elastic thickness structure of South America estimated using wavelets and satellite-derived gravity data. *Earth and Planetary Science Letters*, **253**, 17–36.

TUNIK, M. 2003. Interpretación paleoambiental de la Formación Saldeño (Cretácico superior), en la Alta Cordillera de Mendoza, Argentina. *Revista de la Asociación Geológica Argentina*, **58**, 417–433.

TUNIK, M., FOLGUERA, A., NAIPAUER, M., PIMENTEL, M. & RAMOS, V. A. 2010. Early uplift and orogenic deformation in the Neuquen basin: constraints on the Andean uplift from U–Pb and Hf isotopic data of detrital zircons. *Tectonophysics*, **489**, 258–273.

TURCOTTE, D. L. & SCHUBERT, G. 2002. *Geodynamics*. Cambridge University Press, Cambridge, New York.

VERGARA, M., CHARRIER, F., MUNIZAGA, F., RIVANO, S., SEPÚLVEDA, P., THIELE, R. & DRAKE, M. 1988. Miocene vulcanism in the central Chilean Andes (31°30′S-34°35′S). *Journal of South American Earth Sciences*, **1**, 199–209.

VERGARA, M., MORATA, D., VILLARROEL, R., NYSTRÖM, J. & AGUIRRE, L. 1999. Ar/Ar ages, very low-grade metamorphism and geochemistry of the volcanic rocks from 'Cerro El Abanico', Santiago Andean Cordillera (33°30′S-70°30′–70°25′W). *In: Proceedings of the IV International Symposium on Andean Geodynamics*, Göttingen, Germany, 785–788.

WALL, R., GANA, P. & GUTIÉRREZ, A. 1996. Mapa Geológico del area de San Antonio-Melipilla. Regiones de Valparaíso y Metropolitana. Sernageomin, Santiago, Mapas Geológicos N°1, 1:100.000.

WALL, R., SELLÉS, D. & GANA, P. 1999. Hoja Tiltil-Santiago, Area Metropolitana. 1: 100 000. Sernageomin.

WANG, L., SHUM, C. K. *ET AL*. 2012. Coseismic slip of the 2010 Mw 8.8 Great Maule, Chile, earthquake quantified by the inversion of GRACE observation. *Earth and Planetary Science Letters*, **335**, 167–179.

YAÑEZ, G., CEMBRANO, J., PARDO, M., RANERO, C. & SELLES, D. 2002. The Challenger-Juan Fernández-Maipo major tectonic transition of the Nazca – Andean subduction system at 33–34°S: geodynamic evidence and implications. *Journal of South American Earth Sciences*, **15**, 23–38.

YRIGOYEN, M. R. 1993. Los depósitos sinorogénicos terciarios. *In*: RAMOS, V. A. (ed.) *Geología y recursos naturales de la Provincia de Mendoza*, Mendoza, Argentina, 123–148.

ZURITA, E., MUÑOZ, N., CHARRIER, R., HARAMBOUR, S. & ELGUETA, S. 2000. Madurez termal de la materia orgánica de la Formación Abanico = Coya Machalí, Cordillera Principal, Chile Central: resultados e interpretación. *In*: *Proceedings of the IX Congreso Geológico Chileno*, Puerto Varas, Chile, 726–730.

Role of basin width variation in tectonic inversion: insight from analogue modelling and implications for the tectonic inversion of the Abanico Basin, 32°–34°S, Central Andes

P. JARA[1,2]*, J. LIKERMAN[3,4], D. WINOCUR[5], M. C. GHIGLIONE[3,5], E. O. CRISTALLINI[3,4], L. PINTO[2] & R. CHARRIER[2,6]

[1]*Departamento de Ingeniería en Minas, Facultad de Ingeniería, Universidad de Santiago, Chile*

[2]*Departamento de Geología, Facultad de Ciencias Físicas y Matemáticas, Universidad de Chile, Chile*

[3]*Consejo Nacional de Investigaciones Científicas y Técnicas (CONICET), Avda. Rivadavia 1917, CP C1033AAJ, Ciudad de Buenos Aires, Argentina*

[4]*Laboratorio de Modelado Geológico (LaMoGe), Instituto de Estudios Andinos Don Pablo Groeber. Universidad de Buenos Aires, Ciudad Universitaria, C1428EHA, Buenos Aires, Argentina*

[5]*Departamento de Ciencias Geológicas, Laboratorio de Tectónica Andina, Instituto de Estudios Andinos Don Pablo Groeber, Universidad de Buenos Aires, Buenos Aires, Argentina*

[6]*Escuela de Ciencias de la Tierra, Universidad Andres Bello, Santiago, Chile*

*Corresponding author (e-mail: pamela.jara@usach.cl)

Abstract: We use analogue modelling to investigate the response of compressional deformation superimposed on an extensional basin with along-strike changes in width. Parameters described include extension and shortening distribution and directions, orientation of structures and degree of basin inversion. Two types of model are presented: in the first (Type I), an extensional basin is constructed with variable width (applying differential extension) and subsequently inverted by homogeneous shortening; in the second (Type II), an extensional basin with constant width is subsequently inverted by inhomogeneous shortening (differential compression). From our observations, we compare both types of model to structural patterns observed in some natural cases from the Central Andes. Both models generate oblique structures, but in the Type II model a significant rotation is characteristic. Our results suggest that in the Central Andes region between 32° and 33°S, the Abanico Basin may correspond to a basin of smaller area compared to the larger basin south of 33°S. Our Type I model further explains some patterns observed there, from which we conclude that the control exercised by the width of a pre-existing basin should be considered when interpreting the geological evolution of that area of the Andes.

The complexity of tectonic inversion systems involves a number of controlling factors such as orientation of pre-existing structures, reactivation angle and syntectonic sedimentation, among others. Rift zones associated with extensional regimes are generally controlled by the presence of lithospheric weaknesses or by previous fabrics of the upper-crust. Extensional development along rift basins may be heterogeneous and is generally characterized by along-axis segmentation in a series of single sub-basins with sedimentary and structural differences (Corti 2003). Furthermore, subsequent compressional phases may induce the reactivation of pre-existing normal faults and the inversion of these extensional systems. A large amount of new forming structures can be also generated, producing substantial changes with respect to the original extensional pattern, so different trending structures can coexist (Cooper & Williams 1989; Gillcrist et al. 1989; Coward et al. 1991; Turner & Williams 2004).

Changes in the structural trend of an arc-shaped orogen (map-view curve) can be attributed to various factors, including the architecture of the pre-deformational sedimentary basin, inversion of pre-existing extensional structures (Vergés & Muñoz 1990; Burbank et al. 1992), along-strike displacement gradients (e.g. Elliott 1976), interaction of a thrust belt with foreland obstacles or promontories, salients and recesses (oblique and lateral ramps) in the sole thrust (Macedo & Marshak 1999),

From: Sepúlveda, S. A., Giambiagi, L. B., Moreiras, S. M., Pinto, L., Tunik, M., Hoke, G. D. & Farías, M. (eds) 2015. *Geodynamic Processes in the Andes of Central Chile and Argentina*. Geological Society, London, Special Publications, **399**, 83–107. First published online February 27, 2014, http://dx.doi.org/10.1144/SP399.7
© 2015 The Geological Society of London. For permissions: http://www.geolsoc.org.uk/permissions.
Publishing disclaimer: www.geolsoc.org.uk/pub_ethics

hinterland collision of an indenter (Ghiglione & Cristallini 2007; Reiter *et al.* 2011), interaction with strike-slip faults and warping of the downgoing (underthrust) plate (Marshak 2004), along-strike variations of the frictional properties of the décollement level (Cotton & Koyi 2000), rheological or thickness changes in the detached cover (Thomas 1990; Calassou *et al.* 1993; Corrado *et al.* 1998), lateral variations of thickness in the sedimentary wedge (Soto *et al.* 2002) and/or along-strike variations of syntectonic sedimentation and erosion rates (Nalpas *et al.* 1995; Dubois *et al.* 2002; Panien *et al.* 2005). Some curved orogens involve continental-scale rotation of segments of the fold–thrust belt and are called 'oroclines', assuming that the curve was due to the rotational bending of an originally straight orogeny (Carey 1958; Marshak 2004; Ghiglione & Cristallini 2007).

Analogue modelling is a powerful tool that helps interpret geodynamic environments, because it allows the simplification and investigation of isolated parameters of the studied processes. When scaled experiments are carried out (Hubbert 1937; Ramberg 1981; Weijermars & Schmeling 1986), the results can then be compared with natural examples. Previous analogue models study the mechanics and geometry of tectonic inversion (Koopman *et al.* 1987; McClay 1989; Buchanan & McClay 1991; Krantz 1991; Mitra & Islam 1994; Keller & McClay 1995; Brun & Nalpas 1996; Yamada & McClay 2003; Gartrell *et al.* 2005; Panien *et al.* 2005; Del Ventisette *et al.* 2006; Sandiford *et al.* 2006; Konstantinovskaya *et al.* 2007; Amilibia *et al.* 2008; Yagupsky *et al.* 2008; Pinto *et al.* 2010).

Our modelling is inspired by contributing to a better understanding of factors that control some latitudinal variation in strikes of structures and the evolution of the Central Andes. Experimental results are compared with the Eocene–Miocene Abanico Basin developed between *c.* 28° and 39°S (Charrier *et al.* 2002, 2007, and references therein), which has a history of extension followed by compression and inversion of previous structures. The complexity of the inversion process caused subsequent heterogeneous latitudinal deformation, which produces latitudinal changes in the style of deformation and strike of some structures (Jara *et al.* 2009; Jara & Charrier in press). Within the context of the Abanico Basin, the study region corresponds to the High Central Andean between 32° and 34°S. South of 33°S, many studies recognize that the Abanico Formation's volcanic rocks were deposited in an extensional basin over 50 km wide, with an approximately north–south orientation and high subsidence rates, with some depocentres reaching more than 3.5 km in depth (Charrier *et al.* 2002, 2007; Fock *et al.* 2005; Farías 2007; Farías *et al.* 2010). North of 33°S, new field-based data indicate that distal facies of this formation, which crop out in the eastern side of the Principal Andean Cordillera at this latitude, accumulated in a basin less than 10 km wide with an estimated thickness of *c.* 3.0–3.5 km (Jara & Charrier in press).

In order to study the influence of extensional basin width on subsequent superimposed compressive structures and their trend, two main types of model (I and II) were configured to study the effect of two principal variables: homogeneous shortening with variable basin width (Type I) and inhomogeneous shortening (differential) with a constant-width pre-existing basin (Type II). In both studies we distinguish whether or not the resulting structures are linked with positive inversion (Cooper *et al.* 1989; Coward 1994).

Analogue models

Strategy

Analogue modelling was carried out at the Laboratorio de Modelado Geológico (LaMoGe) at the Universidad de Buenos Aires to investigate the effects of inherited differential basin width on deformation patterns, such as plan view strike, number of structures and structural patterns during basin inversion. The selected experiments are representative of two possible tectonic configurations. In the Type I model, we investigate the influence of along-strike variations produced by differential extension in a wedge-shaped rift basin where extension increased from a pivot point, followed by homogeneous shortening in response to orthogonal convergence. In the Type II model, a homogeneous orthogonal extension was followed by a rotational convergence, where differential shortening increased from a pivot point.

In referring to the sectors of our experiments we will consider that a region affected by differential extension may develop main depocentres or sub-basins in a wider zone, whereas areas with less extension can generate isolated sub-basins in a narrow zone (Fig. 1a). During differential compressional tectonics, the most shortened region will generate greater uplift and compressional structures than the less shortened region (Fig. 1b). These conceptual models do not consider previous structural arrangements, but they are the starting point for our study of deformation width control in a tectonic inversion process.

Set-up and experimental procedure

The experiments were performed without boundaries in a platform of size 70 × 50 cm. Subtraction of edge effects in the final process allowed us to

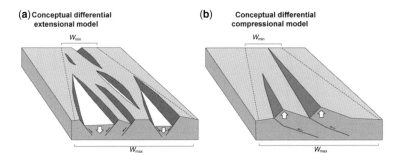

Fig. 1. Conceptual models of (**a**) differential extension and (**b**) differential compression. The vergence of structures is schematic and only referential. In both conceptual models there are no predeformation structures. The white arrows show the subsided (down) and uplifted (up) zones. The dotted lines correspond to the main deformation limits interpreted. W_{min} and W_{max} indicate the minimal and maximal width, respectively.

analyse a deformed sandbox of with dimensions of $55 \times 40 \times 3.5$ cm (Fig. 2). The models were deformed by pulling/pushing a backstop attached to a mobile basal plate, generating a velocity discontinuity (VD) below the sandbox (Fig. 2a). The mobile basal plate produced extension along a basal silicon layer and was disconnected during compression, similarly to other experimental set-ups (Yagupsky et al. 2008; Likerman et al. 2013). In this way, compression was applied from the mobile wall, similar to a natural assembly, because compression through a mobile basal plate would generate structures rooted in the VD.

The backstops used in our experiments allow differential movements through a pivot point located at one end acting as a vertical axis of rotation (Fig. 2b). The backstop rotation was produced by pulling/pushing the free edge of the backstop connected to a worm screw driven by a stepmotor at a constant linear velocity of 4 cm/h in all models. The angular velocity imposed on the backstop was $4°/h$, and can be considered constant throughout the experiment (for more details see Soto et al. 2006). A maximum backstop rotation of $10°$ was reached, for both extensional and compressional phases, equivalent to a maximum extension/shortening of 10 cm achieved in the area furthest from the pivot point, and progressively diminishing toward the pivot end where the deformation was null. For non-rotational deformation a maximum 10 cm shortening was achieved. Deformation phases applied to different sets (Type I and Type II models) of experiments are simplified in Fig. 3.

It is recognized that there is a minor strike-slip component in the model when generating rotational movement of the basal plate (extensional differential phase for the Type I model). This component was measured and is considered negligible in relation to the components perpendicular to the strike. In this study we focus only on the normal and reverse components.

Materials

Brittle layers are represented by well-sorted dry quartz sand with well-rounded grains (finer than

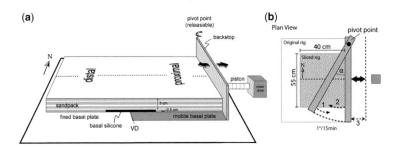

Fig. 2. (**a**) Model set-up and zones for the description of models. Proximal and distal are related to the distance to the backstop. (**b**) Plan view and different movements applied to the backstop: α, angle of backstop rotation; 1, differential extensional movement applying rotating backstop; 2, differential compressional movement applying rotating backstop; 3, homogeneous (orthogonal) movement of backstop by releasing the pivot.

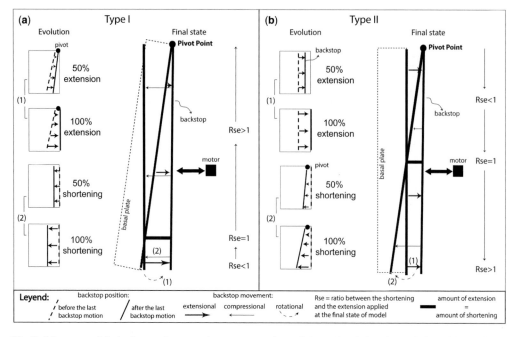

Fig. 3. Movement of the backstop schemes and percent extension and shortening during the evolution of the models, (**a**) Type I and (**b**) Type II, at different stages.

500 μm), with behaviour similar to that of the brittle upper-crustal rocks (Hubbert 1951). In such materials, frictional contacts between the grains result in a Mohr–Coulomb-type failure envelope, with small cohesion (<100 Pa) and an angle of internal friction close to 32.7° (Yagupsky et al. 2008), as measured with a modified Hubbert-type shear apparatus (Hubbert 1951). The material's density ρ is c. 1400 kg m^{-3} and the density ratio between the granular materials used and the rocks is $\rho^* \approx 0.5$. The stretching at the base of the brittle upper crust was modelled with a 0.5-cm-thick layer of SGM 36 polymer. SGM 36, a Newtonian viscous material manufactured by Dow Corning, has a density of 965 kg m^{-3} with an effective viscosity of 5×10^4 Pa s at room temperature (20 °C). A fine silicone layer (0.5 mm) was added above the VD. In these experiments the silicone does not simulate any character of the crust and has only a geometrical purpose (Brun & Nalpas 1996; Pinto et al. 2010), allowing the strain to be distributed and creating a wider deformation zone. The use of SGM 36 in analogue modelling is well documented (Weijermars 1986); it forms an analogue for materials with viscosities between 10^{18} and 10^{21} Pa s, which is sufficiently similar to those commonly used for high décollement viscosities (Van Keken et al. 1993; Weijermars et al. 1993). The models are thus broadly scaled to represent a strong natural décollement.

Scaling

To contrast analogue models with natural examples, the experiments required proper scaling of the model parameters (Hubbert 1937; Ramberg 1981). In the present experiments, the length ratio between model and nature was $L = 10^{-5}$ (so 1 cm in the model corresponds to c. 1 km in nature), and the gravity ratio between model and nature is $g^* = 1$, as both the prototype and the model are subject to the same gravitational acceleration. The corresponding stress ratio between model and nature is $\sigma^* = \rho^* g^* L^* \approx 6 \times 10^{-6}$. The scaling parameters for extension and compression are provided in Table 1.

Although the experimental set-up lacked the prerift brittle structures that may appear in nature, which can play an important role in the interaction and development of extensional structures (Sibson 1985; Huyghe & Mugnier 1992; Faccenna et al. 1995; Ranalli 2000), emphasis was placed on the interaction between rift structures and subsequent shortening.

Data collection and three-dimensional reconstruction of deformed models

A fixed high-resolution digital camera was used to photograph the top surface of the models at regular time intervals in order to study the time–space

Table 1. *Type I and Type II models' scaling parameters in both extensional and compressional phases*

	λ (μm)	g (m s^{-2})	ρ (kg m^{-3})	μ (Pa s)	V (m s^{-1})	σ (Pa)	ε (s^{-1})
Nature	1000	9.81	2300	1.8×10^{21}	5×10^{-13}	2.2×10^{7}	Oct 14
Model	0.01	9.81	1400	5×10^{4}	1.1×10^{-5}	1.4×10^{2}	2.2×10^{-3}
Model/nature(*)	10 May	1	0.6	2.7×10^{-17}	2.2×10^{7}	6×10^{-6}	2.2×10^{11}

evolution of the structures. At the end of the experiments, dry sand was sieved onto the model's surface to preserve the final topography. To reconstruct a three-dimensional geometry, the models were impregnated with hot gelatin solution and, once cooled and slightly hardened, were sliced (c. 5 mm) perpendicular to the principal structure's trend. In this way, internal structural arrangements of the final state of the models could be analysed.

The top surfaces of the models were scanned at regular time intervals with a laser with c. 0.135 mm of vertical accuracy. The data obtained were processed to remove spurious data. Subsidence of the top surface was calculated from the incremental difference between successive gridded data. In the shortening phase, incremental topographic maps were obtained, revealing the development of contractional structures. Model evolution was visualized by measuring diagnostic parameters along strike positions representative of wide and narrow deformed sectors close to (north) or away from (south) the pivot (Fig. 2). This methodology, together with photography, allowed a more rigorous and detailed interpretation of the development of the structures.

Synrift sand thickness was used to identify inverted faults. Synrift sand deposits on the hanging wall of normal faults are thicker than on the equivalent footwall, so inversion will show thicker packages of synrift lifted on the hanging wall of positively inverted faults. Moreover, the presence of faults planes with normal and inverse movement was used to identify incomplete reversal movement.

Statistical fault analysis

In the final stages, fault traces were mapped over the plan view. Fault traces intercepted by a 0.5 cm-wide belt in two representative sections were analysed quantitatively in terms of length and azimuth distribution. To analyse the fault's azimuth distribution, the polyline features used to map the fault traces were transformed into tip-to-tip lines (Agostini et al. 2011). The azimuth data were weighted for the length of the corresponding fault. The weighting factor for each fault was the ratio between its length and the minimum fault length intercepting the belt, such that long faults have higher ratios than short ones. The frequency of the azimuth of a tip-to-tip fault is directly related to this ratio; the longer the fault the higher its frequency. This type of fault pattern outcome has often been coupled with analogue models, providing the basis for a well-established approach that has been successfully applied to the determination of rift kinematics in oblique extension settings (Dauteuil & Brun 1996) and continental rifts (Brun & Tron 1993; Bonini et al. 1997).

Analogue modelling results

Six experiments were performed to test the control that axial width variation in an extensional basin has on the strike of subsequent inversion structures. The structural configuration obtained was relatively similar for each type of experiment. Because of limited space, only two are described here, one for each type. In order to simplify the description of the results, in the following we will use a series of acronyms to refer to the different sectors of both models (Fig. 2). Sequential photographs of the tops of the three-dimensional (3D) models during the evolution of the experiment, together with isopach maps highlighting the differential subsidence or uplift between each stage, are presented. Additionally, internal structure interpretation is shown in serial vertical sections for the end of this phase. For descriptive purposes, the terms proximal and distal are used relative to the advancing moving wall (Fig. 2a), and forethrusts for those faults dipping towards the moving wall and backthrusts for those dipping in the opposite direction.

In the Type I model, maximum extension (100% extension) was reached after applying 10 cm of movement to the backstop, as measured on the zone further away from the pivot (Fig. 3). During the ensuing compressional phase the mobile wall was released from the pivot and was moved until completing 10 cm of homogeneous compression (100% shortening) perpendicular to the last extensional movement, producing total inversion of the extensional basin (Fig. 3a). In the final state of the Type I model there is a variable amount of extension and shortening for a given point, so we have included a factor ('Rse') relating the amount of applied shortening to the amount of previous extension in sections at different distances from the pivot (Fig. 3).

In Type II models, extension was homogeneous in applying a 5 cm motion to the backstop (100% extension, Fig. 3b). The result was a 15-cm-wide basin, a width similar to that of the central zone of the basin in the Type I model. During the compressional phase, the pivot was connected to achieve a differential shortening. The region further away from the pivot reached a maximum 10 cm, or 100% shortening (Fig. 3b). As with the Type I model, in the final stage we use the same factor ('Rse') to compare the amount of extension and shortening applied in different sections according to their distance from the pivot (Fig. 3).

Type I: Along-strike width variation of basin and subsequent inversion

Extensional phase evolution. The initial stages of the extensional phase (Fig. 4a) produced a symmetric basin defined by two sets of normal faults with opposite dips. Subsequently, with 25% extension, the deformation concentrated in a direction parallel to the backstop. The isopach map (Fig. 4a) highlights the differential subsidence between the non-deformation stage and the 100% extension after every 25% displacement of the backstop; deepening of the basin shows progressively more activity and a large basin width focused on the southern zone.

From 25% to 50% of extension, structures in the proximal zone generated deepening of the basin, mainly in the southern zone (Fig. 4a). Subsequently, subsidence ends in the middle part of the model, generating some horsts between sub-basins or graben (see Fig. 4c and non-coloured areas within the main basin in Fig. 4a). The high activity (subsidence and active faults) is seen mainly in the southern and proximal zones. This becomes most evident between 50% and 75% of extensional displacement (Fig. 4a).

Between 50% and 75% extension (Fig. 4a), the activity of the distal zone of the main basin is almost null, while in the proximal zone the basin is active and sub-basins coexist (note subsidence symbology colours in Fig. 4a). Sub-basins in the northern zone are oriented approximately north–south, while in the southern proximal zone they have a general approximately N10°E strike (although one reaches a maximum of 30° obliquity) (Fig. 4a).

Extension ended at 10 cm (100% extension) of displacement in the southern zone, and in this last stage the distal basin zone is reactivated (Fig. 4a) but with much lower activity compared to the proximal zone, which preserves the features described above, but with further deepening of depocentres in the basin (note subsidence symbology colours in Fig. 4a).

Extensional phase final state. The proximal and distal limits of the basin have oblique (<15°) orientations, which generate its triangular shape, but its main depocentres have a north–south orientation (Fig. 4a). In the northern zone, two graben separated by a horst were generated, both bounded by normal faults with opposing dips (Fig. 4c, profile I). Deformation and faulting were localized above the VD characterized by a set of normal faults with dipping angles from 50° to 65° (Fig. 4c, profiles I–II), but towards the southern zone (Fig. 4c, profile III) some normal faults are very flat with dips up to 20°. The deepest sub-basin occurs from the middle of the model to the southern zone (Fig. 5) and is mainly concentrated in the proximal zone of the basin, as seen in the plan view evolution (Fig. 4a) and in the cross-sections of the final result (Fig. 4c).

Compressional phase evolution. An important topographic increase developed homogeneously along the entire model, mainly in the internal region (in the location of the previous extensional basin) and in the proximal zone, as shown by the isopach maps highlighting the differential uplift between each stage (Fig. 4b). With 25% of shortening, there is uplift in the proximal region of the northern zone quite near the mobile wall and an important uplift at the proximal inherited graben in the southern zone (Fig. 4b). Structures that limit these uplifted areas have approximately north–south strike in the southern zone, and approximately N10°W in the proximal zone in the north (Fig. 4b).

Between 25% and 50% of shortening, deformation is mainly concentrated in north–south backthrusts in the central part of the extensional basin (Fig. 4a,b). The activity of these north–south backthrusts decreases towards the north where shortening is transferred to the proximal zone (Fig. 4b). In this region away from the pre-existing basin, the main forethrust with N10°W strike is better developed, generating large uplift (note the increase in topography bordering the structure in Fig. 4b). In the northern zone, uplift also occurs in the inherited basin generated by approximately north–south to N10°E strike backthrusts. Towards the southern zone, shortening is mostly concentrated in the region previously occupied by the basin, although some rising is generated out of the basin, mainly in the proximal zone (Fig. 4b).

With 75% of accumulated shortening, deformation concentrates in two major regions. In the northern region it is concentrated in the N10°W structure, leading to an important rising of the proximal zone, while in the southern region the main structures correspond to approximately

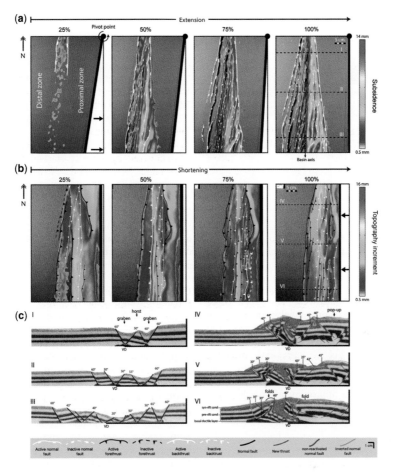

Fig. 4. Type I model evolution. (**a**) Extensional phase: active subsidence at different extension percentages. (**b**) Compressional phase: active topography increment at different shortening percentages. (**c**) I, II, II: profiles of the final result after differential extension. IV, V, VI: profiles of the final result after homogeneous shortening.

north–south strike forethrusts and backthrusts, bounding the uplifted proximal area of the inherited basin. Towards the distal zone, part of the shortening occurs in the inherited basin, as indicated by the 'triangular' shape of the uplift in the northern zone. In the northern zone there is also activity (topographic increase in Fig. 4b) produced by backthrusts in the proximal area (centre of the inherited basin), but this is lower than that in the south.

Between 75% and the last stage of shortening (Fig. 4b), the features described above are preserved, but the activity of the main approximately north–south strike structure (topographic increase at the centre of the inherited basin, Fig. 4b) becomes heterogeneous between north and south. The central portion of this north–south structure is completely inactivated at c. 90% of shortening, while in the northern zone this structure is reactivated between 90% and 100% of shortening. The oblique N10°W structure in the northern zone maintains its activity and generates increasing uplift throughout the course of the experiment (Fig. 4b).

Compressional phase final state. The final state (Fig. 4c, profiles IV, V and VI) shows that (1) faults inherited from the extensional stage have a heterogeneous behaviour (some were not reactivated and others were reactivated), (2) the southern zone has a greater density of normal faults, some preserved within folds of large amplitude that mainly concentrate the shortening in this region (Fig. 4c, profile VI), and (3) towards the northern zone a pop-up structure (uplifted area bordered by two new opposite dipping thrusts) of smaller amplitude and increasing uplift (Fig. 4c, profile IV)

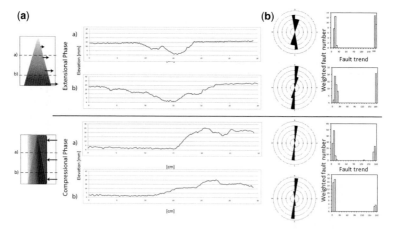

Fig. 5. (a) Topography of the final stages of the Type I model in the northern and southern zones. (b) Rose diagram and weighted fault number trend for the Type I model extensive and compressive end stage.

absorbs part of the shortening in the proximal zone where the basin was narrowest. The major topographic increase (elevation) developed along the entire model mainly in the proximal region and mostly in the northern zone (Fig. 5).

In the northern zone, compressive structures have an oblique strike in both the proximal and distal zones, while inactive normal or inverted normal faults have an approximately north–south orientation (Fig. 4) at the centre of the active zone (evidenced by a topographic increment).

Type II: Homogeneous extension and subsequent differential compression

Extensional phase evolution. Elongated north–south basins were generated (Fig. 6a). Distal to proximal migration of the deformation occurs, while the main activity (see the subsidence maps in Fig. 6a) moves towards the proximal zone at the centre of the basin (Fig. 6a). Normal faults dip towards the basin's centre. In some places, sub-basins are separated by horsts (note uncoloured areas with no subsidence between two adjacent normal faults dipping opposite in Fig. 6a). With more than 75% of extensional displacement, normal faults located in the distal zone became inactive. This did not occur on the basin's axis, above the VD, where the major activity was concentrated.

The extension ended at 5 cm of displacement (100% extension in Fig. 6a). At this last stage, the development of normal faulting along the whole model was emphasized, with approximately north–south strikes and the development of relay ramps between the extensional structures (note areas of lower subsidence between two normal faults displaced in the east–west direction in Fig 6a). Intrarift normal faults grew closely spaced and the rift border faults are well defined.

Extensional phase final state. After applying 5 cm of extension (100%), a 15 cm-wide basin was formed (Fig. 7a) that has a similar structural pattern to the central profile of the basin of the Type I model (Fig. 4), but with near north–south strike (Fig. 7b). The basin is formed by sub-basins, with the central region deeper than the edges (Figs 6c & 7a). The basin is relatively homogeneous along the course, with graben limited by normal faults dipping between c. 50° and 70° (Fig. 6c). In the centre of the basin, the deepest area presents low-angle normal faults, similar to those observed in profile III at the extensional phase of the Type I model (Fig. 4c). Its general structure can also be observed in the less deformed profile in the compressive stage of the Type II model, which shows that all faults are rooted in the ductile layer (Fig. 6c, profile IV).

Compressional phase evolution. During the initial shortening phase (0–25% shortening) the deformation is concentrated mainly in the southern zone (Fig. 6b), between the distal boundary of the pre-existent basin and the area proximal to the backstop. Uplift in the northern zone is less than in the southern zone and it concentrates in the proximal region, as shown in the isopach maps in Fig. 6b. With increasing shortening (50% in Fig. 6b) the elevated region expanded; some deformation is transferred towards the northern distal zone, but within the limits of the pre-existent basin.

From 75% to 100% shortening (Fig. 6b) it is clear that in the distal zone of the model the structures exhibit an approximately north–south strike, similar to the structures of the pre-existent basin,

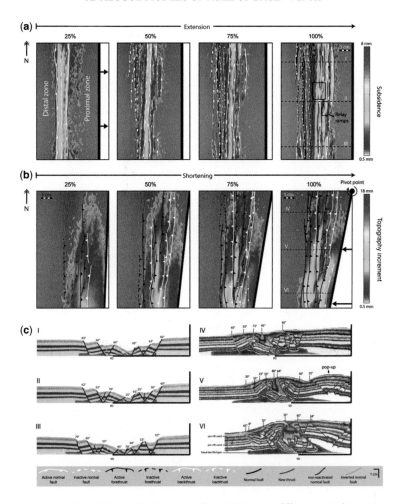

Fig. 6. Type II model evolution. (**a**) Extensional phase: active subsidence at different extension percentages. (**b**) Compressional phase: active topography increment at different shortening percentages. (**c**) I, II, II: profiles of the final result after homogeneous extension. IV, V, VI: profiles of the final result after differential shortening.

whereas structures in the proximal zone occur with an obliquity similar to that of the mobile wall, with a N10°E maximum strike. Note that this obliquity occurs in backthrusts that were already generated before 75% shortening, which means that they have been rotated and are not new oblique structures.

The deformed area does not vary significantly during the evolution of the model, but the amount of uplift increases from the beginning of compression until it reaches 100% (Fig. 6b), indicating that the region is progressively uplifted and rotated until it absorbs 100% of the applied shortening.

Compressional phase final state. In the northern zone (Fig. 6c, profile IV), a principal forethrust absorbs part of the deformation uplifting the proximal zone above the pre-existent basin, while a backthrust in the distal zone produces the same effect with opposite vergence. It is possible to follow this structures to the profiles of the southern zone (Fig. 6c, profiles V and VI), given its approximately north–south strike. Some normal faults were reactivated; however, most of the normal faults were not reactivated. It is remarkable how the number of thrusts increased along the strike with increasing shortening, while their spacing decreased (Fig. 6c). These structures generated greater rise into the southern zone compared to the north (compressional phase in Fig. 7).

In profile IV (Fig. 6c), all deformation is concentrated in the area occupied by the previous basin. The proximal forethrust preserves some non-reactivated normal faults in its hanging wall, while

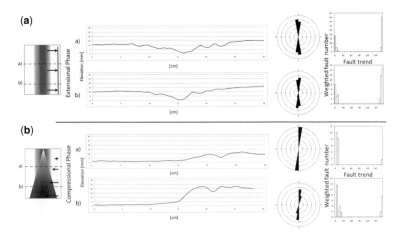

Fig. 7. (a) Topography of the final stages of the Type I model in the northern and southern zones. (b) Rose diagram and weighted fault number trend for the Type I model extensive and compressive end stages.

some inverted normal faults are recognized in the hanging wall of the distal backthrust (Fig. 6c, profile IV). Most of the normal structures are preserved non-reactivated at the proximal zone, where they generated the major subsidence in the previous phase.

In the centre of the model (Fig. 6c, profile V), where the amount of shortening was equal to the amount of applied extension, the proximal forethrust and the distal backthrust into the inherited basin had greater activity and produced two opposite verging folds, closing the basin towards its axis. An anticline fold, consisting of a pop-up structure, was created out of the basin in the proximal region (Fig. 6c, profile V). Towards the south, this anticline fold rises as a result of the activity of the forethrust mentioned above (Fig. 6c, profile VI), generating greater rise into the southern compared to the northern zone (compressional phase in Fig. 7a). Throughout the region, several normal faults are preserved, while some normal faults dipping towards the proximal zone are slightly reactivated.

Discussion

Summary and interpretation of results

It has been reported that initially high-angle normal faults may rotate to gentler dips (more favourable for reactivation) during extension through 'domino' rotation of successive normal fault sets (Jackson & McKenzie 1983; McClay et al. 1989; Buchanan & McClay 1991; Knott et al. 1995; Mandal & Chattopadhyay 1995; Del Ventisette et al. 2006; Bonini et al. 2012). In our models, normal faults distinguished by their low dip angle occur in the central part of the most extended zone in the Type I model (southern zone) and at the centre of the basin of the Type II model. From the observation of profiles in plan evolution, and 3D building blocks shedding light on the north–south continuity of some structures, we interpreted a 'domino' block rotation mechanism (Buchanan & McClay 1992; McClay & Buchanan 1992; Mitra 1993) of initially steep normal faults formed with c. 60° dip in the early stages of extension to minor inclination normal faults at the final stages, presenting angles even less than 20°.

Moreover, the compression in the southern zone of the Type I model is highly concentrated inside the basin (Fig. 8a), so reverse faults are formed mainly within the limits of the previous extensional basin, while in the northern zone, where the pre-existing basin was narrow, new compressive structures developed out of the limits of the basin (Figs 8a & 9). Positive inversion is concentrated in the southern zone of the Type I model, where pre-existing faults prone to positive reactivation due to the low angle are located (Figs 8a & 9).

In both types of model proposed for this study, some normal faults are preserved after inversion of the basin (Figs 4c, 6c & 9). Reactivated faults generally have high dipping angles of 50°–60° (Fig. 9), whereas in the central profiles of both models (profile V in Figs 4 & 6), the dipping of reactivated normal faults decreases toward the surface. Similarly, Likerman et al. (2013) and Konstantinovskaya et al. (2007) noted that, for asymmetric rifts, reactivated normal faults rotated to gentler dips when cutting off to the surface.

Additionally, in the region where the rift basin was poorly developed (northern zone in the Type I model), the deformation is accommodated by

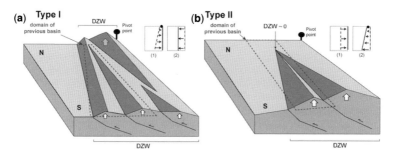

Fig. 8. Conceptual models of main features of the final results of (**a**) the Type I model and (**b**) the Type II model. The vergence of structures is schematic and only referential. White arrows show the uplifted zones. DZW, deformational zone width during compression. (1) and (2) correspond to the extensional and compressional phases respectively.

structures that are capable of absorbing large amounts of shortening by raising a narrower area outside the rift basin (Figs 5 & 8a). This process causes the main forethrust, which generates this higher rise in this narrow region, to have an oblique strike (Fig. 4a) to connect with the wider area (south), where it generates a minor uplift area closest (proximal) to the backstop (Figs 8a & 9).

In the Type II model, compression by a rotating pivot generated a differential deformation zone. Faults and folding were concentrated in the area of maximum shortening (southern zone in Figs 8b & 9). The deformational plan view sequence and the structural pattern of the Type II model (Fig. 6b) have some similarities with compressional rotational models that do not involve the basin inversion process (Soto *et al.* 2006), because both have a progression that starts in the most compressed region, and the rotation is in the same direction as the rotation axis of the backstop. Similar to the Type II model (Figs 8b & 9), Ghiglione & Cristallini (2007) showed that rotation of a rigid indenter against a soft sedimentary cover produces a strong variation in orogenic shortening and width of the thrust belt along the strike of the pivoted edge. Deformation in the southern zone of the Type II model is concentrated in new thrusts, some normal reactivated faults and low wavelength folds (Fig. 9). In the northern zone there is little shortening and therefore most normal faults are preserved (Fig. 9). As compression progresses, deformation in the southern zone produces the rotation of structures (Fig. 6) and closing of the basin through main folds. Distal structures develop parallel to the basin and proximal thrust gradually turns in a more oblique direction (Fig. 6b), parallel to the shortening vector. In the more compressed region we can recognize some similarities with complex inversion models described in previous papers, such as decapitating early normal faults, shortcuts and buttressing (Coward *et al.* 1991; Scisciani 2009; Bonini *et al.* 2012).

Figures 5, 7, 8 and 9 summarize some of the characteristics of the developed models. The Type I model presents many normal structures in the southern zone and fewer in the northern zone, as expected (Fig. 5b). The principal structures are north–south oriented, but some structures show a direction linked to the imposed edges of the silicone basal layer, which forced the direction of the basin limits. During compression, the oblique NNW direction is not directly linked with the limits of the basal silicone; this is made evident because the main forethrust in the northern zone is not generated from the previous proximal limit of the basin, but far closer to the mobile wall (Figs 8 & 9). This NNW orientation of the proximal structures is interpreted as being produced by the greater uplift of the northern compared to the southern zone, which is the result after compressing both regions equally, but with different pre-existing basin widths.

In the Type II model (Fig. 7), a similar number of normal faults are created in the northern and southern zones, with a trend very close to north–south. Compressing this model produces fewer structures in the northern zone than in the Type I model (Fig. 9).

Our modelling technique included complete filling of the basin before inversion. Sediment load makes difficult or delays the reactivation of previous normal faults (Nalpas & Brun 1993; Brun & Nalpas 1996; Dubois *et al.* 2002; Panien *et al.* 2005; Del Ventisette *et al.* 2006; Pinto *et al.* 2010), while in models with load inside and outside the basin, reactivation is facilitated, allowing the interpretation that the dominant factor in this case is the relative difference in load (Pinto *et al.* 2010). In our models, the largest load difference occurs where the basin is deeper and wider, so we expect the reactivation to be difficult during inversion tectonics in these sectors. This is consistent, because, regardless of the width of the basin, inverted normal faults are generally located near the edges of the pre-existing basin (Fig. 9), where the filling (synrift) is thinner.

It is worth noting one important difference between the models mentioned above and ours. In our Type I models the greater width of the basin means that shortening occurs inside it, while in other models (Pinto et al. 2010), an increase in sediment load generates further development of thrust faults outside the basin, similar to what occurred in our filled narrow basin (Fig. 9). Although the sedimentary basin is greater in our wide basin than in our narrow basin, the amount of shortening is equal in both areas, leading to the ratio (Rse, Fig. 3) between the amount of shortening and the amount of extension (and the resulting basin width difference) to be the predominant factor in this case.

As the methodology used in the Type I models allowed extensional structures to be generated with the same strike in both wide and narrow basin areas (approximately north–south faults and basin edges with equal obliquity across the region), the orientation of these structures is not a predominant factor in discerning between reactivated faults among narrower and widest basin zones (Fig. 9).

At stages with differential deformation (rotating the pivot), a maximum of 10° of obliquity was applied to the mobile wall. This generated, for the Type I model, extensional structure strikes between 0° and 15°, leading to the subsequent applied direction of principal compressive stress (perpendicular to the main direction of the extensional basin) being inefficient for reactivation. Analogue model experiments showed that, without a change in the inclinaton of the previously formed normal structures to gentler dips, an optimum angle close to 15° between the direction of extension and compression is necessary to generate subsequent reversal of high-angle faults (Brun & Nalpas 1996). We obtained considerably higher angles between the compressive and extensive strain axes, and there is reactivation of structures in our models. We think this might be caused by flattening of some normal faults before compression.

The Type II models have a similar relation: the direction of compressive stress rotates together with backstop rotation (0–10°), so the angle between the two principal stresses is never as close

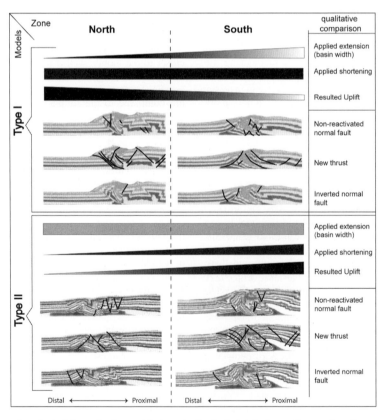

Fig. 9. Comparative chart of the main observed features in the final state in the northern and southern zones of the developed Type I and Type II models.

to 15°, and not enough to generate an efficient inversion process.

The north–south direction is a preferred orientation for major structures in extensional deformation phases of our models. However, in the Type I model this orientation is preferential for the major depocentres or sub-basins, whereas in the Type II model all extensional structures had this orientation prior to the shortening phase, and the obliquity of later structures is generated by the rotational movement of the pivot. Summarizing, the Type II model can explain the development of oblique structures, but it needs an important rotation. In the Type I model, latitudinal difference between uplift and width of the deformation zone play an important role in the obliquity of compressive structures.

The main results of our model are as follows:

(1) The preservation of non-reactivated normal faults, and positive inversion and generation of new compressive oblique structures, occur in both types of model.
(2) A significant rotation is required to generate obliquity of structures in models with no difference in the width of the pre-existing basin, as well as subsequent differential compression (Type II).
(3) In the Type I model, the obliquity of structures is generated due to a different width along the pre-existing basin.
(4) The northern zone in the Type I model and the southern zone in the Type II model are the areas that predominantly exhibit compressive structures (Fig. 9), but in the Type I model they are concentrated in a narrower and raised region due to the absence of a pre-existing wide basin.

Comparison with inverted natural basins

The elongated Abanico Basin, located in the Principal Cordillera of the Central Andes (Fig. 10a), presents significant latitudinal variations in the distribution and width of the Cenozoic outcrops in the strike of the major structures of the eastern Principal Cordillera, and the presence or absence of tectonic inversion structures, those that can be observed in latitudinal profiles (Jara et al. 2009; Godoy 2011, 2012). There is an observed south to north decrease in width of the Abanico Formation between 32°S and 33°S (Fig. 10c, d), which may correspond to an original extensional setting or to a subsequent narrowing due to a northward increased in shortening (Jara et al. 2009). In this comparison we will focus on the major structural features (strike of principal structures) and width of the deformational zone and outcrop of affected deposits (Fig. 11) in order to interpret the origin of the latitudinal changes based on the results of both developed models. Together with a comparison with the applied models, a brief background that will allow us to interpret the origin of these main features is further discussed below.

The Abanico Formation (Fig. 10) is represented by a set of volcanic-sedimentary rocks 3500 m thick (Thiele 1980; Fock 2005; Charrier et al. 2007; Rauld 2011), interpreted to have been deposited in an intra-arc extensional basin with active volcanism during the Late Eocene to Early Miocene, and tectonically inverted between c. 21 and c. 16 Ma (see Charrier et al. 2007, and references therein) and exposed in the Principal Cordillera of central Chile from 33° to 36°S (Aguirre 1960; Klohn 1960; González & Vergara 1962; Charrier 1973; Vergara & Drake 1979; Thiele 1980). Two stages are distinguished in the evolution of this basin: (1) a Late Eocene to Early Miocene extensional stage, where thick deposits of lava, volcaniclastics and minor sediments accumulated, and (2) an Early to Middle Miocene (Charrier et al. 2002, 2007, 2009) compressional stage. During this second stage, the Abanico Basin would have undergone partial tectonic inversion, probably contemporaneous with uninterrupted volcanism and accumulation of material included in the Farellones Formation (Middle–Late Miocene) (Charrier et al. 2002, 2005, 2007; Fock 2005; Fock et al. 2006; Pinto et al. 2010; Muñoz et al. in press). Charrier et al. (2005) showed evidence for the presence of the Abanico extensional basin from at least 33° and 36°S, and suggest its probable continuity north of 31°S and south of 37°S (Fig. 10a).

New data indicate that, between 32° and 32°30′S, and near the 'Las Llaretas' river (Figs 10b & 12), Oligocene–Miocene rocks are deformed by approximately north–south trending structures in the easternmost sector of the Principal Andean Cordillera (Jara & Charrier in press). These rocks are located in areas where intense compressional deformation is superimposed on Late Oligocene depocentres bounded by approximately north–south normal faults that were later inverted. Some of these faults had extensional movements accommodating clastic deposits at least 21 Ma ago. Post-18 Ma, these structures were reversed, deforming clastic deposits of the basin (synrift) as well as overlying discordant accumulated cenozoic strata (post-rift) (Fig. 12).

At these latitudes (32°–32°30′S), eastern units of the Abanico and Farellones formations are affected by NNW-oriented out-of-sequence faults of the La Ramada fold-and-thrust belt (Fig. 11). Although the La Ramada fold-and-thrust belt has been described as a thick-skinned fold-and-thrust belt (Mosquera 1990; Zapata 1990; Cristallini & Cangini 1993; Cristallini et al. 1994; Alvarez

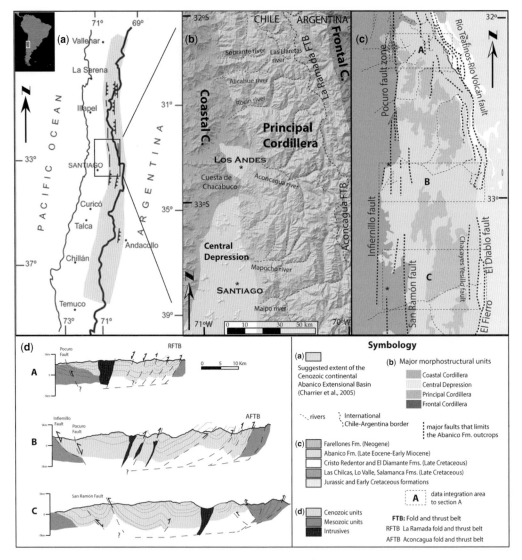

Fig. 10. (a) Tentative suggested extent of the Cenozoic continental Abanico extensional basin along the Principal Cordillera (Charrier et al. 2005). (b) Major morphostructural units and localities at c. 32°–34°S. (c) Simplified geological map and major structures that limit the Abanico Formation outcrops. (d) Schematic sections of deformed units of the Principal Cordillera in central Chile between c. 32° and 34°S.

et al. 1995; Pérez 1995; Cristallini 1996), the out-of-sequence structures in this region only affect Cenozoic levels (Jara & Charrier in press).

The eastern NNW-strike structural pattern in this region has been explained previously (Rivera & Yáñez 2007) as a result of an originally segmented basin with sub-basins bounded by oblique structures controlled by heterogeneities of the Triassic–Jurassic basement under the Mesozoic to Cenozoic cover. The influence of Triassic–Jurassic rift structures on the deformation style in Andean structures has been studied from various points of view by several authors (Giambiagi et al. 2003a, b, 2005, 2009a, b; Yagupsky et al. 2008). In particular, the structural control that these basement structures exerted on Tertiary tectonic inversion has been studied and recognized in the fold-and-thrust belt of La Ramada (RFTB), Aconcagua (AFTB) and Malargüe (MFTB) (Cristallini & Cangini 1993; Cristallini et al. 1994; Maceda & Figueroa 1995; Giambiagi & Ramos 2002; Giambiagi et al. 2005; Yagupsky et al. 2008). Hovewer, the mainly NNW orientation of the main

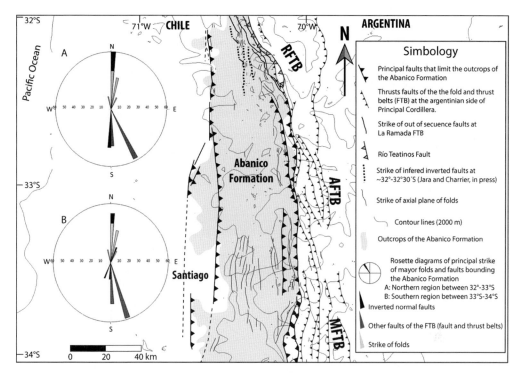

Fig. 11. Abanico Formation outcrops and major structure trends along the Principal Cordillera between c. 32° and 34°S.

structures in the eastern Principal Cordillera in this region would be associated with Tertiary deformation and not necessarily with the Abanico Basin original boundaries, or even older basins. We propose a not previously considered control, variable basin width, to explain in part some of the characteristics of this region, including distribution and width of the Cenozoic outcrops, the strike of the major structures and the presence or absence of tectonic inversion structures.

Cristallini (1996) indicates that the structural evolution of the La Ramada fold-and-thrust belt would have begun around 19–20 Ma ago, before flat-slab development in this region. The first stage of this structural evolution developed the first folding and faulting in Argentinian territory, and the first rise of the Principal Cordillera in the Chile–Argentina bordering region, controlled by approximately north–south preferential trending faults (Cristallini 1996). At this stage, in the eastern sector of the Principal Cordillera between 32° and 33°S (Jara & Charrier in press), some of the normal faults that previously controlled the accumulation of distal deposits of the Abanico Formation were reversed (Fig. 13).

The subsequent stage of deformation in the La Ramada fold-and-thrust belt, developed between c. 14 and 12.7 Ma (Pérez 1996), was controlled by the reactivation and inversion of ancient Triassic rift normal faults, which have an essentially NNW strike and raised basement blocks (Cristallini 1996). Basement uplift by high-angle reverse faults produced a 'sticking point' in the foreland propagation of the thrust belt, responsible for the third stage of deformation characterized by thin-skinned NNW out-of-sequence thrusts developed in the westernmost sector of La Ramada fold-and-thrust belt (Cristallini & Ramos 2000). The NNW-trending faults in the eastern part of the Principal Cordillera between 32° and 33°S, in Chilean territory, correspond to the westernmost of these out-of-sequence structures and are controlled by a shallower detachment level than those that would have inverted the basement of the La Ramada fold-and-thrust belt (Cristallini 1996; Cristallini & Ramos 2000; Ramos 2006).

In the southern region (south of 33°S), the structures are generally north–south oriented and, towards the eastern sector of the Principal Cordillera, the structure is characterized by a series of east-verging thrusts repeating Mesozoic sequences in a typical thin-skinned fold-and-thrust belt: the Aconcagua fold-and-thrust belt (Ramos et al. 1996; Giambiagi & Ramos 2002; Giambiagi et al. 2003a, b; Armijo et al. 2010; Farías et al. 2010). This belt would have been active since c. 20 Ma

Fig. 12. Normal faulting affecting: (**a**) Early Miocene deposits at Las Llaretas; (**b**) the Cuartitos Unit at Tres Quebradas Creek.

(Cegarra & Ramos 1996). The early stages of deformation in this belt would be characterized by the inversion of Jurassic graben prior to the deposition of the volcanic levels of the Farellones Formation and then reactivated in the Late Miocene where migration of deformation to the east would have allowed the development of the Frontal Cordillera (Cegarra & Ramos 1996). As in the north (at the La Ramada fold-and-thrust belt), there is evidence of old normal faults inversion, but here these structures do not involve the basement significantly (Cegarra & Ramos 1996). Furthermore, like the

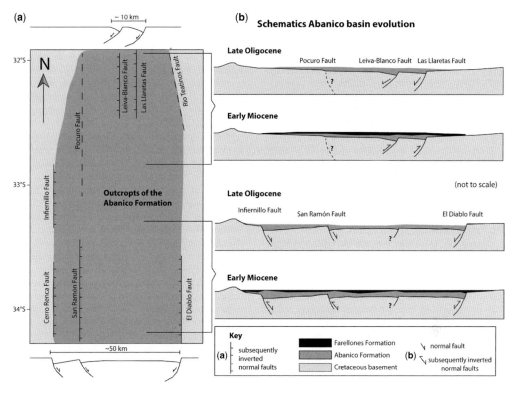

Fig. 13. (a) Schematic Abanico Formation outcrops and major structures that limit them in the Principal Cordillera between c. 32° and 34°S. We interpret the width of the basin where the deposits of the Abanico Formation accumulated. (b) Schematic of the Abanico Basin evolution between the Late Oligocene and Early Miocene at two generalized sections between 32°–33°S and 33°–34°S.

area of the La Ramada fold-and-thrust belt, the Chilean sector of the Principal Cordillera would have been affected by out-of-sequence thrusting, the development of which is not well constrained chronologically due to the structural complexity and scarcity of radiometric dating of some levels, so it is uncertain whether the out-of-sequence thrusts occurred before or after 8.6 Ma (Ramos et al. 1996).

For comparison purposes, we will focus on the strike of major structures and the width of the deformation zones. Our Type I model may explain in part the NNW orientation of compressive structures north of 33°S (Fig. 11), because the NNW-oriented out-of-sequence structures of the La Ramada fold-and-thrust belt, affecting the Abanico and Farellones formations in the eastern sector of the Principal Cordillera between 32° and c. 32°30′S, do not involve the basement and also do not correspond to a preferential orientation of the inverted normal faults that controlled the Abanico Basin (Fig. 13). The model shows that NNW-trending structures are generated in the region of lower basin width (northern zone), because when the same amount of shortening is induced throughout the region, the structures have an oblique orientation to accommodate more uplift in the north with respect to minor uplift on a wider region in the south (Fig. 8), where the structures are generated mainly within the pre-existing basin.

Moreover, the Type II model produces oblique structures generated by a progressive rotation process during the compressional phase. Because at c. 32°–32°30′S the Abanico extensional basin structures present approximately north–south strike and 21 Ma synrift associated strata, and given that post-21 Ma deformation affects both Abanico and Farellones formations (post-18 Ma) with NNW-oriented structures, a rotation process for developing these NNW-trending structures is discarded, because the process would also have rotated previous 18 Ma structures, which is not observed.

South of 33°S, the 'El Diablo' fault (Figs 10c, 11 & 13), located on the eastern boundary of the Chilean territory Principal Cordillera, corresponds to the westernmost structure of the east-verging Aconcagua fold-and-thrust belt and the fault that would have limited the basin to the east (Fock 2005; Fock et al. 2006; Farías 2007). The Aconcagua fold-and-thrust belt developed folds with approximately NNW–SSE-oriented axes and approximately north–south-oriented faults (Giambiagi et al. 2003a, b). Some of these faults cut early developed folds and are therefore out-of-sequence faults. At this latitude, the eastern part of the Abanico Basin is deformed by backthrusts of the 'El Diablo' fault (Fock et al. 2006). The principal approximately north–south orientation of the 'El Diablo' fault and its associated backthrusts and the western part of the Aconcagua fold-and-thrust belt are consistent with the approximately north–south strike of the forethrusts and backthrusts that deformed the proximal-southern zone (wider basin zone) in the Type I model.

Based on evidence for the deposition of the Abanico Formation in a basin more than c. 50 km wide between 33° and 34°S (Fig. 13), controlled by normal faults (see Fock 2005), and the new data indicating that north of 33°S only distal deposits of this formation were accumulated in depocentres controlled by normal faults in a region c. 10 km wide (Jara & Charrier, in press), we believe that the Type I model might be representative of the palaeogeography existing between 32° and 34°S, where the space that accommodated clastic and volcaniclastic deposits is progressively narrower towards the north. Based on the above and on the results of the Type I model (Fig. 8), we can suggest that the curvature of the out-of-sequence faults of the La Ramada fold-and-thrust belt may be accounted for by a higher uplifting accommodation north of 33°S, compared to minor uplift accommodated in the larger width basin south of 33°S, regardless of or as a complement to the hypothesis that major inherited Pre-Andean normal structures would directly cause this oblique direction.

The evolution of the Type I model also allows us to make some interpretations. As deformation progresses, shortening in the southern zone is accommodated within the pre-existing basin, and the activity of the north–south elongated structures (Fig. 4b) is maintained throughout the process until the end of the compression. Instead, where shortening exceeds the amount of extension, the principal north–south structure (Fig. 4b) has less activity and has almost no activity in the centre of the model. If we make a comparison between the evolution of the main north–south elongated fault of the Type I model and our study area, we can illustrate this with the traces of the San Ramón and Pocuro fault zones (Fig. 10c), with the latter showing its main activity prior to the Early Miocene, deforming the Cretaceous outcrops intensely (Mpodozis et al. 2009; Jara & Charrier in press), with Miocene outcrops only slightly deformed. In fact, at 32°10'S latitude, west of the Pocuro Fault, almost flat-lying Oligocene–Miocene rocks cover unconformably deformed Cretaceous deposits of c. 82 Ma (Jara & Charrier, in press). In contrast, south of 33°S the San Ramón Fault shows evidence of important activity post-20 Ma (deforming the Abanico and Farellones formations' deposits) and even showing Quaternary activity (Rauld 2002; Armijo et al. 2008).

Further north, in the High Andes between 28°45'S and 30°30'S in the Norte Chico region of Chile, Early Oligocene to Late Miocene volcanic deposits have been designated as the Doña Ana Group (Martin et al. 1997), which consists of the Tilito (27–23 Ma) and Escabroso (21.5–17.5 Ma) formations (Maksaev et al. 1984; Mpodozis & Cornejo 1988; Kay et al. 1988, 1991; Bissig et al. 2001; Litvak 2009).

The older parts of the Doña Ana Group (Kay & Abbruzzi 1996) were probably deposited in an extensional basin subsequently inverted in Miocene times. The Doña Ana Group is partially coeval with the Abanico Formation further south (see Charrier et al. 2005, 2007). New studies show evidence of an extensional period in the Valle del Cura region located in the eastern Andean flank of the Pampean Flat Slab segment during the Oligocene. This corresponds to the intra-arc basin volcanic deposits of the Tilito Formation. This extensional basin affected the region between 29° and 30°S (Winocur 2010; Winocur & Ramos 2008, 2012; Winocur et al. 2014).

In this region, there is a series of sedimentary deposits of Jurassic age named the Lautaro Formation (Segerstrom 1959) and Cuartitos Unit (Martin et al. 1995), which are affected by normal faults formed in Late Triassic to Early Jurassic times (Winocur 2010). Some of these faults are preserved with a normal displacement recorded by a synsedimentary extensional structural setting (Fig. 12b). These normal structures, with mainly north–south strike, are almost completely preserved in this region after the Miocene compression. The structural cross-section (Fig. 14) shows the palinspastic restoration in Jurassic times at 29°30'S in the Norte Chico Region (Fig. 14c). Some of the normal faults that affected the Cuartitos Unit in Jurassic times are preserved in the structural cross-section (Fig. 14b). Some Jurassic normal faults were inverted in Miocene times, but there are also new faults created under compression in this period (Fig. 14b).

The north–south strike of normal structures preserved in the Type I model is consistent with the north–south orientation of normal faults of Jurassic age preserved in the 29°–30°S region (Winocur 2010). In the latter, it can be noted further that inverted Oligocene normal faults are concentrated

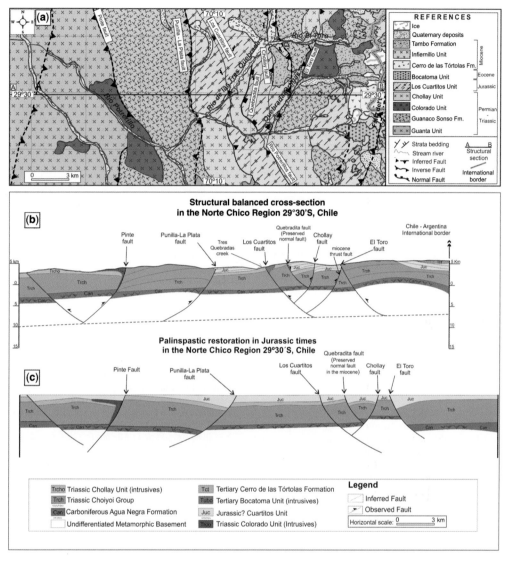

Fig. 14. (a) Geological map at c. 29°30′S. (b) Structural balanced cross-section. (c) Palinspastic restoration in the Norte Chico region, 29°30′S, Chile.

in the eastern region, with preserved Jurassic normal faults to the west (Winocur 2010), similar to what is observed in our and other analogue models (Bonini et al. 2012) where new inverse structures are concentrated near the mobile wall and inverted normal faults in the distal zone, away from the backstop.

Concluding remarks

The Type I model, reproducing differential extension and subsequent homogeneous compression, shows similar patterns to those of the natural field examples exposed here. In this model, compressive structures with uplift concentrated on a narrow area developed on the sector with smaller amounts of previous extension. In the region with greater extension, the compressive deformation was accommodated between the boundaries of the extensional basin, occupying a wider area, with consequent lower rising structures and topography. Furthermore, in the Type I model, these latitudinal differences of uplift between the northern and southern zones played an important role in the obliquity of compressive structures. This is because, in the wider basin, zone shortening is concentrated within it, while to the north the structure turns its strike to accommodate more uplift close to the boundaries of the narrow basin.

Our Type I model shows that NNW-trending structures are generated in the region of smaller basin width (north), which is consistent with the orientantion of structures in the eastern sector of the Abanico Basin at c. 32°–33°S, where the NNW-oriented out-of-sequence structures of the La Ramada fold-and-thrust belt, affecting the Abanico and Farellones formations, do not involve the basement and do not correspond to a preferential orientation of the inverted normal faults that controlled the Abanico Basin.

The model shows that NNW-trending structures are generated in the narrower basin region to the north, which is consistent with a c. 10 km wide area of active depocentres during the Oligocene–Early Miocene at c. 32°10′S, in contrast to the region south of 33°S, where the basin would have been over 50 km wide. This result suggests that the curvature of the La Ramada fold-and-thrust belt may be influenced by a stronger and narrower uplifting north of 33°S compared to the smaller and wider uplift south of 33°S.

In the Type II model, with equal basin width and subsequent superimposed differential compression, the basin did not control the strike of the compressive structures. Conversely, regional structure is controlled by rotation during the shortening phase, which generates the greatest amount of folds and faults and greater uplift in the most compressed area. Thus, the Type II model can explain the development of oblique structures, but it needs an important oroclinal rotation, which is not consistent with geological observations of the studied region.

The experimental models developed show that the control exerted by the width of a pre-existing basin, which was subsequently reversed, has important implications in the resulting structural pattern and should be considered when interpreting the geological evolution of an inverted natural basin.

J. Likerman and E. O. Cristallini acknowledge funding from Laboratorio de Modelado Geológico (LaMoGe), Instituto de Estudios Andinos "Don Pablo Groeber" (IDEAN), Departamento de Ciencias Geológicas, FCEN, Universidad de Buenos Aires, C1428EGA, Argentina. D. Winocur and M. C. Ghiglione acknowledge funding from Laboratorio de Tectónica Andina, Instituto de Estudios Andinos "Don Pablo Groeber" (IDEAN), Departamento de Ciencias Geológicas, FCEN, Universidad de Buenos Aires, C1428EGA, Argentina. Funding for this project was provided by 'Ayuda para estadías cortas de investigación', Vicerrectoria de asuntos académicos, Departamento de postgrado y postítulo de la Universidad de Chile. We also acknowledge funding from Agencia de Promoción Científica, from Consejo Nacional de Investigaciones Científicas (CONICET, Argentina) and from LaMoGe (Universidad de Buenos Aires, UBA), and thank the colleagues and investigators working there for their generous contributions and valuable discussions regarding this research. The assistance of the Laboratorio de Modelamiento Analógico del Departamento de Geología de la Universidad de Chile, of the IGCP-586Y Unesco Proyect, and of L. Giambiagi (IANIGLA, Argentina) and L. Pinto (Universidad de Chile), for their collaboration in this contribution, is gratefully acknowledged. We also acknowledge the valuable ideas of T. Nalpas (Université de Rennes) and C. Jara (Universidad de Chile) in generating the experimental device and conducting the first test that led to the present study. This article is a contribution to FONDECYT Project 1120272: Extension, inversion and propagation: key tectonic styles on the development of the Andean Cordillera of Central Chile–Argentina (32°–36°S).

References

AGOSTINI, A., BONINI, M., CORTI, G., SANI, F. & MAZZARINI, F. 2011. Fault architecture in the Main Ethiopian Rift and comparison with experimental models: implications for rift evolution and Nubia–Somalia kinematics. *Earth and Planetary Science Letters*, **301**, 479–492, http://dx.doi.org/10.1016/j.epsl.2010.11.024

AGUIRRE, L. 1960. Geología de Los Andes de Chile Central. *Instituto de Investigaciones Geológicas*, **9**.

ALVAREZ, P. P., BENOIT, S. & OTTONE, E. G. 1995. Las Formaciones Rancho de Lata, Los Patillos y otras unidades mesozoicas de la Alta Cordillera Principal de San Juan. *Revista de la Asociación Geológica Argentina*, **50**, 123–142.

AMILIBIA, A., SABAT, F., MCCLAY, K. R., MUNOZ, J., ROCA, E. & CHONG, G. 2008. The role of inherited tectono-sedimentary architecture in the development of the central Andean mountain belt: insights from the Cordillera de Domeyko. *Journal of Structural Geology*, **30**, 1520–1539, http://dx.doi.org/10.1016/j.jsg.2008.08.005

ARMIJO, R., RAULD, R., THIELE, R., VARGAS, G., CAMPOS, J., LACASSIN, R. & KAUSEL, E. 2008. An Andean megathrust synthetic to subduction? The San Ramón fault and associated seismic hazard for Santiago (Chile). *In*: *4th Alexander von Humboldt International Conference*. Santiago de Chile, 164.

ARMIJO, R., RAULD, R., THIELE, R., VARGAS, G., CAMPOS, J., LACASSIN, R. & KAUSEL, E. 2010. The West Andean Thrust, the San Ramón Fault, and the seismic hazard for Santiago, Chile. *Tectonics*, **29**, 1–34, http://dx.doi.org/10.1029/2008TC002427

BISSIG, T., CLARK, A. H., LEE, J. K. W. & HEATHER, K. B. 2001. The Cenozoic history of volcanism and hydrothermal alteration in the Central Andean flat-slab region: new ^{40}Ar–^{39}Ar constraints from the El Indio–Pascua Au (−Ag, Cu) Belt, 29°20′–30°30′ S. *International Geology Review*, **43**, 312–340, http://dx.doi.org/10.1080/00206810109465016

BONINI, M., SOURIOT, T., BOCCALETTI, M. & BRUN, J. P. 1997. Successive orthogonal and oblique extension episodes in a rift zone: laboratory experiments with application to the Ethiopian Rift. *Tectonics*, **16**, 347, http://dx.doi.org/10.1029/96TC03935

BONINI, M., SANI, F. & ANTONIELLI, B. 2012. Basin inversion and contractional reactivation of inherited normal faults: a review based on previous and new experimental models. *Tectonophysics*, **522–523**, 55–88, http://dx.doi.org/10.1016/j.tecto.2011.11.014

BRUN, J. P. & TRON, V. 1993. Development of the North Viking graben: inferences from laboratory modelling. *Sedimentary Geology*, **86**, 31–51, http://dx.doi.org/10.1016/0037–0738(93)90132-O

BRUN, J. P. & NALPAS, T. 1996. Graben inversion in nature and experiments. *Tectonics*, **15**, 677–687, http://dx.doi.org/10.1029/95TC03853

BUCHANAN, P. G. & MCCLAY, K. R. 1991. Sandbox experiments of inverted listric and planar fault systems. *Tectonophysics*, **188**, 97–115, http://dx.doi.org/10.1016/0040-1951(91)90317-L

BUCHANAN, P. G. & MCCLAY, K. R. 1992. Experiments on basin inversion above reactivated domino faults. *Marine and Petroleum Geology*, **9**, 486–500, http://dx.doi.org/10.1016/0264-8172(92)90061-I

BURBANK, D. W., VERGÉS, J., MUÑOZ, J.-A. & BENTHAM, P. 1992. Coeval hindward- and forward-imbricating thrusting in the south-central Pyrenees, Spain: timing and rates of shortening and deposition. *Geological Society of America Bulletin*, **104**, 3–17, http://dx.doi.org/10.1130/0016-7606(1992)104<0003:CHAFIT>2.3.CO;2

CALASSOU, S., LARROQUE, C. & MALAVIEILLE, J. 1993. Transfer zones of deformation in thrust wedges: an experimental study. *Tectonophysics*, **221**, 325–344, http://dx.doi.org/10.1016/0040-1951(93)90165-G

CAREY, S. 1958. The tectonic approach to continental drift. *In*: *Continental Drift: A Symposium*. Geology Department, University of Tasmania, Hobart, Tasmania, 177–355.

CEGARRA, M. & RAMOS, V. A. 1996. La Faja Plegada y Corrida del Aconcagua. *In*: RAMOS, V. A. (ed.) *Geología de la región del Aconcagua, provincias de San Juan y Mendoza*. Subsecretaría de Minería de la Nación, Dirección Nacional del Servicio Geológico, Buenos Aires, 387–422.

CHARRIER, R. 1973. Interruptions of spreading and the compressive tectonic phases of the meridional andes. *Earth and Planetary Science Letters*, **20**, 242–249, http://dx.doi.org/10.1016/0012-821X(73)90164-7

CHARRIER, R., BAEZA, O. ET AL. 2002. Evidence for Cenozoic extensional basin development and tectonic inversion south of the flat-slab segment, southern Central Andes, Chile (33°–36° S.L.). *Journal of South American Earth Sciences*, **15**, 117–139, http://dx.doi.org/10.1016/S0895-9811(02)00009-3

CHARRIER, R., BUSTAMANTE, M. ET AL. 2005. The Abanico extensional basin: regional extension, chronology of tectonic inversion and relation to shallow seismic activity and Andean uplift. *Neues Jahrbuch für Geologie und Paläontologie Abh*, **236**, 43–77.

CHARRIER, R., PINTO, L. & RODRÍGUEZ, M. 2007. Tectonostratigraphic evolution of the Andean Orogen in Chile. *In*: MORENO, W. & GIBBONS, T. (eds) *The Geology of Chile*. The Geological Society, London, Special Publications, 21–116.

CHARRIER, R., FARÍAS, M. & MAKSAEV, V. 2009. Evolución tectónica, paleogeográfica y metalogénica durante el Cenozoico en los Andes de Chile norte y central e implicaciones para las regiones adyacentes de Bolivia y Argentina. *Revista de la Asociación Geológica Argentina*, **65**, 5–35.

COOPER, M. & WILLIAMS, G. D. 1989. *Inversion tectonics*. Geological Society, London, Special Publications, **42**, i–vi, http://dx.doi.org/10.1006/jare.1999.0574

COOPER, M. A., WILLIAMS, G. D. ET AL. 1989. Inversion tectonics – a discussion. *In*: COOPER, M. A. & WILLIAMS, G. D. (eds) *Inversion Tectonics*. Geological Society, London, Special Publications, **44**, 335–347.

CORRADO, S., DI BUCCI, D., NASO, G. & FACCENNA, C. 1998. Influence of palaeogeography on thrust system geometries: an analogue modelling approach for the Abruzzi–Molise (Italy) case history. *Tectonophysics*, **296**, 437–453, http://dx.doi.org/10.1016/S0040-1951(98)00147-4

CORTI, G. 2003. Transition from continental break-up to punctiform seafloor spreading: how fast, symmetric and magmatic. *Geophysical Research Letters*, **30**, 1–4, http://dx.doi.org/10.1029/2003GL017374

COTTON, J. T. & KOYI, H. A. 2000. Modeling of thrust fronts above ductile and frictional detachments: application to structures in the Salt Range and Potwar Plateau, Pakistan. *Geological Society of America Bulletin*, **112**, 351–363, http://dx.doi.org/10.1130/0016-7606(2000)112<351:MOTFAD>2.0.CO;2

COWARD, M. P. 1994. Inversion tectonics. *In*: HANCOCK, P. L. (ed.) *Continental Deformation*. Pergamon Press, Oxford, 289–304.

COWARD, M. P., GILLCRIST, R. & TRUDGILL, B. 1991. Extensional structures and their tectonic inversion in the Western Alps. *In*: ROBERTS, A., YIELDING, G. & FREEMAN, B. (eds) *The Geometry of Normal Faults*. Geological Society, London, Special Publication, **56**, 93–112.

CRISTALLINI, E. O. 1996. La faja plegada y corrida de la Ramada. *In*: RAMOS, V. A. (ed.) *Geología de la región del Aconcagua, provincias de San Juan y Mendoza*. Subsecretaría de Minería de la Nación, Dirección Nacional del Servicio Geológico, Buenos Aires, 349–385.

CRISTALLINI, E. O. & CANGINI, A. 1993. Estratigrafía y estructura de las nacientes del río Volcfin, Alta Cordillera de San Juan. *In*: *XII Congreso Geológico Argentino*, Buenos Aires, 85–92.

CRISTALLINI, E. O. & RAMOS, V. A. 2000. Thick-skinned and thin-skinned thrusting in the La Ramada fold and thrust belt: crustal evolution of the High Andes of San Juan, Argentina (32° SL). *Tectonophysics*, **317**, 205–235, http://dx.doi.org/10.1016/S0040-1951(99)00276-0

CRISTALLINI, E. O., MOSQUERA, A. & RAMOS, V. A. 1994. Estructura de la Alta Cordillera de San Juan. *Revista de la Asociación Geológica Argentina*, **49**, 165–183.

DAUTEUIL, O. & BRUN, J. P. 1996. Deformation partitioning in a slow spreading ridge undergoing oblique extension: Mohns Ridge, Norwegian Sea. *Tectonics*, **15**, 870, http://dx.doi.org/10.1029/95TC03682

DEL VENTISETTE, C., MONTANARI, D., SANI, F. & BONINI, M. 2006. Basin inversion and fault reactivation in laboratory experiments. *Journal of Structural*

Geology, **28**, 2067–2083, http://dx.doi.org/10.1016/j.jsg.2006.07.012

DUBOIS, A., ODONNE, F., MASSONNAT, G., LEBOURG, T. & FABRE, R. 2002. Analogue modelling of fault reactivation: tectonic inversion and oblique remobilisation of grabens. *Journal of Structural Geology*, **24**, 1741–1752, http://dx.doi.org/10.1016/S0191-8141(01)00129-8

ELLIOTT, D. 1976. The motion of thrust sheets. *Journal of Geophysical Research*, **81**, 949, http://dx.doi.org/10.1029/JB081i005p00949

FACCENNA, C., NALPAS, T., BRUN, J. P., DAVY, P. & BOSI, V. 1995. The influence of pre-existing thrust faults on normal fault geometry in nature and in experiments. *Journal of Structural Geology*, **17**, 1139–1149, http://dx.doi.org/10.1016/0191-8141(95)00008-2

FARÍAS, M. 2007. *Tectónica y erosión en la evolución del relieve de los Andes de Chile Central durante el Neogeno*. PhD Thesis, Universidad de Chile, Santiago.

FARÍAS, M., COMTE, D. ET AL. 2010. Crustal-scale structural architecture in central Chile based on seismicity and surface geology: implications for Andean mountain building. *Tectonics*, **29**, TC3006, http://dx.doi.org/10.1029/2009TC002480

FOCK, A. 2005. *Cronología y tectónica de la exhumación en el Neógeno de los Andes de Chile central entre los 33° y los 34°S*. MSc (unpublished), Departamento de Geología, Universidad de Chile, Santiago.

FOCK, A., CHARRIER, R., FARÍAS, M., MAKSAEV, V., FANNING, C. M. & ALVAREZ, R. 2005. Deformation and uplift of the western Main Cordillera between 33° and 34°S. *In*: *Proceedings 6th International Symposium on Andean Geodynamics (ISAG)*, Barcelona, Editions IRD (Institut de recherche pour le développement), 273–276.

FOCK, A., CHARRIER, R., MAKSAEV, V. & FARIAS, M. 2006. Neogene exhumation and uplift of the Andean Main Cordillera from apatite fission tracks between 33°30′ and 34°00′ S. *In*: MENDOZA, (ed.) *Backbone of the Americas – Patagonia to Alaska. Geological Society of America Meeting*, Mendoza, 102.

GARTRELL, A., HUDSON, C. & EVANS, B. 2005. The influence of basement faults during extension and oblique inversion of the Makassar Straits rift system: insights from analog models. *American Association of Petroleum Geologists Bulletin*, **89**, 495–506, http://dx.doi.org/10.1306/12010404018

GHIGLIONE, M. C. & CRISTALLINI, E. O. 2007. Have the southernmost Andes been curved since Late Cretaceous time? An analog test for the Patagonian Orocline. *Geology*, **35**, 13, http://dx.doi.org/10.1130/G22770A.1

GIAMBIAGI, L. B. & RAMOS, V. A. 2002. Structural evolution of the Andes in a transitional zone between flat and normal subduction (33°30′–33°45′S), Argentina and Chile. *Journal of South American Earth Sciences*, **15**, 101–116, http://dx.doi.org/10.1016/S0895-9811(02)00008-1

GIAMBIAGI, L. B., ALVAREZ, P. P., GODOY, E. & RAMOS, V. A. 2003a. The control of pre-existing extensional structures on the evolution of the southern sector of the Aconcagua fold and thrust belt, southern Andes. *Tectonophysics*, **369**, 1–19, http://dx.doi.org/10.1016/S0040-1951(03)00171-9

GIAMBIAGI, L. B., RAMOS, V. A., GODOY, E., ALVAREZ, P. P. & ORTS, S. 2003b. Cenozoic deformation and tectonic style of the Andes, between 33° and 34° south latitude. *Tectonics*, **22**, http://dx.doi.org/10.1029/2001TC001354

GIAMBIAGI, L. B., ALVAREZ, P. P., BECHIS, F. & TUNIK, M. A. 2005. Influencia de las estructuras de rift triásico-jurásicas sobre el estilo de deformación en las fajas plegadas y corridas de Aconcagua y Malargüe, Mendoza. *Revista de la Asociación Geológica Argentina*, **60**, 662–671.

GIAMBIAGI, L. B., GHIGLIONE, M., CRISTALLINI, E. O. & BOTTESI, G. 2009a. Kinematic models of basement/cover interaction: insights from the Malargüe fold and thrust belt, Mendoza, Argentina. *Journal of Structural Geology*, **31**, 1443–1457, http://dx.doi.org/10.1016/j.jsg.2009.10.006

GIAMBIAGI, L., GHIGLIONE, M. C., CRISTALLINI, E. O. & BOTTESI, G. 2009b. Características estructurales del sector sur de la faja plegada y corrida de Malargüe (35°–36°s): distribución del acortamiento e influencia de estructuras previas. *Revista de la Asociación Geológica Argentina*, **65**, 140–153.

GILLCRIST, R., COWARD, M. P., TRUDGILL, B., PECHER, A. & MUGNIER, J. L. 1989. Structural inversion in the external French Alps. *In*: COOPER, M. A. & WILLIAMS, G. D. (eds) *Inversion Tectonics*. Geological Society, London, Special Publications, **44**, 354–354.

GODOY, E. 2011. Structural setting and diachronism in the Central Andean Eocene to Miocene volcano-tectonic basins. *In*: SALFITY, J. A. & MARQUILLAS, R. A. (eds) *Cenozoic Geology of the Central Andes of Argentina*. Instituto del Cenozoico, Universidad Nacional de Salta, Salta, 155–167.

GODOY, E. 2012. Sobre el variable marco geotectónico de las formarciones Abanico y Farellones y sus equivalentes al ser de los 35°LS. *Revista de la Asociación Geológica Argentina*, **69**, 570–577.

GONZÁLEZ, O. & VERGARA, M. 1962. Reconocimiento Geológico de la cordillera de los Andes entre los paralelos 35° y 38°S. Universidad de Chile, Instituto de Geología Publicacion **24**.

HUBBERT, M. K. 1937. Theory of scaled models as applied to the study of geological structures. *Geological Society of America Bulletin*, **48**, 1459–1519.

HUBBERT, M. K. 1951. The mechanical basis for certain familiar geologic structures. *Geological Society of America Bulletin*, **62**, 355.

HUYGHE, P. & MUGNIER, J.-L. 1992. The influence of depth on reactivation in normal faulting. *Journal of Structural Geology*, **14**, 991–998, http://dx.doi.org/10.1016/0191-8141(92)90030-Z

JACKSON, J. & MCKENZIE, D. 1983. The geometrical evolution of normal fault systems. *Journal of Structural Geology*, **5**, 471–482, http://dx.doi.org/10.1016/0191-8141(83)90053-6

JARA, P. & CHARRIER, R. In press. Nuevos antecedentes geocronológicos y estratigráficos para el Cenozoico de la Cordillera Principal de Chile entre 32° y 32°30 S. Implicancias paleogeográficas y estructurales. *Andean Geology*. Online.

JARA, P., PIQUER, J., PINTO, L. & ARRIAGADA, C. 2009. Perfiles estructurales de la Cordillera Principal de

Chile Central: resultados preliminares. *In*: *Congreso Geológico Chileno*, Santiago, no. 12, S9–038.

KAY, S. M. & ABBRUZZI, J. M. 1996. Magmatic evidence for Neogene lithospheric evolution of the central Andean 'flat-slab' between 30°S and 32°S. *Tectonophysics*, **259**, 15–28, http://dx.doi.org/10.1016/0040-1951(96)00032-7

KAY, S. M., MAKSAEV, V., MOSCOSO, R., MPODOZIS, C., NASI, C. & GORDILLO, C. E. 1988. Tertiary Andean magmatism in Chile and Argentina between 28°S and 33°S: correlation of magmatic chemistry with a changing Benioff zone. *Journal of South American Earth Sciences*, **1**, 21–38, http://dx.doi.org/10.1016/0895-9811(88)90013-2

KAY, S. M., MPODOZIS, C., RAMOS, V. A. & MUNIZAGA, F. 1991. Magma source variations for mid–late Tertiary magmatic rocks associated with a shallowing subduction zone and a thickening crust in the central Andes (28 to 33 S). *In*: HARMON, R. S. & RAPELA, C. W. (eds) *Andean Magmatism and its Tectonic Setting*. Geological Society of America, Boulder, 113–138, http://dx.doi.org/10.1130/SPE265-p113

KELLER, J. V. A. & MCCLAY, K. R. 1995. 3D sandbox models of positive inversion. *In*: BUCHANAN, J. G. B. & BUCHANAN, P. G. (eds) *Basin Inversion*. Geological Society, London, Special Publications, **88**, 137–146.

KLOHN, C. 1960. Geología de la Cordillera de los Andes de Chile Central, provincias de Santiago, O'Higgins, Colchagua y Curicó. *Instituto de Investigaciones Geológicas*, **8**, 95.

KNOTT, S. D., BEACH, A., WELBON, A. I. & BROCKBANK, P. J. 1995. Basin inversion in the Gulf of Suez: implications for exploration and development in failed rifts. *In*: BUCHANAN, J. G. B. & BUCHANAN, P. G. (eds) *Basin Inversion*. Geological Society, London, Special Publications, **88**, 59–81.

KONSTANTINOVSKAYA, E. A., HARRIS, L. B., POULIN, J. & IVANOV, G. M. 2007. Transfer zones and fault reactivation in inverted rift basins: insights from physical modelling. *Tectonophysics*, **441**, 1–26, http://dx.doi.org/10.1016/j.tecto.2007.06.002

KOOPMAN, A., SPEKSNIJDER, A. & HORSFIELD, W. T. 1987. Sandbox model studies of inversion tectonics. *Tectonophysics*, **137**, 379–388, http://dx.doi.org/10.1016/0040-1951(87)90329-5

KRANTZ, R. W. 1991. Measurement of friction coefficients and cohesion for faulting and fault reactivation in laboratory models using sand and sand mixtures. *In*: COBBOLD, P. R. (ed.) *Experimental and Numerical Modelling of Continental Deformation*. Tectonophysics, **188**, 203–207.

LIKERMAN, J., BURLANDO, J. F., CRISTALLINI, E. O. & GHIGLIONE, M. C. 2013. Along-strike structural variations in the Southern Patagonian Andes: insights from physical modeling. *Tectonophysics*, **590**, 106–120, http://dx.doi.org/10.1016/j.tecto.2013.01.018

LITVAK, V. D. 2009. El volcanismo Oligoceno superior–Mioceno inferior del Grupo Doña Ana en la Alta Cordillera de San Juan. *Revista de la Asociación Geológica Argentina*, **64**, 201–213.

MACEDA, R. & FIGUEROA, D. 1995. Inversion of the Mesozoic Neuquén Rift in the Malargüe fold and thrust belt, Mendoza, Argentina. *In*: TANKARD, A., SUAREZ, R. & WELSINK, H. (eds) *Petroleum Basins of South America*. American Association of Petroleum Geologists, **62**, 369–382.

MACEDO, J. & MARSHAK, S. 1999. Controls on the geometry of fold–thrust belt salients. *Geological Society of America Bulletin*, **111**, 1808–1822, http://dx.doi.org/10.1130/0016-7606(1999)111<1808:COTGOF>2.3.CO;2

MAKSAEV, V., MOSCOSO, R., MPODOZIS, C. & NASI, C. 1984. Las unidades volcánicas y plutónicas del Cenozoico superior en la alta cordillera del Norte Chico (29°–31° S): geología, alteración hidrotermal y mineralización. *Revista Geológica de Chile*, **21**, 11–51.

MANDAL, N. & CHATTOPADHYAY, A. 1995. Modes of reverse reactivation of domino-type normal faults: experimental and theoretical approach. *Journal of Structural Geology*, **17**, 1151–1163, http://dx.doi.org/10.1016/0191-8141(95)00015-6

MARSHAK, S. 2004. Salients, recesses, arcs, oroclines, and syntaxes – a review of ideas concerning the formation of map-view curves in fold–thrust belts. *In*: MCCLAY, K. R. (ed.) *Thrust Tectonics and Hydrocarbon Systems*. American Association of Petroleum Geologists, Memoirs, **82**, 131–156.

MARTIN, M., CLAVERO, J., MPODOZIS, C. & CUITIÑO, L. 1995. *Estudio geológico de la Franja El Indio, Cordillera de Coquimbo*. Servicio Nacional de Geología y Minería, Santiago, Informe Registrado IR-95-6, **1**, 1–238.

MARTIN, M., KATO, T. & CAMPOS, A. 1997. Stratigraphic, structural, metamorphic and timing constraints for the assembly of late Palaeozoic to Triassic rocks in the lake-district, Chile (40 S). *Proceedings VIII Congreso Geológico Chileno*, Antofagasta, 154–158.

MCCLAY, K. R. 1989. Analogue models of inversion tectonics. *In*: COOPER, M. A. & WILLIAMS, G. D. (eds) *Inversion Tectonics*. Geological Society, London, Special Publications, **44**, 41–59.

MCCLAY, K. & BUCHANAN, P. G. 1992. Thrust faults in inverted extensional basins. *In*: MCCLAY, K. R. (ed.) *Thrust Tectonics*. Chapman and Hall, London, 93–104.

MCCLAY, K. R., INSLEY, M. W. & ANDERTON, R. 1989. Inversion of the Kechika Trough, Northeastern British Columbia, Canada. *In*: COOPER, M. A. & WILLIAMS, G. D. (eds) *Inversion Tectonics*. Geological Society, London, Special Publications, **44**, 235–257.

MITRA, S. 1993. Geometry and kinematic evolution of inversion structures. *American Association of Petroleum Geologists Bulletin*, **77**, 1159–1191.

MITRA, S. & ISLAM, Q. 1994. Experimental (clay) models of inversion structures. *Tectonophysics*, **230**, 211–222, http://dx.doi.org/10.1016/0040-1951(94)90136-8

MOSQUERA, A. 1990. *Estudio geológico del extremo sur de la cordillera del Medio y Valle del río Mercedario, provincia de San Juan*. MSc (unpublished), Universidad de Buenos Aires, Departamento de Geología, Buenos Aires.

MPODOZIS, C. & CORNEJO, P. 1988. *Hoja Pisco Elqui, Región de Coquimbo*, scale 1:250000. Servicio Nacional de Geología y Minería, Chile.

MPODOZIS, C., BROCKWAY, H., MARQUARDT, C. & PERELLÓ, J. 2009. Geocronología U/Pb y tectónica de la región Los Pelambres-Cerro Mercedario: implicancias para la evolución cenozoica de los Andes del centro de Chile y Argentina. In: XII Congreso Geológico Chileno, Santiago de Chile.

Muñoz, C., Pinto, L., Charrier, R. & Nalpas, T. In press. Miocene Abanico Basin inversion, Central Chile (33°–35°S): the importance of volcanic load and shortcut faults. Andean Geology.

Nalpas, T. & Brun, J. P. 1993. Salt flow and diapirism related to extension at crustal scale. Tectonophysics, 228, 349–362, http://dx.doi.org/10.1016/0040-1951(93)90348-N

Nalpas, T., Le Douaran, S., Brun, J. P., Unternehr, P. & Richert, J. P. 1995. Inversion of the broad Fourteens Basin (offshore Netherlands), a small-scale model investigation. Sedimentary Geology, 95, 237–250, http://dx.doi.org/10.1016/0037-0738(94)00113-9

Panien, M., Schreurs, G. & Pfiffner, A. 2005. Sandbox experiments on basin inversion: testing the influence of basin orientation and basin fill. Journal of Structural Geology, 27, 433–445, http://dx.doi.org/10.1016/j.jsg.2004.11.001

Pérez, D. J. 1995. Estudio geológico del cordón del Espinacito y regiones adyacentes, provincia de San Juan. PhD thesis, Universidad de Buenos Aires.

Pérez, D. J. 1996. Los depósitos sinorogénicos. In: Ramos, V. A. (ed.) Geología de la región del Aconcagua, provincias de San Juan y Mendoza. Subsecretaría de Minería de la Nación, Dirección Nacional del Servicio Geológico, Buenos Aires, 317–341.

Pinto, L., Muñoz, C., Nalpas, T. & Charrier, R. 2010. Role of sedimentation during basin inversion in analogue modelling. Journal of Structural Geology, 32, 554–565, http://dx.doi.org/10.1016/j.jsg.2010.03.001

Ramberg, H. 1981. Gravity, Deformation and the Earth's Crust: In Theory, Experiments and Geological Application. Academic Press, London/New York.

Ramos, V. A. 2006. Overview of the tectonic evolution of the southern Central Andes of Mendoza and Neuquén (35°–39° S latitude). In: Kay, S. M. & Ramos, V. A. (eds) Evolution of an Andean Margin: A Tectonic and Magmatic View from the Andes to the Neuquén Basin (35°–39°S latitude). Geological Society of America, USA, Special Papers, 1–17, http://dx.doi.org/10.1130/2006.2407

Ramos, V. A., Cegarra, M. & Cristallini, E. O. 1996. Cenozoic tectonics of the high Andes of west–central Argentina (30–36 S latitude). Tectonophysics, 259, 185–200.

Ranalli, G. 2000. Rheology of the crust and its role in tectonic reactivation. Journal of Geodynamics, 30, 3–15, http://dx.doi.org/10.1016/S0264-3707(99)00024-1

Rauld, R. 2002. Análisis morfoestructural del frente cordillerano: Santiago oriente entre el río Mapocho y Quebrada de Macul. Thesis, Departamento de Geología., Universidad de Chile, Santiago.

Rauld, R. 2011. Deformación cortical y peligro sísmico asociado a la falla San Ramón en el frente cordillerano de Santiago, Chile Central (33°S). PhD thesis, Universidad de Chile, Santiago.

Reiter, K., Kukowski, N. & Ratschbacher, L. 2011. The interaction of two indenters in analogue experiments and implications for curved fold-and-thrust belts. Earth and Planetary Science Letters, 302, 132–146, http://dx.doi.org/10.1016/j.epsl.2010.12.002

Rivera, O. & Yañez, G. 2007. Geotectonic evolution of the central Chile Oligo-Miocene volcanic arc, 33–34°S: Towards a multidisciplinary re-interpretation of the inherited lithospheric structures. In: GeoSur, Abstracts. GeoSur, Santiago, Chile, 138.

Sandiford, M., Hansen, D. L. & McLaren, S. N. 2006. Lower crustal rheological expression in inverted basins. In: Buiter, S. & Schreurs, G. (eds) Analogue and Numerical Modelling of Crustal Scale Processes. Geological Society, London, Special Publications, 253, 271–283.

Scisciani, V. 2009. Styles of positive inversion tectonics in the Central Apennines and in the Adriatic foreland: implications for the evolution of the Apennine chain (Italy). Journal of Structural Geology, 31, 1276–1294, http://dx.doi.org/10.1016/j.jsg.2009.02.004

Segerstrom, K. 1959. Cuadrángulo Los Loros: provincia de Atacama. Instituto de Investigaciones Geológicas, Santiago, Carta 1, 33.

Sibson, R. H. 1985. A note on fault reactivation. Journal of Structural Geology, 7, 751–754, http://dx.doi.org/10.1016/0191-8141(85)90150-6

Soto, R., Casas, A. M., Storti, F. & Faccenna, C. 2002. Role of lateral thickness variations on the development of oblique structures at the western end of the South Pyrenean Central Unit. Tectonophysics, 350, 215–235, http://dx.doi.org/10.1016/S0040-1951(02)00116-6

Soto, R., Casas-Sainz, A. M. & Pueyo, E. L. 2006. Along-strike variation of orogenic wedges associated with vertical axis rotations. Journal of Geophysical Research, 111, B10402, http://dx.doi.org/10.1029/2005JB004201

Thiele, R. 1980. Hoja Santiago. Región Metropolitana. Carta Geológica de Chile, scale 1:250.000. Instituto de Investigaciones Geológicas, Santiago, 29, 21.

Thomas, W. 1990. Controls on locations of transverse zones in thrust belts. Ecologae Geologicae Helvetiae, 83, 727–744.

Turner, J. P. & Williams, G. 2004. Sedimentary basin inversion and intra-plate shortening. Earth-Science Reviews, 65, 277–304, http://dx.doi.org/10.1016/j.earscirev.2003.10.002

Van Keken, P. E., Spiers, C. J., Van den Berg, A. P. & Muyzert, E. J. 1993. The effective viscosity of rocksalt: implementation of steady-state creep laws in numerical models of salt diapirism. Tectonophysics, 225, 457–476, http://dx.doi.org/10.1016/0040-1951(93)90310-G

Vergara, M. & Drake, R. E. 1979. Eventos magmáticos–plutónicos en Los Andes de Chile Central. In: II Congreso Geológico Chileno. 2., Arica, 6-11 Agosto, F19–F30.

Vergés, J. & Muñoz, J. A. 1990. Thrust sequences in the southern central Pyrenees. Bulletin de la Société Géologique de Franc, VI, 265–271.

Weijermars, R. 1986. Finite strain of laminar flows can be visualized in SGM36-polymer. Naturwissenschaften, 73, 33–34, http://dx.doi.org/10.1007/BF01168803

Weijermars, R. & Schmeling, H. 1986. Scaling of Newtonian and non-Newtonian fluid dynamics without inertia for quantitative modelling of rock flow due to gravity (including the concept of rheological similarity). *Physics of the Earth and Planetary Interiors*, **43**, 316–330, http://dx.doi.org/10.1016/0031-9201(86)90021-X

WEIJERMARS, R., JACKSON, M. P. A. & VENDEVILLE, B. 1993. Rheological and tectonic modeling of salt provinces. *Tectonophysics*, **217**, 143–174, http://dx.doi.org/10.1016/0040-1951(93)90208-2

WINOCUR, D. 2010. *Geología y estructura del Valle del Cura y el sector central del Norte Chico, provincia de San Juan y IV Región de Coquimbo, Argentina y Chile*. PhD thesis, Universidad de Buenos Aires.

WINOCUR, D. & RAMOS, V. A. 2008. Geología y Estructura del sector norte de la Alta Cordillera de la provincia de San Juan. *In*: *XVII Congreso Geológico Argentino*, Jujuy, 166–167.

WINOCUR, D. & RAMOS, V. A. 2012. Oligocene extensional tectonics at the Main Andes, Valle del Cura Basin, San Juan Province, Argentina. *In*: *XIII Congreso Geológico Chileno*, Antofagasta, 253–255.

WINOCUR, D. A., LITVAK, V. Y. & RAMOS, V. A. 2014. Magmatic and tectonic evolution of the Oligocene Valle del Cura basin, main Andes of Argentina and Chile: evidence for generalized extension. *In*: SEPÚLVEDA, S., GIAMBIAGI, L. B., MOREIRAS, S. M., PINTO, L., TUNIK, M., HOKE, G. D. & FARÍAS, M. (eds) *Geodynamic Processes in the Andes of Central Chile and Argentina*, Geological Society, London, Special Publications, **399**. First published online February 17, 2014, http://dx.doi.org/10.1144/SP399.2

YAGUPSKY, D. L., CRISTALLINI, E. O., FANTIN, J., VALCARCE, G., BOTTESI, G. & VARADE, R. 2008. Oblique half-graben inversion of the Mesozoic Neuquén Rift in the Malargüe fold and thrust belt, Mendoza, Argentina: new insights from analogue models. *Journal of Structural Geology*, **30**, 839–853, http://dx.doi.org/10.1016/j.jsg.2008.03.007

YAMADA, Y. & MCCLAY, K. R. 2003. Application of geometric models to inverted listric fault systems in sandbox experiments. Paper 2: insights for possible along strike migration of material during 3D hanging wall deformation. *Journal of Structural Geology*, **25**, 1331–1336, http://dx.doi.org/10.1016/S0191-8141(02)00160-8

ZAPATA, T. R. 1990. *Estudio geológico de la cordillera Casa de Piedra y del cordón Valle Hermoso*. Unpublished, Master Thesis. Departamento de Geología, Universidad de Buenos Aires.

Magmatic and tectonic evolution of the Oligocene Valle del Cura basin, main Andes of Argentina and Chile: evidence for generalized extension

D. A. WINOCUR*, V. D. LITVAK & V. A. RAMOS

Laboratorio de Tectonica Andina, IDEAN, Instituto de Estudios Andinos Don Pablo Groeber (UBA-CONICET), Departamento de Ciencias Geológicas, FCEyN, Pabellón II, Ciudad Universitaria, Bs, As, Argentina

*Corresponding author (e-mail: Winocur@gl.fcen.uba.ar)

Abstract: The magmatic history and tectonic evolution of the Valle del Cura region has received the attention of several studies in recent years, particularly as part of a larger area of interest named the Indio Belt. These studies have suggested an Eocene volcanic sequence known as the Valle del Cura Formation. The present study, based on extensive field work, robust geochronological and geochemical datasets, shows an Oligocene to early Miocene age for this unit, similar to the Doña Ana Group. The tectonic setting that controlled the volcanism of the Valle del Cura Formation was extensional and corresponds to a retro-arc position of the main arc volcanism of the Doña Ana Group. The field evidence combined with radiometric and geochemical data demonstrate the synextensional characteristic of the volcanic sequence of the Valle del Cura Formation and the Doña Ana Group at these latitudes. This characteristic was dominant at the central part of the Pampean Flat Slab ($29°-30°S$).

The Argentine Frontal Cordillera between $29°-30°S$ and $60°30'-70°00'W$ is located in the middle of the Pampean Flat Slab subduction zone of the Central Andes (Fig. 1). The structure presents great complexity due to changes in the stress regime that have occurred during the Andean evolution, which are partially controlled by pre-Andean basement fabrics (see Ramos *et al.* 2002). The dominant structural style is thick-skinned due to tectonic inversion of previous extensional faults. Thick volcanic sequences of Tertiary age show a complex magmatic history in the study area.

The main objective of this paper is to present a new Oligocene stratigraphy of the region, which improves the understanding of the tectonic evolution and is based on new geochemical, radiometric and geological data of the magmatic rocks and the new structural data obtained by field mapping of the area.

Different magmatic stratigraphies have been proposed by many authors, such as Thiele (1964), who named the Doña Ana Formation and defined it as a series of andesitic and basaltic sequences exposed in the Chilean slopes of the Cordillera.

Maksaev *et al.* (1984) divided the Doña Ana Formation into the Tilito and Escabroso members with ages between 27 and 18.9 Ma. Later, Martin *et al.* (1995) observed an angular unconformity and proposed that the Doña Ana Formation be classified as a group consisting of the Tilito and Escabroso Formations. Nullo (1988), Marín & Nullo (1989) and Ramos *et al.* (1990) recognized similar units in the region of the Cerro de las Tortolas and La Ortiga on the Argentinian side. These authors described the volcaniclastic Río La Sal Formation, which they correlated with a similar sequence on the Chilean side that had been recognized by Reutter (1974). Limarino *et al.* (1999) identified volcanic successions that they grouped into the Eocene Valle del Cura Formation. These volcanic and volcaniclastic deposits are distributed along the foreland side of the main Andes (Litvak & Poma 2005). This study gives new insight into the age of these volcanic sequences and their tectonic setting.

Stratigraphy

The Cenozoic stratigraphy in the Valle del Cura region was discussed by Nullo (1988), Marín & Nullo (1989), Ramos *et al.* (1989), Kay *et al.* (1991, 1999), Limarino *et al.* (1999), Litvak *et al.* (2007) and Litvak & Poma (2005), with several of these authors proposing a Tertiary stratigraphy. However, the superposition of volcanic processes coupled with widespread hydrothermal activity and structural complexity have complicated the understanding of the stratigraphy and structure of the region.

In this study, we carried out a survey of radiometric ages in both the published literature and

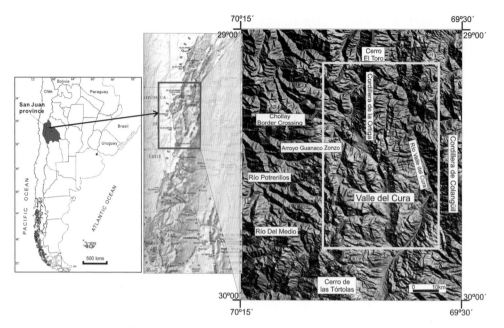

Fig. 1. Location of the Valle del Cura region in the Central Andes of Argentina and Chile.

unpublished reports and obtained new Ar/Ar ages in order to evaluate the stratigraphic setting and regional geological context of these units.

This evaluation pointed to a series of inconsistencies and uncertainties in the ages, which required re-examination in the field. Particularly, many K–Ar ages used to establish the volcanic succession may not be reliable crystallization ages as they were obtained on hydrothermally altered rocks. These ages are also difficult to use in assigning cooling ages to the volcanic sequence, as they may represent different posthumous alteration episodes. In some cases these ages were reanalyzed in recent studies with robust methods like Ar/Ar or U–Pb, the conclusions being that they are not as reliable as crystallization ages as they were obtained on hydrothermally altered rocks.

As noted above, the Oligocene Doña Ana Formation defined by Thiele (1964) was subdivided by Maksaev et al. (1984) and Martin et al. (1995) into two units on the basis of K–Ar ages and fieldwork in the High Cordillera of Chile. The Tilito Formation consists of rhyolitic ignimbrite tuffs and lavas interbedded with dacitic flows with ages from 27 to 22.1 Ma. The upper Escabroso Formation is composed of andesitic to basaltic lavas, volcanic agglomerates and breccias with ages between 26 and 18.9 Ma. Martin et al. (1995) suggested an angular unconformity between them and separated the Tilito Formation (27–22.1 Ma) from the Escabroso Formation (21–17 Ma) and suggested a hiatus of 1–3 million years.

Ramos et al. (1987) were the first to describe these units on the Argentinian side. They recognized outcrops of basaltic necks, grouped in the Las Máquinas basalts with K–Ar ages of 22.8 ± 1.1 Ma (Kay et al. 1991) related to a back-arc activity co-eval with the main volcanic arc of the Doña Ana Group. Nullo (1988) and Marín & Nullo (1989) recognized the Doña Ana Formation on the Argentinian side further north along the Cordillera de la Ortiga.

Bissig et al. (2001) dated the volcanic rocks of the Doña Ana Group using the Ar/Ar method. The range of volcanic activity for the Tilito Formation was constrained between 25.1 ± 0.4 and 23 Ma and for the Escabroso Formation between 21.9 ± 0.9 and 17.6 ± 0.5 Ma. Charchaflié et al. (2007) also reported Ar/Ar dating, which confirmed existing ages and more recently Litvak (2009) provided two new K–Ar and Ar/Ar Oligocene dates in the Cordillera de Zancarrón of the Valle del Cura, and assigned them to the Tilito Formation.

Against this background we present an integrated geological map of the Valle del Cura region, shown in Figure 2a, with the SE portion enlarged in Figure 2b. Here, the study area has been divided into an eastern and western sector to better define the Cenozoic volcanic stratigraphy. Some units are represented in both sectors, while others are restricted to one. This division is critical to understanding the evolution of the region as shown in the revised stratigraphic column in Figure 3.

Oligocene–early Miocene sedimentation

Río La Sal Formation

This oldest unit of the region was defined by Reutter (1974) as a series of red volcanic breccias, conglomerates, sandstones, gypsum and limestones that crop out along the Río de la Sal valley in the Chilean sector. Nullo & Marín (1990) identified the main outcrops of this sequence in the Cordillera de la Ortiga and along the Río de la Sal valley in Argentina. These authors included coarse to fine red sandstones and conglomerates that are 70–150 m thick in this unit. Malizia et al. (1997) recognized additional outcrops and noted that the Río La Sal Formation was unconformably deposited over the Late Palaeozoic to Triassic volcanic Choiyoi Group. This Group was dated by Bissig et al. (2001) in the Valle del Cura between 261 and 259 Ma.

The age of the Río La Sal Formation has been a topic of discussion. Reutter (1974) assigned it to the Oligocene–Miocene interval based on correlation with the Calchaquense beds, which implies Oligocene ages, whereas Nullo & Marín (1990) suggested a late Miocene age. Malizia et al. (1997) suggested a Cretaceous to Palaeogene age based on stratigraphic relations.

The Río La Sal Formation is composed of breccias, conglomerates, sandstones, mudstones and dacitic tuffs, red and brown in colour. It has a thickness of up to 300 m and interbedded in its sediments are finely laminated dacitic tuffs up to 60 m thick. A sample was obtained from a tuffaceous layer interbedded with the red sandstones in the Cordillera de la Ortiga, 4 km to the SE of the Cerro del Toro (Fig. 4). This yellow-brown volcanic rock (sample 39612) has a very fine texture visible to the naked eye. Under the microscope fresh biotite crystals were observed that were used for mineral dating. The matrix is quite heterogeneous, but generally very fine grained, consisting of plagioclase microliths and devitrified glass (Fig. 4a). The biotite (Fig. 4b), which was dated by the Ar/Ar method at the Geochronological Laboratory of Dalhousie University in Canada, yielded a plateau age of 23.2 ± 0.3 Ma, indicating a late Oligocene–early Miocene age (Fig. 4c and Table 1). Based on stratigraphic relationships and this biotite age, an interval between 24 and 22 Ma is proposed for this unit. The regional distribution and tectonic significance of this unit, and its correlation with the Doña Ana Group, are discussed below.

Oligocene–early Miocene volcanism

Valle del Cura Formation

Minera (1968) assigned the name Valle del Cura Formation to a series of tuffs, breccias, conglomerates and volcanic rocks that included the volcanic tuffs associated with multicoloured deposits of Cordillera de la Brea. Aparicio (1975) used the name Tobas Multicolores Formation, for the series of thick conglomerates, sandstones and tuffs that rests unconformably over the volcanic Choiyoi Group in the Cordillera de la Brea. In addition, Nullo & Marín (1990) proposed including some of these sequences in the La Ollita Formation, which were assigned to the Miocene. They recognized a lower member corresponding with the Tobas Multicolores of Aparicio (1975) and an upper member for the La Ollita Formation (sensu Minera 1968; Aparicio 1975).

Malizia et al. (1997) described these units as a sequence of red sandy conglomerates containing smaller proportions of shales associated with gypsum. Limarino et al. (1999) obtained the first K–Ar whole rock ages for these units. Two ages, of 44 ± 2 and 45 ± 2 Ma, came from mesosilicic lavas that crop out on the western slope of the Cordillera de la Brea. They also obtained an age of 36 ± 1 Ma from volcanic rocks that crop out in the Bañitos Valley, and a 34 ± 1 Ma age from a crystal tuff exposed in the Cordillera of Zancarrón. Based on these data, these authors assigned these sequences an age between the middle Eocene and the Eocene–Oligocene limit.

However, whole rock K–Ar dating may misrepresent ages according to Faure (1977). This author explains that argon diffuses within minerals easily, and moreover, is easily removed by leaching of rock; argon could therefore accumulate in anomalous quantities on the surface and be present in excess, thus leading to a rock sample being assigned an age older than it really is. In addition some rocks retain argon to a greater or lesser extent, resulting in uncertain dating using Ar/Ar dating. Some of these samples show high levels of atmospheric argon, between 38.3 and 79.1% (Limarino et al. 1999).

The new radiometric dates in this paper rule out the Eocene age for these units, an issue to be addressed in subsequent descriptions.

More comprehensive studies conducted by Litvak & Poma (2005) and Litvak et al. (2007), in the Valle del Cura Formation provide detailed stratigraphic profiles and geochemical analyses. The unit is shown to have a large lithological range with facies distributions that include mainly volcanic and sedimentary components. Litvak & Poma 2005 present a new stratigraphic division based on identification of separate facies in distinct geographical areas. As pointed out by Malizia et al. (1997) and Limarino et al. (1999), the Valle del Cura Formation is composed of pyroclastic, volcanic and volcaniclastic deposits that mainly occur in two sites along the Valle del Cura. No such deposits occur on the Chilean side, showing this unit is restricted

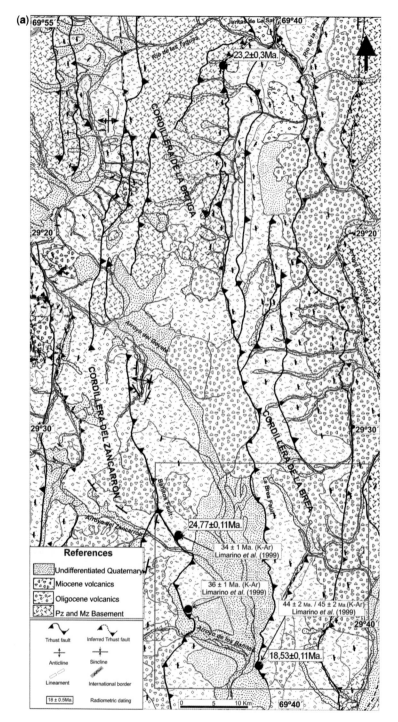

Fig. 2. (a) Geological map of the Valle del Cura Region. (b) Enlargement of selected area, showing a detailed geological map of the southern and eastern sectors of the Valle del Cura region, with K–Ar ages determined by Limarino *et al.* (1999), and Ar/Ar ages determined during this study.

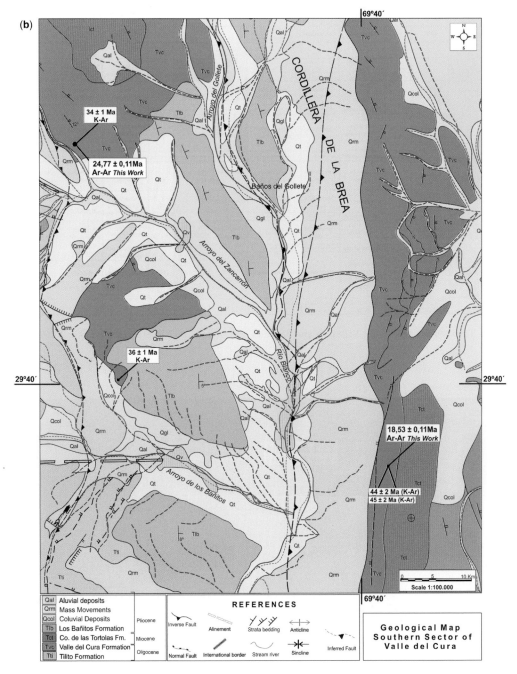

Fig. 2. *Continued*

exclusively to the Argentinian side. Two belts can be clearly distinguished. One, located in the eastern side to the south of the present study, occurs in the Cordillera de la Brea, where it extends north to the eastern slope of the Cordillera de la Ortiga, reaching the Cordillera de San Guillermo. This sector is composed mainly of volcaniclastic and clastic deposits interbedded with ash-fall tuffs. The second belt of outcrops is to the west along the Río Blanco valley. The deposits here are pyroclastic,

Tertiary Unified Stratigraphy at Valle del Cura and Norte Chico Region 29° - 30°S (ARGENTINA AND CHILE)

EPOCH	GROUP	WESTERN SECTOR UNIT	Ma	LITHOLOGY	DESCRIPTION	TECTONIC REGIME	EASTERN SECTOR UNIT West - East		Ma	LITHOLOGY	DESCRIPTION	TECTONIC REGIME
Pliocene		X	X	X	X	X	Cerro de Vidrio Fm.	X	2 Ma	Vitrophyry	Relict magmatism	Relict magmatism
		X	X	X	X	X	Los Bañitos Fm.		5 - 2 ?	Red conglomerates	Braided fluvial systems	Synorogenic compresional deposits
Miocene		Vallecito / Pascua Fms.	7 - 6 Ma	Dacitic Ignimbrites and dikes	Calcalkaline suites	Compressive magmatic arc	Vacas Heladas Ignimbrites	X	6 Ma	Dacitic Ignimbrites	Calcalkaline suites	Compressive magmatic arc
		~~Unconformity~~~~										
		Tambo Formation	13 - 11 Ma	Dacitic Ignimbrites, domes, dacites and hydrotermal breccias	Calcalkaline suites	Compressive magmatic arc	Tambo Formation		13 - 11 Ma	Dacitic Ignimbrites, domes and hydrotermal breccias	Calcalkaline suites	Compressive magmatic arc
		~~Unconformity~~~~										
		Cerro de las Tortolas Formation	16 - 11,5 Ma	Andesites-dacites	Calcalkaline suites	Compressive magmatic arc	Cerro de las Tortolas Formation		16 - 11,5 Ma	Andesites-dacites	Calcalkaline suites	Compressive magmatic arc
		Infiernillo Unit	17 - 14 Ma	Diorites	Calcalkaline suite	Compressive magmatic arc	Infiernillo U.	La Ollita Fm	18? - 16 Ma	Sandstones and gypsum facies	Foreland basin clastic sequences	Foreland synorogenic basin
		~~Unconformity~~~~										
		Escabroso Formation	21 - 17 Ma	Andesites	Calcalkaline suites	Transitional arc	Escabroso Formation	Valle del Cura Formation	26 - 18 Ma	Dacitic ignimbrites, tuff, volcanic sandstones and conglomerates	Pyroclastic, volcanoclastic and clastic deposits	Extensional back arc
Oligocene	Doña Ana Group	~~~~Unconformity~~~~										
		Tilito Formation	27 - 23 Ma	Dacitic and rhyolithic Ignimbrites, lava flows.	Calcalkaline suites	Extensional intra arc basin	Tilito Formation	Rio la Sal Fm.	24 - 22? Ma	Red conglomerates	Volcaniclastic deposits	Extensional back arc
								Las Máquinas Basalts	23 - 22 Ma	Olivinic basalts	Alkaline suites	Extensional back arc
Eocene		Bocatoma Unit	39-31 Ma	Granodiorites and diorites	Calcalkaline suites	Compressional Arc Granitoids	X	X	X	X	X	X

Fig. 3. Tertiary stratigraphy of the Valle del Cura (Argentina) and Norte Chico (Chile) regions (Winocur 2010).

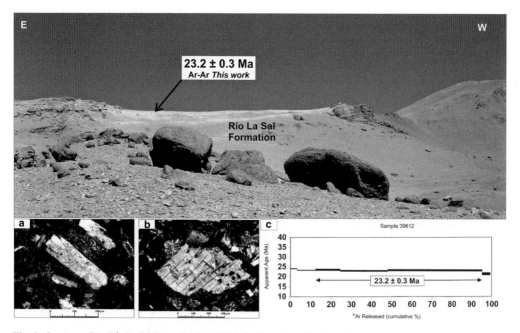

Fig. 4. Outcrops of the Río La Sal Formation interbedded with the dated dacitic tuff, where sample 39612 was obtained. Location is 5 km SW of Juntas de la Sal. (**a**) Crystals of plagioclase in the pyroclastic dacitic tuff of sample 39612. The matrix is formed by an extremely fine aggregate (c. 10 μm) of plagioclase and glass. Cross-polarized light. (**b**) Fresh biotite crystal of c. 0.5 mm was dated. (**c**) Age spectra for sample 39612 of the Río La Sal Formation. Note the plateau age of 23.2 + 0.3 Ma.

Table 1. Heating stages of Ar/Ar dating for sample 39612, 5 km SW of Juntas de la Sal

$T°$ (C)	% ^{39}Ar	Age (Ma)	Error (Ma)
Sample 39612			
600	0.6	5.8	6
700	2.5	17.7	6
800	9.1	21.1	2
900	12	23.6	0.4
1000	23.5	22.9	0.2
1100	47.2	23.3	0.2
1200	4	21.3	0.6
1300	0.4	0	9
1450	0.2	0	50
Plateau age (Ma)		23.2	0.3

with ignimbrites and volcaniclastic beds, as seen in the Los Bañitos and Zancarrón valleys (Fig. 5).

Limarino *et al.* (1999) and Malizia *et al.* (1997) identified four sections that characterize the main features of the unit. The environment is characterized by interlocking river systems and alluvial fans, some of which were drowned by ash-falls. These authors proposed that sequences west of the Río Blanco valley have experienced more pyroclastic flows and ash-fall tuffs due to the proximity to the magmatic vents. These deposits continue with a north trend along the arroyos Deidad, Zancarrón and Bañitos.

The deposits east of the Río Blanco valley have the same orientation, but have a mostly volcaniclastic composition with less pyroclastic material. All outcrops have ash-fall tuffs interbedded but to a lesser extent than the western sector. The main outcrops correspond to tuffaceous sandstone. The deposits of the Valle del Cura Formation in the Cordillera de la Brea continue downhill to the eastern slope of the arroyo de la Ortiga, as far as the Cordillera de San Guillermo.

Radiometric dating and correlations

As discussed above, previous K–Ar dating of the volcanic sequences in the Valle del Cura Formation indicated an Eocene age (Limarino *et al.* 1999). This age is consistent with that of an overlying Doña Ana Group between 27 and 17 Ma but is inconsistent with an Oligocene–early Miocene (24–22 Ma) age for the underlying Río La Sal Formation. Due to this inconsistency, two new Ar/Ar radiometric ages were obtained in the Sernageomin Laboratory, Santiago de Chile, one for each of the two belts of the Valle del Cura Formation.

For the first age determination, sample 07–170 was taken from the ignimbritic levels exposed in the arroyo Zancarrón valley at the same layer as the sample dated by Limarino *et al.* (1999) (Fig. 6a). The sample is from a 20 m thick volcanic agglomerate containing rounded clasts of andesitic composition. Under the microscope, sample 07–170 is seen to be a porphyry with phenocrysts of plagioclase up to 1.8 mm in size in an array of hyalopilitic texture composed of plagioclase in a glassy groundmass (Fig. 6b). The matrix of this sample yielded an Ar/Ar plateau age of 24.77 ± 0.11 Ma. (Fig. 6c and Table 2). The age of this unit ranges between the upper Oligocene and the middle Miocene, which is consistent with stratigraphic relationships and field observations.

For the second age determination, we analysed a sample from the Valle del Cura Formation at

Fig. 5. Pyroclastic flow with a dacitic lapilli tuff of the Valle del Cura Formation at arroyo de los Bañitos.

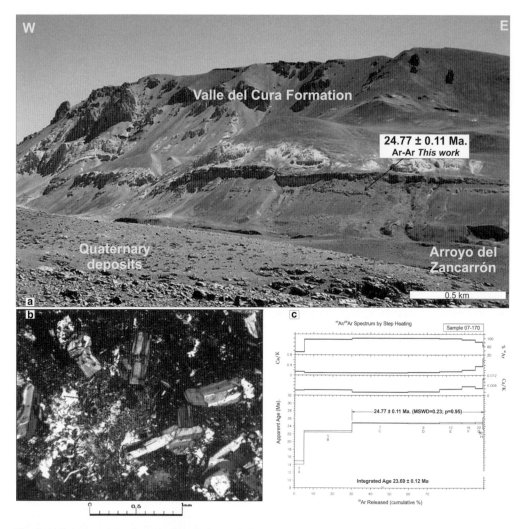

Fig. 6. (a) Location of dated sample (07-170) in the Valle del Cura Formation at arroyo del Zancarrón Valley. (b) Microphotograph showing plagioclase phenocrysts embedded in an intersertal matrix of sample 07-170. (c) Age spectra for sample 07-170 of the Valle del Cura Formation.

the Cordillera de la Brea, in the Río Blanco valley (sample DAW-01), where two Eocene K–Ar ages were obtained by Limarino et al. (1999). The sample comes from close to the base of the Cordillera de la Brea, and corresponds to the oldest deposit (Fig. 7a). It is a tuff of andesitic glass, and most of its components are plagioclase and biotite. The dating was performed by the Ar/Ar method by step-heating the biotite. We obtained a plateau age of 18.53 ± 0.11 Ma, which corresponds to the lower Miocene, and is significantly younger than the previously reported ages (Fig. 7b and Table 3). Litvak (2004) obtained a K–Ar age of 14.5 ± 0.9 Ma at the top of the Cordillera de la Brea, in the Cerro de las Tórtolas Formation, above the unconformity with the Valle del Cura Formation.

One of the samples dated in the study by Limarino et al. (1999) was obtained from a location 5 km to the south of sample 07–170 (Fig. 8), along arroyo de los Bañitos. Based on these new age estimates of samples taken from sites for which older K–Ar age measurements had been made, and the stratigraphic relationships of these sequences along the arroyos del Zancarrón and de los Bañitos valleys, we conclude that this is the same volcanic sequence that was dated in sample 07-170. Due to the possibility of there being an excess of argon in this sample, the Eocene age of 36 ± 1 Ma obtained from the K–Ar radiometric method should be discarded.

Table 2. *Heating stages for sample 07-170 from the Valle del Cura Formation*

Heating stages	%^{39}Ar released by stage	% cumulative	%^{40}Ar	Age (Ma)	Error (Ma)
Sample 07-170					
A	5.2	5.2	33.3	14.41	0.22
B	25.1	30.3	95.1	22.60	0.08
C	29.4	59.7	99.6	24.76	0.03
D	16.7	76.4	99.5	24.73	0.05
E	12.2	88.6	98.7	24.78	0.05
F	7.1	95.7	93.4	24.80	0.07
G	3.8	99.5	81.7	24.80	0.09
H	0.5	100	61.6	24.68	0.73
Integrated age (Ma)				23.69	0.12
Plateau age (Ma)				24.77	0.11

These two new radiometric dates with highly reliable results obtained from the Ar/Ar stepheating method and with excellent plateau ages with errors close to 100 000 years, lead us to reject the previous whole rock K–Ar ages for the Valle del Cura Formation. Based on a combination of these new age determinations and the stratigraphic relationships, we propose an age range of between 26 and 18 Ma for this unit.

Oligocene structural setting

The Valle del Cura structural style corresponds to a tectonic inversion of the previous normal faults. Some of these structures are partially inverted, some are fully inverted and others are preserved in their initial extensional state. These structures are principally north–south trending.

The photos in Figure 9 show several outcrops with extensional sets of faults. The faults in Figure 9a are located along the abandoned mining road at the southern margin of the arroyo Guanaco Zonzo. The dacitic tuff outcrops of the Tilito Formation can be seen to be affected by two normal north-trending faults with c. 1 m displacement (see hammer for scale). These volcanic sequences have been dated c. 2 km to the north where they yielded U–Pb zircon ages of 24.5 ± 0.2 Ma (Charchaflié et al. 2007). This outcrop shows similar extensional structures to those in Figure 9a, affecting a series of volcaniclastic deposits of the Tilito Formation. The strikes are similar to the faults in Figure 9a, but displacements are less than 1 m (Fig. 9b). This sequence also corresponds to the Tilito Formation, and was dated by Bissig et al. (2001) as 23.9 ± 0.3 Ma using the Ar/Ar method in biotite.

Similar extensional structures are seen in the co-eval Oligocene Río La Sal and Valle del Cura Formations. The Río La Sal Formation (24–22 Ma), located at the junction of the Río de la Sal and the Río de las Taguas, presents extensional growth strata in various sectors (Fig. 10).

A dacitic tuff intercalated in the clastic sequence has also been dated, yielding an age of 23.2 ± 0.3 Ma by the Ar/Ar method (sample 39612), showing that the fault was active during the Oligocene sedimentation. Details of the progressive unconformities associated with these deposits are observed along the east bank of Río de la Sal valley, where similar synextensional features are observed. The outcrops correspond to thick sequences of red volcaniclastic conglomerates and sandstones interbedded with dacitic tuffs. The field data show progressive unconformities between the top and the base of the outcrops with dips ranging from about 15 to 25 degrees in the Río La Sal Formation (Fig. 10).

Some Valle del Cura Formation outcrops at the Cordillera de la Ortiga are structurally controlled by normal faults. These volcaniclastic deposits, located on the eastern side of the region, are also Oligocene in age (Fig. 11) and synextensional (Fig. 12). The surveyed area shows many sectors with different units that exhibit robust evidence of synextensional sedimentation during the Tilito Formation deposition in Oligocene times.

A structural cross section and palinspastic restoration for Oligocene times are shown at 29°30′S in Figure 13. This section shows the structural style of the Valle del Cura region, dominated by tectonic inversion of previously normal faults. Some of these faults preserve the synextensional sedimentation.

The orogenic shortening of 7.25 km was calculated from an initial non-deformed length of 51.57 km and a final deformed length of 44.33 km. This represents 8 per cent of contraction, congruent with the structural style proposed. The inverted normal faults imply little shortening and great topographic relief.

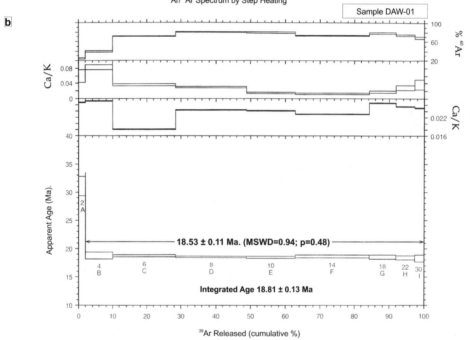

Fig. 7. (a) Location of dated sample (DAW-01) of the Valle del Cura Formation at Cordillera de la Brea, and the 14.5 ± 0.9 Ma rock dated by Litvak (2004). (b) Age spectra for sample DAW-01 of the Valle del Cura Formation.

Geochemistry of late Oligocene–lower Miocene volcanism

The Tertiary volcanic rocks of Valle del Cura were largely analysed in previous studies in the context of the geodynamic evolution of the Pampean Flat Slab segment (Kay et al. 1987, 1988, 1991, 1999; Bissig et al. 2003; Litvak et al. 2007). Overall, the chemical features for the late Oligocene–late Miocene Valle del Cura volcanic rocks indicate a calk-alkaline affinity and an arc-related behavior for the sequences, with changes in the residual mineral assemblages and variations in isotopic signatures that were related to the shallowing of the downgoing slab (Kay et al. 1991, 1999; Kay & Mpodozis 2002; Bissig et al. 2003; Litvak et al. 2007).

New data on major, trace and rare earth elements (REE) of the Oligocene–early Miocene magmatism are presented here: the two dated samples

Table 3. *Heating stages for sample DAW-01 from the Valle del Cura Formation*

Heating stages	% ^{39}Ar released by stage	% cumulative	% ^{40}Ar	Age (Ma)	Error (Ma)
Sample DAW-01					
A	2	2	23.3	31.03	0.86
B	8.1	10.2	38.7	18.72	0.31
C	18.4	28.6	71.2	18.69	0.09
D	20.3	48.9	79.5	18.5	0.06
E	14.2	63.1	78.5	18.44	0.1
F	21.2	84.3	72.5	18.58	0.1
G	7.7	91.9	76.3	18.47	0.18
H	5.5	97.5	72.9	18.31	0.17
I	7.7	100	67.6	18.26	0.31
Integrated age (Ma)				18.81	0.13
Plateau age (Ma)				18.53	0.11

from the Valle del Cura Formation and seven samples from the co-eval Escabroso and Tilito Formations, all of which are now included within the Doña Ana Group (Table 4). The main focus of the discussion here is the difference seen in chemical composition within the Doña Ana Group volcanism and, particularly, the review of geochemical features of the Valle del Cura Formation in the light of the revised stratigraphic position for this unit. Chemical analyses were done in the

Fig. 8. Location of the dated samples of the Valle del Cura Formation. Note the same volcanic sequence dated in this work as Oligocene, in contrast with the Eocene age previously determined by Limarino *et al.* (1999).

Fig. 9. Volcaniclastic sequences of the Tilito Formation affected by extensional faults: (**a**) at the arroyo Guanaco Zonzo; (**b**) at Río Valle del Cura. See location in Figure 12.

Fig. 10. View to the south of the Río de la Sal. Synextensional growth strata composed of red conglomerates interbedded with dacitic tuffs of the Río La Sal Formation. See location in Figure 12.

Sernageomin Laboratory in Chile with major elements analysed by X-ray fluorescence (XRF) and trace elements by inductively coupled plasma mass spectrometry (ICP-MS). New data were completed with a compilation of previous data from Oligocene–early Miocene magmatism, as reported by

Fig. 11. View to the north from Cordillera de la Ortiga. Synextensional growth strata composed of volcanic ash tuffs and volcaniclastic deposits in the Valle del Cura Formation. See location in Figure 12.

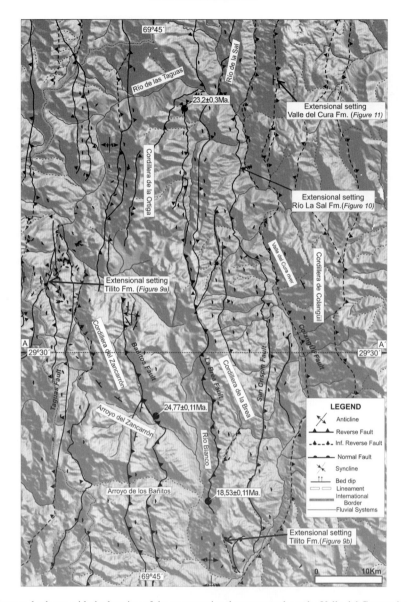

Fig. 12. Structural scheme with the location of the synextensional structures along the Valle del Cura region.

Kay et al. (1987, 1991, 1999), Bissig et al. (2003) and Litvak et al. (2007).

Major element classification, based on new and compiled data, is shown in Figure 14. New samples show a moderate grade of hydrothermal alteration, particularly evidenced by their moderate to high loss on ignition (LOI) content (c. 3–4%), so all of the data were recalculated on an anhydrous basis (Table 4). In a plot of wt% $Na_2O + K_2O$ v. SiO_2, all of these volcanic rocks fall in the sub-alkaline field of Irvine & Baragar (1971) (Fig. 14a).

The K_2O v. SiO_2 diagrams show the different compositions of both samples from the Valle del Cura Formation (Fig. 14b). One, sample 07-710, dated at 24.77 ± 0.11 Ma plots in the high-K dacitic and rhyolitic field, and correlates with early Miocene Tilito Formation samples. The younger sample from Valle del Cura Formation, DAW-01, dated at 18.53 ± 0.11 Ma, is classified as a basaltic andesite, consistent with the andesites and basaltic andesites of the younger Miocene Escabroso Formation. New samples from the Escabroso and Tilito

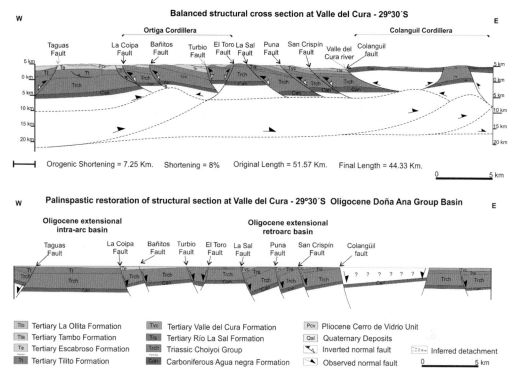

Fig. 13. Structural cross section and palinspastic restoration of the Valle del Cura region at 29°30′S. Note the extensional intra-arc basin of the Doña Ana Group and the extensional retro-arc composed of the Río La Sal and Valle del Cura Formations. See location in Figure 12.

Formations correlate with previous classification fields of the Doña Ana Group volcanic rocks; only one sample from the Tilito Formation is a silicified rhyolite with lower alkali concentrations (sample 07-209).

Mantle normalized trace elements and REE diagrams for new data of the Oligocene–early Miocene volcanic rocks are shown in Figure 15. Escabroso and Tilito Formation samples were plotted separately due to the difference in their silica content. Both samples of the Valle del Cura Formation were plotted within the Tilito and Escabroso diagrams for comparison purposes, according also to their silica content: the dacite (63.98% SiO_2) correlates with Tilito, while the andesitic tuff (52.06% SiO_2) with Escabroso. No significant difference is seen in the mantle normalized trace element pattern between the Tilito and Escabroso Formations with their co-eval Valle del Cura Formation. Despite the difference in silica content, all of the volcanic rocks of the Tilito Formation, as well as of the Escabroso Formation, show an arc-type trace element pattern, given by the enrichment in the large ion lithophile elements (LILE) relative to the high field strength elements (HFSE), which is also seen in the Valle del Cura Formation rocks (Fig. 15a, b).

New data were compared with compiled data of the Oligocene–early Miocene volcanism of the Valle del Cura region, according to their silica content (Fig. 15a,b). It is important to remark that no previous chemical data for andesitic samples assigned to the Valle del Cura Formation have been reported. New results are consistent with previously reported trace element patterns of the Doña Ana Group being indicative of arc mantle sources (Kay *et al.* 1987, 1988, 1991, 1999; Ramos *et al.* 1989; Otamendi *et al.* 1994; Bissig *et al.* 2003). In particular, Valle del Cura Formation samples reveal the same type of arc-related pattern when compare to previous data, as was originally reported by Litvak *et al.* (2007), who mentioned that these volcanic rocks show enrichment of the LILE relative to HFSE and REE, moderate negative Eu anomalies, and light and middle REE enrichment relative to heavy REE (Fig. 15a–d).

Some differences are seen in the dacite 07-170, from the Valle del Cura Formation, and the dacitic tuff 07.53, from the Tilito Formation, which show a positive Ta anomaly that differs

Table 4. Whole rock geochemical analyses*

Formation	Valle del Cura	Valle del Cura	Tilito	Tilito	Escabroso	Escabroso	Escabroso	Escabroso	Escabroso
Sample	07-170	DAW-01	07.53	07.209	07.179	RPW-8	RPW-10	07.178	07.195
SiO_2	63,98	52,06	69,37	77,89	63,65	57,62	57,87	63,5	53,76
TiO_2	0,18	0,5	0,33	0,29	0,61	0,95	1,08	0,59	1,13
Al_2O_3	14,55	20,98	17,88	9,84	14,88	16,85	17,95	16,02	18,18
Fe_2O_3	2,72	3,69	1,85	1,32	4,68	6,26	5,48	3,83	6,63
MnO	0,14	0,35	0,03	0,05	0,09	0,14	0,14	0,14	0,28
MgO	0,52	2,06	0,6	1,2	2,15	2,41	2,62	0,99	2,51
CaO	5,59	10,64	0,99	1,89	4,45	4,67	5,45	4,81	8,58
Na_2O	3,79	4,22	2,7	1,25	2,77	3,7	3,6	3,57	4
K_2O	3,58	1,25	3,92	1,48	3,04	3,01	2,45	2,84	0,56
P_2O_5	0,12	0,04	0,14	0,12	0,15	0,12	0,13	0,22	0,22
LOI	4,72	4,12	2,13	4,44	3,28	4,01	2,94	3,2	4,06
Total	99,89	99,91	99,94	99,77	99,75	99,74	99,71	99,71	99,91
La	52,30	17,40	43,9	27,6	34,7	29,5	29,2	32,1	15,6
Ce	99,90	32,80	86,8	56,9	73,8	59,7	60	64,4	35,1
Nd	45,20	13,60	40,3	25,5	34,4	29,1	29,9	29,4	20,1
Sm	7,37	2,25	6,68	4,09	5,76	5,25	5,39	5,06	4,21
Eu	1,150	1,340	1,22	0,658	1,05	1,33	1,38	1,17	1,24
Tb	0,930	0,277	0,869	0,523	0,74	0,693	0,721	0,62	0,603
Yb	3,44	1,17	3,27	1,95	2,59	2,27	2,32	2,09	1,87
Lu	0,545	0,179	0,52	0,298	0,383	0,338	0,346	0,321	0,276
Cs	3,08	6,76	4,33	9,23	8,18	11,2	0,67	4,87	1,98
U	2,97	3,65	4,07	1,87	6,58	2,17	2,06	3,63	0,363
Th	14,20	7,33	17,2	14,7	24,64	8,16	7,96	13,2	1,8
Hf	9,28	2,32	8,77	4,3	9,01	5,16	4,91	7,34	4,46
Ta	4,04	0,17	3,67	0,85	2,59	0,55	0,35	2,51	0,654
Y	32,2	10,6	34,4	20,2	26,3	26,8	27,7	22,6	20
Pr	10,90	3,47	9,65	6,23	8,34	6,89	7	7,02	4,53
Gd	6,21	1,86	5,91	3,28	5,03	4,56	4,82	4,29	3,94
Dy	5,49	1,69	5,05	3,12	4,34	4,16	4,33	3,66	3,54
Ho	1,12	0,37	1,12	0,669	0,905	0,888	0,895	0,769	0,743
Er	3,24	1,05	3,25	1,93	2,58	2,35	2,48	2,18	1,98
Tm	0,502	0,159	0,496	0,295	0,377	0,343	0,35	0,314	0,279
Nb	20,5	4,3	18,3	8,08	10,2	9,42	9,03	11,3	4,56
La/Sm	7,1	7,7	6,6	6,7	6,0	5,6	5,4	6,3	3,7
La/Yb	15,2	14,9	13,4	14,2	13,4	13,0	12,6	15,4	8,3
Sm/Yb	2,1	1,9	2,0	2,1	2,2	2,3	2,3	2,4	2,3
La/Ta	12,9	102,4	12,0	32,5	13,4	53,6	83,4	12,8	23,9
Th/U	4,8	2,0	4,2	7,9	3,7	3,8	3,9	3,6	5,0
Rock type	Dacite	Andesitic Tuff	Dacitic Tuff	Dacitic Tuff	Dacite	Andesite	Andesite	Andesite	Andesite
Coordinates	29°36' 08 69°47'23	29°42'29 69°42'37	29°25'15 69°54'43	29°24'41 69° 57'31	29° 26'51 69° 55'53	29° 25'56 69° 58'16	29° 25'34 69° 57'56	29°27'10 69°55'28	29°24'34 69°57'47
Location	Arroyo del Zancarrón	Cordillera de la Brea	Río de las Taguas	Vistas del Toro	Cerro Pupa	Arroyo Guanaco Zonzo	Arroyo Guanaco Zonzo	Arroyo Guanaco Zonzo	Río de las Taguas

*Major elements are normalized to 100% totals (LOI). Geochemical whole rock analyses were done by XRF and trace elements by alkaline fusion (ICP-MS). The data were recalculated on an anhydrous basis, particularly due to their high LOI content. Datum of coordinates is UTM-WGS1984.

from the other Tilito dacitic tuff, 07.209. The same is seen in the more silicic sample of the Escabroso Formation (sample 07.178). This could be related to the concentration of an accessory phase, such as apatite, which may contain Ta. This feature is also seen in the compiled data of Oligocene–early Miocene volcanism, so the overall pattern of the new data correlates with that of the compiled data (Fig. 15a, b).

One main difference in trace element content between Doña Ana Group volcanic rocks and the Valle del Cura Formation is seen in the Ba/Ta, Ba/La, and La/Ta ratios, especially between the Valle del Cura Formation samples and those of the Tilito Formation, which have the same silica content (Fig. 16a). As discussed by Litvak et al. (2007), Ba/Ta and La/Ta ratios for these units are mostly arc-like in that their Ba/Ta ratios exceed 450 and La/Ta ratios exceed 25, in contrast to the typical mid-ocean rich basalt (MORB) and ocean island basalt (OIB) rocks. The difference is that the Valle del Cura Formation dacites and rhyolites have higher La/Ta and relatively lower Ba/Ta ratios than the silicic Tilito samples.

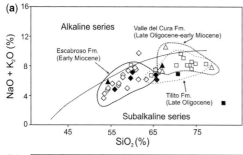

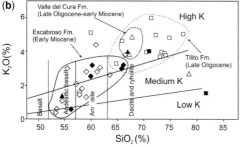

Fig. 14. Major element classification for late Oligocene–early Miocene volcanic rocks in the Valle del Cura area. Squares: Tilito Fm, diamonds: Escabroso Fm, triangles: Valle del Cura Fm, filled symbols: data from this study, unfilled symbols: data compiled by Bissig et al. (2003), Kay et al. (1987, 1991, 1999) and Litvak et al. (2007). (a) Na$_2$O + K$_2$O v. SiO$_2$ diagram with subalkaline field of Irvine & Baragar (1971). (b) K$_2$O v. SiO$_2$ with discrimination fields from Le Maitre et al. (1989).

Figure 15c, d shows REE patterns for samples whose analyses are in Table 4. These data are consistent with previous REE behaviour reported for the Oligocene–early Miocene volcanism of the Doña Ana Group and the Valle del Cura Formation (Litvak et al. 2007). Tilito, Escabroso and Valle del Cura Formation volcanic rocks show a smooth REE pattern (La/Yb c. 13–14) and a low heavy REE (HREE) slope (Sm/Yb c. 2). However, a difference is seen between the silicic and the mafic samples of the Valle del Cura Formation. The dacite shows a small negative Eu anomaly, which is consistent not only with the new and compiled data of Tilito Formation rocks, but also with previous data for the Valle del Cura Formation rocks; in particular, Litvak et al. (2007) reported that the silicic samples of the Valle del Cura Formation show the highest Eu anomalies relative to Tilito rocks. On the other hand, the andesitic tuff of the Valle del Cura Formation shows a positive Eu anomaly not seen in the other Escabroso samples (Fig. 15d); this could be a particular feature related to an accumulation of plagioclase during crystallization. This sample also shows a lesser degree of enrichment in REE when compared to the Escabroso Formation samples, particularly for medium to HREE.

As pointed out by several authors, Doña Ana magmatism has variably equilibrated with low to medium pressure residual mineral assemblages, including plagioclase and pyroxene, according to their REE patterns (Kay et al. 1987, 1988, 1991, 1999; Bissig et al. 2003; Litvak et al. 2007). Its co-eval Valle del Cura volcanism resembles that of the Doña Ana Group, in its low and constant HREE slope (Sm/Yb c. 2), along with higher concentrations of light and intermediate REEs with variable light REE (LREE) slopes.

Finally, the isotopic data of Oligocene–early Miocene volcanic rocks are shown in Figure 16b. In order to evaluate the isotopic behaviour, all of the available data for the Palaeogene to Neogene volcanism of the Valle del Cura region were plotted, included data for Paleocene alkali basalts (Kay et al. 1987, 1988, 1991, 1999; Kay & Abruzzi 1996; Bissig et al. 2003; Litvak et al. 2007; Litvak & Poma 2010).

The plot shows that the $^{87}Sr/^{86}Sr$ ratios for the Tertiary volcanic sequence range from 0.70367 to 0.70552 and the €Nd values range from −4 to 2, according to the compiled data. Paleocene and early Miocene basalts show an isotopically depleted nature, whereas the more enriched values of the silicic magmas are interpreted to reflect crustal contamination. The gradual increase in $^{87}Sr/^{86}Sr$ ratios and decrease in €Nd values from the Oligocene to late Miocene noted by Kay et al. (1991) and Kay & Abruzzi (1996) indicate an increase in the importance of a radiogenic crustal component with time. The Valle del Cura Formation sample from the arroyo del Zancarrón valley (which corresponds to the dated sample 07–170) has a $^{87}Sr/^{86}Sr$ ratio of 0.70533 and €Nd of −0.4 (reported by Litvak et al. 2007); data for this rock follow a Tertiary trend, but plot near data for middle to late Miocene volcanic rocks (see Litvak et al. 2007).

Discussion

Oligocene–early Miocene magmatism

Late Oligocene–early Miocene volcanism in the Valle del Cura area is grouped into three magmatic units within the Doña Ana Group, the Tilito, Escabroso and Valle del Cura Formations, while minor volcanic rocks are interbedded within the Río La Sal Formation. The geochemical analysis of Tilito, Escabroso and Valle del Cura rocks show that they share similar chemical features regarding their subalkaline tendency and arc-related signature. New samples from the Valle del Cura Formation can be correlated with Escabroso and Tilito, according to

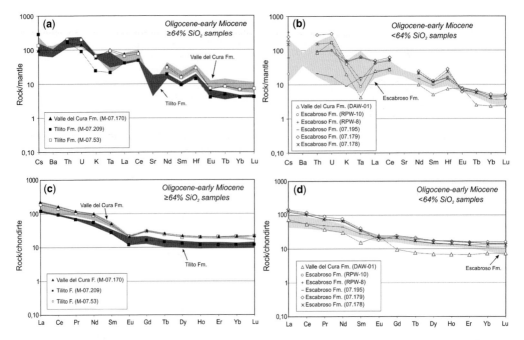

Fig. 15. (**a, b**) Mantle normalized diagrams for Tilito, Escabroso and Valle del Cura Formations rocks, according to their silica content (normalized values of McDonough & Sun 1995). (**c, d**) Chondrite REE normalized diagrams for Tilito, Escabroso and Valle del Cura Formations rocks (normalized chondrite values of Sun & McDonough 1989). Shadow patterns are data compiled by Bissig *et al.* (2003), Kay *et al.* (1987, 1988, 1991, 1999) and Litvak *et al.* (2007).

their age and chemical features: an older sample, of Oligocene age, is consistent with the age and a more intermediate-to-silicic composition of the Tilito Formation; a younger sample, of early Miocene age, is correlated with the timescale of the Escabroso Formation and its more andesitic-to-basaltic composition. Dacitic tuffs within the Río La Sal Formation represent syneruptive deposits related to the main volcanic events of the area, interbedded within a mainly epiclastic unit.

From a petrogenesis point of view, the late Oligocene–early Miocene Doña Ana magmas are the consequence of melts that originated in the mantle wedge and interacted in the crust (Kay *et al.* 1987, 1991). Amphibole and pyroxene are the principal mafic residual mineral phases that indicate magma equilibration in low-to-intermediate pressure conditions. Isotopic ratios of Nd and Sr show that these volcanic rocks were influenced by less radiogenic crustal components than the younger Miocene volcanic sequence.

In comparison, the Valle del Cura Formation rocks have low La/Yb and Sm/Yb ratios, along with their negative Eu anomalies, which provide evidence that they last equilibrated with a low pressure residual mineral assemblage. Evidence of a strong influence of a crustal component in their genesis is registered in their enriched isotopic signatures relative to Paleocene and early Miocene basalts, which do not follow the expected trend for Tertiary volcanism of the Valle del Cura area.

Originally, Litvak *et al.* (2007) and later on Alonso *et al.* (2011), who considered the Valle del Cura Formation to be of Eocene age, suggested that its location east of the Eocene arc-like Bocatoma stocks in Chile was consistent with its origin in a retro-arc foreland basin, east of the volcanic arc front in Chile. These authors proposed that Eocene Valle del Cura silicic rocks contain a large fraction of a melt from an arc-like crust, consistent with their high $^{87}Sr/^{86}Sr$ and low ϵNd ratios, as well as their eruption in a back-arc setting. In this context, magmatic activity in the retro-arc produced andesitic to rhyolitic lavas and tuffs represented by the Valle del Cura Formation, which developed in an extensional back-arc basin (Winocur & Ramos 2008, 2011). Thus, asthenospheric wedge-derived melts were able to ascend through a thin crust, where they were contaminated with crust-derived products. Petrogenesis and the chemical signature of this magmatic unit are independent of the entire and typical arc-type suite of volcanism in Valle del Cura area, of upper Oligocene–late Miocene ages, which follows a different trend. The co-eval

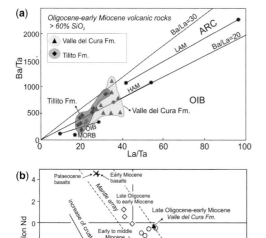

Fig. 16. (a) Ba/Ta v. La/Ta ratios for Valle del Cura and Tilito Formation rocks, data corresponds to Litvak *et al.* (2007); mid ocean ridge basalts (MORB), ocean island basalts (OIB), and high and low abundance (LAM and HAM) arc volcanic rocks for the SVZ defined by Hickey *et al.* (1986) are also included. (b) Isotopic data for the whole volcanic sequence from the Valle del Cura area. Isotopic evolution shows a trend of increasing crustal-derived components from late Oligocene to early Miocene; note that the Valle del Cura Formation shows a distinct isotopic behaviour relative to the rest of the Tertiary sequence (modified from Litvak *et al.* 2007).

late Oligocene–early Miocene Doña Ana volcanic rocks show a more arc-like chemical signature consistent with their evolution in a mildly extensional intra-arc setting.

Oligocene–early Miocene structural and tectonic setting

Based on an exhaustive examination of the outcrops corresponding to the Río La Sal and Valle del Cura Formations, combined with the petrographical features and new radiometric dates and geochemical data obtained, we propose a new stratigraphy for the region, shown in Figure 3. In this scheme, the Río La Sal Formation (24–22 Ma) which lies unconformably on the volcanic Choiyoi Group (Permian–Triassic) is covered by the Valle del Cura Formation (24.7–18.5 Ma). Both the Río La Sal and the Valle del Cura Formations are composed of volcanic and pyroclastic products, as well as clastic and volcaniclastic deposits and both are located east of the Oligocene magmatic intra-arc basin of the Doña Ana Group. The new Ar/Ar ages constrain both of these formations to be deposited between 24.7 and 18.5 Ma, co-eval with the magmatic activity in the intra-arc basin, which was active further to the west.

We thus propose that the Valle del Cura and Río La Sal Formations are part of the Doña Ana Group, taking into account the longitudinal and latitudinal extension of these units. The westernmost area is represented by the Tilito and Escabroso Formations and associated intrusive rocks. The retroarc includes the Las Máquinas basalts and the Río La Sal and Valle del Cura Formations (Fig. 17).

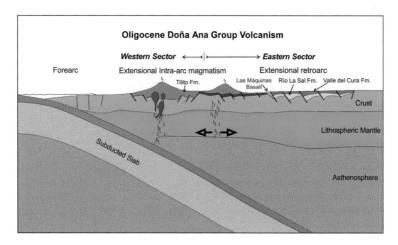

Fig. 17. Schematic tectonic setting for Oligocene times at 29° to 30°S. The Doña Ana Group corresponds to a magmatic arc dominated by extensional tectonics. The Oligocene Río La Sal and Valle del Cura Formations show evidence of extension in the retro-arc.

The observations presented in the study area, together with the new radiometric dating and geochemical information, show a distinctive tectonic evolution in this segment of the Andes (Fig. 17). The region during the Oligocene period was characterized by a generalized extension co-eval with an active calc-alkaline intra-arc volcanic basin in an extensional Andean-type subduction (Mpodozis & Ramos 1989; Ramos 2010).

A number of authors have proposed an extensional regime for the Abanico and Coya Machalí basins in regions further south during the Oligocene, for example, Kay et al. (1999), Godoy et al. (1999) and Charrier et al. (2002). However, at these latitudes Ramos et al. (1989) were the first who proposed the extensional regime for the Las Máquinas basalts in the retro-arc. Using structural arguments, Winocur & Ramos (2008, 2011) also proposed extension for arc and retro-arc regions at Oligocene times at these latitudes. This period was characterized by a generalized extension co-eval with an active calc-alkaline intra-arc volcanic basin in an extensional Andean type subduction (Winocur & Ramos 2008, 2011). In this context, dacitic to rhyolitic volcanic rocks of Valle del Cura Formation represent the retro-arc magmatic activity developed in this extensional basin.

The evidence described here indicates that an extensional regime was widespread, not only in the volcanic arc, but also in the adjacent foreland regions. The extension in the arc and the retro-arc is recognized in the foreland, as far as 100 km away from the volcanic front.

The palinspastic restoration of the structural section at 29°30′S illustrated in Figure 18 shows three main sectors. Near the Chilean border is an extensional basin filled with volcaniclastic sediments of the Tilito Formation. Two extensional faults were mapped in these sectors that were active during sedimentation. The central sector of the section is occupied by a topographic high with no evidence of deposition of the Tilito Formation. The eastern part of the section corresponds to the retro-arc basin where the Río La Sal and Valle del Cura Formations show evidence for concurrent synextensional sedimentation. There are four regions where retro-arc sedimentation located more than 100 km to the east of the main arc shows evidence of extension.

Some of the Oligocene normal faults have reactivated previous extensional faults of Permian–Triassic age (Rodríguez Fernández et al. 1999; Winocur 2004). The normal faults of the Permian–Triassic Choiyoi Volcanic Group basement were reactivated during the deposition of the Oligocene Doña Ana Group, which was inverted during Miocene times (Winocur & Ramos 2008, 2011).

Conclusions

The Oligocene was characterized not only by a low convergence rate normal to the trench (Pardo Casas & Molnar 1987) but also by a very low absolute velocity of the South American plate (Silver et al. 1998). As a result of this setting, a widespread extensional regime dominates the Oligocene. The Valle del Cura region was characterized at this time by a generalized extension co-eval with an active calc-alkaline intra-arc volcanic basin in an extensional Andean-type subduction (Kay et al. 1991; Kay & Abruzzi 1996; Mpodozis & Ramos 1989; Jordan et al. 2001; Ramos 2010). Based on the present observations complemented by new geochronological and geochemical data, we conclude that during Oligocene times a magmatic intra-arc basin occupied the axial part of the main Andes between 29° and 30°S under an extensional tectonic regime (Winocur & Ramos 2008, 2011). This intra-arc basin is represented by the Doña Ana Group, composed of the Tilito and Escabroso Formations, and the retro-arc is occupied by the Valle del Cura and Río La Sal Formations, assigned in this work to the Oligocene. Structural observations demonstrate synextensional sedimentation in all the volcanic sequences of Oligocene ages. The Tilito Formation in the main volcanic arc and Valle del Cura and Río La Sal Formations in the retro-arc have been affected by extensional faults. The palinspastic restoration of a structural cross section at 29°30′S (Fig. 18) shows the principal faults affecting and controlling the extensional deposits of the Oligocene volcanic and volcaniclastic units in arc and retro-arc positions.

The new geochemical data reported here for the Valle del Cura and Tilito Formations confirm the extensional characteristics in the arc as the retro-arc.

The authors are indebted to Dr Suzanne Kay for fruitful discussion and critical reading of this manuscript, and to Dr Maydagan for her encouragement, and review of this manuscript. The members of the Laboratorio de Tectónica Andina of the Universidad de Buenos Aires are also thanked for many years of fruitful discussion and collaborative work. This is the contribution R76 of the Instituto de Estudios Andinos Don Pablo Groeber (UBA-CONICET).

References

ALONSO, S., LIMARINO, C. O., LITVAK, V., POMA, S., SURIANO, J. & REMESAL, M. 2011. Palaeogeographic, magmatic and palaeoenviromental scenarios at 30°SL during the Andean Orogeny: cross sections from the volcanic-arc to the orogenic front (San Juan province, Argentina). In: SALFITY, J. A. & MARQUILLAS, R. A.

(eds) *Cenozoic Geology of the Central Andes of Argentine*. Tucumán, SCS Publisher, Salta, 23–45.

APARICIO, E. P. 1975. *Mapa geológico de San Juan*. Editorial Universitaria. Universidad Nacional de San Juan, San Juan.

BISSIG, T., CLARK, A. H., LEE, J. K. W. & HEATHER, K. B. 2001. The Cenozoic history of volcanism and hydrothermal alteration in the Central Andean flat-slab region: new ^{40}Ar-^{39}Ar constraints from the El Indio-Pascua Au(-Ag, Cu) belt, 29°20′–30°30′ S. *International Geology Review*, **43**, 312–340.

BISSIG, T., CLARK, A. H., LEE, J. K. W. & Y VON QUADT, A. 2003. Petrogenetic and metallogenetic responses to miocene slab flattenig: New constrains from the El Indio-Pascua Au-Ag-Cu Belt, Chile/Argentina. *Mineralium Deposita*, **38**, 844–862.

CHARCHAFLIÉ, D., TOSDAL, R. M. & MORTENSEN, J. K. 2007. Geologic framework of the Veladero high-sulfidation epithermal deposit area, Cordillera Frontal, Argentina. *Economic Geology*, **102**, 171–192.

CHARRIER, R., BAEZA, O. *ET AL*. 2002. Evidence for Cenozoic extensional basin development and tectonic inversion south of the flat-slab segment, southern Central Andes, Chile (33°–36°S.L.). *Journal of South American Earth Sciences*, **15**, 117–139.

FAURE, G. 1977. *Principles of Isotope Geology*. John Wiley & Sons, New York.

GODOY, E., YAÑEZ, G. & VERA, E. 1999. Inversion of an Oligocene volcano-tectonic basin and uplift of its superimposed Miocene magmatic arc in the Chilean Central Andes: first seismic and gravity evidences. *Tectonophysics*, **306**, 217–236.

HICKEY, R. L., FREY, F. A., GERLACH, D. C. & LÓPEZ ESCOBAR, L. 1986. Multiple sources for basaltic arc rocks from the Southern Volcanic Zone of the Andes (34–41S): trace elements and isotopic evidence for contributions from subducted oceanic crust, mantle and continental crust. *Journal of Geophysical Research*, **91**, 5963–5983.

IRVINE, T. N. & BARAGAR, W. R. A. 1971. A guide to the chemical classification of the common volcanic rocks. *Canadian Journal of Earth Sciences*, **8**, 523–548.

JORDAN, T. E., MATTHEW BURNS, W., VEIGA, R., PÁNGARO, F., COPELAND, P., KELLEY, S. & MPODOZIS, C. 2001. Extension and basin formation in the southern Andes caused by increased convergence rate: a mid-Cenozoic trigger for the Andes. *Tectonics*, **20**, 308–324.

KAY, M. S. & ABRUZZI, J. M. 1996. Magmatic evidence for Neogene lithospheric evolution of the Central Andes "flat slab" between 30°S and 32°S. *Tectonophysics*, **259**, 15–28.

KAY, S. M. & MPODOZIS, C. 2002. Magmatism as a probe to Neogene shalowing of the Nazca plate beneath the modern Chilean flat-slab. *Journal of South American Earth Science*, **15**, 39–57.

KAY, M. S., MAKSAEV, V. A., MOSCOSO, R., MPODOZIS, C. & NASI, C. 1987. Probing the evolving Andean lithosphere: Mid–Late Tertiary Magmatism in Chile (29°–30°30′) over the modern zone of subhorizontal subduction. *Journal of Geophysical Research*, **92**, 6173–6189.

KAY, M. S., MAKSAEV, V. A., MOSCOSO, R., MPODOZIS, C., NASI, C. & GORDILLO, C. E. 1988. Tertiary Andean magmatism in Chile and Argentina between 28°S and 33°S: correlation of magmatic chemistry with changing Benioff zone. *Journal of South American Earth Sciences*, **1**, 21–38.

KAY, S. M., MPODOZIS, C., RAMOS, V. A. & MUNIZAGA, F. 1991. Magma source variations for mid late Tertiary magmatic rocks associated with shallowing zone and thickening crust in the central Andes (28° to 33°S). *In*: HARMON, R. S. & RAPELA, C. W. (eds) *Andean Magmatism and its Tectonic Setting*. Geological Society of America, Special Paper, Boulder, **265**, 113–137.

KAY, S. M., MPODOZIS, C. & COIRA, B. 1999. Neogene magmatism, tectonism and mineral deposits of the Central Andes (22°–23°S Latitude). *In*: SKINNER, B. J. (ed.) *Geology and Ore Deposits of the Central Andes*. Society of Economic Geologists, Special Publications, 7, 27–59.

LE MAITRE, R. W., BATEMAN, P. *ET AL*. 1989. *A Classification of Igneous Rocks and Glossary of Terms*. Blackwell, Oxford.

LIMARINO, C. O., GUTIÉRREZ, P. R., MALIZIA, D., BARREDA, V., PAGE, S., OSTERA, H. & LINARES, E. 1999. Edad de las secuencias paleógenas y neógenas de las cordilleras de la Brea y Zancarrón, Valle del Cura, San Juan. *Revista de la Asociación Geológica Argentina*, **54**, 177–181.

LITVAK, V. D. 2004. Evolución del volcanismo terciario en el Valle del Cura sobre el segmento de subducción horizontal pampeano, provincia de San Juan. PhD thesis (unpublished), Universidad de Buenos Aires.

LITVAK, V. D. 2009. El volcanismo Oligoceno superior-Mioceno inferior del Grupo Doña Ana en la Alta Cordillera de San Juan. *Revista de la Asociación Geológica Argentina*, **64**, 201–213.

LITVAK, V. D. & POMA, S. 2005. Estratigrafía y facies volcánicas y volcaniclásticas de la Formación Valle del Cura: magmatismo paleógeno en la Cordillera Frontal de San Juan. *Revista de la Asociación Geológica Argentina*, **60**, 402–416.

LITVAK, V. D. & POMA, S. 2010. Geochemistry of mafic Paleocene volcanic rocks in the Valle del Cura region: Implications for the petrogenesis of primary mantle-derived melts over the Pampean Flat Slab. *Journal of South American Earth Sciencies*, **29**, 705–716.

LITVAK, V. D., POMA, S. & KAY, S. M. 2007. Palaeogene and Neogene magmatism in the Valle del Cura region: a new perspective on the evolution of the Pampean flat slab, San Juan province, Argentina. *Journal of South American Earth Sciences*, **24**, 117–137.

MCDONOUGH, W. F. & SUN, S. S. 1995. The composition of the Earth. *Chemical Geology*, **120**, 223–253.

MAKSAEV, V., MOSCOSO, R., MPODOZIS, C. & NASI, C. 1984. Las unidades volcánicas y plutónicas del Cenozoico superior en la Alta Cordillera del Norte Chico (29°–31°S), Geología, alteración hidrotermal y mineralización. *Revista Geológica de Chile*, **21**, 11–51.

MARÍN, G. & NULLO, F. 1989. Geología y estructura del oeste de la Cordillera de la Ortiga, San Juan. *Revista de la Asociación Geológica Argentina*, **43**, 153–163.

MALIZIA, D., LIMARINO, C. O., SOSA-GOMEZ, J., KOKOT, R., NULLO, F. & GUTIÉRREZ, P. 1997. *Descripción* de la Hoja Geológica Cordillera del Zancarrón, escala 1: 100 000. Secretaría de Minería de la Nación (Inédito), Buenos Aires.

MARTIN, M. W., CLAVERO, J., MPODOZIS, C. & Y CUITIÑO, L. 1995 Estudio Geológico de la Franja El Indio, Cordillera de Coquimbo: Servicio Nacional de Geología y Minería, Informe Registrado IR-95-6, **1**: 1–238, Santiago.

MARTIN, M., CLAVERO, R. & MPODOZIS, C. 1995. Eocene to late Miocene magmatic development of El Indio belt, 30°S, North Central Chile. *In*: Actas del *VIII Congreso Geológico Chileno*, Antofagasta, **1**, 149–153.

MARTIN, M. W., CLAVERO, J. R. & MPODOZIS, C. 1999. Late Palaeozoic to Early Jurassic tectonic development of the high Andean Principal Cordillera, El Indio Region, Chile (29°–30°S). *Journal of South American Earth Sciences*, **12**, 33–49.

MINERA, T. E. A. 1968. *Geología de la Alta Cordillera de San Juan. Su prospección y áreas con posibilidades mineras*. Departamento de Minería de San Juan (inédito), San Juan.

MPODOZIS, C. & RAMOS, V. A. 1989. The Andes of Chile & Argentina. *In*: ERICKSEN, G. E., CAÑAS PINOCHET, M. T. & REINEMUD, J. A. (eds) *Geology of the Andes and its Relation to Hydrocarbon and Mineral Resources*. Circumpacific Council for Energy and Mineral Resources, Earth Sciences Series, **11**, 59–90.

NULLO, F. 1988. Geología y estructura del área de Guanaco Zonzo y Veladero, oeste de la Cordillera de Zancarrón, San Juan. *In*: *Actas del III Congreso Nacional de Geología Económica*, Olavarría, **2**, 501–515.

NULLO, F. & MARÍN, G. 1990. Geología y estructura de las quebradas de la Sal y de la Ortiga, San Juan. *Revista de la Asociación Geológica Argentina*, **45**, 323–335.

OTAMENDI, J., NULLO, F., GODEAS, M. & PEZZUTTI, N. 1994. Petrogenesis del volcanismo terciario del Valle del Cura, San Juan, Argentina. *In*: *Actas del VII Congreso Geológico Chileno*, Concepción, **2**, 1130–1135.

PARDO CASAS, F. & MOLNAR, P. 1987. Relative motion of the Nazca (Farallón) and South America plates since Late Cretaceous time. *Tectonics*, **6**, 233–248.

RAMOS, A. R., CRISTALLINI, E. O. & PÉREZ, D. J. 2002. The Pampean flat-slab of the Central Andes. *Journal of South American Earth Sciences*, **15**, 59–78.

RAMOS, V. A. 2010. The tectonic regime along the Andes: present-day and Mesozoic regimes. *Geological Journal*, **45**, 2–25.

RAMOS, V. A., KAY, S. M., PAGE, R. & MUNIZAGA, F. 1989. La ignimbrita Vacas Heladas y el cese del volcanismo en el Valle del Cura, provincia de San Juan. *Revista de la Asociación Geológica Argentina*, **44**, 336–352.

RAMOS, V. A., KAY, S. M., PAGE, R. & MUNIZAGA, F. 1990. La ignimbrita Vacas Heladas y el cese del volcanismo en el Valle del Cura, provincia de San Juan. *Revista de la Asociación Geológica Argentina*, **44**, 336–352.

REUTTER, K. J. 1974. Entwicklung und Bauplan der chilenischen Hanchkordillere im Bereich 29° sudlicher Breite. *Neues Jahrbuch derGeologie und Paläntologie, Abhlandulgen*, **146**, 153–178. Stuttgart.

RODRÍGUEZ FERNÁNDEZ, L. R., HEREDIA, N., ESPINA, R. G. & CEGARRA, M. I. 1999. Estratigrafía y estructura de los Andes centrales argentinos entre los 30° y 31° de latitud sur. *In*: BUSQUETS, P., COLOMBO, F., PÉREZ, E. A. & RODRÍGUEZ, F. R. (eds) *Geología de los Andes centrales argentino-chilenos*. Instituto Nacional de Geología, Acta Geológica Hispánica, Barcelona, **32**, 51–75.

SILVER, P. G., RUSSO, R. M. & LITHGOW-BERTELLONI, C. 1998. Coupling of South American and African plate motion and plate deformation. *Science*, **279**, 60–63.

SUN, S. S. & MCDONOUGH, W. F. 1989. Chemical and isotopic systematics of oceanic basalts: implications for mantle composition and processes. *In*: SAUNDERS, A. D. & NORRY, M. J. (eds) *Magmatism in the Ocean Basins*. Geological Society, London, Special Publications, **42**, 313–345.

THIELE, R. 1964. *Reconocimiento geológico de la Alta Cordillera de Elqui*. Universidad de Chile, Departamento de Geología, Publicaciones, Santiago, **27**.

WINOCUR, D. A. 2004. Geología y estructura de la región de los Despoblados, Valle del Cura, Provincia de San Juan. *Trabajo Final de Licenciatura*, Universidad de Buenos Aires, inédito, Buenos Aires.

WINOCUR, D. A. 2010. *Geología y estructura del Valle del Cura y el sector central del Norte Chico, provincia de San Juan y IV Región de Coquimbo, Argentina y Chile*. PhD thesis (unpublished), Universidad de Buenos Aires.

WINOCUR, D. A. & RAMOS, V. A. 2008. Geología y Estructura del sector norte de la Alta Cordillera de la provincia de San Juan. *In*: *Actas del XVII Congreso Geológico Argentino*, Jujuy, **3**, 166–167.

WINOCUR, D. A. & RAMOS, V. A. 2011. La Formación Valle del Cura: Su edad y ambiente tectónico. *In*: *Actas del XVIII° Congreso Geológico Argentino*, Neuquén, CD-ROM, 2.

U–Pb detrital zircon ages of Upper Jurassic continental successions: implications for the provenance and absolute age of the Jurassic–Cretaceous boundary in the Neuquén Basin

MAXIMILIANO NAIPAUER[1,2]*, MAISA TUNIK[2,3], JULIANA C. MARQUES[4], EMILIO A. ROJAS VERA[1,2], GRACIELA I. VUJOVICH[1,2], MARCIO M. PIMENTEL[4] & VICTOR A. RAMOS[1,2]

[1]*Instituto de Estudios Andinos 'Don Pablo Groeber', Departamento de Ciencias Geológicas, FCEN – Universidad de Buenos Aires, Argentina*

[2]*CONICET*

[3]*Instituto de Investigación en Paleobiología y Geología, Sede Alto Valle, Universidad Nacional de Río Negro*

[4]*Laboratorio de Geología Isotópica, Universidade Federal do Rio Grande do Sul, Brasil*

*Corresponding author (e-mail: maxinaipauer@gl.fcen.uba.ar)

Abstract: New U–Pb detrital zircon ages are presented for the Tordillo Formation. The ages indicate that the most important source region of sediment supply was the Jurassic Andean arc (peaks at c. 144, 153 and 178 Ma), although two secondary sources were defined at c. 218 and 275 Ma. Temporal variation in the provenance indicates that at the beginning of the sedimentation, Carboniferous to Lower Jurassic magmatic rocks and Lower Palaeozoic metamorphic rocks were the most important sources. Towards the top, the data suggest that the Andean arc becomes the main source region. The comparison between provenance patterns of the Tordillo Formation and of the Avilé Member (Agrio Formation) showed some differences. In the former, the arc region played a considerable role as a source region, but this is not identified in the latter. The results permit a statistically robust estimation of the maximum deposition age for the Tordillo Formation at c. 144 Ma. This younger age represents a discrepancy of at least 7 Ma from the absolute age of the Kimmeridgian and Tithonian boundary (from the chronostratigraphic timescale accepted by the International Commission of Stratigraphy, IUGS), and has strong implications for the absolute age of the Jurassic–Cretaceous boundary.

Supplementary material: Sample coordinates, values of the sandstone compositional framework and U–Pb (LAM-MC-ICP-MS) age measurements of zircons grains are available at http://www.geolsoc.org.uk/SUP18718

Upper Jurassic continental sequences deposited on the southwestern proto-Pacific margin of Gondwana represent an important component of the Neuquén Basin between 31° and 41° South latitude in western central Argentina. These continental sediments are mainly grouped into the Tordillo Formation (Groeber 1946) and form one of the most characteristic lowstand wedges of the basin (Spalletti & Veiga 2007; Veiga & Spalletti 2007). This lowstand wedge is the main hydrocarbon reservoir and contains the largest conventional oil reserves in Argentina (Maretto *et al.* 2002). The Neuquén Basin is located in a retroarc setting (Vergani *et al.* 1995; Ramos 1999; Howell *et al.* 2005) where thousands of metres of marine and continental sediments were accumulated along the eastern side of the Andean volcanic arc from the Early Jurassic (Ramos 2010; and references therein) (Fig. 1a). Because of its retroarc position, important input from volcanic sources to the detrital sediments should be expected. However, the detrital input from the Andean arc was not constant. It varied over time and throughout the basin and was mixed with other pre-Andean volcanic materials. Previous analyses of sedimentary provenance in the Tordillo Formation were mainly based on petrographic and geochemical studies. These demonstrated the participation of the volcanic supply, but were not able to identify the ages of the sources (Marchese 1971; Gulisano 1988; Eppinger & Rosenfeld 1996; Spalletti *et al.* 2008). Systematic studies based on sedimentary provenance in the Neuquén Basin, including U–Pb zircon dating, remain scarce (Tunik *et al.* 2010; Naipauer *et al.* 2012; Di Giulio *et al.* 2012).

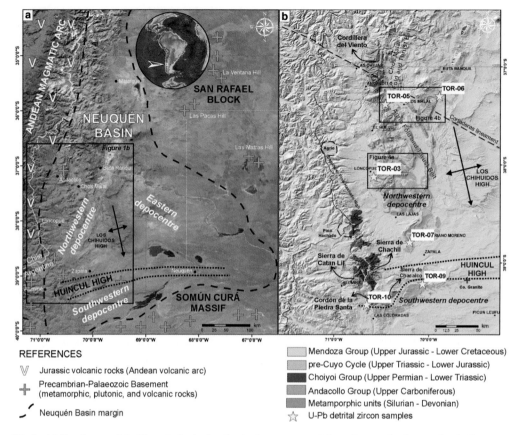

Fig. 1. (a) Sketch map of the Neuquén Basin showing its depocentres and the main tectonic elements for the Late Jurassic (based on Spalletti & Colombo Piñol 2005). (b) Geological map with distribution of the outcrops of the Mendoza Group (Late Jurassic–Early Cretaceous) and the main pre-Middle Jurassic basement rocks. Sample localities for U–Pb work for the Tordillo Formation are also shown.

The present study presents new U–Pb data (LAM-MC-ICP-MS; Laser Ablation Microprobe-Multi-Collector-Inductively Coupled Plasma-Mass Spectrometer) on detrital zircons from outcrops of the northwestern region of the retroarc Neuquén Basin with a view to defining the age of the sediment sources. We investigate how the provenance patterns along a north–south transect through the western margin of the basin differ. The new data permit a more precise assessment of the maximum age of deposition of the Tordillo Formation and the description of regional and temporal changes in the sedimentary provenance. We also compare the pattern of detrital zircon ages obtained in this study with those obtained previously by Tunik *et al.* (2010) in a lowstand wedge developed in the Early Cretaceous within the Agrio Formation. The absence of provenance from the Andean arc is remarkable in this lowstand wedge. We therefore describe and discuss a major change in the palaeogeography and tectonic evolution between these lowstands within the Neuquén Basin, which is important in order to understand the mechanics of subsidence and the sedimentation control in a typical Andean retroarc basin. Finally, we discuss the maximum depositional age obtained in the Upper Jurassic succession and its relationship to the absolute age of the Jurassic–Cretaceous boundary.

Geological and stratigraphic framework

The geological history of the Neuquén Basin is intimately linked to the evolution of the southern Central Andes and was mainly affected by (1) changes in tectonic conditions of the proto-Pacific margin, (2) the installation of the Jurassic magmatic arc, and (3) eustatic global sea-level changes (Legarreta & Gulisano 1989; Legarreta & Uliana 1991; Vergani *et al.* 1995; Ramos 1999; Howell *et al.* 2005; Mpodozis & Ramos 2008).

Additionally, it is important to point out the tectonic control exerted by the structural fabrics of the Palaeozoic basement (Franzese & Spalletti 2001). Although poorly exposed, this basement largely conditioned the evolution of the southern region of the basin, especially along the Huincul deformation zone (Vergani et al. 1995; Mosquera & Ramos 2006; Mpodozis & Ramos 2008; Ramos et al. 2011; Naipauer et al. 2012) (Fig. 1a, b).

The basement of the basin is composed of three main components (Fig. 2): (1) low- to high-grade metamorphic rocks of the Early Palaeozoic (Piedra Santa, Guaraco Norte, and Colohuincul formations; Cingolani et al. 2011, and references therein); (2) volcanic and volcaniclastic rocks with marine sediments of the Late Carboniferous (Andacollo Group; Llambías et al. 2007), and (3) plutonic and volcanic rocks grouped into the Upper Palaeozoic to Lower Triassic Choiyoi Group (Llambías & Sato 2011, and references therein). These basement components are mainly exposed in the Cordillera del Viento range, NW of the basin and in the western portion of the Huincul High in the Sierra de Chachil, Sierra de Chacaico, and Cordon de la Piedra Santa, south of the basin (see Fig. 1b). The Andacollo Group is only exposed on the western side of the Cordillera del Viento range (Fig. 1b). The basement was exhumed and eroded during the Early Triassic extension prior to the beginning of the filling of the Neuquén Basin; this extension phase produced a regional erosion surface on which the volcanic units of the Late Triassic and Early Jurassic were later deposited (Llambías et al. 2007).

During the complex tectonic evolution of the basin, three main stages are identified, which are also recognized in the stratigraphy (Fig. 2): the first Late Triassic–Early Jurassic rift stage is seen in the pre-Cuyo Cycle (e.g. Ñireco, Lapa, Cordillera del Viento, and Remoredo formations; Gulisano 1981; Franzese & Spalletti 2001; Schiuma & Llambías 2008; Carbone et al. 2011), followed by a long period of thermal sag (Mid Jurassic to Early Cretaceous) comprising the Cuyo, Lotena, Mendoza and Bajada del Agrio groups and finally a Late Cretaceous–Cenozoic foreland stage being included in the Neuquén and Malargüe groups (Gulisano et al. 1984; Vergani et al. 1995; Franzese & Spalletti 2001; Howell et al. 2005). This last stage coincides with the beginning of the compressive Andean deformation at these latitudes (Tunik et al. 2010, and references therein).

The stratigraphy of the Neuquén Basin is complex, not only because of its large regional extension but also because of the diversity of local names. A summary of the general stratigraphy, tectonic evolution, biostratigraphic resolution and available U–Pb absolute ages is shown in Figure 2. As shown there, good biostratigraphic control is well documented in many units of the Cuyo, Lotena and Mendoza groups. However, it highlights the low geochronological control that exists throughout the whole stratigraphic column (Fig. 2).

The Mendoza Group (Stipanicic et al. 1968) represents a sedimentary cycle developed during the sag phase (Howell et al. 2005). Sedimentation started with typical continental red facies (Tordillo Formation), which are the subject of this study. These sediments are covered by marine dark shales and limestones of the Vaca Muerta Formation (Weaver 1931), which represent a widespread marine transgression, well documented by ammonite faunas throughout the entire basin (Leanza 1981; Gulisano & Gutiérrez Pleimling 1995; Vergani et al. 1995; Riccardi 2008a, b). Finally, mixed continental and marine facies are increasingly restricted and are included in the Mulichinco and Agrio formations (Weaver 1931).

The base of the Mendoza Group is assigned to the Kimmeridgian due to the stratigraphic position of the Tordillo Formation (Fig. 2). This unit unconformably overlies the Lotena Group (Leanza 1992). In this group the La Manga Formation (Stipanicic 1966) is characterized by its poor ammonoid fauna of mid Oxfordian age (Stipanicic 1951; Riccardi 1984, 2008a, b). Below the Lotena Group, the Cuyo Group was developed (see Fig. 2). It is important to point out that, in the eastern sector of the Cordillera del Viento (Fig. 1b), a tuff layer intercalated in the Chacay Melehue Formation, at the top of the Cuyo Group, yielded the ID-TIMS (Isotope Dilution-Thermal Ionization Mass Spectrometry) zircon age of 164.6 ± 0.2 Ma (Kamo & Riccardi 2009). This absolute age is in agreement with well-documented ammonoid fauna present in the sequence, coinciding with the Bathonian–Callovian boundary (late Mid Jurassic) (Kamo & Riccardi 2009) (Fig. 2).

The age of the top of the Mendoza Group is well constrained through the study of the upper Agrio Formation. The fossiliferous content of this unit indicates a late Hauterivian age, which was corroborated by a U–Pb absolute zircon age (SHRIMP) at c. 132 Ma (Aguirre-Urreta et al. 2008). The zircons analysed correspond to a tuff layer in the Agua de la Mula Member, extending the age of the Mendoza Group to the early Barremian (Early Cretaceous) (Aguirre-Urreta et al. 2008; Aguirre-Urreta & Rawson 2012) (Fig. 2).

The Tordillo Formation and its equivalents

Previous studies on the Tordillo Formation focused on its lithological composition and stratigraphic relationships (Groeber 1946; Stipanicic 1966), as well as on its main sedimentological aspects

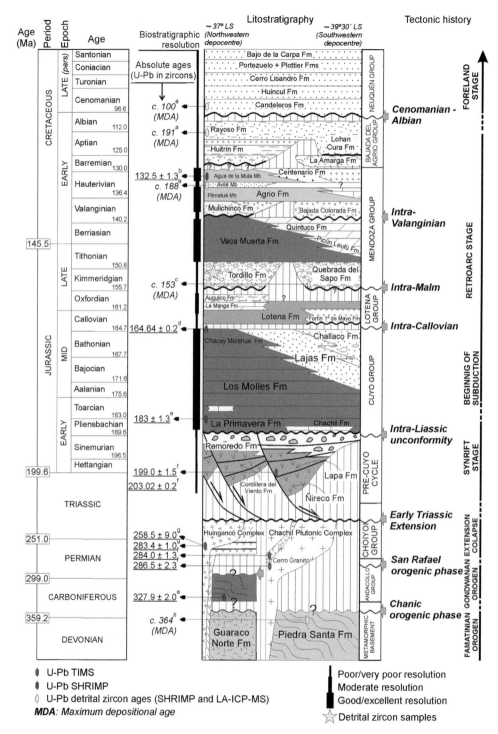

Fig. 2. Tectonostratigraphic chart of the Neuquén Basin (modified from Howell *et al.* 2005). The table shows the chronostratigraphy, U–Pb zircon ages (a, Tunik *et al.* 2010; b, Aguirre-Urreta *et al.* 2008; c, Naipauer *et al.* 2012; d, Kamo & Riccardi 2009; e, Suárez *et al.* 2008; f, Schiuma & Llambías 2008; g, Sato *et al.* 2008; h, Ramos *et al.* 2010), biostratigraphic resolution, lithostratigraphic, tectonic setting and stratigraphic location of the studied samples.

(Marchese 1971; Legarreta & Gulisano 1989; Legarreta & Uliana 1999; Zavala et al. 2005; Spalletti & Colombo Piñol 2005; Veiga & Spalletti 2007; Spalletti et al. 2008, 2011; López-Gómez et al. 2009). Rocks of equivalent age exposed along the Main Cordillera of Argentina and Chile from nearly 31°S to 37°S are known as the Rio Damas Formation (Vicente 2005, 2006; Charrier et al. 2007). South of 37°S, where the outcrops extend eastward and become part of the Neuquén Embayment up to c. 41°S (Fig. 1a) the Tordillo Formation is composed of conglomerates, sandstones, shales and tuffs (Fig. 3). In the southern and western areas of the basin, coarse facies including conglomerates and sandstones are dominant, while in the northern and eastern parts distal facies predominate, where pelites, siltstones and tuffs are common (Marchese 1971). Thickness is also variable within the basin, where three main depocentres have been commonly identified (Spalletti & Colombo Piñol 2005; Spalletti & Veiga 2007) (Fig. 1a). The northwestern and eastern depocentres (Spalletti & Colombo Piñol 2005) are located north of the Huincul High and separated by the Los Chihuidos High (Ramos 1978) (Fig. 1a). The Tordillo Formation is mostly known from outcrops of the northwestern depocentre, whereas in the subsurface of the Neuquén Embayment, the rocks of the eastern depocentre are known as the Sierras Blancas and Catriel formations (Digregorio 1972; Spalletti & Veiga 2007; Spalletti et al. 2011). South of the Huincul High the Kimmeridgian deposits are located in a southwestern depocentre and correspond to the Quebrada del Sapo Formation (Zavala et al. 2005; Spalletti & Veiga 2007) (Fig. 1a). The maximum thickness is up to 900 m in the northwestern depocentre, decreasing considerably towards the east to 90 m on the edge of the Los Chihuidos High, while south of the Huincul High the thickness is less than 100 m (Marchese 1971; Spalletti & Veiga 2007; Spalletti et al. 2011).

The Tordillo Formation and equivalent units display highly variable depositional systems in the three different depocentres (Marchese 1971; Gulisano 1988; Legarreta & Gulisano 1989; Legarreta & Uliana 1999; Spalletti & Colombo Piñol 2005; Spalletti & Veiga 2007). Ephemeral fluvial systems and playa lakes are the main deposits in the north (Spalletti & Colombo Piñol 2005; Spalletti & Veiga 2007) (Fig. 3a–e). Particularly at the top of the Río Neuquén and Loncopué sections, primary deposits of tuff and volcaniclastic rocks are present (Spalletti & Colombo Piñol 2005) (Fig. 3f). Proximal alluvial fan and fluvial deposits are dominant in the southern sector of the northwestern depocentre, for example the Río Covunco section (Fig. 3g). In the southwestern and eastern depocentres, the depositional systems are characterized by the transition from ephemeral fluvial to aeolian systems (Gulisano 1988; Zavala et al. 2005; Veiga & Spalletti 2007) (Fig. 3h).

The Kimmeridgian to earliest Tithonian age of these deposits was historically constrained by their stratigraphic position due to the lack of fossils with biostratigraphic value and radiometric dates. They rest on an angular unconformity with the Lotena Group (Callovian–Oxfordian) and are concordant with, and transitional to, the Vaca Muerta Formation. The age of the latter is late Early Tithonian to Early Valanginian based on its ammonite faunas (Leanza 1981; Aguirre-Urreta & Rawson 1999). The first U–Pb absolute ages on detrital zircons of the Tordillo and Quebrada del Sapo formations showed that the maximum age of deposition is c. 153 Ma (Kimmeridgian), at least for the proximal fluvial facies from the southern sector of the basin (Naipauer et al. 2012). These authors mentioned, however, that it is noteworthy that in all samples analysed a significant number of detrital zircon ages (c. 145 Ma) are younger than the currently accepted age for the Kimmeridgian/Tithonian boundary.

Provenance data background

The Neuquén Basin is bounded to the NE and SW by the San Rafael Block and the Somún Curá Massif, respectively, and its western margin is defined by an almost continuous Andean volcanic arc (Digregorio et al. 1984) (Fig. 1a). Previous provenance studies in the Tordillo Formation concluded that the main source of sediment supply comprised basic to acid volcanic rocks (Marchese 1971; Gulisano 1988; Mescua et al. 2008; Spalletti et al. 2008). Gulisano (1988) was the first to suggest that the source area of clastic, pyroclastic and volcanic material was the Andean magmatic arc. Direct evidence of the activity of this arc is observed in the Rio Damas Formation, where andesitic volcanic rocks are interlayered in the clastic sequence (Charrier et al. 2007). The petrographic and geochemical studies agree with an Andean arc source, but provenance from basement acid volcanics of the Choiyoi Group was also identified (Mescua et al. 2008; Spalletti et al. 2008).

As well as the significant volcanic component, igneous–metamorphic rocks are also represented in a lesser proportion. U–Pb ages on detrital zircons, separated from conglomerates from the southern region of the basin, indicate a dominant contribution from the Andean Jurassic arc with ages between c. 178 and 153 Ma and from synrift Late Triassic volcanism between 220–200 Ma. They also show an input from Upper Permian (280–260 Ma) and Upper Devonian (c. 360 Ma)

sources of igneous–metamorphic basement (Naipauer *et al.* 2012).

Samples and analytical methods

The analysed samples were taken from different localities within the Tordillo Formation with well-known stratigraphic control. The samples are distributed along the western part of the Neuquén Basin (Fig. 1b). The provenance analysis included detailed petrographic studies and zircon morphology analyses to define the main populations in conjunction with U–Pb geochronological ages (LAM-MC-ICP-MS).

Petrographic studies were performed on samples of sandstone and conglomerate matrix from six classic profiles of the basin. Detailed stratigraphic columns and sedimentological aspects have been surveyed previously (see Gulisano 1988; Gulisano & Gutiérrez Preimling 1995; Spalletti & Colombo Piñol 2005; Zavala *et al.* 2005; Veiga & Spalletti 2007; Spalletti & Veiga 2007). The samples in the northwestern depocentre were taken from the classic sections: (1) Loncopué, TOR-01, TOR-02 and TOR-03; (2) Río Neuquén, TOR-04 and TOR-05; (3) Pampa Tril, TOR-06; and (4) Arroyo Covunco, TOR-07 (Fig. 1b). For the southwestern depocentre, the samples correspond to the following profiles: (5) Río Picún Leufú, TOR-08 and TOR-09 and (6) Fortín Primero de Mayo, TOR-10 (Fig. 1b). The sandstones were petrographically classified according to the diagram of Folk *et al.* (1970), and were also plotted in the provenance diagram of Dickinson *et al.* (1983). The petrography was also used to select samples for U–Pb dating.

We analysed the U–Pb (LAM-MC-ICP-MS) ages of three samples (TOR-03, TOR-05 and TOR-06) of the Tordillo Formation from the northwestern depocentre. These results are compared with data from the southern region (samples TOR-07, TOR-09 and TOR-10; Naipauer *et al.* 2012). The analysed samples correspond to a fine-grained red sandstone located at the base of the Loncopué profile (TOR-03; Fig. 4a), a fine-grained green volcaniclastic sandstone from the top of the Río Neuquén profile (TOR-05; Fig. 4b) and a green siltstone at the top the Pampa Tril profile (TOR-06;

Fig. 4b). The total results complete a set of ^{237}U–Pb concordant ages of detrital zircons.

For U–Pb LAM-MC-ICP-MS zircon analyses, 5 kg of each rock sample were crushed. They were then powdered and sieved to fractions between 75 and 212 μm. Heavy mineral concentrates were obtained by panning and were subsequently purified by an electromagnetic (Frantz) method to eliminate only the highly magnetic fraction. Zircon grains were selected and set in epoxy resin mounts. The mount surface was then polished to expose the grain interiors. Backscattered electron (BSE) images of zircons were obtained using an SEM (Scanning Electron Microscope) JEOL JSM 5800 at Universidade Federal do Rio Grande do Sul (UFRGS), Brazil. The U–Pb LAM-MC-ICP-MS zircon analyses were carried out using a Finnigan Neptune coupled to a UP-213nm Nd-YAG laser ablation system (New Wave Research, USA) installed in the Isotope Geology Laboratory of the Geoscience Institute of UFRGS.

The analyses were performed as a single spot of 30 μm using the following laser parameters: repetition rate of 10 Hz, energy of 0.5–1.1 mJ cm^{-2}, 40 s ablation time and 1 s integration time. Faraday cup configuration of the MC-ICP-MS was ^{208}Pb, ^{232}Th and ^{238}U, and ion counters on cup L4 with ^{202}Hg, ^{204}Pb, ^{206}Pb and ^{207}Pb. Main gas flow was 15 l/min Ar, auxiliary gas flow 0.8 l m^{-1}, while the sample was carried with 0.75 l/min Ar plus 0.42 l/min He. Unknown analyses were bracketed for internal instrumental fractionation control and mass bias following the Albarede *et al.* (2004) method by measurements of the international standard GJ-1 (Jackson *et al.* 2004) and blanks at every set of four zircon spots.

The raw data were reduced using an Excel worksheet (Buhn *et al.* 2009). The ^{206}Pb–^{238}U ratio was recalculated using the linear regression method of Košler *et al.* (2002) when laser fractionation correction was necessary. The ^{204}Pb (common lead) interference and background was observed using ^{202}Hg and (^{204}Hg + ^{204}Pb) masses during analyses. The analysed grains have low common lead, and the usual ^{204}Pb correction using the Stacey & Kramers (1975) model was not necessary. Corrected ^{206}Pb/^{207}Pb and ^{206}Pb–^{238}U ratios were shifted into absolute age data using ISOPLOT/Ex (Ludwig 2003). The probability

---

Fig. 3. (a–g) Photographs of the Tordillo Formation, some of which are the localities of the U–Pb samples. (a, b) Red fluvial deposits at Loncopué, a fine red sandstone located at the base of the profile was sampled from U–Pb analysis (sample TOR-03). (c, d) Overview of the deposits and the fine green volcaniclastic sandstone from the top of the Río Neuquén profile (TOR-05). (e, f) View in detail of the transitional contact between Tordillo and Vaca Muerta formations in the Pampa Tril area; see the sample locality at the top the Pampa Tril profile (sample TOR-6). (g) Fine-grained conglomerate and sandstone levels at the Arroyo Covunco locality (sample TOR-07). (**h**) Aeolian deposits of the Quebrada del Sapo Formation at Arroyo Picún Leufú, southwestern depocentre.

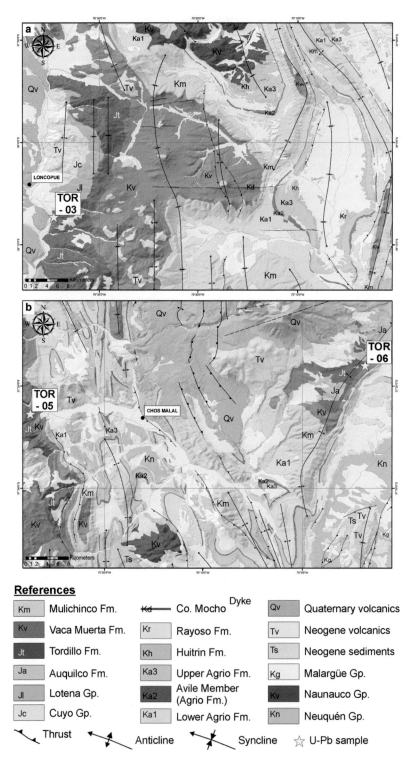

Fig. 4. Geological map with sample locations of the Tordillo Formation. Detail of geology of (a) the Loncopué region (Cerro Mocho) and (b) the Chos Malal region.

diagrams were constructed using the same software. The reported errors were propagated by quadratic addition of external reproducibility and within-run precision.

Results

Sandstone petrography provenance

The samples were studied under a polarized microscope and classified following the scheme of Dott (1964) separating sandstones and wackes (TOR-01 and TOR-06). The sandstones are poorly sorted and most of the detrital fragments are sub-angulose, although some subrounded clasts are also identified (Fig. 5). There are shards and pumice fragments immersed in a carbonate cement (TOR-02; Fig. 5b). Other types of cements are argillaceous and zeolitic.

The Gazzi–Dickinson method was used for the provenance studies as suggested by Zuffa (1985). Some samples were discarded due to their fine grain size (TOR-02, TOR-04 and TOR-08). Depending on the size of the thin section, 400 or 500 points were counted. The five analysed samples were classified as lithoarenite ($n = 1$) Q15F14L71 and feldspatic lithoarenites ($n = 4$) Q20F30L50 after Folk et al. (1970) (Fig. 6).

The sandstones are classified in order of abundance of lithic fragments, feldspar and quartz. The monocrystalline quartz (Fig. 5c) is more abundant than the polycrystalline in samples TOR-03, TOR-05 and TOR-07. In samples TOR-09 and TOR-10 the relation is reversed; the polycrystalline quartz is more abundant than the monocrystalline. The relative abundances of plagioclase and K-feldspar are variable. In general, plagioclase is more abundant, but in sample TOR-09 the relationship is inverted. This can be explained by the large amount of plutonic rock fragments and metamorphic lithic fragments (Fig. 5a).

The analyses of the rock fragments indicate that volcanic rocks are dominant (Fig. 5e). However, the composition of the rock fragments in the southwestern and northwestern depocentres is quite different. In the SW, the plutonic and metamorphic rock fragments (Fig. 5c) are remarkable, whereas in the NW, sedimentary rock and angular quartz fragments are abundant (Fig. 5d). The presence of altered lithic fragments (Fig. 5f) and pseudomatrix is common in the northwestern samples and absent or very rare in the samples from the SW.

Zircon morphology analysis

The zircons were separated by conventional techniques and their external morphologies were studied under a binocular microscope. For each sample, the main zircon populations were recognized and described according to their shape, colour, size, inclusions and fractures.

Sample TOR-03. The zircon crystals belong to the size fraction of <150 μm. The predominant morphological features are prismatic habit and subrounded to subidiomorphic form (P1). Many crystals are characterized by their elongation ($>5:1$) and represent c. 20% of the total grains (P2). They are mostly transparent or have a yellow colour. A third population (P3) is represented by a few crystals that have rounded form. In general, the zircons investigated have few inclusions and fractures (Fig. 7a).

Sample TOR-06. The small numbers of grains separated are smaller than 100 μm. The largest population of zircons (P1) is characterized by crystals with short prismatic habit, although some zircon grains forming long prisms (P2) are also present. The forms are subrounded and subidiomorphic, although a few crystals have idiomorphic form and some of them present bipyramidal crystal forms. They are transparent and have some inclusions. Some less abundant crystals are rounded (P3) (Fig. 7b).

Sample TOR-05. Two groups of zircons with different sizes are recognized. Crystals forming population P1 are the most abundant grains and are smaller than 100 μm, with short prismatic habit (elongation $<3:1$) and subrounded to rounded form. Few crystals were observed with subidiomorphic forms. They are mostly transparent. Some are yellow in colour, with inclusions, and many are broken. The second group (P2) is characterized by grain size larger than 200 μm; the crystals have subrounded to subidiomorphic form. Some have a short prismatic habit, but others have a long prismatic habit and are broken. They are mostly transparent and there are some grains with a pink colour. Inclusions were observed in several grains (Fig. 7c).

U–Pb data

The U–Pb geochronological analysis of detrital zircon grains is a powerful tool with which to obtain valuable information on the provenance and maximum age of sedimentary deposition (Mueller et al. 1994; Morton & Hallsworth 1999; Fedo et al. 2003; Bahlburg et al. 2009; Dickinson & Gehrels 2009; Bahlburg et al. 2010). A fraction of zircon crystals were randomly separated and mounted in epoxy sections. The minerals were then analysed using the U–Pb method (LAM-MC-ICP-MS) to obtain absolute ages.

In sample TOR-03, a total of 44 zircon grains were analysed, but eight analyses were rejected because of high discordance (Fig. 8a). The spectra

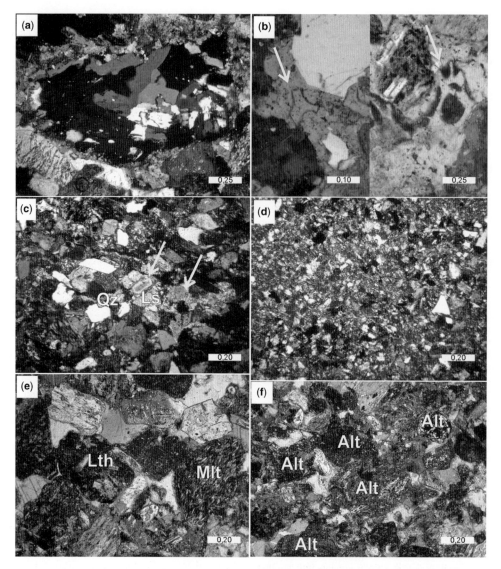

Fig. 5. Photomicrographs of the Tordillo Formation. (**a**) Metamorphic lithic fragments are common in the base of the column (sample TOR-02, Loncopué section). (**b**) Cuspate and pumice shards (arrows) provide evidence of active volcanism during sedimentation (sample TOR-02). (**c**) Monocrystalline quartz (Qz) and sedimentary lithic fragments (Ls and arrows) distinguish the samples from the northern depocentre from those of the southern depocentres. The southern samples are characterized by the presence of polycrystalline quartz and plutonic lithic fragments (see fig. 8c in Naipauer *et al.* 2012) (sample TOR-09, Picún Leufú section). (**d**) Angular quartz on the sublithic wacke (Dott 1964) suggests a volcanic provenance (sample TOR-06, Pampa Tril section). (**e**) Microlithic (Mlt) and lathwork textures (Lth) of volcanic lithic fragments suggest the provenance from the Jurassic volcanic arc, clearly different from the provenance from the Choiyoi volcanics (see fig. 8d in Naipauer *et al.* 2012) (sample TOR-05, Río Neuquén section). (**f**) The presence of altered lithic fragments (Alt) is remarkable; much of them are probably of volcanic and sedimentary origin (sample TOR-05, Río Neuquén section).

of the 36 concordant ages are between 145 Ma and 777 Ma. The distribution of the ages is characterized by multimodal maximum peaks at *c.* 145, *c.* 160, *c.* 191, *c.* 207 and *c.* 280 Ma. There are two isolated ages at *c.* 684 Ma and *c.* 777 Ma (Fig. 9a). The inset of Figure 9a shows some BSE images of the most typical detrital zircon populations from sample TOR-03.

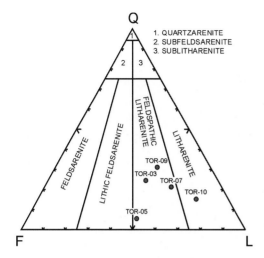

Fig. 6. Sandstone classification based on the scheme of Folk et al. (1970).

A total of 36 ages were obtained from sample TOR-06, but two grains were rejected, one because of high discordance and another for its high uncertainty (Fig. 8c). The concordant ages plot between 142 and 2114 Ma but most of the ages are Late Jurassic to Early Carboniferous. The curve of relative probability shows several maximum peaks at c. 143, c. 153, c. 170 and c. 268 Ma, plus isolated ages at c. 334, c. 348 and c. 2114 Ma (Fig. 9b). Some BSE images of the detrital zircons analysed are included in Figure 9b; these show the internal structure, which is characterized by oscillatory zoning that indicates a magmatic origin for these grains. The idiomorphic form and prismatic shape of the crystals may suggest a volcanic origin.

Sample TOR-05 had 36 zircons analysed, five of which were rejected due to high discordance (Fig. 8e) and another due to the high uncertainty. The remaining grains yielded 30 concordant ages in the relatively narrow interval between 138 and 213 Ma, producing a unimodal distribution (Fig. 9c). The most important maximum peak occurs at c. 144 Ma, although there are also minor peaks at c. 170 Ma and a single age at c. 213 Ma (Fig. 9c). The BSE images of the zircon grains (inset in Fig. 9c) show faint internal structures, but oscillatory zoning can be recognized, so a magmatic origin is also postulated for these grains.

Integration of the data and discussion

Source regions along the Neuquén Basin

The source regions defined by the provenance analysis are in accordance with the sediment transport directions as shown by palaeocurrent studies in the basin. These are characterized by axial systems parallel to the Andean chain with north to NE trends (Gulisano 1988; Spalletti & Colombo 2005; Spalletti et al. 2008). The petrographic provenance studies using the Dickinson et al. (1983) diagrams indicate that all samples belong to the arc fields (Fig. 10a, b), which is not surprising considering that the Tordillo Formation was deposited near the Andean arc in a retroarc position. However, they could also have received sediments from the pre-Cuyo Cycle volcanism and from the Choiyoi magmatic rocks, as mentioned by Spalletti et al. (2008). The zircon morphology and internal textures also indicate a magmatic origin for the main detrital zircons for the northwestern depocentre. The dominance of prismatic and long prismatic crystal habits (elongation 5:1) suggests a volcanic origin for these zircons (P1 and P2). The absence of zircon populations characterized by idiomorphic, multifaceted and pink coloured grains in the northwestern depocentre is remarkable; these kinds of grains are present in samples from the southern part of the basin and were probably derived from basement areas (Naipauer et al. 2012).

The summary U–Pb age probability curve for all six analysed samples indicates that the most important source of sediments was the Jurassic Andean magmatic arc (Fig. 11). The maximum peaks of zircon ages at c. 144, c. 153 and c. 178 Ma coincide with the main magmatic pulses in the Andean arc (Castro et al. 2011, and references therein). Also, two secondary source regions are defined according to the maximum frequency peaks at c. 218 Ma and c. 275 Ma (Fig. 11). The younger peaks are consistent with the ages of synrift volcanism represented in the pre-Cuyo Cycle, which has ages varying between c. 182 and c. 219 Ma (Rapela et al. 1983; Pángaro et al. 2002; Franzese et al. 2006; Schiuma & Llambías 2008). Rocks with these ages are well exposed along the Huincul High and in the Cordillera del Viento. The Late Palaeozoic peaks are consistent with the c. 258 and c. 286 Ma Choiyoi magmatic province (Llambías & Sato 2011, and references therein). These rocks form the core of large basement anticlines in the Huincul High and the Cordillera del Viento, but they are also exposed in the San Rafael Block, in the northeastern margin of the basin (Fig. 1a, b). The smaller amount of ages from the Early Palaeozoic and many minor Precambrian ages might represent evidence of metamorphic basement sources at the western margin of the basin (Fig. 1b). In the southern part, this basement is extensively exposed and corresponds to the Piedra Santa Formation (Franzese 1995) with detrital zircon ages of c. 364 Ma (Ramos et al. 2010). Alternatively, the Colohuincul Complex may also be mentioned

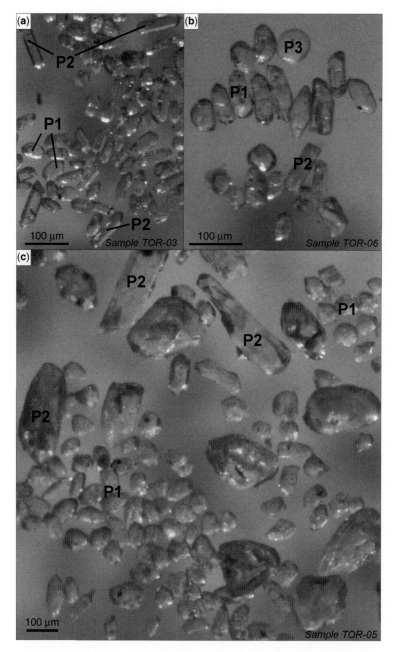

Fig. 7. Stereo microscope images of the zircons in the studied samples: (**a**) TOR-03, (**b**) TOR-06 and (**c**) TOR-05.

(Lucassen et al. 2004). To the north, in the Cordillera del Viento, there is also a small outcrop of metamorphic basement known as the Guaraco Norte Formation (Zappettini et al. 1987) that could have acted as a source region at this latitude (Figs 1b & 2). Further outcrops of the metamorphic basements grouped in the Huechulaufquen Formations are discontinuously exposed along the Patagonian Cordillera (Pankhurst et al. 2006; Ramos 2008).

To the SE of the basin, Palaeozoic, Triassic and Lower Jurassic magmatic rocks are widely

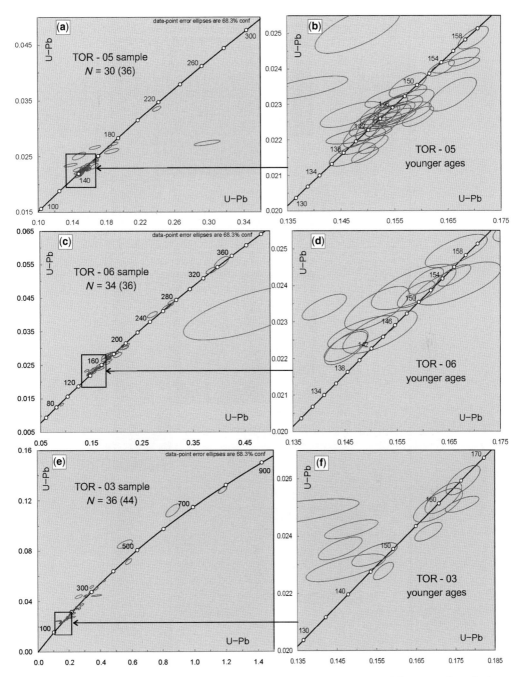

Fig. 8. U–Pb concordia plots: (**a, b**) TOR-05, (**c, d**) TOR-06, (**e, f**) TOR-03. The detrital zircon ages shown in the concordia plots are the concordant ages (in green) and discordant ages (in red) discussed in the text. For interpretation of the references to colour in this figure legend, the reader is referred to the web version of this article.

exposed in the Somún Curá Massif (Cingolani *et al.* 1991; Rapela *et al.* 1991, 2005; Varela *et al.* 2005; Pankhurst *et al.* 2006) (Fig. 1a). Several authors have interpreted that the detrital influx in the southwestern depocentre was supplied by this region (Spalletti & Veiga 2007; Veiga & Spalletti

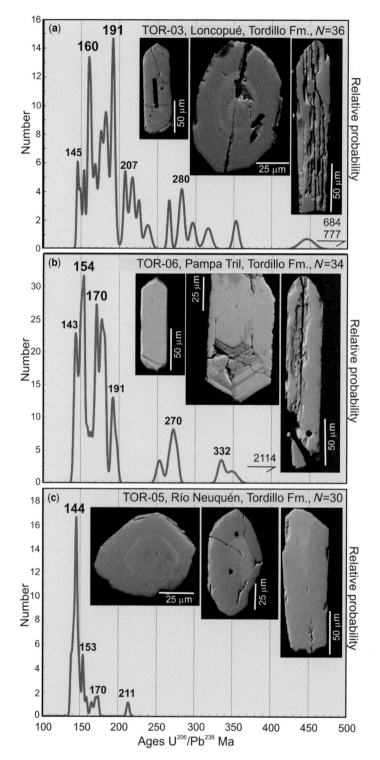

Fig. 9. Relative probability curves of Pb^{206}–U^{238} ages on detrital zircons from the Tordillo Formation: (a) TOR-03, (b) TOR-06 and (c) TOR-05. Inset are BSE images from the study samples.

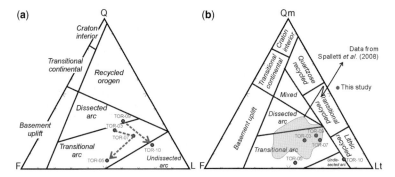

Fig. 10. Provenance data from petrography of the Tordillo Formation using the schemes of Dickinson *et al.* (1983): (**a**) quartz, feldspar and lithic fragments (Q, F, L); (**b**) monocrystalline quartz, feldspar, total lithic grains (Qm, F, Lt).

2007). However, palaeocurrent measurements in the southwestern depocentre indicate an orientation of drainage opposed to the Somún Curá Massif (Zavala *et al.* 2005). Consequently, the source region of detrital material supplied should be the Huincul High. This basement high might also have acted as a barrier to sediment provenance from further south (Naipauer *et al.* 2012).

Regional and temporal changes in the sedimentary provenance of the Tordillo Formation

Detailed analysis of the sandstone petrography and patterns of the detrital zircon ages suggests important regional and temporal changes in the sedimentary provenance for the Tordillo Formation.

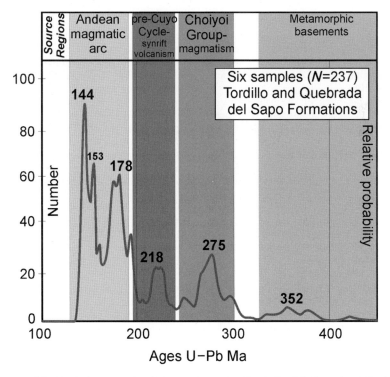

Fig. 11. Summary of the U–Pb zircon ages for all the samples analysed in the Tordillo Formation: TOR-03, TOR-05, TOR-06, TOR-07, TOR-09 and TOR-10. Within the main peaks of ages, the possible source areas are interpreted.

From an initial examination, two different source areas are distinguished based on the amount of polycrystalline quartz and differences in the lithic composition. The samples from the southwestern depocentre have higher amounts of polycrystalline quartz, probably related to the presence of a metamorphic source. This metamorphic source is also seen in the detailed composition of the rock fragment framework. On the other hand, samples from the northwestern depocentre contain a large amount of monocrystalline quartz together with the larger amount of sedimentary and altered rock fragments. Although the proportion and composition of volcanic lithic fragments are similar in both depocentres, it is clear that in the southwestern depocentre the presence of metamorphic and vitric volcanic rock fragments indicates sources within the Huincul High or from the Patagonian Cordillera.

Analysing separately the volcanic lithic fragments in both depocentres, it becomes clear that there is a change in the proportion of acid v. basic textures, the acid textures (granular and seriate) being more common at the base of the unit and the basic textures (microlithic and lathwork) more common at the top. This temporal variation could be interpreted as a change of the volcanic rock source from the Choiyoi magmatic province at the base to the Andean magmatic arc at the top. If we analyse the southwestern depocentre, the provenance changes from a dissected to an undissected arc (Fig. 10b). Analysing the northwestern depocentre, a similar pattern is observed, with sources of sediments changing from a dissected to a transitional arc (Fig. 10b).

The detailed analyses of the patterns of the detrital zircon ages are consistent with petrographic studies. Samples TOR-03, TOR-07 and TOR-09 taken from the base of the sequence are characterized by the predominance of Carboniferous to Early Jurassic zircon ages, representing more than 50% of the total number of grains for each sample (Fig. 12). The Late Triassic to Early Jurassic ages (pre-Cuyo Cycle) are represented by 40% of the zircon grains, and the relative abundances of the less frequent Late Palaeozoic ages (Choiyoi Group) are between 14 and 36%. The Precambrian to Early Palaeozoic ages (<11%) indicate less important metamorphic sources (Fig. 12). These samples also have Middle to Upper Jurassic detrital zircon grains with abundance of less than 40%. Samples from the top of the sequence have a predominance of zircon ages from the Mid to Early Jurassic (TOR-06, 62%; TOR-05, 97%), and the older ages that characterize the basal units disappear (Fig. 12). This temporal variation in the provenance of the detrital zircons suggests that, at the beginning of the sedimentation, Carboniferous to Lower Jurassic magmatic rocks and minor Precambrian to Lower Palaeozoic metamorphic areas were the most important sources. Towards the top of the sequence the basement rocks seem to have been eroded or were covered, while the Andean magmatic arc becomes much more important, predominating as the main source region (Fig. 12).

There are also some variations in the provenance patterns according to sample location (Fig. 12). In the northwestern depocentre (samples TOR-03, TOR-05, TOR-6 and TOR-07) the Andean arc sources are predominant, while in the southwestern depocentre (samples TOR-09 and TOR-10) basement sources (pre-Cuyo Cycle, Choiyoi Group and metamorphic complexes) are more prominent, but there are also maximum peaks indicating sources in the Andean arc (Fig. 12). The complete U–Pb dataset confirms the relevance of the Huincul deformation zone as a positive topographic region in the southern part of the basin, which acted as a source area during the Late Jurassic and also divided the Neuquén Basin into two depocentres (Naipauer *et al.* 2012).

Changes in the provenance patterns and the tectonic evolution of the Neuquén Basin

Several second-order lowstand wedges were recognized in the Neuquén Basin during the retroarc stage (Jurassic to Early Cretaceous) (Howell *et al.* 2005; Spalletti & Veiga 2007). The patterns of detrital zircon ages in the Kimmeridgian (Tordillo Formation) and the Hauterivian (Avilé Member of the Agrio Formation; Fig. 2) lowstands show significant differences. These are shown by the variations in the relative frequency peaks corresponding to maximum contributions from the Andean magmatic arc, from the basements of the basin or from the eastern foreland regions (Fig. 13).

A lowstand wedge is developed during the Late Jurassic in the Kimmeridgian to early Tithonian, represented by the Tordillo Formation (Spalletti & Veiga 2007). This stage is characterized by detrital zircon supply from the Andean magmatic arc, which is coeval or a little older than the sedimentation. Local variations in the patterns of detrital zircon ages occur principally at the beginning of sedimentation and in the Huincul deformation zone south of the basin (Fig. 11).

A major inundation of the Neuquén Basin occurred after the Kimmeridgian lowstand wedge that marked the beginning of the marine deposition of the Mendoza Group (Legarreta & Uliana 1991). However, another important lowstand wedge was developed in the Hauterivian during the deposition

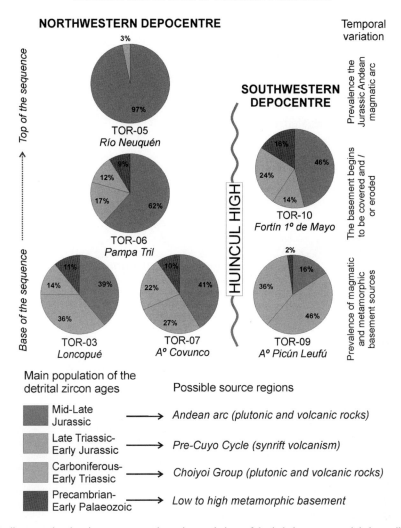

Fig. 12. Pie diagrams showing, in percentages, the main populations of detrital zircon ages and their possible source areas in the Neuquén Basin.

of the Agrio Formation. This is represented by the Avilé Member (Legarreta & Gulisano 1989; Veiga et al. 2007, 2011). Two fluvial sandstones at the base of this member were dated by U–Pb SHRIMP on detrital zircons (Tunik et al. 2010); these samples were taken in Tricao Malal and Pichaihue in the western central part of the basin (Fig. 13), where fluvial depositional systems dominate (Veiga et al. 2011). Detrital zircons from this member show a significant change in the age pattern in comparison with the Tordillo Formation. The Hauterivian lowstand wedge has a decrease in the relative abundance of the Late Jurassic and Cretaceous zircon ages, as well as an increase in the abundance of the older ages (Tunik et al.

2010). The prominent age peaks show provenance from the synrift volcanism (pre-Cuyo Cycle), from the basement of the basin (Choiyoi Group) and from the eastern foreland cratonic region (Grenvillian, Pampean and Famatinian magmatic arcs) (Fig. 13). This provenance pattern is probably linked with the growth of topographic barriers such as the Huincul High in the south or the Cordillera del Viento range in the NW. Although there are no Late Jurassic and Early Cretaceous detrital zircon ages in the Avilé Member, petrographic studies have demonstrated that the magmatic arc supplied volcanic material into the basin during the Early Cretaceous (Eppinger & Rosenfeld 1996). This source is only detected in the fall tuff

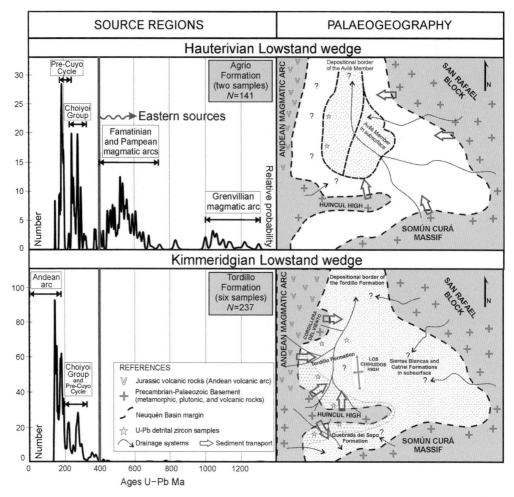

Fig. 13. Summary of the U–Pb zircon ages for all the samples from the Tordillo Formation and the Avilé Member of the Agrio Formation with the Kimmeridgian and Hauterivian lowstand wedges palaeogeography in the Neuquén Basin. Note the potential source areas for the basin: a western magmatic arc and emerged areas along the east–west oriented Huincul deformational zone as well as northeastern and southeastern basement margins.

layers of the Agrio Formation (Aguirre-Urreta et al. 2008). Consequently, the present data suggest that the major changes in the patterns of detrital zircon ages during the Late Jurassic and Early Cretaceous possibly reflect global sea-level changes and could also be a result of positive topographic sediment barriers.

Our zircon data confirm previous studies by Legarreta & Uliana (1991) for the Kimmeridgian lowstand wedge. They interpreted that the arc played a considerable role as a source region during sea-level lowstands, but this has not been recognized in the less significant Hauterivian lowstand wedge where the foreland cratonic area acted as the main source region.

Maximum depositional and minimum biostratigraphic ages

The youngest U–Pb ages of detrital zircons are commonly used to constrain the maximum depositional ages of Precambrian and metamorphic sedimentary rocks where the fossiliferous content is inconclusive (Fedo et al. 2003). It is also used in unfossiliferous Palaeozoic, Mesozoic and Cenozoic sequences. Dickinson & Gehrels (2009) have proposed and discussed four alternative measures to constrain the maximum depositional age using thousands of U–Pb data from Mesozoic sedimentary units with biostratigraphic age control. We discuss our data in terms of the two most statistically robust alternatives proposed by these authors.

One of the measures of the maximum depositional age is the youngest single grain age, although this age might be altered due to Pb loss (Dickinson & Gehrels 2009). In the analysed samples, the youngest single grain ages are 145.2 Ma (TOR-03), 142.2 Ma (TOR-06) and 137.6 Ma (TOR-05). The samples analysed from the southern part of the basin have younger single grain ages at c. 143.7 Ma (TOR-07), 147.5 Ma (TOR-09) and 140.2 Ma (TOR-10). Therefore, there are small variations among the analysed samples, and the younger ages (i.e. 137.6 Ma) are probably due to Pb loss, because their discordances are slightly higher (see tables in the Supplementary Material).

The second most statistically robust approach is to considerer the youngest graphical age peak as the most reliable maximum age (Dickinson & Gehrels 2009), which in the Tordillo Formation is at c. 145 Ma (TOR-03), c. 143 Ma (TOR-06) and c. 144 Ma (TOR-05). Samples from the southern part of the basin have similar maximum age peaks, but they are a little older: c. 144 Ma (TOR-07) for the Tordillo Formation and c. 151 Ma (TOR-09) and c. 150 Ma (TOR-10) for the Quebrada del Sapo Formation. Moreover, the summary U–Pb age probability curves for all six analysed samples (237 concordant ages) shows a statistically robust youngest age peak at c. 144 Ma (Fig. 11). This result is in agreement with the individual youngest graphical age peak for each sample analysed and we consider it the most reliable value for the maximum depositional age for the Tordillo and Quebrada del Sapo formations.

The minimum relative age of the Tordillo Formation is well constrained by the concordant deposition of the overlying Vaca Muerta Formation. The age of the base of this unit is late early Tithonian while the top is early Valanginian, based on its abundant ammonoid fauna and other biostratigraphic markers (Leanza 1981; Riccardi 2008a, b; Leanza et al. 2012; Aguirre-Urreta et al. 2011). According to the chronostratigraphic Mesozoic timescale of Ogg (2004), the boundaries of the Tithonian are, at the base, 150.8 ± 4 Ma and, at the top, 145.5 ± 4 Ma (Fig. 2), the latter being the present absolute age for the Jurassic–Cretaceous boundary accepted by the International Commission of Stratigraphy of the IUGS (Ogg 2004). Recent studies by Ogg & Hinnov (2012), moved the base of the Tithonian to an older age, and propose a limit at 152.1 Ma. However, in the Tordillo Formation, the U–Pb detrital zircon ages indicate a statistically robust measure of the maximum depositional age at c. 144 Ma. This absolute age is at least 7 Ma younger than the value in the chronostratigraphic Mesozoic timescale.

It should be noted that U–Pb detrital zircon ages separated from the Lagunillas Formation, which is the temporal and facies equivalent to the Tordillo Formation in the Tarapacá Basin, northern Chile, yielded a group of younger ages between 148.5 ± 5.1 and 144.6 ± 3.5 Ma (Oliveros et al. 2012). Furthermore, volcanic zircons separate from ignimbrite and tuff levels interlayered in this volcaniclastic sequence yielded similar ages between c. 144 and 151 Ma (Labbé et al. 2012, and references therein). This age group is the equivalent to that found in the younger population of zircons from the Tordillo Formation.

Moreover, south of the Neuquén Basin, between 43°S and 55°S in Chilean and Argentinian Patagonia, another retroarc basin known as the Austral Basin was developed. In the northwestern region a marine transgression started in the Tithonian until the early Hauterivian with the deposition of volcanic sandstone and pelitic successions with ammonoid and bivalve faunas (De La Cruz et al. 1996). The marine sediments cover a thick sequence of volcanic rocks named the Ibáñez Formation from the Mid to Late Jurassic. These volcanic rocks were recently dated by U–Pb SHRIMP on zircons at c. 136–140 Ma (Suárez et al. 2009). These ages also show a discrepancy between the radiometric ages and the biostratigraphy for the Late Jurassic to Early Cretaceous in the Austral Basin. This discrepancy seems to extend through several basins in the southwestern Andean margin.

Pálfy et al. (2000) also clearly exposed the discrepancies in the absolute ages for the Jurassic stage boundaries and they presented new U–Pb and Ar/Ar ages. The proposed upper Tithonian boundary was at c. 141.8 Ma, which is more in accordance with the ages obtained in the Tordillo Formation from the Neuquén Basin. Geochronological work in progress on U–Pb ages (TIMS and SHRIMP) in zircons separated from tuffs interlayered in fossiliferous sequences of the Vaca Muerta Formation will shed further light on this discussion.

Implications

(1) The summary of U–Pb detrital zircon age probability curves for all analysed samples for the Tordillo Formation indicates that the most important sediment source was the Jurassic Andean magmatic arc (peaks at c. 144, c. 153 and c. 178 Ma). Also, two secondary source regions are defined according to the prominent peaks at c. 218 and c. 275 Ma, which are consistent with ages from the pre-Cuyo Cycle and the Choiyoi magmatic province, respectively. Less abundant age

groups from the Early Palaeozoic to Precambrian could represent evidence of metamorphic basement sources for the basin.

(2) Temporal and regional changes in the provenance patterns were defined in the Tordillo Formation. The temporal variation indicates that, at the beginning of sedimentation, Carboniferous to Lower Jurassic magmatic rocks and Precambrian to Lower Palaeozoic metamorphic basement rocks were the most important sources of sediment supply. Towards the top of the unit the basement rocks were eroded and/or covered, and the Andean magmatic arc becomes the main source region. Some geographic variations in the provenance patterns throughout the basin were also indentified; in the northwestern depocentre the Jurassic arc is the predominant source, but in the southwestern depocentre the basement sources (pre-Cuyo Cycle, Choiyoi Group and metamorphic complex) are the most important, although there are also significant zircon ages from the Late Jurassic arc. This confirms the relevance of the Huincul deformation zone as a positive element and source region in the southern part of the Neuquén Basin for the Jurassic.

(3) The comparison between the provenance patterns from the Kimmeridgian and the Hauterivian lowstand wedges showed that the major changes are possibly related to global sea-level changes and the growth of topographic barriers such as the Huincul High in the south or the Cordillera del Viento range in the NW. Our zircon data confirm that the arc region played a considerable role as a source region for the Tordillo Formation during these sea-level lowstands. However, it is not possible to recognize this arc source in the smaller Hauterivian lowstand wedge. In the Avilé Member of the Agrio Formation, the Late Jurassic and Cretaceous zircon ages are not present and therefore the foreland area acted as the main source. Possibly, the lack of the Andean detrital zircons in the Avilé Member indicates that the volcanic arc was retracted to the trench at this time.

(4) The results obtained show a statistically robust measure of the maximum deposition age for the Tordillo Formation. We used the youngest graphical age peak and this is defined by a prominent peak composed of 37 concordant zircon ages at *c.* 144 Ma. This younger age indicates a discrepancy of at least 7 Ma with the absolute ages for the Kimmeridgian and Tithonian boundaries from the chronostratigraphic timescale of Ogg & Hinnov (2012). This also raises questions about the true absolute age of the Jurassic–Cretaceous boundary. However, these younger ages are more in accordance with the ages obtained by Pálfy *et al.* (2000) where the upper Tithonian boundary was proposed at 141.8 Ma.

Maximiliano Naipauer acknowledges the financial support of CONICET, the Bunge & Born Foundation and ANPCyT PICT-2010-2099, Argentina. Marcio Pimentel acknowledges continuous financial support from CNPq, Brazil. B. Aguirre-Urreta and E. García Morabito are thanked for their comments and suggestions for the final version of the manuscript. We would like to thank the two anonymous referees for providing us with constructive comments and suggestions. This is contribution **R-81** of the Instituto de Estudios Andinos '*Don Pablo Groeber*' (CONICET-UBA).

References

AGUIRRE-URRETA, B. & RAWSON, P. F. 2012. Lower Cretaceous ammonites from the Neuquén Basin, Argentina: a new heteromorph fauna from the uppermost Agrio Formation. *Cretaceous Research*, **35**, 208–216.

AGUIRRE-URRETA, B., LAZO, D. G. ET AL. 2011. Megainvertebrados del Cretácico y su importancia bioestratigráfica. *In*: LEANZA, H. A., ARREGUI, C., CARBONE, O., DANIELI, J. C. & VALLÉS, J. M. (eds) *Geología y Recursos Naturales de la Provincia de Neuquén*. Relatorio del VXIII Congreso Geológico Argentino, Buenos Aires, 465–488.

AGUIRRE-URRETA, M. B. & RAWSON, P. F. 1999. Stratigraphic position of *Valanginites*, *Lissonia* and *Acantholissonia* in the Lower Valanginian (Lower Cretaceous) sequence of the Neuquén Basin, Argentina. *In*: OLÓRIZ, F. & RODRIGUEZ-TOVAR, F. J. (eds) *Advancing Research on Living and Fossil Cephalopods*. Plenum Press, New York, 521–529.

AGUIRRE-URRETA, M. B., PAZOS, P. J., LAZO, D. G., FANNING, C. M. & LITVAK, V. D. 2008. First U–Pb SHRIMP age of the Hauterivian stage, Neuquén Basin, Argentina. *Journal of South American Earth Sciences*, **26**, 91–99.

ALBAREDE, F., TELOUK, P., BLICHERT-TOFT, J., BOYET, M., AGRANIER, A. & NELSON, B. 2004. Precise and accurate isotopic measurements using multiple-collector ICPMS. *Geochimica et Cosmochimica Acta*, **68**, 2725–2744.

BAHLBURG, H., VERVOORT, J. D., DU FRANE, S. A., BOCK, B., AUGUSTSSON, C. & REIMANN, C. 2009. Timing of crust formation and recycling in accretionary orogens: insights learned from the western margin of South America. *Earth-Science Reviews*, **97**, 215–241.

BAHLBURG, H., VERVOORT, J. D. & DU FRANE, S. A. 2010. Plate tectonic significance of Middle Cambrian and Ordovician siliciclastic rocks of the Bavarian Facies, Armorican Terrane Assemblage, Germany – U–Pb and Hf isotope evidence from detrital zircons. *Gondwana Research*, **17**, 223–235.

BUHN, B., PIMENTEL, M. M., MATTEINI, M. & DANTAS, E. 2009. High spatial resolution analysis of Pb and U isotopes for geochronology by laser ablation multi-collector inductively coupled plasma mass spectrometry (LA-MC-ICP-MS). *Anais da Academia Brasileira de Ciências*, **81**, 99–114.

CARBONE, O., FRANZESE, J., LIMERES, M., DELPINO, D. & MARTÍNEZ, R. 2011. El Ciclo Precuyano (Triásico Tardío – Jurásico Temprano) en la Cuenca Neuquina. *In*: LEANZA, H. A., ARREGUI, C., CARBONE, O., DANIELI, J. C. & VALLÉS, J. M. (eds) *Geología y Recursos Naturales de la Provincia de Neuquén. Relatorio del VXIII Congreso Geológico Argentino*, Buenos Aires, 63–75.

CASTRO, A., MORENO-VENTAS, I. ET AL. 2011. Petrology and SHRIMP U–Pb zircon geochronology of Cordilleran granitoids of the Bariloche area, Argentina. *Journal of South American Earth Sciences*, **32**, 508–530.

CHARRIER, R., PINTO, L. & RODRÍGUEZ, M. 2007. Tectonostratigraphic evolution of the Andean Orogen in Chile. *In*: GIBBONDS, W. & MORENO, T. (eds) *The Geology of Chile*. Geological Society, London, Special Publications, 21–116.

CINGOLANI, C., DALLA SALDA, L., HERVÉ, F., MUNIZAGA, F., PANKHURST, R. J., PARADA, M. A. & RAPELA, C. W. 1991. The magmatic evolution of northern Patagonia; new impressions of pre-Andean and Andean tectonics. *In*: HARMON, R. S. & RAPELA, C. W. (eds) *Andean Magmatism and its Tectonic Setting*. Geological Society of America, Boulder, CO, Special Papers, **265**, 29–44.

CINGOLANI, C. A., ZANETTINI, J. C. M. & LEANZA, H. A. 2011. El basamento ígneo metamórfico. *In*: LEANZA, H. A., ARREGUI, C., CARBONE, O., DANIELI, J. C. & VALLÉS, J. M. (eds) *Geología y Recursos Naturales de la Provincia de Neuquén. Relatorio del VXIII Congreso Geológico Argentino*, Buenos Aires, 37–47.

DE LA CRUZ, R., SUÁREZ, M., COVACEVICH, V. & QUIROZ, D. 1996. Estratigrafía de la zona de Palena y Futaleufú (43°15′–43°45′ Latitud S), X Región, Chile. *XIII Congreso Geológico Argentino and III Congreso de Exploración de Hidrocarburos*, Buenos Aires, Actas, **I**, 417–424.

DICKINSON, W. R. & GEHRELS, G. E. 2009. Use of U–Pb ages of detrital zircons to infer maximum depositional ages of strata: a test against a Colorado Plateau Mesozoic database. *Earth and Planetary Science Letters*, **288**, 115–125.

DICKINSON, W. R., BEARD, L. S. ET AL. 1983. Provenance of North American Phanerozoic sandstone in relation to tectonic setting. *Geological Society of American Bulletin*, **94**, 222–235.

DI GIULIO, A., RONCHI, A., SANFILIPPO, A., TIEPOLO, M., PIMENTEL, M. & RAMOS, V. A. 2012. Detrital zircon provenance from the Neuquén Basin (South Central Andes): Cretaceous geodynamic evolution and sedimentary response in a retroarc–foreland basin. *Geology*, **40**, 559–562.

DIGREGORIO, J. H. 1972. Neuquén. *In*: LEANZA, H. F. (ed.) *Geología Regional Argentina*. Academia Nacional de Ciencias, Córdoba, 439–505.

DIGREGORIO, R. E., GULISANO, C. A., GUTIÉRREZ-PLEIMLING, A. R. & MINITTI, S. A. 1984. Esquema de la evolución geodinámica de la Cuenca Neuquina y sus implicancias paleogeográficas. *IX Congreso Geológico Argentino*, San Carlos de Bariloche, Actas, **II**, 147–162.

DOTT, R. H. 1964. Wacke, greywacke and matrix – what approach to immature sandstone classification? *Journal of Sedimentary Petrology*, **34**, 625–632.

EPPINGER, K. J. & ROSENFELD, U. 1996. Western margin and provenance of sediments of the Neuquén Basin (Argentina) in the Late Jurassic and Early Cretaceous. *Tectonophysics*, **259**, 229–244.

FEDO, C. M., SIRCOMBE, K. N. & RAINBIRD, R. H. 2003. Detrital zircon analysis of the sedimentary record. *In*: HANCHAR, J. M. & HOSKIN, P. W. O. (eds) *Zircon, Reviews in Mineralogy and Geochemistry*. Mineralogical Society of America, Washington DC, **53**, 277–303.

FOLK, R. L., ANDREWS, P. B. & LEWIS, D. W. 1970. Detrital sedimentary rock classification and nomenclature for use in New Zealand. *New Zealand Journal of Geology and Geophysics*, **13**, 937–968.

FRANZESE, J. R. 1995. El Complejo Piedra Santa (Neuquén, Argentina): parte de un cinturón metamórfico neopalaeozoico del Gondwana suroccidental. *Revista Geológica de Chile*, **22**, 193–202.

FRANZESE, J. R. & SPALLETTI, L. A. 2001. Late Triassic–early Jurassic continental extension in southwestern Gondwana: tectonic segmentation and pre-break-up rifting. *Journal of South American Earth Sciences*, **14**, 257–270.

FRANZESE, J. R., VEIGA, G. D., SCHWARZ, E. & GÓMEZ-PÉREZ, I. 2006. Tectono-stratigraphic evolution of a Mesozoic graben border system: the Chachil depocentre, southern Neuquén Basin, Argentina. *Journal of the Geological Society, London*, **163**, 207–221.

GROEBER, P. 1946. Observaciones geológicas a lo largo del meridiano 70. 1. Hoja Chos Malal. *Revista de la Sociedad Geológica Argentina*, **1**, 177–208.

GULISANO, C. A. 1981. El ciclo cuyano en el norte de Neuquén y sur de Mendoza. *XVIII Congreso Geológico Argentino*, San Juan, Actas, **III**, 573–592.

GULISANO, C. A. 1988. *Análisis estratigráfico y sedimentológico de la Formación Tordillo en el oeste de la Provincia del Neuquén, Cuenca Neuquina, Argentina*. PhD thesis, Facultad de Ciencias Exactas y Naturales, Universidad de Buenos Aires.

GULISANO, C. A. & GUTIÉRREZ PLEIMLING, A. R. 1995. Field guide: the Jurassic of the Neuquén Basin. a) Neuquén province. Asociación Geológica Argentina, Buenos Aires, Serie E, 1–111.

GULISANO, C. A., GUTIÉRREZ PLEIMLING, A. R. & DIGREGORIO, R. E. 1984. Esquema estratigráfico de la secuencia jurásica del oeste de la Provincia de Neuquén. *IX Congreso Geológico Argentino*, San Carlos de Bariloche, Actas, **I**, 236–259.

HOWELL, J. A., SCHWARZ, E., SPALLETTI, L. A. & VEIGA, G. D. 2005. The Neuquén Basin: an overview. *In*: VEIGA, G. D., SPALLETTI, L. A., HOWELL, J. A. & SCHWARZ, E. (eds) *The Neuquén Basin, Argentina: A Case Study in Sequence Stratigraphy and Basin Dynamics*. Geological Society, London, Special Publications, **252**, 1–14.

JACKSON, S. E., PEARSON, N. J., GRIFFIN, W. L. & BELOUSOVA, E. A. 2004. The application of laser ablation-inductively coupled plasma-mass spectrometry to *in* situ U–Pb zircon geochronology. *Chemical Geology*, **211**, 47–69.

KAMO, S. L. & RICCARDI, A. C. 2009. A new U–Pb zircon age for an ash layer at the Bathonian–Callovian boundary, Argentina. *GFF*, **131**, 177–182.

KOŠLER, J., FONNELAND, H., SYLVESTER, P., TUBRETT, M. & PEDERSEN, R. B. 2002. U–Pb dating of detrital zircons for sediment provenance studies – a comparison of laser ablation ICP-MS and SIMS techniques. *Chemical Geology*, **182**, 605–618.

LABBÉ, M., SALAZAR, E., ROSSEL, P., MERINO, R. & OLIVEROS, V. 2012. Variaciones laterales en la arquitectura estratigráfica del Jurásico Superior en el valle del Tránsito: ¿Evidencias del desarrollo de un rift continental? *XIII Congreso Geológico Chileno*, Antofagasta, Actas, 718–720.

LEANZA, H. A. 1981. The Jurassic–Cretaceous boundary beds in West Central Argentina and their ammonite zones. *Neues Jahrbuch für Geologie und Paläontologie, Abhandlungen*, **161**, 62–92.

LEANZA, H. A. 1992. Estratigrafía del Paleozoico y Mesozoico anterior a los Movimientos Intermálmicos en la comarca del cerro Chachil, provincia del Neuquén. *Revista de la Asociación Geológica Argentina*, **45**, 272–299.

LEANZA, H. A., SATTLER, F., MARTINEZ, R. S. & CARBONE, O. 2012. La Formación Vaca Muerta y equivalentes (Jurásico Tardío–Cretácico Temprano) en la Cuenca Neuquina. *In*: LEANZA, H. A., ARREGUI, C., CARBONE, O., DANIELI, J. C. & VALLÉS, J. M. (eds) *Geología y Recursos Naturales de la Provincia de Neuquén. Relatorio del VXIII Congreso Geológico Argentino*, Buenos Aires, 113–129.

LEGARRETA, L. & GULISANO, C. A. 1989. Análisis estratigráfico secuencial de la Cuenca Neuquina (Triásico superior–Terciario inferior, Argentina). *In*: CHEBLI, G. & SPALLETTI, L. (eds) *Cuencas Sedimentarias Argentinas*. Universidad Nacional de Tucumán, Tucumán, Serie Correlación Geológica, **6**, 221–243.

LEGARRETA, L. & ULIANA, M. A. 1991. Jurassic–Cretaceous marine oscillations and geometry of backarc basin fill, central Argentine Andes. *In*: MACDONALD, D. I. M. (ed.) *Sedimentation, Tectonics and Eustasy, Sea Level Changes at Active Plate Margins*. International Association of Sedimentologists, Oxford, Special Publications, **12**, 429–450.

LEGARRETA, L. & ULIANA, M. A. 1999. EL Jurásico y Cretácico de la Cordillera Principal y Cuenca Neuquina. *In*: CAMINOS, R. (ed.) *Geología Argentina*. Instituto de Geología y Recursos Minerales, Buenos Aires, Anales, **29**, 399–432.

LLAMBÍAS, E. J. & SATO, A. M. 2011. Ciclo Gondwánico: la provincia magmática Choiyoi en Neuquén. *In*: LEANZA, H. A., ARREGUI, C., CARBONE, O., DANIELI, J. C. & VALLÉS, J. M. (eds) *Geología y Recursos Naturales de la Provincia de Neuquén*. Relatorio del VXIII Congreso Geológico Argentino, Buenos Aires, 53–62.

LLAMBÍAS, E. J., LEANZA, H. A. & CARBONE, O. 2007. Evolución tectono-magmática durante el Pérmico al Jurásico Temprano en la Cordillera del Viento (37°05′S–37°15′S): nuevas evidencias geológicas y geoquímicas del inicio de la cuenca Neuquina. *Revista de la Asociación Geológica Argentina*, **62**, 217–235.

LÓPEZ-GÓMEZ, J., MARTÍN-CHIVELET, J. & PALMA, R. M. 2009. Architecture and development of the alluvial sediments of the Upper Jurassic Tordillo Formation in the Cañada Ancha Valley, northern Neuquén Basin, Argentina. *Sedimentary Geology*, **219**, 180–195.

LUCASSEN, F., TRUMBULL, R., FRANZ, G., CREIXELL, C., VÁSQUEZ, P., ROMER, R. L. & FIGUEROA, O. 2004. Distinguishing crustal recycling and juvenile additions at active continental margins: the Paleozoic to Recent compositional evolution of the Chilean Pacific margin (36–41°S). *Journal of South American Earth Sciences*, **17**, 103–119.

LUDWIG, K. R. 2003. *Isoplot 3.00: a geochronological toolkit for Microsoft Excel*. Berkeley Geochronological Center, Berkeley, CA, Special Publications, **4**, 70.

MARCHESE, H. G. 1971. Litoestratigrafía y variaciones faciales de las sedimentitas mesozoicas de la Cuenca Neuquina, provincia de Neuquén, República Argentina. *Revista de la Asociación Geológica Argentina*, **26**, 343–410.

MARETTO, H., CARBONE, O., GAZZERA, C. & SCHIUMA, M. 2002. Los reservorios de la Formación Tordillo. *In*: SCHIUMA, M., HINTERWIMMER, G. & VERGANI, G. (eds) *Rocas Reservorio de las Cuencas Productivas de la Argentinas*. V Congreso de Exploración y Desarrollo de Hidrocarburos, Actas CD, Mar del Plata, 335–358.

MESCUA, J. F., GIAMBIAGI, L. B. & BECHIS, F. 2008. Evidencias de tectónica extensional en el Jurásico Tardío (Kimmeridgiano) del suroeste de la provincia de Mendoza. *Revista de la Asociación Geológica Argentina*, **63**, 512–519.

MORTON, A. C. & HALLSWORTH, C. R. 1999. Processes controlling the composition of heavy mineral assemblages in sandstones. *Sedimentary Geology*, **124**, 3–29.

MOSQUERA, A. & RAMOS, V. A. 2006. Intraplate deformation in the Neuquén Embayment. *In*: KAY, S. M. & RAMOS, V. A. (eds) *Evolution of an Andean Margin: A Tectonic and Magmatic View from the Andes to the Neuquén Basin (35°–39° lat)*. Geological Society of America, Boulder, Special Papers, **407**, 97–124.

MPODOZIS, C. & RAMOS, V. A. 2008. Tectónica Jurásica en Argentina y Chile: extensión, subducción oblicua, rifting, deriva y colisiones? *Revista de Asociación Geológica Argentina*, **63**, 481–497.

MUELLER, P. A., HEATHERINGTON, A. L., WOODEN, J. L., SHUSTER, R. D., NUTMAN, A. P. & WILLIAMS, I. S. 1994. Precambrian zircons from the Florida basement: a Gondwana connection. *Geology*, **22**, 119–122.

NAIPAUER, M., GARCÍA MORABITO, E. ET AL. 2012. Intraplate Late Jurassic deformation and exhumation in western central Argentina: constraints from surface data and U–Pb detrital zircon ages. *Tectonophysics*, **524–525**, 59–75.

OGG, J. G. 2004. The Jurassic period. *In*: GRADSTEIN, F., OGG, J. & SMITH, A. (eds) *A Geologic Time Scale*. Cambridge University Press, Cambridge, 307–343.

OGG, J. G. & HINNOV, L. A. 2012. The Jurassic Period. *In*: GRADSTEIN, F. M., OGG, J. G., SCHMITZ, M. D. & OGG, G. M. (eds) *The Geologic Time Scale 2012*. Elsevier, China, 731–791.

OLIVEROS, V., LABBÉ, M., ROSSEL, P., CHARRIER, R. & ENCINAS, A. 2012. Late Jurassic paleogeographic evolution of the Andean back-arc basin: new constrains from the Lagunillas Formation, northern Chile (27°30′–38°30′S). *Journal of South American Earth Sciences*, **37**, 25–40.

PÁLFY, J., SMITH, P. L. & MORTENSEN, J. K. 2000. A U–Pb and ^{40}Ar/^{39}Ar time scale for the Jurassic. *Canadian Journal of Earth Sciences*, **37**, 923–944.

PÁNGARO, F., VEIGA, R. & VERGANI, G. 2002. Evolución tecto-sedimentaria del área de Cerro Bandera, Cuenca Neuquina, Argentina. *V Congreso de Exploración y Desarrollo de Hidrocarburos*, CD-ROM, 16.

PANKHURST, R., RAPELA, C., FANNING, C. & MÁRQUEZ, M. 2006. Gondwanide continental collision and the origin of Patagonia. *Earth-Science Reviews*, **76**, 235–257.

RAMOS, V. A. 1978. Estructura. *VII Congreso Geológico Argentino*, Relatorio, Neuquén, 99–118.

RAMOS, V. A. 1999. Evolución tectónica de la Argentina. *In*: CAMINOS, R. (ed.) *Geología Argentina*. Instituto de Geología y Recursos Minerales, Buenos Aires, Anales, **29**, 715–784.

RAMOS, V. A. 2008. Patagonia: A Paleozoic continent adrift? *Journal of South American Earth Sciences*, **26**, 235–251.

RAMOS, V. A. 2010. The tectonic regime along the Andes: present-day and Mesozoic regimes. *Geological Journal*, **45**, 2–25.

RAMOS, V. A., GARCÍA MORABITO, E., HERVÉ, F. & FANNING, C. M. 2010. Grenville-age sources in Cuesta de Rahue, northern Patagonia: constrains from U–Pb/SHRIMP ages from detrital zircons. *International Geological Congress on the Southern Hemisphere. Bollettino de Geofisica*, **51**, 42–44.

RAMOS, V. A., MOSQUERA, A., FOLGUERA, A. & GARCÍA MORABITO, E. 2011. Evolución tectónica de los Andes y del Engolfamiento Neuquino adyacente. *In*: LEANZA, H. A., ARREGUI, C., CARBONE, O., DANIELI, J. C. & VALLÉS, J. M. (eds) *Geología y Recursos Naturales de la Provincia de Neuquén. Relatorio del VXIII Congreso Geológico Argentino*, Buenos Aires, 335–348.

RAPELA, C. W., SPALLETTI, L. & MERODIO, J. 1983. Evolución magmática y geotectónica de la 'Serie Andesitica' andina (Paleoceno–Eoceno) en la Cordillera Norpatagónica. *Revista de la Asociación Geológica Argentina*, **38**, 469–484.

RAPELA, C. W., DIAS, G., FRANZESE, J., ALONSO, G. & BENVENUTO, A. 1991. El Batolito de la Patagonia Central: evidencias de un magmatismo triásico–jurásico asociado a fallas transcurrentes. *Revista Geológica de Chile*, **18**, 121–138.

RAPELA, C. W., PANKHURST, R. J., FANNING, C. M. & HERVÉ, F. 2005. Pacific subduction coeval with the Karoo mantle plume: the Early Jurassic Subcordilleran belt of northwestern Patagonia. *In*: VAUGHAN, A. P. M., LEAT, P. T. & PANKHURST, R. J. (eds) *Terrane Processes at the Margins of Gondwana*. Geological Society, London, Special Publications, **246**, 217–240.

RICCARDI, A. C. 1984. Las asociaciones de Amonites del Jurásico y Cretácico de Argentina. *IX Congreso Geológico Argentino*, San Carlos de Bariloche, Actas, **IV**, 569–595.

RICCARDI, A. C. 2008a. The marine Jurassic of Argentina: a biostratigraphic framework. *Episodes*, **31**, 326–335.

RICCARDI, A. C. 2008b. El Jurásico de la Argentina y sus amonites. *Revista de la Asociación Geológica Argentina*, **63**, 625–643.

SATO, A. M., LLAMBÍAS, E. J., BASEI, M. A. S. & LEANZA, H. A. 2008. The Permian Choiyoi Cycle in Cordillera del Viento (Principal Cordillera, Argentina): Over 25 Ma of magmatic activity. *VI South American Symposium on Isotope Geology*, San Carlos de Bariloche, Proceedings in CD-ROM, 4.

SCHIUMA, M. & LLAMBÍAS, E. J. 2008. New ages and chemical analysis on Lower Jurassic volcanism close to the Huincul High, Neuquén. *Revista de la Asociación Geológica Argentina*, **63**, 644–652.

SPALLETTI, L. A. & COLOMBO PIÑOL, F. 2005. From alluvial fan to playa: an Upper Jurassic ephemeral fluvial system, Neuquén Basin, Argentina. *Gondwana Research*, **8**, 363–383.

SPALLETTI, L. A. & VEIGA, G. D. 2007. Variability of continental depositional systems during lowstand sedimentation: an example from the Kimmeridgian of the Neuquén Basin, Argentina. *Latin American Journal of Sedimentology and Basin Analysis*, **14**, 85–104.

SPALLETTI, L. A., QUERALT, I., MATHEOS, S. D., COLOMBO, F. & MAGGI, J. 2008. Sedimentary petrology and geochemistry of siliciclastic rocks from the upper Jurassic Tordillo Formation (Neuquén Basin, western Argentina): implications for provenance and tectonic setting. *Journal South American Earth Sciences*, **25**, 440–463.

SPALLETTI, L. A., ARREGUI, C. D. & VEIGA, G. D. 2011. La Formación Tordillo y equivalentes (Jurásico Tardío) en la Cuenca Neuquina. *In*: LEANZA, H. A., ARREGUI, C., CARBONE, O., DANIELI, J. C. & VALLÉS, J. M. (eds) *Geología y Recursos Naturales de la Provincia de Neuquén. Relatorio del VXIII Congreso Geológico Argentino*, Buenos Aires, 99–111.

STACEY, J. S. & KRAMERS, J. D. 1975. Approximation of terrestrial lead isotope evolution by a two-stage model. *Earth and Planetary Science Letters*, **26**, 207–221.

STIPANICIC, P. N. 1951. Sobre la presencia del Oxfordense superior en el arroyo de La Manga. *Revista de la Asociación Geológica Argentina*, **6**, 213–239.

STIPANICIC, P. N. 1966. El Jurásico en Vega de la Veranda (Neuquén), el Oxfordense y el diastrofismo Divesiano (Agassiz-Yaila) en Argentina. *Revista de la Asociación Geológica Argentina*, **20**, 403–478.

STIPANICIC, P. N., RODRIGO, F., BAULIES, O. L. & MARTÍNEZ, C. G. 1968. Las Formaciones presenonianas en el denominado Macizo Nordpatagónico y regiones adyacentes. *Revista de la Asociación Geológica Argentina*, **23**, 367–388.

SUÁREZ, M., DE LA CRUZ, R., FANNING, M. & ETCHART, H. 2008. Carboniferous, Permian and Toarcian magmatism in Cordillera del Viento, Neuquén, Argentina: first U–Pb SHRIMP dates and tectonic implications. *XVII Congreso Geológico Argentino*, San Salvador de Jujuy, Actas, 906–907.

SUÁREZ, M., DE LA CRUZ, R., AGUIRRE-URRETA, B. & FANNING, M. 2009. Relationship between volcanism and marine sedimentation in northern Austral (Aisén) Basin, central Patagonia: stratigraphic, U–Pb SHRIMP and paleontologic evidence. *Journal of South American Earth Sciences*, **27**, 309–325.

TUNIK, M., FOLGUERA, A., NAIPAUER, M., PIMENTEL, M. M. & RAMOS, V. A. 2010. Early uplift and orogenic deformation in the Neuquén Basin: constraints on the Andean uplift from U–Pb and Hf isotopic data of detrital zircons. *Tectonophysics*, **489**, 258–273.

VARELA, R., BASEI, M. A. S., CINGOLANI, C. A., SIGA, O., JR. & PASSARELLI, C. R. 2005. El basamento cristalino de los Andes Norpatagónicos en Argentina: geocronología e interpretación tectónica. *Revista Geológica de Chile*, **32**, 167–187.

VEIGA, G. D. & SPALLETTI, L. A. 2007. The Upper Jurassic (Kimmeridgian) fluvial/aeolian systems of the southern Neuquén Basin, Argentina. *Gondwana Research*, **11**, 286–302.

VEIGA, G. D., SPALLETTI, L. A. & FLINT, S. S. 2007. Anatomy of a fluvial lowstand wedge: the Avilé Member of the Agrio Formation (Hauterivian) in central Neuquén Basin (NW Neuquén province), Argentina. *In*: NICHOLS, G., WILLIAMS, E. & PAOLA, C. (eds) *Sedimentary Environments, Processes and Basins, A Tribute to Peter Friend*. International Association of Sedimentologists, New Jersey, Special Publications, **38**, 341–365.

VEIGA, G. D., SPALLETTI, L. A. & SCHWARZ, E. 2011. El Miembro Avilé de la Formaicón Agrio (Cretácico Temprano). *In*: LEANZA, H. A., ARREGUI, C., CARBONE, O., DANIELI, J. C. & VALLÉS, J. M. (eds) *Geología y Recursos Naturales de la Provincia de Neuquén. Relatorio del VXIII Congreso Geológico Argentino*, Buenos Aires, 161–173.

VERGANI, G. D., TANKARD, A. J., BELOTTI, H. J. & WELSINK, H. J. 1995. Tectonic evolution and paleogeography of the Neuquén Basin, Argentina. *In*: TANKARD, A. J., SUÁREZ SORUCO, R. & WELSINK, H. J. (eds) *Petroleum Basins of South America*. American Association of Petroleum Geologists, Tulsa, Memoirs, **62**, 383–402.

VICENTE, J. C. 2005. Dynamic paleogeography of the Jurassic Andean Basin: pattern of transgression and localization of main straits through the magmatic arc. *Revista de la Asociación Geológica Argentina*, **60**, 221–250.

VICENTE, J. C. 2006. Dynamic paleogeography of the Jurassic Andean Basin: pattern of regression and general considerations on main features. *Revista de la Asociación Geológica Argentina*, **61**, 408–437.

WEAVER, C. E. 1931. Palaeontology of the Jurassic and Cretaceous of west central Argentina. *Memoir of the University of Washington*, **1**, 594.

ZAPPETTINI, E. O., MÉNDEZ, V. & ZANETTINI, J. C. M. 1987. Metasedimentitas mesopaleozoicas en el noroeste de la provincia del Neuquén. *Revista de Asociación Geológica Argentina*, **42**, 206–207.

ZAVALA, C., MARETTO, H. & DI MEGLIO, M. 2005. Hierarchy of bounding surfaces in Aeolian sandstones of the Tordillo Formation (Jurassic). *Neuquén Basin, Argentina. Geologica Acta*, **3**, 133–145.

ZUFFA, G. 1985. Optical analyses of arenites: influence of methodology on compositional results. *In*: ZUFFA, G. G. (ed.) *Provenance of Arenites: NATO-ASI*. Reidel Publication Company, Dordrecht, **148**, 165–189.

The north-western margin of the Neuquén Basin in the headwater region of the Maipo drainage, Chile

ESTANISLAO GODOY

Tehema Consultores Geológicos/Ministerio de Obras Públicas, Morandé 59 Santiago, Chile (e-mail: estanislao.godoy@mop.gov.cl)

Abstract: Volcanic and pyroclastic rocks of intermediate composition, dating from Kimmeridgian time with peak ages during both the early–middle Tithonian and the late Hauterivian, characterize several localities of the north-western margin of the Neuquén basin along the main Andean Range at 33–34°S. The latter is thought responsible for the desiccation of the basin during Aptian–Albian time. The thick Tithonian submarine lavas at the Volcán valley (33°30′S), on the other hand, coincide spatially with the only pass-way recognized at that time in the Chilean Coast Range. This coincidence may be related to an isolated relic rift-like structure that coexisted with the regional intra-arc extensional setting during Tithonian time. Continental-scale rifting in the southern Andes has been recognized only for the Late Triassic–Early Jurassic period.

The Early Cretaceous rocks inside the western-most thrust sheet of the Neuquén Neogene fold-and-thrust belt at Volcán valley are intruded by Oligocene andesitic sills. They have been considered as flows from a stratigraphic unit of that age, the Abanico Formation, and to unconformably cover Late Cretaceous red beds. At this latitude, however, no K–T (Cretaceous–Tertiary) orogenic event is recognized in the Andean Main Range.

Middle Jurassic to uppermost Early Cretaceous sedimentary and volcanic rocks crop out in the headwater region of the Maipo and its tributaries (Fig. 1). They represent the NW margin of the Neuquén basin, a major structure that may be described as a Late Jurassic–Late Cretaceous NNW-oriented back-arc basin superimposed on Late Triassic–Early Jurassic episodic rifting. The rocks of this basin crop out mainly east of the continental divide, a feature that here coincides with the Main Range. At these latitudes (33°20′–34°20′S) however, three Pleistocene–Holocene volcanic complexes (the Tupungato, Marmolejo-San José and the Diamante Caldera, the latter hosting the Maipo strato-cone) have shifted the continental divide eastwards and left most of the basin outcrops in the Pacific watershed.

This paper reviews the available literature on the western margin while adding structural and stratigraphic data collected by students of the Geology Department, University of Chile during the past 30 years in Tithonian–Late Cretaceous rocks from the headwater region of the Maipo drainage. Most of this data are available only in the proceedings of national meetings or unpublished university theses. The area is characterized by rugged topography and many along-strike facies changes involving thick volcanic successions. Most of the information regarding the evolution of the Neuquén basin in the Argentinian Neuquén and Mendoza provinces has been taken from Legarreta & Uliana (1996, 1999).

Because Miocene–Holocene volcanic and pyroclastic rocks cover much of the Main Range, palaeogeographic reconstructions such as Figure 2a for the Tithonian (which shows the western margin of the basin in white) refer to a volcanic arc lying in the present Chilean Coast Range. The outcrops we describe in this paper may help to better understand small areas of that poorly known western margin.

Geological setting

The Yeso and Volcán valleys lie directly west of the Middle Jurassic Yeguas Muertas and Nieves Negras depocentres (Álvarez et al. 1997). Because the inner basinal black pelites of those localities crop out only along the core of broad NNW-oriented anticlines, the westward extension of the depocentres is uncertain. In the profiles presented by Giambiagi & Ramos (2002), for example, covered Triassic–Jurassic rift faults controlling deposition have been extended 30 km to the west. North of the Nieves Negras depocentre, the Neuquén basin was separated during the Middle Jurassic from the Ramada basin by the Alto del Tigre High (Álvarez 1996), an area where a Tithonian transgression occurred above Permian–Triassic rocks of the Cordillera Frontal.

This paper deals mainly with the Kimmeridgian–Late Cretaceous rocks of the Neuquén basin that crop out in the western-most thrust sheet of the northern, basement involved fold-and-thrust belt, excluding the hanging wall of the Fierro thrust. As shown by Godoy et al. (1999) this Late Miocene out-of-sequence thrust juxtaposes at 35°S Oligocene lavas and volcanoclastics (Abanico

From: SEPÚLVEDA, S. A., GIAMBIAGI, L. B., MOREIRAS, S. M., PINTO, L., TUNIK, M., HOKE, G. D. & FARÍAS, M. (eds) 2015. *Geodynamic Processes in the Andes of Central Chile and Argentina*. Geological Society, London, Special Publications, **399**, 155–165. First published online February 27, 2014, updated December 24, 2014, http://dx.doi.org/10.1144/SP399.13 © 2015 The Geological Society of London.
For permissions: http://www.geolsoc.org.uk/permissions. Publishing disclaimer: www.geolsoc.org.uk/pub_ethics

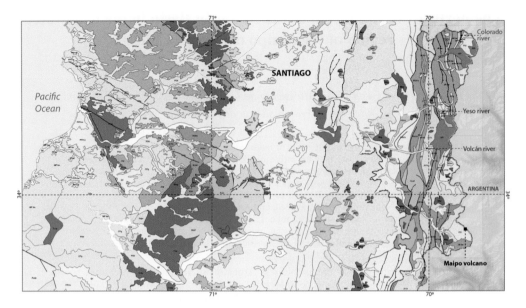

Fig. 1. 1 000 000 Geological Map of Chile (Sernageomin 2002) blown up to twice the scale. Units referred to in the text: Jsg: Middle–Late Jurassic plutonic rocks; Js2c: middle–late continental sedimentary and pyroclastic rocks (Horqueta Formation); OM2c: Oligocene–Miocene volcanic and pyroclastic rocks (Abanico Formation); Kia1c: upper Early Cretaceous continental sediments (Colimapu Formation, here extended to the Turonian); Kia3: Early Cretaceous volcanic rocks; JK1m: Late Jurassic–Early Cretaceous marine sedimentary rocks (Lo Valdés Formation); Js1c: Late Jurassic continental sedimentary rocks (Río Damas Formation).

Formation), first on top of Tithonian limestones and then on Kimmeridgian red conglomerates and sandstones. As it shallows northwards, at the leading edge (33°30′S) it thrusts Miocene lavas on top of the Abanico Formation.

The Kimmeridgian margin

According to Legarreta & Uliana (1996), following the Oxfordian Messinian-style event that desiccated a large tract of the basin and favoured massive precipitation of anhydrite, during Kimmeridgian time 'the basin was largely flooded by siliciclastics and turned into a broad mudflat-salina complex linked to an ephemeral drainage system'. Near the Argentina–Chile boundary and further west, the succession is represented by a '... crudely stratified mixture of immature volcanic debris, tuffs and breccias, lava flows, and locally derived volcanic conglomerates. These rocks record the presence of an emergent-arc and intra-arc terrain, irregularly segmented by axial graben systems (Groeber 1918; Charrier 1984).' No definite western margin for the basin may be established at this stage, unless we consider the Coastal Range Jurassic batholith (see Figs 1, 2a) as such. The stratigraphically well bracketed age of these deposits has been confirmed by 151-Ma-old zircons reported in sandstones of the Rio Damas Formation (Tordillo in Argentina) at the headwater region of the Volcán valley by Aguirre et al. (2009).

The fact that Kimmeridgian volcanism is better developed north of the area discussed in this paper coincides with the emplacement of voluminous Middle–Late Jurassic plutons at that latitude. As pointed out by Charrier et al. (2007), thick Kimmeridgian volcanic and volcano-clastic rocks known as the Horqueta Formation also crop out directly west of those plutons along the Coast Range.

As already noted by Klohn (1960) who defined the Río Damas Formation, this continental unit shows remarkable thickness and lithology variations. At 35°S (the type locality) he reports 5500 m thickness of section, which includes 1400 m of andesitic lavas. On the other hand, 2500 m of mainly conglomerates are present at 34°20′S, while only 2000 m of lavas and conglomerates are estimated at 34°S. Biro-Bagoczky (1964, 1980) reports 3-km-thick continental conglomerates and minor andesitic lavas underlying early Tithonian pelites in the headwaters of the Volcán river valley. According to Thiele (1980), this conglomerate-rich formation is close to 3000 m thick in the middle section of the Colorado valley. At this locality these matrix-rich conglomerates underlie 250 m of

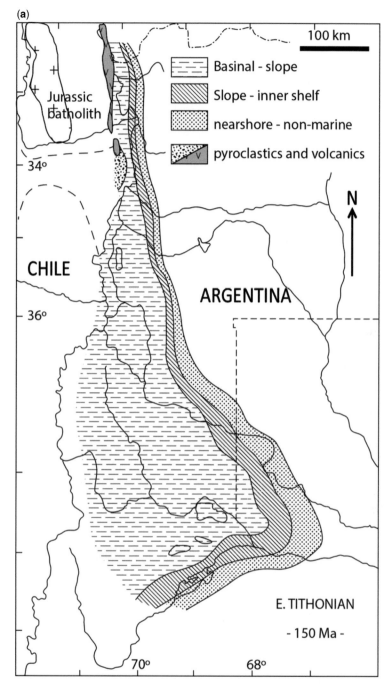

Fig. 2. (a) Early Tithonian palaeogeography of the Neuquén basin. The volcanic rocks that Legarreta & Uliana (1996) restrict to an area around 34°S have been extended northwards and a sea-way to the Pacific sketched at the same latitude. Absolute age taken from the 2012 IUGS chart (also in Fig. 7). (b) Stratigraphic column of Lo Valdés Formation at its type locality (legend translated from Biro-Bagoczky 1980).

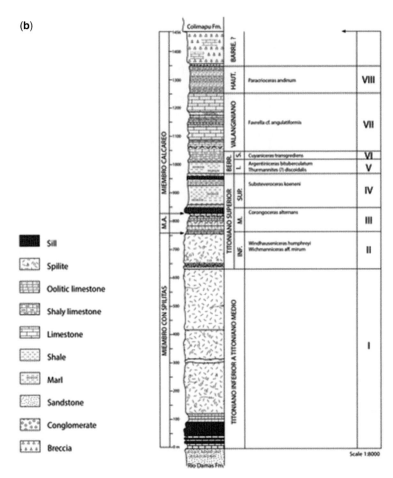

Fig. 2. *Continued.*

Tithonian–Hauterivian sediments that compare well with the Quintuco, Mulichinco and Agrio formations inside the basin (Álvarez *et al.* 2000). The conglomerates are described as having been thrust eastward on top of 2000-m-thick subvertical andesitic lavas of 'uncertain age' that wedge down to 200 m south of the valley. On the other hand, Ramos *et al.* (1991) prefer to interpret the whole conglomerate–lava package as lithologies belonging to a magmatic arc that developed here during Kimmeridgian–early Tithonian time. This assignment is supported by the fact that these rocks lie directly north of the volcanic- and conglomerate-rich eastern section included in the Lo Valdés Formation along the Yeso valley.

Legarreta & Uliana (1996) interpret the different formational names as the result of a setting where a volcanic arc (Río Damas) interfingers with its erosional products (Tordillo). Further north at 32°–33°S, Sanguinetti & Ramos (1993) report that the Tordillo Formation shows local multiple composite centres, cinder cones and up to 400 m of porphyritic basaltic andesite lava flows in its lower sections (Fig. 3). Subaerial exposure promoted by a low-base-level stand probably favoured rapid incision and erosion of the volcanic cores.

Klohn (1960) proposes that the variable thicknesses in the southern part of the area described above may be compared to those in a graben setting. Field evidence for what has been interpreted as a Late Jurassic synrift period has been found in eastern thrust sheets of the fold-and-thrust belt (Pángaro *et al.* 1996; Giambiagi *et al.* 2003).

The Tithonian–Hauterivian margin

During early Tithonian time, extreme base level rise extended the Neuquén Embayment shoreline to locations updip of any previous Jurassic

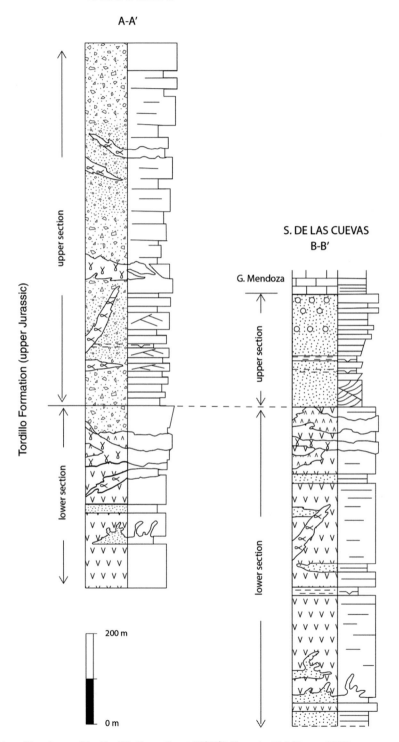

Fig. 3. Stratigraphic column of the Tordillo Formation at 32°50′S (Sanguinetti & Ramos 1993).

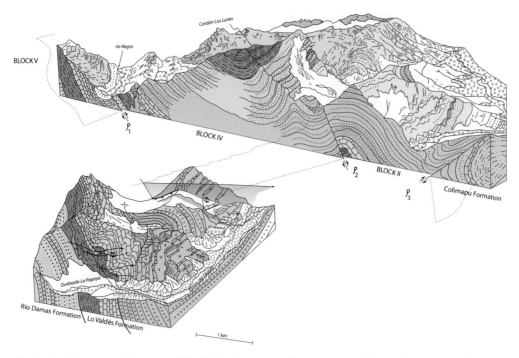

Fig. 4. Block diagrams looking south of Río Volcán (slightly modified from Palma 1991). Notice the sinistral transfer fault separating the northern from the southern blocks. Lithologies as referred to in the text and in Figure 5.

deposition (Legarreta *et al.* 1993). Both the long trend of mainly submarine volcanism developed along the margin north of 34°S as well as the probable location of the passage to the open Pacific at that time are highlighted in Figure 2a (modified from Legarreta & Uliana 1996). The Coast Range localities bearing Tithonian ammonites that support this connection are discussed by Godoy *et al.* (2009). The sea covered a wider area of the present Chilean Coast Range north of 35°S only from Valanginian time.

As previously recognized in the early 1960s (Biro-Bagoczky 1964, 1980), the basal 762-m-thick member of the subvertical Lo Valdés Formation (González 1963) in the Volcán valley mainly comprises submarine mafic metavolcanic lavas, previously referred to as spilites. He reported the presence of scarce lower Tithonian *Virgatosphinctes andesensis* in thin limestone and marl beds that overlie the Río Damas Formation (Fig. 2b). Grooves found in the marls underlying the spilites indicate transport towards the ESE.

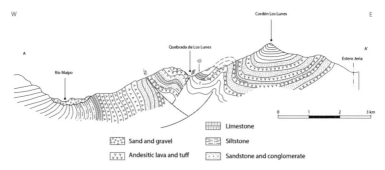

Fig. 5. Profile across the north face of the southern blocks in Figure 4, showing the location of the late Hauterivian ammonite-bearing limestone (slightly modified from Palma 1991).

The overlying late Tithonian–Hauterivian limestones and sandstones are conformably covered, after an event of shelfal exposure and incision, by red siltstones and shales of the Colimapu Formation (Klohn 1960), the age of which is discussed in the next section.

Godoy & Vela (1985) and Palma (1991) have shown that Biro's thick succession of submarine

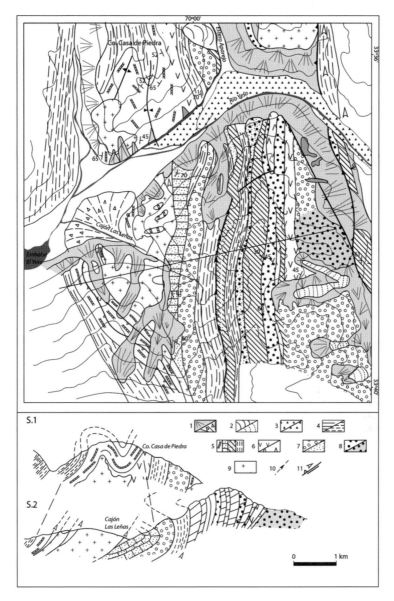

Fig. 6. Geological map and profiles directly south of the Río Yeso reservoir. 1: Debris and Cajón Las Leñas alluvial cone deposits; 2: Landslide and moraine deposits; 3: Matrix/clast rich facies of the Marmolejo debris avalanche; 4: Colimapu/Lo Valdés Formation shales and siltstones; 5: Upper limestones (Agrio Formation)/intermediate limestones and carbonate-rich sandstones/lower limestones (Tithonian); 6: Andesite flows (Tithonian); 7: Upper conglomerates and volcanic-rich sandstones (Tithonian–Hauterivian); 8: Tithonian–Kimmeridgian conglomerates; 9: Actinolite microdiorite (Miocene?); 10: Tithonian gypsum and anhydrite; 11: Diapiric gypsum (Aptian in the west/ Oxfordian in the east). The thrusts associated with the actinolite granodiorite in the southern section transition into folds in the northern profile.

lavas quickly gives way southwards to green sandstones, which then wedge out under complexly folded black shales (lower block diagrams in Fig. 4, taken from Palma's thesis).

In this area located south of the Volcán valley the upper Hauterivian limestones of the Lo Valdés Formation are replaced by abundant andesitic lavas with thin interbedded limestones bearing the late Hauterivian ammonite *Crioceratites diamantensis* (Palma 1991). These andesites are coloured brown in the middle southern block diagram of Figure 4 and shown in the centre of the profile in Figure 5. The red beds he found both underlying and overlying the andesites induced him, following Klohn (1960) and Godoy *et al.* (1988), to wrongly assign the whole package to the Colimapu Formation. (I regret coaxing my student at that time into such an interpretation.)

Red beds and thin evaporites have however been developed at several time intervals in different margins of the Neuquén basin. As pointed out by Legarreta & Uliana (1996), red mudstones fringe the late Tithonian eastern margin of the basin. In the western slopes of the Barroso (southern tributary of the Maipo at 34°10′S), Charrier (1981) recognizes abundant dacitic pyroclastics and lavas deposited on top of the Río Damas Formation. On the other hand, Ramos *et al.* (1990) report early Hauterivian *Olcostephanus* sp. as present in thin limestones at the base of a thick andesitic lava and red tuffaceous sandstone succession that mark the western margin at 32°50°S, the headwater region of the Mendoza River, Argentina at that time.

A special reference must be made to a 100-m-thick unit of red siltstones, marls, sandstones and conglomerates that crop out on the southern slopes of the Cajón del Morado, sandwiched between limestones with interbedded green hyaloclastites of the basal Lo Valdés Formation and 300 m of mafic lavas and tuffs (Godoy & Vela 1985). These red beds taper out southwards to 16 m in the northern slopes of the Volcán valley and widen northwards to 500 m in the southern slopes of the Yeso valley. In the southern slopes of that valley, they overlie a 50-m-thick anhydrite bed that covers a volcanic- and conglomerate-rich limestone section of the Lo Valdés Formation (Fig. 6). The red pelites underlie 300 m of conglomerates, sandstones and andesitic tuffs that grade north of the valley into andesite flows. A more detailed description of the succession is available in Godoy *et al.* (1988). The map and profiles in Figure 6 were drawn together with J.C. Castelli on the same year; they remained unpublished, however.

The Valanginian–Hauterivian palaeogeographic reconstruction of Legarreta & Uliana (1999) has been modified in Figure 7 in order to highlight late Hauterivian widespread volcanism along the western margin mentioned in this paper. According to the figure, what they interpret as a new yet smaller Messinian-style desiccation may be explained by the effect of volcanoes that closed the passage to open sea.

Tunik *et al.* (2010) report a 130 Ma age for the youngest detrital zircons from the base of the Neuquén Group directly south of the area considered in this paper. They believe that an uplifted Coast Range may be their source area. The sequence of andesites (green in Fig. 4) may well be a closer and therefore more probable source area for their zircons. A younger age (102 Ma) is reported by Di Giulio *et al.* (2012) for detrital zircons from that base, which fits better with the Albian–Cenomanian unconformity recognized for that time in the central Neuquén Basin. Di Giulio

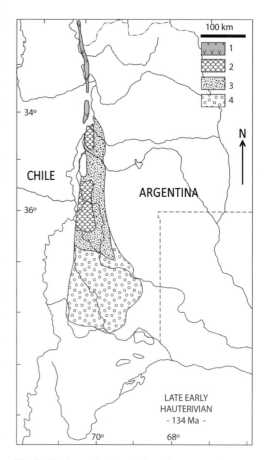

Fig. 7. The late early Hauterivian palaeogeography, modified from Legarreta & Uliana (1999) in order to show the northwards extension of a volcanic chain. 1 = andesitic lavas 2 = evaporates 3 = mudflat pelites and sandstones 4 = fluvial conglomerates and sandstones.

et al. (2012) however link their source 'to the exhumation of magmatic granitoids and volcanic rocks in the Andean Cordillera, to the west'. The belt of Late Cretaceous granitoids show in their location map close to the border does not exist (they are Miocene in age). The source rock must therefore be restricted to the volcanic units mentioned in this paper.

The Late Cretaceous margin

As already stated, continental red sediments of the Colimapu Formation overlie Hauterivian limestones in the Volcán valley as a talus breccia, recording an event of shelfal exposure and incision. They resemble some of the successions that constitute the late-stage Neuquén Group in Argentina.

For a long time carofites assigned to Aptian–Albian time was the only available palaeontological age control (Martinez & Osorio 1963). The dating of detrital zircon has recently provided new ages. Detrital zircons of age 90 Ma have been reported in sandstones from both the headings of the Tinguirírica valley at 34°30' and from the Volcán profile by Aguirre et al. (2009). The youngest zircons from the second locality come from sandstones close to the top of the formation. Based on these first radiometric ages it now seems that the western margin of the Neuquén basin accommodated continental sediments mainly during Turonian time. At the moment, no rocks have been found that record the Maastrichtian Atlantic transgression, well preserved in the Neuquén basin.

The contact relation of the Colimapu Formation with the overlying Abanico Formation (Eocene–Early Miocene lavas and pyroclastics) has recently been a matter of controversy. The early authors (González 1963; Thiele 1980) describe it as conformable, a relation that agreed with the Late Cretaceous age accepted for the base of Abanico Formation at that time. Giambiagi et al. (2014), Godoy (2011) and Godoy et al. (1999) agree on the structure of their profiles. The latter authors point out that the fact that Oligocene lavas conformably overlie Late Cretaceous red beds implies a long hiatus (Fig. 8b). North of the Aconcagua valley (32°S) and south of 36°S, however, several authors have mapped a clear-cut unconformity. Its age is K–T in the northern section and intra-Late Cretaceous in the southern section (see references in Di Giulio et al. 2012). The wide northern half of the coastal Late Palaeozoic batholith may have acted as a rigid block between both latitudes, resisting shortening until Early Miocene (18 Ma) time, as

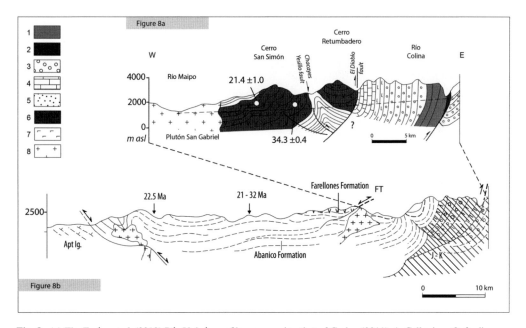

Fig. 8. (a) The Farías et al. (2010) Río Volcán profile compared to that of Godoy (2011). 1: Callovian–Oxfordian marine and continental sedimentary rocks; 2: Diapiric gypsum; 3: Kimmeridgian conglomerates; 4: Tithonian–Hauterivian spilites, marls and sandstones; 5: ?Barremian–Turonian red beds; 6: Oligocene lavas and pyroclastic rocks; 7: Early Miocene epiclastic and pyroclastic rocks; 8: Late Miocene granodiorite. (b) Apt Ig: Aptian Ignimbrites from the eastern slopes of the Coast Range; FT: northern end of the northwards-shallowing Fierro Thrust. The contact between the hatched (Mesozoic rocks) and unhatched units represents a 45 Ma long hiatus.

recorded by (Giambiagi et al. 2001) in the foreland earliest synorogenic sediments. The Main Andes may have been passively transported eastwards before Miocene time at the latitude of concern of this paper.

Both Fock (2005) as well as Farías et al. (2005) consider the contact as faulted (which is the case further south; see Fig. 4) and add a new Cenozoic formation that locally unconformably covers the Colimapu red beds (Fig. 8a). Their evidence for such a new unit is mainly geochronological however, and refers to an all-too-well-defined 29.39 Ma U–Pb average age on 25 zircons from a rock they classify as a tuff (M. Farías, pers. comm., 2008). In a recent field trip organized by the IGCP Project 586Y we have recognized that andesites with Miocene Ar/Ar ages that Aguirre et al. (2009) reported from the base of the Colimapu Formation, also in the Volcán profile, were emplaced as sills. It seems appropriate to focus on detailed mapping of the area, rather than on isolated radiometric ages.

Conclusions

Marked lithological variations occur along-strike in the north-western margin of the Neuquén basin during the Kimmeridgian and up to the Hauterivian, sometimes in very short distances. They are controlled mainly by the construction of volcanic edifices and their erosional continental to littoral aprons. During early Tithonian time, however, the seaway that developed close to 34°S and accommodated over 700 m of submarine lavas and sills may have been related to a local rift-like event.

Thick sections of Hauterivian andesitic flows are described in the high drainage of the Maipo system and by other authors north of 33°S. The fact that the Neuquén basin remained closed during Aptian–Albian time, despite the highstand recorded in much of the planet, may be a long-lasting local effect of this Hauterivian volcanism.

The author is deeply indebted to the many University of Chile students who participated in the Field Geology courses led by the author and P. Alvarez, held in the area from 1985 to 1999. V. Ramos and B. Aguirre-Urreta kindly helped pointing to important references. The former, together with T. Kato, also kindly revised the paper. References to Chilean Geological meetings may be down-loaded as congreso geologico chileno files from http://biblioserver.sernageomin.cl/opac/index.asp?Op=17.

References

AGUIRRE, L., CALDERÓN, S., VERGARA, M., OLIVEROS, V., MORATA, D. & BELMAR, M. 2009. Edades isotópicas de rocas de los valles Volcán y Tinguiririca, Chile central. *Proceedings 12th Congreso Geológico Chileno*, Santiago, S8-001.

ÁLVAREZ, P. 1996. Los depósitos triásicos y jurásicos de la Alta Cordillera de San Juan. *In*: RAMOS, V. (ed.) *Geología de la región del Aconcagua, provincias de San Juan y Mendoza*. Subsecretaría de Minería, Dirección Nacional del Servicio Geológico, Buenos Aires, Anales, **24**, 59–137.

ÁLVAREZ, P., AGUIRRE URRETA, B., GODOY, E. & RAMOS, V. 1997. Estratigrafía del Jurásico de la Cordillera Principal de Argentina y Chile (33°45′–34°S). *Proceedings 8th Congreso Geológico Chileno*, Antofagasta, **1**, 425–429.

ÁLVAREZ, P., GODOY, E. & SELLÉS, D. 2000. Geología de la región del río Colorado, Andes Principales de Chile (33°25′S). *Proceedings IX Congreso Geológico Chileno*. Puerto Varas, **I**, 736–740.

BIRO-BAGOCZKY, L. 1964. *Estudio sobre el límite Titoniano y el Neocomiano en la Formación Lo Valdés, Provincia de Santiago (33°50′ l.s.), Chile, principalmente en base a ammonoideos*. Thesis, Departamento de Geología, Universidad de Chile.

BIRO-BAGOCZKY, L. 1980. Estudio sobre el límite Titoniano y el Neocomiano en la Formación Lo Valdés, Provincia de Santiago (33°50′ l.s.), Chile, principalmente en base a ammonoideos. *Proceedings 2° Congreso Argentino de Paleontología y Bioestratigrafía y 1er Congreso Latinoamericano de Paleontología, 1978*, Buenos Aires, **5**, 137–152.

CHARRIER, R. 1981. Geologie der chilenischen Hauptkordillere zwischen 34°30′ südlicher Breite und ihre tektonische, magmatische und paleogeographische Entwicklung. *Berliner Geowissenschaftliche Abhandlungen (A)*, **36**, 270.

CHARRIER, R. 1984. Áreas subsidentes en el borde occidental de la cuenca tras-arco jurásico~cretácica, Cordillera Principal Chilena entre 34″ y 34″30′S. *In*: *Actas Noveno Congreso Geológico Argentino*, Buenos Aires, **2**, 107–124.

CHARRIER, R., PINTO, L. & RODRIGUEZ, M. 2007. Tectonostratigraphic evolution of the Andean Orogen. *In*: MORENO, T. & GIBBONS, W. (eds) *The Geology of Chile*. Geological Society, London, 21–116.

DI GIULIO, A., RONCHI, A., SANFILIPPO, A., TIEPOLO, M., PIMENTE, M. & RAMOS, V. 2012. Detrital zircon provenance from the Neuquén Basin (south-central Andes): Cretaceous geodynamic evolution and sedimentary response in a retroarc-foreland basin. *Geology*, **40**, 559–562.

FARÍAS, M., CHARRIER, R., FOCK, A., CAMPBELL, D. & MARTINOD Y D. COMTE, J. 2005. Rapid Late Cenozoic surface uplift of the central Chile Andes (33°–35°S). VI International Symposium on Andean Geodynamics, Barcelona. Expanded abstracts, 261–265.

FARÍAS, M., COMTE, D. ET AL. 2010. Crustal-scale structural architecture in central Chile based on seismicity and surface geology: implications for Andean mountain building. *Tectonics*, **29**, TC3006, http://dx.doi.org/10.1029/2009TC002480

FOCK, A. 2005. *Cronología y tectónica de la exhumación en el Neógeno de los Andes de Chile central entre los 33° y los 34°S*. Thesis, Departamento de Geología, Universidad de Chile.

GIAMBIAGI, L. & RAMOS, V. 2002. Structural evolution of the Andes between 33°30′ and 33°45′ S, above the transition zone between the flat and normal subduction segment, Argentina and Chile. *Journal of South American Earth Sciences*, **15**, 101–116, http://dx.doi.org/10.1016/S0895-9811(02)00008-1

GIAMBIAGI, L., TUNIK, M. & GIHIGLIONE, M. 2001. Cenozoic tectonic evolution of the Alto Tunuyan foreland basin above the transition zone between the flat and normal subduction segment (33°30′–34°S), western Argentina. *Journal of South American Earth Sciences*, **14**, 707–724.

GIAMBIAGI, L., ÁLVAREZ, P., GODOY, E. & RAMOS, V. 2003. The control of pre-existing extensional structures in the evolution of the southern sector of the Aconcagua fold and thrust belt. *Tectonophysics*, **369**, 1–19, http://dx.doi.org/10.1016/S0040-1951(03)00171-9

GIAMBIAGI, L., TASSARA, A. *ET AL*. 2014. Evolution of shallow and deep structures along the Maipo–Tunuyán transect (33°40′S): from the Pacific coast to the Andean foreland. *In*: SEPÚLVEDA, S. A., GIAMBIAGI, L. B., MOREIRAS, S. M., PINTO, L., TUNIK, M., HOKE, G. D. & FARÍAS, M. (eds) *Geodynamic Processes in the Andes of Central Chile and Argentina*. Geological Society, London, Special Publications, **399**, First published online February 27, 2014, http://dx.doi.org/10.1144.SP399.14

GODOY, E. 2011. Structural setting and diachronism in the Central Andean Eocene to Miocene volcano-tectonic basins. *In*: SALFITY, J. A. & MARQUILLAS, R. A. (eds) *Cenozoic Geology of the Central Andes of Argentina*. Instituto del Cenozoico, Universidad Nacional de Salta, Salta, 155–167.

GODOY, E. & VELA, I. 1985. Consideraciones sobre la Formación Colimapu en la alta cordillera de Santiago y el control paleogeográfico de la estructura. *In*: *IV Congreso Geológico Chileno*, Antofagasta, **4**, 370–384.

GODOY, E., CASTELLI, J. C., LOPEZ, M. C. & RIVERA, O. 1988. ... y Klohn tenía razón: la Formación Colimapu recupera sus miembros basales. *In*: *V Congreso Geológico Chileno*, Santiago, **3**, H101–H120.

GODOY, E., YAÑEZ, G. & VERA, E. 1999. Inversion of an Oligocene volcano-tectonic basin and uplifting of its superimposed Miocene magmatic arc in the Chilean Central Andes: first seismic and gravity evidences. *Tectonophysics*, **306**, 217–236.

GODOY, E., SCHILLING, M., SOLARI, M. & FOCK, A. 2009. *Carta Rancagua. Escala 1:100.000 Sernageomin*. Serie Geología Básica, 118, 50 p.

GONZÁLEZ, O. 1963. Observaciones geológicas en el valle del Río Volcán. *Revista Minerales*, Santiago, **17**, 20–61.

GROEBER, P. 1918. Estratigrafia del Dogger en la República Argentina. *Boletín Dirección General de Minas, Geología e Hidrogía. B*, Buenos Aires, 18.

KLOHN, C. 1960. *Geología de la Cordillera de los Andes de Chile Central*. Instituto de Investigaciones Geológicas, Boletín, **8**.

LEGARRETA, L. & ULIANA, M. A. 1996. The Jurassic succession in west-central Argentina: stratal patterns, sequences and paleogeographic evolution. *Palaeogeography, Palaeoclimatology, Palaeoecology*, **120**, 303–330.

LEGARRETA, L. & ULIANA, M. A. 1999. El Jurásico y Cretácico de la Cordillera Principal y la Cuenca Neuquina. *In*: CAMINOS, R. (ed.) *Geología Argentina*. Anales 29 Segemar, Buenos Aires, 399–416.

LEGARRETA, L., GULISANO, C. A. & ULIANA, M. A. 1993. Las secuencias sedimentarias jurasicas-cretacicas. *In*: RAMOS, V. A. (ed.) *Geologia y Recursos Naturales de Mendoza. Relat. 12th Congr. Geol. Argent.*, 1, Geol., Buenos Aires, 87–114.

MARTINEZ, R. & OSORIO, R. 1963. Consideraciones preliminares sobre la presencia de carófitas fósiles en la Formación Colimapu. *Revista Minerales*, **82**, 28–43.

PALMA, W. 1991. *Estratigrafía y estructura de la Formación Colimapu entre el Estero del Diablo y el cordón Los Lunes, región Metropolitana*. Thesis, Departamento de Geología, Universidad de Chile.

PANGARO, F., RAMOS, V. & GODOY, E. 1996. La Faja plegada y corrida de la Cordillera Principal de Argentina y Chile a la latitud del Cerro Palomares (33°35′S). *In*: *XIII Congreso Geológico Argentino y III Congreso Exploración de Hidrocarburos*, Buenos Aires, Actas II, 315–324.

RAMOS, V., RIVANO, S., AGUIRRE-URRETA, B., GODOY, E. & LO FORTE, G. 1990. El Mesozoico del Cordón del límite entre portezuelo Navarro y Monos de Agua (Chile–Argentina). *In*: *XI Congreso Geológico Argentino*, San Juan, **2**, 43–46.

RAMOS, V., GODOY, E., LO FORTE, G. & AGUIRRE-URRETA, B. 1991. La franja plegada y corrida del norte del río Colorado, región Metropolitana, Chile central. *Proceedings VI Congreso Geológico Chileno*, Santiago, 323–327.

SANGUINETTI, A. & RAMOS, V. 1993. El volcanismo de arco Mesozoico. *In*: RAMOS, V. (ed.) *Geología y recursos naturales de Mendoza. Relatorio 12° Congreso Geológico Argentino*, Buenos Aires, 115–122.

SERNAGEOMIN 2002. *Servicio Nacional de Geología y Minería*, Santiago. Carta Geológica de Chile, escala 1:1.000.000.

THIELE, R. 1980. *Hoja Santiago, Región Metropolitana*. Instituto de Investigaciones Geológicas Carta Geológica de Chile, Santiago, **29**, 21p.

TUNIK, M., FOLGUERA, A., NAIPAUER, M., PIMENTEL, M. & RAMOS, V. 2010. Early uplift and orogenic deformation in the Neuquén Basin: constraints on the Andean uplift from U–Pb and Hf isotopic data of detrital zircons. *Tectonophysics*, **489**, 258–273.

Geophysical characterization of the upper crust in the transitional zone between the Pampean flat slab and the normal subduction segment to the south (32–34°S): Andes of the Frontal Cordillera to the Sierras Pampeanas

M. SÁNCHEZ[1]*, F. LINCE KLINGER[1], M. P. MARTINEZ[1,2], O. ALVAREZ[1], F. RUIZ[2], C. WEIDMANN[1] & A. FOLGUERA[3]

[1]*CONICET. Instituto Geofísico y Sismológico Ing. Volponi, Universidad Nacional de San Juan, Ruta 12, km. 17, CP 5407, San Juan, Argentina*

[2]*Instituto Geofísico y Sismológico Ing. Volponi, Universidad Nacional de San Juan, Ruta 12, km. 17, CP 5407, San Juan, Argentina*

[3]*Department Cs. Geol. FCEN. U.B.A. Buenos Aires, Inst. Estudios Andinos 'Don Pablo Groeber', Argentina*

Corresponding author (e-mail: 1marcossanchez@gmail.com)

Abstract: The Nazca Plate subducting beneath the South American Plate has strongly influenced Cenozoic mountain growth in western Argentina and Chile sectors (32–34°S; 70–66°W). At these latitudes, the Pampean flat slab has induced the development of prominent mountain systems such as the Frontal Cordillera, the Precordillera, and the associated Sierras Pampeanas in the eastwards foreland region. Through a gravity study from the Frontal Cordillera to the Sierras Pampeanas region between 32 and 34°S, we delimit a series of geological structures that are accommodating shortening in the upper crust and others of regional and subsurface development, without any clearly defined mechanics of deformation. Additionally, through an isostatic residual anomaly map based on the Airy-Heiskanen local compensation model, we obtain a decompensative gravity anomaly map that highlights anomalous gravity sources emplaced in the upper crust, related to known geological structures. In particular, by applying the Tilt method which enhances the gravity anomalies, the NW-trending Tunuyan Lineament is depicted south of 33.4°S following previous proposals. Using the decompensative gravity anomaly, two profiles were modelled through the northern sector of the study area using deep seismic refraction lines, borehole data and geological information as constraints. These density models of the upper crust of this structurally complex area accurately represent basin geometries and basement topography and constitute a framework for future geological analysis.

At the transitional zone between the Pampean flat subduction zone and the normal subduction system located immediately to the south, the area between the Frontal Cordillera and the Sierras Pampeanas is of great interest due to the existence of crustal discontinuities, attributed to sutures between different allochthonous and parautochthonous Palaeozoic terranes (Ramos *et al.* 2002). These lithospheric blocks are known as the Pampia, Cuyania and Chilenia terranes and constitute the Andean basement at these latitudes (latitude 27–33°S). Basin development is thought to have been influenced by reactivation of these basement discontinuities in Late Triassic and Early Cretaceous times; understanding the anatomy of this system is important for geological analyses.

Located in the north-central region of Argentina, the Pampia terrane collided (according to certain hypotheses) with the Rio de la Plata craton during latest Proterozoic–Early Cambrian time (Ramos & Vujovich 1993; Brito Neves *et al.* 1999; Almeida *et al.* 2000). This led to deformation, metamorphism and magmatism (González Bonorino 1950; Caminos 1979; Casquet *et al.* 2008) exposed in the Pampean Orogen beyond the area of interest.

To the west, the Cuyania exotic microcontinent detached from Laurentia during Cambrian time collided in Late Ordovician time with western Gondwana (Thomas & Astini 2003). Its definition is based on fauna, palaeomagnetic position and isotopic and geochemical composition (Abbruzzi *et al.* 1993; Benedetto & Astini 1993; Kay *et al.* 1996; Rapalini & Astini 1997; Benedetto *et al.* 1999; Keller 1999; Ramos 2004). However, a parautochthonous origin was proposed for it in other studies (Aceñolaza *et al.* 2002; Finney *et al.* 2003;

Peralta *et al.* 2003). During the amalgamation of Cuyania – which potentially occurred during the Famatinian orogenic cycle (Late Ordovician) (Rapela *et al.* 2001) – magmatism affected the western sector of the Sierras Pampeanas in La Rioja province (Pankhurst *et al.* 1998; Quenardelle & Ramos 1999) and to the south in the Sierra de San Luis (Llambías *et al.* 1996).

Mesoproterozoic basement, attributed to the Cuyania exotic microcontinent, is exposed in the Western Sierras Pampeanas. In particular, in the Sierra Pie de Palo the basement is thrust on top of an Early Cambrian carbonate platform in metamorphic facies, considered laterally equivalent to the Precordillera sedimentary facies (Ramos *et al.* 1986; Ramos 1988, 1995; Vujovich *et al.* 2004). According to these hypotheses, Mesoproterozoic rocks exposed in this region would have had a complex evolution prior to its final amalgamation with the Gondwana supercontinent, in which a suprasubduction system would have accreted in the Greenvillian Orogen showing typical mid-ocean ridge and island-arc basalt (MORB and IAB) components. Alternative models have considered Sierra de Pie de Palo basement as part of the proto-Andean margin (Galindo *et al.* 2004) and, even as an independent block, amalgamated to Cuyania before the Ordovician (Mulcahy *et al.* 2007).

In the Precordillera, limestones and associated siliciclastic successions belonging to this terrane are exposed as a thin-skinned fold-and-thrust belt that to the south merges in a thick-skinned belt (Ramos 1995; Ramos *et al.* 1998).

Most of these crustal discontinuities, potentially related to sutures between different terranes, have suffered extensional reactivations in Late Triassic and Early Cretaceous times. The Cuyo basin constitutes one of the Late Triassic rift systems that intersect the Andean trend obliquely with a predominant NW orientation. Eastern-most depocentres are sparse over the foreland area beneath thick sections of Cenozoic synorogenic strata (Uliana & Biddle 1988; Legarreta *et al.* 1993), while those at the orogenic wedge in the west were selectively incorporated during Andean growth stages. These Late Triassic depocentres are related to the initial stages of the western Pangea break-up, while the Cretaceous depocentres are spatially and temporally associated with the Paraná hotspot that affected the early Atlantic margin (Boll & de la Colina 1993; Lagorio 2008).

In particular, the south-western section of the Sierras Pampeanas corresponding to the Sierra del Gigante and Sierra de las Quijadas has been the locus of a Cretaceous rift system which led to the accumulation of within-plate volcanic rocks interfingered with continental facies (Rivarola & Spalletti 2006; Martinez *et al.* 2012).

Between the eastern Precordillera and Sierras Pampeanas, the Jocoli basin registers Andean exhumation as a frontal foredeep due to the flexure of the eastern Cuyania block (Fig. 1a).

The western and central sectors of the Precordillera in the Pampean flat slab segment (Baldis & Chebli 1969; Baldis *et al.* 1982) are composed of typical east-vergent thin-skinned thrust sheets (Allmendinger *et al.* 1990; von Gosen 1992). Contrastingly, the eastern Precordillera and Sierras Pampeanas to the east are thick-skinned blocks with opposite vergences (Bracaccini 1946; Rolleri & Baldis 1969; Ortiz & Zambrano 1981; von Gosen 1992; Zapata & Allmendinger 1994; Zapata & Allmendinger 1997).

These two systems interfere at a triangular thick-skinned zone (Zapata & Allmendinger 1994, 1996). South of Mendoza city, this triangular zone disappears at the zone where the inclination of the Wadati–Benioff zone increases, passing to a normal subduction zone (Figueroa & Ferraris 1989).

The Jocoli basin is partially located on that triangular zone (Cominguez & Ramos 1991). Its infill records the beginning of the uplift of the Frontal Cordillera that began at c. 15 Ma (Bercowski *et al.* 1993). Afterwards, the basin records the uplift of the Sierra de Pedernal in the Precordillera simultaneously to the uplift of the Western Sierras Pampeanas to the east (Ramos *et al.* 1997).

The aim of this paper is to make a density model that reflects the structure and basin geometry in this transition zone from a flat to a normal subduction zone (32–34°S and 70–66°W; Fig. 1a), using a large database of high-quality gravity data, boreholes and deep seismic refraction lines, constrained with pre-existing geological models in the area. The model will describe Andean anatomy at this key area, a valuable tool in structural, basin and tectonic studies.

Methodology

Database

This study is based on a dataset which comprises 23 680 gravity stations that are the property of the Geophysical Seismological Institute of the National University of San Juan (IGSV). The dataset covers the central region of Argentina in an area between 27.5° and 36.5°S and from 71° to 65°W, extending outside of the boundaries of our study area in order to avoid border effects. The spatial distribution of gravity stations from the original sources is shown in Figure 1b.

This database was measured using geodetic gravimeters with precisions of ± 0.1 mGal. In order to ensure the accuracy of the measurements and to homogenize all stations obtained on different

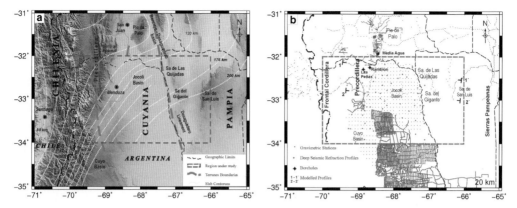

Fig. 1. (**a**) Digital elevation model (DEM 90 m × 90 m) that indicates the study area (red box) in a complex zone characterized by the transition from a flat subduction zone of the Nazca plate in the north, to a normal subduction segment. Nazca plate isodepth contours obtained by Alvarado *et al.* (2007) are represented by white dotted lines. The proposed boundaries between Palaeozoic terranes are indicated by grey dashed lines (see text for references). (**b**) Geophysical databases available in the region of study. Green dashed line: area under study; red dots: gravity data courtesy of Instituto Geofísico Sismológico Volponi (IGSV); light blue dots: deep seismic refraction data available in the area by Yacimientos Petroliferos Fiscales (YPF) boreholes. 1–1′: Modelled profiles in Figures 9 and 10.

campaigns, each gravity station has been linked to the national altimetry network. This process allows any possible artefact due to problems in the levelling of the different sources to be avoided, as all were referred to the IGSN 71 network (International Gravity Standardization Net 1971) and linked to the fundamental station Miguelete (Buenos Aires) through the nodal 145 City of San Juan and PF9 into the N24 line (Morelli *et al.* 1974). More details about the homogenization and reductions to the altimetry network are provided in Villella & Pacino (2010). This methodology allows a proper data reduction to be performed for anomaly calculation using classical corrections detailed below (Blakely 1995; Hinze *et al.* 2005).

The theoretical or normal gravity accounting for the mass, shape and rotation of the Earth is the predicted gravitational acceleration on the best-fitting terrestrial ellipsoidal surface. We have used the latest ellipsoid recommended by the International Union of Geodesy and Geophysics: the 1980 Geodetic Reference System (GRS80) (Moritz 1980). The Somigliana closed-form formula (Somigliana 1930) for theoretical gravity g_T on this ellipsoid at latitude (south or north) φ is:

$$g_T = \frac{g_e(1 + k\sin^2\varphi)}{\sqrt{1 - e^2\sin^2\varphi}}, \quad (1)$$

where the GRS80 reference ellipsoid has the value g_e of 978 032.67715 mGal, where g_e is normal gravity at the equator; k of 0.001931851353 where k is a derived constant; and e^2 of 0.0066943 800229, where e is the first numerical eccentricity.

The height correction is called the free-air correction and is based on the elevation (or orthometric height) above the geoid (sea level) rather than height above the ellipsoid. The revised standards use the ellipsoid as the vertical datum rather than sea level. Conventionally, the first-order approximation formula of Δgh in milligal, where gravity anolamy $\Delta g = 0.3086$, is used for this correction.

The Bouguer correction accounts for the gravitational attraction of the layer of the Earth between the vertical datum (i.e. the ellipsoid) and the station. This correction, ΔgB in milligals, is traditionally calculated assuming that the Earth between the vertical datum and the station can be represented by an infinite horizontal slab with equation:

$$\Delta gB = 2\pi G \sigma h = 4.193 \times 10^{-5} \sigma h \quad (2)$$

where G the gravitational constant is $6.673 \pm 0.001 \times 10^{-11}$ m^3 kg^{-1} s^{-2} (Mohr & Taylor 2001) and σ is the density of the horizontal slab (kg m^{-3}). The mean used density is 2670 kg m^{-3} (Hinze 2003), and h is the height of the station (m) relative to the ellipsoid in the revised procedure or relative to sea level in the conventional procedure.

The terrain correction adjusts the gravity effect produced by a mass excess (mountain) or deficit (valley) with respect to the elevation of the observation point. The terrain correction was obtained using two digital elevation models, a local and a regional DEM. Software program OASIS v.7.2 combines the algorithms developed by Kane (1962) and Nagy (1966), where elevation models were obtained from the Shuttle Radar Topography

Mission (SRTM) of the United States Geological Survey (USGS). Through the use of a sampling procedure, the corresponding topographic correction value was assigned to each gravity station. The resulting maximum error for this correction was ±1.8 mGal. Finally, the complete Bouguer anomaly values (Fig. 2) were calculated on a regular grid cell size of 5 × 5 km using the minimum curvature method (Briggs 1974).

Isostatic anomaly

The 'regional' flexural compensation models proposed by Watts et al. (1995), Tassara et al. (2007), Wienecke et al. (2007), Pérez-Gussinyé et al. (2008) and Tassara & Echaurren (2012) for the Central Andes have enabled the determination of equivalent elastic thickness variables that are progressively higher eastwards in the foreland area, beyond the magmatic arc locus associated with the subduction process in the west. However, crustal thickening in the hinterland of the Central Andes, where the arc is located and was hosted (Introcaso et al. 1992b), can be explained by relative low values of the effective elastic thickness which justify the use of a 'local' compensation model (Airy–Heiskanen) in the study area.

This assumption has already been applied in this region by several authors with fairly good results (Götze & Evans 1979; Introcaso et al. 1992a; Chapin 1996; Götze & Kirchner 1997; Whitman 1999; Introcaso et al. 2000; Tassara & Yáñez 2003).

In order to obtain the isostatic root, we performed the calculation of the Airy–Heiskanen model frame using different parameters (see Table 1). We have taken into consideration other models integrating (1) gravity with seismic data; (2) the global model presented by Assumpção et al. (2013); and (3) the Moho depths determined by Gans et al. (2011) using receiver functions (see Fig. 3). Finally, the isostatic root that best fits with the above models was calculated using the following parameters: normal thickness crust $T_n = 40$ km; contrast density $\rho = 400$ kg m^{-3} and density crust $\rho = 2670$ kg m^{-3}. This T_n value was found by Christensen & Mooney (1995) as the mean global crustal thickness. The adopted crust–mantle density contrast ($\rho_{cm} = 400$ kg m^{-3})

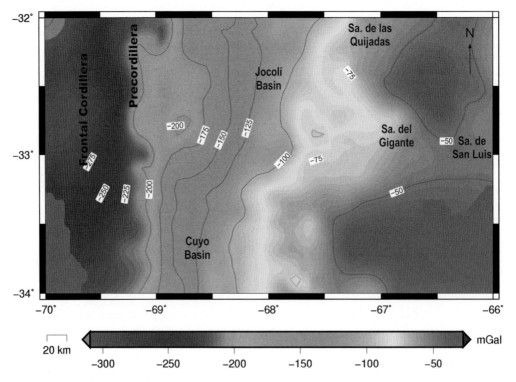

Fig. 2. Bouguer anomaly map obtained from 23 680 stations of terrestrial gravity data. The Bouguer anomaly values were calculated on a regular grid cell size of 5 × 5 km, using the minimum curvature method. To the west at Frontal Cordillera, the negative effect of the Andean root has a clear influence on the gravity signal; the effect of this decreases towards the east at Sierras Pampeanas.

Table 1. *Parameters used in the isostatic root calculation*

Crustal thickness (km)	T_{n1}	30
	T_{n2}	35
	T_{n3}	40
Contrasts of crust–mantle interface densities (kg m^{-3})	ρ_{cm1}	330
	ρ_{cm2}	350
	ρ_{cm3}	400
Topographic load density (kg m^{-3})	ρ_{ct}	2670

was used previously by Martinez *et al.* (2006) and Gimenez *et al.* (2009).

With the exception of the south-western edge of Figure 3 (where there is a lack of data), there exists a good correspondence between the depths determined by the hydrostatic Moho model with the solutions (coloured dots) obtained by Gans *et al.* (2011). Likewise, our results are consistent with depths found by the global model presented by Assumpção *et al.* (2013). The comparison is plotted with dashed white lines on Figure 3. For this case, we observed a trend deepening to the west (with the exception of isoline 40*).

Using the Moho hydrostatic geometry (Fig. 3), the isostatic gravity root effect was calculated (Fig. 4). This map presents a high negative gravity gradient ranging from −25 to −275 mGal. By subtracting the gravity effect of the isostatic root from the Bouguer anomaly (Figs 2 & 4), an isostatic residual anomaly is obtained (Fig. 5). These residual anomalies could mask short-wavelength anomalies, generated from more superficial sources (Simpson *et al.* 1986). The isostatic corrections could therefore be used to partially remove the effect of the crustal roots produced by the topographic elevations and depressions. Nevertheless, they fail to resolve the problem when the crustal roots are derived from high-density crustal regions with or without topographic expression. To overcome this drawback, we apply the decompensative gravity anomaly technique as proposed by Cordell *et al.* (1991).

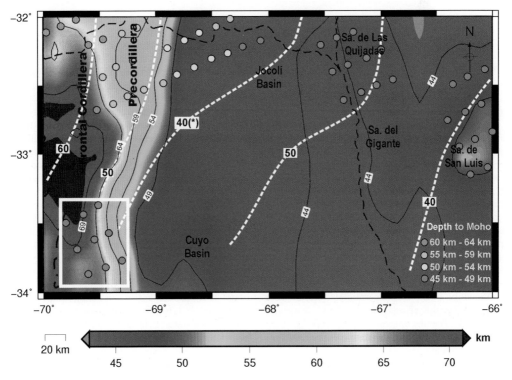

Fig. 3. Depth of mantle–crust interface corresponding to the geometry of the computed hydrostatic Moho in the region of study. Circular dots represent the differences between the Moho depth obtained by Gans *et al.* (2011) and the gravimetric results, and dashed white lines represent the differences between a global model by Assumpção *et al.* (2013) and the gravimetric results. Our results present a general good fit with the Gans *et al.* (2011) results, with the exception of the region within the white rectangle. Assumpção *et al.* (2013) results are also consistent with our results, with a deeper root under the Andes and a shallowing trend towards the east (with the exception of 40*).

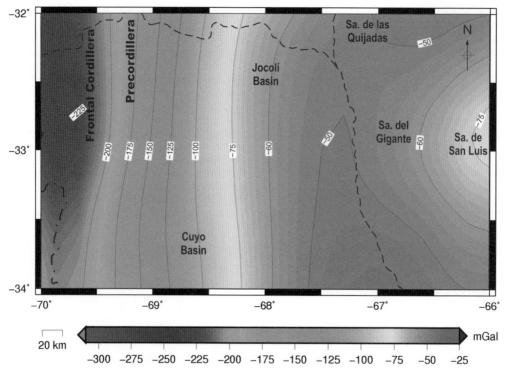

Fig. 4. Gravity response of isostatic root. Isoanomalies decreasing to the west indicate the importance of the horizontal component of the Andean root. Recall that the gravity effect of the root is equal and opposite to the isostatic correction.

Decompensative gravity anomaly

A 'decompensative' correction to the isostatic gravity anomaly also allows us to remove the gravity effect of isostatic compensation of geological loads. Under a hypothesis of local (as opposed to regional) compensation, the gravity effect of a shallow mass can be separated from the effect of gravity of a geological body emplaced on upper crust by its deep compensating root. The decompensative anomaly is the Bouguer gravity anomaly with isostatic and decompensative corrections added. The decompensative correction (Cordell et al. 1991) is calculated based on the upwards continuation of the isostatic anomaly (AI) in order to minimize the effects of the short-wavelength structures. The decompensative anomaly is then obtained by subtracting the regional anomaly (upwards continuation) from the isostatic anomaly (AI).

In this work we apply a variant of this procedure (Cordell et al. 1991), working with the power spectrum method of the isostatic anomaly instead of performing an upwards continuation in order to separate the different wavelengths (Fig. 6) through the proposed transformation by Mishra & Naidu (1974).

The above-mentioned improvement allows a depth value of the sources causing anomalies to be estimated; average depths of the anomalous sources are then determined, as proposed by Spector & Grant (1970). Additionally, spectral analysis determines the cut-off frequency for long wavelengths and applies the filter more efficiently.

As indicated, the lower frequencies of the power spectrum indicate an approximate depth of 20 km. This value coincides with the décollement of the main contractional structures determined for this region by Cominguez & Ramos (1990) using deep seismic reflection data. These results were confirmed by Gilbert et al. (2006) using receiver functions. The décollement of the fold-and-thrust belt are located in a brittle-ductile transition that could constitute a natural discontinuity, where magmas associated with Late Triassic–Early Cretaceous rifting stages reside temporally or permanently, constituting an anomalous mass in the middle–upper crust.

To fulfil the objective of studying structures shallower than 20 km above this mass anomaly area, we used a high-ass filter with a cut-off wave number k_c of 0.022 cyc km^{-1} (Fig. 6). Using this cut-off parameter, we obtained the decompensative isostatic residual anomaly map (Fig. 7).

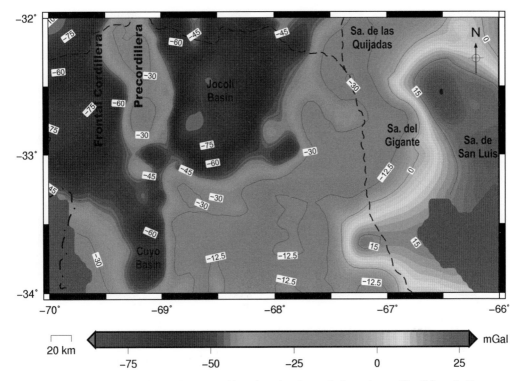

Fig. 5. Map of isostatic residual anomaly computed by subtracting the gravity isostatic root (Fig. 4) from the Bouguer anomaly map (Fig. 2).

Enhancement of anomalies: Tilt method

The Tilt method is effective in enhancing the edges of bodies that generate anomalous effects in the gravity field, assuming a vertical contact model. This method uses the horizontal and vertical gradients of the gravity field, and does not require previous knowledge about the geometry. The Tilt

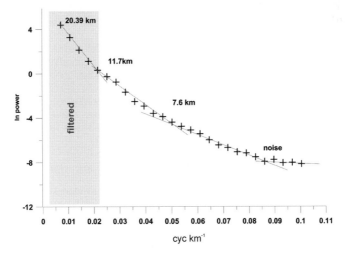

Fig. 6. Power spectrum of the radial isostatic anomaly obtained from the proposed transformation of Mishra & Naidu (1974). The isostatic anomaly input signal is used. Numbers in kilometres indicate the average depths obtained for each line. The shaded area indicates the filtered frequency range.

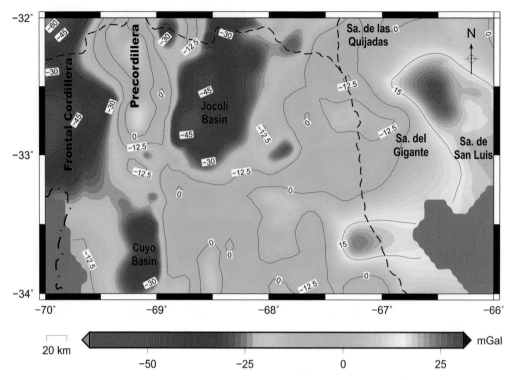

Fig. 7. Decompensative gravity anomaly map. Even although this map is morphologically similar to Figure 5, the range of values is different and physically represents residual anomalies of shallow sources.

method has therefore been applied with success to highlight edges and shapes of geological structures (e.g. Salem et al. 2008; García Torrejón et al. 2011; Oruç & Selim 2011). The tilt angle (Miller & Singh 1994; Thurston & Smith 1997; Verduzco et al. 2004) is defined:

$$\text{Tilt} = \arctan\left(\frac{T_{zz}}{\sqrt{T_{xz}^2 + T_{yz}^2}}\right) \quad (3)$$

where the denominator represents the horizontal gravity gradient and T_{zz} is the vertical gravity gradient. Figure 8 depicts the results obtained by applying this method. From this analysis two regional structures with an unknown direct significance are identified: the San Pedro Ridge and the Tunuyan Lineament that are subsequently described.

2D models

In order to relate the gravity signal to the potential basement segmentation observed in the decompensative gravity anomaly map and the Tilt map, we constructed two density models over selected NW-trending profiles (Fig. 7). Constraints used in these models are deep seismic refraction lines, borehole data and geological information (Fig. 1b).

The initial reference model is a simple two-layer model which includes depth of the basement and sediments, which fill the basins. We then improved the model by introducing potential lateral density variations between the basements of Cuyania and Pampia terrains. Variable sediment densities are assigned for each basin based on log constraints (see following section). From these models we obtained the direct gravity response. Finally, the difference between observed data and the calculated response was minimized varying the geometry of the polygons from the original model using a 2D forward modelling algorithm (Talwani et al. 1959; Marquardt 1963).

Geophysical constraints

Densities used to model profiles have been taken from oil exploration borehole data, corresponding to Las Peñas log (YPF.SJ.Sp.EP.-1) located within the Jocoli basin (Fig. 1b). To determine the mean density of the sedimentary basin filling, we obtained the weighted averaged densities of each lithological unit (by estimating its thickness). We found a mean density value of 2390 kg m^{-3}, which

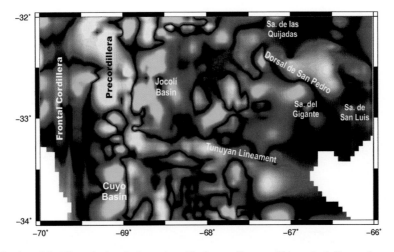

Fig. 8. Application of the Tilt method on the isostatic residual anomalies map. This method allows enhancement of the edges and shapes of geological structures that generate anomalous effects in the gravity field, assuming a vertical contact model from the horizontal and vertical gradients of the gravity field. The high gradient located south of 33.4°S latitude trending approximately westwards is associated with the Tunuyan Lineament, identified by Kostadinoff & Gregori (2004). This feature, which has no surface expression, has been associated with the boundary between two regions with different seismotectonic behaviour, possibly associated with the change in the Nazca plate subduction angle (Perucca & Bastias 2005).

is similar to the density value of 2400 kg m^{-3} proposed by Azeglio et al. (2010) towards the east of the neighbouring Las Salinas basin. In order to obtain the crystalline basement density, we used the velocities obtained from deep seismic refraction data available in the area from Yacimientos Petroliferos Fiscales (YPF). By means of the Gardner equation (Gardner et al. 1974) we determined a range of density values from 2676 to 2716 kg m^{-3} corresponding to velocities from 5600 to 6000 m s^{-1}. These velocity values are similar to those of the crystalline basement in Western Sierras Pampeanas found by Snyder et al. (1990).

The geometry of the crystalline basement along the profiles was adjusted using information obtained from: (1) seismic refraction data (YPF); (2) seismic sections interpreted by Cominguez & Ramos (1990) and Vergés et al. (2007) between Precordillera and Sierras Pampeanas; and (3) surficial data from Western Sierras Pampeanas (Snyder et al. 1990; Zapata & Allmendinger 1996).

Profile 1

This profile was traced in the northern region over 340 km (see location on Fig. 1b). This section cut across the Eastern Precordillera, the Jocoli basin and the northern part of Sierra de las Quijadas (Western Sierras Pampeanas). The profile crosses the Salinas basin up to the NW border of the Sierra de San Luis. The gravity model obtained along this section was adjusted through deep seismic refraction profiles, borehole records and geological information available on this zone (Fig. 9). The western border was modelled using a triangular structural geometry determined by overthrusts with opposite vergences proposed previously (Figueroa & Ferraris 1989; Ramos et al. 1997). In Cerro Salinas, a series of thrust sheets are detached from the contact between the sedimentary section and the basement producing a west-vergent structure as described by Vergés et al. (2007).

The Jocoli basement geometry ($\rho_{cb} = 2670$ kg m^{-3}) is characterized by faulted blocks that have a strong gravimetric contrast with the basin fill ($\rho_{sed} = 2390$ kg m^{-3}). The highest sedimentary infill (c. 3 000 m) is located to the east of Cerro Salinas, diminishing towards the Sierra de las Quijadas (up to 1500 m). Eastwards in the Western Sierras Pampeanas a pronounced increase in the gravity residual anomaly is observed (Fig. 7). This anomaly has been linked to potential mafic bodies buried at the contact zone between the Cuyania and Pampia terrains (Introcaso & Pacino 1987; Ramos 1994; Miranda & Introcaso 1999; Gimenez et al. 2000; Martinez et al. 2007). This was resolved by modelling the crust assigned to the Pampia terrain with a higher density ($\rho = 2720$ kg m^{-3}). In this model the Salinas basin with a sedimentary thickness of 1 km and a mean density of 2400 kg m^{-3} therefore has a basement constituted by the Pampia terrain (Azeglio et al. 2010).

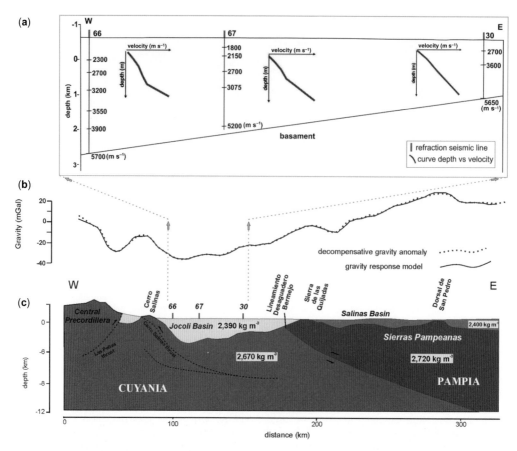

Fig. 9. Modelled Profile 1 from western Precordillera to eastern Sierras Pampeanas through the Sierra de las Quijadas. (**a**) The location of deep seismic refraction lines is shown in red and the velocities used to calculate densities from the Gardner equation are indicated in blue. (**b**) Fit between the decompensative gravity anomaly and the modelled profile. (**c**) Density model of the upper crust. The highest sedimentary thickness (c. 3000 m) is located in the Jocoli basin ($\rho_{fill} = 2390$ kg m^{-3}). Eastwards in the Western Sierras Pampeanas, a pronounced increase in the decompensative gravity anomaly has been linked to the contact zone between the basements of the Cuyania and Pampia terrains ($\rho = 2670$ kg m^{-3} v. $\rho = 2720$ kg m^{-3}). Pampia terrain constitutes the basement of the Salinas basin which has a sedimentary thickness of 1 km and a mean density of 2400 kg m^{-3} (Azeglio et al. 2010). See the text for more details.

Profile 2

This section extends for 300 km from the Eastern Precordillera to the western flank of the Sierra de San Luis in the Western Sierras Pampeanas, passing through the southern Jocoli basin and Sierra del Gigante (see location on Fig. 1b). The maximum sedimentary thickness of this basin is located to the eastern border of Central Precordillera as described in Profile 1, modelled with a density of 2670 kg m^{-3}. In this case the sediment thickness reaches up to 8000 m. The increasing depth of the Jocoli basin towards the Precordillera deformational front has been determined by Vergés et al. (2007) from a seismic section a few kilometres to the north (Fig. 10).

On the eastern-most region of this profile a high gravity value, potentially related to a density variation between the Cuyania and Pampia terrains, is observed in the Sierra del Gigante area. We do not discard a certain contribution to this density contrast derived from Early Cretaceous mafic bodies.

Results

The isostatic root found on the basis of the Airy–Heiskanen isostatic model (Fig. 3) shows a good correspondence with the Moho obtained by Gans et al. (2011) based on receiver functions and by Assumpção et al. (2013) based on active source experiments (deep seismic reflection surveys) and

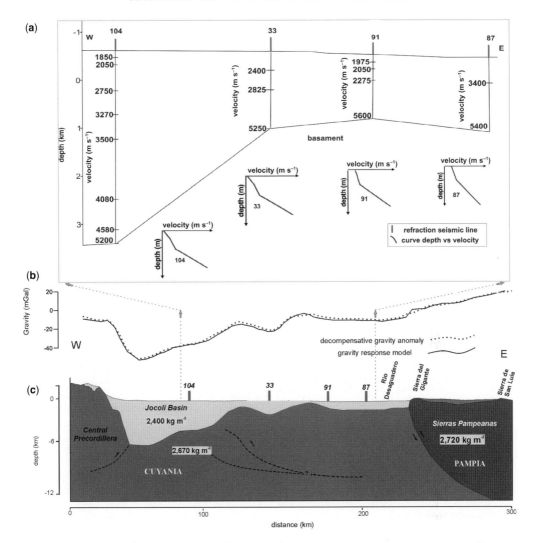

Fig. 10. Modelled Profile 2 from Precordillera to the western flank of the Sierra de San Luis in the Western Sierras Pampeanas. (**a**) The location of deep seismic refraction lines used as constraints is shown in red and velocities used to calculate densities from the Gardner equation are indicated in blue. (**b**) Fit between the decompensative gravity anomaly and the modelled profile. (**c**) Density model of the upper crust, where maximum sedimentary thicknesses reach up to 8000 m. On the easternmost region of this profile in the Sierra del Gigante area, a high gravity value is related to a density variation between the basements of the Cuyania and Pampia terrains.

receiver functions. The minimum values of the Bouguer anomaly (−300 mGal) found in Frontal Cordillera (Fig. 2) are related to the crustal thickening of c. 65 km obtained by the application of the Airy–Haiskenan local compensation model. These results are consistent with previous work. The gravity signal increases eastwards up to c. −50 mGal at Sierras Pampeanas, in relation to a progressive decrease in the crustal thickness to the foreland zone (45 km) (see Fig. 3). The gravity effect of the hydrostatic Moho (Fig. 4) behaves as a warped plane in which the eastern part corresponds to isoanomalies of the order −75 mGal, descreasing towards the west up to −225 mGal.

The isostatic anomaly map (Fig. 5) and the decompensative gravity anomaly (Fig. 7) are consistent with previous studies that are detailed below. Even though the figures are morphologically similar, the absolute scale and consequently the corresponding values vary. In the decompensative gravity anomaly map (Fig. 7), we identified different regions related to the main geological features:

(1) Frontal Cordillera (Introcaso et al. 1992b; Miranda & Robles 2002).
(2) The relative positive gravity response of the Precordillera within the general negative trend marked by the Andean root is clearly depicted (Introcaso et al. 1992a; Gimenez et al. 2000, 2009; Alvarez et al. 2012).
(3) The Jocoli basin, which is limited to the east by high anomalous values, is possibly associated with the continuity of the basement of Pie de Palo underneath (Martinez et al. 2008). The lowest values (-45 mGal) are localized towards the west, constituting evidence of the increase in sedimentary thickness towards the Precordillera deformational front.
(4) To the north of Sierra de San Luis, a high gravity value of more than 15 mGal is associated with a belt of mafic and ultra-mafic rocks with densities of the order 3000 kg m^{-3} (see Kostadinoff et al. 2010).
(5) The Cuyo basin with values of -30 mGal is linked to a sedimentary fill with density estimated as 2320 kg m^{-3} and thicknesses of 5 km (Miranda & Robles 2002).

Additionally, to enhance the anomalies related to density contrasts, we applied the Tilt method (Fig. 8). Areas of high gradients are indicated with warm colours, while cold colours indicate the sedimentary basins. Analysing the Tilt map, we can distinguish the following features.

(1) The Precordillera region is contoured by the contrast between the areas of warm and cold colours. This is in agreement with the decompensative gravity anomaly results.
(2) The Jocoli basin edges are depicted by the Tilt method, achieving a better resolution than in previous work (Fig. 7).
(3) The Bermejo–Desaguadero basin, an extensional basin associated with the reactivation of the boundary between Cuyania and Pampia terranes to the north, can be traced in the studied zone bordered by the Sierra de las Quijadas to the east. The Tilt method also highlights the San Pedro Ridge, a NW-oriented lineament of regional development (Kostadinoff et al. 2003; Kostadinoff & Gregori 2004; Azeglio et al. 2009). Even although this lineament cannot be related to a geological body with superficial expression, its presence is gravimetrically constrained; further work is necessary to understand its significance.
(4) Another important feature in this map is the high gradient located south of 33.4°S latitude, in a roughly NW direction. This is associated with the continuation of the Tunuyan lineament, identified east of 66.5°W by Kostadinoff & Gregori (2004). The Tunuyan lineament has been associated with a boundary between two regions with different seismotectonic behaviour (to the north, the Precordillera and to the south, the Southern Mendoza area) by Perucca & Bastias (2005). The seismotectonic behavioir has been associated with the transition zone between the flat subduction of the Nazca plate to the north and the segment of normal angle to the south (Perucca & Bastias 2005). Its real significance again merits further work, since it is transversal to the structural and gravimetric grain. Hypothetically, this feature could be related to some kind of crustal structure associated with the change in angle of the subduction zone or to an inherited basement structure.

Finally, results obtained by applying the Tilt method are consistent with the solutions found by Gimenez et al. (2008) when implementing the 3D Euler deconvolution method using gravity data.

Conclusions

The computed isostatic anomalies (Fig. 5) indicate that in the potential cessation of the dynamic forces that are sustaining the topography the Frontal Cordillera and Jocoli basin would ascend isostatically, while the Precordillera would ascend to a lesser extent. Contrastingly, the Sierra de San Luis and the Sierras Pampeanas to the foreland area should descend in order to reach isostatic equilibrium.

The decompensative gravity anomaly allows us to identify geological features such as Frontal Cordillera, Precordillera, Sierras Pampeanas, and the Cuyo and Jocoli basins. When applying the Tilt method, a technique that is used to highlight gravity anomalies associated with sources emplaced in the upper crust, the previously mentioned results are enhanced. Two regional lineaments are clearly delineated: San Pedro Ridge and the Tunuyan Lineament. The latter, located south of 33.4°S with an approximately NW orientation, coincides with morphological changes that occur in the subducted Nazca plate from flat to normal subduction angle.

The gravity models obtained from the decompensative gravity anomaly and constrained by the existing geological and geophysical data reveal the geometry of a segmented basement in the north of this region. Results obtained from both modelled profiles reveal that the basement deepens towards the Central Precordillera region. The depths of the modelled depocentres reach up to 8 km in the Jocoli basin and 2 km for the Salinas basin. To the foreland area, a significant increase in the residual gravity anomaly could be due to the presence of high-density rocks associated with the closure of Palaeozoic ocean basins.

The authors acknowledge the use of the GMT-mapping software of Wessel & Smith (1998). We are also grateful to CONICET, Consejo Nacional de Investigaciones Científicas y Técnicas, Project PIP 6044, CICITCA, Project N°21 E901 and the Agencia Nacional de Promoción Científica y Tecnológica, Project PICT- 2007-01903 for their financial support.

References

ABBRUZZI, J. M., KAY, S. M. & BICKFORD, M. E. 1993. Implications for the nature of the Precordilleran basement from the geochemistry and age of Precambrian xenoliths in Miocene volcanic rocks, San Juan province. *In*: *12° Congreso Geológico Argentino y 2° Congreso de Exploración de Hidrocarburos*, Buenos Aires, Actas, **3**, 331–339.

ACEÑOLAZA, F. G., MILLER, H. & TOSELLI, A. J. 2002. Proterozoic-Early Paleozoic evolution in western South America: a discussion. *Tectonophysics*, **354**, 121–137.

ALLMENDINGER, R. W., FIGUEROA, D., SNYDER, D., BEER, J., MPODOZIS, C. & ISACKS, B. L. 1990. Foreland shortening and crustal balancing in the Andes at 30° S Latitude. *Tectonics*, **9**, 789–809.

ALMEIDA, F. F. M., BRITO NEVES, B. B. & CARNEIRO, C. D. R. 2000. The origin and evolution of the South American Platform. *Earth-Science Reviews*, **50**, 77–111.

ALVARADO, P., BECK, S. & ZANDT, G. 2007. Crustal structure of the south-central Andes Cordillera and backarc region from regional waveform modelling. *Geophysical Journal International*, **170**, 858–875.

ALVAREZ, O., GIMENEZ, M., BRAITENBERG, C. & FOLGUERA, A. 2012. GOCE satellite derived gravity and gravity gradient corrected for topographic effect in the South Central Andes region. *Geophysical Journal International*, **190**, 941–941.

ASSUMPÇÃO, M., BIANCHI, M. ET AL. 2013. Crustal thickness map of Brazil: data compilation and main features. *Journal of South American Earth Sciences*, **43**, 74e85.

AZEGLIO, E. A., GIMÉNEZ, M. E. & INTROCASO, A. 2009. La aplicación de técnicas de Señal Analítica a las estructuras de la cuenca de Las Salinas. *Revista de la Asociación Geológica Argentina*, **64**, 183–193.

AZEGLIO, E. A., GIMÉNEZ, M. E. & INTROCASO, A. 2010. Análisis estructural de la cuenca de Las Salinas y su relación con la acumulación de sedimentos eólicos en el área Médanos Negros. *Revista Asociación Geológica Argentina*, **67**, 2.

BALDIS, B. A. & CHEBLI, G. 1969. Estructura profunda del área central de la Precordillera sanjuanina. *IV Jornadas Geologia Argentina*, **1**, 47–66. Argentina.

BALDIS, B., BERESI, M., BORDONARO, O. & VACA, A. 1982. Síntesis evolutiva de la Precordillera Argentina. *In*: *V Congreso Latinoamericano de Geología*, Buenos Aires, **IV**, 399445.

BENEDETTO, J. L. & ASTINI, R. A. 1993. A Collisional model for the stratigraphic evolution of the Argentine Precordillera during the Early Paleozoic. *In*: *2nd International Symposium on Andean Geodynamics*, Oxford, 501–504.

BENEDETTO, J. L., SÁNCHEZ, T. M., CARRERA, M. G., BRUSSA, E. D. & SALAS, M. J. 1999. Paleontological constraints on successive paleogeographic positions of Precordillera terrane during the early Paleozoic. *In*: RAMOS, V. A. & KEPPIE, J. D. (eds) *Laurentia-Gondwana Connections before Pangea*. Geological Society, America, Special Paper, **336**, 21–42.

BERCOWSKI, F., RUZYCKI, L., JORDAN, T., ZEITLER, P., CABALLERO, M. M. & PEREZ, I. 1993. Litofacies y edad isotópica de la secuencia La Chilca y su significado paleogeográfico para el Neógeno de Precordillera. *In*: *Actas del 12° Congreso Geológico Argentino y 2° Congreso de Exploración de Hidrocarburos*, Mendoza, **1**, 212–217.

BLAKELY, R. 1995. *Potential Theory in Gravity and Magnetic Applications*. Cambridge University Press, Cambridge.

BOLL, A. & DE LA COLINA, J. 1993. Armazón estratigráfico del Triásico- Jurásico en Atamisqui-Cuenca Cuyana-Mendoza. *In*: *Actas XII Congreso Geológico Argentino y II Congreso de Exploración de Hidrocarburos*, Mendoza, Acta, **I**, 33–40.

BRACACCINI, O. I. 1946. Contribución al conocimiento geológico de la Precordillera Sanjuanino-mendocina. *In*: *Boletín de Informaciones Petroleras*, Buenos Aires, 258–264.

BRIGGS, I. 1974. Machine contouring using minimum curvature. *Geophysics*, **39**, 39–48.

BRITO NEVES, B. B., CAMPOS NETO, M. C. & FUCK, R. A. 1999. From Rodinia to Western Gondwana: an approach to the Brasiliano-Pan African cycle and orogenic collage. *Episodes*, **22**, 155–166.

CAMINOS, R. 1979. Sierras Pampeanas Noroccidentales Salta, Tucumán, Catamarca, La Rioja y San Juan. *Geología Regional Argentina*, Academia Nacional de Ciencias, Córdoba, **1**, 225–291.

CASQUET, C., PANKHURST, R. J. ET AL. 2008. The Mesoproteroic Maz Terrane in the Western Sierras Pampeanas, Argentina, equivalent to the Arequipa-Antofalla block of southern Peru? Implications for West Gondwana margin evolution. *Gondwana Research*, **13**, 163–175.

CHAPIN, D. A. 1996. A deterministic approach toward isostatic gravity residuals-A case study from South America. *Geophysics*, **61**, 1022–1033.

CHRISTENSEN, N. & MOONEY, W. 1995. Seismic velocity structure and composition of the continental crust: a global view. *Journal of Geophysical Research*, **100**, 9761–9788.

COMINGUEZ, A. H. & RAMOS, V. A. 1990. Sísmica de reflección profunda entre Precordillera y Sierras Pampeanas. *In*: *XI Congreso Geológico Argentino*. San Juan. Actas, **II**, 311314.

COMINGUEZ, A. H. & RAMOS, V. A. 1991. La estructura profunda entre Precordillera y Sierras Pampeanas de la Argentina: evidencias de la sísmica de reflexión profunda. *Revista Geológica de Chile*, **18**, 3–14.

CORDELL, L., ZORIN, Y. A. & KELLER, G. R. 1991. The decompensative gravity anomaly and deep structure of the region of the Rio Grande rift. *Journal of Geophysical Research*, **96**, 6557–6568.

FIGUEROA, D. E. & FERRARIS, O. R. 1989. Estructura del margen oriental de la Precordillera Mendocino-sanjuanina. *In: 1° Congreso Nacional de Exploración de Hidrocarburos*, Mar del Plata, Buenos Aires, **1**, 515–529.

FINNEY, S. C., GLEASON, J. D., GEHRELS, G. G., PERALTA, S. H. & ACEÑOLAZA, G. 2003. Early Gondwana connection for the Argentina Precordillere Terrane. *Earth and Planetary Sciences Letters*, **205**, 349–359.

GALINDO, C., CASQUET, C., RAPELA, C. W., PANKRURST, R. J., BALDO, E. & SAAVEDRA, J. 2004. Sr, C, and O isotope geochemistry and stratigraphy of Precambrian and lower Paleozoic carbonate sequences from the western Sierras Pampeanas of Argentina: tectonic implications. *Precambrian Research*, **131**, 55–71.

GANS, C. R., BECK, S. L., ZANDT, G., GILBERT, H., ALVARADO, P., ANDERSON, M. & LINKIMER, L. 2011. Continental and oceanic crustal structure of the Pampean flat slab region, western Argentina, using receiver function analysis: new high-resolution results. *Geophysical Journal International*, **186**, 45–58.

GARCÍA TORREJÓN, M., ALVAREZ PONTORIERO, O. ET AL. 2011. Evidencias de la zona de contacto entre los terrenos de Precordillera y Pie de Palo, Provincias de San Juan y Mendoza. *Revista de la Asociación Geológica Argentina*, **68**, 502–506.

GARDNER, G. H. F., GARDNER, L. W. & GREGORY, R. 1974. Formation velocity and density – the diagnostic basis for stratigraphic traps. *Geophysics*, **39**, 770–780.

GILBERT, H., BECK, S. & ZANDT, G. 2006. Lithosperic and upper mantle structure of Central Chile and Argentina. *Geophysical Journal International*, **165**, 383–398.

GIMENEZ, M. E., MARTINEZ, M. P. & INTROCASO, A. 2000. A Crustal Model based mainly on Gravity data in the Area between the Bermejo basin and the Sierras de Valle Fértil- Argentina. *Journal of South American Earth Sciences*, **13**, 275–286.

GIMENEZ, M., MARTINEZ, P. & INTROCASO, A. 2008. Determinaciones de lineamientos regionales del basamento cristalino a partir de un análisis gravimétrico. *Revista de la Asociación Geológica Argentina*, **63**, 288–296.

GIMENEZ, M. E., BRAITENBERG, C., MARTINEZ, M. P. & INTROCASO, A. 2009. A comparative analysis of Seismological and Gravimetric Crustal Thicknesses below the Andean Region with flat Subduction of the Nazca Plate. Hindawi Publishing Corporation International. *Journal of Geophysics*. Article ID 607458, http://dx.doi.org/10.1155/2009/607458

GONZÁLEZ BONORINO, F. 1950. Algunos problemas geológicos de las Sierras Pampeanas. *Revista de la Asociación Geológica Argentina*, **49**, 81110.

GÖTZE, C. & EVANS, B. 1979. Stress and temperature in the bending lithosphere as constrained by experimental rock mechanism. *Geophysical Journal of the Royal Astronomical Society*, **59**, 463–478.

GÖTZE, H. J. & KIRCHNER, A. 1997. Interpretation of gravity and geoid in the Central Andes between 20° and 29° S. *Journal of South American Earth Sciences*, **10**, 179–188.

HINZE, W. J. 2003. Bouguer reduction density, why 2.67. *Geophysics*, **68**, 1559–1560.

HINZE, W. J., AIKEN, C. ET AL. 2005. New standards for reducing gravity data: the North American gravity database. *Geophysics*, **70**, 4, J25–J32.

INTROCASO, A. & PACINO, M. C. 1987. Gravity Andean model associated with subduction near 24°25′ south latitude. *Revista Geofísica*, **2**, 43.

INTROCASO, A., GUSPÍ, F., ROBLES, A., MARTINEZ, P. & MIRANDA, S. 1992a. Carta gravimétrica de Precordillera y Sierras Pampeanas entre 30° y 32° de Latitud Sur. *In: 17° Reunión de la Asociación Argentina de Geofísicos y Geodesias*, Buenos Aires.

INTROCASO, A., PACINO, M. C. & FRAGA, H. 1992b. Gravity, isostasy and Andean crustal shortening between latitudes 30° and 35° S. *Tectonophysics*, **205**, 31–48.

INTROCASO, A., PACINO, M. C. & GUSPI, F. 2000. The Andes of Argentina and Chile: Crustal configuration, Isostasy, Shortening and Tectonic features from Gravity Data. *Temas de Geociencia*, **5**, 31.

KANE, M. F. 1962. A comprehensive system of terrain corrections using a digital computer. *Geophysics*, **27**, 455–462.

KAY, S. M., ORRELL, S. & ABBRUZZI, J. M. 1996. Zircon and whole rock Nd-Pb isotopic evidence for a Grenville age and a Laurentian origin for the basement of the Precordilleran terrane in Argentina. *Journal of Geology*, **104**, 637–648.

KELLER, M. 1999. Argentine Precordillera: Sedimentary and plate tectonic history of a Laurentian crustal fragment in South America. *Geological Society of America, Special Paper*, **341**, 134.

KOSTADINOFF, J. & GREGORI, D. 2004. La Cuenca de Mercedes, provincia de San Luis. *Revista de la Asociación Geológica Argentina*, **59**, 488–494.

KOSTADINOFF, J., BJERG, E., RANIOLO, A. & SANTIAGO, E. 2003. Anomalías del campo gravitatorio y magnético terrestre en la sierra de Socoscora, provincia de San Luis. *Revista de la Asociación Geológica Argentina*, **58**, 505–510.

KOSTADINOFF, J., FERRACUTTI, G. R. & BJERG, E. A. 2010. Interpretación de una sección gravi-magnetométrica sobre la Pampa de las Invernadas, Sierra Grande de San Luis. *Revista de la Asociación Geológica Argentina*, **67**, 349–353.

LAGORIO, S. L. 2008. Early Cretaceous alkaline volcanism of the Sierra Chica de Córdoba (Argentina): mineralogy, geochemistry and petrogénesis. *Journal of South American Earth Sciences*, **26**, 152–171.

LEGARRETA, L., KOKOGIAN, D. & DELLAPE, D. 1993. Estructuración Terciaria de la Cuenca Cuyana. Cuánto de inversión tectónica? *Revista de la Asociación Geológica Argentina*, **47**, 83–86.

LLAMBÍAS, E. J. 1966. Geología y petrografía del volcán Payún Matru. *Acta Geológica Lilloana*, **8**, 265–310.

MARQUARDT, D. 1963. An Algorithm for Least-Squares Estimation of Nonlinear Parameters. *SIAM Journal of Applied Mathematics*, **11**, 431–441.

MARTINEZ, A., RIVAROLA, D., STRASSER, E., GIAMBIAGI, L., ROQUET, M. B., TOBARES, M. L. & MERLO, M. 2012. Petrografía y geoquímica preliminar de los basaltos cretácicos de la sierra de Las Quijadas y Cerrillada. *Serie Correlación Geológica*, **28**, 9–22.

MARTINEZ, M. P., GIMENEZ, M. E., BUSTOS, G., LINCE KLINGER, F., MALLEA, M. & JORDAN, T. 2006. Detección de Saltos de Basamento de la Cuenca del Valle de la Rioja-Argentina a partir de un Modelo Hidrostático. *GEOACTA*, **31**, 1–9.

MARTINEZ, M. P., GIMENEZ, M. E., INTROCASO, A. & RUIZ, F. 2007. Preliminary Geophysic Results in The Calingasta Bolson, Province of San Juan, Argentina. *In*: *U51B-01, AGU Meetings 2007*, Acapulco.

MARTINEZ, M. P., PERUCA, P. L., GIMENEZ, M. E. & RUIZ, F. 2008. Manifestaciones Geomorfológicas y Geoísicas de una Estructura Geológica al Sur de La Sierra de Pie de Palo. *Revista de la Asociación Geológica Argentina*, **63**, 104–111.

MILLER, H. G. & SINGH, V. 1994. Potential field tilt – A new concept for location of potential field sources. *Journal of Applied Geophysics*, **32**, 213–217.

MIRANDA, S. & INTROCASO, A. 1999. Cartas gravimétricas de la provincia de Córdoba, República Argentina: interpretación de la estructura profunda de la Sierra de Córdoba: Argentina, Universidad Nacional de Rosario. *Temas de Geociencia*, **1**, 45.

MIRANDA, S. & ROBLES, J. A. 2002. Posibilidades de atenuación cortical en la cuenca Cuyana a partir del análisis de datos de gravedad. *Revista de la Asociación Geológica Argentina*, **5**, 271–279.

MISHRA, D. C. & NAIDU, P. S. 1974. Two-dimensional power spectral analysis of aeromagnetic fields. *Geophysical Prospecting*, **22**, 345–353.

MOHR, P. J. & TAYLOR, B. N. 2001. The fundamental physical constant. *Physics Today*, **54**, 6–16.

MORELLI, C., GANTAR, C. ET AL. 1974. *The International Gravity Standardization Net 1971 (IGSN71)*. IUGG-IAG International Union of Geodesy and Geophysics, Special Publications, Paris, **4**.

MORITZ, H. 1980. Geodetic Reference System 1980. *Journal of Geodesy*, **54**, 395–405.

MULCAHY, S., ROESKE, S., MCCLELLAND, W., NOMADE, S. & RENNE, P. 2007. Cambrian initiation of the Las Pirquitas thrust of the western Sierras Pampeanas, Argentina: implications for the tectonic evolution of the proto-Andean margin of South America. *Geology*, **35**, 443–446.

NAGY, D. 1966. The gravitational attraction of a right rectangular prism. *Geophysics*, **31**, 362–371.

ORTIZ, A. & ZAMBRANO, J. J. 1981. La Provincia Geológica Precordillera Oriental. *In*: *VIII Congreso Geológico Argentino*, San Luis, Argentina, Actas, **III**.

ORUÇ, B. & SELIM, H. 2011. Interpretation of magnetic data in the Sinop area of Mid Black Sea, Turkey, using tilt derivative, Euler deconvolution, and discrete wavelet transform. *Journal of Applied Geophysics*, **74**, 194–204.

PANKHURST, R. J., RAPELA, C. W., SAAVEDRA, J., BALDO, E., DAHLQUIST, J., PASCUA, I. & FANNING, C. M. 1998. The Famatinian magmatic arc in the central Sierras Pampeanas: an Early to Mid- Ordovician continental arc on the Gondwana margin. *In*: PANKHURST, R. J. & RAPELA, C. W. (eds) *The Proto-Andean Margin of Gondwana*. Geological Society, London, Special Publication, **142**, 343–367.

PERALTA, S., PÖTHE DE BALDIS, E., LEÓN, L. & PEREYRA, M. 2003. Silurian of the San Juan Precordillera, western Argentina: stratigraphic framework. *In*: ORTEGA, G. & ACEÑOLAZA, G. F. (eds) *Proceedings of the 7° International Reappraisal of the Silurian Stratigraphy at Cerro del Fuerte section*, San Juan, Argentina INSUGEO, Serie Correlaciones Geológicas, **18**, 151–155.

PÉREZ-GUSSINYÉ, M., LOWRY, A. R., PHIPPS MORGAN, J. & TASSARA, A. 2008. Effective elastic thickness variations along the Andean margin and their relationship to subduction geometry. *Geochemistry Geophysics Geosystems*, **9**, Q02003, http://dx.doi.org/10.1029/2007GC001786

PERUCCA, L. & BASTIAS, H. 2005. El terremoto argentino de 1894: Fenómenos de licuefacción asociados a Sismos. Simp. Bodenender. INSUGEO. *Serie de Correlación Geológica*, **19**, 55–70.

QUENARDELLE, S. M. & RAMOS, V. A. 1999. Ordovician western Sierras Pampeanas magmatic belt: record of Precordillera accretion in Argentina. *In*: RAMOS, V. A. & KEPPIE, J. D. (eds) *Laurentia-Gondwana Connections before Pangea*. Geological Society of America, Boulder, Special Paper, **336**, 63–86.

RAMOS, V. A. 1988. The tectonics of the Central Andes; 30° to 33° S latitude. *In*: CLARK, S. & BURCHFIEL, D. (eds) *Processes in Continental Lithospheric Deformation*. Geological Society of America, Boulder, Special Paper, **218**, 31–54.

RAMOS, V. A. 1994. Terranes in Southern Gondwana land and their control in the Andean Structure (30°–33° latitude). *In*: REUTTER, K. J., SCHEUBER, E. & WIGGER, P. J. (eds) *Tectonics of the Southern Central Andes*. Springer-Verlag, New York, 249–261.

RAMOS, V. A. 1995. Sudamérica: un mosaico de continentes y océanos. *Ciencia Hoy*, **6**, 24–29.

RAMOS, V. A. 2004. Cuyania, an exotic block to Gondwana: review of a historical success and the present problems. *Gondwana Research*, **7**, 1009–1026.

RAMOS, V. A. & VUJOVICH, A. G. 1993. *Laurentia – Gondwana Connection: A South American Perspective*. GSA, Abstracts with Programs, Boston, **1**.

RAMOS, V. A., JORDAN, T. E., ALLMENDINGER, R., MPODOZIS, C., KAY, S. M., CORTES, J. M. & PALMA, M. 1986. Paleozoic terranes of the Central Argentine-Chilean Andes. *Tectonics*, **5**, 855–880.

RAMOS, V. A., CEGARRA, M. I., LO FORTE, G. & COMINGUEZ, A. 1997. El frente orogénico de la sierra de Pedernal (San Juan, Argentina): su migración a través de los depósitos sinorogénicos. *In*: *Actas 8° Congreso Geológico Chileno*, Chile, **3**, 1709–1713.

RAMOS, V. A., DALLMEYER, R. D. & VUJOVICH, G. 1998. Time constraints on the Early Palaeozoic docking of the Precordillera, central Argentina. *In*: *The Proto-Andean Margin of Gondwana*. Geological Society, London, Special Publications, **142**, 143–158.

RAMOS, V. A., CRISTALLINI, E. O. & PÉREZ, D. J. 2002. The Pampean Flat-Slab of the Central Andes. *Journal of South American Earth Sciences*, **15**, 59–78.

RAPALINI, A. E. & ASTINI, R. A. 1997. First Paleomagnetic Evidence for the Laurentian Origin of the Argentine Precordillera. *In*: *8th Scientific Assembly IAGA*, Uppsala, Sweden, Abstracts, 56.

RAPELA, C. W., PANKHURST, R. J. & FANNING, C. M. 2001. U-Pb SHRIMP ages of basement rocks from Sierra de la Ventana (Buenos Aires, Argentina).

In: *3rd South American Symposium on Isotope Geology*, Extended Abstracts (CD edition), Sociedad Geológica de Chile, Santiago, 225–228.

RIVAROLA, D. & SPALLETTI, L. 2006. Modelo de sedimentación continental para el rift cretácico de la Argentina central. Ejemplo de la Sierra de las Quijadas, San Luis. *Revista de la Asociación Geológica Argentina*, **61**, 63–80.

ROLLERI, E. & BALDIS, B. 1969. *Paleography and distribution of Carboniferous deposits in the Precordillera, Argentina*. La estratigrafía del Gondwana, Ciencias de la Tierra-2, Coloquio U.I.C.G. UNESCO, París (France).

SALEM, A., WILLIAMS, S., FAIRHEAD, D., SMITH, R. & RAVAT, D. 2008. Interpretation of magnetic data using tilt-angle derivatives. *Geophysics*, **73**, 1, doi: 10.1190/1.2799992.

SIMPSON, R. W., JACHENS, R. C., BLAKELY, R. J. & SALTUS, R. W. 1986. A new isostatic residual gravity map of the conterminous United States with a discussion on the significance of isostatic residual anomalies. *Journal of Geophysical Research*, **91**, 8348–8372.

SNYDER, D. B., RAMOS, V. A. & ALLMENDINGER, R. W. 1990. Thick-skinned deformation observed on deep seismic reflection profiles in western Argentina. *Tectonics*, **9**, 773–788.

SOMIGLIANA, C. 1930. Geofisica – Sul campo gravitazionale esterno del geoide ellissoidico: Atti della Accademia nazionale dei Lincei. Rendiconti. *Classe di Scienze fisiche, Matematiche e Naturali*, **6**, 237–240.

SPECTOR, A. & GRANT, F. S. 1970. Statistical models for interpreting aeromagnetic data. *Geophysics*, **35**, 283–302.

TALWANI, M., WORZEL, J. L. & LANDISMAN, M. 1959. Rapid gravity computations for two dimensional bodies with application to the Mendocino Submarine Fracture zone. *Journal of Geophysical Research*, **64**, 49–58.

TASSARA, A. & ECHAURREN, A. 2012. Anatomy of the Andean subduction zone: three-dimensional density model upgraded and compared against global-scale models. *Geophysical Journal International*, **189**, 161–168.

TASSARA, A. & YÁÑEZ, G. 2003. Relación entre el espesor elástico de la litósfera y la segmentación tectónica del margen andino (15–478S). *Revista Geológica de Chile*, **30**, 159–186.

TASSARA, A., SWAIN, C., HACKNEY, R. & KIRBY, J. 2007. Elastic thickness structure of South America estimated using wavelets and satellite-derived gravity data. *Earth and Planetary Science Letters*, **253**, 17–36.

THOMAS, W. A. & ASTINI, R. A. 2003. Ordovician accretion of the Argentine Precordillera terrane to Gondwana: a review. *Journal of South American Earth Sciences*, **16**, 67–79.

THURSTON, J. B. & SMITH, R. S. 1997. Automatic conversion of magnetic data to depth, dip and susceptibility contrast using the SPITM method. *Geophysics*, **62**, 807–813.

ULIANA, M. A. & BIDDLE, K. T. 1988. Mesozoic-Cenozoic paleogeographic and geodynamic evolution of Southern South America. *Revista Brasileira de Geociencias*, **18**, 172–190.

VERDUZCO, B., FAIRHEAD, J. D., GREEN, C. M. & MACKENZIE, C. 2004. New insights into magnetic derivatives for structural mapping. *The Leading Edge*, **23**, 116–119.

VERGÉS, J., RAMOS, V. A., MEIGS, A., CRISTALLINI, E., BETTINI, F. H. & CORTÉS, J. M. 2007. Crustal wedging triggering recent deformation in the Andean thrust front between 31°S and 33°S: Sierras Pampeanas-Precordillera interaction. *Journal of Geophysical Research*, **112**, http://dx.doi.org/10.1029/2006JB004287

VILLELLA, J. C. & PACINO, M. C. 2010. Interpolación gravimétrica para el cálculo de los números geopotenciales de la red altimétrica de Argentina en zonas de alta montaña. *GEOACTA*, **35**, 13–26.

VON GOSEN, W. 1992. Structural evolution of the Argentine Precordillera: the Río San Juan section. *Journal of Structural Geology*, **14**, 643–667.

VUJOVICH, G., VAN STAAL, C. R. & DAVIS, W. 2004. Age Constraints on the Tectonic Evolution and Provenance of the Pie de Palo Complex, Cuyania Composite Terrane, and the Famatinian Orogeny in the Sierra de Pie de Palo, San Juan, Argentina. *Gondwana Research*, **7**, 1041–1056.

WATTS, A. B., LAMB, S. H., FAIRHEAD, J. D. & DEWEY, J. F. 1995. Lithospheric flexure and bending of the Central Andes. *Earth and Planetary Science Letters*, **134**, 9–20.

WESSEL, P. & SMITH, W. H. F. 1998. *New, improved version of the Generic Mapping Tools released*, Eos Trans. AGU, **79**, 579.

WHITMAN, D. 1999. Isostatic residual gravity anomaly in the Central Andes: 12 to 29 deg. S: A guide to interpreting crustal structure and deeper lithospheric processes. *International Geology Review*, **41**, 457–475.

WIENECKE, S., BRAITENBERG, C. & GÖTZE, H. J. 2007. A new analytical solution estimating the flexural rigidity in the Central Andes. *Geophysics Journal International*, **169**, 789–794.

ZAPATA, T. R. & ALLMENDINGER, R. W. 1994. 'Thick-Skinned' Triangular zone of the Precordillera Thrust Belt, Argentina. *In*: *Canadian Society of Exploration Geophysicist and Canadian Society of Petroleum Geologist Joint National Convention. Program with Expanded Abstracts and Bibliographies*, 72–73.

ZAPATA, T. R. & ALLMENDINGER, R. W. 1996. Thrust-front zone of the Precordillera, Argentina: a thick-skinned triangle zone. *AAPG Bulletin*, **80**, 359–381.

ZAPATA, T. R. & ALLMENDINGER, R. W. 1997. Evolución de la deformación del frente de corrimiento de Precordillera, provincia de San Juan. *Revista de la Asociación Geológica Argentina*, **52**, 115–131.

New insights into the Andean crustal structure between 32° and 34°S from GOCE satellite gravity data and EGM2008 model

O. ALVAREZ[1,2]*, M. E. GIMENEZ[1,2], M. P. MARTINEZ[1,2], F. LINCEKLINGER[1,2] & C. BRAITENBERG[3]

[1]*Instituto Geofísico y Sismológico Ing. Volponi, Universidad Nacional de San Juan, Ruta 12, Km 17, CP 5407, San Juan, Argentina*

[2]*Consejo Nacional de Investigaciones Científicas y Tecnicas, CONICET, Argentina*

[3]*Dipartimento di Geoscienze, Universita di Trieste, Via Weiss 1, 34100 Trieste, Italy*

Corresponding author (e-mail: orlando_a_p@yahoo.com.ar)

Abstract: The subduction of the Nazca oceanic plate under the South American plate in the south-central Andes region is characterized by the oblique collision of the Juan Fernandez Ridge against the continental margin. The upper plate is characterized by a broken foreland, a thrust-and-fold belt and eastward migration of the volcanic arc promoted by the flattening of the slab. Topographic load, thermal state and plate rheology determine the isostatic state of the continental plate. We calculated the vertical gravity gradient from GOCE satellite data in order to delineate the main tectonic features related to density variations resulting from internal and external loads. Then, using the Bouguer anomaly, we calculated the crust–mantle discontinuity and the elastic thickness in the frame of the isostatic lithospheric flexure model applying the convolution method approach. The results obtained show substantial variations in the structure of the continental lithosphere related to variations in the subduction angle of the Nazca plate. These variations are reflected in the varying Moho depths and in the plate rigidity, presenting a distinct behaviour in the southern zone, where the oceanic plate subducts with an approximate 'normal' angle with respect to the northern zone of the study area where the flat slab occurs.

The south-central Andes region is characterized by the oblique subduction of the Nazca plate beneath the South American plate. The processes associated with the subduction of the oceanic slab beneath the continental plate, such as flattening, shortening, volcanism and heating, gave rise to the Andes mountains uplift. The shallowing of the Nazca plate in the southern part of the Pampean flat subduction zone has been connected to the collision of the Juan Fernández ridge (JFR) and is based on its potential subducted geometry inferred from hot-spot trajectories conserved on the western Pacific (Yañez *et al.* 2001; Kay & Coira 2009). Deformation of the overriding plate margin under the flat slab region has been associated with the subduction of this elevated oceanic feature (Yañez *et al.* 2001; Yañez & Cembrano 2004; among others). The flat slab can be tracked beneath the continent 500 km from the trench (Sacks 1983; McGeary *et al.* 1985). This segment is associated with vast continental regions elevated above 4000 m and a wide deformational zone that extends beyond 700 km east of the trench.

Gravity field gradients obtained from global gravity field models and from satellite-only data represent an innovative tool for regional gravity modelling. This becomes more meaningful in regions where terrestrial data are sparse or unavailable. Improvements are significant where geological structures are concealed by sediments (Braitenberg *et al.* 2011a) or in mountainous areas (Alvarez *et al.* 2012) where terrestrial access is difficult, and also to overcome problems related to non-unified height measurements from different terrestrial campaigns (Reguzzoni & Sampietro 2010; Gatti *et al.* 2013). Derivatives of the gravity field such as the gravity anomaly (Ga) and the vertical gravity gradient (Tzz) highlight equivalent geological features related to density differences in a different and complementary way, and are very useful for geological mapping. The vertical gravity gradient is appropriate for detecting mass heterogeneities located in the upper crust where high- and low-density rocks are faced.

The flexural rigidity of the crust is a measure of the lithospheric strength (thickness and viscosity), which in turn depends strongly on lithospheric thermal state and composition (i.e. rheological properties; Lowry *et al.* 2000). The flexural rigidity can be interpreted in terms of the elastic thickness (Te) by making some assumptions regarding Poisson's ratio and Young's modulus. The spatial distribution of Te is useful to understand the processes related to the isostatic state and deformation of the upper crust, its variation can be explained by temperature distribution and a change of the Young's modulus.

From: SEPÚLVEDA, S. A., GIAMBIAGI, L. B., MOREIRAS, S. M., PINTO, L., TUNIK, M., HOKE, G. D. & FARÍAS, M. (eds) 2015. *Geodynamic Processes in the Andes of Central Chile and Argentina.* Geological Society, London, Special Publications, **399**, 183–202. First published online February 6, 2014, http://dx.doi.org/10.1144/SP399.3
© 2015 The Geological Society of London. For permissions: http://www.geolsoc.org.uk/permissions.
Publishing disclaimer: www.geolsoc.org.uk/pub_ethics

The value of the Te is equivalent to the thickness of a corresponding plate with a constant Young's modulus (Wienecke 2006). The base of the mechanical lithosphere for oceanic areas is marked by the approximate depth of the isotherm of 600 °C and presents a good correspondence with the estimated Te values. For the continental lithosphere, the relation between specific geological and physical boundaries and the results of Te is not so evident (Watts 2001; Wienecke 2006). Different authors (e.g. Goetze & Evans 1979; Lyon-Caen & Molnar 1983; Burov & Diament 1995; Hackney et al. 2006) defined a dependence between Te and the composition and geometry of the plate, external forces and thermal structure.

In the present study, we examine the interplay between the subduction of the JFR, the subducting Nazca plate and the deformation in the overriding plate on the basis of gravity field modelling. In order to improve the tectonic knowledge of the area under study, our aim is to analyse the geological structure of the crust through gravimetric satellite data. To achieve this goal the new data of the satellite GOCE (Floberghagen et al. 2011; Pail et al. 2011) was used to calculate the vertical gravity gradient for the south-central Andes area. The same calculation was performed using the global gravity field model EGM2008 (Pavlis et al. 2008, 2012), which has higher spatial resolution. Both were corrected for the topographic effect, in a spherical approximation, by using a one arc-minute global relief model of the Earth's surface that integrates land topography and ocean bathymetry (Amante & Eakins 2009). By inverse calculation of the Bouguer anomaly from GOCE data (http://icgem.gfz-potsdam.de/ICGEM/) we obtained the crustal–mantle discontinuity. We later obtained Te by flexural modelling. Results were mapped and compared to a schematic geological map of the south-central Andes region, which includes the main geological features with regional dimensions, possibly associated with crustal density variations.

Tectonic setting

The central Chilean margin is marked by the subduction of the JFR (Fig. 1), a hot spot chain formed by intraplate volcanism c. 900 km west of the trench, which first collided with the Chilean margin in the north (c. 20°) at c. 22 Ma, and then moved progressively southwards to the current collision point located at c. 32–33°S (Yañez et al. 2001). The seamounts are aligned in a chain that trends c. 85°E, but changes its strike to a NE direction when approaching the trench (Yañez et al. 2001). The collision of the JFR affects the tectonic margin, producing erosion, extensive deformation (von Huene et al. 1997) and local uplift where it collides with the continent affecting the entire continental slope (Ranero et al. 2006). The subducting plate descends with a maximum dip of 30° from the trench to a depth of c. 100–120 km and then flattens underneath the overriding lithosphere for

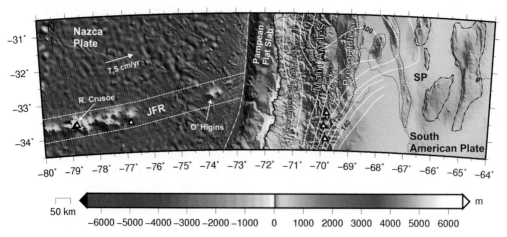

Fig. 1. Shaded DEM of the southern-central Andes region. The JFR is indicated (white dotted line) colliding against the Chilean trench. Robinson Crusoe Island and the O'Higgins seamount are indicated in the figure. The trench is outlined with white dotted and dashed lines. The Precordillera is depicted by a black dashed line, Western Sierras Pampeanas by a black dotted line, and Eastern Sierras Pampeanas by a black dotted and dashed line. These mountain systems were related to the development of the flat slab in the last 17 Ma. A volcanic arc gap is associated with the shallowing of the subducted Nazca plate north of 33°S. SP, Sierras Pampeanas. Triangles indicate the current position of the active volcanic arc (Siebert & Simkin 2002). White solid lines indicate contours of the subducting slab in the Pampean flat-slab zone (Anderson et al. 2007). Nazca–South American plates' convergence rate and azimuth are from DeMets et al. (2010).

several hundreds of kilometres (c. 300 km, Cahill & Isacks 1992; Gutscher et al. 2000; Anderson et al. 2007) before sinking into the upper-mantle asthenosphere. The asthenospheric wedge is repelled as far as 600 km away from the trench (Brooker et al. 2004; Martinod et al. 2010).

The southern limit of the Pampean flat-slab segment has been extensively analysed (Jordan et al. 1983a, b; Jordan & Allmendinger 1986; Cahill & Isacks 1992; Anderson et al. 2007; among others). Several studies based on hypocentre datasets indicate that the shallowest portion of the flat slab is associated with the inferred location of the subducting JFR at 31°S and that the slab deepens both to the south and north of this region. The gross interpreted structure of the Wadati–Benioff zone of Cahill & Isacks (1992) suggests that the shallowest portion of the slab is spatially correlated with the JFR. Anderson et al. (2007) have proposed that the subducting slab geometry is consistent with a buoyant ridge hypothesis for slab-flattening.

The Central Chilean flat-slab segment is expressed superficially by a volcanic gap between 28°S and 32°S, a deformed and faulted foreland zone (Fig. 1) and an expanded Neogene to Quaternary arc volcanism that reached the Sierras Pampeanas (Stauder 1973; Barazangui & Isacks 1976, 1979; Pilger 1981; Jordan et al. 1983a, b; Smalley & Isacks 1987; Kay et al. 1988, 1991; Allmendinger et al. 1990; Ramos et al. 1991, 2002; Cahill & Isacks 1992; Kay & Abbruzzi 1996; Yañez et al. 2001). Multiple authors (Allmendinger et al. 1997; Kay et al. 1999; Gutscher 2000; Kay & Mpodozis 2002; Ramos et al. 2002) have linked the eastward expansion and subsequent extinction of the Miocene to Quaternary volcanic arc and contemporary migration of the compressive strain towards the foreland with the gradual flattening of the subducted slab produced from c. 15 Ma. The associated change in the thermal structure and consequent shallow fragile–ductile transitions (James & Sacks 1999; Ramos & Folguera 2009) led to the uplift of the Precordillera expressed by a thin-skinned fold and thrust belt and the thick-skinned Sierras Pampeanas, a set of Laramide-style crystalline basement blocks uplifted during the shallowing of the slab from the late Miocene (Fig. 1) (Ramos et al. 2002; Kay & Coira 2009).

Global gravity field models

Global Earth gravity field models like EGM2008 (Pavlis et al. 2008, 2012) that combine satellite and terrestrial data present a higher spatial resolution (e.g. $N = 2159/\lambda = 19$ km, for EGM2008) than models obtained from satellite-only data (e.g. the preliminary model derived from data of the GOCE mission (Pail et al. 2011) with $N = 250/\lambda = 160$ km). The relation between the degree/order of the spherical harmonic development and the spatial resolution is given by $\lambda_{min} \approx 2\pi R/N_{max}$ (Barthelmes 2009), where R is Earth's radius and N_{max} is the maximum degree and order of the expansion. The GOCE model used in this work was obtained from pure satellite GOCE data (e.g. Pail et al. 2011; GO_CONS_GCF_2_TIM_R3, http://icgem.gfz-potsdam.de/ICGEM/). The TIM (Time Wise Model) solution is a GOCE-only solution in a rigorous sense as no external gravity field information is used (neither as reference model, nor for constraining the solution). The TIM models have been externally validated by independent GPS/levelling observations for Germany (875 stations) and Japan (873). Results indicate that the lasts pure GOCE-model TIM performs significantly better than EGM2008, even though the latter also contains terrestrial gravity data (see GO_CONS_GCF_2_TIM_xx datasheets, http://icgem.gfz-potsdam.de/ICGEM/). Thus, the GOCE-only gravitational model is useful to evaluate the quality of the terrestrial data entering the EGM2008 by a comparison analysis up to degree $N = 250$ (for degrees greater than $N = 120$, EGM2008 relies entirely on terrestrial data; see Appendix I). Therefore GOCE is a remarkably important independent quality assessment tool for EGM2008. In a recent paper, Braitenberg et al. (2011b) showed in detail how errors at high degree enter the error of a downscaled EGM2008.

We utilized the convolution approach (Braitenberg et al. 2002) to calculate the flexural strength. This method requires the gravity field data on a much smaller scale (on the order of 100 km length) than when using spectral methods (Wienecke 2006). Only the topography must be known over an extensive scale, which depends on the Te and therefore the radius of convolution, as explained by Wienecke (2006). The statistical analyses presented (Appendix I) show that the EGM2008 model present errors with respect to the satellite-only model of GOCE, especially in the Andes region. Based on this, we used the satellite-only model of GOCE for calculation of the different quantities derived from the gravity field (e.g. Moho, Te, Bouguer, Tzz), as it gives greater accuracy. Despite this, the higher spatial resolution of the EGM2008 model was exploited to find the main tectonic features in the Tzz map, which then was compared with the Tzz map obtained from GOCE (as made in Alvarez et al. 2012).

Flexural strength

When calculating the flexural strength of the lithosphere using spectral methods (coherence and admittance) a large spatial window is required over the study area and the method becomes unstable if the input topography is smooth. Both methods require an averaging process, so the variation in rigidity may be retrieved only to a limited extent (Wienecke 2006). For this reason, these techniques have been questioned when applied to the continental lithosphere. The convolution approach (Braitenberg et al. 2002) and the use of a newly derived analytical solution for the fourth-order differential equation that describes the flexure of a thin plate, a concept introduced by Vening-Meinesz in 1939, allow these problems to be overcome (to calculate analytically the deflection of a thin plate for any irregular shape of topography; see Wienecke (2006, and references therein) for a more detailed discussion). This method calculates the flexure parameters by the best fit of the observed crust–mantle interface (e.g. Moho by gravity inversion) and a crust–mantle interface computed with a flexure model. The gravity inversion method and the convolution approach have been tested extensively in synthetic models and in different geographical areas (Braitenberg et al. 1997, 2002; Braitenberg & Drigo 1997; Zadro & Braitenberg 1997; Braitenberg & Zadro 1999; Ebbing et al. 2001; Wienecke 2002, 2006; Pérez-Gussinyé et al. 2004; Bratfisch et al. 2010; Steffen et al. 2011; Ferraccioli et al. 2011).

Methodology

To accomplish the inverse modelling of flexural rigidity we used the Lithoflex software package (www.lithoflex.org) (Braitenberg et al. 2007; Wienecke et al. 2007). This tool fulfils a series of different functions that are concerned with studying the gravity field as well as the isostatic state, and combines forward and inverse calculation for gravity and flexural rigidity. The evaluation method used in these calculations allows a relatively high spatial resolution, superior to spectral methods (see Braitenberg et al. (2007) for more detail). Isostatic modelling adopts the isostatic thin plate flexure model (e.g. Watts 2001). To perform isostatic calculations, that is, to estimate the elastic properties of the plate for a known load, a crustal load and the crust–mantle interface need to be used as a reference surface (Wienecke 2006). The load acting on the crust is provided as the combination of the overlying topography and a density model (Braitenberg et al. 2007). A density variation within the crust represents a variation in the load and must be reflected in the isostatic response (Ebbing et al. 2007). The topographic load was calculated using topo/bathymetry data from ETOPO1 (Amante & Eakins 2009), while the densities used for calculation were 1.03 g cm^{-3} for water and 2.8 g cm^{-3} for the crust.

The undulating boundary or discontinuity corresponding to the Moho was calculated from observed gravity data by gravity inversion. The Bouguer anomaly field used for the gravity inverse calculations was obtained from the Calculation Service of the International Centre for Global Earth Models (ICGEM, http://icgem.gfz-potsdam.de). The Bouguer anomaly was calculated using the GOCE satellite data (Pail et al. 2011) up to degree/order $N = 250$. Long-wavelength information of the gravity field mainly corresponds to the crust–mantle density contrast, but sedimentary basins can also produce a long-wavelength signal, thus influencing the correct gravity crust–mantle interface estimation by the gravity inversion process (Wienecke 2006). Therefore, the gravity effect of sediments was calculated in order to reduce the gravity data.

The forward calculation of the gravity effect for the sediment package (Fig. 2) was calculated taking into account a linear variation of density with depth. To perform this calculation we defined a two-layer reference model of the continental crust with the following densities: upper crustal density, 2.7 g cm^{-3}; lower crustal density, 2.9 g cm^{-3}. The relation density/depth was defined using a linear variation (see Braitenberg et al. 2007). To perform this operation we used the bathymetry from ETOPO1 (Amante & Eakins 2009) and off-shore sediment thickness from Divins (2003). On-shore basins were modelled using depths to top basement from gravimetric studies and the seismic lines of Yacimientos Petroliferos Fiscales (YPF), Texaco, Repsol YPF, YPF S.A. and OIL M&S, and from Kokogian et al. (1993), Milana & Alcober (1994), Fernandez Seveso & Tankard (1995), Miranda & Robles (2002), Rosello et al. (2005) and Barredo et al. (2008). The correction amounts were up to -40 mGal for the main onshore basins and up to a few mGal for oceanic sediments, reaching their maximum over the Chilean trench.

From this reduced Bouguer anomaly (Fig. 3) we estimated the gravimetric crust–mantle discontinuity (Fig. 4) by gravity inversion. This method uses an iterative algorithm that alternates downward continuation with direct forward modelling (Braitenberg & Zadro 1999) and is somewhat analogous to the Oldenburg–Parker inversion approach (Parker 1972; Oldenburg 1974; see Braitenberg et al. (2007) for a detailed explanation). This method requires two input parameters: density contrast and reference depth. The density contrast between crust and mantle is unknown and has to be assumed as a constant value. Standard parameters such as normal crust thickness $Tn = 35$ km, and a crust–mantle density contrast of -0.4 g cm^{-3} were used.

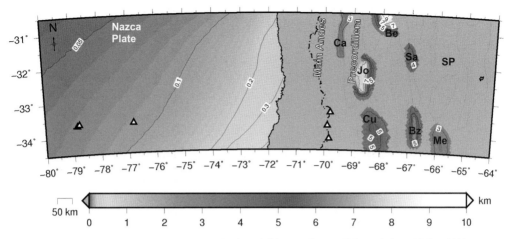

Fig. 2. Sediment thickness used to reduce the gravity data. Offshore sediment thickness is from Divins (2003). The basins over the South American plate were approximated using gravity databases and basement depths from seismic lines (YPF S.A. unpublished report).

For the inverse flexure calculation, the crustal load (obtained from topo/bathymetry data and the density model) and the Moho undulations (obtained by inversion of the reduced Bouguer anomaly) were utilized. The flexural rigidity is inverted in order to match the known loads with the known crustal thickness model (i.e. to model the gravity Moho in terms of an isostatic model). The elastic thickness was allowed to vary in the range $1 < Te < 50$ km and was iteratively estimated over moving windows of size 80 km × 80 km. The model parameters are given in Table 1, where the adopted densities are standard values already used by Introcaso et al. (2000), Gimenez et al. (2000), Miranda & Robles (2002) and Gimenez et al. (2009).

The difference between the Moho from gravity inversion and the flexure Moho is the residual Moho (gravity Moho minus flexure Moho) (Fig. 5). The Moho undulations obtained from gravity inversion agree with the CMI undulations expected for the flexural model, c. 90% within 3 km of difference (Fig. 6). Positive values of the residual Moho indicate that the gravity Moho is shallower than the flexure Moho, which is the case when high-density rocks are present in the crust,

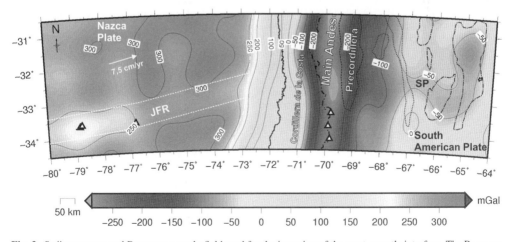

Fig. 3. Sediment corrected Bouguer anomaly field used for the inversion of the crust–mantle interface. The Bouguer anomaly was obtained from GOCE up to degree and order 250 (Pail et al. 2011). The JFR can be tracked by a well-defined gravity signal. The main Andes presents a low gravity signal representative of the Andean root.

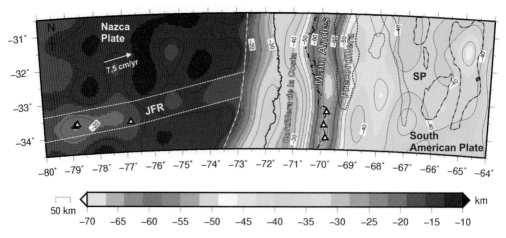

Fig. 4. Moho undulations obtained by inversion of the sediment corrected Bouguer anomaly. The Moho depths indicate the existence of an over-thickened oceanic crust in the JFR path. East of the trench, the contours exhibit a distinct behaviour north and south of the ridge. The Andean root presents depths of more than -60 km.

as in the Cordillera de la Costa (Figs 7 & 8). Correction of the gravity and load effect of the sediments at sedimentary basins allows gravity Moho to comply with load changes: once the negative effect on the Bouguer from the sediments is removed, the Moho from gravity inversion will be shallower and will follow the flexure Moho. When at sedimentary basins there are positive values of residual Moho this is an indication of increased crustal densities below the basin. Negative values of the residual Moho in the Main Andes indicate that the gravity Moho is deeper than the flexure Moho, as the Bouguer anomaly is strongly influenced by the negative effect of the Andean root. The flexural model used in this work is a simplification and would be influenced by the stress of the downgoing plate. Thus, the Te solutions along the active subduction margin could be distorted (as explained by Braitenberg et al. 2006). Gimenez et al. (2000) found that the effect of the flattened Nazca plate has a positive gravimetric effect (max. of 100 mGal) in the flat-slab region with very long wavelengths. This effect over the Bouguer anomaly then produces an increase in the Moho depth of c. 7–10 km (max.), again with very long wavelengths. Not accounting for this effect in our calculations does not change the conclusions.

Vertical gravity gradient Tzz

As previously explained, the lithosphere deforms in response to topographic and internal loads. To delineate the geological structures related to density variations at a regional scale, we calculated Tzz in terms of the spherical harmonic coefficients up to degree/order $N = 250$ for GOCE (Pail et al. 2011) and up to degree/order $N = 2159$ for EGM2008 (Pavlis et al. 2008), on a regular grid with a cell size of $0.05°$. The need for higher resolution to perform this task justifies the calculation with the EGM2008 model, while considering that some areas exhibit differences with respect to GOCE. For geological mapping, the vertical gravity gradient is ideal, as it highlights the centre of the anomalous mass (Braitenberg et al. 2011a). This allows unknown geological structures to be revealed that are either concealed by sediments or that have not been mapped previously.

The topographic effect is removed from the potential field derivatives to eliminate the correlation with the topography (see Alvarez et al. (2012) for more details). The calculation height is 7000 m to ensure that all values are above the topography, and topographic mass elements are approximated with spherical prisms to take into account the Earth's curvature (Uieda et al. 2010). The topographic correction amounts to tens of Eötvös for the vertical gradient and up to a few hundred mGal for gravity and is greatest above the highest topographic elevations and above the lower topographic depressions, as in the Chilean trench.

Table 1. *Parameters used in the flexural modelling*

Masses above sea-level	ρ_s	2.67 g cm^{-3}
Upper crustal density	ρ_{uc}	2.7 g cm^{-3}
Lower crustal density	ρ_{lc}	2.9 g cm^{-3}
Upper mantle density	ρ_m	3.3 g cm^{-3}
Young modulus	E	10^{11} N m^{-2}
Poisson ratio	Σ	0.25

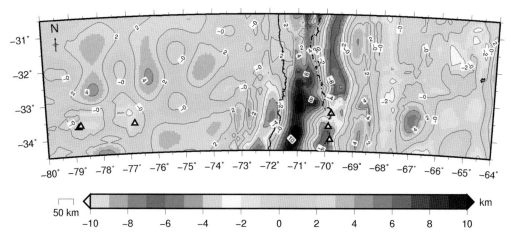

Fig. 5. Residual obtained by subtracting the crust–mantle interface obtained by gravity inversion of the Bouguer anomaly minus the crust–mantle interface obtained from the flexural model.

Results

To perform interpretation and determine the relation of the main geological units in a regional dimension, the T_{zz} and Bouguer anomaly were compared. A comparison of the fields reveals that the locations of the anomalies are well-correlated, but the T_{zz} highlights more detail than the gravity, as explained by Braitenberg *et al.* (2011*a*). We first analysed the T_{zz} obtained with the EGM2008 model as it presents the greatest spatial resolution available to date, taking into account the presence of some differences with respect to the pure satellite GOCE data (see Appendix I). The T_{zz} obtained with GOCE was then analysed and contrasted with the results obtained with EGM2008. This allows us to determine the location and morphology of the geological structures related to density variations in an optimal way, as done by Braitenberg *et al.* (2011*a*) and Alvarez *et al.* (2012).

The track of the JFR is delimited by a well-defined gravimetric signal lower than the surrounding plate, reaching its minimum expression over the seamount chain and islands with less than +250 mGal (Fig. 3). The T_{zz} obtained with GOCE (Fig. 8) presents values less than −5 Eötvös in the vicinity of the Robinson Crusoe Island, and increases its values as it approaches the trench. The T_{zz} obtained with EGM2008 (Fig. 7) exhibits different positive anomalies of more than +25 Eötvös, an expression of the numerous small volcanic edifices and hot-spot derived seamounts over the JFR (e.g. the O'Higgins seamount is expressed by a high T_{zz} value). These positive anomalies are not detectable in the GOCE T_{zz} map (Fig. 8) as a consequence of the lower spatial resolution; that is, the high-wavelength character of the GOCE signal does not resolve the high-frequency anomalies.

Seaward of the trench, the flexural bulge of the downgoing Nazca plate is marked by a positive anomaly in the T_{zz} signal greater than +20 Eötvös (Fig. 7). This is not apparent in the T_{zz} obtained with GOCE, although it is also detectable (Fig. 8) by the +15 Eötvös contour, which is segmented by the ridge path. Eastward from the trench to the coastline, the values of the T_{zz} anomaly reflect the differences of the trench sediment infill south and

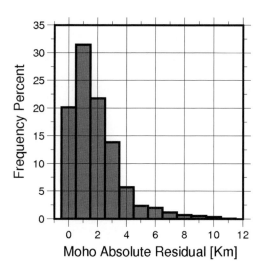

Fig. 6. Histogram of the residual Moho between the crust–mantle interface obtained by gravity inversion and the crust–mantle interface from the flexural model. More than 90% of the error is less than 4 km.

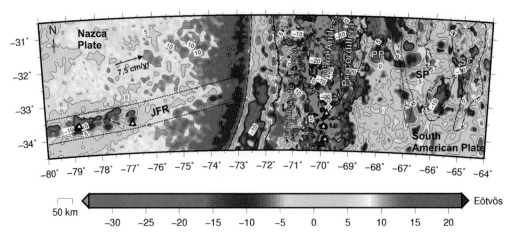

Fig. 7. Map of the vertical gravity gradient (EGM2008 up to degree and order 2159) corrected by topography. The seamounts of the JFR (black dotted line) can be tracked by a high gradient signal. The flexural bulge exhibits a high gradient signal that parallels the Chilean trench (black dotted and dashed line). The main Andes presents a low gradient signal representative of the Andean root. The Precordillera (black dashed line) exhibits high gradient values. Western Sierras Pampeanas is depicted by a black dotted line and Eastern Sierras Pampeanas by a black dotted and dashed line. White triangles indicate the current position of the active volcanic arc (Siebert & Simkin 2002).

north of the collision point of the JFR. To the south, the T_{zz} values are lesser than −10 Eötvös, and to the north the T_{zz} signal increases, indicating the abrupt decease in the sediment infill in this region (Figs 7 & 8).

Inland, the negative gravimetric effect of the Andes Cordillera presents a smaller amplitude in the southern direction, reflecting the lower Andean elevations and the consequent reduction of the Andean root. This can be observed in the Bouguer anomaly, which has smaller values to the north (<−300 mGal) than to the south (Fig. 3), whereas the T_{zz} signal has values of <−30 Eötvös (Fig. 8). The T_{zz} obtained with the model EGM2008 exhibits some positive anomalies higher than +25 Eötvös within this negative response (Fig. 7). West to the

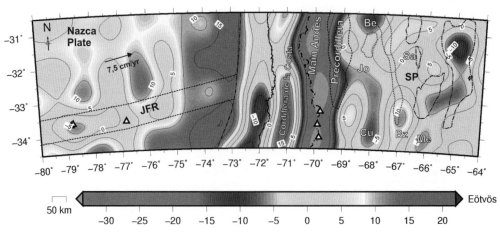

Fig. 8. Map of the vertical gravity gradient (GOCE up to degree and order 250) corrected by topography. The path of the JFR can be tracked by a low gradient signal relative to the surrounding plate. The effect of the Andean root over the gradient signal is notorious. A higher gradient signal is located in the Precordillera and in Western Sierras Pampeanas. Sedimentary basins: Be, Bermejo; Jo, Jocoli; Sa, Salinas; Cu, Cuyana; Bz, Beazley; Me, Mercedes.

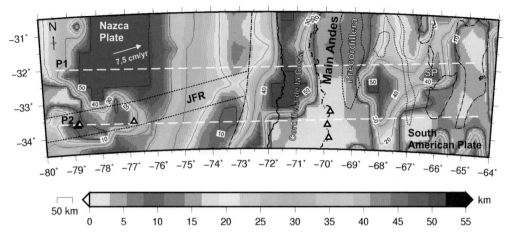

Fig. 9. Elastic thickness values obtained for the region. Note the weakening of the oceanic lithosphere in the ridge path and over the flexural bulge. A region of great rigidity is located in the forearc over the flat-slab region. The compensation to the topographic load of the main Andes is expressed by low Te values. To the west of the active volcanic arc, the low Te values are the expression of a weakened crust due to heating of the asthenospheric wedge. A more rigid plate is located to the east of the main Andes up to the location of Eastern Sierras Pampeanas.

Main Andes and to the coast line, the Cordillera de la Costa presents a positive gravimetric response and high Tzz values.

The positive gravimetric response of the Precordillera within the general negative trend marked by the Andean root is clearly depicted in both Tzz maps, with higher resolution over the EGM2008 Tzz map (Fig. 7), and is marked by the 0 Eötvös contour in the GOCE Tzz map (Fig. 8). The gravity anomaly map presented by Alvarez et al. (2012) also exhibits a different response of the Precordillera with respect to the Main Andes, presenting higher values. Other works (Introcaso et al. 1992; Gimenez et al. 2000, among others) also report a high gravimetric signal for the Precordillera based on terrestrial gravimetric data. These gravimetric highs in the northern Precordillera could be related to denser bodies generated by the fusion of the continental lithosphere through intrusion of asthenospheric magmas (Astini et al. 2009; Dahlquist et al. 2010).

Sedimentary basins such as Jocoli, Cuyana, Mercedes, Bermejo and Beazley basins exhibit low Tzz values, while the Salinas basin exhibits more intermediate values (Fig. 8). North of Beazley Basin we observe a high Tzz value greater than +10 Eötvös (Figs 7 & 8). This anomaly was also detected in the Bouguer anomaly map (more than +50 mGal) obtained with GOCE and corresponds to the Quijadas range, the southernmost expression of the Western Sierras Pampeanas (SP).

The Pie de Palo (PP) mountain range, an exposure of Mesoproterozoic crystalline basement and the westernmost expression of the western SP, was identified in the EGM2008 map (Fig. 7) by a high Tzz signal of more than +70 Eötvös. This is undetectable with GOCE because its topographic signal has a magnitude on the order of the spatial resolution of the model derived from GOCE. East of PP are the Western SP, which are mainly composed of Ordovician plutonic rocks. These mountains, which form part of the Famatinian arc (Fig. 10b) within the SP, exhibit high Tzz values greater than +20 Eötvös (Fig. 7). The easternmost region of the SP is characterized by the Sierras de Cordoba (SC), basement-cored uplifts that present low Tzz values because the density of granitic rocks is lower than 2.67 g cm^{-3}.

Once the spatial locations of the main anomalies related to density variations were determined, we analysed their relation with plate strength variations. Figure 12 shows the relation between elastic thickness Te (Fig. 9), the gravimetric crust–mantle discontinuity (Fig. 4) obtained from the inversion of the sediment corrected Bouguer anomaly (Fig. 3), and Tzz (Fig. 8).

Above the Nazca Oceanic plate, the Moho undulations delineate the track of the JFR, which reaches its maximum depth ($c.$ −30 km) over Robinson Crusoe Island (Fig. 4). The Te over the seamount chain presents low values (Figs 9 & 10a) indicative of the bending of the oceanic plate in this area (consequence of the topographic load and a young, warm oceanic lithosphere). This is consistent with the Bouguer anomaly (Fig. 3), which presents lower values over the ridge than over the surrounding plate. Previous studies (Wienecke 2006) have also shown a good correlation between lower Te values and the occurrence of seamounts, especially in the

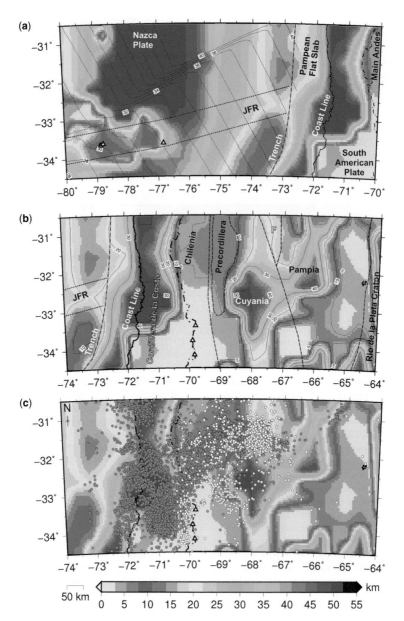

Fig. 10. (a) Elastic thickness values obtained for the Nazca plate. Superimposed age plates from Müller et al. (2008). (b) Elastic thickness values obtained for the South American plate. Superimposed terranes from Ramos (2009) and Ramos et al. (2010). F, Famatina. (c) Hypocentres of seismicity (EHB Catalog) for crustal earthquakes (white circles) and subducted Nazca plate earthquakes (grey circles).

area of the Sala y Gomez Ridge, Nazca Ridge and Juan Fernandez Ridge. Other work has reported an over-thickened oceanic crust beneath the JFR based on wide-angle seismic data (von Huene et al. 1997) and also related negative satellite-derived gravity anomalies to a crustal root indicative of crustal flexure from loading associated with seamounts (Sandwell & Smith 1997).

Towards the trench the Moho undulations, expression of the ridge path, are shallower and the Te decreases (to 30 km). An inflexion in the Te signal occurs where the ridge path intersects the bent oceanic plate prior to subduction (Figs 9 & 10a). We obtained higher values of *Te* over the ridge track and smaller values south and north of it over the fore bulge (less than +15 km) where the astenosphere is more superficial. These minimum values of *Te* at the flexural bulge are coincident with high plate curvatures, strong bending moment, fracturing and faulting of the oceanic basement and a reduction in crustal and mantle seismic velocities in this region, as mentioned by Contreras-Reyes & Osses (2010). These authors also reported a reduction in flexural rigidity towards the trench and related this to a weakening of the oceanic lithosphere.

Another factor to consider is the age of the subducting plate. Conductive cooling exerts a primary control over the strength of the oceanic lithosphere (Watts 2001), so its strength is greater where it is cooler and older. A cooler subducted lithosphere may reduce heat from the base of the upper plate (Yañez & Cembrano 2004; Tassara 2005). Before subduction, the Nazca plate ages range from 38 to 42 Ma north of the JFR, while to the south it is younger than 38 Ma (Fig. 10a). These slight differences in age may also contribute to the greater strength estimated in the flexural-bulge region north of the JFR. Note that the flexural bulge presents lower *Te* values south of the JFR where the plate is younger. However, this is not conclusive, as other factors must be taken into account such as the loading history of the plate prior to subduction, as explained by Contreras-Reyes & Osses (2010). The JFR is an important barrier, restraining the transport of sediment along the trench axis separating a heavily sedimented trench to the south from a trench north of 32.5°S that is starved of sediment or contains less than 1 km thickness of turbidites confined to a narrow axial zone (Bangs & Cande 1997, among others). Taking this into account, the lower *Te* values over the fore bulge to the south of the JFR when compared to the north should be a consequence of the combined effect of a younger and warmer lithosphere and to the sediment load.

East to the trench, the *Te* exhibits distinct behaviour north and south of the JFR trend. To the north, in the flat-slab region, the oceanic Moho deepens more gradually than to the south. In the forearc region, the subducting slab reverses its flexural curvature to flatten and travels subhorizontally (see profiles from Anderson *et al.* (2007), Fig. 11). The flattening of the slab repels the asthenospheric wedge far inland, hundreds of kilometres east, generating a conductive cooling of the continent, which is reflected by its increased strength. In this area the *Te* values increase towards the coastline, indicating an increasing rigidity eastwards (Figs 9 & 10b). Previous work has also found high *Te* values over the flat-slab regions (Stewart & Watts 1997; Tassara 2005). Interplate seismicity (Engdahl *et al.* 1998; EHB 2009) is well correlated with high *Te* values in the forearc region (Fig. 10c).

In a previous work, Pérez-Gussinyé *et al.* (2008) proposed that the flat slabs are characterized by high *Te* values, high shear wave velocity, thick thermal lithosphere and low heat flow, indicating that the continental lithosphere is thicker and cooler. The estimated *Te* in this area, where the upper plate and the subducted slab are in contact, can be expected to have contributions from both plates (see Pérez-Gussinyé *et al.* (2008) for a detailed discussion) and the *Te* values without other constraints cannot distinguish if the high flexural rigidity is also reflecting a thicker and stronger continental lithosphere prior to subduction.

To the south of the collision of the JFR, the oceanic plate presents a 'normal' dip angle of *c.* 30° and the active volcanic arc is located roughly above the 110 km isodepth contour of the slab (Fig. 1). Here, the oceanic Moho deepens more rapidly than in the northern zone and exhibits a depth of more than 40 km in the vicinity of the coastline. Anderson *et al.* (2007) obtained *c.* 50 km of depth to Moho from seismicity data. The higher values of *Te* (40 km) east to the trench may be interpreted as upper plate cooling by the underlying slab, which displaces the thermal structure downwards. The opposite occurs at the forearc region where the upper plate and the slab decouple, where melting ascends in the volcanic arc and basal heating by the asthenospheric wedge flow produces advective heating of the upper plate. Thus, low *Te* values estimated in this area reflect a weakened upper plate in the forearc to the south of the ridge collision point (Figs 9 & 10b).

The Moho depths obtained for the main Andes reach more than 66 km over the Andean axis and *Te* values are lower than 5 km (Figs 9 & 10b). Moho depths are consistent with those obtained by Gans *et al.* (2011) (70 km) for the main Cordillera, based on 'receiver functions'. In the backarc, the low gravity anomalies and low *Te* values affect the flexural root of the main Andes. This low *Te* value corresponds to a flexural model where the plate has no strength, the classical Airy model for local compensation. Similar results have been obtained previously by other authors, such as Introcaso *et al.* (1992), who explained that the Andes would be near the isostatic balance. The results are also consistent with other previous work for the Andean region (Stewart & Watts 1997; Tassara & Yañez 2003; Tassara *et al.* 2007; Pérez-Gussinyé *et al.* 2008; Sacek & Ussami 2009).

The Precordillera exhibits a different behaviour from that of the main Andes; the crust–mantle interface is more superficial, and higher values of *Te* are

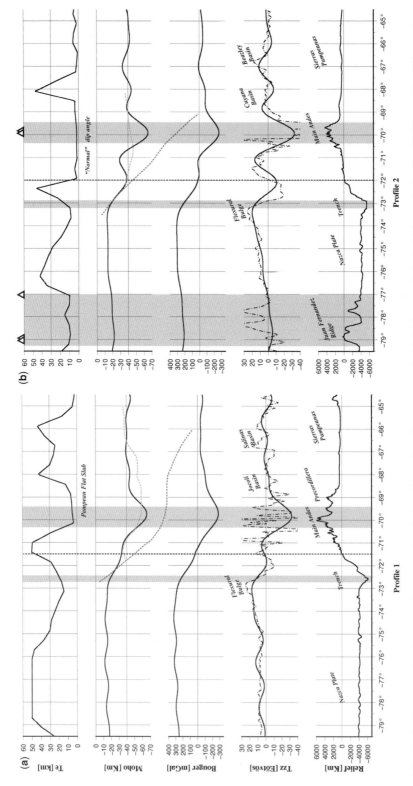

Fig. 11. Profiles comparing the topography corrected vertical gravity gradient, Bouguer anomaly, Moho depths and elastic thickness obtained by GOCE up to $N = 250$, over the northern region. (a) Profile 1: grey shaded area depicts the trench and the main Andes. (b) Profile 2: grey shaded area depicts the trench, the main Andes and the JFR. Dotted and dashed line indicates T_{zz}-EGM2008 corrected by topography; dashed line indicates the coastline; red dashed line indicates profiles of the Wadati–Benioff zone; blue dashed line indicates Moho depths from Anderson et al. (2007).

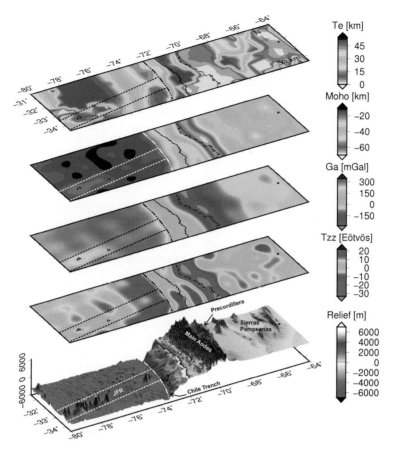

Fig. 12. Relation between the elastic thickness (uppermost and downwards), Moho, Bouguer, vertical gravity gradient and relief. Note the relation between the ridge path, which can be delineated in all signals, and the differences north and south of it in the region east of the trench to the volcanic arc. The effect of the Andean root is also well correlated in all quantities. The high gravity gradient signal obtained in the southernmost region of the Sierras Pampeanas is also observed in the Bouguer anomaly, in the Moho depths and in the elastic thickness, although there is no evident topographic signal in this region.

detected in this area. The Western SP are characterized by more intermediate Moho depths (between 40 and 50 km) and intermediate Te values (c. 20 km) indicative of a more global compensation (Figs 9 & 10b). Moho depths are consistent with those obtained by Gans et al. (2011). In this area the Tzz map indicates the presence of high-density bodies (Fig. 7). Weidmann et al. (2013) also found high-density areas with high flexural rigidity in the Western SP, based on terrestrial data. This is consistent with the tectonic evolution of these mountains, which were elevated by listric faults and thus do not present a local compensation. Quite the opposite occurs at the southern ending of the Western SP. In this region a low value of Te is detected (Figs 9 & 10b), which is well matched with a shallowing of the Moho (Figs 4 & 12). In this area there is no significant topographic expression, but there are high Tzz values (Figs 7 & 8) and high Bouguer anomalies (Fig. 3) related to the Sierra de la Quijadas, as previously stated. This low plate rigidity is well correlated with an absence of intraplate seismic activity in the area (Fig. 10c).

The SC are the easternmost expression of the SP and form the eastern border with the Rio de la Plata Craton (Fig. 10b). These mountain ranges are characterized by shallower Moho depths (c. 40 km), consistent with previous work (Gans et al. 2011). Te values indicate low rigidity under these ranges and increasing values to the east where the Rio de la Plata Craton is located (Fig. 10b). The Sierra de San Luis, the southwesternmost expression of the eastern SP, presents Te values ranging from 5 km to 30 km, increasing in the northwestern direction. Sedimentary basins such as

Jocoly and Mercedes basins exhibit high rigidity values, while the Bermejo, Cuyana and Salinas basins have more intermediate values.

An approximate correlation between different terranes and plate rigidity was found (Fig. 10b). The Precordillera Terrane exhibits an homogeneous Te value of $c.$ 7 km. Cuyania presents high rigidity, which is diminished by the asthenopheric wedge to the south. Famatina has intermediate Te values. An inflection in the Te is observed on the border between the Pampia Terrane and the Rio de la Plata Craton where the values increase indicating more plate rigidity.

Profiles across the region

Two different profiles were traced over the study region in the west–east direction (for location of profiles see Fig. 9). Profile 1 (Fig. 11a) was traced in the northern region at 32°S where the flattening of the slab occurs. Profile 2 (Fig. 11b) was traced at 33.5°S and cut along the JFR to the active volcanic arc.

North of the JFR the Te values (Fig. 11a) are more than +50 km, indicating a more rigid part in this area. Tebbens & Cande (1997) proposed that these rigid parts surrounded by areas of lower Te values, as well as between the Nazca Ridge and the JFR (see Fig. 9), might indicate the existence of microplates. Towards the trench the plate rigidity decreases, reaching up to +15 km in the flexural bulge, which is well defined by a high Tzz signal (Fig. 11a). The Te values over the JFR (Fig. 11b) are lower than to the north (less than +20km), indicative of weakened oceanic crust over the ridge (Fig. 9). The minimum Te values in this area are well correlated with active volcanoes over the Nazca plate. Eastward, where the ridge does not present a 'topographic' signal, the plate becomes more rigid and then weakens again in the vicinity of the flexural bulge, where it reaches a constant value of $c.$ 10 km (Fig. 11b).

East to the trench, the Te values rise up to $c.$ +50 km. The velocity of growth for this region is greater in profile 2 than in profile 1, but once it reaches its maximum it decays abruptly in profile 2. The opposite occurs to the north (profile 1), where the transition is more subtle, reaching the maximum Te under the coastline. This value is maintained for a few kilometres to the east (Fig. 11a) and is characteristic of the flat-slab segments as exposed previously (Fig. 9). The rigidity then decays continuously to almost 0 km under the main Andes along both profiles.

The high topographic load of the main Andes is expressed by a minimum in the Bouguer anomaly, a minimum in the Tzz and a minimum in the Moho depths (Fig. 11). Based on the above we can say that the low values of Te in this section are related to the local compensation of the Andes root (Fig. 11). In profile 1, the Te values increases eastwards to the Precordillera, while in profile 2 the low Te values are maintained eastwards due to the presence of the asthenospheric wedge.

The SP exhibit distinct behaviour to the south and north. The Western SP exhibit intermediate Te values (more than 30 km), which then decay eastwards approaching the Eastern SP (profile 1). The behaviour to the south is rather distinct (profile 2). Once the Te values reach their minimum values they are maintained, except at the Cuyana Basin where they become more rigid. Low Te values are maintained over the southern ending of the SP. Shallower Moho depths and smaller Te values at the southern region of the SP indicate a weakened continental crust in this region compared to that in the north (Fig. 12). This region exhibits a low seismic activity (Fig. 10c). The opposite occurs in the northern region, where hypocentral seismicity (Fig. 10c) shows greater activity related to the JFR (as explained by Anderson et al. (2007), among others).

Conclusions

Deformation in the overriding plate, volcanism and discontinuities in the pattern of seismicity reflect anomalies through the Benioff zones that are caused by flat subduction. Previous work (e.g. Pérez-Gussinyé et al. (2008) has shown that subduction-related processes, as variations in the subduction angle, are related to variations in the compositional and thermal structure of the continental lithosphere modifying its flexural strength. The new global Earth gravity field models and exclusive satellite-only data offer a tool for studies on a regional scale, for example by determining the connections between anomalously thickened subducted oceanic crust (as is the case for the JFR) and the associated deformational effects in the overriding plate.

Calculation of the Tzz was performed with both EGM2008 and GOCE, allowing optimization with the higher resolution of EGM2008 (but reduced quality over the Andes) and the uniform quality of the GOCE data (but reduced spatial resolution). From these we delineated the main tectonic features, as intruded denser bodies in the region defined by 31°–34°S and 66–80°W. Then, using the Bouguer anomaly obtained from GOCE satellite data, we calculated the crustal and upper mantle discontinuity and the Te using the convolution approach. This method makes use of a newly derived analytical solution for the fourth-order differential equation that describes the flexure of a thin plate.

The results obtained show a weakening of the oceanic plate over the track of the JFR and over the flexural bulge. We obtained substantial variations in the crustal structure of the continental lithosphere in the northern zone when compared to the south. To the north, where the flat slab occurs, plate rigidity is greater than in the southern zone where the oceanic plate subducts with an approximately normal angle. Here, the plate rigidity appears to be equivalent to zero, reflecting a weakened continental lithosphere due to the ascent of the magmas from the asthenophere. We obtained a gross correlation between the estimated Te values and some terranes, such as Precordillera, Cuyania, Famatina and Rio de la Plata Craton. Chilenia appears to be greatly affected by the differences in subduction angle, whereas the SP presents a distinct behaviour representing more rigidity in the northern region, and a weakened crust to the south and to the east. The results for Te indicate that the crustal structure in the region under study is far from homogeneous, with contrasting differences in strength properties.

The authors acknowledge the use of the GMT-mapping software of Wessel & Smith (1998). The authors also thank the Ministerio de Ciencia y Tecnica–Agencia de Promocion Científica y Tecnologica, PICT07–1903, Agenzia Spaziale Italiana for the GOCE–Italy Project, the Ministero dell'Istruzione, dell'Universita' e della Ricerca (MIUR) under project PRIN, 2008CR4455_003 for financial support, and ESA for granting AO_GOCE_proposal_4323_Braitenberg.

Appendix I: Statistical analysis

We calculated the gravity anomaly derived from the EGM2008 model (Pavlis et al. 2008) and from the GOCE satellite (Pail et al. 2011) up to $N = 250$. The absolute value of the difference field (EGM2008-GOCE) is shown in Fig. A1. Statistical parameters for the difference between the two fields are shown in Table A1. A high-quality region is compared with a low-quality region in terms of the residual histogram. The white square in Fig. A1 marks a $1° \times 1°$ area with relatively high quality, which is compared to a square of equal size (black) of degraded quality. The histograms of the residuals (Fig. A2) illustrate the higher values for the black square.

The root-mean-square deviation was calculated from the mean on sliding windows measuring $0.5° \times 0.5°$ as a statistical measure of EGM2008 quality. The result is shown on Fig. A3. The most frequent value of the root-mean-square deviation is 3 mGal as shown in Fig. A4. The locations where the terrestrial data have problems reflects greatly increased values (up to -48 mGal).

The sparseness of terrestrial data in large regions, especially in areas of difficult access, and a non-unified height system used in different terrestrial studies resulted in these differences. The precision of the height measurements directly affects the accuracy of the (inland) gravity observations and their derivatives, and greater inconsistencies arise when considering large areas (Reguzzoni & Sampietro 2010). This highlights the usefulness of satellite-only derived data in mountainous areas that are difficult to access, as in the central to eastern region under study (see Braitenberg et al. (2011a, b) and Alvarez et al. (2012) for a more detailed explanation).

Table A1. *Statistical parameters for the difference*

Average difference	0.173 mGal
Standard deviation	15.208 mGal
Maximal value of difference	−48.109 mGal

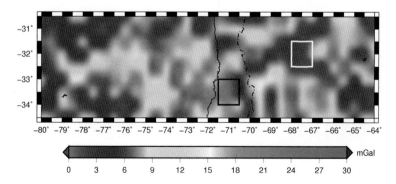

Fig. A1. Absolute difference between the gravity anomalies from EGM2008 and GOCE. The black square over the Andes indicates an area with erroneous data. The white square indicates an area over the Sierras Pampeanas with better data. The national border is given by a dotted and dashed line, and the coastal border by a solid line. Erroneous terrestrial data or lack of it in the EGM2008 model generates these differences between the two fields.

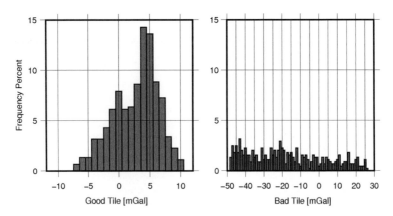

Fig. A2. Histogram of the residual gravity anomaly between EGM2008 and GOCE (up to degree and order $N = 250$). Left (good tile): white square of Figure A1. Right (bad tile): black square of Figure A1.

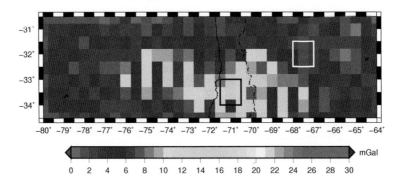

Fig. A3. Root mean square of the gravity anomaly residual on $0.5° \times 0.5°$ tiles.

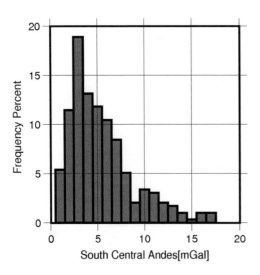

Fig. A4. Histogram of the root-mean-square deviations on $0.5° \times 0.5°$ tiles.

References

ALLMENDINGER, R. W., FIGUEROA, D., SNYDER, D., BEER, J., MPODOZIS, C. & ISACKS, B. L. 1990. Foreland shortening and crustal balancing in the Andes at 30°S Latitude. *Tectonics*, **9**, 789–809.

ALLMENDINGER, R. W., ISACKS, B. L., JORDAN, T. E. & KAY, S. M. 1997. The evolution of the Altiplano–Puna plateau of the Central Andes. *Annual Review of the Earth and Planetary Science*, **25**, 139–174.

ALVAREZ, O., GIMENEZ, M. E., BRAITENBERG, C. & FOLGUERA, A. 2012. GOCE satellite derived gravity and gravity gradient corrected for topographic effect in the South Central Andes region. *Geophysical Journal International*, **190**, 941–959, http://dx.doi.org/10.1111/j.1365-246X.2012.05556.x

AMANTE, C. & EAKINS, B. W. 2009. *ETOPO1 1 Arc-Minute Global Relief Model: Procedures, Data Sources and Analysis*. National Oceanic and Atmospheric Administration Technical Memorandum NESDIS NGDC-24, 19 March 2009.

ANDERSON, M. L., ALVARADO, P., BECK, S. & ZANDT, G. 2007. Geometry and brittle deformation of the subducting Nazca plate, central Chile and Argentina. *Geophysical Journal International*, **171**, 419–434, http://dx.doi.org/10.1111/j.1365-246X.2007.03483.x

ASTINI, R. A., MARTINA, F., EZPELETA, M., DÁVILA, F. M. & CAWOOD, P. A. 2009. Chronology from rifting to foreland basin in the Paganzo Basin (Argentina), and a reappraisal on the 'Eo- and Neohercynian' tectonics along Western Gondwana. *In*: *XII Congreso Geológico Chileno*, Santiago, 22–26 November 2009.

BANGS, N. L. & CANDE, S. C. 1997. Episodic development of a convergent margin inferred from structures and processes along the southern Chile margin. *Tectonics*, **16**, 489–505.

BARAZANGUI, M. & ISACKS, B. 1976. Spatial distribution of earthquakes and subduction of the Nazca Plate beneath South America. *Geology*, **4**, 686–692.

BARAZANGUI, M. & ISACKS, B. 1979. Subduction of the Nazca plate beneath Peru – evidence from spatial distribution of earthquakes. *Geophysical Journal of the Royal Astronomical Society*, **57**, 537–555.

BARREDO, S., CRISTALLINI, E., ZAMBRANO, O., PANDO, G. & GARCÍA, R. 2008. Análisis tectono-sedimentario del relleno de edad precuyana y cuyana inferior de la region del alto Kaufmann, Cuenca Neuquina. *In*: *VII Congreso de Exploración y Desarrollo de Hidrocarburos*, Mar del Plata, Argentina, November 2008, Actas, 443–446.

BARTHELMES, F. 2009. Definition of Functionals of the Geopotential and their Calculation from Spherical Harmonic Models. Theory and formulas used by the calculation service of the International Centre for Global Earth Models (ICGEM). GFZ German Research Centre for Geosciences Scientific Technical Report **STR09/02**, http://icgem.gfz-postdam.de

BRAITENBERG, C. & DRIGO, R. 1997. A crustal model from gravity inversion in Karakorum. *In*: *International Symposium on Current Crustal Movement and Hazard Reduction in East Asia and South-East Asia*, Wuhan, 4–7 November, Symposium Procedures, 325–341.

BRAITENBERG, C. & ZADRO, M. 1999. Iterative 3D gravity inversion with integration of seismology data. Bollettino di Geophisica Teorica ed Applicata. *Proceedings of the 28th Joint Meeting IAG*, Trieste, Italy, **40**, 469–476.

BRAITENBERG, C., PETTENATI, F. & ZADRO, M. 1997. Spectral and classical methods in the evaluation of Moho undulations from gravity data: the NE Italian Alps and isostasy. *Journal of Geodynamics*, **23**, 5–22.

BRAITENBERG, C., EBBING, J. & GÖTZE, H. J. 2002. Inverse modelling of elastic thickness by convolution method – the Eastern Alps as a case example. *Earth Planetary Science Letters*, **202**, 387–404.

BRAITENBERG, C., WIENECKE, S. & WANG, Y. 2006. Basement structures from satellite-derived gravity field: South China Sea ridge. *Journal of Geophysical Research*, **111**, B05407, http://dx.doi.org/10.1029/2005JB003938

BRAITENBERG, C., WIENECKE, S., EBBING, J., BOM, W. & REDFIELD, T. 2007. Joint gravity and isostatic analysis for basement studies – a novel tool. *In*: *EGM 2007 International Wokshop, Innovation in EM, Grav and Mag Methods: A New Perspective for Exploration*, Villa Orlandi, Capri, Extended Abstracts.

BRAITENBERG, C., MARIANI, P., EBBING, J. & SPRLAK, M. 2011*a*. The enigmatic Chad lineament revisited with global gravity and gravity-gradient fields. *In*: VAN HINSBERGEN, D. J. J., BUITER, S. J. H., TORSVIK, T. H., GAINA, C. & WEBB, S. J. (eds) *The Formation and Evolution of Africa: A Synopsis of 3.8 Ga of Earth History*. Geological Society, London, Special Publications, **357**, 329–341, http://dx.doi.org/10.1144/SP357.18

BRAITENBERG, C., MARIANI, P. & PIVETTA, T. 2011*b*. GOCE observations in exploration geophysics. *In*: *Proceedings of 4th International GOCE User Workshop*, Munich, 31 March–1 April 2011, ESA SP-696.

BRATFISCH, R., JENTZSCH, G. & STEFFEN, H. 2010. A 3D Moho depth model for the Tien Shan from EGM2008 gravity data. *In*: *7th EGU General Assembly*, Vienna, 5 February 2010. Geophysical Research Abstracts 12, EGU2010-442. Poster presentation. Winner of the EGU Young Scientists Outstanding Poster Paper (YSOPP) Award 2010 in the Seismology division.

BROOKER, J. R., FAVETTO, A. & POMPOSIELLO, M. C. 2004. Low electrical resistivity associated with plunging of the Nazca flat slab beneath Argentina. *Nature*, **429**, 399–403.

BUROV, E. B. & DIAMENT, M. 1995. The effective elastic thickness (Te) of continental lithosphere: what does it really mean? *Journal of Geophysical Research*, **100**, 3905–3927.

CAHILL, T. & ISACKS, B. 1992. Seismicity and shape of the subducted Nazca plate. *Journal of Geophysical Research*, **97**, 17 503–17 529.

CONTRERAS-REYES, E. & OSSES, A. 2010. Lithospheric flexure modelling seaward of the Chile trench: implications for oceanic plate weakening in the Trench Outer Rise region. *Geophysical Journal International*, **182**, 97–112.

DAHLQUIST, J., ALASINO, P. H., EBY, N. E., GALINDO, C. & CASQUET, C. 2010. Fault controlled Carboniferous A-type magmatism in the proto-Andean foreland (Sierras Pampeanas, Argentina): geochemical constraints and petrogenesis. *Lithos*, **115**, 65–81.

DEMETS, C., GORDON, R. G. & ARGUS, D. F. 2010. Geologically current plate motions. *Geophysical Journal International*, **181**, 1–80, http://dx.doi.org/10.1111/j.1365-246X.2009.04491.x

DIVINS, D. L. 2003. *Total Sediment Thickness of the World's Oceans & Marginal Seas*. NOAA National Geophysical Data Center, Boulder.

EBBING, J., BRAITENBERG, C. & GÖTZE, H. J. 2001. Forward and inverse modeling of gravity revealing insights into crustal structures of the Eastern Alps. *Tectonophysics*, **337**, 191–208.

EBBING, J., BRAITENBERG, C. & WIENECKE, S. 2007. Insights into the lithospheric structure and tectonic setting of the Barents Sea region from isostatic considerations. *Geophysical Journal International*, **171**, 1390–1403.

EHB-CATALOG 2009. *International Seismological Centre EHB Bulletin*. International Seismological Center, Thatcham, http://www.isc.ac.uk

ENGDAHL, E. R., VAN DER HILST, R. & BULAND, R. 1998. Global teleseismic earthquake relocation with improved travel times and procedures for depth determination. *Bulletin of the Seismological Society of America*, **88**, 722–743.

FERNANDEZ SEVESO, F. & TANKARD, A. 1995. Tectonics and stratigraphy of the late Paleozoic Paganzo Basin of western Argentina and its regional implications. *In*: TANKARD, A. J., SUÁREZ SORUCO, R. & WELSINK, H. J. (eds) *Petroleum Basins of South America*. American Association of Petroleum Geologists, Tulsa, Memoirs, 285–301.

FERRACCIOLI, F., FINN, C. A., JORDAN, T. A., BELL, R. E., ANDERSON, L. M. & DAMASKE, D. 2011. East Antarctic rifting triggers uplift of the Gamburtsev Mountains. *Nature*, **479**, 388–392, http://dx.doi.org/10.1038/nature10566

FLOBERGHAGEN, R., FEHRINGER, M. ET AL. 2011. Mission design, operation and exploitation of the gravity field and steady-state ocean circulation explorer mission. *Journal of Geodesy*, **85**, 749–758.

GANS, C., BECK, S., ZANDT, G., GILBERT, H., ALVARADO, P., ANDERSON, M. & LINKIMER, L. 2011. Continental and oceanic crustal structure of the Pampean flat slab region, western Argentina, using receiver function analysis: new high-resolution results. *Geophysical Journal International*, **186**, 45–58.

GATTI, A., REGUZZONI, M. & VENUTI, G. 2013. The height datum problem and the role of satellite gravity models. *Journal of Geodesy*, **87**, 15–22. http://dx.doi.org/10.1007/s00190-012-0574-3

GIMENEZ, M. E., MARTÍNEZ, M. P. & INTROCASO, A. 2000. A Crustal Model based mainly on Gravity data in the Area between the Bermejo Basin and the Sierras de Valle Fértil–Argentina. *Journal of South American Earth Sciences*, **13**, 275–286.

GIMENEZ, M. E., BRAITENBERG, C., MARTINEZ, M. P. & INTROCASO, A. 2009. A comparative analysis of seismological and gravimetric crustal thicknesses below the Andean Region with flat subduction of the Nazca Plate. *Journal of Geophysics*, **2009**, 607458, http://dx.doi.org/10.1155/2009/607458

GOETZE, C. & EVANS, B. 1979. Stress and temperature in the bending lithosphere as constrained by experimental rock mechanism. *Geophysical Journal of the Royal Astronomical Society*, **59**, 463–478.

GUTSCHER, M. A. 2000. An Andean model of interplate coupling and strain partitioning applied to the flat subduction of SW Japan (Nankai Trough). *Tectonophysics*, **333**, 95–109.

GUTSCHER, M. A., SPAKMAN, W., BIJWAARD, H. & ENGDAHL, E. R. 2000. Geodynamics of flat subduction: seismicity and tomographic constraints from the Andean margin. *Tectonics*, **19**, 814–833.

HACKNEY, R. I., ECHTLER, H. P., FRANZ, G., GÖTZE, H. J., LUCASSEN, F., MARCHENKO, D., MELNICK, D., MEYER, U., SCHMIDT, S., TAŠÁROVÁ, Z., TASSARA, A. & WIENECKE, S. 2006. The segmented overriding plate and coupling at the south-central Chile margin (36–42°S). *In*: ONCKEN, O., CHONG, G., FRANZ, G., GIESE, P., RAMOS, V. A., STRECKER, M. R. & WIGGER, P. (eds) *The Andes: Active Subduction Orogen*. Springer Verlag, Berlin, Frontiers in Earth Sciences, **1**, 355–374.

INTROCASO, A., PACINO, M. C. & FRAGA, H. 1992. Gravity, isostasy and Andean crustal shortening between latitudes 30°S and 35°S. *Tectonophysics*, **205**, 31–48.

INTROCASO, A., PACINO, M. C. & GUSPI, F. 2000. The Andes of Argentina and Chile: crustal configuration, isostasy, shortening and tectonic features from gravity data. *Temas de Geociencia*, **5**, 31.

JAMES, D. E. & SACKS, S. 1999. Cenozoic formation of the Central Andes: a geophysical perspective. *In*: SKINNER, B. (ed) *Geology and Mineral Deposits of Central Andes*. Society of Economic Geology, London, Special Publications, **7**, 1–25.

JORDAN, T. E. & ALLMENDINGER, R. 1986. The Sierras Pampean of Argentina: a modern analogue of Rocky Mountain foreland deformation. *American Journal of Science*, **286**, 737–764.

JORDAN, T. E., ISACKS, B., ALLMENDINGER, R., BREWER, J., RAMOS, V. A. & ANDO, C. J. 1983a. Andean tectonics related to geometry of the subducted Nazca Plate. *Geological Society of America Bulletin*, **94**, 341–361.

JORDAN, T. E., ISACKS, B., RAMOS, V. A. & ALLMENDINGER, R. 1983b. Mountain building in the Central Andes. *Episodes*, **3**, 20–26.

KAY, S. M. & ABBRUZZI, J. M. 1996. Magmatic evidence for Neogene lithospheric evolution of the central Andean 'flat-slab' between 30°S and 32°S. *Tectonophysics*, **259**, 15–28.

KAY, S. M. & COIRA, B. 2009. Shallowing and steepening subduction zones, continental lithospheric loss, magmatism, and crustal flow under the Central Andean Altiplano–Puna Plateau. *In*: KAY, S., RAMOS, V. & DICKINSON, W. (eds) *Backbone of the Americas: Shallow Subduction Plateau Uplift and Ridge and Terrane Collision*. Geological Society of America, Boulder, Colorado, USA, Memoirs, **204**, 229–259.

KAY, S. M. & MPODOZIS, C. 2002. Magmatism as a probe to the Neogene shallowing of the Nazca plate beneath the modern Chilean flat-slab. *Journal of South American Earth Science*, **15**, 39–57.

KAY, S. M., MAKSAEV, V., MOSCOSO, R., MPODOZIS, C., NASI, C. & GORDILLO, C. E. 1988. Tertiary Andean magmatism in Chile and Argentina between 28 and 33°S: correlation of magmatic chemistry with a changing Benioff zone. *Journal of South American Earth Science*, **1**, 21–38.

KAY, S. M., MPODOZIS, C., RAMOS, V. A. & MUNIZAGA, F. 1991. Magma source variations for mid–late Tertiary magmatic rocks associated with a shallowing subduction zone and thickening crust in the Central Andes (28–33°S). *In*: HARMON, R. S. & RAPELA, C. W. (eds) *Andean Magmatism and Its Tectonic Setting*. Geological Society of America, Boulder, Colorado, USA, Special Papers, **26**, 113–137.

KAY, S. M., MPODOZIS, C. & COIRA, B. 1999. Neogene magmatism, tectonism and mineral deposits of the Central Andes (22°–33°S latitude). *In*: SKINNER, B. (ed) *Geology and Ore Deposits of the Central Andes*. Society of Economic Geology, London, Special Publications, **7**, 27–59.

KOKOGIAN, D. A., SEVESO, F. F. & MOSQUERA, A. 1993. Las secuencias sedimentarias triásicas. *In*: RAMOS, V. A. (ed.) *Geología y Recursos Naturales de Mendoza*. XII Congreso Geología Argentina y II Congreso de Exploración de Hidrocarburos, Relatorio, Mendoza, **I**, 65–78.

LOWRY, A. R., RIBE, N. M. & SMITH, R. B. 2000. Dynamic elevation of the Cordillera, western United States. *Journal of Geophysical Research*, **105**, 23371–23390.

LYON-CAEN, H. & MOLNAR, P. 1983. Constraints on the structure of the Himalaya from an analysis of gravity anomalies and a flexural model of the lithosphere. *Journal of Geophysical Research*, **88**, 8171–8191.

MARTINOD, J., HUSSON, L., ROPERCH, P., GUILLAUME, B. & ESPURT, N. 2010. Horizontal subduction zones, convergence velocity and the building of the Andes. *Earth and Planetary Science Letters*, **299**, 299–309.

MCGEARY, S., NUR, A. & BEN-AVRAHAM, Z. 1985. Spatial gaps in arc volcanism: the effect of collision or subduction of oceanic plateaus. *Tectonophysics*, **119**, 195–221.

MILANA, J. P. & ALCOBER, O. 1994. Modelo tectosedimentario de la cuenca triásica de Ischigualasto (San Juan, Argentina). *Revista de la Asociación Geológica Argentina*, **49**, 217–235.

MIRANDA, S. & ROBLES, J. A. 2002. Posibilidades de atenuación cortical en la cuenca Cuyana a partir del análisis de datos de gravedad. *Revista de la Asociación Geológica Argentina*, **57**, 271–279.

MÜLLER, R. D., SDROLIAS, M., GAINA, C. & ROEST, W. R. 2008. Age, spreading rates and spreading symmetry of the world's ocean crust. *Geochemistry Geophysics Geosystems*, **9**, Q04006, http://dx.doi.org/10.1029/2007GC001743

OLDENBURG, D. 1974. The inversion and interpretation of gravity anomalies. *Geophysics*, **39**, 526–536.

PAIL, R., BRUISMA, S. ET AL. 2011. First GOCE gravity field models derived by three different approaches. *Journal of Geodesy*, **85**, 819–843.

PARKER, R. L. 1972. The rapid calculation of potential anomalies. *Geophysical Journal of the Royal Astronomical Society*, **31**, 447–455.

PAVLIS, N. K., HOLMES, S. A., KENYON, S. C. & FACTOR, J. K. 2008. An Earth Gravitational Model to degree 2160: EGM2008. Paper presented at the *2008 General Assembly of the European Geosciences Union*, Vienna.

PAVLIS, N. K., HOLMES, S. A., KENYON, S. C. & FACTOR, J. K. 2012. The development and evaluation of the Earth Gravitational Model 2008. *Journal of Geophysical Research*, **117**, B04406.

PÉREZ-GUSSINYÉ, M., LOWRY, A. R., WATTS, A. B. & VELICOGNA, I. 2004. On the recovery of the effective elastic thickness using spectral methods: examples from synthetic data and the Fennoscandian shield. *Journal of Geophysical Research*, **109**, 409.

PÉREZ-GUSSINYÉ, M., LOWRY, A. R., PHIPPS MORGAN, J. & TASSARA, A. 2008. Effective elastic thickness variations along the Andean margin and their relationship to subduction geometry. *Geochemistry Geophysics Geosystems*, **9**, Q02003.

PILGER, R. H. 1981. Plate reconstructions, aseismic ridges, and low-angle subduction beneath the Andes. *Geological Society of America Bulletin*, **92**, 448–456.

RAMOS, V. A. 2009. Anatomy and global context of the Andes: main geologic features and the Andean orogenic cycle. *In*: KAY, S., RAMOS, V. A. & DICKINSON, W. (eds) *Backbone of the Americas: Shallow Subduction, Plateau Uplift, and Ridge and Terrane Collision*. Geological Society of America, Boulder, Colorado, USA, Memoirs, **204**, 31–65.

RAMOS, V. A. & FOLGUERA, A. 2009. Andean flat subduction through time. *In*: MURPHY, B., KEPPIE, J. & HYNES, A. (eds) *Ancient Orogens and Modern Analogues*. Geological Society, London, Special Publications, **327**, 31–54.

RAMOS, V. A., MUNIZAGA, F. & KAY, S. M. 1991. El magmatismo cenozoico a los 33°S de latitud: Geocronologia y relaciones tectónicas. Paper presented at *6° Congreso Geológico Chileno*, Viña del Mar, Chile, Actas, **1**, 892–896.

RAMOS, V. A., CRISTALLINI, E. O. & PEREZ, D. 2002. The Pampean flat-slab of the Central Andes. *Journal of South American Earth Science*, **15**, 59–78.

RAMOS, V. A., VUJOVICH, G., MARTINO, R. & OTAMENDI, J. 2010. Pampia: a large cratonic block missing in the Rodinia supercontinent. *Journal of Geodynamics*, **50**, 243–255.

RANERO, C., VON HUENE, R., WEINREBE, W. & REICHERT, C. 2006. Tectonic processes along the Chile convergent margin. *In*: ONCKEN, O., CHONG, G. ET AL. (eds) *The Andes, Active Subduction Orogeny*. Springer-Verlag, Berlin/Heidelberg/New York, Frontiers in Earth Science Series, 91–121.

REGUZZONI, M. & SAMPIETRO, D. 2010. An inverse gravimetric problem with GOCE data. *International Association of Geodesy Symposia, 'Gravity, Geoid and Earth Observation'*, **135**, 451–456, http://dx.doi.org/10.1007/978-3-642-10634-7_60

ROSSELLO, E. A., LIMARINO, C. O., ORTIZ, A. & HERNÁNDEZ, N. 2005. Cuencas de los Bolsones de San Juan y La Rioja. *In*: CHEBILI, G., CORTIÑAS, J. S., SPALLETTI, L. A., LEGARRETA, L. & VALLEJO, E. L. (eds) *Simposio Frontera Exploratoria de la Argentina*. VI Congreso de Exploración y Desarrollo de Hidrocarburos, Mar del Plata, Argentina, 147–173.

SACEK, V. & USSAMI, N. 2009. Reappraisal of the effective elastic thickness for the sub-Andes using 3-D finite element fexural modelling, gravity and geological constraints. *Geophysical Journal International*, **179**, 778–786.

SACKS, I. S. 1983. The subduction of young lithosphere. *Journal of Geophysical Research*, **88**, 3355–3366.

SANDWELL, D. T. & SMITH, W. H. F. 1997. Marine gravity anomaly from Geosat and ERS-1 satellite altimetry. *Journal of Geophysical Research*, **102**, 10 039–10 050.

SIEBERT, L. & SIMKIN, T. 2002. *Volcanoes of the World: An Illustrated Catalog of Holocene Volcanoes and Their Eruptions*. Smithsonian Institution, Washington DC, USA, Global Volcanism Program Digital Information Series, GVP-3, http://www.volcano.si.edu/world/

SMALLEY, R. & ISACKS, B. 1987. A high resolution local network study of the Nazca Plate Wadati–Benioff Zone under Western Argentina. *Journal of Geophysical Research*, **92**, 13903–13912.

STAUDER, W. 1973. Mechanism and spatial distribution of Chilean earthquakes with relation to subduction of the oceanic plate. *Journal of Geophysical Research*, **78**, 5033–5061.

STEFFEN, R., STEFFEN, H. & JENTZSCH, G. 2011. A three-dimensional Moho depth model for the Tien Shan from EGM2008 gravity data. *Tectonics*, **30**, TC5019, http://dx.doi.org/10.1029/2011TC002886

STEWART, J. & WATTS, A. B. 1997. Gravity anomalies and spatial variations of flexural rigidity at mountain ranges. *Journal of Geophysical Research*, **102**, 5327–5352.

TASSARA, A. 2005. Interaction between the Nazca and South American plates and formation of the Altiplano–Puna plateau: review of a flexural analysis along the Andean margin (15°–34°S). *Tectonophysics*, **399**, 39–57.

TASSARA, A. & YAÑEZ, G. 2003. Relación entre el espesor elástico de la litósfera y la segmentación tectónica del margen Andino (15–478S). *Revista Geológica de Chile*, **30**, 159–186.

TASSARA, A., SWAIN, C., HACKNEY, R. & KIRBY, J. 2007. Elastic thickness structure of South America estimated using wavelets and satellite-derived gravity data. *Earth and Planetary Science Letters*, **253**, 17–36.

TEBBENS, S. F. & CANDE, S. C. 1997. Southeast Pacific tectonic evolution from early Oligocene to present. *Journal of Geophysical Research*, **102**, 12 061–12 084.

UIEDA, L., USSAMI, N. & BRAITENBERG, C. F. 2010. Computation of the gravity gradient tensor due to topographic masses using tesseroids. *EOS, Transactions AGU*, **91** Meeting of the Americas Supplement, Abstract G22A-04, http://leouieda.github.io/tesseroids/

VON HUENE, R., CORVALÁN, J., FLUEH, E. R., HINZ, K., KORSTGARD, J., RANERO, C. R., WEINREBE, W. & CONDOR SCIENTISTS 1997. Tectonic control of the subducting Juan Fernández Ridge on the Andean margin near Valparaiso, Chile. *Tectonics*, **16**, 474–488.

WATTS, A. 2001. *Isostasy and Flexure of the Lithosphere*. Cambridge University Press, Cambridge, UK.

WEIDMANN, C., SPAGNOTTO, S., GIMENEZ, M. E., MARTINEZ, P., ÁLVAREZ, O., SANCHEZ, M. & LINCEKLINGER, F. 2013. Crustal structure and tectonic setting of the south central Andes from gravimetric analysis. *Geofisica Internacional*, **52**, 121–133.

WESSEL, P. & SMITH, W. H. F. 1998. New, improved version of the generic mapping tools released. *EOS, Transactions of the American Geophysical Union*, **79**, 579.

WIENECKE, S. 2002. *Homogenisierung und Interpretation des Schwerefeldesentlang der SALTTraversézwischen 36°–42°S*. Unpublished diploma thesis, FreieUniversität, Berlin.

WIENECKE, S. 2006. *A new analytical solution for the calculation of flexural rigidity: significance and applications*. PhD thesis, Free University Berlin, http://www.diss.fu-berlin.de/2006/42

WIENECKE, S., BRAITENBERG, C. & GÖETZE, H. J. 2007. A new analytical solution estimating the flexural rigidity in the Central Andes. *Geophysics Journal International*, **169**, 789–794.

YAÑEZ, G. & CEMBRANO, J. 2004. Role of viscous plate coupling in thelate Tertiary Andean tectonics. *Journal of Geophysical Research*, **109**, http://dx.doi.org/10.1029/2003JB002494

YAÑEZ, G. A., RANERO, C. R., VON HUENE, R. & DIAZ, J. 2001. Magnetic anomaly interpretation across the southern Central Andes (32°–34° S): the role of the Juan Fernandez Ridge in the late Tertiary evolution of the margin. *Journal of Geophysical Research, Solid Earth*, **106**, 6325–6345.

ZADRO, M. & BRAITENBERG, C. 1997. Spectral methods in gravity inversion: the geopotential field and its derivatives. *Annali di geofisica XL*, **5**, 1433–1443.

The Neocomian of Chachahuén (Mendoza, Argentina): evidence of a broken foreland associated with the Payenia flat-slab

LUCIA SAGRIPANTI*, BEATRIZ AGUIRRE-URRETA, ANDRÉS FOLGUERA & VICTOR A. RAMOS

Instituto de Estudios Andinos Don Pablo Groeber (UBA – CONICET), Universidad de Buenos Aires, Buenos Aires, Argentina

*Corresponding author (e-mail: lusagripanti@gmail.com)

Abstract: Isolated marine sedimentary Lower Cretaceous deposits crop out in the foreland of the Neuquén Basin, west-central Argentina. They are the result of an anomalous uplift of the Sierra de Chachahuén in the far foreland region. These outcrops are assigned to the Agrio Formation based on their rich fossil contents. In particular, the study reveals a unique outcrop of continental facies along the eastern proximal margin of the basin that were known only from core wells, and constitutes the first exposed evidence at the surface. These deformed deposits are 70 km from the Andean orogenic front and present 2 km of local uplift produced by high-angle basement reverse faults that reactivated a previous Early Mesozoic rift system. The increase in compression was related to the decrease in the subduction angle. This fact, together with the expansion of the magmatic arc, controlled the Chachahuén calc-alkaline Late Miocene volcanic centre and the uplift of the Mesozoic deposits in the foreland. This broken foreland was associated with localized heating of the Miocene volcanic centre that produced the rising of the brittle-ductile transitions. This fact weakens the foreland area, which was broken by compression during the development of the Payenia flat-slab.

The large Miocene–Holocene volcanic centres behind the present volcanic arc in the southern Central Andes in northern Neuquén and southern Mendoza provinces, Argentina, hide some remarkable features along the eastern edge of the Neuquén Basin. The eastern flank of the basin is characterized by almost undeformed Tithoneocomian sequences far away from the thrust front located at 69°40′W, several kilometres west of the basin margin (Fig. 1). Patchy exposures of Mesozoic marine deposits near the Chachahuén volcanic centre have been known of for many years, but without a detailed examination. This study partly analyses why these Lower Cretaceous sequences are exposed in this particular region of the retroarc zone. Two alternatives have been proposed in previous studies to explain their abnormal positions: either they were some kind of large roof-pendant associated with the emplacement of the large volcanic domes that conforms the Chachahuén volcanic complex, or they represent some atypical exposures of a tectonically uplifted block of unusual behaviour. To evaluate these alternatives, detailed field work was carried out and seismic lines and borehole data of previous work were examined. We made a detailed analysis and description of the Lower Cretaceous (Agrio Formation) with the aim of explaining the tectonic evolution of the area. We present a detailed map, a sedimentary profile of the Agrio Formation and a structural cross-section. The aims of this contribution are first to identify and date these Mesozoic sequences, and second to understand the local structure and evaluate its location based on the available subsurface data in the eastern flank of the Neuquén Basin. The general geology of the area has been known since the early oil exploration conducted by Padula (1948) with descriptions of the stratigraphy of the area. Since this study, the outcrops of the upper Agrio Formation in the study area have been recognized as marine and conglomeratic continental facies. However, the peculiar conglomeratic facies were only known at the subsurface based on borehole data (Vergani *et al.* 2001; Barrionuevo & Pérez 2002). This is the first contribution that presents a detailed description of the outcrops of this continental facies. The identification of this deposit shows the real north extension of this facies along the eastern margin of the basin, previously only known in the Neuquén province further south. This contribution therefore represents a re-evaluation of the Early Cretaceous palaeogeography of the Neuquén Basin.

The first reconnaissance study of the Chachahuén volcanic complex, intruding the Early Cretaceous sections, was a regional-scale geological mapping of the Sierra de Chachahuén (Holmberg 1962). As the Sierra de Chachahuén became a target for oil exploration during the 1990s, Yacimientos Petroliferos Fiscales (YPF) produced a detailed map of the volcanic rocks (Perez & Condat

From: SEPÚLVEDA, S. A., GIAMBIAGI, L. B., MOREIRAS, S. M., PINTO, L., TUNIK, M., HOKE, G. D. & FARÍAS, M. (eds) 2015. *Geodynamic Processes in the Andes of Central Chile and Argentina*. Geological Society, London, Special Publications, **399**, 203–219. First published online February 5, 2014, http://dx.doi.org/10.1144/SP399.9
© 2015 The Geological Society of London. For permissions: http://www.geolsoc.org.uk/permissions.
Publishing disclaimer: www.geolsoc.org.uk/pub_ethics

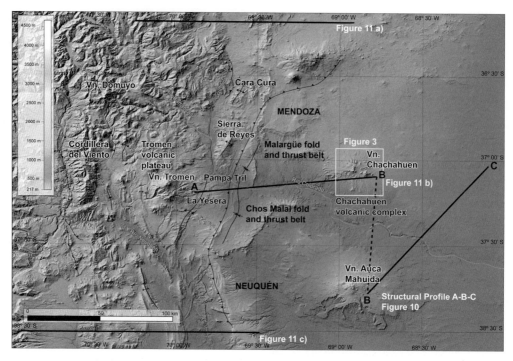

Fig. 1. Digital elevation model of the study area. Note the thrust front east of the Sierras de Reyes, Cara Cura, Pampa Tril and La Yesera. The Chachahuén volcanic complex is located to the east of the thrust front in the foreland area.

1996) and proposed a volcanic stratigraphy by correlating the unconformities between measured sections and from 24 new Late Miocene K–Ar ages. This information was later analysed by Kay et al. (2006) who proposed a coherent stratigraphic scheme. Kay (2001) and Kay & Mancilla (2001) emphasized the petrological features of the Chachahuén volcanic centre that make it unique among the Neogene volcanic centres in this part of the Neuquén Basin. Finally, Kay et al. (2006) produced a detailed description of the geochemistry and geochronology of these volcanic centres located through the eastern Mendoza and Neuquén foreland zone, among which the Chachahuén volcanic centre is located.

Tectonic setting

The study area is located in the south of the Mendoza province and in the north-eastern part of the Neuquén Basin. The geology of the Neuquén Basin can be divided into three general stages: (1) a Triassic–Lower Jurassic pre-rift and rift stage, which started as a rift system with non-connected depocentres and accumulated more than 7000 m of continental, marine and volcaniclastic components (Vergani et al. 1995; Legarreta & Uliana 1996); (2) an Upper Jurassic–Cretaceous stage associated with thermal subsidence; in particular, the Upper Cretaceous records synorogenic deposits of the Neuquén Group in a foreland basin; and (3) a Paleocene–Holocene stage which includes sedimentary deposits associated with important Cenozoic volcanic sequences (Vergani et al. 1995). In the eastern part of the Neuquén Basin, where the study area is located, Miocene–Holocene magmatic rocks are widespread in the retroarc region. The large Quaternary volcanic province of basaltic composition, known as the Payenia volcanic province, constitutes the main association at these latitudes. Here, the main depocentres of the Mesozoic rifting are at the subsurface in marginal positions.

The Andean structure is characterized by tectonic inversion of Mesozoic normal faults and new reverse structures produced during Late Cretaceous–Neogene time. This deformation took place in the western side of the basin with the development of basement structures that transferred shortening to the east (Giambiagi et al. 2009). The orogenic front at these latitudes is located at 69°40′W, coincident with the eastern margin of the Sierras de Cara Cura, Reyes and Pampa Tril structural trend (Fig. 1). The only structure developed in the foreland area, to the east of this front, is the Sierra de Chachahuén, where some minor structures are partially covered by the products of the Chachahuén volcanic complex. This volcanic centre creates a topographic anomaly north of the Río Colorado in southern Mendoza and dominates the landscape of this part of the retroarc region. This complex extends over an area of 2000 km^2 and consists of a broad volcanic dome, a large caldera, the Chachahuén Caldera and at least 12 other minor calderas (Kay et al. 2006). Llambías et al. (2010) described more than 30 younger monogenetic cones and one phreatomagmatic volcano, Los Loros (Németh et al. 2012), that cover the older volcanic structures in this area. It is interesting to note that in this maar exotic pieces of the Agrio Formtion are recognized (M. Barrionuevo, pers. comm. 2013).

Evidence of contractional deformation is found near the Sierra de Chachahuén. In the Cerro Corrales in particular, there are exposures of the marine Agrio Formation (Hauterivian), continental red sandstones of the Rayoso Formation (Aptian–Albian) and of the Neuquén Group (Late Cretaceous) that have been tectonically emplaced (Holmberg 1962; Kay et al. 2006) (Fig. 2). Note that these marine Lower Cretaceous outcrops are the only rocks of that age exposed in the foreland area, located at least 70 km eastwards from the orogenic front (Fig. 2).

The Chachahuén volcanic complex is part of the Payenia volcanic province (Ramos & Folguera 2011). This province is exposed in the retroarc of the Andes, between 33°40′S and 38°00′S over an area of almost 40 000 km^2, parallel to the active volcanic arc of the Southern Volcanic Zone (Stern 2004). It has an estimated volcanic volume of about 8387 km^3 erupted through more than 800 volcanic centres in the last c. 2 Ma (Risso et al. 2008; Folguera et al. 2009; Llambías et al. 2010; Ramos & Folguera 2011), being one of the densest volcanic provinces of South America. It is a typical retroarc assemblage with two peaks of activity, one between 12 and 4 Ma and another younger than 2 Ma (Fig. 3).

The study of the distribution in time and space of the different eruptions, combined with geophysical data, indicates an important crustal attenuation beneath the Payenia volcanic province associated with a hot sublithosphere (Wagner et al. 2005; Folguera et al. 2012; Burd et al. 2008; Søager et al. 2013). The Early Pleistocene collapse of the Payenia region affected the San Rafael Block which was uplifted during Late Miocene time. The Quaternary Payenia volcanic province is explained as a consequence of the steepening of the subducted slab and the injection of hot asthenosphere (Burd et al. 2008). This magnetotelluric study shows that the Payún Matrú volcanic caldera coincides with the area of impact of this asthenospheric flow currently, associated with a deep source melt that rises from a depth of 400 km and connects with the upper mantle. This justifies the ocean-island basalt (OIB) geochemical signature of the Payún Matrú caldera (Fig. 3), corresponding to an enriched mantle source typical of hotspots (Llambías et al. 2010; Søager et al. 2013).

Early Cretaceous deposits in the foreland area

The oldest sedimentary rocks of the area are the Lower Cretaceous sequences of the Agrio and Rayoso formations. The gently folded Late Cretaceous Neuquén Group is exposed south of the study area near the Río Colorado valley.

The Agrio Formation, defined by Weaver (1931) near Bajada del Agrio in central Neuquén (location in Fig. 2), reaches nearly 1200 m in its type locality and is divided into three distinctive members: Pilmatué, Avilé and Agua de la Mula (Leanza & Hugo 2001). Both the Pilmatué and the Agua de la Mula members represent marine facies and are largely composed of massive dark shales and mudstones interbedded with packstones, wackestones and sandstones containing abundant molluscs and other fossil invertebrates. The Avilé Member, 30–40 m thick, is composed of yellowish coarse sandstones of fluvial and aeolian origin in the central part of the basin (Weaver 1931; Gulisano & Gutiérrez Pleimling 1988); towards the north, coeval deposits are represented by greenish claystones with gypsum nodules and halite hoppers (Legarreta & Uliana 1991). This member represents a lowstand wedge produced by a rapid relative sea-level drop during middle Hauterivian time (Legarreta & Gulisano 1989; Veiga et al. 2002).

The Agrio Formation has been interpreted as a storm-dominated shallow-marine environment with mixed siliciclastic and carbonate sedimentation (Spalletti et al. 2001). The abundance of ammonoids has allowed the Agrio Formation to be accurately dated as being of late Early Valanginian–Late Hauterivian–Early Barremian age (Aguirre-Urreta et al. 2007).

In the study area, the outcrops of the Agrio Formation are located in the southern part of the Cerro Corrales in an area slightly over 1 km^2, some 500 m west of the Puesto Ojo de Agua (the only permanent settlement in the area; Fig. 4).

A detailed stratigraphic profile of the Agrio Formation was surveyed (A–A′ in Fig. 4) and illustrated in Figure 5.

(a) The base of the section is an intrusive contact with a Late Miocene andesitic sill. There are 15 m of black to grey shales, interbedded with massive mudstones and a coquina. The

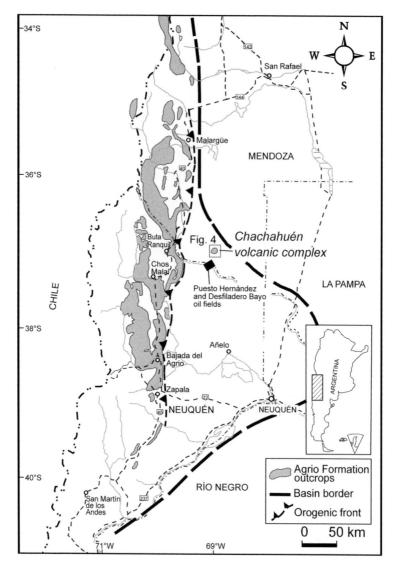

Fig. 2. Agrio Formation outcrops in the Neuquén Basin. Most of the outcrops are restricted to the Andean belt limited by the orogenic front; the only exposure in the foreland area is located near the Chachahuén volcanic centre.

70-cm-thick coquina has tabular structure and records abundant oysters, thin-shelled bivalves and the ammonite *Hoplitocrioceras gentilii* Giovine (Fig. 6a). Above, another level with *Hoplitocrioceras gentilii* was found. This part of the section ends with a metre-thick massive mudstone, with ammonites assigned to *Weavericeras vacaense* (Weaver).

(b) Upsection there is a 5-m-thick Late Miocene andesitic sill with large amphibole xenocrysts. The sedimentary sequence, more than 120 m thick, continues with laminated black shales that grade upwards into massive mudstones, wackestones and coquinas towards the top (Fig. 7). The lower 3 m of black shales have the ammonite *Spitidiscus* sp., Discinidae brachiopods (Fig. 8e, f) and aggregates of small serpulids. Upwards, yellowish thin massive wackestones have well-preserved fossils, including *Neocomiceramus curacaensis* (Weaver) (Figs 6b, 8d) and *Crioceratites* sp. below and *Amphidonte (Ceratostreon)* sp., *Aetostreon* sp. (Fig. 8g), *Neocomiceramus curacoensis* and *Crioceratites* sp. above. In the top of the sequence the coquinas have abundant fossils, including: *Neocomiceramus*

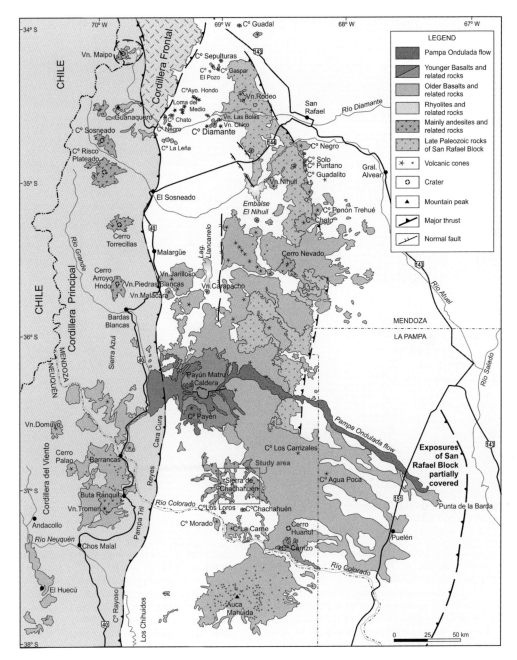

Fig. 3. Main volcanic centres and basaltic flows of the Payenia volcanic province. The Sierra de Chachahuén is the biggest volcanic centre with andesites and related rocks in the Payenia (Ramos & Folguera 2011).

curacoensis, *Gervillaria alatior* (Imlay), *Pinna* sp., *Aetostreon* sp., *Amphidonte* (*Ceratostreon*) sp., *Mimachlamys* sp. and *Crioceratites* sp. (Fig. 6c).

(c) Another 10-m-thick andesitic sill is emplaced in this section with similar lithology to the previous sills (Fig. 6d). Over the andesites, there are 5 m of medium- to coarse-grained cross-bedded calcareous sandstones with a coquina near the base. The next 5 m are matrix-supported, light grey conglomerates with calcareous matrix (Fig. 6e). The clasts are rounded, poorly selected

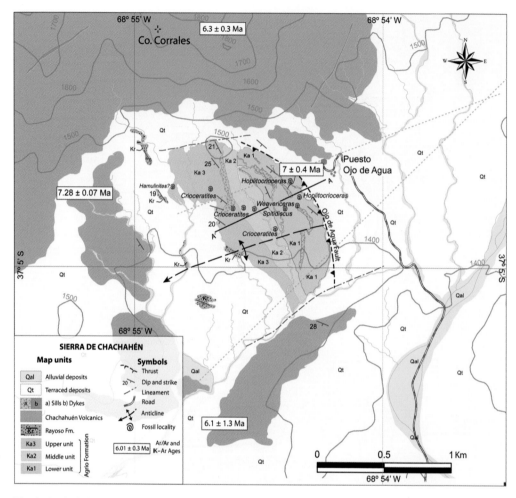

Fig. 4. Geological map of the southern part of the Sierra del Chachahuén (note the outcrops of the Agrio Formation south of Cerro Corrales).

and their granulometry varies from 1 to 3 cm, predominantly of rhyolitic porphyries. The nature of the clasts indicates a second cycle reworking, and are probably derived from the Choiyoi Group. From cores from the PH-1044, PH-1143 and PH-1147 wells of the Puesto Hernandez field to the south of the study area, Vergani et al. (2001) describe similar conglomeratic facies in the same interval of the upper Agrio Formation. They explain the occurrence of these conglomerates as a conspicuous spontaneous event related to a forced regression (Vergani et al. 2001).

(d) Above the conglomerates there are grainstones and 20 m of amalgamated coquinas formed mostly by large and small oysters in a calcareous matrix (Fig. 6f), also preserving ammonites. There are several levels where the dominant ammonite is *Crioceratites diamantensis* (Gerth), and most of the specimens are colonized by oysters (Figs 6g, 8a). In the next 20 m the coquinas, each 2–3 m thick, are interbedded with massive yellowish mudstones and wackestones.

(e) The succession continues with 10 m of medium- to coarse-grained cross-bedded sandstones with carbonate cement, which bear a few levels with internal moulds of *Neocomiceramus* sp.

(f) Above the sandstones there are more than 25 m of coquinas interbedded with mudstones. The fauna includes *Myoconcha transatlantica* Burckhardt, *Pterotrigonia coihuicoensis* (Weaver), *Ptychomya koeneni* Beherendsen, *Cucullaea gabrielis* Deshayes, *Amphidonte (Ceratostreon)* sp., *Eriphyla* sp., other

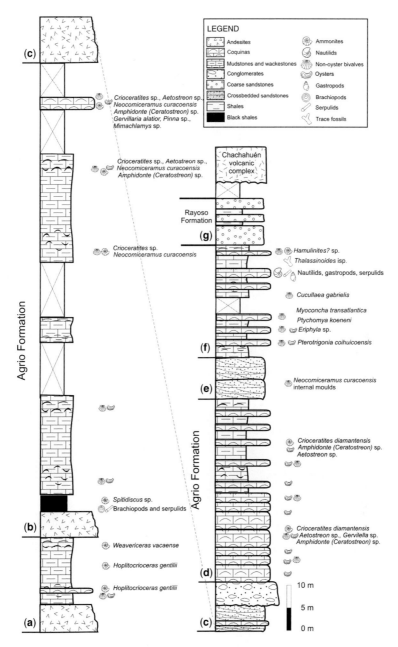

Fig. 5. Sedimentary profile of the Agrio Formation exposed in the southern part of Cerro Corrales underlying the Chachahuén volcanic complex.

indeterminate bivalves, nautiloids, serpulids, gastropods and the ammonite *Hamulinites?* sp. (Fig. 8b, c). The mudstones preserve horizontal networks of *Thalassinoides* isp. (Fig. 6h).

(g) Paraconformably lying on top of the section are poorly exposed coarse sandstones interbedded with shales of the Rayoso Formation of Aptian–Albian age.

In the study section, the black–grey shales interbedded with massive mudstones and coquinas represent the most common and thickest lithofacies. Coquinas occur with loose to dense shell packing

Fig. 7. View to the east of the *Spitidiscus* shales, mudstones and wackestones of the Agrio Formation above the second sill in the Cerro Corrales section.

and the marine benthic fauna is mostly represented by diverse bivalves, with less-common gastropods and serpulids. Ammonites are also common, while nautilids are very rare. Thalassinoid trace fossils appear in the mudstones. These rocks are interpreted as representing fair-weather suspension deposits of an outer ramp. Some firm substrates were also developed as indicated by the presence of trace fossil networks with distinct sedimentary filling. The coquinas are the result of accumulation of shells reworked by storms events (Lazo *et al.* 2005).

The laminated black shales with *Spitidiscus* point to a high organic content typical of dysoxic or anoxic seafloors (Tyson *et al.* 2005) where the low-diversity small-sized benthic elements, mostly represented by discinid brachiopods and small serpulids, are concentrated in pavements.

Conglomeratic fluvial facies are interbedded with massive coarse carbonatic grainstones of shore facies. This indicates the presence of incise valleys in a coastal zone, similar to the conglomerates and sandstones described in the subsurface by Barrionuevo & Pérez (2002) a few kilometres to the south of the study area.

The amalgamated coquinas are interpreted to have been deposited in a mid-ramp setting during reduced input of siliciclastic material, and record the stacking and amalgamation of multiple storm events (Lazo *et al.* 2005).

The presence of several ammonites in different levels of the study section allows the age of the Agrio Formation to be determined as between the early Hauterivian (*Hoplitocrioceras gentilii* zone) and the latest Hauterivian–early Barremian (*Sabaudiella riverorum* zone) (Aguirre-Urreta & Rawson 2001, 2012).

When a comparison between the measured section near Cerro Corrales and the typical development of the Agrio Formation is made, some striking differences arise.

- *The base of the exposed section*: In the study area the lowermost exposures correspond to the upper third of the Pilmatué Member evidenced by the presence of the ammonite *Hoplitocrioceras gentilii*. The initial succession in the area was described by Padula (1948) as bearing the ammonite *Hatchericeras* sp., but the presence of this Austral genus is ruled out as it is most probably a misidentification of the large *Hoplitocrioceras*. The base of the Agrio Formation in the Cerro Corrales locality is not exposed as it is tectonically truncated by the Ojo de Agua thrust (Fig. 4).
- *The lack of the Avilé Member*: There is no equivalent of the Avilé Member either in its typical continental sandstone facies of the Neuquén embayment or in the shallow evaporitic playa lake deposits exposed towards the north in southern Mendoza. It is clear in our section that the black shales with *Spitidiscus* (ammonite of the base of the Agua de la Mula Member) cover with a sharp contact the massive mudstones with *Weavericeras vacaense* (ammonite that defines the last biozone of the Pilmatué Member).
- *The unique development of conglomerates in the middle of the Agua de la Mula Member*: The upper member of the Agrio Formation depicts

---

Fig. 6. (**a**) Large *Hoplitocrioceras gentilii* (Giovine) at the base of the succession; (**b**) wackestone with *Neocomiceramus curacoensis* (Weaver); (**c**) coquina with abundant *Mimachlamys* sp. and other bivalves; (**d**) andesitic sill; (**e**) fluvial conglomerates; (**f**) oyster coquina; (**g**) coquina with bivalves and *Crioceratites diamantensis* (Gerth); and (**h**) massive mudstone bedding plane with networks of *Thalassinoides* isp.

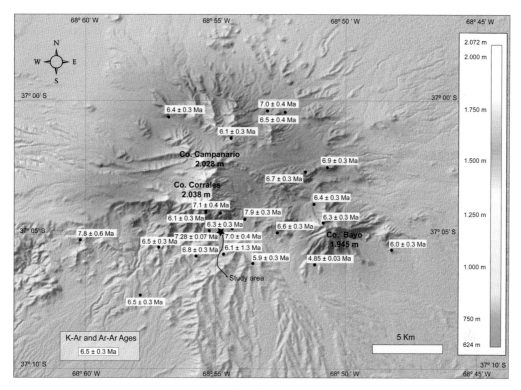

Fig. 9. Digital elevation model with the K–Ar and $^{40}Ar/^{39}Ar$ ages compiled from Kay *et al.* (2006).

different marine facies throughout its development all along the Neuquén Basin. This is the only exposed section where fluvial conglomerates are intercalated within the mudstones, wackestones and coquinas bearing *Crioceratites diamantense* in the middle part of the Agua de la Mula Member. These conglomerates have been studied in well cores by Barrionuevo & Pérez (2002, figs 24–26) as they are important reservoirs in several oil fields such as Puesto Hernández and Desfiladero Bayo in NE Neuquén and SE Mendoza (location in Fig. 2). These conglomerates can also be correlated towards the east with the upper member of Centenario Formation, a subsurface unit of continental origin (Cabaleiro *et al.* 2002).

- *The total thickness of the Agrio Formation*: The measured section of c. 230 m is much more reduced when compared to the equivalent development in central Neuquén. This is partially due to the fact that the lower part of the Pilmatué Member is tectonically truncated, but also because towards the basin margin the sequences are stratigraphically condensed. Nevertheless, this thickness is similar to the 200 m known from the subsurface in the eastern border of the basin (Barrionuevo & Pérez 2002).

The Chachahuén volcanic complex

The Chachahuén volcanic complex is the only Late Miocene volcanic centre in the extra-Andean region of the central Neuquén Basin at these latitudes. It has been recently studied by Kay *et al.* (2006), who established the geochemical, geochronological and isotopic characteristics of these volcanic rocks. The complex had been divided previously into three groups by Perez & Condat (1996), criteria followed by Kay *et al.* (2006): the Late Miocene Vizcachas Group, the Early Chachahuén Group, and the Late Chachahuén Group.

Fig. 8. (**a**) *Crioceratites diamantensis* (Gerth), CPBA 21491; (**b, c**) *Hamulinites*? sp. CPBA 18217; (**d**) *Neocomiceramus curacoensis* (Weaver) CPBA 21495; (**e, f**) Discinidae indet. CPBA 21497.1–2; (**g**) *Aetostreon* sp. CPBA 21493. All specimens are coated with ammonium chloride. Black bars are 10 mm.

The base corresponds to the Vizcachas Group, which includes porphyritic diorites, andesitic and dacitic to rhyodacitic lava flows, and dacitic dykes. These dykes cut the diorites as well as the Cretaceous sedimentary rocks. The age of this unit ranges from 7.90 ± 0.3 Ma and 7.28 ± 0.07 Ma ($^{40}Ar/^{39}Ar$) to 7.1 ± 0.4 Ma (K–Ar) in Cerro Corrales (Kay et al. 2006).

The Early Chachahuén Group consists of sandstones and pyroclastic deposits that include ignimbrites, lava flows, and volcanic conglomerates. Ages reported by Perez & Condat (1996), vary from 6.9 ± 0.3 to 6.3 ± 0.3 Ma (K–Ar).

The Late Chachahuén Group comprises a prominent andesitic to dacitic ignimbrite-pyroclastic unit overlaid by mafic andesitic lava flows interbedded with thin pyroclastic layers. Ages reported by Perez & Condat (1996) of this group range from 6.4 ± 0.4 to 5.6 ± 0.3 Ma (K–Ar). A new $^{40}Ar/^{39}Ar$ age of 4.85 ± 0.03 Ma on a mafic andesitic lava flow on the SW side of Cerro Chachahuén (Kay et al. 2006) extends the younger age limit into the earliest Pliocene (Fig. 9).

The chemical and isotopic data show that the magma evolved from a dacitic to rhyodacitic Vizcachas Group with intraplate-like chemical tendencies, to a basaltic and dacitic composition in the early and late Chachahuén groups with arc-like signature. These data are consistent with a model in which the magma of the Chachahuén volcanic complex contains crustal- and mantle-derived components that have changed through time due to the increasing influence of subducted components above a shallowing subduction zone (Kay et al. 2006).

The Chachahuén volcanic complex is unconformably covered by Pleistocene–recent mafic volcanic flows and sedimentary units. The geochemistry of these Quaternary basaltic flows, dominated by an oceanic-island basalt signature, contrasts with the calc-alkaline subduction-related signature of the Miocene rocks (Kay et al. 2006).

Based on the previous model, the occurrence of this andesitic to dacitic centre far from the volcanic front was interpreted by Ramos & Kay (2006) as controlled by an expansion of the magmatic arc during a period of shallow subduction. The location of this volcanic complex on a regional scale, like other Neogene magmatic centres, was controlled by previous extensional structures. These structures correspond to the Mesozoic rifting of the basin (Vergani et al. 1995; Zapata et al. 1999; Cristallini et al. 2006).

Structural setting

The Lower Cretaceous rocks exposed west of the Puesto Ojo de Agua (Fig. 4) are gently dipping to the southwest between 15° and 29°. This homoclinal structure defines a smooth anticline nose plunging to the SW with a half-wavelength of 2 km. This structural nose is truncated by the SW-dipping Ojo de Agua fault, a reverse fault that uplifted the sedimentary sequence. The uplifted Early Cretaceous block is bound to the north and south by two lineaments, a fact which was recognized by Padula (1948). These lineaments are interpreted as previous normal faults that segmented the structure at both ends because, outside of the central block, there are younger deposits disposed subparallel to the lineaments. This fact was also described by Padula (1948), who interpreted the northern lineament as a normal fault.

The integration of the local features with the regional structure as illustrated in Figure 10

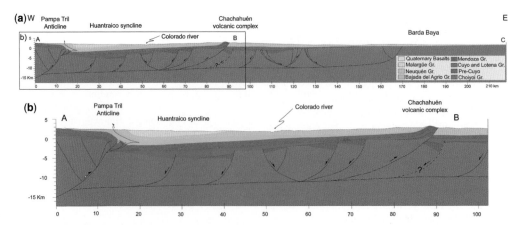

Fig. 10. (a) Regional structural cross-section from the Pampa Tril anticline to the non-deformed area (see Fig. 1 for location). (b) Detail of the previous section from the Pampa Tril anticline to the Chachahuén volcanic complex (modified from Zamora Valcarce & Zapata 2005; Cristallini et al. 2006; Rojas Vera et al. 2010).

indicates that a series of Early Mesozoic normal faults were partially inverted during the Andean deformation. Limited available seismic information shows that immediately to the SW the polarity of the half-graben is opposite to that interpreted in the section of Figure 10 (Vergani et al. 2001). Additionally, in the nearby sectors (such as in Puesto Hernandez field where 3D seismic lines are available) the contractional structures are not produced by reactivation of previous normal fault. In this area, the structure is controlled by strike-slip offsets (Vergani et al. 2001). The lack of seismic information in the Chachauén structure only allows a basement reactivation to be interpreted, with no clear evidence of reactivation of normal faults. In the structural section of Figure 10 the Chachahuén structure has therefore been interpreted as a basement fault that does not reactivate a previous normal fault, cutting through the synrift filling probably due to its inappropriate orientation in relation to Andean contraction.

Most of the extensional structures are characterized by high-angle NW-trending grabens and half-grabens (Zamora Valcarce & Zapata 2005; Cristallini et al. 2006). The main normal faults were active until the deposition of lower Upper Jurassic sequences. Subsequently, only a few faults with favourable orientation were reactivated during the Andean contractional deformation.

The section shows that most of the half-graben structures were not modified during the Andean deformation. The orogenic front coincides at these latitudes with the Pampa Tril anticline. There, a triangle zone developed in the gypsum levels of the Huitrín Formation at the base of the Bajada del Agrio Group, detached through a back-thrust in Upper Cretaceous sequences of the Neuquén Group from the older deposits. In many areas substantial lateral migration of the Huitrín evaporites occupied the triangle zone with important local thickening of this unit. Some younger extensional structures produced subtle lows and highs in the foreland area (Cristallini et al. 2006), mainly due to differential compaction. As seen in the section of Figure 10b, Chachahuén constitutes a structural anomaly in the foreland. There is a striking coincidence between the unique Chachahuén volcanic centre, the only calc-alkaline volcanic complex at these latitudes, and the possible inversion of a normal fault during the Andean deformation. Some ENE-trending faults that truncate the Lower Cretaceous outcrops (see Fig. 4) could be interpreted as extensional Andean faults, concomitant with the contraction. It is interesting to note that three sills derived from this volcanic centre are emplaced in the Lower Cretaceous sequences. As seen in many areas of the Main Andes (see Ramos et al. 1996) during maximum compression associated with the thrusting, $\sigma 1$ is horizontal while $\sigma 3$ is vertical. The positive dilatancy associated with $\sigma 3$ controls the emplacement of the sills during contraction in these settings (Ramsay & Huber 1983). As the age of the sills are constrained in 7 ± 0.4 Ma in the Chachahuén volcanic complex by Kay et al. (2006), we can bracket the age of deformation and therefore the exposition of the Early Cretaceous sections during Late Miocene time.

The comparison between different structural cross-sections at different latitudes, both north and south of the study area involving the thrust fronts and the foreland regions, illustrates some interesting characteristics (Zamora Valcarce et al. 2011; Giambiagi et al. 2012). The northern-most section cuts the Malargüe fold-and-thrust belt at 36°S (Fig. 11a) east of the eastern back-thrust of the Palauco oil field, showing no deformation immediately to the east. In the far foreland region, the San Rafael block was uplifted at the time of the development of the Payenia shallow subduction zone beyond the limits of this analysis. The southern section across the Agrio fold-and-thrust belt at 38°S has no deformation east of the thrust front. The Chihuidos High, a gentle uplift produced in during Late Miocene time, has been interpreted as a peripheral bulge associated with the loading of the Andean fold-and-thrust belt by Sigismondi (2013). Even although several half-grabens are observed in this sector, no normal faults seem to have been reactivated at depth.

The comparison between the three cross-sections highlights three preliminary conclusions: (1) the Chachahuén section is the only section developed across an important Late Miocene cal-calkaline volcanic field; (2) the detachment of the half-graben system, located in the three sections at about 12 km depth, seems to be an effective décollement to transfer the displacement towards the foreland (Zamora Valcarce et al. 2011; Giambiagi et al. 2012) and (3) the heating produced by the important volcanic activity that lasted almost three million years in the area (from 7.8 ± 0.6 Ma to 4.85 ± 0.03 Ma) could be an important factor in the development of brittle-ductile transitions far away from the orogenic front. Similar scenarios have been described in the Sierras Pampeanas further north (Ramos et al. 2002).

Discussion

As a result of this field study we were able to precisely correlate the exposed Lower Cretaceous deposits with the middle and upper part of the Agrio Formation based on the ammonoids found in the study area. The importance of this correlation

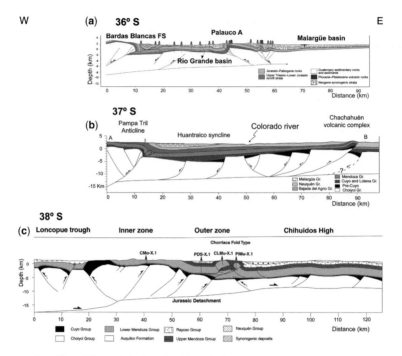

Fig. 11. A structural profile at different latitudes. The Chachahuén volcanic complex is the only structure in the three profiles at the foreland area: (**a**) cross-section of the Malargüe fold-and-thrust belt and the Cenozoic Río Grande Basin taken from Giambiagi et al. (2012); (**b**) cross-section showing the broken foreland associated with the Chachahuén volcanic complex (this work); and (**c**) cross-section of the Agrio fold-and-thrust belt and the Chihuidos High taken from Zamora Valcarce et al. (2006).

is that this unit has been uplifted at least 2 km from the regional depth at this part of the eastern flank of the Neuquén Basin.

The age of deformation has been studied by several authors in the Malargüe and Agrio fold-and-thrust belts. The main episodes of faulting in the Malargüe fold-and-thrust belt occurred during Early Miocene–Late Pliocene time, when the thrusting related to the Neogene deformation migrated towards the foreland in a series of successive steps. These steps, precisely depicted by Giambiagi et al. (2009) and Silvestro & Atencio (2009), were prior to reaching the present position at the thrust front. The orogenic front at latitude 36°S coincides with the Bardas Blancas anticline and its associated frontal back-thrust, which characterizes a triangle zone. Close to the orogenic front, the tectonic inversion of the basement produced the west-verging Palauco anticline during Miocene times (Giambiagi et al. 2012). East of this anticline there are no structures exposed in the eastern flank of the basin.

The orogenic front at latitude 38°S has a complex structure formed by the Pichi Mula and La Mula–Naunauco detached anticlines. These thin-skinned structures were produced during Late Cretaceous deformation. The reactivation of basement structures was finally shaped during Miocene time, as proposed by Zamora Valcarce et al. (2006). East of that region no other structure is developed in the eastern flank of the basin with the only exception being the Chihuidos High, a wide ridge interpreted as a peripheral bulge by Sigismondi (2013).

The orogenic front at 37°S has the Pampa Tril triangle zone defined by Ploszkiewicz et al. (1984) and Zamora Valcarce et al. (2011); 70 km from the thrust front is the anomalous broken foreland of Sierra de Chachahuén.

The comparison between the three sections highlights the unusual deformation of Sierra de Chachahuén far away from the emergent orogenic front. The central section at 37°S is the only one where a large calc-alkaline centre (the Chachahuén volcanic complex) was emplaced (Fig. 9). Even although most of the Payenia region is covered by volcanic rocks as illustrated in Figure 3, there is an important difference between a thin cover of basaltic flows and the long-lasting dacitic to andesitic centre of Chachahuén in terms of thermal processes. The first episode is a short lasting and surficial activity of a few thousand years, in comparison to the large calc-alkaline centre, one of the largest of its type in the extra-Andean region (see Kay *et al.* 2006). There, the residence time of a magmatic chamber of 4–5 million years was enough to heat the upper–middle crust. As seen in different areas, heating produced by this kind of calc-alkaline volcanic complex may have developed the brittle-ductile transition that favoured the décollement and the subsequent deformation. That centre, formed as far as 500 km away from the trench, can only be explained by a shallowing of the subduction zone as proposed by Kay (2001). This shallowing not only explains the expansion of the magmatism, but also the increased compression related to the decrease in the angle of subduction that has broken the foreland as observed in different flat-slab settings (Jordan *et al.* 1983; Gutscher *et al.* 2000). Based on this correlation, we propose that the Chachahuén basement inversion is related to the shallowing of the subducted slab, the heating related to the expansion of the calc-alkaline magmatism and the subsequent contraction.

Conclusions

The combination of detailed fieldwork and subsurface seismic data from previous work, combined with a comprehensive analysis of neighbouring areas, has shown that the Chachahuén volcanic centre is unique in the foreland region where the Payenia volcanic province developed. This large volcanic complex produced abnormal heat conditions that triggered basement deformation far from the orogenic front. The transfer of deformation was facilitated by the development of brittle-ductile transitions in the basement, which localized the deformation in the Sierra de Chachahuén.

The topographic relief of this range is explained by uplift through high-angle faults related to basement reactivation; only those of certain orientation have led to the inversion of previous normal faults at a décollement located at *c.* 12 km depth, in coincidence with the previous potential brittle-ductile transition as depicted in the Sierras Pampeanas flat-slab region.

The stratigraphic study of these isolated Lower Cretaceous exposures, combined with the structural evaluation of the region, yielded a solid explanation for this unusual setting in the foreland of the Neuquén Basin. This fact exposes atypical facies of the upper Agrio Formation, only known by subsurface data along the basin border. The occurrence of fluvial conglomerates incised in typical marine carbonatic ramp facies of the *Crioceratites diamantensis* biozone characterizes an important reservoir facies and highlights the strong dynamics of the eastern basinal margin in Late Hauterivian times.

This study was carried out with the financial assistance of Grant ANPCyT Pict 2142/2008 to VAR. The assistance of Francisco Rusconi in the field is highly appreciated. Dario Lazo, Leticia Luci and Cecilia Cataldo (Universidad de Buenos Aires) helped with annelid, bivalve and gastropod identifications. The authors acknowledge the critical comments of Marcelo Barrionuevo and an anonymous reviewer which contributed to an improved manuscript. This is contribution R-86 of the Instituto de Estudios Andinos Don Pablo Groeber.

References

AGUIRRE-URRETA, M. B. & RAWSON, P. F. 2001. Lower Cretaceous ammonites from the Neuquén Basin, Argentina: the Hauterivian neocomitid genus *Hoplitocrioceras* (Giovine, 1950). *Cretaceous Research*, **22**, 201–218.

AGUIRRE-URRETA, M. B. & RAWSON, P. F. 2012. Lower Cretaceous ammonites from the Neuquén Basin, Argentina: a new heteromorph fauna from the uppermost Agrio Formation. *Cretaceous Research*, **35**, 208–216.

AGUIRRE-URRETA, M. B., MOURGUES, F. A., RAWSON, P. F., BULOT, L. G. & JAILLARD, E. 2007. The Lower Cretaceous Chañarcillo and Neuquén Andean basins: ammonoid biostratigraphy and correlations. *Geological Journal*, **42**, 143–173.

BARRIONUEVO, M. & PÉREZ, D. E. 2002. Reservorios del Miembro Agrio Superior de la Formación Agrio. *In*: SCHIUMA, M., HINTERWIMMER, G. & VERGANI, G. (eds) *Rocas Reservorio de las Cuencas Productivas de la Argentina*. Instituto Argentino del Petróleo, Buenos Aires, 447–456.

BURD, A., BOOKER, J. R., POMPOSIELLO, M. C., FAVETTO, A., LARSEN, J., GIORDANENGO, G. & BERNAL, L. O. 2008. Electrical conductivity beneath the Payún Matrú Volcanic field in the Andean back-arc of Argentina near 36.5°S: insights into the magma source. *VII° International Symposium on Andean Geodynamics*. Nice, *Extended Abstract*, 90–93.

CABALEIRO, A., CAZAU, L., LASALLE, D., PENNA, E. & ROBLES, D. 2002. Los reservorios de la Formación Centenario. *In*: SCHIUMA, M., HINTERWIMMER, G. & VERGANI, G. (eds) *Rocas Reservorio de las Cuencas Productivas de la Argentina*. Instituto Argentino del Petróleo, Buenos Aires, 407–425.

CRISTALLINI, E., BOTTESI, G., GAVARRINO, A., RODRIGUEZ, L., TOMEZZOLI, R. & COMERON, R. 2006. Synrift geometry of the Neuquén Basin in northeastern Neuquén Province, Argentina. *In*: KAY, S. M. & RAMOS, V. A. (eds) *Evolution of an Andean Margin: A Tectonic and Magmatic View from the Andes to the Neuquén Basin (35°–39° S lat)*. Geological Society of America, Boulder, Special Paper, **407**, 147–162.

FOLGUERA, A., NARANJO, J. A., ORIHASHI, Y., SUMINO, H., NAGAO, K., POLANCO, E. & RAMOS, V. A. 2009. Retroarc volcanism in the northern San Rafael block (34°–35° 30′S), southern Central Andes: occurrence, age, and tectonic setting. *Journal of Volcanology and Geothermal Research*, **189**, 69–185.

FOLGUERA, A., ALASONATI TAŠÁROVA, Z., GÖTZE, H. J., ROJAS VERA, E., GIMÉNEZ, M. & RAMOS, V. A. 2012.

Retroarc extension in the last 6 Ma in the South-Central Andes (36°S–40°S) evaluated through a 3-D gravity modelling. *Journal of South American Earth Sciences*, **40**, 23–37.

GIAMBIAGI, L., GHIGLIONE, M., CRISTALLINI, E. & BOTTESI, G. 2009. Kinematic models of basement/cover interaction: insights from the Malargüe fold and thrust belt, Mendoza, Argentina. *Journal of Structural Geology*, **31**, 1443–1457.

GIAMBIAGI, L., MESCUA, J., BECHIS, F., TASSARA, A. & HOKE, G. 2012. Thrust belts of the southern Central Andes: along-strike variations in shortening, topography, crustal geometry, and denudation. *Geological Society of America, Bulletin*, **124**, 1339–1351.

GULISANO, C. A. & GUTIÉRREZ PLEIMLING, A. 1988. Depósitos eólicos del Miembro Avilé (Formación Agrio, Cretácico inferior) en el norte del Neuquén, Argentina. *2 Reunión Argentina de Sedimentología*. Buenos Aires, Actas, 120–124.

GUTSCHER, M. A., SPAKMAN, W., BIJWAARD, H. & ENGDAHL, E. R. 2000. Geodynamic of flat subduction: seismicity and tomographic constraints from the Andean margin. *Tectonics*, **19**, 814–833.

HOLMBERG, E. 1962. Descripción geológica de la hoja 32-d, Chachahuén, Provincias de Neuquén y Mendoza. *Dirección Nacional de Geología y Minería, Boletín*, **91**, 1–72.

JORDAN, T., ISACKS, B., RAMOS, V. A. & ALLMENDINGER, R. W. 1983. Mountain building model: the Central Andes. *Episodes*, **1983**, 20–26.

KAY, S. M. 2001. Geochemical evidence for a late Miocene shallow subduction zone in the Andean Southern Volcanic Zone near 37°S latitude. *EOS (Transactions, American Geophysical Union)*, Abstract, 81 V12C-099.

KAY, S. M. & MANCILLA, O. 2001. Neogene shallow subduction segment in the Chilean/Argentine Andes and Andean-type margins. *Geological Society of America*, Abstract with programs, **34**, 6 156, abs. 63-0.

KAY, S. M., MANCILLA, O. & COPELAND, P. 2006. Evolution of the late Miocene Chachahuén volcanic complex at 37°S over a transient shallow subduction zone under the Neuquén Andes. *In*: KAY, S. M. & RAMOS, V. A. (eds) *Evolution of an Andean Margin: A Tectonic and Magmatic View from the Andes to the Neuquén Basin (35°–39°S lat)*. Geological Society of America, Special Paper, **407**, 215–246.

LAZO, D., CICHOWOLSKI, M., RODRÍGUEZ, D. & AGUIRRE-URRETA, M. B. 2005. Lithofacies, palaeocology and palaeoenvironments of the Agrio Formation, Lower Cretaceous of the Neuquén basin, Argentina. *In*: VEIGA, G. D., SPALLETTI, L. A., HOWELL, J. A. & SCHWARZ, E. (eds) *The Neuquén Basin: A Case Study in Sequence Stratigraphy and Basin Dynamics*. The Geological Society, London, Special Publications, **252**, 295–315.

LEANZA, H. A. & HUGO, C. 2001. *Hoja Geológica Zapala, Hoja 3969-I, 1:250 000*. Instituto de Geología y Recursos Minerales, Boletín, **275**, 1–128.

LEGARRETA, L. & GULISANO, C. A. 1989. Análisis estratigráfico secuencial de la cuenca Neuquina (Triásico Superior-Terciario Inferior). *In*: CHEBLI, G. A. & SPALLETTI, L. A. (eds) *Cuencas Sedimentarias Argentinas*. Facultad de Ciencias Naturales, Universidad Nacional de Tucumán, Tucumán, Serie Correlación Geológica, **6**, 221–243.

LEGARRETA, L. & ULIANA, M. A. 1991. Jurassic–Cretaceous marine oscillations and geometry of back-arc basin fill, central Argentine Andes. *In*: MACDONALD, D. I. (ed.) *Sedimentation, Tectonics and Eustasy: Sea Level Changes at Active Plate Margins*. International Association of Sedimentologists, Oxford, Special Publication, **12**, 429–450.

LEGARRETA, L. & ULIANA, M. A. 1996. The Jurassic succession in west-central Argentina; Stratal patterns, sequences and palaeogeographic evolution. *Palaeogeography, Palaeoclimatology, Palaeoecology*, **120**, 303–330.

LLAMBÍAS, E. J., BERTOTTO, G. W., RISSO, C. & HERNANDO, I. 2010. El Volcanismo cuaternario en el retroarco de Payenia: una revisión. *Revista de la Asociación Geológica Argentina*, **67**, 278–300.

NÉMETH, K., RISSO, C., NULLO, F., SMITH, I. E. M. & PÉCSKAY, Z. 2012. Facies architecture of an isolated long-lived, nested polygenetic silicic tuff ring erupted in a braided river system: The Los Loros volcano, Mendoza, Argentina. *Journal of Volcanology and Geothermal Research*, **239–240**, 33–48.

PADULA, L. A. 1948. Sobre la presencia del Hauterivense marino en la Sierra de Chachahuén, Provincia de Mendoza. *Boletín de Informaciones Petroleras*, **287**, 49–53.

PEREZ, M. A. & CONDAT, P. 1996. *Geología de la Sierra de Chachahuén, Área CNQ-23, Puelen: Buenos Aires, Argentina*. Geólogos Asociados, S.A., Report to YPF, 82.

PLOSZKIEWICZ, J. V., ORCHUELA, I. A., VAILLARD, J. C. & VIÑES, R. F. 1984. Compresión y desplazamiento lateral en la zona de Falla Huincul: estructuras asociadas, provincia de Neuquén. *IX Congreso Geológico Argentino*. Buenos Aires, Actas, **2**, 163–169.

RAMOS, V. A. & FOLGUERA, A. 2011. Payenia volcanic province in the Southern Andes: an appraisal of an exceptional Quaternary tectonic setting. *Journal of Volcanology and Geothermal Research*, **201**, 53–64.

RAMOS, V. A. & KAY, S. M. 2006. Overview of the tectonic evolution of the souther Central Andes of Mendoza and Neuquén (35°–39° latitude). *In*: KAY, S. M. & RAMOS, V. A. (eds) *Evolution of an Andean Margin: A Tectonic and Magmatic View from the Andes to the Neuquén Basin (35°–39°S lat)*. Geological Society of America, Special Paper, **407**, 1–17.

RAMOS, V. A., CEGARRA, M. & CRISTALLINI, E. 1996. Cenozoic tectonics of the High Andes of west-central Argentina (30–36°S latitude). *Tectonophysics*, **259**, 185–200.

RAMOS, V. A., CRISTALLINI, E. & PÉREZ, D. J. 2002. The Pampean flat-slab of the Central Andes. *Journal of South American Earth Sciences*, **15**, 59–78.

RAMSAY, J. G. & HUBER, M. I. 1983. *The Techniques of Modern Structural Geology*. Academic Press, London.

RISSO, C., NÉMETH, K., COMBINA, A. M., NULLO, F. & DROSINA, M. 2008. The role of phreatomagmatism in a Plio-Pleistocene high-density scoria cone field: Llancanelo Volcanic Field (Mendoza), Argentina. *Journal of Volcanology and Geothermal Research*, **169**, 61–86.

ROJAS VERA, E., FOLGUERA, A. ET AL. 2010. Neogene to Quaternary extensional reactivation of a fold and thrust belt: the Agrio belt in the southern Central Andes and its relation to the Loncopué Trough (38°–39°S). *Tectonophysics*, **492**, 279–294.

SIGISMOSNDI, M. E. 2013. *Estudio de la deformación litosférica de la cuenca neuquina: Estructura termal, datos de gravedad y sísmica de reflexión*. PhD thesis, Universidad de Buenos Aires.

SILVESTRO, J. & ATENCIO, M. 2009. La cuenca Cenozoica del Río Grande y Palauco: Edad, evolución y control estructural, faja plegada de Malargüe (36°S). *Revista de la Asociación Geológica Argentina*, **65**, 154–169.

SØAGER, N., MARTIN HOLM, O. & LLAMBÍAS, E. J. 2013. Payenia volcanic province, southern Mendoza, Argentina: OIB mantle upwelling in a backarc environment. *Chemical Geology*, **349–350**, 36–53.

SPALLETTI, L. A., POIRÉ, D., PIRRIE, D., MATHEOS, S. & DOYLE, P. 2001. Respuesta sedimentológica a cambios en el nivel de base en una secuencia mixta clástica-carbonática del Cretácico Inferior de la cuenca Neuquina, Argentina. *Revista de la Sociedad Geológica de España*, **14**, 57–74.

STERN, C. R. 2004. Active Andean volcanism: its geologic and tectonic setting. *Revista Geologica de Chile*, **31**, 161–208.

TYSON, R. V., ESHERWOOD, P. & PATTISON, K. A. 2005. Organic facies variations in the Valanginian-mid-Hauterivian interval of the Agrio Formation (Chos Malal area, Neuquén, Argenina): local significance and global content. *In*: VEIGA, G. D., SPALLETTI, L. A., HOWELL, J. A. & SCHWARZ, E. (eds) *The Neuquén Basin: A Case Study in Sequence Stratigraphy and Basin Dynamics*. The Geological Society, London, Special Publications, **252**, 251–266.

VEIGA, G. D., SPALLETTI, L. A. & FLINT, S. 2002. Aeolian/fluvial interactions and high-resolution sequence stratigraphy of a non-marine lowstand wedge: the Avilé Member of the Agrio Formation (Lower Cretaceous), central Neuquén Basin, Argentina. *Sedimentology*, **49**, 1001–1019.

VERGANI, G. D., TANKARD, J., BELOTTI, J. & WELSINK, J. 1995. Tectonic evolution and Paleogeography of the Neuquén Basin, Argentina. *In*: TANKARD, A. J., SUAREZ, R. & WELSINK, H. J. (eds) *Petroleum Basins of South America*. American Association of Petroleum Geologists, Memoir, **62**, 383–402.

VERGANI, G. D., BARRIONUEVO, M., SOSA, H. & PEDRAZZINI, M. 2001. Análisis estratigráfico secuencial de alta resolución en las formaciones Agrio y Huitrín del Yacimiento Puesto Hernández. *BIP (Boletín de infomaciones Petroleras)*, **67**, 76–87.

WAGNER, L. S., BECK, S. & ZANDT, G. 2005. Upper mantle structure in the south central chilean subduction zone (30°–36°S). *Journal of Geophysical Research*, **110**, http://dx.doi.org/10.1029/2004JB003238

WEAVER, C. 1931. Paleontology of the Jurassic and Cretaceous of West Central Argentina. *Memoirs of the University of Washington*, **1**, 1–595.

ZAMORA VALCARCE, G. & ZAPATA, T. B. 2005. Estilo estructural del frente de la faja plegada neuquina a los 37°S. *VI Congreso de Exploración y Desarrollo de Hidrocarburos*, Mar del Plata, **16**, Electronic files.

ZAMORA VALCARCE, G., ZAPATA, T. B., DEL PINO, D. & ANSA, A. 2006. Structural evolution and magmatic characteristics of the Agrio fold-and-thrust belt. *In*: KAY, S. M. & RAMOS, V. A. (eds) *Evolution of an Andean Margin: A Tectonic and Magmatic View from the Andes to the Neuquén Basin (35°–39°S lat)*. Geological Society of America, Boulder, Special Paper, **407**, 125–146.

ZAMORA VALCARCE, G., ZAPATA, T. & RAMOS, V. A. 2011. La Faja Plegada y corrida del Agrio. *In*: LEANZA, H. A., ARREGUI, C., CARBONE, O., DANIELI, J. C. & VALLÉS, J. M. (eds) *Geología y Recursos Naturales de la Provincia del Neuquén*. Asociación Geológica Argentina, 367–374.

ZAPATA, T. R., BRISSÓN, I. & DZELALIJA, F. 1999. La estructura de la faja plegada y corrida Andina en relación con el control del basamento de la cuenca Neuquina. *BIP (Boletín de Informaciones Petroleras)*, **60**, 142–164.

Sedimentation model of piggyback basins: Cenozoic examples of San Juan Precordillera, Argentina

J. SURIANO[1]*, C. O. LIMARINO[1], A. M. TEDESCO[1,2] & M. S. ALONSO[1]

[1]*IGeBA-Departamento de Ciencias Geológicas, Facultad de Ciencias Exactas y Naturales, Universidad de Buenos Aires-CONICET, Intendente Güiraldes 2160 – Ciudad Universitaria-Pab.II-CABA, C1428EGA, Buenos-Aires, Argentina*

[2]*Present Address: SEGEMAR (Servicio Geológico Minero Argentino). Av. General Paz 5445 Edificio 14 y Edificio 25 San Martín (B1650 WAB), Buenos-Aires, Argentina*

*Corresponding author (e-mail: jsuriano@gl.fcen.uba.ar)

Abstract: Piggyback basins are one of the most important sediment storage systems for foredeep basins within foreland basin systems, so understanding the dynamics of sediment accumulation and allocyclic changes is essential. Three alluvial systems are proposed here to depict sediment movement along the piggyback basin: piedmont, axial and transference systems. We propose differentiation between open continental piggyback basins that include a transference system that is able to deliver sediment to the foredeep and closed piggyback basins that are isolated. Two idealized models of sedimentation in piggyback basins are proposed. For open piggyback basins we identify four stages: (a) the incision stage; (b) the confined low accommodation system tract; (c) the high accommodation system tract; and (d) the unconfined low accommodation system tract. Meanwhile two stages are proposed for closed ones: (a) the high accommodation system tract; and (b) the low accommodation system tract. To test these models, Quaternary deposits and a Miocene unit are analysed. The first one is controlled by climatic changes, and the second is related to tectonic activity in the Precordillera.

Most of the research in foreland basin systems (DeCelles & Giles 1996) has been focused on the evolution of the foredeep region and its relation to the orogenic wedge and the forebulge areas (Jordan 1981, 1995; Beaumont et al. 1988; Flemings & Jordan 1989; Mitrovica et al. 1989; Holt & Stern 1994; Johnson & Beaumont 1995; DeCelles & Giles 1996; Limarino et al. 2001; DeCelles & Horton 2003). These models are useful in understanding the sedimentary patterns in the foredeep as well as in establishing the close relationship between tectonism and sedimentary filling.

Even though they comprise a major storage system of sediments that can feed the foredeep, much less attention has been paid to piggyback basins, developed between thrust sheets, particularly the continental ones. Therefore, understanding the sedimentary dynamics of piggyback areas is important because: (a) piggyback basins are a significant reservoir of sediments exported periodically to the foreland (DeCelles & Giles 1996; Horton & DeCelles 2001; Suriano & Limarino 2006); (b) the volume of exported sediments from piggyback basins may even exceed the volume supplied by the orogenic front to the foredeep; (c) the stratigraphic record gives direct information about the uplifting of the different thrusts sheets and the advance of the clastic wedge; and (d) part of the foreland basin record that is eroded as the orogenic front advances can be preserved in the piggyback basins.

In this contribution the piggyback basins located along the Precordillera in the Argentinian Andes (San Juan Province, 30–31°S, Fig. 1) are analysed. This study intends to characterize the sedimentation patterns and dynamics of piggyback basins as well as to propose a new classification and to establish a sequence stratigraphic model. In order to accomplish these goals, the work is divided in two sections. First, a stratigraphic model based on the changes in the accommodation space is built, taking into account the analysis of piggyback basins, their definition, main sedimentological features, sedimentary environments, stacking patterns and relationship with allocyclic controls. In the second part, the proposed model is tested, analysing two stratigraphic successions (Quaternary deposits and Miocene Cuesta del Viento Formation) of the Jáchal River area, controlled by climatic and tectonic changes.

Piggyback basin concept

The piggyback basins were defined by Ori & Friend (1984) as 'basins that are formed and filled up while

From: SEPÚLVEDA, S. A., GIAMBIAGI, L. B., MOREIRAS, S. M., PINTO, L., TUNIK, M., HOKE, G. D. & FARÍAS, M. (eds) 2015. *Geodynamic Processes in the Andes of Central Chile and Argentina.* Geological Society, London, Special Publications, **399**, 221–244. First published online February 28, 2014, http://dx.doi.org/10.1144/SP399.17
© 2015 The Geological Society of London. For permissions: http://www.geolsoc.org.uk/permissions.
Publishing disclaimer: www.geolsoc.org.uk/pub_ethics

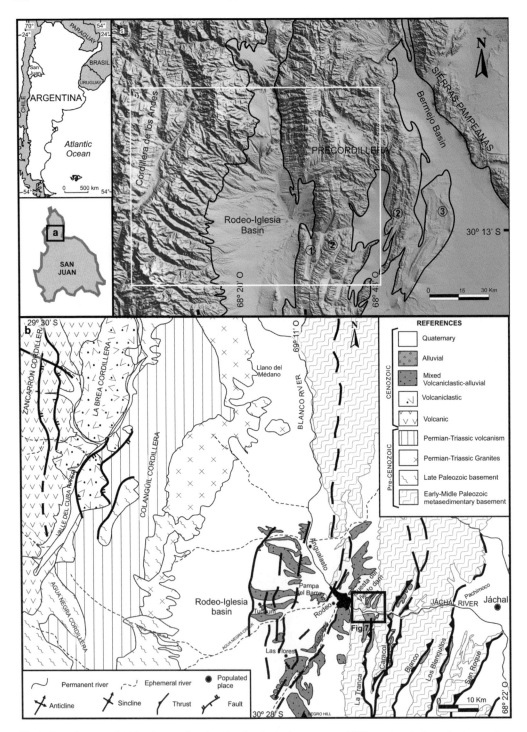

Fig. 1. Location maps. (**a**) Morphotectonic units related to the Andean orogen at 30°S; numbers indicate the three major sectors of the Precordillera (1, western; 2, central and 3, eastern), the white frame shows the location of (b).
(**b**) Geological map, where two main domains can be recognized: mainly granitic and volcanic (Cordillera) in the west and a metasedimentary one in the east (Precordillera); the frame shows the location of Figure 7.

being carried on top of thrust sheets'. This term is similar to 'satellite basins' proposed by Ricci Lucchi (1986). Thrust-top basins or parched basins (Turner 1990; Butler & Grasso 1993; Butler *et al.* 1995; Bonini *et al.* 1999; Doglioni *et al.* 1999; Hesse *et al.* 2010) have a related meaning, although they indicate a more specific arrangement associated with basins that are formed and filled up while a portion of the foreland basin is dismantled. According to the above, piggyback and satellite basins are synonymous, and they are related to thrust-top and parched basins. Years later, piggyback basins were included by DeCelles & Giles (1996) in their wedge-top depozone within the foreland system. Gugliotta (2012) divides the wedge-top depozone in inner and outer sequences. In this paper, we follow this terminology, using the term inner-wedge-top depozone for the piggyback area, which designates basins separated from the foreland by thrust sheets or anticlines, while the outer-wedge-top depozone is where blind thrust structures have no superficial expression. This last zone transitionally passes into the foredeep.

Sedimentary dynamics of piggyback basins

To understand the auto and allocyclic factors that could affect sedimentation in piggyback basins, it is essential to inspect the sediment movement pattern all along the basin. Based on the studied examples, the pattern of sediment circulation in continental piggyback basins is here related to three alluvial systems (Fig. 2). The first one corresponds to the piedmont systems, which convey poorly sorted breccias and conglomerates directly from the mountain front to the floor of the basin. This environment includes taluses, different kinds of colluvial fans (Bilkra & Nemec 1998), alluvial fans and alluvial slopes. In the case studied here, four piedmont associations related to accommodation space were recognized (Fig. 3; Suriano & Limarino 2009). As the piedmont area has permanent availability of material from the mountain front, the movement of sediment is regulated by axial systems and, if present, by the transference system.

Sediment transport along the elongate axis of the piggyback basins is represented by longitudinal, frequently braided channels (the collector river–conoid systems of Suriano & Limarino 2009). As shown by Suriano & Limarino (2009), these systems are defined as conoids and the associated collector river as axial distributary systems.

Finally, the transference systems (Fig. 2) connect piggyback basins among them and to the foreland basin. These transference systems can be represented by different kinds of rivers or even by open lacustrine arrangements.

Regarding this occurrence, we propose that continental piggyback basins (Type 2 of Wagerich 2001) can be classified into two main types, open and closed (Fig. 4). The open piggyback basins are those including a transference system that connects them and is able to export sediment to the foreland. Closed piggyback basins do not have a transference system, which makes them endorreic. A piggyback basin can shift from open to closed and vice versa owing to allocyclic factors, not only tectonics, directly related to the evolution of the fold and thrust belt, but also climatic

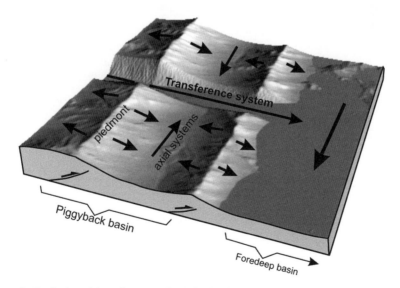

Fig. 2. Schematic distribution of the sedimentary circulation in piggyback basins.

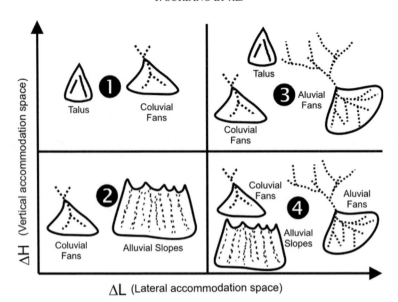

Fig. 3. Arid piedmont associations for the Precordillera (modified from Suriano & Limarino 2009). Highest energy corresponds to association 1, composed of taluses and colluvial fans with steeper slopes; if there is enough lateral space this association could evolve into association 2. Association 3 is related to gentle slope mountain fronts as much as association 4, which develops in wider valleys and evolved drainages.

modifications that affect the erosion capability of the transference system.

Open piggyback basins can change from highly efficient to inefficient (or vice versa) according to the efficacy of the major sediment exporter system formed by the fluvial systems inside the piggyback basin and transference system. Highly efficient transference systems are those able to export all the sediment delivered by piggyback basins to the foredeep (Fig. 4). In contrast, a transference system is inefficient if it is not able to transport the material from the piggyback to the foredeep

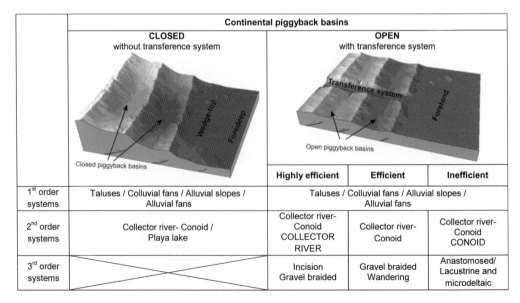

Fig. 4. Piggyback basin classification proposed in this work.

(Fig. 4), which leads to the storage of sediment inside the piggyback basin.

Before building a sedimentary model for piggyback basins it is important to examine the factors that control their patterns of sedimentation. They can be local if they affect each basin individually or regional if they have influence over the entire orogenic wedge. Another equally significant aspect to take into account is if those factors affect the accommodation space (A) and/or the sediment supply (S). These factors and their effects are shown in Figure 5.

A local factor that increases the relationship A/S is the upstream thrust movement (Fig. 5a). This factor produces an increase in the accommodation space in the footwall as well a higher sediment supply from the hanging wall. Another way of increasing accommodation space is a local base level rise (Fig. 5b, c). This could be generated by a sediment dam within the basin (Fig. 5b) or by activity of the downstream thrust (Fig. 5c). Efficacy of this last factor is limited as major movements produce narrowing and shallowing of the basins, which decreases the accommodation space (Fig. 5d; Talling et al. 1995). Lower local A/S ratio could also be generated by a downstream river capture (Fig. 5e).

There are also regional factors affecting the A/S relationship. Increase of the A/S ratio owing to subsidence is related to the structural arrangement and dynamics of the orogenic wedge, with regard to tectonic loading and to slab pull (Fig. 5f). Decrease in the accommodation space is instead associated with periods of tectonic quietness and isostatic rebound (Fig. 5h). Fluvial equilibrium profiles, independently of their driving force, could rise (Fig. 5g) or fall (Fig. 5i), causing an increase or decrease in the accommodation space. Finally, dramatic changes in the amount of sediment production, owing to climatic factors or the

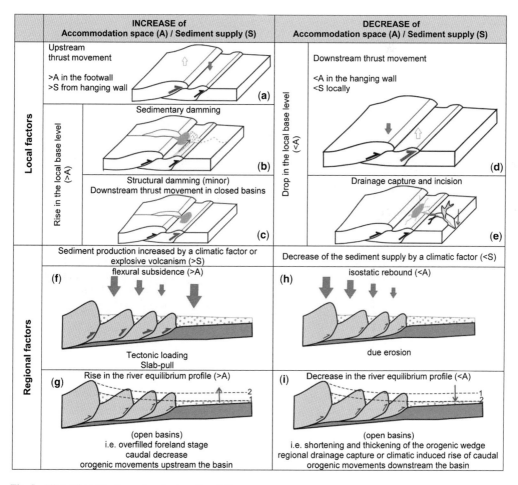

Fig. 5. Allocyclic controls on piggyback sedimentation.

occurrence of explosive volcanism, should be taken into account.

The piggyback sedimentary models presented here are based on the principles of sequence stratigraphy in continental environments (e.g. Shanley & McCabe 1994; Quirk 1996; Dahle et al. 1997; Dalrymple et al. 1998; Blum & Törnqvist 2000; Miall 2002). According to Dahle et al. (1997), two major stages can be recognized in continental basins disconnected of the sea-level: (a) the high accommodation system tract; and (b) the low accommodation system tract. This proposal detaches system tracts from sea-level and relates them to changes in accommodation space (Catuneanu 2006). Following Shanley & McCabe (1994) and Dalrymple et al. (1998), discontinuities generated by fluvial incision are used here as sequence boundaries. Two kinds of erosive bounding surfaces were recognized according to their degree of confinement (confined or unconfined). Finally the existence of sedimentological evidence of the presence of a fluvial transference system can be used to differentiate open from closed piggyback basins, a concept that is also incorporated in the proposed model.

Model for piggyback basins

Using the elements mentioned above we constructed a sedimentary model for continental piggyback basins that is based on the temporal variations of accommodation space and sediment supply. These changes are based in four supporting elements: (a) basin subsidence; (b) tectonic activity in the fold and thrust belt; and (c) changes in sediment amount by climatic changes and/or major explosive volcanism.

Subsidence in piggyback basins has been considered short-lived and of much smaller magnitude than that at the associated foredeep (Xie & Heller 2009). In fact, subsidence owing to tectonic loading, a major mechanism in foreland basins (Jordan 1981; Naylor & Sinclair 2008), has lower values in piggyback areas (Zoetemeijer et al. 1993; Xie & Heller 2009). The key factor in the creation of local accommodation space in most piggyback basins is the tectonic activity in the upstream thrust sheet (tectonically induced accommodation, Fig. 5a), which in turn produces migration of the depocentre (Bonini et al. 1999; Zoetemeijer et al. 1993). In this sense, migration of the depocentres in piggyback basins has been associated with passive rotation of the internal thrusts owing to the successive external thrust activation (Bonini et al. 1999; Roure 2008) or active thrusting. In this last case, migration of the depocentre towards the foredeep has been interpreted as in-phase thrusting

sequences while hinterland-directed depocentre migration seems to be related to out-of-phase thrusting (Zoetemeijer et al. 1993). In the same way Bonini et al. (1999) showed that the depocentre migration is controlled by the thrust having the highest displacement rate.

Sequence deposition is essentially ruled by the above factors, which affect the A/S relationship. On this basis we propose a sequence stratigraphy model for open piggyback basins involving: (a) the incision stage; (b) the confined low accommodation system tract; (c) the high accommodation system tract; and (d) the unconfined low accommodation system tract or overfill stage (Fig. 6).

The first stage or incision stage begins with an important drop in the equilibrium profile (a in Fig. 6) that produces incision in the transference system and generates the entrenchment of the piedmont system and the development of an incision surface across the basin (incision stage, a in Fig. 6). As Shanley & McCabe (1994) pointed out, even in high-incision events some remnant deposits can be found in the sides of the valley owing to short moments of stabilization during this mainly erosive phase.

The confined low accommodation system tract can be separated into two substages. The first one takes place following the incision stage when there is a time of relative instability during which the river profile is close to equilibrium, generating the infilling of incised transference system deposits (b in Fig. 6). In this stage deposition of a coarsening-up incised channel complex can be observed (Catuneanu 2006). The amount of sediment involved during this stage relates to the time span of this unstable situation and therefore it is not always recognized. The next stage occurs when the river profile reaches or is slightly above the equilibrium profile. At that point slow aggradation takes place in the previously eroded piggyback area (c in Fig. 6). This stage is denoted by the presence of amalgamated high-energy braided channels, generated by the transference system and also by limited aggradation in the piggyback area.

If the rise of the equilibrium profile continues, the transference system produces high aggradation in its alluvial plain, promoting high equilibrium profiles in the axial and piedmont systems and the consequent filling of the piggyback basins (high accommodation system tract, d in Fig. 6). As the sediment supply from the mountain front is more or less continuous, if no new accommodation space is created, the basin tends to fill up, reaching the unconfined low accommodation system tract (or overfill stage, e in Fig. 6).

For closed piggyback basins, only two stages are proposed. Initially, when the basin is created, the A/S ratio is high, so there is also high

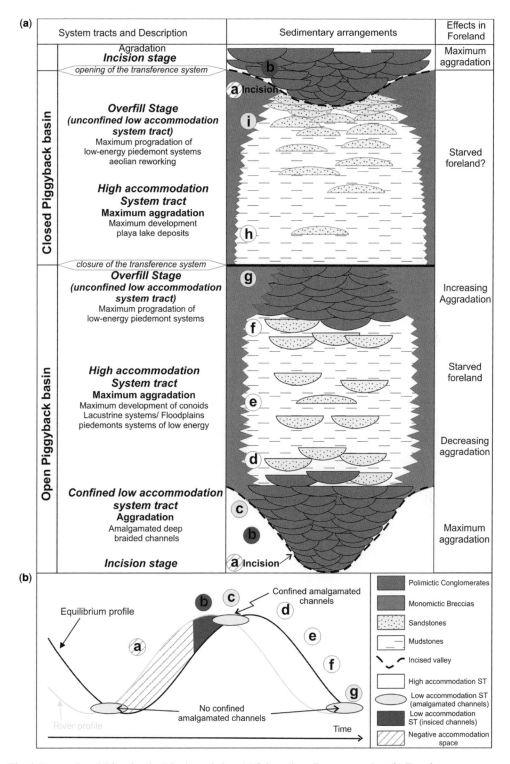

Fig. 6. Proposed model for piggyback basin evolution. (a) Schematic sedimentary section. (b) Transference system profile and its equilibrium profile through time. Numbers relate to time.

accommodation space. This stage is dominated by fine-grained playa lake deposits (f in Fig. 6) with an external ring of coarse piedmont facies. Again, if there is no newly created accommodation space, the basin tends to fill up, and the overfill stage is developed. This stage is represented by the progradation of low-energy piedmont associations (2 in Fig. 3) like alluvial slopes and fluid flow-dominated colluvial fans and a marked decrease in the area previously occupied by playa lake deposits. The late stage of this system tract is characterized by a decrease in sediment supply associated with an extensive aeolian reworking and development of a desert pavement.

Stratigraphic context

Between 30° and 31°S, the Andean orogen is made up of five morphotectonic units (Fig. 1a): (a) western and higher Cordillera de los Andes; (b) the wide intermontane Rodeo–Iglesia Basin; (c) the fold and fault thrust belt of Precordillera; (d) the Bermejo Basin; and (e) the Pampean Ranges. In this area the Cordillera Frontal (part of (a)) is composed of a Carboniferous metasedimentary rocks (Cerro de Agua Negra Formation), Triassic batholiths (Llambías & Sato 1990) and Triassic to Miocene volcanic rocks (Ramos 1999). To the east, the wide Rodeo-Iglesia Basin is developed, where Cenozoic volcaniclastic rocks are exposed.

The Precordillera belt is divided into three structural provinces (Fig. 1a): western, central and eastern (Ortiz & Zambrano 1981). The studied area is located within the western Precordillera (Fig. 1b), which has been built since the Early Miocene (Jordan et al. 1993, 2001; Cardozo & Jordan 2001; Alonso et al. 2011) by eastward migration of the orogenic front. The thin-skinned fold and thrust belt of the western Precordillera expose Early Palaeozoic rock sheets that contain the Miocene to Holocene piggyback sedimentation. In particular, we studied the Quaternary piggyback basins along the Jáchal River (Figs 1b & 7) and the Cuesta del Viento Formation, a Miocene unit deposited within the same stratigraphic context (Fig. 7). The western Precordillera is characterized by its west vergence of thick-skinned structures (Zapata & Allmendinger 1996). Finally, Bermejo Basin separates the Precordillera from the basement uplifted blocks of Pampean Ranges.

Dynamics of the Quaternary piggyback basins of Jáchal River area

In order to apply the models described to the Jáchal River area, the Quaternary filling of the piggyback basins was analysed in detail. Taking into account the presence of incision surfaces, resulting in decametric- to metric-scale relief, the sedimentary infill of the four Jáchal River piggyback basins (La Tranca, Caracol, Zanja Honda and Pachimoco) was divided into four depositional sequences (Fig. 8). The Pleistocene–Holocene sequences studied here do not show evidence of significant tectonic disturbance. This suggests that tectonic uplift of thrust sheets was not a main mechanism to produce changes in the accommodation space during Pleistocene–Holocene times. Moreover, the lack of evidence of volcanic activity (Kay et al. 1988) sets climate change as a major allocyclic control for sedimentation in these basins. In addition, the sequence's age fits with regional climate changes (Iriondo 1999).

Depositional sequences

Sequence I: pediment. The base of this unit is formed by a low-relief erosive pedimentation surface on Miocene and Palaeozoic deformed rocks. Over this unconformity a relatively thin layer (up to 10 m thick) of Quaternary deposits is found (Fig. 8). These deposits are dominated by monomictic breccias of piedmont accumulations with minor participation of lenses of polimictic conglomerates.

The monomictic breccias are fine-grained and occur mainly as clast-supported tabular, massive or horizontally stratified beds. Lenticular bodies of massive mud-supported breccias interbedded with clast-supported breccias are scarce. The latter show planar cross-bedding or clast imbrications, and it is worth mentioning that their scarcity does not allow a statistical measurement of palaeocurrents. Clasts are pebbles, mainly slates (more than 95%) and basic volcanic rocks derived from the Yerba Loca Formation, which indicates local provenance as it is the sole unit outcropping in the Precordillera in the area. Deposits are interpreted as having formed in alluvial slopes (Smith 2000) with minor participation of colluvial fans (Bilkra & Nemec 1998), so they can be interpreted as a low-energy piedmont association (2 in Fig. 3).

The infrequent polimictic conglomerates are coarse, well-rounded and clast-supported. They show imbrication, horizontal and planar cross-bedding. They appear as lenticular to lentiform beds and are interpreted as gravel bars in a multi-channelized fluvial system. Cobbles are dominated by granites and acid volcanics fragments. These lithologies are widely present at the Colangüil Cordillera located to the west (Fig. 1b), indicating an external provenance. An ancient transference system (ancient Jáchal River) draining the Precordillera from west (Cordillera de los Andes) to east

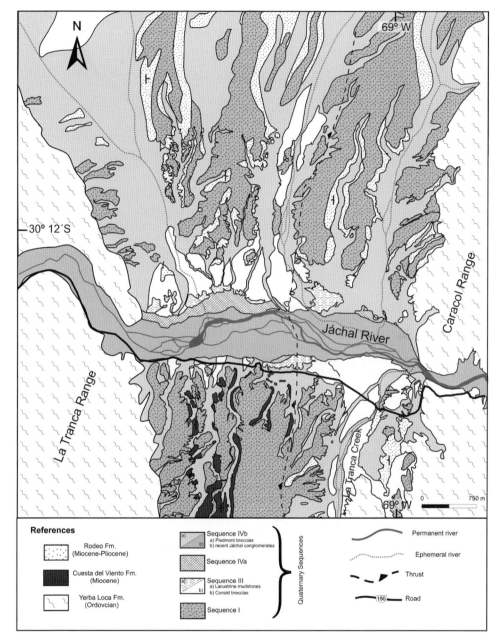

Fig. 7. Detailed geological map of La Tranca Valley. See Figure 1b for general location.

(present day foredeep) can be interpreted from the clast composition of this conglomerate unit.

Sequence II: Quaternary incised valley. The lower boundary of Sequence II is a high relief surface carved on Quaternary deposits of Sequence I, Miocene or Palaeozoic rocks (Figs 8 & 9a). Deposits are few and scattered, with relatively small thicknesses (up to 25 m). This unit is dominated by polimitic conglomerates with scarce intercalations of gravelly sandstones.

The polimictic conglomerates are coarse to very coarse and clast-supported, with erosive base, and occur as massive or planar cross-bedded beds.

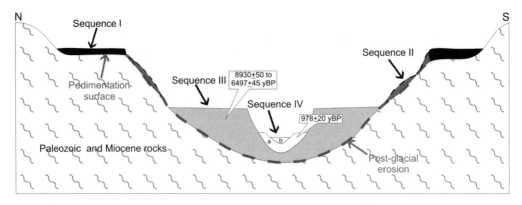

Fig. 8. Schematic distribution of the Quaternary sequences and ages in the Jáchal Valley.

The clasts (sized up to boulders) are rounded to well-rounded, compositions are granite and acid-volcanic (reflecting external provenance), with little participation of greenish grey and yellow sandstones, quartz and greenish slates. The gravelly sandstones have planar or trough cross-bedding. The described facies are interpreted as ancient Jáchal River deposits, as they are similar to the present Jáchal River, sharing the provenance, lithology and sedimentary structures. Occasionally, breccias associated with taluses, coluvial fans and collector rivers (piedmont and axial systems) are interbedded within the fluvial conglomerate.

We correlate this sequence with an important erosive event responsible for the generation of the lowest Quaternary valley floor, which can be traced up to Llano del Médano, an upstream site (Fig. 1b), where the incision is evidenced by several hundred metres of relief between the oldest Quaternary deposits and the valley floor (Suriano & Limarino 2006). The eroded material was probably exported from the piggyback area and formed two megafans (Damanti 1993; Horton & DeCelles 2001) that can currently be seen at the mouth of Huaco and Jáchal Rivers in the Bermejo Valley (Fig. 10).

Sequence III: sedimentary dams. The lower boundary of this sequence largely fits the incision marking the base of Sequence II (Fig. 8). Above this surface, there is one of the thickest and best preserved Quaternary units in the area, characterized by the presence of conspicuous intermontane shallow lacustrine sediments (Colombo et al. 2000, 2005, 2009; Suriano & Limarino 2005, 2008, 2009). Deposits of these natural dams can be found in Rodeo, La Tranca, Caracol and Zanja Honda basins, all of them with similar characteristics. Three main facies associations could be recognized in the lacustrine system: (a) siltstones and fine sandstones; (b) monomictic roughly stratified breccias; and (c) polimictic conglomerates.

The fine-grained association is dominant and is formed mainly by muddy and sandy lithofacies (1 in Fig. 11a, c) with scarce breccias. Tabular beds of whitish grey mudstones with horizontal lamination or massive are abundant. Mudstones commonly show root marks and mudcracks and there are also some levels with shells of freshwater gastropods (Colombo et al. 2005) and others with plant remains. Brown shales with high contents of organic matter are present, though scarce. Gypsum levels are present towards the top. Light brown sandstones are massive or with ripple lamination, and some planar cross-bedded structures are also found.

Fine-grained lithofacies are interpreted as produced by both decantation and underflows, which according to May et al. (1999) can take place in shallow lacustrine environments. The ephemeral nature of the lake system is inferred by the root marks, mudcracks and evaporite deposits at the top of the association (May et al. 1999; Colombo et al. 2009). Massive sandstones represent rapid deposition from sheet floods. Horizontally laminated sandy lithofacies record the migration of subaqueous micro- and mesoforms and represent the coastal zone of the lake when dominant.

The palynological assemblage of these lacustrine deposits has been studied by Colombo et al. (2009), who proposed periods of heavy rain, alternating with episodes of extreme aridity in a warm climate for the base of the section, turning arid towards the top. Colombo et al. (2005) dated gastropods shells and organic matter by ^{14}C technique obtaining ages from 8930 ± 50 (near the base of the lacustrine deposits) to 6497 ± 45 years BP (in the upper part). These ages indicate that the lacustrine system persisted for at least 2700 years.

Fig. 9. (**a**) Quaternary valley carved on Miocene aeolian units. (**b**) Deposits of Sequence IVa in the Pachimoco area; the white arrow shows the abundant bioturbation in the floodplain deposits; the red line shows the scale. (**c**) Massive matrix-supported chaotic breccias (Bmm) of facies association 1, Cuesta del Viento Formation. (**d**) Mudstones with polimictic conglomerates in coarsening-up arrangements of facies association 4, Cuesta del Viento Formation; the red bar is 1.5 m long. (**e**) Mudstones with monomictic breccias in coarsening-up arrangements belonging to facies association 5, Cuesta del Viento Formation. (**f**) Close view of the coarsening-up cycles of facies association 5; the red bar is 1.5 m long.

This fine-grained association alternates with breccia levels. Roughly stratified matrix- and clast-supported monomictic breccias appear as tabular massive beds. Some of them also show imbricated clasts, normal gradation or planar cross-bedded stratification. Massive and laminated mudstones occur to the top of some bodies.

Breccias are clearly dominant downstream (2 in Fig. 11c), and are interpreted as conoids (axial distributaries systems). As they always occur downstream from the lacustrine deposits, these breccias are interpreted as resulting from the dam closure (Fig. 11d, Colombo *et al.* 2000, 2009 Suriano & Limarino 2005).

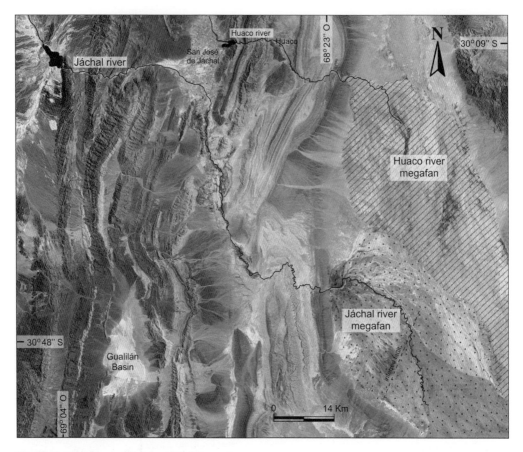

Fig. 10. Satellite image of Huaco and Jáchal megafans.

While downstream the muddy lacustrine association pinches out into conoid facies, upstream intercalations of polimictic conglomerates become frequent (at left in Fig. 11c), forming the last facies association. The beds of polimictic conglomerate form biconvex beds from 2 to 7 m in thickness. They show horizontal, planar cross-bedded or massive structures. Between the conglomeratic beds, sandy facies are common, forming beds up to 1 m with a coarsening upward arrangement (3 in Fig. 11b).

The interpretation of these deposits and their lateral variations is shown in Figure 12. Fine-grained sediments are found at the centre zone of the lake (Fig. 11a). In the upstream area, the transference system (ancient-Jáchal River) flowing into the lake generated a microdeltaic arrangement (Fig. 11b). Figure 11c shows the natural dam deposits represented by the light grey fine lacustrine sediments that laterally pinch out to the downstream conoid breccias (dark grey) responsible for the dam closure. The above-described arrangement is related to a high equilibrium profile in the transference system owing to the damming that also caused a local level rise in the whole piggyback basin, so that aggradation was favoured. This led to the deposition of thick units in conoids and collector rivers (axial systems) as well as in the piedmont areas.

Sequence IV Jáchal River. This stage is developed over an important incision surface that corresponds to the present Jáchal River Valley (Fig. 8) and includes two subsequences. The first subsequence (IVa) involves the highest terraced levels dated by Colombo *et al.* (2005) in La Tranca area between 978 ± 20 and 525 ± 32 years BP. This unit is composed of two types of deposits, the first characterized by clast-supported conglomerates and sandstones with lenticular to lentiform geometry, which represents channel facies, and the second corresponding to tabular mudstones floodplain beds.

The conglomerates of the channel facies have clasts of granite and acid volcanic rocks, indicating a foreign provenance similar to those of the Stage II,

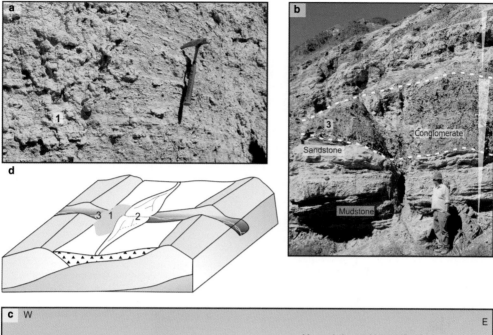

Fig. 11. Scheme and pictures from the Quaternary lacustrine arrangement. (**a**) Dam–lacustrine fine-grained association. (**b**) Coarsening-up cycles from a micodeltaic system. (**c**) Closure of a dam represented by conoid facies, lateraly related to fine dam–lacustrine facies. (**d**) Scheme of the dam facies; numbers in the scheme correspond to those shown in the photos.

but clasts have a smaller mean diameter (minor to 30 cm). There are two types of channel arrangements according to their hierarchy; the first one is dominated by polimictic clast-supported conglomerates. Some of the beds begin with massive or imbricate conglomerates, but overlying clast-supported conglomerates are a conspicuous lithofacies, with planar cross-bedding and horizontal stratification. On top of beds, gravelly sandstones and sandstones with planar cross-bedded and horizontal lamination are found. This arrangement corresponds to migrating transversal and longitudinal gravelly bars (Hein & Walker 1977). Considering the geometry of the beds, they are interpreted as migrating in a conglomeratic channel of low to medium sinuosity.

The second type of channel facies (Fig. 9b) comprises gravelly sandstone or sandstone lentiform to lenticular beds. Massive or horizontally laminated fine-grained conglomerates or gravelly sandstones are found at the base. Above them, sandstones with planar and trough cross-bedded stratification occur. This association is interpreted as having been deposited by sinuous channels in which longitudinal gravel bars and 2D and 3D sand dunes migrated.

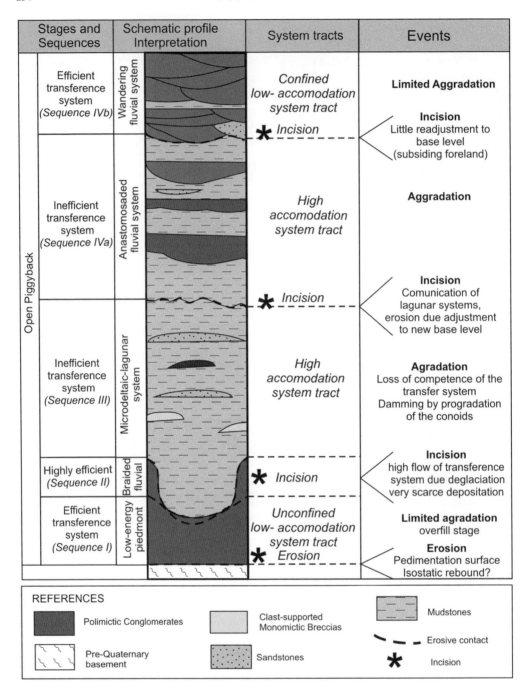

Fig. 12. Schematic section of the Quaternary sequences, with system tract interpretation and geological events that caused them (right column).

Floodplain beds are the dominant facies of the Subsequence IVa. They are formed by mudstones and subordinated sandstones. Mudstones show horizontal lamination, ripple cross lamination or massive structures. They are greenish or light brown and have intense root bioturbation. Sporadic beds of planar cross-bedded, horizontally laminated or massive sandstones are intercalated within mudstones. The fine-grained lithofacies were formed by decantation or low-energy fluid flows in the waning of floods. These floodplains (Fig. 9b) show different degrees of maturity (according to Reinfelds & Nanson 1993), which range from incipient to well evolved, this last one with important development of palaeosols. During the flood events, the sandstones reach these areas with the migration of megaripples, plane bed or sudden deposition by competence loss.

In summary, this association is characterized by different hierarchy channels with low to intermediate sinuosity and extensive development of confined floodplains (Nadon 1994) or cohesive low-energy floodplains (Nanson & Croke 1992). Therefore we interpret the subsequence IVa as an anastomosed fluvial system (Smith & Smith 1980; Nadon 1994), which represents the Early Holocene dynamics of the Jáchal River.

Subsequence IVa is best developed in Pachimoco area (Fig. 2), where a wider piggyback basin allowed for the deposition of large amounts of fine-grained deposits (Pachimoco Formation after Furque 1979) of ancient Jáchal River. Subsequence IVb includes the recent and present deposits of the Jáchal River, younger than 525 ± 32 years BP. Its lower limit is marked by a low relief surface (up to 5 m high) carved into the Subsequence IVa deposits (Fig. 8). Above this surface, the youngest terraces and active channels with scarce floodplains of the Jáchal River are found. In addition to the incision surface between them, the difference among Subsequences IVa and b is the minor participation of fine-grained floodplain deposits in Subsequence IVb that could be described as a bed load-dominated system intermediate between braided and wandering (Miall 1996).

Accommodation stages in the Quaternary Precordillera's piggyback basins

In this section, we present the evolution of the Quaternary sequences in the context of temporal changes in relative accommodation space and sediment supply. The scenario before the deposition of Sequence I is a regional-scale erosive surface (pediment) carved into Palaeozoic to Miocene rocks (Figs 9a, 12 & 13). This low-relief surface indicates a long erosive period. Its origin cannot be stated unequivocally but may be associated with a period of relative tectonic quiescence of the fold and fault thrust belt. Tectonic quiescence could have been related to an isostatic rebound phase owing to erosion, during which a slow but continuous fall in the equilibrium profile (Fig. 13) led to an important erosive event throughout the entire orogenic wedge, generating a pedimentation surface all over the piggyback area. That situation was followed by a small rise in the equilibrium profile that produced a relatively thin layer of breccias and conglomerates of Sequence I deposited in

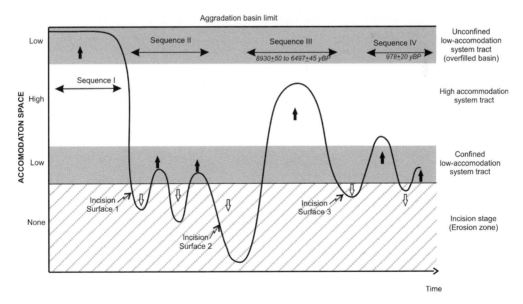

Fig. 13. Changes in the accommodation space through time, and their relationship with different system tracts (at right) and with the Quaternary sequences. The black arrows show an increase in accommodation space and the white ones a decrease.

an unconfined low accommodation system tract (overfill stage; Fig. 12).

Then, long-term degradation in the piggyback area, which generated the present valley of the Jáchal River, took place. This phase had probably begun in late Pleistocene and involved an intensive erosion period, with a low degree of discontinuous stages of deposition in isolated patches which were gathered within Sequence II (incision stage; Figs 12–14a).

The origin of Sequence II is related to deglaciation times that took place after the late glacial maximum (isotopic stage 2; Iriondo & Kröling 1996; Iriondo 1999), when the Cordillera de los Andes was glaciated (above 3500 m). During the deglaciation (between 15 500 and 16 500 years BP; Iriondo & Kröling 1996) melted ice generated high discharge in rivers that drained the Andes, including the transference systems across the Precordillera (i.e. Huaco and Jáchal Rivers), all of them with catchment areas including large portions of glaciated regions. The climate amelioration caused a higher flow of the transference fluvial systems that increased its efficiency and produced the fall of the equilibrium profile (Fig. 13). As a result of this, huge volumes of sediments eroded, promoting the deposition of large megafans (DeCelles & Cavazza 1999; Horton & DeCelles 2001) in the foredeep. According to Damanti (1993), the megafans of Huaco and Jáchal reach areas of 700 and 1400 km^2 with catchment areas of 7100 and 27 700 km^2, respectively. Owing to the huge amount of sediment delivered to the foredeep, Bermejo River had to divert its course next to the Pampean range.

Sequence III deposits represent the end of the deglaciation phase little before 8930 ± 50 to beyond 6.497 ± 45 years BP (Colombo et al. 2005), when a dramatic fall in the flow rates of the transference rivers occurred. This situation favoured high rates of aggradation (up to 150 m thickness) and the occurrence of the high accommodation system tract (Figs 12 & 13) in the piggyback basins. Note that these ages, according to our interpretation, are related to climatic amelioration proposed by Iriondo & Kröling (1996). During this period, the loss of efficiency of the transference system together with an increase in sediment supply from piggyback basins could have produced the advance of these axial systems (conoid-collector

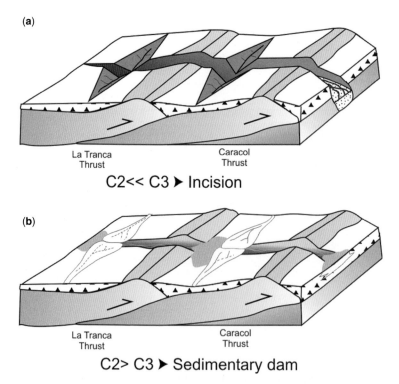

Fig. 14. Block diagrams of the Jáchal River-related basins. C2 = sediment transport efficiency of the longitudinal fluvial system; C3 = sediment transport efficiency of the transference system. (**a**) Sequence II deposition arrangement scheme (mainly erosive phase, few related isolated deposits). (**b**) Sequence III deposition scheme.

rivers) over the transference system valley. In addition, it would have led to the formation of the numerous sedimentary dams (La Tranca, Caracol and Zanja Honda), recorded at the margins of the present Jáchal River Valley (Figs 11c & 14b). The increase of sediment supply can be related to the humid period associated with the deglaciation proposed by García et al. (1999) for the southern Precordillera. It is also consistent with the pollen record of the base of the lacustrine sediments in the area studied by Colombo et al. (2009). Although thick monomictic breccias, from piedmont and axial systems, were deposited in the piggyback area, it is important to point out that, despite the local high sediment supply and the low efficiency of the transference system, the overfill stage was never reached. This was probably due to the large accommodation space created during the previous stage (deglaciation; Figs 12 & 13).

A new incision surface, which contains terraces and the present day alluvial belt of the Jáchal River, was carved into the lacustrine deposits of the Sequence III, which form the lower boundary of Sequence VIa. The origin of the change in the accommodation space between sequences III and IVa is not clear, but it may have resulted from the adjustment of the transference system to its regional base-level fall (the Bermejo River). This was associated with the end of lacustrine system sedimentation (the local-base level of the piggyback basins in Sequence III), which could be related to the breaking of dams by connection with the foredeep (regional-base level) and/or the readjustment to the new base level (probably falling by a continued subsidence in the foredeep). For this reason the Jáchal River produced an incision in its valley in order to catch up with its new base level (degradation stage in the base of Sequence IV; Figs 8, 12 & 13). This period of degradation, with very limited or no aggradation, also reached the piggyback area. From 978 ± 20 to 525 ± 32 years BP (Colombo et al. 2005) the Jáchal River had a short aggradation period recorded by Sequence IVa, corresponding to a high accommodation system tract (Fig. 12). A minor incision surface (Figs 8 & 12) separates Sequence IVa from the present-day fluvial deposits of the Jáchal River, only interrupted by short episodes of aggradation limited to the transference system that form Sequence IVb.

Ancient record of piggyback basin sedimentation in La Tranca

The proposed model for piggyback sedimentation can also be applied to ancient settings. As an example, we analysed the Cuesta del Viento Formation, a Miocene unit, described and proposed by Suriano et al. (2011), located within the same geographic context as the previous Quaternary basins. These are the oldest synorogenic strata in the area and record the transition from outer-wedge-top to inner-wedge-top (piggyback basin) sedimentation. The upper section was dated as Lower Miocene (19.5 ± 1.1 and 19.1 ± 1.3 Ma) by Jordan et al. (1993).

Sedimentary environments

This unit was studied in detail by Suriano et al. (2011), who have defined six facies association that are summarized here.

Facies association 1 (chaotic breccias). This facies association comprises monomitic breccias, with metamorphic clasts from the Yerba Loca Formation. They are dominated by coarse-grained chaotic matrix-supported massive breccias (Fig. 9c) in beds up to 60 cm thick. Massive and imbricated clast-supported breccias in lenticular amalgamated beds also appear. This is interpreted as a high energy piedmont arrangement composed of hyperconcentrated flow-dominated colluvial fans, with some participation of taluses (1 in Fig. 3).

Facies association 2 (breccias). This facies association is composed of breccias with the same provenance as facies association 1, sandstones and mudstones (Fig. 15). It is dominated by clast-supported massive and imbricated breccias. Matrix-supported breccias appear as tabular bodies with massive or strong parallel fabric deposited by hyperconcentrated flows (Bilkra & Nemec 1998). Sandstones and gravelly sandstones locally appear with massive, planar cross-bedded or horizontal lamination. These lithofacies are more common to the top of the section (Fig. 15). Massive and laminated mudstones also appear as milimetric mud drapes or as continuous centimetre-thick levels.

This facies association is interpreted as braided river deposits, in particular a transitional system between gravel-bed braided with sediment-gravity-flow and shallow gravel-bed braided types of Miall (1996). The full interpretation of this facies association depends upon its geomorphic context (Suriano et al. 2011). At both the bottom and the top of the section (Fig. 15) it appears as braided channels developed in piedmonts, colluvial fans and alluvial slopes, associated with proximal facies. In contrast, in the middle part of the unit an alternation of transference and dam facies occurs, allowing their interpretation as collector river deposits (axial systems). These interpretations are supported by the scarce clast imbrications observed (Fig. 15), which are not abundant enough to make a statistical analysis.

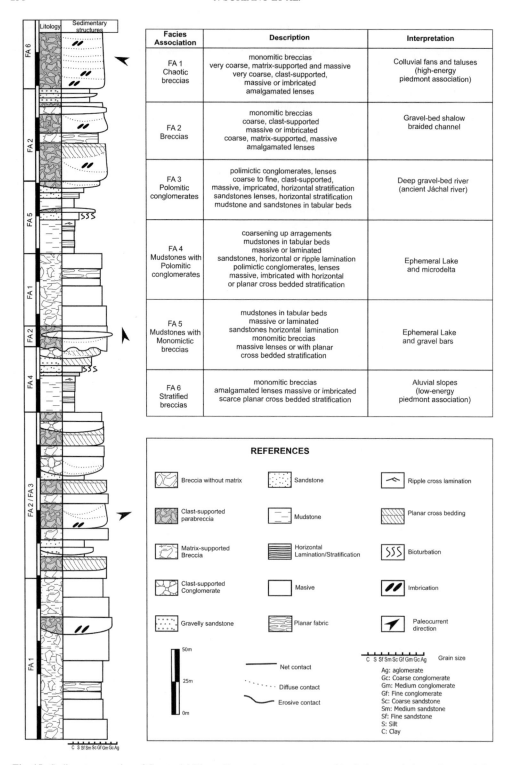

Fig. 15. Sedimentary section of Cuesta del Viento Formation and summary of its facies associations, characteristics and interpretation.

Facies association 3 (polimictic conglomerates). This unit comprises lenses of coarse- to fine-grained clast-supported conglomerates and sandstones. The conglomerates bearing granites, mudstones, slates, acid and ultrabasic volcanic clasts record a mixture of local and foreign provenance (Suriano *et al.* 2011). The conglomeratic lenses occur as massive beds or with imbricated or horizontally stratified structures, thus indicating the deposition of transversal gravel bars (Hein & Walker 1977). Subordinately massive or horizontally laminated sandstones appear as bar tops. Tabular beds of sandstones and mudstones, massive or laminated, deposited in a floodplain environment are also present.

This facies association is interpreted as a low-sinuosity river with channels of multiple hierarchies, similar to the deep gravel-bed braided river of Miall (1996) or the Donjek type of Williams & Rust (1969). Owing to its lithology and environmental interpretation, this association is envisaged as a transference system, which represents the ancient Jáchal River.

Facies association 4 (mudstones with polimictic conglomerates). Two fine-grained facies associations were recognized in this formation. The facies association 4; Fig. 9d) is formed from coarsening-up cycles up to 20 m thick. The cycles are dominated by massive or slightly laminated light brown mudstones, intercalated with fine-grained sandstones. Deposition took place owing to alternating processes of sheetfloods (Davis 1938) and decantation. These facies are overlaid by sandstones with ripple cross-lamination and planar cross-bedded, which are covered in turn by lenses of polimictic conglomerates. Conglomerate composition is similar to that described in the facies association 2, highlighting a mixing of local and external provenance. Conglomerate beds have an erosive base and appear as massive or planar cross stratification, deposited by channelized fluid flows.

In summary, facies association 4 is interpreted as the occurrence of clastic ephemeral lake (natural damming) associated with microdeltaic bars facies related to ancient Jáchal River. The fine-grained tabular facies are related to lacustrine deposits and the sandstones and conglomerates in coarsening arrangements are interpreted as microdeltaic bars facies in the ancient Jáchal channel mouths.

Facies association 5 (mudstones with monomictic breccias). This facies association (Fig. 9e, f) is formed from thick tabular beds of light greenish mudstones, dominantly laminated. Coarsening-up arrangements of laminated sands and massive and trough cross-bedded breccias, with local provenance, appear in a minor proportion. This association is interpreted as closed ephemeral clastic lacustrine system deposits, dominated by decantation processes and the prograding lobes of sandstone sheetfloods or thin monomictic gravel bars with local provenance (microdeltaic facies with local provenance).

Facies association 6 (stratified breccias). This is made up of tabular beds of 10–20 cm of thickness composed of fine to medium clast-supported breccias, imbricated or massive. Locally, some lenses of planar cross-bedded stratification are found. Some palaeocurrent measurements indicate east to west direction of flow. This unit is interpreted as shallow channels with broad interchannel areas dominated by gravel sheetfloods, corresponding to alluvial slopes (Smith 2000) developed in a low-energy piedmont environment (2 in Fig. 5).

Basin evolution

The Cuesta del Viento Formation records the earliest stages of the Bermejo Foreland Basin (Jordan *et al.* 1993, 2001; Alonso *et al.* 2011; Suriano *et al.* 2011), related to the uplift of the Sierra de La Tranca. The base of the unit (Figs 15 & 16, facies association 1 and 2) corresponds to wedge-top deposition in the inner wedge-top depocentre, represented by the development of high-energy and local provenance piedmont (Figs 16 & 17a). Overlying this basal unit, lower-energy piedmont facies were deposited, represented by braided channels of colluvial fans. Finally, the last facies associations deposited in this basin correspond to the ancient Jáchal River, interbedded with colluvial fans (Fig. 16, facies associations 3 and 2).

As the Andean deformation advanced to the east, the uplift of the Caracol Range started. This phase represents the onset of the La Tranca area as a piggyback basin configuration, and the deposition of lacustrine and microdeltaic facies recording this change (Fig. 16, facies association 4). The dam was, in this case, related to an increase in the local accommodation space owing to the Caracol range uplift (Fig. 17b). Afterwards, the base level was readjusted to the foreland, probably because the damming was broken, an incision surface was carved (incision stage, Fig. 16) and axial systems prograded in response to a low equilibrium profile on the transference system (confined low accommodation system tract; Fig. 17c). Finally, the major uplift of the Caracol range took place. La Tranca area was isolated, forming a closed piggyback basin (Figs 16 & 17d), recording a high accommodation system tract, registered by muddy lacustrine deposits and a high-energy piedmont association (Fig. 16). In the latter phase, low-energy piedmont systems (facies association 3) prograded onto playa lake facies, producing the filling of

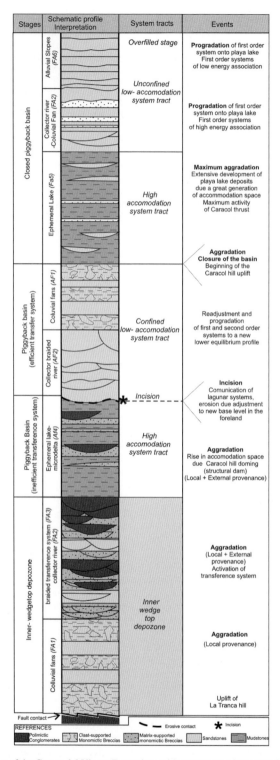

Fig. 16. Schematic section of the Cuesta del Viento Formation, with system tract interpretation and geological events that caused them (right column).

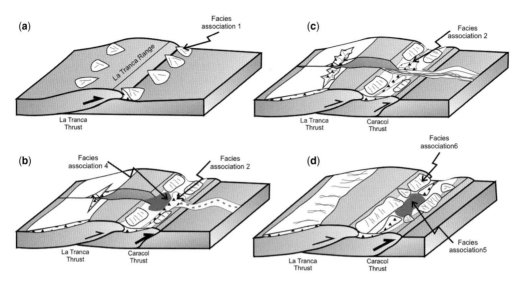

Fig. 17. Block diagrams showing the evolution of Cuesta del Viento Formation and its relationship with the facies association (thick arrows represent active thrusting).

the basin (unconfined low accommodation system tract, Fig. 16).

Conclusions

The movement of sediments within continental piggyback basins can be sketched as driven by three different types of alluvial–fluvial systems: (a) piedmont accumulations (talus, different types of colluvial fans and, in wide enough basins, alluvial fans) forming alluvial aprons during mountain denudation; (b) axial fluvial systems which collect sediments derived from the piedmont area and transport them to transversal rivers; and (c) transference systems formed by large rivers draining the fold and thrust belt and transferring sediments from the piggyback basins to the foredeep region.

Despite the fact that this model is a simplification of the alluvial–fluvial systems in the Precordillera, it is useful to explain the transference of sediments not only within the piggyback basins but also from the piggyback area to the foreland. In addition, the existence of transference fluvial systems (in our case the Jáchal River) enables the recognition of two different types of piggyback: open and closed piggyback basins (when the lack of transference fluvial systems prevents the transport of sediments from the piggyback to the foreland, turning the piggyback into an endorheic isolated basin). The evidence obtained from Cenozoic deposits as analysed in this paper clearly shows that open piggyback basins can undergo temporary closures and, in the same way, closed piggyback areas can be open when a new transference fluvial system develops in the thrust and fold belt.

The stratigraphic record of open piggyback basins has been divided in this paper into four major theoretical stages: (a) the incision stage; (b) a confined low accommodation system tract; (c) a high accommodation system tract; and (d) an unconfined low accommodation system tract or overfill stage. The incision stage is reached when long-lived periods of erosion within the piggyback allow/produce a massive transference of sediments to the foreland. Degradation stages are recorded by incision surfaces. These surfaces can be related not only to changes in subsidence or tectonism in the foreland or in the fold and thrust belt, but also to climatic fluctuations significant enough to produce an increase in the sediment transport capacity of the third-order fluvial systems. The high accommodation system tract is characterized by strong aggradation in the piggyback basins mainly related to either low rates of subsidence in the foredeep or inefficiency of the transference fluvial systems. Low-accommodation stages reflect limited capacity of the piggyback to store sediments. The confined stage occurs in subsequent periods of erosion and transference of sediments towards the foreland, recorded as the first fill over the incision surface. Finally, the unconfined stage is associated with basin overfill.

The deposits of close piggyback basins have been divided in two end members stages: (a) high accommodation system tracks; and (b) low accommodation system tracts. In order to examine the utility of the proposed model, two case studies in the Precordillera of San Juan province, Argentina were analysed: the Pleistocene–Holocene deposits of the Jáchal River piggyback basins and the Oligocene–Miocene Cuesta del Viento Formation.

The first example associated with relative relatively quiescent tectonic shows that changes in subsidence rates along the foreland under these conditions can exert a strong control in the equilibrium profiles of the transference systems, thus promoting periods of aggradation and degradation of the piggyback. A good example of this situation appears in Sequence I, which illustrates how the isostatic rebound in the thrust and fold belt controlled the enhancement of aggradation phases. On the other hand, changes in climatic conditions should not be overlooked, In fact, the widespread development of intermontane lakes (Sequence III) and the large progradation of megafans (Sequence II) into the foreland are linked to postglacial conditions and result in good examples of how climate plays an important role in the accumulation patterns in both piggyback and foreland basins.

The second case study records the transition between the outer-wedge-top depozone and piggyback basin as well as different types of piggyback basin (open to close). We conclude that the model proposed here is useful to analyse the sediments of the Jáchal River area deposited during limited or absent tectonic activity with climatic controls, as well as the Cuesta del Viento Formation deposited in an active fold and thrust belt. In both cases, the model provides a framework in which piggyback basin evolution can be described and related to the factors that control the dynamics of sedimentation.

This study was supported by grant BID 1728 OC/AR PICT 20752 of the Agencia de Promoción Científica y Tecnológica (Argentina), UBACyT 20020110300025 and the Departamento de Ciencias Geológicas of the Universidad de Buenos Aires. We would like to thank J. Mescua for critical comments that helped to improve our manuscript. The authors acknowledge the critical reading and the valuable proposals suggested by the reviewers (M. Tunik andan anonymousreviwer).

References

ALONSO, S., LIMARINO, C. O., LITVAK, V., POMA, S. M., SURIANO, J. & REMESAL, M. 2011. Paleogeographic, magmatic and paleoenviremontal scenarios at 30°SL during the andean orogeny: cross sections from the volcanic-arc to the orogenic front (San Juan Province, Argentina). In: SALFITY, J. A. & MARQUILLAS, R. A. (eds) Cenozoic Geology of the Central Andes of Argentina. SCS, Salta, 23–45.

BEAUMONT, C., QUINLAN, G. M. & HAMILTON, J. 1988. Orogeny and stratigraphy: numerical models of the Paleozoic in the eastern interior of North America. Tectonics, 7, 389–416.

BILKRA, L. H. & NEMEC, W. 1998. Postglacial colluvium in western Norway: depositional process, facies and paleoclimatic record. Sedimentology, 45, 909–959.

BLUM, M. D. & TÖRNQVIST, T. E. 2000. Fluvial responses to climate and sea-level chamge: a review and look forward. Sedimentology, 47, 2–48.

BONINI, M., MORATTI, G. & SANIB, F. 1999. Evolution and depocentre migration in thrust-top basins: inferences from the Messinian Velona Basin (Northern Apennines, Italy). Tectonophysics, 304, 95–108.

BUTLER, R. W. H. & GRASSO, M. 1993. Tectonic controls on base-level variations and depositional sequences within thrust-top and foredeep basins: examples from the Neogene thrust belt of central Sicily. Basin Research, 5, 137–151.

BUTLER, R. W. H., LICKORISH, W. H., GRASSO, M., PEDLEY, H. M. & RAMBERTI, L. 1995. Tectonics and sequence stratigraphy in Messinian basins, Sicily: constraints on the initiation and termination of the Mediterranean 'salinity crisis'. Geological Society of America, Bulletin, 107, 425–439.

CARDOZO, N. & JORDAN, T. 2001. Causes of spatially variable tectonic subsidence in the Miocene Bermejo Foreland Basin, Argentina. Basin Research, 13, 335–357.

CATUNEANU, O. 2006. Principles of Sequence Stratigraphy. Elsevier, Amsterdam.

COLOMBO, F., BUSQUETS, P., RAMOS, E., VERGÉS, J. & RAGONA, D. 2000. Quaternary alluvial terraces in an active tectonic region: the San Juan River Valley, Andean Ranges, San Juan Province, Argentina. Journal of South American Earth Sciences, 13, 611–626.

COLOMBO, F., LIMARINO, C., BUSQUETS, P., SOLÉ DE PORTA, N., HEREDIA, N., RODRÍGUEZ FERNÁNDEZ, R. & ALVAREZ MARRÓN, J. 2005. Primeras edades absolutas de los depósitos lacustres holocenos del río Jáchal, Precordillera de San Juan. Revista de la Asociación Geológica Argentina, 60, 605–608.

COLOMBO, F. P., BUSQUETS BUEZO, P., SOLÉ DE PORTA, N., LIMARINO, C. O., HEREDIA, N., RODRÍGUEZ FERNÁNDEZ, L. R. & ÁLVAREZ MARRÓN, J. 2009. Holocene intramontane lake development: a new model in the Jáchal River Valley, Andean Precordillera, San Juan, Argentina. Journal of South American Earth Sciences, 28, 228–238.

DAHLE, K., FLESJA, K., TALBOT, M. R. & DREYER, T. 1997. Correlation of fluvial deposits by the use of Sm-Nd isotope analysis and mapping of sedimentary architecture in the Escanilla Formation (Ainsa Basin, Spain) and the Statfjord Formation (Norwegian North Sea), Sixth International Conference on Fluvial Sedimentology, Cape Town, South Africa, 46.

DALRYMPLE, M., POSSER, J. & WILLIAMS, B. 1998. A dynamic systems approach to the regional controls on deposition and architecture of alluvial sequences, illustrated in the Stafjord Formation (United Kingdom, northern North Sea). In: SHANLEY, K. W. & MCCABE, P. J. (eds) Relative Role of Eustasy, Climate and Tectonism in Continental Rocks. Society of Economic Paleontologists and Mineralogists, Tulsa, OK, Special Publications, 65–81.

DAMANTI, J. F. 1993. Geomorphic and structural controls on facies patterns and sediment composition in a modern foreland basin. *In*: MARZO, M. & PUIGDEFÁBREGAS, C. (eds) *Alluvial Sedimentation*. International Association of Sedimentologists, Special Publications, Blackwell, Oxford, **17**, 221–233.

DAVIS, W. M. 1938. Sheetfloods and streamfloods. *Bulletin of the Geological Society of America*, **49**, 1337–1416.

DECELLES, P. G. & CAVAZZA, W. 1999. A comparison of fluvial megafans in the Cordilleran (Upper Cretaceous) and modern Himalayan foreland basin systems. *Geological Society of America, Bulletin*, **111**, 1315–1334.

DECELLES, P. G. & GILES, K. A. 1996. Foreland basin systems. *Basin Research*, **8**, 105–123.

DECELLES, P. G. & HORTON, B. K. 2003. Early to middle Tertiary foreland basin development and the history of Andean crustal shortening in Bolivia. *Geological Society of America Bulletin*, **115**, 58–77.

DOGLIONI, C., MERLINI, S. & CANTARELLA, G. 1999. Foredeep geometries at the front of the Apennines in the Ionian Sea (central Mediterranean). *Earth and Planetary Science Letters*, **168**, 243–254.

FLEMINGS, P. B. & JORDAN, T. E. 1989. A synthetic stratigraphic model of foreland basin development. *Journal of Geophysical Research*, **94**, 3851–3866.

FURQUE, G. 1979. *Descripción de la Hoja Geológica 18 c, Jáchal, Provincia de San Juan*. Carta Geológico-Económica de la República Argentina Escala 1:200 000. Servicio Geológico Nacional, Boletín 164, Buenos Aires.

GARCÍA, A., ZÁRATE, M. & PÁEZ, M. 1999. The Pleistocene/Holocene transition and the human occupation in the Central Andes of Argentina: Agua de la Cueva Locality. *Quaternary International*, **53/54**, 43–52.

GUGLIOTTA, C. 2012. Inner v. outer wedge-top depozone 'sequences' in the Late Miocene (late Tortonian–early Messinian) Sicilian Foreland Basin System: new data from the Terravecchia Formation of NW Sicily. *Journal of Geodynamics*, **55**, 41–55.

HEIN, F. J. & WALKER, R. G. 1977. Bar evolution and development of stratification in the gravelly, braided, Kicking Horse River, British Columbia. *Canadian Journal of Earth Sciences*, **14**, 562–570.

HESSE, S., BACK, S. & FRANKE, D. 2010. The structural evolution of folds in a deepwater fold and thrust belt – a case study from the Sabah continental margin offshore NW Borneo, SE Asia. *Marine and Petroleum Geology*, **27**, 442–454.

HOLT, W. E. & STERN, T. A. 1994. Subduction, platform subsidence, and foreland thrust loading: the late Tertiary development of Taranaki basin, New Zealand. *Tectonics*, **13**, 1068–1092.

HORTON, B. K. & DECELLES, P. G. 2001. Modern and ancient fluvial megafans in the foreland basin system of the central Andes, southern Bolivia: implications for drainage network evolution in fold-thrust belts. *Basin Research*, **13**, 43–63.

IRIONDO, M. 1999. Climatic changes in the South American plains: records of a continent-scale oscillation. *Quaternary International*, **57/58**, 93–112.

IRIONDO, M. & KRÖLING, D. M. 1996. Sedimentos eólicos del noreste de la llanura pampeana (Cuaternario superior), *XIII Congreso Geológico Argentino y III Congreso de Exploración de Hidrocarburos*, Buenos Aires, 27–48.

JOHNSON, D. D. & BEAUMONT, C. 1995. *Preliminary Results from a Planform Kinematic Model of Orogen Evolution, Surface Processes and the Development of Clastic Foreland Basin Stratigraphy*. Society of Economic Paleontologists and Mineralogists, Tulsa, OK, Special Publications, **52**, 3–24.

JORDAN, T. E. 1981. Thrust loads and foreland basin evolution, Cretaceous,Western United-States. *American Association of Petroleum Geologists Bulletin*, **65**, 2506–2520.

JORDAN, T. E. 1995. Retroarc Foreland and related basins. *In*: BUSBY, C. J. & INGERSOL, R. V. (eds) *Tectonics of Sedimentary Basins*. Blackwell Science, Oxford, 331–362.

JORDAN, T. E., DRAKE, R. E. & NAESER, C. W. 1993. Estratigrafía del Cenozoico Medio en la Precordillera a la latitud del Río Jáchal, San Juan, Argentina. *XII Congreso Geológico Argentino y II Congreso de Exploración de Hidrocarburos*, Mendoza, Argentina, 132–141.

JORDAN, T., SCHLUNEGGER, F. & CARDOZO, N. 2001. Unsteady and spatially variable evolution of the Neogene Andean Bermejo foreland basin, Argentina. *Journal of South American Earth Sciences*, **14**, 775–798.

KAY, S M., MAKSAEV, V., MOSCOSO, R., MPODOZIS, C., NASI, C. & GORDILLO, C. E. 1988. Tertiary Andean magmatism in Chile and Argentina between 28°S and 33°S: correlation of magmatic chemistry with changing Benioff zone. *Journal of South American Earth Sciences*, **1**, 21–38.

LIMARINO, C., TRIPALDI, A., MARENSSI, S., NET, L., RÉ, G. & CASELLI, A. 2001. Tectonic control on the evolution of the fluvial systems of the Vinchina Formation (Miocene), Northwestern Argentina. *Journal of South American Earth Sciences*, **14**, 751–762.

LLAMBÍAS, E. J. & SATO, A. M. 1990. El batolito de Colangüil (29°–31°S), Cordillera Frontal de Argentina: estructura y marco tectónico. *Revista Geológica de Chile*, **17**, 89–108.

MAY, G., HARTLEY, A. J., STUART, F. M. & CHONG, G. 1999. Tectonic signatures in arid continental basins: an example from the Upper Miocene–Pleistocene, Calama Basin, Andean forearc, northern Chile. *Palaeogeography, Palaeoclimatology, Palaeoecology*, **151**, 55–77.

MIALL, A. D. 1996. *The Geology of Fluvial Deposits, Sedimentary Facies, Basin Analysis and Petroleum Geology*. Springer, Berlin.

MIALL, A. D. 2002. Architecture and sequence stratigraphy of Pleistocene fluvial systems in the Malay Basin, based on seismic time-slice analysis. *American Association of Petroleum Geologists Bulletin*, **86**, 1201–1216.

MITROVICA, J. X., BEAUMONT, C. & JARVIS, G. T. 1989. Tilting of continental interiors by dynamical effects of subduction. *Tectonics*, **8**, 1079–1094.

NADON, G. C. 1994. The genesis and recognition of anastomosed fluvial deposits: data from the St. Mary River Formation, southwestern Alberta, Canada. *Journal of Sedimentary Research*, **64**, 451–463.

NANSON, G. C. & CROKE, J. C. 1992. A genetic classification of floodplains. *Geomorphology*, **4**, 459–486.

NAYLOR, M. & SINCLAIR, H. D. 2008. Pro- v. retro-foreland basins. *Basin Research*, **20**, 285–303.

ORI, G. G. & FRIEND, P. F. 1984. Sedimentary basins formed and carried piggyback on active thrust sheets. *Geology*, **12**, 475–478.

ORTIZ, A. & ZAMBRANO, J. J. 1981. La provincia geológica Precordillera Oriental, *VII Congreso Geológico Argentino*, Actas, San Luis, **3**, 59–74.

QUIRK, D. G. 1996. 'Base profile': a unifying concept in alluvial sequence stratigraphy. *In*: HOWELL, J. A. & AITKEN, J. F. (eds) *High Resolution Sequence Stratigraphy*. Geological Society, London, Special Publications, **104**, 37–49.

RAMOS, V. A. 1999. Rasgos estructurales del territorio argentino. I. evolucion tectónica de la Argentina. *In*: *Geología Argentina*. Anales, Buenos Aires, **29**, 715–784.

REINFELDS, I. & NANSON, G. 1993. Formation of braided river floodplains, Waimakariri River, New Zealand. *Sedimentology*, **40**, 1113–1127.

RICCI LUCCHI, F. 1986. The Oligocene to recent foreland basins of the northern Apennines. *In*: ALLEN, P. A. & HOMEWOOD, P. (eds) *Foreland Basins*. International Association of Sedimentologists, Special Publications, Blackwell, Oxford, **8**, 105–139.

ROURE, F. 2008. Foreland and hinterland basins: what controls their evolution? *Swiss Journal of Geoscience*, **101**(Suppl. 1), S5–S29.

SHANLEY, K. W. & MCCABE, P. J. 1994. Perspectives on the sequence stratigraphy of continental strata. *American Association of Petroleum Geologists Bulletin*, **78**, 544–568.

SMITH, D. G. & SMITH, N. D. 1980. Sedimentation in anastomosed river systems: examples from alluvial valleys near Banff, Alberta. *Canadian Journal of Earth Sciences*, **17**, 1396–1406.

SMITH, G. A. 2000. Recognition and significance of streamflow-dominated piedmont facies in extensional basins. *Basin Research*, **12**, 399–411.

SURIANO, J. & LIMARINO, C. O. 2005. Depósitos de endicamiento en la sección superior del Río Jáchal, quebradas de La Tranca y Caracol (Precordillera de San Juan), *XVI Congreso Geológico Argentino*, La Plata, Buenos Aires, Argentina, 223–230.

SURIANO, J. & LIMARINO, C. O. 2006. Modelo para la generación de valles incisos en cuencas intermontanas: controles climáticos y subsidencia, *IV Congreso Latinoamericano de Geología y XI Reunión Argentina de sedimentología*, Bariloche, Río Negro, Argentina, 218.

SURIANO, J. & LIMARINO, C. O. 2008. Balance sedimentario entre sistemas fluviales colectores y de transferencia, su implicancia en los ambientes sedimentarios intermontanos, *XII Reunión Argentina de Sedimentología*, Ciudad de Buenos Aires, Argentina, 171.

SURIANO, J. & LIMARINO, C. O. 2009. Sedimentación pedemontana las naciente del Río Jáchal y Pampa de Gualilán, Precordillera de San Juan. *Revista de la Asociación Geológica Argentina*, **65**, 516–532.

SURIANO, J., ALONSO, M. S., LIMARINO, C. O. & TEDESCO, A. M. 2011. La Formación Cuesta del Viento (*nov. nom.*): una nueva unidad litoestratigráfica en la evolución del orógeno precordillerano. *Revista de la Asociación Geológica Argentina*, **68**, 246–260.

TALLING, P. J., LAWTON, T. F., BURBANK, D. W. & HOBBS, R. S. 1995. Evolution of latest Cretaceous–Eocene nonmarine deposystems in Axhandle piggyback basin of central Utah. *Bulletin of the Geological Society of America*, **107**, 297–315.

TURNER, J. P. 1990. Structural and stratigraphic evolution of the West Jaca thrust-top basin, Spanish Pyrenees. *Journal of the Geological Society*, **147**, 177–184.

WAGERICH, M. 2001. A 400-km-long piggyback basin (Upper Aptian–Lower Cenomanian) in the Eastern Alps. *Terra Nova*, **13**, 401–406.

WILLIAMS, P. F. & RUST, B. R. 1969. The sedimentology of a braided river. *Journal of Sedimentary Petrology*, **39**, 649–679.

XIE, X. & HELLER, P. L. 2009. Plate tectonics and basin subsidence history. *Geological Society of America Bulletin*, **121**, 55–64.

ZAPATA, T. R. & ALLMENDINGER, R. W. 1996. The thrust front zone of the Precordillera, Argentina: a Thick-skinned triangle zone. *American Association of Petroleum Geologists Bulletin*, **80**, 359–381.

ZOETEMEIJER, R., CLOETINGH, S., SASSI, W. & ROURE, F. 1993. Modelling of piggyback-basin stratigraphy; record of tectonic evolution. *Tectonophysics*, **226**, 253–269.

Quaternary shortening at the orogenic front of the Central Andes of Argentina: the Las Peñas Thrust System

CARLOS H. COSTA[1]*, EMILIO A. AHUMADA[1], CARLOS E. GARDINI[1], FABRICIO R. VÁZQUEZ[1,2] & HANS DIEDERIX[3]

[1]*Departamento de Geología, Universidad Nacional de San Luis, Ej. de los Andes 950, Bloque II, 5700 San Luis, Argentina*
[2]*CONICET*
[3]*Servicio Geológico Colombiano, Diagonal 53 N° 34-53, Bogotá, D.C, Colombia*
*Corresponding author (e-mail: costa@unsl.edu.ar)

Abstract: The NNW-trending Las Peñas Thrust System is one of the key structures along the Andean orogenic front between 32°15′ and 32°40′S in the Southern Precordillera of Argentina. This east-verging structure crops out over a distance of c. 40 km and provides one of the best opportunities for a detailed field survey of Quaternary thrusting in the Andean frontal deformation zone. We present a systematic description of the geometry and geomorphic signatures of the main thrust deformation zone, which emplaces Neogene rocks over Quaternary alluvium, and usually behaves as a blind propagating thrust into the youngest (Late Pleistocene–Holocene) alluvial deposits. The Las Peñas Thrust System is understood to represent the latest stage of the eastward migration of an imbricated fan structure, which has driven the neotectonic uplift of the Las Peñas–Las Higueras range. Excellent outcrops provided by well-incised creek outlets reveal that the thrust system is made up either by a single fault surface or by two or more frontal splays. Several sections along its length can be differentiated on the basis of thrust geometries and/or morphotectonic features. The northern sections are characterized by isolated outcrops of Neogene rocks in the hanging wall, surrounded by alluvial bajadas. Remnants of fold limbs scarps depict the geomorphic signature of the thrust propagation into the Quaternary layers, although the preserved topographic relief always underestimates the cumulated thrust slip during the Quaternary. The southern part of this thrust system is defined by a frontal range, cored by a transposed south-plunging anticline in bedrock. Our observations suggest a dynamic and unsteady interaction between thrust propagation and sedimentation/erosion processes along the thrust trace during deposition of the Quaternary alluvial layers.

Descriptions of the geometry and neotectonics of the Quaternary thrust front along the eastern Andean foothills only exist in a very few places along its near 8000 km length. Its study has been approached by tectonic geomorphology analysis at a regional scale (Dumont 1996; Audemard 1999; Horton 1999; Mora *et al.* 2009; 2010) and through the interpretation of seismic reflection profiles or punctual field descriptions (Baby *et al.* 1992, 1997; Ramos *et al.* 2004, 2006; Mora *et al.* 2006; 2009; Parra *et al.* 2009). Space geodesy indicates that a significant part of the interseismic elastic deformation of the Andean back-arc is being accommodated at the frontal deformation zone of the orogen (Kendrick *et al.* 1999, 2001, 2003, 2006; Norabuena *et al.* 1999; Brooks *et al.* 2003, 2011). Accordingly, description and analysis of Quaternary deformation may help to link the present-day kinematic data provided by space geodesy with the long-term framework modelled by geological studies.

However, field surveys of Quaternary deformation along the Andean front have commonly been hampered by adverse morphoclimatic conditions (i.e. dense vegetation cover and high erosion/deposition rates), with a resulting scarcity of exposures and the preservation of only non-diagnostic geomorphic signatures of neotectonic thrusts and fold-related phenomena in Quaternary sediments. As a result, there are only a few places along the entire chain where a proper examination of the geometry and kinematics of these structures can be conducted at the outcrop scale.

At the southern Precordillera in central western Argentina (Fig. 1), an arid environment in combination with numerous well-incised creeks across the Las Peñas–Las Higueras range provide probably the best opportunity along the entire chain for analysing the surface geometry of the Quaternary thrust front of the Andes at outcrop scale. This morphostructure is bounded for the most part by the

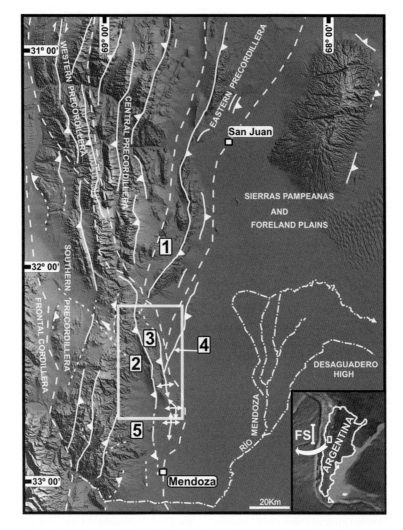

Fig. 1. Main tectonic settings in the Precordillera region (central western Argentina; white box in inset) between 31°S and 33°S, showing the major neotectonic faults and folds in white trace. FS, flat slab subduction segment (in inset). The white rectangle corresponds to the Sierra de Las Peñas–Las Higueras area (see Fig. 2). 1, Pedernal thrust system; 2, Las Higueras thrust system; 3, Las Peñas thrust system; 4, Cerro Salinas–Montecito thrust system; 5, La Cal fault. Continuous white lines, main thrusts; dashed white lines, major morphostructural units boundaries; dashed and dotted lines, recent and active courses of the Mendoza River.

east-directed Las Peñas Thrust System, which has concentrated significant Quaternary shortening at these latitudes (Cortés & Costa 1996; Costa et al. 2000a, 2006a, b; Vergés et al. 2007; Ahumada & Costa 2009; Schmidt et al. 2011a).

This contribution aims to provides the first systematic description of the geometries, geomorphic signatures and styles of Quaternary deformation related to this thrust system. It also aims to contribute to a better understanding of the interaction between Quaternary thrusting and surface processes involved in current mountain building.

Tectonic setting

The Nazca Plate subducts with a subhorizontal angle beneath the South American plate between latitudes 27°S and 33°30′S, resulting in significant geological differences in comparison with those segments, which exhibit normal subduction angles to the north and south of it (Barazangi & Isacks 1976; Jordan et al. 1983; Ramos 1988; Gutscher et al. 2000; Ramos et al. 2002; Yañez et al. 2001, among others). Along this flat-lying subduction segment the main mountain-building processes from the Pliocene to the present, and notably the current tectonic activity along the Andean frontal deformation zone in Argentina, have been located along the eastern foothills of the Precordillera (Fig. 1). Here, active tectonic processes are indicated by a belt of concentrated shallow crustal seismicity, which has generated the most significant earthquakes of the last two centuries. Space geodesy data also point out that ongoing shortening is being concentrated in the area between the Precordillera and the Sierras Pampeanas broken foreland, with shortening rates ranging from 2 to 4 mm/a (Kendrick et al. 1999, 2001, 2006; Brooks et al. 2003).

The Precordillera is a Palaeozoic orogen whose Neogene evolution has resulted in different neotectonic deformation styles, controlled by inheritances of pre-Cenozoic structures. In the San Juan Province, north of 32°15′S, the western and central Precordillera behave as a thin-skinned fold-and-thrust belt with eastern vergence and a general NNE trend. In contrast, the west-directed (Pampean-type) thrusts of the eastern Precordillera have been interpreted to be rooted and controlled by the fabric of the underlying crystalline basement of the Sierras Pampeanas (Fig. 1) (Rolleri 1969; Ortiz & Zambrano 1981; Baldis et al. 1982; Comínguez & Ramos 1991; von Gosen 1992; Zapata & Allmendinger 1996; Siame et al. 2002; Alvarado et al. 2009). These Pampean-type thrusts have also been understood to be the result of crustal wedging (Meigs et al. 2006; Vergés et al. 2007; Meigs & Nabelek 2010).

The orogenic front at the Southern Precordillera (Figs 1 & 2) is characterized by NNW-trending east-verging (Andean-type) thrusts, controlled by former normal faults of an inverted Triassic hemigraben (Bettini 1980; Ramos & Kay 1991; Dellapé & Hegedus 1995; Cortés et al. 2005; Giambiagi et al. 2011). This inheritance is responsible for the general NNW structural trend, which allows the resulting neotectonic structures of the Las Peñas–Las Higueras range to be separated from those of the Precordilleran domains in the San Juan Province to the north, which follow a general NNE trend (Fig. 1).

Along most of the San Juan Precordillera, the NNE-trending Andean frontal deformation zone is characterized by the interaction of these two thrust systems with opposite vergence (Ahumada & Costa 2009; Ahumada 2010). This tectonic array has been defined north of San Juan city as a thick-skinned triangle zone (Zapata & Allmendinger 1996). At the NNW-trending Southern Precordillera thrust front, this antithetic linkage pattern and the main thrust trends determine two oblique junctions where opposite-verging thrusts converge (Fig. 1) (Costa et al. 2006b; Ahumada & Costa 2009). The northern junction is between the Las Higueras (LHTS) and Pedernal (PTS) thrusts systems (Ahumada & Costa 2009; Ahumada 2010), whereas the southern one is between the Las Peñas (LPTS) and Cerro Salinas-Montecito (CSMT) thrusts systems (Costa et al. 2000b; Vergés et al. 2007; Bohon 2008; Enderlin et al. 2009; Cisneros et al. 2010). The latter junction has not been the subject of detailed study because of its burial beneath a blanket of alluvial piedmont cover sediments. In both cases, the Pampean-type thrusts (PTS and CSMT) converge obliquely with the Andean-type structures (LHTS and LPTS).

This latitude section of the orogenic front is the result of the overimposed Neogene shortening of the Barreal–Las Peñas Block of the Southern Precordillera (Cortés et al. 2005), whose southeastern part at the Las Peñas–Las Higueras range became fully involved in the Andean frontal deformation zone. The oblique (NNW) trend of the main structures regarding the regional frontal deformation zone, as well as the key role of inherited extensional structures in the Neogene contraction, turn the orogenic front along the Southern Precordillera into an area with particular characteristics along the entire Andean front.

Geological and tectonic outline of the Las Peñas–Las Higueras range

The Las Peñas–Las Higueras range constitutes a morphological unit that covers the southeastern corner of the Southern Precordillera (Figs 1 & 2).

The western part of the range is composed of Palaeozoic and Mesozoic sedimentary and volcanic sequences, thrusted over Mesozoic and Cenozoic continental clastic rocks (sandstones, conglomerates and siltstones) (Fossa Mancini 1942; Harrington 1971; Ahumada 2004; Ahumada et al. 2006). The Cenozoic (Miocene and Pliocene) deposits have been correlated with the Mariño and Río de Los Pozos formations (Cortés & Costa 1996; Sepúlveda & López 2001), which form the bedrock involved in the Las Peñas thrust deformation zone throughout its trace. At the southeastern part of the range, these stratigraphic units are seen to be in conformable sequence with the overlying Plio-Pleistocene strata of the Mogotes Formation (Costa et al. 2000a; Ahumada 2004; Ahumada et al. 2006). The latter unit is made up of psephitic proximal sediments (coarse grey conglomerates) and is considered to record the onset of the Precordilleran uplift (Yrigoyen 1993; Irigoyen et al. 2000).

Farther to the north, continued tectonic activity and uplift has led to the erosion of the Cenozoic rocks and created a palaeo-erosion surface (Fig. 2). This surface has subsequently been covered by Quaternary fanglomerates, remnants of which are

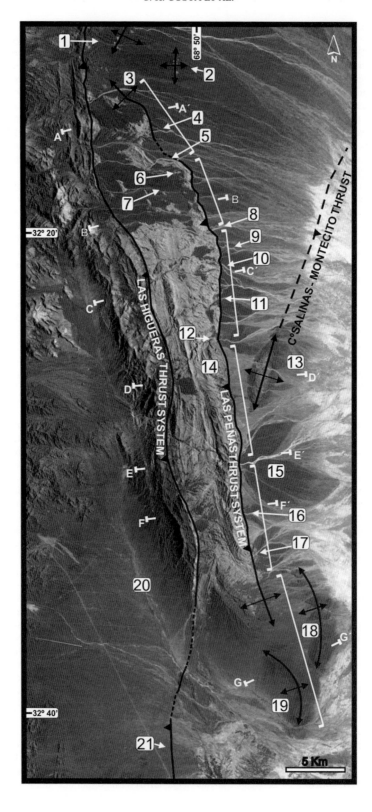

numerous and increase their coverage to the north where the range relief diminishes. The composition of the clasts of the fanglomerates indicates the source rocks to be Palaeozoic–Triassic age formations that constitute the western rim of the range, but a mixture of volcanic and intrusive rock fragments point to a source farther to the west and/or possibly reworking of the Mogotes Formation.

The eastern flank of the range borders an extensive piedmont plain made up of fanglomerates in a typical alluvial fan environment. Fluvial processes of deposition and erosion have given rise to the development of stairs of fluvial terraces. The real thickness of these alluvial deposits could not be established because stream incision has not been sufficiently deep to expose the entire sedimentary column. Stream downcutting and stream widening has also resulted in characteristic wing-like shapes that stand out remarkably on space imagery because of their typical shape and dark tones (Figs 1 & 2).

The range hosts two east-verging thrusts systems, the LHTS in the west and the LPTS in the east (Figs 1 & 2). The thrust sheet located between the LHTS and LPTS exhibits, for the most part, a homocline structure where several imbrications at a macroscopic scale can be recognized or suspected from terrain analysis. However, the Cenozoic strata appear intensively folded in the Cerros Colorados area (Fig. 2).

The LHTS system runs along the centre line of the Sierra, thrusting Palaeozoic/Mesozoic rocks over Mesozoic/Neogene rocks (Harrington 1971; Ahumada et al. 2006; Ahumada & Costa 2009). It is expressed by a prominent scarp that separates the topographically high ground of the Sierra in the west from the relatively flat morphology of the Mesozoic/Cenozoic rocks and the Quaternary alluvial cover in the east (Fig. 2). The fault trace shows a gently sinuous pattern with a general 340° trend. The northern section marks the main mountain-piedmont junction at the northern end of the Las Peñas–Las Higueras range, as well as the west-bounding structure of the linkage zone between the southern and eastern Precordillera (Ahumada & Costa 2009; Ahumada 2010) (Figs 1 & 2). The thrust surface dips 30° to 45°W, with a general NNW direction, and exhibits several bends in plan view (Fig. 2). The central section of the thrust is in that part of the range with the highest elevations. Here, the identification of diagnostic neotectonic evidence along its length is hampered by elusive exposures due to a coarse colluvial–fluvial cover overlying the thrust trace. The dip of the thrust surface ranges between 45° and 80°E and can even reach an overturned position (Bea 2000; Ahumada 2004). The southern section of the LHTS (south of 32°35′) comprises a string of low hills that stand out from the surrounding piedmont to the south of the Sierra (La Bomba and La Cal hills, Fig. 2). In this extensive piedmont area there is almost no pre-Quaternary substratum cropping out in the hanging wall of the LHTS (Ahumada et al. 2006) (Fig. 1). Evidence of thrust-related Holocene tectonic activity has been reported both at the northern (Ahumada & Costa 2009; Ahumada 2010) and southern ends of the LHTS (Bastías et al. 1993; Ahumada et al. 2006; Mingorance 2006; Schmidt et al. 2011a, b, 2012; Salomon et al. 2013). The continuation of this thrust system farther to the south can be traced to the urban areas of Mendoza city (Fig. 1).

Quaternary activity along the LPTS

This NNW-trending Andean-type thrust bounds the eastern foothills of the Las Peñas–Las Higueras range over a distance of at least 40 km. It causes the Neogene to override the Quaternary deposits and is considered to be the main Quaternary thrust front of the Southern Precordillera between 32°16′ and 32°34′S (Cortés & Costa 1996; Costa et al. 2000a, 2005; Ahumada et al. 2006; Schmidt et al. 2011a) (Figs 1 & 2). The strike of the thrust trace fluctuates between 320° and 350°, with dip angles ranging from 15° to 55°W. However, orientations close to an east–west trend of the thrust trace have been measured at different places, suggesting significant local variation in attitude of the thrust surface.

This thrust cannot be regarded as just one single structure because of the identification of several thrust branches and geometric complexities. Thus, it is designated as the LPTS.

Fig. 2. Structural sketch of the Las Peñas–Las Higueras range highlighting the observed (black solid line) and interpreted (black dashed line) traces of the Las Peñas and Las Higueras thrust systems. 1, Las Trancas Creek; 2, Loma de Los Burros; 3, Riquiliponche Creek; 4, Los Loros Creek; 5, Las Chacras Creek; 6, YPF-SJ.SP es-1; 7, Las Chilcas Creek; 8, Los Guanacos Creek; 9, El Jarillal Creek; 10, Agua Las Muñeras Creek; 11, El Infiernillo Creek; 12, El Cóndor Creek; 13, Montecito anticline; 14, Cerros Colorados; 15, Las Peñas Creek; 16, Escondida Creek; 17, Baños Colorados Creek; 18, Lomas de Jocolí; 19, Cordón de Barda Negra; 20, Río Seco de Las Higueras; 21, La Cal Fault section. Thrust sections are indicated by white solid lines. From north to south: Riquiliponche–Las Chacras; Las Chacras–Los Guanacos; Los Guanacos–El Cóndor; El Cóndor–Las Peñas; Las Peñas–Baños Colorados, south of Baños Colorados. Cross-sections depicted in Fig. 3 are indicated as A–A′ to G–G′.

The Quaternary cumulated slip of the LPTS based on surface geology data remains unknown, and only minima values can be estimated. At the Las Peñas Creek outlet, the exposed thrust surface cored by Neogene bedrock exhibits c. 25 m of dip slip over Quaternary alluvial fan deposits.

The LPTS commonly exhibits frontal splays where bedrock is thrust over Quaternary alluvial fan deposits (Cortés & Costa 1996; Costa et al. 2000a, 2005). The westerly hanging-wall splays can only be documented reliably where there is a cover of alluvial deposits. Thus, it is likely that more thrust splays may exist in the Cenozoic sequence and remain unnoticed. The main thrust zone constitutes a deformation belt with very good exposures in the main creeks that usually cut across them at almost right angles.

Footwall strata have been thrusted in sharp and clean contacts and usually show no internal deformation, the flat-lying attitude of the strata being preserved in most cases. Only submetric deformation and drag folding have been observed in a few exposures.

The LPTS is not a homogeneous structure in terms of its geometry and morphotectonic assemblages. Two main parts can be distinguished that are separated by the Los Guanacos Creek (Fig. 2). To the north of it the topographic relief related to the LPTS Quaternary uplift is not very significant, generally being less than 30 m. Also, its trace exhibits some east–west-striking bends. Exposures of the thrust plane are rare, and the geomorphic expression of thrusting diminishes northward, where it eventually disappears completely under the piedmont alluvial cover. South of the Los Guanacos Creek, the thrust-related uplift is much more pronounced, with strong topographic and structural relief controlled by a transposed south-plunging anticline. It imposes the structural fabric of intensely deformed Cenozoic bedrock on the Quaternary deformation. Several exposures of the main thrust surface can be observed in the valleys of steeply incised streams that cut the mountain front. Fold-limb scarps are the expression of fault propagation folds in alluvial terrace deposits. The LPTS also seems to be related or connected to other folds and blind faults described south of this range (Figueroa & Ferraris 1989; Olgiati & Ramos 2003; Ahumada et al. 2006).

The many along-strike variations along the length of the thrust front warrant a partitioning of the range front in the following sections (Figs 2 & 3 and Table 1).

Riquiliponche–Las Chacras creeks section

The main mountain-piedmont slope break of the Las Peñas–Las Higueras range in this northern section coincides with the LHTS trace at a distance of c. 5 km to the west of the northward projection of the LPTS, marked by a steep bedrock scarp (Figs 1 & 2). North of the Riquiliponche River, an inselberg-like outcrop of bedrock stands out within the bajada environment of the piedmont plains (Fig. 2). These inselbergs have east-facing bedrock scarps presumably related to the LPTS, although its exposed surface has not been found in this sector.

South of the Riquiliponche River, the Quaternary alluvial surfaces appear to be interrupted by a string of low hills (up to 20 m above the current stream profile) cored by Cenozoic rocks with remnants of old alluvial surfaces perched on top (Fig. 2). The LPTS propagation has here resulted in back-tilting and folding of the alluvial cover in the hanging wall, with maximum inclinations of 11°W in the oldest levels that gradually decrease towards the younger beds (Ahumada 2010).

Between the Riquiliponche and the Los Loros creeks, surface evidence of thrust propagation is partly preserved in the form of a broad asymmetric fold scarp (c. 10 m high). Dip angles of the monoclinal front limb in alluvial deposits range from 15° to 26°E, whereas the underlying bedrock exhibits an average attitude of 350°/45°SW (Ahumada 2010).

Gently folded and thrusted Quaternary alluvial deposits have been observed in the north wall of Los Loros Creek (Figs 2–4), where the thrust propagates all the way through to the surface, overriding Quaternary alluvium (Ahumada 2010).

South of the Las Chacritas Creek (32°17.022′S–68°50.488′W), the LPTS fault surface (345°/35°SW) cored by the Cenozoic Mariño Formation deforms alluvial sediments with an estimated minimum dip slip of 3 m (Ahumada 2010).

Between the Los Loros and the Las Chacras creeks (Fig. 2), the thrust trace bends almost 90° to an east–west trend over a distance of 1.5 km, accompanied by the bedding attitude of the Mariño Formation. South of this latitude the piedmont alluvial cover located between the LHTS and LPTS has been eroded for the most part, suggesting an increase in the exhumation and uplift of this area (Fig. 2).

Las Chacras–Los Guanacos creeks section

This section is bounded by two sharp left bends of the LPTS trace. The northern bend increases the distance between the LHTS–LPTS traces to the south of it, whereas the southern bend, which produces a similar but smaller left lateral bend, marks the southern limit of the almost continuous alluvial cover preserved between both thrusts to the north of it (Figs 2 & 3).

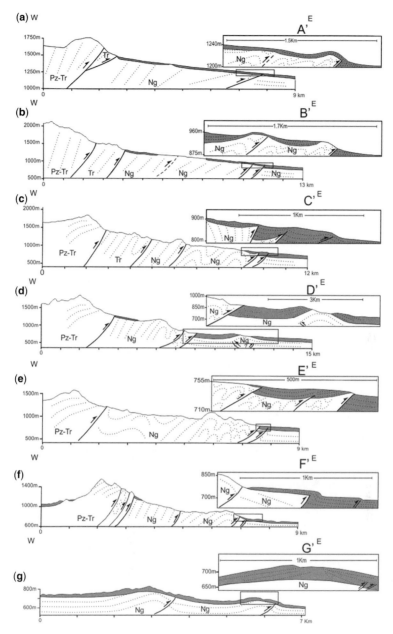

Fig. 3. Simplified cross-sections across the Las Peñas–Las Higueras range, highlighting in the insets the main representative geometries at the LPTS deformation zone. Location of cross-sections are as follows: (**a**) A–A′, north of Los Loros Creek; (**b**) B–B′, between Las Chilcas and Divisadero Bayo creeks (Fig. 5); (**c**) C–C′, north of El Cóndor Creek (Fig. 11); (**d**) D–D′, between El Cóndor and Las Peñas creeks; (**e**) E–E′, Las Peñas Creek; (**f**) F–F′, La Escondida Creek; (**g**) G–G′, Barda Negra-Lomas de Jocolí (Fig. 11).

The LPTS exhibits a sinuous trace with a general NNW trend, coincident with the bedrock bedding attitude (Figs 2 & 5). Evidence of Quaternary tectonic activity is related to a gentle coaxial folding of the older alluvial surfaces in the hanging wall, where a gentle syncline flanked by two east-verging asymmetric anticlines is preserved in the Quaternary beds (Fig. 3). The western limb of the syncline dips up to 35°E in the oldest Quaternary beds, in a close relationship with the bedrock attitude

Table 1. *Main sections identified along the Las Peñas Thrust System. See location in Figure 2.*

Section	Latitude	Thrust geometry	Surface expression
Riquiliponche–Las Chacras	32°13'42"S 32°16'30"S	Single thrust, left bends of the thrust trace	Fold scarps
Las Chacras–Los Guanacos	32°16'30"S 32°19'30"S	Hanging-wall syncline piggy-back style, several thrust splays and related anticlines	Fold scarps, drainage difluences
Los Guanacos–El Condor	32°19'30"S 32°24'02"S	Hanging-wall syncline piggy-back style, several thrust splays and related anticlines	Fold scarps, bedrock cropping out in the hanging wall barely covered by Quaternary alluvium
El Cóndor–Las Peñas	32°24'02"S 32°29'10"S	Local rejoining splays	Fold scarps with higher heights in older terrace deposits, bedrock scarps, upturned forelimb beds
Las Peñas–Baños Colorados	32°29'10"S 32°33'32"S	Parallel/rejoining splays	Fold scarps, bedrock scarps
South of Baños Colorados	32°33'32"S 32°40'08"S	Blind thrust with related anticlines	Piedmont hills, no diagnostic morphologies

(Ahumada 2010) (Fig. 5). Most of the anticline hinges in the Quaternary cover deposits have been eroded away. Only the lower parts of their frontal limb remain at the main thrust deformation zone. It is understood that such a piggy-back style of folding results from the propagation of two or more parallel thrust splays (Figs 3 & 5). The development and/or preservation of the interpreted western thrust splay and overlying fault propagation fold seems to be restricted to the Las Chacras–Los Guanacos creeks area (Fig. 2).

A flat iron-like morphology is preserved in east-tilted alluvium as remains of the fold front limb of the 'bulldozed' scarp (Costa *et al.* 2000a). In the vicinity of Los Guanacos, this fold forelimb shows attitudes of 315°/67°E in the Quaternary layers. Bedding-parallel shear phenomena in the bedrock are present near the projection of the thrust trace (335°–350°/33°–50°NW).

At a right bend of the trace (32°17.337'S–68°50.458'W), a NE-trending thrust surface (40–50°/45°NW) exhibits slickenlines with a rake angle of 70°SW. Curiously, even at orientations closely perpendicular to the thrust average trend, the strike–slip component is almost negligible (Ahumada 2010).

The almost continuous bedrock outcrops in this section constitute a topographic barrier to drainage flow, giving rise to local drainage reorganization near these hillocks, in particular, a stream difluence on reaching the western bound of the folded alluvium at Las Chilcas Creek (Figs 2 & 5). This drainage anomaly is thought to have been induced by the Late Pleistocene (or younger) backtilt of the LPTS hanging wall.

Los Guanacos–El Cóndor creeks section

A short sharp left bend of the thrust trace in the Los Guanacos Creek, marks the northern boundary of this section and the eastward shift of the mountain front from the LHTS trace to the LPTS trace to the south of it (Figs 2 & 3). In this section bedrock crops-out continuously in the thrust hanging wall, exposing isolated relics of an exhumed erosion surface. The LPTS is here characterized by a fold-related scarp affecting old alluvial deposits with different degrees of forelimb preservation.

The distribution of perched patches of Quaternary alluvial cover preserved on top of the palaeo-erosion surface permit the outlining of a gentle asymmetric syncline in the LPTS hanging wall, similar to the one described in the previous section. Between the Las Chilcas and Divisadero Bayo creeks (Figs 2 & 5), the syncline is bounded on both flanks by thrust-related anticlines.

The partly preserved alluvial cover sediments in the frontal limb unconformably overly the Mariño Formation (bedding strike *c.* 315–320°) in the Los Guanacos Creek area. The fold scarp here is *c.* 5 m high, and the tilt of Quaternary deposits ranges from 4°E in the youngest beds up to 17°E in the older beds of the middle slope, outlining a bedding fabric fanning upwards. Higher dip angles have been measured (340°/40°NE–327°/57°NE) in the oldest sediments close to the bedrock contact. This illustrates that syntectonic sedimentation has taken place over an extended period.

In the south wall of the El Jarillal Creek outlet (Figs 2 & 6), two thrust splays exposed in Quaternary deposits present a submeridional trend (*c.* 5°)

Fig. 4. The LPTS at the north wall of the Los Loros Creek outlet (see Fig. 2 for location). Arrows indicate the thrust surface (350°/25°W). Dotted lines show the bedding attitude of the Quaternary alluvium involved in thrusting. The orientative scale (bottom left) refers to the base of the outcrop.

and a related shear zone with an average thickness of 30 cm. A thin alluvial cover (c. 1.5 m) is preserved in the hanging wall, which contrasts sharply with the exposed thicknesses in the footwall alluvium (c. 11 m). Fault propagation of the western splay has led to the development of a monoclinal scarp, whereas the eastern splay does not seem to deform the near-surface sediments. A fanning-up geometry with an overlap array between both thrust surfaces suggests a contemporaneity of thrust propagation and alluvial sedimentation and permits more recent activity and/or higher activity rates for the western splay to be postulated. Also, the tendency of fold scarp height to be greater in older terrace deposits suggests a sustained activity of the thrust slip during sedimentation (Fig. 6).

Fig. 5. Oblique aerial view (NNE-looking) from the Las Chacras–Los Guanacos section (see Fig. 2 for location). Arrows indicate the LPTS main trace. Hanging-wall topography outlines a gentle coaxial syncline (solid white line), where Neogene bedrock outcrops are recognized by their light colours. Drainage difluence is within the dashed circle.

Fig. 6. South-looking view of the LPTS surfaces at the El Jarillal Creek, highlighting a fanning-up geometry recorded in the footwall strata. A topographic saddle results at the eroded hinge area, as observed in the background. Such a fold-limb scarp (reconstructed with black dashed lines) depicts more cumulated deformation than the thrust-related fold scarp related to the outcrops exposed in the forefront. The black solid line bounds the bedrock/alluvium boundary at the hanging wall. Q, Quaternary alluvial sediments; NG, Neogene clastic rocks.

The estimated minimum vertical slip of both thrust splays based on the exposed thickness of the footwall alluvial deposits is c. 15 m. This value is greater than the height of the fold-limb scarp, indicating that erosion of the hanging wall took place during thrust propagation, at least in the case of the western splay. A similar deformation pattern was also found in the south wall of an unnamed creek, 130 m south of the El Jarillal Creek (32°20.538′S–68°48.957′W).

The height of the fold-limb scarp tends to increase southwards (Fig. 6) where older Quaternary units are apparently better preserved. East-tilted Quaternary alluvial deposits in the front limb scarps give rise to a flatiron-like pattern, with the triangular shape of the dark-coloured alluvial deposits standing out in the landscape. Erosion of the front limb scarp generates topographic saddles between the Cenozoic bedrock and the Quaternary alluvium. Their alignment along the thrust scarp may give the misleading impression of a fault lineament marked by a fault saddle.

A lobate pattern in bedding (and possibly a left bend) in the thrust trace is located at Agua Las Muñeras Creek (Fig. 2). The bedrock here shows a dominant west-dipping attitude (310°/55°W average) with overturned chevron-type folds. The internal structure of a fold-limb scarp, exposed in the southern wall of this creek, reveals the presence of two main propagating thrust splays in the alluvial cover (Costa et al. 2000a) (Fig. 7). Minor splays with less cumulated deformation are also present. The eastern splay flattens at the surface (from 36° to 5°W), and the east tilt of the monocline forelimb is thought to be related to bulldozing due to the propagation of this thrust with a surface concave downwards. Clear examples of growth strata geometry are preserved in Quaternary sediments between both thrusts splays (Fig. 7), pointing once again to the recurrent interaction between aggradational and thrusting events (Costa et al. 2000a). This deformation affects only the older Quaternary deposits, whereas younger terrace beds preserved in the north side of the outlet do not seem to have been deformed by thrust propagation.

Between the El Infiernillo and El Cóndor creeks, a gentle syncline preserved in the hanging wall is best visible in the bedding attitude of

Fig. 7. South-looking view at the Agua Las Muñeras Creek outlet (see Fig. 2 for location). Two main propagating splays of the LPTS (arrows) highlighted by the concentration of light-coloured material, show the internal structure of the partly preserved fold-limb scarp. Note the alternation between fine-grained (light-coloured) and coarse-grained (dark-coloured) alluvial deposits, the former highlighted with solid black lines. Black dashed lines indicate bedding attitude (lines do not correlate). Backpack (circled) for scale.

the older alluvial beds (Figs 3, profile C–C′ & 8). The thrust zone is characterized here by at least three splays that are presumably discontinuous. They are best exposed (Figs 2, 8 & 9), where the easternmost splay gives rise to one of the best preserved piedmont scarps (Fig. 9). Preliminary mapping suggests that thrusting does not continue all the way through to the younger alluvium outcropping south of the El Cóndor Creek (Vázquez et al. 2012).

El Cóndor–Las Peñas creeks section

South of the El Cóndor Creek, the bedrock topography rises from 850 m to 1000 m (on average) and defines the Cerros Colorados frontal range, which extends down to the southernmost exposures of the LPTS (Figs 2 & 3, profile D–D′). A significant shortening and intense deformation characterize the bedrock structure of this range. This corresponds to a highly deformed anticline, with a complex geometry where tight to isoclinal flexural-flow folding is dominant in the core zone. These structures are widely exposed in the Las Peñas and nearby creeks (Fig. 2), where Cortés & Costa (1996) interpreted a west-bounding fault for the Cerros Colorados range. The structural complexity of the bedrock here described contrasts with the rather simple homocline fabric prevailing westwards of the Cerros Colorados and to the north of the El Cóndor Creek (Bea 2000).

The Cerros Colorados Range exhibits a rough whale-back morphology, which seems to correlate with the complex double-plunging anticline in the bedrock. Its periclinal closure is more obvious south of the Baños Colorados Creek (Figs 2 & 11). The Cerros Colorados Range is not capped by Quaternary alluvial deposits, although small patches of smooth topography are preserved at higher altitudes, resembling possible remnants of a palaeolandsurface.

The abrupt rise of the Cerros Colorados frontal range occurs just opposite the Montecito Neogene-growing anticline, which crops out in the piedmont area at close distance to the range front (Costa et al. 2000b, 2010; Vergés et al. 2007; Bohon 2008; Enderlin et al. 2009). This double-plunging anticline is cored by deposits of Pliocene age (Río de Los Pozos Formation) and exposes, in its flanks, synorogenic conglomerates of the Mogotes Formation of known Pliocene–Pleistocene age, overlain by younger alluvial deposits. Outcrops of these sediments present excellent examples of progressive unconformities that bear witness to continued co-eval processes of sedimentation and deformation (Costa et al. 2000b, 2010; Bohon 2008).

The dominant piedmont deposits here are light-coloured materials that have their source in the medium- to fine-grained reddish clastics Cenozoic rocks cropping out in the Cerros Colorados Range, and are characterized by a high reflectance in

Fig. 8. North-looking view from the El Cóndor Creek outlet (see Fig. 2 for location) of a western splay of the LPTS (see Fig. 9). The fault surface clearly overrides alluvial fan deposits and a syncline geometry can be recognized at the hanging wall to the upper left of the photo, partly modified by erosion. Dashed lines correspond to the bedrock–Quaternary alluvium contact and dotted lines indicate the internal bedding of the alluvial layers. The thrust exposure that crops out at the north wall of the creek in the forefront corresponds to the central splay indicated by arrows in Figure 9.

Fig. 9. Oblique NW-looking view of the El Cóndor Creek area. At least three thrust surfaces were recognized in the field (vertical arrows). The eastern trace develops one of the best preserved fold-limb scarps. The western trace corresponds to the one shown in Fig. 8. The central splay thrust bedrock over alluvial fan deposits (309°/23°SW), as also seen in Fig. 8. LHTS is displayed as a white solid line in the background.

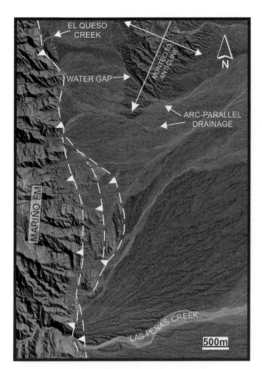

Fig. 10. Google Earth supported image exposing LPTS traces with diagnostic and suspicious evidence of Quaternary activity north of the Las Peñas Creek outlet (see Fig. 2 for location). A water gap and piedmont drainage anomalies, such as an arc-parallel pattern, highlight Quaternary upbulging at the southern closure of the Montecito anticline.

aerial images (Figs 2 & 10). It is understood, however, that older, coarse, dark-coloured alluvial fan deposits lie underneath.

Drainage running down from the range exhibits a distributive pattern on reaching the proximal piedmont plains, but soon shows a tendency to converge into more powerful streams, capable of downcutting across the Montecito anticline. These conditions have given rise to ponded alluvium and a parallel-arc drainage pattern at both periclinal fold closures (Costa et al. 2000b, 2010) (Figs 2 & 10). The local topographic barrier constituted by the Montecito anticline and related anomalies in the longitudinal profiles of streams has resulted in lower gradients on reaching the piedmont plains that explain the reduced power of incision observed at the range outlets. It has been noted that this piedmont section is 20 m lower in average elevation than the piedmont altitudes north of Los Guanacos Creek. This suggests that this section has been an active subsiding area during the Pleistocene.

The absence of Quaternary alluvium in the hanging wall, which would permit the detection and identification of diagnostic neotectonic landforms, as well as the poor stream incision across the thrust-front, have made it difficult to collect enough data to delineate the LPTS geometry along this section. In spite of the limited availability of field evidence, three west-dipping thrust splays (20–28°W) have been observed cropping out in the north wall of an unnamed creek 800 m to the south of El Cóndor Creek (32°24.406′S–68°48.863′W). Also, at El Queso Creek (32°26.348′S–68°48.403′W) (Figs 2 & 10), a small preserved patch of Quaternary conglomerates shows evidence of deformation by thrust propagation, which has given rise to overturned strata (77–84°W) in the poorly preserved frontal limb of the LPT scarp. However, no outcrop of the LPTS has been encountered.

Between the southern nose of the Montecito anticline and the Las Peñas River outlet, the most recognizable LPTS trace at the surface exhibits a remarkable curvature concave to the west (Fig. 10). The interpreted rejoining splays are consistent with a slope break in the hanging wall, and the light-coloured younger alluvial cover sediments prevailing in this section are deformed by the LPTS (Fig. 10).

Las Peñas–Baños Colorados creeks section

Significant stream downcutting at the Las Peñas antecedent river outlet has resulted in this location being the most spectacular exposure of the thrust geometry and related morphologies of the LPTS deformation zone (Fig. 12) (Cortés & Costa 1996; Costa et al. 2000a, 2005). Several sets of unpaired terraces were affected by the thrust propagation, with a maximum cumulated vertical offset in the oldest alluvial surfaces of 9.5 m (Costa et al. 2005). The LPTS here exposes two main NNW-trending splays, with minor faults in between (Fig. 3, profile E–E′). The shear zone in bedrock at the thrust surface is characterized by a 30- to 80 cm-wide belt, where argilic rocks have developed a cataclastic foliation parallel to the main thrust plane (300°/32°W on average). In the footwall, a c. 30 cm-wide shear belt could be recognized through clasts aligned along the fault surface.

Dip angles of the eastern splay (Cerro Colorado Este Fault, sensu Cortés & Costa 1996) range from 30° to 35°W, whereas the surface inclination of the western splay (Cerro Colorado Oeste Fault) varies from 35° to 45°W (Cortés & Costa 1996; Costa et al. 2000a, 2005). The eastern splay is exposed at the 14 m-high cliff in the south wall (Fig. 12), allowing an estimate to be made of an exposed apparent dip slip of c. 25 m. The bedrock is not cropping out in the footwall, making it difficult to quantify the overall Quaternary thrust slip. However,

Fig. 11. Aerial oblique NW-looking view of the plunging anticline at the southern end of the Las Peñas–Las Higueras range. Its western flank is transposed by the emerging trace of the LPTS (solid line).

Fig. 12. Exposure of the main LPTS trace at the south wall of the Las Peñas Creek. Note that the fold scarp amplitude does not correlate with the exposed thrust slip, which is regarded as a minimum value for the thrust Quaternary displacement. White dashed lines highlight bedding surfaces. Geologist for scale at the bottom centre of the image.

this minimum value for the entire Quaternary displacement of the LPTS is higher than the slip value derived from the oldest preserved terraces (c. 10 m).

At the southern domain of the Las Peñas alluvial fan, east-tilted Quaternary conglomerates belonging to the scarp forelimb of the eastern splay can be recognized. Again, clear slope breaks within the range front hillslope may suggest the location of other splays to the west of the main LPTS trace.

As the Cerros Colorados frontal range decreases in altitude towards the south, the morphologic imprint of a south-plunging asymmetric anticline becomes more evident south of the La Escondida Creek, with its eastern limb being transposed by thrusting (Fig. 11). The folding complexity of the Cerros Colorados at a mesoscopic scale also decreases southwards (Ahumada et al. 2006), probably because outcrops only expose the outer anticline core.

Schmidt et al. (2011a, b, 2012) have reported Late Pleistocene–Holocene shortening rates for the LPTS, based on ^{10}Be exposure, optically stimulated luminiscence and radiocarbon ages at the La Escondida and Baños Colorados creeks (Figs 2, 3, profile F–F' & 11). At the first location, estimations were based on a thrust splay scarp affecting terrace deposits, located to the east of the main thrust trace. They suggested a shortening rate of 2.0 ± 0.4 mm/a for this structure during the last c. 16–20 ka. At the Baños Colorados Creek, data also derived from fold scarp profiling resulted in shortening rates averaging 1.2 ± 0.2 mm/a for the last c. 13–16 ka.

South of Baños Colorados creek section

Bedrock outcrops along the mountain front do not continue farther to the south of the Baños Colorados Creek and the appearance of the range front exhibits a notorious change (Figs 2, 3, profile G–G' & 11). Cortés et al. (2011) have interpreted the existence of an oblique ramp at this latitude, which would transfer the thrust-related deformation zone c. 6–7 km to the east. Exposures of mesoscale thrust-related faulting propagating into the Mogotes Formation have been observed in quarries southwards of the Baños Colorados Creek.

At the southernmost range end, two NNW-trending hillocks stand out: from west to east, the Cordón de Barda Negra and the Lomas de Jocolí (CBN and LJ, Fig. 2). They are characterized by dark-coloured synorogenic conglomerates of the Mogotes Formation (Pliocene–Pleistocene) and younger Quaternary alluvium. According to Ahumada (2004) and Ahumada et al. (2006), it is possible to outline anticline geometries from each hillock with axes concave westwards, based on isolated data of bedding attitude. Progressive unconformities with onlap relation have been described at the southern end of these knolls (Costa et al. 2000a; Ahumada et al. 2006).

The Río Seco de las Higueras (Fig. 2) constitutes the southern boundary of the range front outcrops. It exhibits a parallel arc pattern around the CBN and LJ, with dominant fine-grained, light-coloured deposits.

Discussion

Shortening rates

Despite the outstanding exposures described here, there is a regrettable scarcity of numerical dating of the stratigraphic units involved in the Quaternary deformation. This fact has hindered, so far, a proper understanding of the thrusting processes and is the main target for current and ongoing research efforts. Several attempts using different approaches (Cosmogenic Radio Nuclides (CRN), Optical Stimulated Luminescence (OSL) and Ar/Ar) have been unprofitable or troublesome, although a discussion of recent achievements in dating these alluvial surfaces forms the basis of an upcoming paper (Schoenbohm et al. 2013).

Schmidt et al. (2011a, b, 2012) have estimated shortening rates of 2.0 ± 0.4 mm/a for the last 11–13 ka at La Escondida Creek and 1.2 ± 0.2 mm/a for the last 13–16 ka at the Baños Colorados Creek (Figs 2 & 11). The cumulative vertical offset surveyed in these localities (8.1 m) present similar values to those obtained at the Las Peñas Creek (9.5 m) (Costa et al. 2005) and are lower than the scarps surveyed at the El Cóndor Creek (19.23 m) (Vázquez et al. 2012). Chronostratigraphic correlation of alluvial terrace units along the range front remains speculative, but photogeological and morphostratigraphic attributes suggest that the terrace levels profiled through total station and kinematic Global Positioning System (GPS) at the Las Peñas and El Cóndor creeks, are older in age than those dated by Schmidt et al. (2011a, b, 2012). Therefore, it is likely that the LPTS have had higher Holocene shortening rates at the Baños Colorados and La Escondida creeks than north of that area. However, it should be taken into account that the referred topographic surveys were conducted at different thrust splays.

Future studies on Quaternary shortening rates should attempt to include the entire thrust deformation zone in the estimates of the cumulated slip, as well as conduct comparative studies among different thrust splays and different thrust sections.

According to the rates estimated by Schmidt et al. (2011a, b, 2012), recurrence intervals may fall in the range of several thousands of years. Therefore, to arrive at reliable estimates of slip rates, it is necessary to consider a timespan larger than the Holocene in order to avoid the rate values being biased by recent displacements.

Bedrock control

Two main sections can be distinguished along the LPTS bounding thrust regarding the bedrock structure (Figs 2 & 3). North of the Los Guanacos Creek, the bedrock structure is dominated by a homoclinal array, occasionally exhibiting a simple folding geometry close to the LPTS trace. The thrust surface is controlled by the bedding of the Mariño Formation. Evidence of Quaternary thrusting and bedrock faulting were found along the submeridional sections of the thrust trace, whereas along the east–west trending bends folding in the Mariño Formation prevails. These remarkable bends may result from the structural control of inherited structures (lateral ramps?) or other type of secondary faults that may had a tectonic significance in the architecture of the Mesozoic extensional basins (Costa et al. 2000a, 2005).

South of the Los Guanacos Creek (Fig. 2), there is an abrupt change in the hanging-wall morphology. The LPT thrust here transposes a complex asymmetric double-plunging anticline structure in the bedrock that makes up the Cerros Colorados frontal range, which stands out because of its macro-scale whale-back morphology. This remarkable change seems to coincide with some structural control in the cumulated slip of the LPTS.

The sudden topographic rise of the Cerros Colorados in front of the Montecito anticline may be related to the wedge-like closure between the LPTS and CSMT thrust systems. (Figs 1–3, profile D–D' & 10). This thrust convergence defines the southern oblique junction between the two opposite thrust systems referred earlier (Costa et al. 2006b; Ahumada & Costa 2009).

None of the Quaternary alluvial units cropping out west of the Cerros Colorados range has been preserved in the high part of it (Figs 1–3, profile D–D' & 11). This might indicate that uplift along the LPT in this section occurred prior to the deposition of these older alluvial bajadas. However, remnants of deformed Quaternary fanglomerates at the El Queso Creek outlet (Figs 2 & 10) indicate that such alluvial cover was at least partly overlying the Cerros Colorados range. The patches of smooth topography present in some places along the western hillslopes might be regarded as remnants of a deformed erosion surface previously capped by old Quaternary alluvial deposits, as is currently preserved west and north of this range. This alternative interpretation implies that unroofing of the Cerros Colorados took place after the deposition of the LHTS-related bajadas, contemporaneous with and in close interrelation with the eastward LPTS propagation and range uplift. This uplift subsequently led to the exhumation and complete removal of the erstwhile alluvial cover. The coarse lithological facies of these alluvial deposits cropping out between the LHTS and LPTS (Bea 2000) favour interpreting the absence of significant topographic barriers in the eastern part of the range, during the deposition of these coalescing alluvial fans.

Front-limb scarp evolution and thrust splay migration

Most LPTS outcrops do not present a complete picture of the thrust propagation into the Quaternary sediments and related landform development. However, two main geometries representing different evolutionary stages of the thrust propagation can be distinguished from bottom to top.

The thrust surface pushes bedrock up over alluvial beds in the footwall below the tip point, without noticeable mesoscopic-scale disturbance in the footwall beds (Figs 4, 5 & 12). Although a well-developed fold scarp commonly highlights the LPTS trace, the tip point of the thrust fault cannot usually be identified. This leads us to suspect that several erosive and aggradational cycles have modified the present bedrock–Quaternary boundary during thrust propagation. Usually, the thrust surface does not propagate upwards into the youngest deposits, although emerging thrusts do seem to appear due to erosion (Fig. 8).

When thrust splays propagate into the upper Quaternary alluvial cover, beds in the forelimb are bulldozed (Kelson et al. 2001) and upturned eastward, depicting a wedge thrust structure above a blind thrust (Figs 6, 7 & 13). Dissection of the upturned alluvial beds in the forelimb results in landforms that can be classified as flat-irons.

Thrust splays tend to flatten near the surface. This is thought to be controlled by the mechanical condition of less consolidated sediments in the footwall and by a sort of free-border effect, as described in other compressive settings (Ikeda 1983; Kelson et al. 2001; Ishiyama et al. 2007).

It appears that slip activity among propagating thrust surfaces has not been synchronous, so slip events may not have been co-eval along the LPTS. This statement is in line with the geometric complexities described and highlights that caution is required when comparing rates or thrusting ages among different thrust splays or sections.

The LPTS seems to exhibit a forward-breaking rupture sequence in most exposures examined, such as at Las Muñeras, El Cóndor (Vázquez et al. 2012), Las Peñas (Cortés & Costa 1996; Costa et al. 2000a,

2005) and La Escondida creeks (Cortés et al. 2011). However, a breakback sequence was observed at the El Jarillal Creek at the outcrop scale (Fig. 6).

Significant stream incision at the main creek outlets of the central part of the range may point to a blind footwall shortcut located eastwards of the LPTS, which is in agreement with the forward-breaking migration proposed for these Andean-type thrusts. Moreover, epicentres of historic earthquakes and the remarkable northward deflection of the Mendoza River (Fig. 1) may suggest the western margin of the Desaguadero High (Ortiz & Salfity 1989; Perucca 1994; Martínez et al. 2008; Suvires et al. 2012) to be the location of a rising basement structure (west-verging?) in a early stage of landscape imprint.

Thrust activity v. aggradation/erosion processes

Clear examples of progressive unconformities and onlap patterns in association with the thrust-related fold forelimb are exposed at the Agua Las Muñeras Creek (Fig. 7). Also, growth strata and progressive unconformities are present in the backlimb at the Las Peñas (Cortés & Costa 1996; Costa et al. 2005) and El Cóndor creeks (Vázquez et al. 2012). These sedimentary fabrics have also been documented in the Quaternary alluvial deposits outlining the thrust-related anticline at the southern end of the range and constitute evidence of close interaction between tectonism and sedimentation/erosion phenomena during the Quaternary (Costa et al. 2000a, b, 2011; Ahumada 2004; Ahumada et al. 2006).

Relations between thrust slip rate, sedimentation rate, local fluvial erosion, climate variations and local topography should have been too variable to be represented by a simple evolutionary model. However, in an attempt to reconstruct the interactive thrust propagation together with the related landscape, the following stages are envisaged (Fig. 13):

(1) Thrust propagation into the Quaternary alluvial cover gives rise to the development of a gentle monocline fold or flexure above the propagating thrust tip (Fig. 13a). Continued sediment supply determines the development of growth strata geometries, mainly in the hanging wall. The depositional slope in both walls records cumulative tilt with each new pulse of fault activity.

(2) Alternation between fine-grained and coarse-grained sediments suggests that thrust propagation through the surface should have led to anomalous slopes and consequently to ponded alluvium in the hanging wall. This,

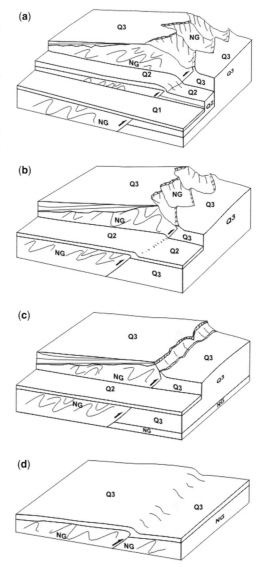

Fig. 13. Simplified sketch for the main evolutionary stages interpreted for interactions between the emerging thrust and co-eval aggradational/erosive processes from (**a**) (oldest stage) to (**d**) (youngest stage). See text for an explanation.

in turn, created rising barriers to antecedent/superimposed stream runoff towards the east. Continued stream down-cutting and sediment supply then gave rise to water gaps and entrenched fluvial terraces, usually carved into bedrock in the hanging wall and into older Quaternary alluvium in the footwall (Fig. 13b).

(3) Erosion of the fold-related scarp, concentrated in the hanging wall and hinge area, takes place in close interaction with continuing folding/tilting of the alluvial cover remnants. Sedimentation, meanwhile, concentrates in the footwall, where the oldest Quaternary sequences continue to be overthrusted. This fact explains the differences in alluvial thicknesses in the thrust hanging wall and footwall (Fig. 13c, d).

(4) Younger monoclinal scarps develop at one or more terrace levels in sections with sustained thrust activity (Fig. 13d) and/or at footwall shortcuts located to the east of the main thrust (i.e. Los Guanacos, Las Peñas, La Escondida creeks).

The imposition of fold limb scarps in the landscape suggests a long-term uplift rate higher than the sedimentation rate. However, the common occurrence of onlap relations in the thrust scarp environment points to unstable relationships between the rates of tectonic activity and sedimentation. To this has to be added the effects of dramatic changes in climate and sediment supply that have to be expected during the late Pleistocene. Thrust activity shifting into different splays might also have played a role in this dynamic scenario.

This evolutionary sketch (Fig. 13) also implies that the fold-limb scarp height should be regarded as the minimum vertical slip of the LPTS, related to a particular thrust splay. Because it is possible that not all the thrust splays were active at the surface at the same time, it may be important to address this issue when estimating activity rates.

Another matter that might lead to an underestimation of the LPTS vertical slip is the difficulty in correlating the alluvial deposits that are found on both the hanging wall and footwall of the thrust. These terrace deposits exhibit very similar or identical lithofacies, but because of the evolutionary stages sketched in Figure 13, their ages are not necessarily the same. In most cases, the alluvial deposits in the footwall are younger than those in the hanging wall. This implies that the actual height of the topographic scarp cannot unequivocally be taken as a measure of the true vertical offset, as it may well result in an underestimate.

Timing and evolution of main thrusting activity

Main Neogene mountain building along the central section of the LHTS appears to have been prior to the thrusting activity at the LPTS, but there is clear evidence of Holocene activity recorded at both its northern and southern ends (Ahumada & Costa 2009; Ahumada 2010; Schmidt et al. 2011a, b; Salomon et al. 2013). Folding of the Mogotes Formation and the underlying Cenozoic sequences are evidence of an important phase of deformation and uplift, which seems to have been coincident with thrust activity of the LHTS and LPTS. This must have taken place during Upper Pliocene–Lower Pleistocene times.

Upthrusting of the Palaeozoic–Triassic rocks along the LHTS has resulted in the topographic high ground that constitutes the entire western half of the sierra. Steep dip angles and even overturning of the Las Higueras thrust plane might suggest a back-tilt due to more recent slip along a listric surface of the LPTS that probably extends at depth towards the LHTS. These considerations point to a forward-breaking sequence of the thrust activity, at least in the central part of the Las Peñas–Las Higueras range, where LPTS splays propagate eastward into late Pleistocene–Holocene beds.

An initial phase of erosion led to unroofing of the sierra high ground and gave rise to the deposition of the Mogotes Formation at the end of the Pliocene (Yrigoyen 1993; Irigoyen et al. 2000). A subsequent period of folding and uplift caused strong deformation of the pre-Quaternary rocks located between the LHTS and LPTS and ended with the cutting of an erosion surface across it. Shortening was controlled by reactivated and inverted anisotropies of the Cuyo Basin hemigraben and by bedding planes, presumably organized in an imbricated fan geometry.

Reactivation of the LHTS, including uplift of the western half of the sierra, caused further unroofing of the Palaeozoic–Triassic formations and the deposition of an apron of coalescing alluvial fan deposits on top of this erosion surface. Shortly afterwards, or simultaneously, the main thrust activity shifted eastwards to the Las Peñas thrust, although some movement along the LHTS continued and became more significant at both the northern and southern thrust ends (Ahumada & Costa 2009). Thus, unroofing of the Paleozoic–Triassic high, deposition of the fanglomerate apron, deformation along the LPTS and dissection of the fanglomerate cover with co-eval fluvial terrace formation are understood to have been interactive processes.

Activity along the central section of the LHTS eventually ceased or decreased sometime during the Upper Pleistocene, when thrusting processes concentrated along the LPTS front at these latitudes, uplifting the Cerros Colorados range. The early stages of this morphostructure could be envisaged as a piedmont foreland (Bull 2007). However, it is difficult to figure out whether the alluvial cover derived from the main range has ever been complete over the entire width of the bedrock belt. It appears that the strongest uplifted parts of Cerros Colorados might have stood out as islands in a sea of fanglomerates.

Conclusions

The LPTS is the easternmost expression at the surface of the Quaternary Andean frontal deformation zone between 32°15′ and 32°45′S, where thrusting activity has been concentrated during the Late Quaternary.

Several sections have been differentiated along the LPTS trace on the basis of its deformation geometry and related landforms. However, a northern and a southern part stand out as a first-order division (Figs 2 & 3). The northern part, defined from the Riquiliponche to Los Guanacos creeks, is characterized by (1) isolated patches of bedrock at the thrust hanging wall, disrupting a rather continuous piedmont alluvial cover; (2) two major bedding-controlled sharp left-bends along the thrust trace; (3) discontinuous remnants of fold-limb scarps; and (4) a gently folded alluvial cover, preserved on top of the hanging wall.

The high topography in the hanging wall in the Cerros Colorados range defines the southern part of the LPTS. Their greater heights coincide with the piedmont upbulge caused by the Montecito anticline (Fig. 2), whereas the southern culmination of the LPTS is characterized by a thrust-bounded asymmetric south-plunging anticline.

Where Quaternary layers are present on the hanging wall, the LPTS is usually characterized by two or more frontal thrust splays. Blind splays not detected by terrain analysis could be present to the east in the piedmont domain.

From bottom to top, the LPTS exposures are usually characterized by a bedrock-cored thrust surface propagating into Quaternary alluvial deposits. The hanging-wall bedrock is highly deformed and the contact between the bedrock and the overthrusted deposits is sharp and well defined. It follows major bedrock bedding surfaces without noticeable perturbation of the footwall sediments.

When the thrust propagates into the upper alluvial deposits, several splays have developed in many places and have a fan-like array, commonly in the forelimb beds. Thrust propagation angle decreases upwards and growth strata are usually related.

The LPTS cannot be regarded as a typical fault propagation fold. Cumulative thrust slip decreases towards the younger alluvium not only as a consequence of slip accommodation upwards, but also because of a discontinuous stratigraphic record due to an unsteady prevalence of either thrust slip or aggradational/erosive processes. Growth strata, progressive unconformities and nested terraces are usually present in the LPTS deformation zone on both sides of the thrust plane. This fact indicates an interaction between thrust activity and sedimentation/erosion processes along the rising range front during the Quaternary. It is understood that cumulative thrust slip is not always completely preserved in the fold-limb scarps, even when old Quaternary morphostratigraphic units are involved in the deformation process.

The dominant thrust-related landform in the Quaternary alluvial units is represented by the fold-limb scarp with different degrees of degradation. Usually, only the forelimb remains and its dissection sometimes gives rise to a flat-iron morphology (Fig. 13d). The complete fold-limb scarp is preserved only in a few places with maximum heights of c. 20 m (El Cóndor Creek).

Our preliminary conclusion is that the main phase of deformation and shortening along the LPTS took place during the middle/late Pleistocene. Sustained Holocene activity has been recorded by Schmidt et al. (2011a, b, 2012), and ongoing shortening could also be distributed to the east of the LPTS (Costa et al. 2006b).

We thank all the colleagues who have shared fruitful discussions over many years. Helpful insights and suggestions provided by two anonymous reviewers have contributed to improving this manuscript. Field works were funded by the Universidad Nacional de San Luis P-340303 Project.

References

AHUMADA, E. A. 2004. *Geología y estructura del extremo sur de la Sierra de Las Peñas, provincia de Mendoza*. Unpublished BSc thesis, Universidad Nacional de San Luis.

AHUMADA, E. A. 2010. *Neotectónica del frente orogénico andino entre los 32°08′S – 32°19′S, provincias de Mendoza y San Juan*. PhD thesis, Universidad Nacional de San Luis.

AHUMADA, E. A. & COSTA, C. H. 2009. Antithetic linkage between oblique Quaternary thrusts at the Andean front, Argentine Precordillera. *Journal of South Amercan Earth Science*, **28**, 207–216, http://dx.doi.org/10.1016/j.jsames.2009.03.008

AHUMADA, E. A., COSTA, C. H., GARDINI, C. E. & DIEDERIX, H. 2006. La estructura del extremo sur de la Sierra de Las Peñas–Las Higueras, Precordillera de Mendoza. *Asociación Geológica Argentina. Serie D, Publicación Especial*, **6**, 11–17.

ALVARADO, P., PARDO, M., GILBERT, H., MIRANDA, S., ANDERSON, M., SAEZ, M. & BECK, S. 2009. Flatslab subduction and crustal models for the seismically active Sierras Pampeanas region of Argentina. *In*: KAY, S. M., RAMOS, V. A. & DICKINSON, W. R. (eds) *Backbone of the Americas: Shallow Subduction, Plateau Uplift, and Ridge and Terrane Collision*. Geological Society of America, Boulder, Memoirs, **204**, 261–278.

AUDEMARD, F. A. 1999. Morpho-structural expression of active thrust fault systems in the humid tropical foothills of Colombia and Venezuela. *Zeitschrift für Geomorphologie*, **118**, 1–18.

BABY, P., HÉRAIL, G., SALINAS, R. & SEMPÉRÉ, T. 1992. Geometry and kinematic evolution of passive roof duplexes deduced from cross section balancing-example from the foreland thrust system of the Southern Bolivian Subandean Zone. *Tectonics*, **11**, 23–536.

BABY, P., ROCHAT, P., MASCLE, G. & HERAIL, G. 1997. Neogene shortening contribution to crustal thickening in the back arc of the Central Andes. *Geology*, **25**, 883–886.

BALDIS, B., BERESI, M., BORDONARO, O. & VACA, A. 1982. Síntesis evolutiva de la Precordillera Argentina. *In: 5° Congreso Latinoamericano de Geología*, Buenos Aires, Actas, **4**, 399–445.

BARAZANGI, M. & ISACKS, B. 1976. Spatial distribution of earthquakes and subduction of the Nazca plate beneath South America. *Geology*, **4**, 686–692.

BASTÍAS, H., TELLO, G., PERUCCA, L. & PAREDES, J. 1993. Peligro sísmico y neotectónica. *In*: RAMOS, V. (ed.) *Geología de la Provincia de Mendoza. 12° Congreso Geológico Argentino Relatorio*, Buenos Aires, 645–658.

BEA, S. 2000. *Geología y estructura de la sierra de Las Peñas, al norte del río homónimo (32°30'S) Precordillera, provincia de Mendoza, Argentina*. Unpublished BSc thesis, Universidad Nacional de San Luis.

BETTINI, F. 1980. Nuevos conceptos tectónicos del centro y borde occidental de la cuenca cuyana. *Revista de la Asociación Geológica Argentina*, **35**, 579–581.

BOHON, W. 2008. *Geomorphic and fluvial analysis of a fault propagation fold, Montecito anticline, Mendoza, Argentina*. Unpublished MSc thesis, Ohio State University.

BROOKS, B., BEVIS, M. *ET AL*. 2003. Crustal motion in the Southern Andes (26°–36°S): Do the Andes behave like a microplate? *Geochemistry, Geophysics, Geosystems*, **4**, http://dx.doi.org/10.1029/2003GC000505

BROOKS, B. A., BEVIS, M. *ET AL*. 2011. Orogenic-wedge deformation and potential for great earthquakes in the central Andean backarc. *Nature Geoscience Letters*, http://dx.doi.org/10.1038/NGEO1143

BULL, W. B. 2007. *Tectonic Geomorphology of Mountains: A New Approach to Paleoseismology*. Blackwell, Malden.

CISNEROS, H., COSTA, C. H. & GARDINI, C. 2010. Análisis neotectónico del área Cerro Salinas, Departamento Sarmiento, Provincia de San Juan. *Revista de la Asociación Geologica Argentina*, **67**, 439–449.

COMÍNGUEZ, A. & RAMOS, V. 1991. La estructura profunda entre Precordillera y Sierras Pampeanas de la Argentina: Evidencia de la sísmica de reflexión profunda. *Revista Geológica de Chile*, **18**, 3–14.

CORTÉS, J. M. & COSTA, C. H. 1996. Tectónica Cuaternaria en la desembocadura del Río de las Peñas, Borde oriental de la Precordillera de Mendoza. *In: 13° Congreso Geológico Argentino*, Buenos Aires, **2**, 225–238.

CORTÉS, J., YAMÍN, M. & PASINI, M. 2005. La Precordillera Sur, provincias de Mendoza y San Juan. *In: 14° Congreso Geológico Argentino*, La Plata, Actas, **1**, 395–402.

CORTÉS, J., PASINI, M. & PRIETO, C. 2011. Propagación y migración de la estructura cuaternaria del frente montañoso precordillerano en la sierra de Las Peñas, Mendoza. *In: 18° Congreso Geológico Argentino*, Buenos Aires, Actas, 723–724.

COSTA, C., AUDEMARD, F., AUDIN, L. & BENAVENTE, C. 2010. Geomorphology as a tool for analysis of seismogenic sources in Latin America and the Caribbean. *In*: LATRUBESSE, E. (ed.) *Natural Hazards and Human-Exacerbated Disasters in Latin America*. Elsevier, Special Volumes of Geomorphology, Developments in Earth Surface Processes, Amsterdam, **13**, 29–47.

COSTA, C. H., GARDINI, C., DIEDERIX, H. & CORTÉS, J. 2000a. The Andean thrust front at Sierra de Las Peñas, Mendoza, Argentina. *Journal of South American Earth Sciences*, **13**, 287–292.

COSTA, C. H., GARDINI, C. & DIEDERIX, H. 2000b. The Montecito anticline: A Quaternary growing structure in the Precordilleran foothills of northern Mendoza, Argentina. *In: 9° Congreso Geológico Chileno*, Santiago, **1**, 758–762.

COSTA, C. H., GARDINI, C. & DIEDERIX, H. 2005. Tectónica v. sedimentación en el Río de Las Peñas, Precordillera de Mendoza. *In: 14° Congreso Geológico Argentino*, La Plata, (CD ROM), **6**, 237–240.

COSTA, C. H., AUDEMARD, F., BECERRA, F. H., LAVENU, A., MACHETTE, M. N. & PARÍS, G. 2006a. An overview of the main Quaternary deformation of South America. *Revista Asociación Geológica Argentina*, **61**, 461–479.

COSTA, C. H., GARDINI, C., DIEDERIX, H., CISNEROS, H. & AHUMADA, E. A. 2006b. The active Andean orogenic front at the Souhternmost Pampean flat-slab. *In: Backbone of the Americas*, Boulder, Abstract with Programs, 15–1.

DELLAPÉ, D. & HEGEDUS, A. 1995. Structural inversion and oil occurrence in the Cuyo Basin of Argentina. *In*: TANKARD, A., SUÁREZ, R. & WELSINK, H. (eds) *Petroleum Basins of South America*. American Association of Petroleum Geologists, Tulsa, Memoirs, **62**, 359–367.

DUMONT, J. F. 1996. Neotectonics of the Subandes–Brazilian craton boundary using geomorphological data – the Marañon and Beni basins. *Tectonophysics*, **257**, 137–151.

ENDERLIN, P. A., SCHOENBOHM, L., BROOKS, B. & COSTA, C. 2009. Identifying blind thrust anticlines in the subsurface using drainage patterns: Andean foreland of Central Argentina. *In: Transactions American Geophysical Union, Fall Meeting*, San Francisco, T43B – 2024.

FIGUEROA, D. & FERRARIS, O. 1989. Estructura del margen oriental de la Precordillera Mendocina–Sanjuanina. *In: 1° Congreso Nacional de Exploración de Hidrocarburos*, Buenos Aires, **1**, 515–529.

FOSSA MANCINI, E. 1942. Algunas particularidades del sinclinal de Salagasta, Mendoza. *Notas del Museo de la Plata*, **7**, 39–68.

GIAMBIAGI, L., MESCUA, J., BECHIS, F., MARTÍNEZ, A. & FOLGUERA, A. 2011. Pre-Andean deformation of the Precordillera southern sector, Southern Central Andes. *Geosphere*, **7**, 1–21.

GUTSCHER, M., SPAAKMAN, W., BIJWAARD, H. & ENGDAHL, R. 2000. Geodynamics of flat-subduction: Seismicity and tomographic constraints from the Andean margin. *Tectonics*, **19**, 814–833.

HARRINGTON, H. 1971. Descripción geológica de la Hoja 22c, 'Ramblón', provincias de Mendoza y San

Juan. *Dirección Nacional de Geología y Minería. Boletín*, **114**.

HORTON, B. K. 1999. Erosional control on the geometry and kinematics of thrust belt development in the central Andes. *Tectonics*, **18**, 1292–1304.

IKEDA, Y. 1983. Thrust-front migration and its mechanism – evolution of intraplate thrust fault systems. *Bulletin of the Department of Geography University of Tokyo*, **15**, 125–159.

IRIGOYEN, M. V., BUCHAN, K. L. & BROWN, R. L. 2000. Magnetostratigraphy of Neogene Andean foreland-basin strata, lat 33°S, Mendoza Province, Argentina. *Geological Society of America Bulletin*, **112**, 803–816.

ISHIYAMA, T., MUELLER, K., SATA, H. & TOGO, M. 2007. Coseismic fault-related folding. Growth structure and the historic multisegment blind thrust earthquake on the basement-involved Yoro thrust, central Japan. *Journal of Geophysical Research*, **112**, B03807, http://dx.doi.org/10.1029/2006JB004320

JORDAN, T., ISACKS, B., ALLMENDINGER, R., BREWER, J., RAMOS, V. & ANDO, C. 1983. Andean tectonics related to geometry of subducted Nazca plate. *Geological Society of America Bulletin*, **94**, 341–361.

KELSON, K., KANG, K., PAGE, W., LEE, C. & CLUFF, L. 2001. Representative styles of deformation along the Chelungpu Fault from the 1999 Chi-Chi (Taiwan) earthquake: Geomorphic characteristics and responses of man-made structures. *Bulletin of the Seismological Society of America*, **91**, 930–952.

KENDRICK, E., BEVIS, M., SMALLEY, R., JR., CIFUENTES, O. & GALBÁN, F. 1999. Current rates of convergence across the Central Andes; estimates from continuous GPS observations. *Geophysical Research Letters*, **26**, 541–544.

KENDRICK, E., BEVIS, M., SMALLEY, R., JR. & BROOKS, B. 2001. An integrated crustal velocity field for the central Andes. *Geochemistry, Geophysics, Geosystems*, **2**. http://dx.doi.org/10.1029/2001GC000191

KENDRICK, E., BEVIS, M., SMALLEY, R., JR., BROOKS, B., VARGAS, R. B., LAURÍA, E. & FORTES, L. P. S. 2003. The Nazca–South America Euler vector and its rate of change. *Journal of South American Earth Sciences*, **16**, 125–131.

KENDRICK, E., BROOKS, B. A., BEVIS, M., SMALLEY, R., JR., LAURÍA, E., ARAUJO, M. & PARRA, H. 2006. Active orogeny of the South-Central Andes studied with GPS geodesy. *Revista de la Asociación Geológica Argentina*, **61**, 555–566.

MARTÍNEZ, M. P., PERUCCA, L. P., GIMÉNEZ, M. E. & RUÍZ, F. 2008. Manifestaciones geomorfológicas y geofísicas de una estructura geológica profunda al sur de la Sierra de Pie de Palo, Sierras Pampeanas. *Revista de la Asociación Geológica Argentina*, **63**, 264–271.

MEIGS, A. J. & NABELEK, J. 2010. Crustal-scale pure shear foreland deformation of western Argentina. *Geophisical Research Letters*, **37**, 11304, http://dx.doi.org/10.1029/2010GL043220

MEIGS, A., KRUGH, W., SCHIFFMAN, C., VERGÉS, J. & RAMOS, V. 2006. Refolding of thin-skinned thrust sheets by active basement-involved thrust faults in the eastern Precordillera of western Argentina. *Revista de la Asociación Geológica Argentina*, **61**, 589–603.

MINGORANCE, F. 2006. Morfometría de la escarpa de falla histórica identificada al norte del Cerro La Cal, zona de falla La Cal, Mendoza. *Revista de la Asociación Geológica Argentina*, **61**, 620–638.

MORA, A., PARRA, M., STRECKER, M. R., KAMMER, A., DIMATÉ, C. & RODRIGUEZ, F. 2006. Cenozoic contractional reactivation of Mesozoic extensional structures in the Eastern Cordillera of Colombia. *Tectonics*, **25**, TC2010, http://dx.doi.org/10.1029/2005TC001854

MORA, A., GAONA, T., KLEY, J., PARRA, M., QUIROZ, L., REYES, G. & STRECKER, M. 2009. The role of inherited extensional fault segmentation and linkage in contractional orogenesis: A reconstruction of Lower Cretaceous inverted rift basins in the Eastern Cordillera of Colombia. *Basin Research*, **21**, 11–137.

MORA, A., PARRA, M. *ET AL.* 2010. The eastern foothills of the Eastern Cordillera of Colombia: An example of multiple factors controlling structural styles and active tectonics. *Geological Society of America Bulletin*, **122**, 1846–1864, http://dx.doi.org/10.1130/B30033.1

NORABUENA, E., DIXON, T., STEIN, S. & HARRISON, C. G. A. 1999. Decelerating Nazca–South America and Nazca–Pacific motions. *Geophysical Research Letters*, **26**, 3405–3408.

OLGIATI, S. & RAMOS, V. 2003. Neotectónica Cuarternaria en el Anticlinal Borbollón, Provincia de Mendoza – Argentina. *In: 10° Congreso Geológico Chileno*, Concepción, **11**.

ORTIZ, A. & SALFITY, J. 1989. Estructuración geológica profunda y expresión superficial: Ejemplos argentinos. *In: 1° Congreso Nacional de Exploración de Hidrocarburos*, **2**, 885.899.

ORTIZ, A. & ZAMBRANO, J. 1981. La provincia geológica de Precordillera Oriental. *In: 8° Congreso Geológico Argentino*, Buenos Aires, **3**, 59–74.

PARRA, M., MORA, A. *ET AL.* 2009. Orogenic wedge advance in the northern Andes: evidence from the Oligocene–Miocene sedimentary record of the Medina Basin, Eastern Cordillera, Colombia. *Geological Society of America Bulletin*, **121**, 780–800.

PERUCCA, J. C. 1994. Imágenes satelitarias y procesos tectónicos. *In: Proceedings of the 3° International Symposium on High-Mountain Remote Sensing Cartography*, Mendoza, 154–163.

RAMOS, V. 1988. The tectonics of the Central Andes; 30° to 33°S latitude. *In:* CLARK, S. & BURCHFIEL, C. (eds) *Processes in Continental Lithospheric Deformation*. Geological Society of America, Boulder, Special Papers, **218**, 31–54.

RAMOS, V. & KAY, S. 1991. Triassic rifting and associated basalts in the Cuyo basin, central Argentina. *In:* HARMON, R. & RAPELA, C. (eds) *Andean Magmatism and its Tectonic Setting*. Geological Society of America, Boulder, Special Papers, **265**, 79–91.

RAMOS, V., CRISTALLINI, E. & PÉREZ, D. 2002. The Pampean flat-slab of the Central Andes. *Journal of South American Earth Sciences*, **15**, 59–78.

RAMOS, V., ALONSO, R. & STRECKER, M. 2006. Estructura y Neotectónica de las Lomas de Olmedo, zona de transición entre los sistemas Subandino y de Santa Bárbara, provincia de Salta. *Revista de la Asociación Geológica Argentina*, **64**, 579–588.

Ramos, V. A., Zapata, T., Cristallini, E. & Introcaso, A. 2004. The Andean thrust system − latitudinal variations in structural styles and orogenic shortening. *In*: McClay, K. R. (ed.) *Thrust Tectonics and Hydrocarbon Systems*. American Association of Petroleum Geologists, Tulsa, Memoirs, **82**, 30–50.

Rolleri, E. 1969. Rasgos tectónicos generales del valle de Matagusanos y de la zona entre San Juan y Jocolí, provincia de San Juan, República Argentina. *Revista de la Asociación Geológica Argentina*, **24**, 408–412.

Salomon, E., Schmidt, S., Hetzel, R., Mingorance, F. & Hampel, A. 2013. Repeated near-surface folding during late Holocene earthquakes on the Cal thrust fault near Mendoza city (Argentina). *Bulletin of the Seismological Society of America*, San Francisco, **103**, http://dx.doi.org/10.1785/0120110335

Schmidt, S., Hetzel, R., Mingorance, F. & Ramos, V. A. 2011*a*. Coseismic displacements and Holocene slip rates for two active thrust faults at the mountain front of the Andean Precordillera (∼33°S). *Tectonics*, **30**, TC5011, http://dx.doi.org/10.1029/2011TC002932

Schmidt, S., Hetzel, R., Kuhlmann, J., Mingorance, F. & Ramos, V. A. 2011*b*. A note of caution on the use of boulders for exposure dating of depositional surfaces. *Earth and Planetary Science Letters*, **302**, 60–70.

Schmidt, S., Tsukamoto, S., Salomon, E., Frechen, M. & Hetzel, R. 2012. Optical dating of alluvial deposits at the orogenic front of the Andean Precordillera (Mendoza, Argentina). *Geochronometria*, **39**, 62–75, http://dx.doi.org/10.2478/s13386-011-0050-5

Schoenbohm, L., Costa, C., Brooks, B., Bohon, W., Gardini, C. & Cisneros, H. 2013. Fault interaction along the Central Andean thrust front: The Las Peñas thrust, Cerro Salinas thrust and the Montecito anticline. *In*: *Transactions American Geophysical Union, Fall Meeting*, San Francisco, Suppl., **94**, T31D–2543.

Sepúlveda, E. & López, H. 2001. *Descripción geológica de la Hoja 3369-II Mendoza, Provincia de Mendoza*. SEGEMAR, Buenos Aires, Boletín 252.

Siame, L., Bellier, O., Sébrier, M., Bourlès, D. L., Leturmy, P., Perez, M. & Araujo, M. 2002. Seismic hazard reappraisal from combined structural geology, geomorphology and cosmic ray exposure dating analyses: The Eastern Precordillera thrust system (NW–Argentina). *Geophysical Journal International*, **150**, 241–260.

Suvires, G., Mon, R. & Gutiérrez, A. A. 2012. Tectonic effects on the drainage disposition in mountain slopes and orogen forelands. A case study: the Central Andes of Argentina. *Revista Brasileira de Geociências*, **42**, 229–239.

Vázquez, F. R., Gardini, C., Montenegro, V., Costa, C. & Ahumada, E. 2012. Nuevas estimaciones de acortamiento cuaternario en el sector central de la Sierra de Las Peñas, Precordillera de Mendoza. *In*: *16° Reunión de Tectónica*, Actas, CD ROM.

Vergés, J., Ramos, V. A., Meigs, A., Cristallini, E., Bettini, F. H. & Cortés, J. M. 2007. Crustal wedging triggering recent deformation in the Andean Thrust front between 31_S and 33_S: Sierras Pampeanas–Precordillera interaction. *Journal of Geophysical Research*, **112**, B03S15, http://dx.doi.org/10.1029/2006JB004287

von Gosen, W. 1992. Structural evolution of the Argentine Precordillera: the Rio San Juan section. *Journal of Structural Geology*, **14**, 643–667.

Yañez, G., Ranero, C., von Huene, R. & Díaz, J. 2001. Magnetic anomaly interpretation across the southern central Andes (32°–34°S): The role of the Juan Fernández Ridge in the late Tertiary evolution of the margin. *Journal of Geophysical Reaserch*, **106**, 6325–6345, http://dx.doi.org/10.1029/2000JB900337

Yrigoyen, M. R. 1993. Los depósitos sinorogénicos terciarios. *In*: Ramos, V. A. (ed.) *Geología y Recursos Naturales de Mendoza. 12 Congreso Geológico Relatorio*, Buenos Aires, 123–148.

Zapata, T. & Allmendinger, R. 1996. Thrust front zone of the Precordillera, Argentina: A thick-skinned triangle zone. *American Association of Petroleum Geologists*, **80**, 359–381.

Quaternary tectonics along oblique deformation zones in the Central Andean retro-wedge between 31°30′S and 35°S

J. M. CORTÉS[1]*, C. M. TERRIZZANO[1], M. M. PASINI[1], M. G. YAMIN[2] & A. L. CASA[2]

[1]*Laboratorio de Neotectónica, IGEBA, Departamento de Ciencias Geológicas, FCEN, Universidad de Buenos Aires. Intendente Güiraldes 2160, Ciudad Universitaria, Pabellón II C1428EHA, Buenos Aires, Argentina*

[2]*SEGEMAR, IGRM, Dirección de Geología Regional, Av. General Paz 5445, Edificio 14, B1650WAB, San Martín, provincia de Buenos Aires, Argentina*

*Corresponding author (e-mail: cortes@gl.fcen.uba.ar)

Abstract: The distribution of the Quaternary deformation in the outer retro-wedge of the Andes (31°30′–35°S) is controlled by the subduction geometry, the position of the structural front, and the location of oblique pre-Cenozoic mechanical anisotropies. In the Southern Precordillera, Quaternary structures tend to group along the Barreal–Las Peñas deformation zone with a NW to NNW trend. This Cenozoic belt (31°30′–32°40′S) developed on the northern segment of the Triassic Cuyo Basin and broadened laterally during the Quaternary. New radiocarbon ages on dam deposits confirm Holocene tectonic activity on the northwestern edge. The oldest ages of dam deposits are 5810 ± 90 a BP (Cabeceras Creek) and 810 ± 50 a BP (Dolores Creek). Palaeoseismological and seismic data suggest active tectonic growth on the NW and SE extremes. Quaternary tectonics has contributed to modifying the relief along this oblique belt. This contribution is evident from the tectonic uplift of blocks (minimum 90–120 m in the Barreal block), the initial development of intermontane basins (Pampa de los Burros Basin), the Quaternary rejuvenation of tectonic depressions (at least 32–37 m of tectonic subsidence in the Vizcacheras half-graben) and the incipient development of low-relief morphotectonic units by soft-linkage of Quaternary structures.

Along the length of the convergent margin of the Central Andes, variations occur in the geometry of subduction and the rate of convergence between plates. The Nazca plate inclines c. 30° to the east, except for two lithospheric segments with subhorizontal subduction, the Peruvian segment (8°S–14°S) and the Pampean segment (27°S–33°S, Fig. 1) (Barazangi & Isacks 1976; Cahill & Isacks 1992). In the upper plate, these changes along the subducted slab are accompanied by significant variations in crustal seismicity (Gutscher et al. 2000), Cenozoic tectonic evolution and physiographic features (Jordan et al. 1983). This article presents information on the Quaternary tectonics of a sector of the Central Andes retro-wedge in Argentina between 31°30′S and 35°S. The sector under consideration straddles the southern portion of the Pampean segment and part of the normal subduction segment located further to the south (Fig. 1b). In the Andean-type orogenic wedges it is possible to identify retro-wedge domains (Ziegler et al. 1998, 2002; McClay et al. 2004) characterized by intraplate compressional structures. They are retro-wedge thrust belts composed of antithetic thrust systems with cratonward vergence. To the east of the highest axial zone of the Central Andes, the retro-wedge and the broken foreland on the flat slab segment (Fig. 2b) contain a great concentration of structures with Quaternary activity (Massabie 1987; Bastías et al. 1990, 1993; Cortés et al. 1999c; Costa et al. 2000a, 2006; Vergés et al. 2007; Schmidt et al. 2011b) (Fig. 3a) extending more than 750 km to east of the oceanic trench.

The hazard and seismic zonation of this region has been determined by INPRES (1977) based on historical and instrumental records. Nevertheless, increased availability of geological and geophysical data related to Quaternary deformation of the Andean retro-wedge in the last 15 years makes it possible to begin exploring what controls the distribution and pattern of Quaternary structures. In addition to the subduction geometry, the rejuvenation of pre-Cenozoic mechanical anisotropies appears to play a significant role in the location of Quaternary deformation. Over the entire broken foreland of the Pampean segment, multiple episodes of pre-Cenozoic fault rejuvenation have been identified, as well as contacts between Palaeozoic terranes that have controlled the architecture and location of Andean morphotectonic units (Comínguez & Ramos 1991; Schmidt et al. 1995; Ramos et al. 2001; Simpson et al. 2001).

From: SEPÚLVEDA, S. A., GIAMBIAGI, L. B., MOREIRAS, S. M., PINTO, L., TUNIK, M., HOKE, G. D. & FARÍAS, M. (eds) 2015. *Geodynamic Processes in the Andes of Central Chile and Argentina.* Geological Society, London, Special Publications, **399**, 267–292. First published online February 5, 2014, http://dx.doi.org/10.1144/SP399.10
© 2015 The Geological Society of London. For permissions: http://www.geolsoc.org.uk/permissions.
Publishing disclaimer: www.geolsoc.org.uk/pub_ethics

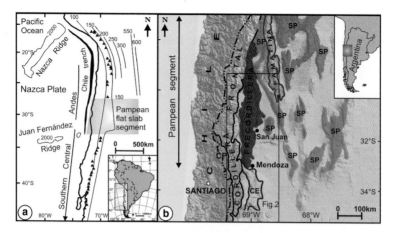

Fig. 1. Tectonic setting and morphotectonic units in the Southern Central Andes. (**a**) Geometry of the subducted Nazca plate under the South America plate. After Barazangi & Isacks (1976). The Pampean flat-slab segment is indicated by the 150 km contour line on top of the Benioff Zone and the interrupted Quaternary volcanic-arc. Modified from Yeats et al. (1997). Black triangles are active volcanoes. (**b**) Morphotectonic units in the Andes of western Argentina between 28°S and 35°S. CP, Cordillera Principal; CE, Cerrilladas Pedemontanas; SP, Sierras Pampeanas.

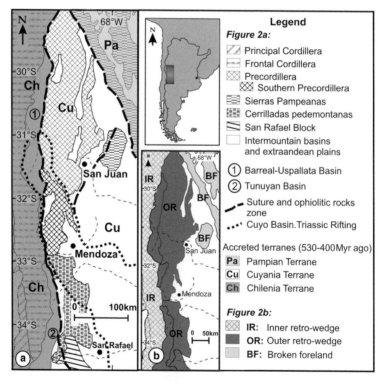

Fig. 2. (**a**) Morphotectonic units in the retro-wedge of the Southern Central Andes between 30°S and 35°S and its relationship with major palaeotectonic features. Modified from Cortés et al. (2006). Accreted terranes and Palaeozoic suture zones are distinguished. After Ramos et al. (2001). (**b**) Inner and outer retro-wedge and broken foreland of the Andes between 30°S and 35°S.

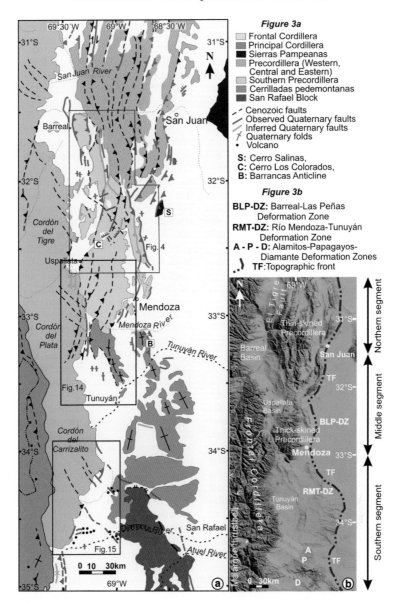

Fig. 3. (a) Distribution of Quaternary structures (post-2.4 Ma of Diaguita Phase) in morphotectonic units of the outer retro-wedge between latitudes 31°S and 35°S. The inferred or subsurface structures have not been taken into account. Modified from Cortés *et al.* (1999c). (b) Structural styles, topographic front and major oblique deformation zones along the northern, middle and southern segments differentiated in the outer retro-wedge between 31°S and 35°S.

Similarly, in the Andean retro-wedge located between 31°30′S and 35°S, numerous discontinuities in the Palaeozoic basement, among which oblique fractures of the Cordilleran belt predominate, have suffered successive Neogene tectonic rejuvenation (Baldis *et al.* 1982; Ramos & Kay 1991; Ramos 1994; Dellapé & Hegedus 1995; Japas & Kleiman 2004; Cortés *et al.* 2006; Giambiagi & Martínez 2008; Giambiagi *et al.* 2012) and Quaternary rupturas (Cortés *et al.* 2005a, 2006; Vergés *et al.* 2007; Terrizzano *et al.* 2010).

One of the aims of this paper is to analyse the distribution patterns of Quaternary deformation affecting different levels of alluvial piedmont deposits located next to the main mountain fronts of the Andean retro-wedge between 31°30'S and 35°S. Many of these structures are concentrated along the Andean oblique deformation zones, which result from reactivation of pre-Cenozoic structures. One of the major oblique structures in that region, the Barreal-Las Peñas deformation zone (31°30'–32°40'S) cuts the southern section of the Precordillera (Cortés et al. 2005a, b) (Fig. 3b). It is a 25-km-wide, 125-km-long belt characterized by large blocks and thrust sheets with a left-stepping arrangement in map view. This geometric pattern is defined by reverse and strike–slip faults striking NNW to NW. The high concentration of Quaternary structures that are present there offer the possibility of examining in this article the geometry, growth and morphotectonic role of Quaternary deformation associated with this tectonic belt.

Tectonic framework

The Southern Central Andes (Ramos 2000) extend between 22°S and 47°S (Fig. 1a). A part of the eastern margin of that orogenic belt may be considered as a retro-wedge domain characterized by an antithetical system of compressional provergent structures. Between 29°S and 35°S (Fig. 2b), the Andean retro-wedge is formed by an inner zone represented by High Andes thrust belts in the eastern margin of Frontal Cordillera and Principal Cordillera. The outer zone of the retro-wedge, characterized by a lower-contrast relief and elevation, is represented by the Precordillera–Cerrilladas Pedemontanas thrust belts and the Barreal-Uspallata and Tunuyán intermontane basins (Fig. 2). East of the retro-wedge, over the flat slab, straddles the broken foreland of Sierras Pampeanas and Famatina, consisting of Proterozoic–Palaeozoic basement fault blocks uplifted during Cenozoic deformation (Fig. 1b, 2b). The outer retro-wedge was formed during the Neogene and Quaternary as a result of the forelandward migration of the structural front. However, the morphotectonic configuration along the strike is non-uniform. Two main tectonic processes control these longitudinal variations: geodynamics boundary conditions imposed by subduction geometry (Jordan et al. 1983) and the influence of pre-Andean mechanical anisotropies on the Cenozoic deformation (Kozlowski et al. 1993; Cortés et al. 2005a, b, 2006; Giambiagi & Martínez 2008).

Between 32°S and 33°S, the subducted oceanic slab presents a first-order geometric irregularity indicated by the 150 km contour line on top of the Beniof zone (Fig. 1a). Through this irregularity it passes from a slightly inclined plate (5–10°) in the north (Pampean segment) to a plate with a normal dip (30°) in the south (Cahill & Isacks 1992; Anderson et al. 2007).

North of 33°S, the outer retro-wedge is represented by the Precordillera fold-and-thrust belt. It was formed during the last 18–20 Ma (Yáñez et al. 2001; Ramos et al. 2002) under a tectonic regime controlled by the progressive southernward collision of the aseismic Juan Fernández ridge (Pilger 1984; Von Huene et al. 1997; Yáñez et al. 2001) and the process of shallowing by the subducted Nazca plate (Barazangi & Isacks 1976; Cahill & Isacks 1992). The highest elevation of the Precordillera varies from c. 4500 m in the north to c. 3300 m in the south. South of 33°S, the physiography of the outer retro-wedge changes abruptly and the maximum elevation falls c. 1000 m. The retro-wedge is represented there by a NNW-striking belt of low hills and a fault block referred to, respectively, as Cerrilladas Pedemontanas and the San Rafael Block (Fig. 2a). At latitude 33°30'S, the uplift of the Frontal Cordillera between 11.7 and 9.0 Ma (Irigoyen et al. 2000) was followed by the thrusting and folding of the clastic foreland sequence in the Cerrilladas Pedemontanas. In a similar manner, at latitude 35°S, the structural front of Principal Cordillera (Malargüe thrust belt) migrated (13–10 Ma) towards the foreland (Kozlowski et al. 1993; Giambiagi et al. 2008), eventually uplifting the San Rafael Block. At these latitudes, Ramos & Folguera (2009) have postulated the hypothesis of a progressive shallow subduction from 15 to 5 Ma followed, in the last 4 Ma, by a subducted slab steepening.

The second boundary condition on the morphotectonic evolution of the retro-wedge was the presence of zones of lithospheric weakness and pre-Andean structures that have acted as mechanical anisotropies during the Cenozoic deformation. These anisotropies are markedly printed in the outer retro-wedge region south of 31°30'S. There, suture zones and ophiolitic belts link a mosaic of accreted terranes on the margin of Gondwana during the Famatinian orogenic cycle between 530 and 400 Ma (Astini & Thomas 1999; Rapela 2000; Ramos 2004). The locations of suture zones that link the terranes of Chilenia, Cuyania and Pampia (Ramos et al. 2001) are indicated in Figure 2a. During the Early Permian, a new orogenic phase (the San Rafael Phase) developed under oblique convergence and flat subduction (Rapalini & Vilas 1991; Kleiman & Japas 2009; Ramos & Folguera 2009) imprinted on the retro-wedge region south of 31°30'S a transpressive system of faults and folds trending predominantly NW and NNW (von Gosen 1995; Cortés et al. 1999b; Cortés & Kleiman 1999; Giambiagi & Martínez 2008). Later, during the late Permian to late Triassic, rifting developed along the Palaeozoic suture zones (Uliana et al. 1989; Ramos & Kay 1991; Franzese & Spalletti 2001) under a regime of oblique extension (Giambiagi & Martínez 2008; Japas et al. 2008). During the Late Permian to the Middle Triassic periods, a post-orogenic extensional arc led to the accumulation of thick volcanic rocks sequences included within the Choiyoi Group. Later, the Late Triassic rifting gave rise to the Cuyo Basin (Fig. 2a).

It is possible to distinguish three different morphotectonic segments in the outer retro-wedge based on the geometry of the subducted slab and pre-Cenozoic anisotropies (Fig. 3b):

(1) *North segment (30°30′S–31°30′S): flat subduction + low concentration of oblique anisotropies.* Along this segment, the Precordillera is a thin-skinned fold-and-thrust belt (Fig. 3b), characterized by an east vergence imbricated thrusts system (Western and Central Precordillera, Ortiz & Zambrano 1981; Baldis et al. 1982) with a detachment to 5–10 km deep (von Gosen 1992). To the east, the Eastern Precordillera (Ortiz & Zambrano 1981) is a west vergent thrust system with a basement decollement depth between 14 and 20 km (Allmendinger et al. 1990). The structural geometry of the eastern margin of Precordillera defines a thick-skinned triangle zone (Zapata & Allmendinger 1996).

(2) *Middle segment (31°30′S–33°S): flat subduction + high concentration of oblique anisotropies.* The Neogene and Quaternary rejuvenation of Palaeozoic anisotropies and the tectonic inversion of Cuyo Basin (Baldis et al. 1982; Ramos & Kay 1991; Kozlowski et al. 1993; Cortés 1998; Cortés et al. 1999b) created the thick-skinned Southern Precordillera morphotectonic subunit (Cortés et al. 2005b) (Fig. 3b). The oblique structures, trending mainly NW to NNW, introduced a simple shear component in the Cenozoic deformation of the Southern Precordillera. The most extensive and complex oblique structure in the Southern Precordillera is the 150 km long, NNW trending Barreal – Las Peñas Deformation Zone (BLP-DZ, Fig. 3b) (Cortés et al. 2005b, 2006). The belt is formed by five imbricated fault-block mountains and associated intermontane basins, laid in a left-stepping pattern (Fig. 4). Its location coincides with the northern portion of the Cuyo Basin and represents the onset of the thick-skinned deformation of the Precordillera.

(3) *South segment (33°S–35°S): Normal subduction + high concentration of oblique anisotropies.* The southern segment of the retro-wedge has a network of anisotropies generated by the same history of deformation that affected the Southern Precordillera. South of Mendoza River, geophysical evidence of Cenozoic tectonic inversion of Triassic halfgrabens has been obtained (Dellapé & Hegedus 1995). There, an association of NNW- to NW-trending thrusts, strike–slip faults and folds define the Río Mendoza–Tupungato Deformation Zone (Cortés et al. 2006; Casa et al. 2010) (Fig. 3b). Across the outer retro-wedge between Frontal Cordillera and San Rafael Block, Quaternary deformation is mainly focused in the Alamitos, Papagayos and Diamante fault zones (Bastías et al. 1993; Tello 1994; Cortés 2000; Casa et al. 2011) (Fig. 3b).

Distribution patterns of Quaternary deformation

Neotectonics studies on a mountain-ridge scale (Costa et al. 2000b, 2006; Cortés et al. 2006; Casa et al. 2010; Terrizzano et al. 2010; Giambiagi et al. 2012), publication of regional geological maps (Cortés et al. 1999a, b; Folguera et al. 2004; Sruoga et al. 2005) and papers on the compilation of Quaternary structures (Cortés et al. 1999c; Costa et al. 2000a) carried out in the last 15 years have allowed us to reconstruct the distribution of evidence of Quaternary deformation in the Central Andean retro-wedge between 31°S and 35°S (Fig. 3a). Nevertheless, the geochronological and palaeoseismological data of these structures remain limited in terms of giving a more complete picture of Quaternary seismic activity and its regional variations. Geological evidence of Quaternary deformation in this region consists of (1) faulted Quaternary deposits, observed in natural sections or trenches, (2) folded or tilted Quaternary strata, (3) rotation of Quaternary layers determined by geophysical methods (geoelectric tomography and palaeomagnetism), (4) landslides and dam deposits generated by tectonic activity, and (5) tectonic geomorphology features such as bedrock and piedmont scarps, fold-limb scarps, pressure ridges, sag basins, pull-apart basins and drainage anomalies. Quaternary faults are for the most part inherited faults, a product of the rejuvenation of Neogene fractures and weakness planes generated by pre-Cenozoic deformation. Neoformed faults cut the sedimentary fill of Quaternary basins and are located mainly in the piedmont environment close to the mountain fronts.

The map in Figure 3a shows a non-uniform distribution of the main Quaternary structures observed on the surface. The location and relative abundance

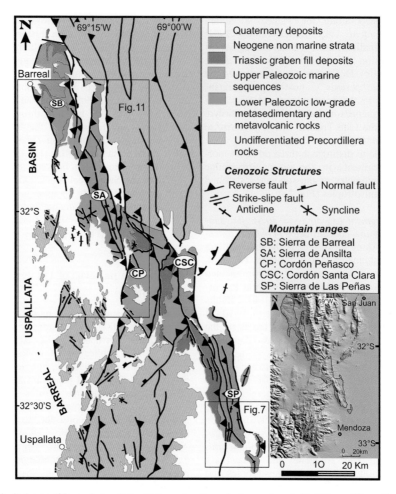

Fig. 4. Geological map of Barreal–Las Peñas Deformation Zone, in the southern half of Precordillera. The major Cenozoic structures and mountain ranges are indicated. Modified from Cortés et al. (2005b).

of Quaternary deformation is controlled by both the tectonic activity generating structures and the potential for preservation in the geological record. The entire region studied shows a broad spatial distribution of Quaternary basins and good exposure of deposits likely to register tectonic activity. In addition, the geological survey of Quaternary structures has fairly homogeneous coverage throughout the outer retro-wedge. For this reason, the distribution of tectonic evidence is assumed to be representative of the distribution of Quaternary deformation.

The map in Figure 3a shows a greater concentration of structures between 31°S and 33°S, in coincidence with the flat-slab subduction Pampean segment. This region forms part of the zone with the greatest seismic hazard on Argentine territory (INPRES 1977). In the Precordillera between 31°S and 33°S, c. 50% of the Quaternary structures are concentrated mainly around the eastern mountain front (Topographic Front, Fig. 3b) and piedmont margin of the retro-wedge. This pattern appears to be consistent with the forward-breaking sequence model of the Precordillera fold-and-thrust belt during the Cenozoic period. South of 33°S, Quaternary structures are concentrated equally on the eastern margin of the Cerilladas Pedemontanas and the San Rafael Block but are also located next to the mountain front of the Frontal Cordillera, further to the west (Fig. 3a).

Another factor controlling the location of Quaternary structures in the region is the distribution of pre-Quaternary mechanical anisotropies. North of c. 31°30′S, inherited Quaternary faults are longitudinal, with north–south to NNE–SSW trends. South of c. 31°30′S, on the other hand, the structure of the Cenozoic morphotectonic units show the influence of orogen-oblique Triassic and Permian anisotropies (Kozlowski et al. 1993; Cortés et al. 2005a, b, 2006; Giambiagi & Martínez 2008). For this reason, the Quaternary rejuvenation of that region has the same geometric pattern. In the interior of the Southern Precordillera, for example, the Quaternary structure distribution shows a high concentration (c. 78%) along the length of the Barreal–Las Peñas Deformation Zone (Fig. 3a). This oblique belt coincides with the northern section of the Triassic Cuyo Basin (Figs 2 & 5) and appears to be a preferential zone of rejuvenation and propagation of mountain range fronts during the Quaternary time (Cortés et al. 2005a, 2006). South of 33°S, there is a broad distribution of Quaternary faults associated with oblique deformation zones lying NNW and NW, such as the Río Mendoza–Tupungato Deformation Zone and the Los Alamitos, Papagayos and Diamante fault zones (Fig. 3b).

This article presents an analysis of Quaternary deformation in the Barreal–Las Peñas Deformation Zone. As a complement, information about other oblique belts located south of the Precordillera will be presented.

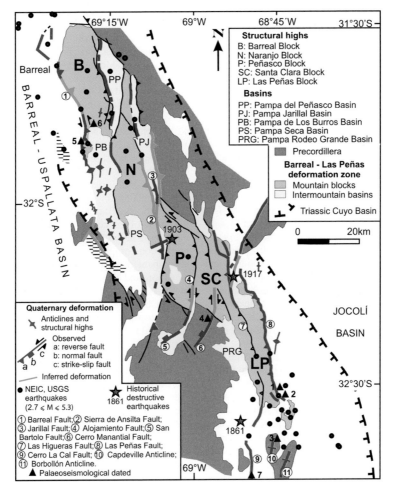

Fig. 5. Observed and inferred evidence of Quaternary deformation in the Barreal–Las Peñas Deformation Zone. The tectonic scheme shows the structural highs, basins and major neotectonic faults of that morphotectonic unit. Modified from Cortés et al. (2005a, 2006). The distribution of earthquakes (INPRES 1995; Perucca et al. 2006; Mingorance 2006, NEIC–USGS Database) is indicated. Dated palaeoseismological evidence from ▲1, Escondida Creek (Schmidt et al. 2011a, b); ▲2, Baños Colorados Creek (Schmidt et al. 2011a, b); ▲3, Capdeville Anticline (Olgiati & Ramos 2003); ▲4, Cerro Manantial Fault (Cortés et al. 2011c); ▲5, Lomas del Inca; ▲6, Cabeceras Creek (this paper); ▲7, La Cal Fault (Salomon et al. 2013).

Quaternary deformation in the Barreal–Las Peñas Deformation Zone (BLP-DZ)

Geometry and growth of Quaternary structures

A geological map and the distribution of Quaternary faults in the BLP-DZ are shown in Figures 4 and 5, respectively. Geological evidence indicates that there has been tectonic activity in all blocks of this deformation zone during the Quaternary. Deformation has been concentrated mainly along the fault zones bordering the mountain blocks. There, the Quaternary rejuvenation of Neogene fronts can be seen from (1) faulting and folding of various levels of Quaternary alluvial sequences close to the piedmont–mountain union (Barreal and Las Peñas blocks), (2) strath terraces and Quaternary alluvial deposits preserved on the slopes of tectonically uplifted blocks, (3) palustrine deposits accumulated from the closure of channels as a consequence of tectonic activity on the mountain–piedmont junction (Barreal Fault) and (4) a combination of geomorphic indices for measuring subtle perturbations along mountain-front stream courses (i.e. stream length–gradient index, mountain-front sinuosity and ratio of valley-floor width to valley height; Bull 2007).

The growth of Quaternary fault systems occurs in two ways. On the one hand, the longitudinal faults that delimit the Barreal, Naranjo and Las Peñas blocks (Fig. 5) have migrated (*sensu* Stewart & Hancock 1991) by means of rejoining, diverging and isolated splays, which uplift piedmont sectors close to the fault. In the Escondida and Baños Colorados creeks at Sierra de las Peñas (Fig. 5 & 7), splays associated with fault-propagation folds show a forward-breaking sequence, as indicated from their association with alluvial fans with apexes that migrate to the east (Schmidt *et al.* 2011a; Cortés *et al.* 2011a).

The other way by which Quaternary fault systems grow is by means of their longitudinal propagation. Because of the en echelon geometry of the BLP-DZ blocks, propagated faults end in the interior of the intermontane basins, where they cause small bedrock scarps or composite piedmont scarps (Stewart & Hancock 1990) some tens of metres high. This relationship can be observed, for example, in the propagation of the Jarillal Fault in the Pampa Jarillal Basin (Figs 5 & 6a, b) or in the propagation to the north of the Barreal Fault as piedmont scarps that penetrate the Barreal–Uspallata basin (Fig. 5). The Quaternary propagation of some BLP-DZ mountain fronts is affected by the synchronous rejuvenation of oblique fractures and anisotropies printed on the substratum. This interference is expressed in interruptions, deflections or changes in structural style, generally accompanied by changes in the relief of the mountain block. The eastern edge of Naranjo Block is bounded by the bedrock escarpment generated by the Sierra de Ansilta Fault. The southern portion of this block ends in an oblique fracture trending NE (Fig. 5). Southward, the Sierra de Ansilta Fault continues as a residual composite piedmont scarp in the Quaternary alluvium of Pampa Seca Basin (Fig. 6c, d). This relationship suggests that the oblique fracture acts as a 'leaky barrier' (Crone & Haller 1991) for larger earthquakes related to the Sierra de Ansilta Fault.

Towards the southern end of Sierra de la Peñas (Fig. 4), Quaternary propagation of the faults shows a distinct deviation towards the SE, with the loss of topographic elevation and transfer of the emergent structural front (topographic front) to the Lomas de Jocolí, 6 km to the east (Cortés *et al.* 2011a) (Fig. 7a). There, the incipient deformation of post-Mogotes Formation deposits is represented by pre-rupture structures (Riedl shears) and minor folds (Fig. 7b, c). In the north of the Las Peñas block, interference relations between NNW-striking thrusts with structures trending NE have been studied by Figueroa & Ferraris (1989), Vergés *et al.* (2007) and Ahumada & Costa (2009).

The southward propagation of the Barreal Fault during the Quaternary has been interfered by oblique structures oriented to the NW (Cortés & Cegarra 2004; Terrizzano *et al.* 2010, 2011a). To the south, the Barreal Fault switches laterally to a system of en echelon structural highs forming the Pampa de los Burros Shear Zone (Fig. 8). The interaction of these two structural systems is accompanied by a reduction in the Barreal Block elevation and the southward-growing significance of strike–slip movements. In the Peñasco and Santa Clara blocks (Fig. 5), the Quaternary southward propagation of the Alojamiento and Cerro Manantial longitudinal faults corresponds to releasing bends controlled by oblique fractures striking NE (Cortés *et al.* 2011b).

The height of the mountain front bounding the Barreal, Naranjo and Las Peñas fault blocks is greater in the central part of those blocks. Towards their northern and southern edges, progressively younger Quaternary strata have been faulted and folded (Yamin 2007; Terrizzano 2010; Cortés *et al.* 2011a). These elements, as well as the higher concentration of Quaternary deformations at the ends of the mountain blocks, and their propagation to adjacent basins, are a sign of gradual widening of the Barreal–Las Peñas Deformation Zone during its late Cenozoic evolution. The along-strike propagation of faults that delimited blocks into the adjacent intermontane basins during the Quaternary varies between a maximum length of 12–13 km (Barreal Fault and Las Peñas Thrust)

Fig. 6. Propagation of Quaternary faults on intermontane adjacent basins. (**a**) Bedrock scarp of Jarillal Fault along the Pampa Jarillal Basin. (**b**) The arrow points to the steep slope of Jarillal Fault scarp. (**c**) Propagation of Sierra de Ansilta Fault into Pampa Seca Basin as a piedmont scarp. (**d**) Northeastern view of degraded upper original surface of piedmont Sierra de Ansilta scarp.

and a minimum of 3–4 km (Sierra de Ansilta and Cerro Manantial faults). The average propagation on both sides of the BLP-DZ is 8–9 km. As a result, there seems to be a replication on a regional scale of the enlargement processes of the mesoscopic shear zones with the deformation, as have been reproduced in analogue laboratory tests and observational models (Naylor et al. 1986; Cox & Scholz 1988; Marrett & Allmendinger 1990).

Late Cenozoic kinematics of the BLP-DZ

The late Cenozoic faults, with NNW to north–south directions, bound blocks in the BLP-DZ and are generally high-angle faults dipping at 45–60°, with a lesser proportion of thrusts dipping at 25–40°. Examples include the Sierra de Ansilta and Las Peñas faults (Cortés et al. 1999a, b; Ahumada & Costa 2009; Schmidt et al. 2011b; Salomon et al. 2013). Kinematic indicators measured to date within the BLP-DZ indicate that they are pure or oblique-slip reverse faults (Cortés & Pasini 2008; Terrizzano 2010). In contrast, in fault zones with NW to WNW orientations, the left-lateral slip component predominates (Fig. 5). Quaternary deformation evidence suggests a similar kinematic pattern. Fault scarps running NNW in the main mountain fronts indicate shortening and dip–slip components with uplifts of several tens of metres. On the other hand, the Peñasco and Santa Clara blocks are controlled by longitudinal faults in a north–south direction. There, releasing bends with normal faults in the southern end of the Alojamiento and Cerro Manantial faults (Fig. 9a) enable the inference of right-lateral slip components in the Quaternary movements along these longitudinal faults.

In addition, Quaternary fault zones with a NW trend are associated with geomorphic features that indicate a left-lateral slip. One excellent example of en echelon geometries associated with such structures is the Pampa de los Burros Fault Zone (Fig. 8a). This structure is defined by structural highs with lengths of 4–8 km, left stepping along a broad belt with a NW strike (Cortés & Cegarra 2004; Terrizzano et al. 2010). Smaller shear zones with stepped pressure ridges (2–3 km long) and trending in the same direction have been recorded

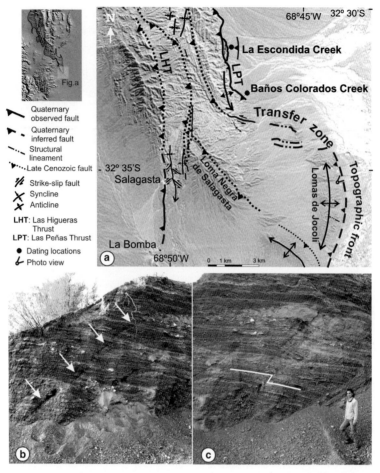

Fig. 7. Quaternary structures at the southern end of Sierra de las Peñas. (**a**) Las Peñas thrust deflected to the SE, transferring the Quaternary deformation to the eastern slope of Lomas de Jocolí. The emergent structural front (topographic front) shifted 6 km to the east through oblique faults and lineaments. Modified from Prieto (2010) and Cortés *et al.* (2011*a*). (**b**) Incipient Quaternary deformation of alluvial sediments to the east of Lomas de Jocolí. Riedel shears indicates a sense of shear in a pre-fault stage along narrow deformation zones. (**c**) Minor fault-propagation folds associated with Riedel shears. This mesoscopic structures indicates blind faults at depth.

within this fault zone (Terrizzano *et al.* 2010). Notable are the Lomas Bayas High, the Pampa de los Burros Anticline and the Lomitas Negras High (Fig. 8b), consisting of folds and uplifted basement blocks that deform Pleistocene alluvial sequences.

The structure comprising left-stepping blocks and an association of reverse and strike–slip faults is characteristic of the Barreal–Las Peñas Deformation Zone. This combination of structures suggests a left transpressive regime in late Cenozoic times (Cortés *et al.* 2005*b*, 2006; Terrizzano *et al.* 2010). Such a tectonic regime is consistent with the directions of compressive stresses obtained by Siame *et al.* (2005) for the Andean Cenozoic deformation of the San Juan Precordillera. It is similarly consistent with the Neogene and Quaternary shortening obtained by Cortés & Pasini (2008) for the Southern Precordillera.

Ages of late Cenozoic tectonics in the BLP-DZ

The Neogene tectonic evolution of Precordillera associated with the shallowing of the subduction zone was noted for the progressive eastward shifting of deformation in the last *c.* 20 Ma (Jordan *et al.* 1993). Neogene deformation was also diachronic along the strike (Beer *et al.* 1990; Jordan *et al.* 1993; Ramos *et al.* 2002), becoming younger

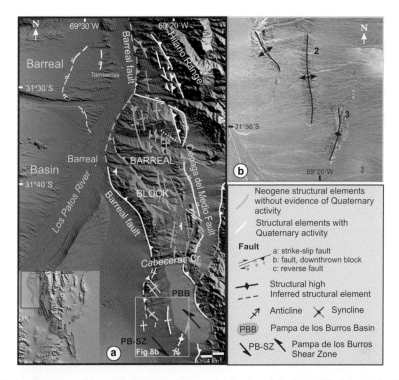

Fig. 8. (a) During Quaternary time, southward propagation and migration of the Barreal Fault is interfered by a NW-trending left-lateral shear zone (Pampa de los Burros Shear Zone). Stepped structural highs of that zone begin to define a western border of a developing Quaternary basin (Pampa de los Burros Basin). Modified from Terrizzano *et al.* (2011*a*). (b) Detail of the en echelon pattern of Quaternary structural highs of Pampa de los Burros Shear Zone. 1, Lomas del Inca Anticline; 2, Pampa de los Burros Anticline; 3, Lomitas Negras High.

towards the south, in agreement with the southward collision of the Juan Fernandez aseismic ridge (Pilger 1984; Von Huene *et al.* 1997). Timing constraints on the initial uplift of the Southern Precordillera are scarce. At the Barreal Block, thrusted foreland sequences of the Lomas del Inca Formation overlie andesitic rocks (Yamin & Cortés 2008) dated by Leveratto (1976) as 20.6 ± 2.5 Ma (K–Ar). Similarly, at the central region of the Southern Precordillera, the Alojamiento thrust cuts volcanic breccias of the Puesto Uno Formation (Massabie *et al.* 1985; Cortés *et al.* 1999*b*) that are dated by Reyna *et al.* (2011) as 19.64 ± 0.54 Ma (Ar/Ar). An equivalent age (K–Ar, 18.4 ± 0.7 Ma) for Puesto Uno Formation volcanic rocks from Cerro Los Colorados was obtained by Kay *et al.* (1991). All these data confirm the post-Early Miocene age of Neogene deformation in the Southern Precordillera.

The initial uplift of the Southern Precordillera would have started at 11 Ma according to Yáñez *et al.* (2001). This date was obtained from the estimated age for the arrival of the Juan Fernández Ridge collision to the 33°S latitude. By means of quantitative geomorphological analysis of palaeolandscape remnants (Walcek & Hoke 2012), the uplift of the southernmost Precordillera is constrained to have initiated at *c.* 10 Ma. Volcanic ash-rich deposits near the base of growth strata of Cerro Salinas Anticline were correlated by Vergés *et al.* (2007) with Tobas Angostura Formation and other Miocene volcanic units of Precordillera. From the Cerro Salinas seismic profile analysis, the shortening and uplift of the eastern piedmont of Precordillera would have been initiated after *c.* 8.5 Ma.

Evidence of the last orogenic phase that substantially modified the morphotectonic configuration of the Southern Precordillera and Cerrilladas Pedemontanas may be obtained along the eastern margin of the retro-wedge in Sierra de las Peñas and Barrancas Anticline (Figs 3 & 4). There, a thick sequence of Miocene to Pliocene continental deposits was folded, faulted and incorporated into the mountain front by the Diaguita Phase (Milana & Zambrano 1996; Cortés *et al.* 1999*c*; Costa *et al.* 2000*b*; Chiaramonte *et al.* 2000; Sepúlveda 2001; Prieto 2010). Time constraints on Diaguita deformation is given by the Mogotes Formation, the

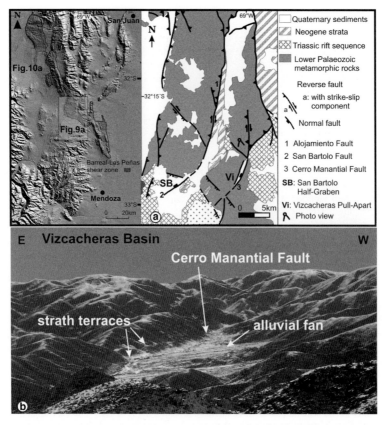

Fig. 9. (a) Releasing bends in the southern end of Alojamiento and Cerro Manantial faults. Tectonic basins are represented there by the San Bartolo half-graben and Vizcacheras pull-apart basin, respectively. (b) Southward view of Vizcacheras Basin. Distal segments of alluvial fans were displaced and uplifted as a result of Late Pleistocene activity of the Cerro Manantial Fault. Strath terraces were preserved on the slope of this bedrock scarp. Modified from Cortés et al. (2011b, c).

youngest unit of the Neogene sequence. From stratigraphical correlations (Yrigoyen 1994) and isotopic dates on tephra horizons preserved in Quaternary terraces (0.39 ± 0.03 Ma and 0.24 ± 0.05 Ma), Irigoyen et al. (2000) suggest that Mogotes Formation is 3–1 Ma old. Nevertheless, the minimum age for this deformed unit is indicated by its cover of pediment with tuffs (2.4 ± 0.7 Ma) and fill terraces (1.9 ± 0.8 Ma) dated by Sarewitz (1988). The Quaternary tectonics considered in this paper corresponds to events that have deformed post-Mogotes Formation alluvial deposits accumulated in a Quaternary landscape already modified by the Diaguita Phase.

Palaeoseismological record in the eastern sector of BLP-DZ

Geochronological data from the palaeoseismological record remain scarce, but provide constraints on the rupture events age for some faults. In the Southern Precordillera the ages of strata deformed by active tectonics published so far are taken from structures of the eastern Precordilleran front. Schmidt et al. (2011a) determined ^{10}Be exposure ages for three fluvial terraces located in the piedmont of the Las Peñas Block, at the mouth of the Escondida Creek and Baños Colorados Creek (Figs 5 & 7a). They obtained an age of 3–5 ka for terrace T2, 11–13 ka for T3 and 16–20 ka for T4. These geoforms are cut by splays connected to the Las Peñas thrust and produced differential displacements on the surface of those terraces. Terrace deposits containing growth strata have been deformed by blind thrusts associated with fault-propagation folds (Cortés et al. 2011a; Schmidt et al. 2011b). Ages and geological evidence indicate Holocene and Late Pleistocene tectonic activity, synchronous with the fluvial sedimentation of the mountain front.

Other palaeoseismological evidence of active tectonics in the southeastern edge of the BLP-DZ is indicated by an association of fault scarps located on the piedmont plain between the Sierra de Las Peñas and the city of Mendoza. These scarps correspond to the Cerro La Cal Fault (Bastías et al. 1993) or La Cal Fault Zone (Mingorance 2006; Salomon et al. 2013) and are the southern continuation of the Las Higueras Fault (Fig. 5). Various morphometric, seismological and historical faulting parameters have enabled Mingorance (2006) to correlate historic ruptures in these scarps with the earthquake in 1861 that destroyed the city of Mendoza. Scarps in this fault zone have been identified in the Mendoza city subsoil and nearby areas (INPRES 1995; Mingorance 2000). Other seismic ruptures of Cerro La Cal Fault were studied 5 km north of Mendoza city by Schmidt et al. (2011b). There, displaced terraces (T2–T4) yield ages of c. 0.8 ka (optical stimulated luminescence, OSL), c. 3.9 ka (^{14}C) and ≤12 ka (^{10}Be), respectively. The youngest terrace indicates well-preserved fault-propagation folding, emphasizing the active character of the La Cal Fault, with at least two earthquakes in the past c. 800 years (Salomon et al. 2013).

East of this fault, the Late Pleistocene deformation is evidenced by the folding of sequences with tuffs dated (^{39}Ar/^{40}Ar) by Olgiati & Ramos (2003) at 27.9 ± 0.6 ka and 16.2 ± 0.6 ka (Capdeville and Borbollón anticlines, Fig. 5).

To the north, Las Higueras Fault has propagated along the mountain front of Santa Clara Block and has deformed unconsolidated deposits of young terraces (Ahumada & Costa 2009). The Santa Clara Block is furthermore segmented by the Cerro Manantial Fault (Cortés et al. 1999c) (Figs 5 & 9a). The southern bend of this fault has been rejuvenated during the Quaternary as a normal fault that affected both strata and Quaternary erosion surfaces (Cortés & Pasini 2008; Cortés et al. 2011b). There, a west-facing bedrock fault scarp originating from the local transtensional regime has been an obstacle to alluvial fans drainage (Fig. 9b), favouring the accumulation of silty dam deposits in the sunken western segment of the fault. Natural trenches excavated by fluvial streams in a hanging wall of the fault show alluvial coarse-grained gravel deposits, dark brown in colour, covered with an upper section of thinly bedded, fine-grained sedimentary strata. This uppermost section, interpreted as dam deposits, comprises mainly silty to fine sand sediments that are pale brown in colour, 1.5–2 m thick, and contains small carbonaceous stalks and scattered fine gravel at the base. Two samples of dam deposits were dated by OSL, giving ages of 2190 ± 400 and 1320 ± 200 a (Cortés et al. 2011c). Dam deposits have, in turn, been gently folded as a consequence of subsequent fault rejuvenation. From these relationships and ages the active nature of the Cerro Manantial Fault has been inferred, as well as from the occurrence of at least two fault rupture events on the surface during the more recent Holocene period (Cortés et al. 2011c).

Palaeoseismological record in the western sector of BLP-DZ: new data

Intermontane basins on the western edge of the Precordillera between 31°S and 32°S, such as the valley of the El Tigre Fault (Fig. 3) and the Pampa del Peñasco and Pampa de los Burros basins (Fig. 5), contain a deformed Quaternary alluvial fill. These deposits make up prograding telescopic alluvial fans and fluvial plains of differing piedmont aggradation levels. On the basis of ^{10}Be exposure ages for selected alluvial surfaces associated with the El Tigre Fault, Siame (1998) showed that the depositional periods ended during successive major interglacial stages. Cosmogenic dating ranges from 20 to 700 ka. To the north of 31°S, trench surveys in the El Tigre Fault and ^{10}Be ages in associated Quaternary surfaces (INPRES 1995; Siame 1998) have confirmed displacements during the Upper Pleistocene and Holocene periods. Nevertheless, until now, there has been no dating of palaeoseismic features proving recent activity in the western half of the BLP-DZ.

Evidence of Quaternary deformation along the Barreal Fault is provided from Cabeceras Creek to the south (Fig. 10a). The oldest Quaternary alluvial deposits (Q1, Cesco Formation; Yamin & Cortés 2008) preserved at the top of Barreal Block (Figs 10 & 11a) were uplifted and tilted as a result of tectonic activity in this fault. Palaeozoic basement in the upthrown block is exposed at the mountain front. Oblique faults and salients on the mountain fronts give rise to geometric and structural segments (Crone & Haller 1991) in the Barreal Fault (Fig. 10a). Geomorphic and stratigraphic evidence of palaeoseismological activity have concentrated on two of these segments (Cabeceras and Lomas del Inca segments, Fig. 10a). Fluvial and palustrine sediments of the Casleo Formation (Yamin & Cortés 2008) were preserved along antecedent fluvial valleys associated with these two segments (Cabeceras and Dolores creeks). Palustrine facies, considered as dam deposits, are composed of laminated silty and clayey sediments.

A wineglass pattern characterizes the drainage basin of Cabeceras Creek. To the east of the mountain front, Cabeceras Creek forms a U-shaped valley, with the valley-floor width being 290 m, on average. The tectonically induced downcutting along the mountain front yields a narrow V-shaped

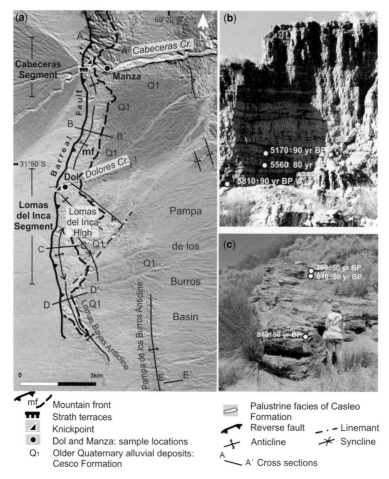

Fig. 10. (a) Quaternary deformation along the Barreal Fault from the Cabeceras Creek to the south. The minimum tectonic uplifts of the older Quaternary deposits (Q1) were measured along five cross-sections. The Cabeceras and Lomas del Inca segments contain stratigraphical and geomorphological evidence of active deformation. (b, c) Sampled locations for radiocarbon dating in dam deposits of the Casleo Formation profile at (b) Cabeceras Creek (Manza) and (c) Dolores Creek (Dol).

valley, where the valley-floor width is 90 m, on average. The recent tectonic uplift of the mountain front is indicated by two strath terraces (Fig. 11b) located c. 12 and c. 25 m above the current channel of the stream, respectively. Several knickpoints have accumulated to create a large step (waterfall) in the streambed. The palustrine facies are restricted to the east of the mountain front (Fig. 10a). The thickness of the Casleo Formation decreases in the mountain front, where it is represented by coarse-grained fluvial facies. Deposits and erosive features resulting from landslides have not been found along the mountain front of the Barreal Block. Dam deposits of the Casleo Formation were also preserved in the Dolores Creek valley on the Lomas del Inca High (Fig. 10a). The

tectonic uplift of the mountain front in this structural high is indicated both by the convexity of the longitudinal profile of its rivers (Fig. 12a, b) and by the presence of hanging valleys on the Barreal Fault scarp (Fig. 12c). The tectonically induced downcutting in the Dolores creek is also supported by the presence of younger alluvial piedmont deposits faulted next to the front. A knickpoint in the bedrock (c. 14 m high) is located at the mouth of this valley. The geomorphic and structural features described above allow us to interpret that the palustrine facies of the Casleo Formation accumulated as a result of an obstruction to drainage and changes in the slope channel as a consequence of tectonic uplift along defined structural segments of the Barreal Fault.

Fig. 11. (a) Western mountain front of the Barreal Block at the Lomas del Inca segment, from the Barreal Basin to the east–SE. The southern portion of Barreal Block is a Quaternary step emerging at the piedmont plain of the higher Sierra de Ansilta (back). (b) Two strath terraces at the southern margin of Cabeceras Creek in the mountain front of Barreal Block.

Three samples of clayey silt with abundant organic material derived from the lower section of the Casleo Formation, cropping out in the Cabeceras Creek valley (Manza site, Fig. 10a & b), were dated (^{14}C) as 5810 ± 90 a BP, 5560 ± 80 a BP and 5170 ± 90 a BP, (Table 1). In addition, a further three samples of palustrine sequences of the Casleo Formation, located in the Dolores Creek on the Lomas del Inca High (Dol site, Fig. 10a, c), yielded the following (^{14}C) ages: 810 ± 50 a BP, 610 ± 50 a BP and 300 ± 50 a BP (Table 1). The obtained ages constrain to c. 5810 a BP and c. 810 a BP the minimum Holocene age for the palaeoearthquakes responsible for the beginning of

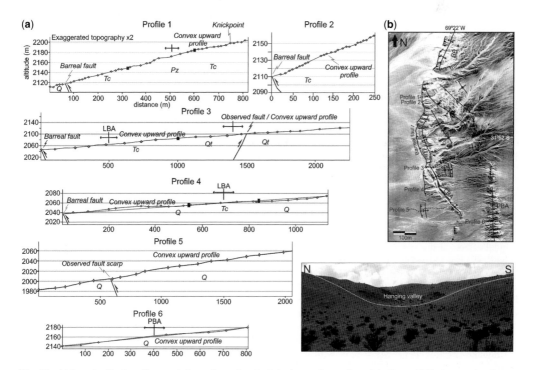

Fig. 12. (a) Longitudinal profiles carried out along river beds in the southern edge of the Barreal Block; note the clear correlation between upward convex profiles and the Quaternary tectonic structures. Distances are expressed in metres, and in all the cases the vertical exaggeration is ×2. LBA, Lomas Bayas Anticline; PBA, Pampa de los Burros Anticline. (b) Location of the longitudinal profiles and structural context of the southern edge of the Barreal Block and its neighbouring piedmont. BB, Barreal Block. (c) Field view of the hanging valley corresponding to Profile 1; note the pickup to the right side of the picture, which is on the current valley floor.

Table 1. *Radiocarbon ages obtained from northwestern BLP-DZ*

Sample site	Coordinate	Sample depth (m)	Lab. no.	Carbon quantity (%)	^{14}C age (a BP)	Relative area under distribution	Calibrated age range (2 Sigma)
Manzanares, Cabeceras Creek	31°47'57"S	7.83	MANZA-1	3.22	5810 ± 90	0.04	6319–6374
						0.96	6390–6748
						0	6769–6771
	69°21'39"W	6.73	MANZA-2	3.96	5560 ± 80	0.01	6023–6048
						0.01	6063–6078
						0.03	6114–6153
						0.95	6175–6483
		6.05	MANZA-3	7.87	5170 ± 90	0	5619–5626
						0.95	5644–6023
						0.01	6049–6063
						0.02	6078–6115
						0.01	6153–6175
Lomas del Inca, Dolores Creek	31°50'28"S	2.4	DOL-1	1.15	810 ± 50	0.03	571–590
						0.97	640–773
	69°22'58"W	1.1	DOL-2	0.64	610 ± 50	0.53	511–577
						0.47	582–650
		0.6	DOL-3	0.62	300 ± 50	0.03	157–164
						0.44	282–331
						0.52	366–442

Measurement method: beta-counting. $^{12}C/^{13}C$ (estimated): -24 ± 2‰. Radiocarbon ages were calibrated using CALIB 6.0.1 Program. South Hemisphere calibration curves: SHCal04 14c (McCormac *et al.* 2004); radiocarbon 46,1087–1092. Material: charcoal from palustrine dam deposits. Samples analysed in Laboratorio de Tritio y Radiocarbono (LATYR).

the mountain front uplift and the closure of the valleys in which were deposited the palustrine facies of the Casleo Formation.

Historical and instrumental record

The active tectonics of the BLP-DZ is indicated primarily by the historical and instrumental record. Three destructive crustal earthquakes with magnitudes from 6 to 7 (Fig. 5) took place in the nineteenth and twentieth centuries (INPRES 1995; Mingorance 2006; Perucca *et al.* 2006). These earthquakes caused damage in the city of Mendoza or the surrounding towns, and all generated co-seismic geological effects at the surface. The most destructive seism was the Mendoza earthquake (20 March 1861, M = 7), which caused 6000 deaths and totally destroyed the city of Mendoza. The earthquake on 12 August 1903 (Ms = 6), with its epicentre in the northern margin of the Peñasco Block, destroyed homes and churches in the urban region of Greater Mendoza. The Panquehua earthquake (27 July 1917, Ms = 6.5), with its epicentre in the far north of the Sierra de las Peñas, was one of the first in Argentina for which there is an instrumental record. It caused hydrogeological disturbances and liquefaction phenomena to the north of the city of Mendoza (Perucca *et al.* 2006).

The NEIC-USGS database records 49 smaller earthquakes (M = 2.7–5.3) in the BLP-DZ region during the 1973–2012 period. Epicentres were mainly concentrated along the two extremes of the deformation zone (Fig. 5), both in the orogenic front of the Precordillera (Sierra de las Peñas) and in its northwestern margin (Barreal and Naranjo blocks).

Accordingly, seismological and palaeoseismological data indicate that Pleistocene tectonics of the BLP-DZ continued in the Holocene. Seismological evidence gathered to date has mainly focused on the SE and NW extremes of the deformation zone, where the faults grow into broad depressions filled with Quaternary deposits (Jocolí and Barreal–Uspallata basins, respectively, Fig. 5). Although data remain limited, the mentioned broadening of the BLP-DZ by means of the fault scarp propagation on the stepped lateral basins seems to be accompanied by a propagation of the deformation towards the NW and SE edges of this deformation zone.

Quaternary deformation and relief construction in the BLP-DZ

Much of the Quaternary deformation in BLP-DZ has been channelled through the principal pre-existing Neogene structures. As these structures control different morphostructural units, their Quaternary rejuvenation has contributed to accentuating the relative relief. This is particularly evident at the ends of the deformation zone in the Las Peñas and Barreal blocks, where the earlier Quaternary deposits have been partially uplifted by growing fault blocks. The Barreal Block, for example, is bounded by bedrock and piedmont fault scarps. Studies of the Late Cenozoic stratigraphy of that block (Yamin & Cortés 2004, 2008) have made it possible to discard the hypothesis that its tectonic uplift was exclusively Quaternary in age, as proposed by Zöllner (1950). On the basis of structural cross-sections through the western mountain front of Barreal Block at the latitude of Cabeceras Creek (sections A–A′ and B–B′, Fig. 10), the minimum tectonic uplifts calculated for the oldest Pleistocene deposits of the piedmont alluvial fans (Q1, Cesco Formation; Yamin & Cortés 2008) located on the top of the block (Fig. 11) and displaced along the Barreal Fault are 115–120 m in section A (31°47′12″S) and 90–95 m in section B (31°49′14″S).

Incipient processes for the formation and interconnection of new morphotectonic units have been detected by means of the combined use of fluvial tectonic geomorphology and detailed cartography on the broad bajadas on the northwestern margin of the BLP-DZ, south and west of the Barreal and Naranjo blocks, respectively. There, structural highs uplifted by Quaternary tectonic activity emerge, apparently disconnected between themselves, on the extended piedmont alluvial plain (Fig. 13a). At Lomas del Inca High–Lomas Bayas Anticline and Pampa de los Burros Anticline (Figs 10a & 11), Pleistocene deposits of the Cesco Formation (Q1) were folded and thrust. In these structural highs, the minimum Quaternary uplift calculated from cross-sections (Fig. 10a) is c. 50 m (section C–C′), c. 55 m (section D–D′) and c. 30 m (section E–E′), respectively. Scarplets, wind gaps, hanging valleys and upward convexity along the longitudinal profile of stream valleys are common in such structural highs (Fig. 12). Detailed mapping of this zone also reveals the presence of drainage anomalies associated with subtle tectonic alterations to the original slope of the alluvial plain (Terrizzano & Cortés 2009; Terrizzano et al. 2011b). These include anomalous changes of patterns and longitudinal profile, sinuosity and stream adjustment to tilting, river diversions, and numerous fold and fault piedmont scarps. Analysis of the distribution of such anomalies shows that they are grouped in zones that link and interconnect the various seemingly isolated structural highs

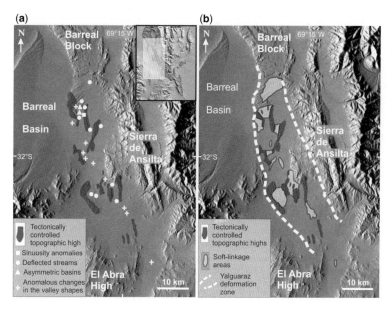

Fig. 13. Quaternary deformation and incipient morphotectonic development along the piedmont of Sierra de Ansilta. (a) Structural highs apparently isolated in the piedmont alluvial plain. The locations of various types of drainage anomalies are indicated. (b) Areas that gather drainage anomalies linking to different structural highs suggest mechanical interactions between them. Together, all these features are interpreted as a soft-linkage structural system (Yalguaraz Deformation Zone) in an early stage of morphotectonic development (modified from Terrizzano et al. 2011b).

(Fig. 13a, b). The geological mapping of these zones therefore suggests mechanical interactions between the emergent structural highs that form part of a broader and more extended soft-linked system of Quaternary structures known as the Yalguaraz Deformation Zone (Terrizzano & Cortés 2009) (Fig. 13b). En echelon folds and stepped pressure ridges on a different scale within this system suggests a left-lateral component. The Yalguaraz Deformation Zone distribution evidences the mechanical interconnection between the BLP-DZ, the major structural highs of the Uspallata Valley (El Abra High) and the eastern mountain front of the Frontal Cordillera. The Yalguaraz Deformation Zone constitutes a defined low-relief Quaternary belt developed on the piedmont plain, so it can be considered as a morphotectonic proto-unit in the Precordillera.

In the Precordillera, most of the Quaternary sedimentation is concentrated in tectonic depressions formed during the Neogene period. North of 31°30'S, Quaternary deposits form part of the longitudinal intermontane basins located between the thrust sheets of the thin-skinned belt (Fig. 3). To the south of that latitude and along the BLP-DZ, Quaternary deposits are lodged in elongated and stepped depressions situated between imbricate blocks (Fig. 5). Some of these basins, such as the Pampa Jarillal or Pampa Seca, have begun to be cannibalized by the propagation of Quaternary splays that uplift the pre-Quaternary bedrock to the surface (Fig. 6). In other cases, the basins are at their initial stages of formation (Pampa de los Burros Basin) as a consequence of the propagation and interconnection of different Quaternary deformation zones in the mountain piedmont (Fig. 8a).

In the interior of the Southern Precordillera, other Quaternary tectonic basins linked to the rejuvenation of faults in the BLP-DZ follow different geometric patterns. In the Sierra de Uspallata, at an altitude of 2100–2700 m, two NE-oriented Quaternary basins (San Bartolo and Vizcacheras basins, Fig. 9) have been identified. These are transtensive basins structured on releasing bends of longitudinal thrusts with dextral strike–slip components (Cortés et al. 2011b). The San Bartolo half-graben is bounded by a belt of piedmont and bedrock scarps 2–20 m high (San Bartolo Fault). Their extensional character has been determined on the basis of electrical resistance tomography profiles (Fazzito et al. 2006, 2009). The Vizcacheras Basin is controlled by the releasing bend of the Cerro Manantial Fault. It is a fusiform pull-apart basin (Fig. 9b) that displays, on the surface, abundant evidence of extensional faulting of successive alluvial and dam Quaternary deposits (Cortés et al. 1999c, 2011b; Cortés & Pasini 2008). This pull-apart basin is a halfgraben controlled by the extensional tectonic activity of the Cerro Manantial Fault (Fig. 9b). Wind gaps and beheaded channels are found on the top of that master-fault bedrock scarp. Three strath terraces covered with Quaternary alluvial sediments are located on the slopes of major antecedent streams that cross through the fault scarpment. The oldest terraces are found 32–37 m above the active stream courses. Both basins are examples of the rejuvenation of tectonic depressions in the mountainous interior of the Precordillera Sur during the Quaternary time.

Quaternary deformation in oblique deformation zones to the south of the Precordillera

Río Mendoza–Tupungato Deformation Zone

The Río Mendoza–Tupungato Deformation Zone (Cortés et al. 2006) is a neotectonic belt with azimuth 150° bounding the southern edge of the Precordillera (Figs 3 & 14). Its main body (southeastern segment) coincides with the western margin of the Cerrilladas Pedemontanas. Towards the NW, it penetrates into the Frontal Cordillera (Cordón del Plata), where it is represented by the La Carrera Fault system (Caminos 1965, Fig. 14). This zone has developed on regional pre-Cenozoic anisotropies such as the western edge of the Triassic Cuyo Basin and the suture zone between accreted Palaeozoic terrains (Fig. 2a).

The southeastern strand of this belt is formed by western vergence system of reverse faults and strike–slip faults associated with en echelon folds (Fig. 14). This structural system is in part the result of Cenozoic tectonic inversion of a Triassic half-graben (Cuyo Basin) under a transpressive regime (Dellapé & Hegedus 1995; Massabie 1998). Cenozoic deformation make up a landscape of low hills (50–200 m high) formed by Triassic and Neogene clastic continental sequences deformed during the Diaguita Phase after the accumulation of the Mogotes Formation at the beginning of the Quaternary (Polanski 1963; Bastías et al. 1993; INPRES 1995; Cortés et al. 1999c; Irigoyen et al. 2000). As yet, no geological evidence has appeared of post-Diaguita deformation, except Quaternary folds to the west of Tupungato city (Fig. 14), which may be linked to the deformation zone. The southern edge of the Precordillera and the Rio Mendoza–Tupungato Zone record a high level of seismicity (Cortés et al. 2006) that declines abruptly to the west. In the NW sector of the belt, on the other hand, Pleistocene rejuvenations of structural segments of the faults controlling the mountain front of the Frontal Cordillera have been identified. They are splays that uplift pediments and give rise

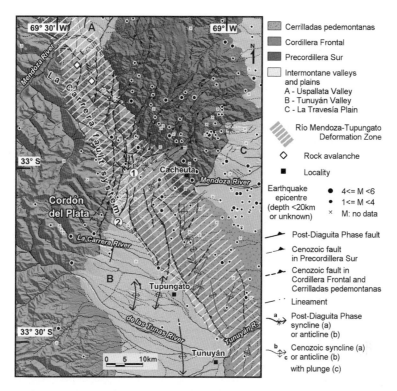

Fig. 14. Major structural lines of the Río Mendoza–Tupungato Shear Zone (modified from Cortés *et al.* 2006). This NW-trending transpressive zone was built in late Cenozoic times over an early Palaeozoic suture zone reactivated later in Permian and Triassic periods (after Cortés *et al.* 2006; Giambiagi & Martínez 2008).

to bedrock and piedmont scarps, like the La Aguadita Fault (Polanski 1963) and the El Salto Fault (Casa *et al.* 2010) (Fig. 14). The latter is associated with anomalous values of morphotectonic indices on the adjacent mountain front. Furthermore, in the northern edge of the Cordón del Plata, large rock-avalanche deposits have breakage scar zones located on the crossed faults of La Carrera system faults (Fauque *et al.* 2000). Exposure ages for block surfaces determined by cosmogenic nuclide (^{36}Cl) in those deposits indicate ages between 46 and 256 ka (Fauque *et al.* 2008).

Alamitos, Papagayos and Diamante deformation zones

Between 34°S and 34°30′S, the piedmont plain of the Cordón Carrizalito in the Frontal Cordillera (Fig. 15) displays Quaternary tectonic and volcanic landforms, trending in a NW–SE direction to form the Papagayos and Diamante fault zones (Bastías *et al.* 1993; Tello 1994; Cortés 2000) and the Los Alamitos Fault Zone (Casa *et al.* 2011). These fault zones are tens of kilometres long and some 5 km wide. They contain piedmont and rock fault scarps, pressure ridges, sag ponds and mono and polygenetic Quaternary volcanic centres. Pressure ridges associated with bends are a characteristic feature (Fig. 15) and reflect a left-lateral strike–slip component in the fault zones. To the NW, fault zones penetrate the mountain front, where they continue as oblique lineaments.

Continuity of these fault zones has been interrupted by the erosive fluvial action of the main Cordilleran streams (the Rosario, Yaucha and Papagallos streams) and the subsequent filling by sudden accumulations of thick volcanic ash flows (Asociación Piroclástica Pumícea, Polanski 1963) during the Middle Pleistocene.

Quaternary tectonic activity in these deformation zones has visibly modified drainage, landforms and sedimentation along the piedmont fluvial valleys, but in competition with exogenous processes it has not been able to develop greater, continuous morphostructural units. Composite piedmont fault scarps in the northwestern reach of Diamante Fault Zone are 6–30 m high (Cortés & Sruoga 1998), while bedrock scarps in Papagayos Fault Zone reach 40 m high. Where streams are

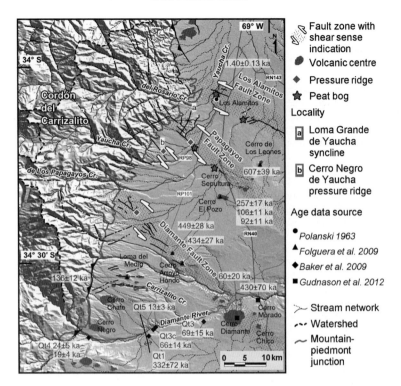

Fig. 15. Quaternary fault zones developed in the piedmont area of Cordón del Carrizalito in Cordillera Frontal. These NW-trending zones are defined by the alignment of fault scarps, pressure ridges and volcanic centres (modified from Casa et al. 2011). Isotopic ages are included.

cut by fault zones, the presence of peat deposits is common in the downthrown block of the scarps and upstream from the courses (Fig. 15). Changes in the channel patterns, such as sinuosity and valley-floor width have been observed by Casa et al. (2011) in the Yaucha Stream valley on both blocks of Alamitos Fault (Fig. 15).

The relationship between the structure and the piedmont stratigraphic units indicates successive rejuvenation during Quaternary time. The principal piedmont scarps next to the Frontal Cordillera predate the ignimbrite of the Asociación Piroclástica Pumícea (Cortés & Sruoga 1998). This unit was correlated with similar deposits in the Chilean side of the Andes dated by fission track at c. 450 ka (Stern et al. 1984) and by U–Th–He at c. 150 ka (Lara et al. 2008). Assuming that the activity of volcanic centres aligned along the Diamante Fault Zone is the result of deformation, Middle to Late Pleistocene tectonic events are revealed by the Ar/Ar ages of 430 ± 70 ka to 60 ± 20 ka at Cerro Diamante (Gudnason et al. 2012) and 434 ± 27 ka and 449 ± 28 ka at Arroyo Hondo (Folguera et al. 2009). Late Pleistocene deformation on the piedmont area is indicated by the gentle folding of younger terraces on Diamante River dated at (^{10}Be) 22 ± 7 ka to 13 ± 3 ka (Baker et al. 2009). Where the Los Alamitos Fault crosses the Yaucha Stream (Fig. 13), Casa et al. (2011) described fine lamination clayey strata bearing organic material that they interpret as palaeoseismological evidence of temporary obstructions of the stream caused by fault activity. The ^{14}C age of 1400 ± 130 a BP (Polanski 1963) and 2622 ± 42 a BP (Mehl & Zárate 2012) of these deposits is evidence of the active nature of this fault. From these data the age of tectonics activity along this system of oblique fault zones varies from the Middle Pleistocene to the Holocene.

Concluding remarks

In the Central Andean retro-wedge between 31°30′S and 35°S, the spatial distribution of geological evidence of Quaternary deformation is not uniform. On the Pampean flat-slab segment (31°30′–33°S), a larger number of Quaternary structures are recorded than on the normal subduction segment, located further to the south. In addition to this geodynamic first-order control, two characteristic distribution patterns have been recognized throughout the region: (1) Quaternary structures tend to concentrate next to the topographic front of the retro-wedge, represented by the eastern edge of both the Precordillera and the edge of the Cerrilladas Pedemontanas–San Rafael Block, which is consistent with the forward-breaking sequence model; (2) Quaternary structures tend to group along deformation belts oblique to the orogen, with a NW and NNW trend. To the south of 31°30′S, these directions coincide with pre-Cenozoic mechanical anisotropies generated mainly by the San Rafael Orogeny (Permian age) and the extensional tectonics that originated the Triassic rift of the Cuyo Basin. The oblique shear zone with the greatest evidence of Quaternary deformation in the region corresponds to the Barreal–Las Peñas Deformation Zone, which largely coincides with the northern segment of the Cuyo Basin. Other oblique deformation zones with Quaternary structures are the Río Mendoza–Tupungato Zone and the Los Alamitos, Papagayos and Diamante fault zones.

The Late Cenozoic Barreal–Las Peñas Deformation Zone forms a morphotectonic unit made up of stepped blocks and basins mainly delimited by reverse and strike–slip faults. During the Quaternary, the reverse faults grew by means of (1) migration of rejoining, diverging and isolated splays that uplifted the piedmont sector contiguous to the mountain front, and (2) propagation in the internal intermontane basins or those adjacent to the unit and through propagation and interference with oblique faults and shear zones. Longitudinal propagation of the faults delimiting the stepped blocks towards the adjacent basin suggests the gradual enlargement of the deformation zone during the Quaternary.

The instrumental seismic record reveals that most of the epicentres of the 49 earthquakes associated with this deformation zone are concentrated at its NW (Barreal Block) and SE (Las Peñas and Santa Clara blocks) ends. Palaeoseismological records, although scarce, seem to confirm the Late Quaternary tectonic activity on its northwestern and southeastern edges. Cosmogenic dating of Quaternary deformed alluvial sequences (Las Peñas Block) and OSL ages for dam deposits generated by fault activity (Santa Clara Block) support the tectonic activity of the southeastern edge of the Barreal–Las Peñas Deformation Zone during the last 20 kyr. In addition, in the extreme NW of the deformation zone, new geochronological data (^{14}C) from dam deposits associated with Barreal Fault activity confirm the Holocene tectonic activity of this fault. As a result, the concurrence of seismological and palaeoseismological data suggest that the NW and SE ends of the deformation zone are active fronts of the deformation. The Las Peñas Block represents the intersection of the structural front of the Southern Precordillera with the Barreal–Las Peñas Deformation Zone. The epicentres of two destructive earthquakes were located close by, one of which (1861, M = 7) completely destroyed the city of Mendoza.

Quaternary post-Diaguita tectonics has contributed to modifying the Precordillera relief along the Barreal–Las Peñas Belt. This modification is evident from the tectonic uplift of blocks (a minimum of $c.$ 100 m in the Barreal Block), the initial development of intermontane basins (Pampa de los Burros Basin), shortening of intermontane basins (southern sector of the Pampa Jarillal Basin), generation of transtensive basins in the mountain sector (at least $c.$ 35 m of subsidence in the Vizcacheras Basin) and the construction of low-relief morphotectonic proto-units by soft-linkage of Quaternary structure systems (Yalguaraz Deformation Zone). On the normal subduction segment, Quaternary tectonic activity in the Los Alamitos, Papagayos and Diamante deformation zones is associated with the creation of pressure ridges, piedmont and bedrock scarps and the modification of fluvial valleys with the development of swamps where they cross active faults.

For a complete picture of the Quaternary tectonic deformation at the Andean retro-wedge front, further studies are still required, especially regarding the number and size of fault displacements during specific time intervals.

We wish to thank the two reviewers, S. Schmidt and C. Costa, and the Managing Editor, G. Hoke, for the valuable suggestions and comments about this article. We express our appreciation to A. Rapalini and S. Fazzito, with whom we have realized geophysical studies about numerous Quaternary structures of the Precordillera. Support for this research was provided by CONICET grant 02295 and UBACYT grants x049 and 0778.

References

AHUMADA, E. A. & COSTA, C. H. 2009. Antithetic linkage between oblique Quaternary thrusts at the Andean front, Argentine Precordillera. *Journal of South American Earth Sciences*, **28**, 207–216.

ALLMENDINGER, R. A., FIGUEROA, D., SNYDER, N. P., BEER, D. J., MPODOZIS, C. & ISACKS, B. L. 1990. Foreland shortening and crustal balancing in the Andes at 30°S latitude. *Tectonics*, **9**, 789–809.

ANDERSON, M., ALVARADO, P., ZANDT, G. & BECK, S. L. 2007. Geometry and brittle deformation of the subducting Nazca Plate, central Chile and Argentina. *Geophysical Journal International*, **171**, 419–434.

ASTINI, R. A. & THOMAS, W. A. 1999. Origin and evolution of the Precordillera terrane of western Argentina: a drifted Laurentian orphan. *In*: RAMOS, V. A. &

KEPPIE, D. (eds) *Laurentia–Gondwana Connections Before Pangea*. Geological Society of America, Special Papers, Boulder, **336**, 1–20.

BALDIS, B. A., BERESI, M. S., BORDONARO, L. O. & VACA, A. 1982. Síntesis evolutiva de la Precordillera Argentina. *V Congreso Latinoamericano de Geología (Buenos Aires)*, Actas, **4**, 399–445.

BAKER, S. E., GOSSE, J. C., MCDONALD, E. V., EVENSON, E. B. & MARTÍNEZ, O. 2009. Quaternary history of the piedmont reach of Río Diamante, Argentina. *Journal of South American Earth Sciences*, **28**, 54–73.

BARAZANGI, M. & ISACKS, B. 1976. Spatial distribution of earthquakes and subduction of the Nazca plate beneath South America. *Geology*, **4**, 686–692.

BASTÍAS, H., ULIARTE, E., PAREDES, J. de D., SANCHEZ, A., BASTÍAS, J. A., RUZYCKI, L. & PERUCCA, L. P. 1990. Neotectónica de la provincia de San Juan. *XI Congreso Geológico Argentino, Relatorio*, San Juan, 228–244.

BASTÍAS, H., TELLO, G. E., PERUCCA, L. P. & PAREDES, J. de D. 1993. Peligro sísmico y Neotectónica. *In*: RAMOS, V. A. (ed.) *Geología y Recursos Naturales de Mendoza. XII Congreso Geológico Argentino y II Congreso de Exploración de Hidrocarburos, Relatorio*, Mendoza, **6**, 645–658.

BEER, J. A., ALLMENDINGER, R. W., FIGUEROA, D. E. & JORDAN, T. 1990. Seismic stratigraphy of a Neogene Piggyback Basin, Argentina. *Bulletin of the American Association of Petroleum Geologists*, **74**, 1183–1202.

BULL, W. B. 2007. *Tectonic Geomorphology of Mountains. A New Approach to Paleoseismology*. Blackwell Publishing, Oxford.

CAHILL, T. & ISACKS, B. L. 1992. Seismicity and the shape of the subducted Nazca plate. *Journal of Geophysical Research*, **97**, 17503–17529.

CAMINOS, R. 1965. Geología de la vertiente oriental del Cordón del Plata, Cordillera Frontal de Mendoza. *Revista de la Asociación Geológica Argentina*, **20**, 351–392.

CASA, A. L., CORTÉS, J. M. & BORGNIA, M. 2010. Evidencias de deformación pleistocena en el sistema de falla de La Carrera (32°40′–33°15′ LS), Cordillera Frontal de Mendoza. *Revista de la Asociación Geológica Argentina*, **67**, 91–104.

CASA, A. L., CORTÉS, J. M. & RAPALINI, A. E. 2011. Fallamiento activo y modificación del drenaje en el piedemonte del cordón del Carrizalito, Mendoza. *XVIII Congreso Geológico Argentino*, **S12**, 712–713.

CHIARAMONTE, L., RAMOS, V. A. & ARAUJO, M. 2000. Estructura y sismotectónica del Anticlinal Barrancas, provincia de Mendoza. *Revista de la Asociación Geológica Argentina*, **55**, 309–336.

COMÍNGUEZ, A. H. & RAMOS, V. A. 1991. La estructura profunda entre Precordillera y Sierras Pampeanas de la Argentina: evidencia de la sísmica de reflexión profunda. *Revista Geológica de Chile*, **18**, 3–14.

CORTÉS, J. M. 1998. Tectónica de desplazamiento de rumbo en el borde sur de la depresión de Yalguaraz, Mendoza, Argentina. *Revista de la Asociación Geológica Argentina*, **53**, 147–157.

CORTÉS, J. M. 2000. Fallas cuaternarias oblicuas al frente montañoso en la Cordillera Frontal de Mendoza (34°–34°30′LS). *Revista de la Asociación Geológica Argentina, Serie D, Publicación Especial*, **4**, 57–62.

CORTÉS, J. M. & CEGARRA, M. 2004. Plegamiento cuaternario transpresivo en el piedemonte suroccidental de la Precordillera sanjuanina. *Revista de la Asociación Geológica Argentina, Serie D, Publicación Especial*, **7**, 68–75.

CORTÉS, J. M. & KLEIMAN, L. E. 1999. La Orogenia Sanrafaélica en los Andes de Mendoza. *XIV Congreso Geológico Argentino*, Salta, Actas, **1**, 31.

CORTÉS, J. M. & PASINI, M. M. 2008. Reorganización cinemática tardío-cenozoica en la Precordillera Sur. *XVII Congreso Geológico Argentino*, Jujuy, Actas, **1**, 91–92.

CORTÉS, J. M. & SRUOGA, P. 1998. Zonas de fracturas cuaternarias y volcanismo asociado en el piedemonte de la Cordillera Frontal (34°30′LS), Argentina. *XIII Congreso Latinoamericano de Geología y III Congreso Iberoamericano de Geología Económica*, Buenos Aires, Actas, **2**, 116–121.

CORTÉS, J. M., GONZALEZ BONORINO, G., KOUKHARSKY, M. L., BRODKORB, A. & PEREYRA, F. 1999a. *Hoja Geológica 3369–03 Yalguaraz, Mendoza. Carta Geológica de la República Argentina Escala 1:100.000*. Servicio Geológico Minero Argentino, Boletín, Buenos Aires, **280**.

CORTÉS, J. M., GONZALEZ BONORINO, G., KOUKHARSKY, M. L., BRODKORB, A. & PEREYRA, F. 1999b. *Hoja Geológica 3369–09 Uspallata, Mendoza. Carta Geológica de la República Argentina Escala 1:100.000*. Servicio Geológico Minero Argentino, Boletín, Buenos Aires, **281**.

CORTÉS, J. M., VINCIGUERRA, P., YAMIN, M. G. & PASINI, M. M. 1999c. Tectónica Cuaternaria de la Región Andina del Nuevo Cuyo (28°–38° LS). *In*: CAMINOS, R. (ed.) *Geología Argentina*. Subsecretaría de Minería de la Nación, Servicio Geológico Minero Argentino, Anales, Buenos Aires, **29**, 760–778.

CORTÉS, J. M., PASINI, M. M. & YAMIN, M. G. 2005a. Paleotectonic control on the distribution of Quaternary deformation in the Southern Precordillera, Central Andes (31°30′–33°SL). *6° International Symposium on Andean Geodynamics (ISAG 2005)*, Barcelona, Extended Abstracts, 186–189.

CORTÉS, J. M., YAMÍN, M. & PASINI, M. 2005b. La Precordillera Sur, provincias de Mendoza y San Juan. *XVI Congreso Geológico Argentino*, La Plata, Actas, **1**, 395–402.

CORTÉS, J. M., CASA, A., PASINI, M. M., YAMIN, M. G. & TERRIZZANO, C. 2006. Fajas oblicuas de deformación neotectónica en Precordillera y Cordillera Frontal (31°30′–33°30′LS). Controles paleotectónicos. *Revista de la Asociación Geológica Argentina*, **61**, 639–646.

CORTÉS, J. M., PASINI, M. M. & PRIETO, M. C. 2011a. Propagación y migración de la estructura cuaternaria del frente montañoso precordillerano en la Sierra de las Peñas, Mendoza. *XVIII Congreso Geológico Argentino*, Neuquén, Actas, **1**, 723–724.

CORTÉS, J. M., PASINI, M. M. & ROSSELLO, J. J. 2011b. Tectónica cuaternaria transtensiva en la región central de la Precordillera Sur, Mendoza. *XVIII Congreso Geológico Argentino*, Neuquén, Actas, **1**, 725–726.

CORTÉS, J. M., HERMANNS, R. L., PASINI, M. M. & ROSAS, M. 2011c. Tectónica activa en la Sierra de Uspallata,

Mendoza. *XVIII Congreso Geológico Argentino*, Neuquén, Actas, **1**, 727–728.

COSTA, C., MACHETTE, M. N. ET AL. 2000a. *Map and Database of Quaternary Faults and Folds in Argentina*. United States Geological Survey Open-File Report **00-108**.

COSTA, C. H., GARDINI, C. E., DIEDERIX, H. & CORTÉS, J. M. 2000b. The Andean orogenic front at Sierra de las Peñas–Las Higueras, Mendoza, Argentina. *Journal of South American Earth Sciences*, **13**, 287–292.

COSTA, C. H., AUDEMARD, F. H. R., BECERRA, M. F. A., LAVENU, A., MACHETTE, M. N. & PARÍS, G. 2006. An overview of the main quaternary deformation of South America. *Revista de la Asociación Geológica Argentina*, **61**, 461–479.

COX, S. J. D. & SCHOLZ, C. H. 1988. On the formation and growth of faults: an experimental study. *Journal of Structural Geology*, **10**, 413–430.

CRONE, A. J. & HALLER, K. M. 1991. Segmentation and coseismic behavior of basin-and-range normal faults; examples from east-central Idaho and southwestern Montana, U.S.A. *Journal of Structural Geology*, **13**, 151–164.

DELLAPÉ, D. & HEGEDUS, A. 1995. Structural inversion and oil occurence in the Cuyo basin of Argentina. *In*: TANKARD, A. J., SUÁREZ-SORUCO, M. & WELSINK, H. (eds) *Petroleum Basins of South America*. American Association of Petroleum Geologists, Boulder, Memoirs, **62**, 359–367.

FAUQUE, L., CORTÉS, J. M., FOLGUERA, A. & ETCHEVERRÍA, M. 2000. Avalanchas de roca asociadas a neotectónica en el valle del río Mendoza, al sur de Uspallata. *Revista de la Asociación Geológica Argentina*, **55**, 419–423.

FAUQUE, L., CORTÉS, J. M. ET AL. 2008. Edades de las avalanchas de roca ubicadas en el valle del río Mendoza, aguas debajo de Uspallata. *XVII Congreso Geológico Argentino*, Neuquén, Actas, **1**, 282–283.

FAZZITO, S., RAPALINI, A. & CORTÉS, J. M. 2006. Tomografía geoeléctrica en zonas de fallas cuaternarias: dos ejemplos en la Precordillera centro-occidental de Mendoza. *Revista de la Asociación Geológica Argentina, Serie D: Publicación Especial*, **9**, 41–47.

FAZZITO, S. Y., RAPALINI, A. E., CORTÉS, J. M. & TERRIZZANO, C. M. 2009. Characterization of Quaternary faults by electric resistivity tomography in the Andean Precordillera of Western Argentina. *Journal of South American Earth Sciences*, **28**, 217–228.

FIGUEROA, D. E. & FERRARIS, O. R. 1989. Estructura del margen oriental de la Precordillera mendocina-sanjuanina. *I Congreso Nacional de Exploración de Hidrocarburos*, **1**, 515–529.

FOLGUERA, A., ETCHEVERRÍA, M. ET AL. 2004. Hoja Geológica 3369-15, Potrerillos, provincia de Mendoza. Carta Geológica de la República Argentina Escala 1:100.000. Servicio Geológico Minero Argentino, Boletín, Buenos Aires, **301**.

FOLGUERA, A., NARANJO, J. A., ORIHASHI, Y., SUMINO, H., NAGAO, K., POLANCO, E. & RAMOS, V. A. 2009. Retroarc volcanism in the northern San Rafael Block (34°–35°30′S), southern Central Andes: occurrence, age, and tectonic setting. *Journal of Volcanology and Geothermal Research*, **186**, 169–185.

FRANZESE, J. R. & SPALLETTI, L. A. 2001. Late Triassic–early Jurassic continental extension in southwestern Gondwana: tectonic segmentation and pre-break-up rifting. *Journal of South American Earth Sciences*, **14**, 257–270.

GIAMBIAGI, L. & MARTÍNEZ, A. N. 2008. Permo-Triassic oblique extensión in the Potrerillos–Uspallata area, western Argentina. *Journal of South American Earth Sciences*, **26**, 252–260.

GIAMBIAGI, L. B., BECHIS, F., GARCÍA, V. & CLARK, A. 2008. Temporal and spatial relationship between thick- and thin-skinned deformation in the Malargüe fold and thrust belt, southern Central Andes. *Tectonophysics*, **459**, 123–139.

GIAMBIAGI, L., MESCUA, J. F., BECHIS, F., TASSARA, A. & HOKE, G. 2012. Thrust belts of the Southern Central Andes: along-strike variations in shortening, topography, crustal geometry, and denudation. *Geological Society of America Bulletin*, **124**, 1339–1351.

GUDNASON, J., HOLM, P. M., SØAGER, N. & LLAMBÍAS, E. J. 2012. Geochronology of the late Pliocene to recent volcanic activity in the Payenia back.arc volcanic province, Mendoza Argentina. *Journal of South American Earth Sciences*, **37**, 191–201.

GUTSCHER, M. A., SPAKMAN, W., BIJWAARD, H. & ENGDAHL, E. R. 2000. Geodynamic of flat subduction: seismicity and tomographic constraints from the Andean margin. *Tectonics*, **19**, 814–833.

INPRES 1977. *Zonificación sísmica de la República Argentina*. Instituto Nacional de Prevención Sísmica Publicación Técnica **5**.

INPRES 1995. *Microzonificación sísmica del Gran Mendoza*. Instituto Nacional de Prevención Sísmica. Resumen Ejecutivo Publicación Técnica **19**.

IRIGOYEN, M. V., BUCHAN, K. L. & BROWN, R. L. 2000. Magnetostratigraphy of Neogene Andean foreland-basin strata, lat 33°S, Mendoza Province, Argentina. *Geological Society of America Bulletin*, **112**, 803–816.

JAPAS, M. S. & KLEIMAN, L. E. 2004. *El Ciclo Choiyoi en el Bloque de San Rafael: de la orogénesis tardía a la relajación mecánica*. Asociación Geológica Argentina, Buenos Aires, Serie D, Publicación Especial, **7**, 89–100.

JAPAS, M. S., CORTÉS, J. M. & PASINI, M. M. 2008. Tectónica extensional triásica en el sector norte de la cuenca Cuyana: Primeros datos cinemáticos. *Revista de la Asociación Geológica Argentina*, **63**, 213–222.

JORDAN, T. E., ISACKS, B. L., ALLMENDINGER, R. W., BREWER, J. A., RAMOS, V. A. & ANDO, C. J. 1983. Andean tectonics related to geometry of subducted Nazca plate. *Geological Society of America Bulletin*, **94**, 341–361.

JORDAN, T. E., ALLMENDINGER, R. W., DAMANTI, J. F. & DRAKE, R. E. 1993. Chronology of motion in a complete thrust belt – Precordillera, 30–31-degrees-S, Andes Mountains. *Journal of Geology*, **101**, 135–156.

KAY, S. M., MPODOZIS, C., RAMOS, V. A. & MUNIZAGA, F. 1991. Magma source variations for mid–late Tertiary magmatic rocks associated with a shallowing subduction zone and a thickening crust in the central Andes (18° to 33°S). *In*: HARMON, R. S. & RAPELA, C. W.

(eds) *Andean Magmatism and Its Tectonic Setting*. Geological Society of America, Boulder, Special Papers, **265**, 113–138.

KLEIMAN, L. E. & JAPAS, M. S. 2009. The Choiyoi volcanic province at 34°S–36°S (San Rafael, Mendoza, Argentina): implications for the Late Paleozoic evolution of the southwestern margin of Gondwana. *Tectonophysics*, **473**, 283–299.

KOZLOWSKI, E. E., MANCEDA, R. & RAMOS, V. A. 1993. Estructura. *In*: RAMOS, V. A. (ed.) *Geología y Recursos Naturales de Mendoza. XII Congreso Geológico Argentino y II Congreso de Exploración de Hidrocarburos*, Relatorio, **1**, 235–256.

LARA, L. E., WALL, R. & STOCKLI, D. 2008. La ignimbrita Pudahuel (Asociación Piroclástica Pumícea) y la caldera Diamante (33°S): nuevas edades U–Th–He. *XVII Congreso Geológico Argentino*, Jujuy, Actas, **3**, 1365.

LEVERATTO, M. A. 1976. Edad de intrusivos cenozoicos en la Precordillera de San Juan y su implicancia estratigráfica. *Revista de la Asociación Geológica Argentina*, **31**, 53–58.

MCCLAY, K. R., WHITEHOUSE, P. S., DOOLEY, T. & RICHARDS, M. 2004. 3D evolution of fold and thrust belts formed by oblique convergence. *Marine and Petroleum Geology*, **21**, 857–877.

MARRETT, R. & ALLMENDINGER, R. W. 1990. Kinematic analysis of fault-slip data. *Journal of Structural Geology*, **12**, 973–986.

MASSABIE, A. C. 1987. Neotectónica y sismicidad en la región de las Sierras Pampeanas Orientales, sierras de Córdoba, Argentina. *X Congreso Geológico Argentino*, San Miguel de Tucumán, Actas, **1**, 271–274.

MASSABIE, A. C. 1998. Reactivación transpresiva del fallamiento triásico durante la inversión tectónica de la Cuenca Cuyana. *Revista de la Asociación Geológica Argentina*, **53**, 261–272.

MASSABIE, A. C., RAPALINI, A. E. & SOTO, J. L. 1985. Estratigrafía del cerro Los Colorados, Paramillo de Uspallata, Mendoza. *Primeras Jornadas sobre Geología de Precordillera*, San Juan, Acta, **1**, 71–76.

MCCORMAC, F. G., HOGG, A. G., BLACKWELL, P. G., BUCK, C. E., HIGHAM, T. F. G. & REIMER, P. J. 2004. SHCal04 Southern Hemisphere calibration, 0–11.0 cal kyr BP. *Radiocarbon*, **46**, 1087–1092.

MEHL, A. E. & ZÁRATE, M. A. 2012. Late Pleistocene and Holocene environmental and climatic conditions in the eastern Andean piedmont of Mendoza (33°–34° S, Argentina). *Journal of South American Earth Sciences*, **37**, 41–59.

MILANA, J. P. & ZAMBRANO, J. J. 1996. La cerrillada pedemontana mendocina: un sistema geológico retrocorrido en vías de desarrollo. *Revista de la Asociación Geológica Argentina*, **51**, 289–303.

MINGORANCE, F. A. 2000. Peligro de desplazamiento superficial de falla en el núcleo urbano del Gran Mendoza, Argentina. *IX Congreso Geológico Chileno*, Puerto Varas, Actas, **1**, 81–85.

MINGORANCE, F. A. 2006. Morfometría de la escarpa de falla histórica identificada al norte del cerro La Cal, zona de falla La Cal, Mendoza. *Revista de la Asociación Geológica Argentina*, **61**, 620–638.

NAYLOR, M. A., MANDL, G. & SIJPESTEIJN, C. H. K. 1986. Fault geometries in basement-induced wrench faulting under different initial stress states. *Journal of Structural Geology*, **8**, 737–752.

OLGIATI, S. & RAMOS, V. A. 2003. Neotectónica cuaternaria en el anticlinal Borbollón, provincia de Mendoza – Argentina. *10° Congreso Geológico Chileno*, Concepción, Actas en CD, 11.

ORTIZ, A. & ZAMBRANO, J. J. 1981. La provincia geológica Precordillera oriental. *VIII Congreso Geológico Argentino*, San Luis, Actas, **3**, 59–74.

PERUCCA, L., PÉREZ, A. & NAVARRO, C. 2006. Fenómenos de licuefacción asociados a terremotos históricos. Su análisis en la evaluación del peligro sísmico en la Argentina. *Revista de la Asociación Geológica Argentina*, **61**, 567–578.

PILGER, R. H. 1984. Cenozoic plate kinematics, subduction and magmatism, South American Andes. *Journal of the Geological Society, London*, **141**, 793–802.

POLANSKI, J. 1963. Estratigrafía, neotectónica y geomorfología del Pleistoceno pedemontano, entre los ríos Diamante y Mendoza. *Revista de la Asociación Geológica Argentina*, **17**, 127–349.

PRIETO, M. C. 2010. *Estratigrafía y Neotectónica del extremo sur de la sierra de Las Peñas–Las Higueras, Precordillera Sur, provincia de Mendoza*. Master thesis (unpublished), Facultad de Ciencias Exactas y Naturales, Universidad de Buenos Aires.

RAMOS, V. A. 1994. Terranes of southern Gondwanaland and their control in the Andean structure (30°–33°S latitude). *In*: REUTTER, K. J., SCHEUBER, E. & WIGGER, P. J. (eds) *Tectonics of the Southern Central Andes*. Springer-Verlag, Berlin, 249–261.

RAMOS, V. A. 2000. The Southern Central Andes. *In*: CORDANI, U. G., MILANI, E. J., THOMAZ FILHO, A. & CAMPOS, D. A. (eds) *Tectonic Evolution of South America*. 31° International Geological Congress, Río de Janeiro, 560–604.

RAMOS, V. A. 2004. Cuyania, an exotic block to Gondwana; review of a historical success and the present problems. *Gondwana Research*, **7**, 1009–1026.

RAMOS, V. A. & FOLGUERA, A. 2009. Andean Flat Slab Subduction Through Time. *In*: MURPHY, B. (ed.) *Ancient Orogens and Modern Analogues*. Geological Society, London, Special Publications, **327**, 31–54.

RAMOS, V. A. & KAY, S. M. 1991. Triassic rifting and associated basalts in the Cuyo basin, Central Argentina. *In*: HARMON, R. S. & RAPELA, C. W. (eds) *Andean Magmatism and its Tectonic Setting*. Geological Society of America, Boulder, Special Papers, **265**, 79–91.

RAMOS, V. A., ESCAYOLA, M., MUTTI, D. & VUJOVICH, G. I. 2001. Proterozoic–Early Paleozoic ophiolites in the Andean basement of southern South America. *In*: DILEK, Y., MOORES, E., ELTHON, D. & NICHOLAS, A. (eds) *Ophiolites and Oceanic Crust: New Insights from Field Studies and Ocean Drilling Program*. Geological Society of America, Boulder, Memoirs, **349**, 331–349.

RAMOS, V. A., CRISTALLINI, E. O. & PÉREZ, D. J. 2002. The Pampean flan-slab of the Central Andes. *Journal of South American Earth Sciences*, **15**, 59–78.

RAPALINI, A. E. & VILAS, J. F. 1991. Tectonic rotations in the Late Paleozoic continental margin of southern South America determined and dated by palaeomagnetism. *Geophysics Journal International*, **107**, 333–351.

RAPELA, C. 2000. The Sierras Pampeanas of Argentina: Paleozoic building of the southern proto-Andes. *In*: CORDANI, U. G., MILANI, E. J., THOMAZ-FILHO, A. & CAMPOS, D. A. (eds) *Tectonic Evolution of South America*, 31° International Geological Congress, Rio de Janeiro, 381–387.

REYNA, G., HOKE, G. D., DÁVILA, F. M. & SUDO, M. 2011. Edades Ar–Ar y correlación de las sucesiones volcaniclásticas expuestas en la Precordillera Normendocina. *XVIII Congreso Geológico Argentino*, S12, Tectónica Andina, Neuquén, 837–838.

SALOMON, E., SCHMIDT, S., HETZEL, R., MINGORANCE, F. & HAMPEL, A. 2013. Repeated folding during Late Holocene earthquakes on the La Cal thrust fault near Mendoza City (Argentina). *Bulletin of the Seismological Society of America*, 103, 936–949.

SAREWITZ, D. 1988. High rates of late Cenozoic crustal shortening in the Andean foreland, Mendoza Province, Argentina. *Geology*, 16, 1138–1142.

SCHMIDT, C. J., ASTINI, R. A., COSTA, C. H., GARDINI, C. E. & KRAEMER, P. E. 1995. Cretaceous rifting, alluvial fan sedimentation, and Neogene inversion, southern Sierras Pampeanas, Argentina. *In*: TANKARD, A., SUÁREZ, R. & HELSINK, H. J. (eds) *Petroleum Basins of South America*, American Association of Petroleum Geologists, Boulder, Memoirs, 62, 341–358.

SCHMIDT, S., HETZEL, R., KUHLMANN, J.,, MINGORANCE, F. & RAMOS, V. A. 2011a. A note of caution on use of boulders for exposure dating of depositional surfaces. *Earth and Planetary Sciences Letters*, 302, 60–70.

SCHMIDT, S., HETZEL, R., KUHLMANN, J., MINGORANCE, F. & RAMOS, V. A. 2011b. Coseismic displacements and Holocene slip rates for two active thrust faults at the mountain front of the Andean Precordillera (∼33°S). *Tectonics*, 30, TC5011, http://dx.doi.org/10.1029/2011TC002932

SEPÚLVEDA, E. G. 2001. Hoja Geológica 3369-II, Mendoza, Provincias de Mendoza y San Juan. Carta Geológica de la República Argentina Escala 1:250.000. *Servicio Geológico Minero Argentino. Boletín*, 252, 55.

SIAME, L. L. 1998. *Cosmonucléide produit in-situ (^{10}Be) et quantification de la déformation active dans les Andes centrales*. Thèse de doctorat, Université de Paris-Sud, Orsay.

SIAME, L. L., BELLIER, O., SÉBRIER, M. & ARAUJO, M. 2005. Deformation partitioning in flat subduction setting: Case of the Andean foreland of western Argentina (28°S–33°S). *Tectonics*, 24, TC5003, http://dx.doi.org/10.1029/2005TC001787

SIMPSON, C., WHITMEYER, S. J., DE PAOR, D. G., GROMET, L. P., MIRO, R., KROL, M. A. & SHORT, H. 2001. Sequential ductile to brittle reactivation of major fault zones along the accretionary margin of Gondwana in Central Argentina. *In*: HOLDSWORTH, R. E., STRACHAN, R. A., MAGLOUGHLIN, R. A. & KNIPE, R. J. (eds) *The Nature and Tectonic Significance of Fault Zone Weakening*. Geological Society, London, Special Publications, 186, 233–255.

SRUOGA, P., ECHEVERRÍA, M., FOLGUERA, A., REPOL, D., CORTÉS, J. M. & ZANETTINI, J. C. 2005. Hoja Geológica 3569-I, Volcán Maipo, provincia de Mendoza Escala 1:250.000. *Servicio Geológico Minero Argentino, Boletín*, 290, 92.

STERN, C. R., AMINI, H., CHARRIER, R., GODOY, E., HERVÉ, F. & VARELA, J. 1984. Petrochemistry and age of rhyolitic pyroclastic flows which occur along the drainage valleys of the Río Maipo and Río Cachapoal (Chile) and the Río Yaucha and Río Papagayos (Argentina). *Revista Geológica de Chile*, 23, 39–52.

STEWART, I. S. & HANCOCK, P. L. 1990. What is a fault scarp? *Episodes*, 13, 256–263.

STEWART, I. S. & HANCOCK, P. L. 1991. Scales of structural heterogeneity within neotectonic fault zones in the Aegean region. *Journal of Structural Geology*, 13, 191–204.

TELLO, G. E. 1994. Fallamiento cuaternario y sismicidad en el piedemonte cordillerano de la provincia de Mendoza. Argentina. *VII Congreso Geológico Chileno*, Concepción, Actas, 1, 380–384.

TERRIZZANO, C. M. 2010. *Neotectónica del extremo noroccidental del cinturón Barreal – Las Peñas, Precordillera Sur*. PhD thesis (unpublished), Facultad de Ciencias Exactas y Naturales, Universidad de Buenos Aires.

TERRIZZANO, C. M. & CORTÉS, J. M. 2009. Analysis of drainage anomalies as a tool to elucidate neotectonic soft-linked deformation zones. *Latein Amerika Kolloquium, Göttingen*, 1, 285–287.

TERRIZZANO, C. M., FAZZITO, S. Y., CORTÉS, J. M. & RAPALINI, A. E. 2010. Studies of Quaternary deformation zones through geomorphic and geophysical evidence: a case in the Precordillera Sur, Central Andes of Argentina. *Tectonophysics*, 490, 184–196.

TERRIZZANO, C. M., YAMIN, M. G. & CORTÉS, J. M. 2011a. Neotectónica y propagación de la deformación cuaternaria del frente serrano occidental del bloque Barreal, Precordillera Sur. *XVIII Congreso Geológico Argentino*, Neuquén, Actas, 873–874.

TERRIZZANO, C. M., FAZZITO, S. Y., CORTÉS, J. M. & RAPALINI, A. E. 2011b. Electrical resistivity tomography applied to the study of neotectonic structures, northwestern Precordillera Sur, Central Andes of Argentina. *Journal of South American Earth Sciences*, 34, 47–60.

ULIANA, M. A., BIDDLE, K. T. & CERDAN, J. 1989. Mesozoic extension and the formation of Argentine sedimentary basins. *In*: TANKARD, A. J. & BALKWILL, H. R. (eds) *Extensional Tectonics and Stratigraphy of North Atlantic Margins*. American Association of Petroleum Geologists, Tulsa, Oklahoma, Memoirs, 46, 599–614.

VERGÉS, J., RAMOS, V. A., MEIGS, A., CRISTALLINI, E., BETTINI, F. & CORTÉS, J. M. 2007. Crustal wedging triggering recent deformation in the Andean thrust front between 31°S and 33°S. Sierras Pampeanas – Precordillera interaction. *Journal of Geophysical Research*, 112, B03S15, 1–22, http://dx.doi.org/10.1029/2006JB004287

VON GOSEN, W. 1992. Structural evolution of the Argentine Precordillera: the Rio San Juan section. *Journal of Structural Geology*, 6, 643–667.

VON GOSEN, W. 1995. Polyphase structural evolution of the southwestern Argentine Precordillera. *Journal of South American Earth Sciences*, 8, 377–404.

VON HUENE, R., CORVALÁN, J., FLUEH, E. R., HINZ, K., KORSTGARD, J., RANERO, C. R. & WEINREBE, W. 1997. Tectonic control of the subducting Juan

Fernández Ridge on the Andean margin near Valparaíso, Chile. *Tectonics*, **16**, 474–488.

WALCEK, A. A. & HOKE, G. D. 2012. Surface uplift and erosion of the southernmost Argentine Precordillera. *Geomorphology*, **153–154**, 156–168.

YAMIN, M. G. 2007. *Neotectónica del Bloque Barreal, borde noroccidental de la Precordillera Sur*. PhD thesis (unpublished), Facultad de Ciencias Exactas y Naturales, Universidad de Buenos Aires.

YAMIN, M. G. & CORTÉS, J. M. 2004. La deformación tardiocenozoica en el interior del bloque Barreal, margen suroeste de la Precordillera de San Juan. Revista de la Asociación Geológica Argentina, Serie D, Publicación Especial, **7**, 137–144.

YAMIN, M. G. & CORTÉS, J. M. 2008. Neotectónica y estratigrafía cenozoica del bloque Barreal, margen noroccidental de la Precordillera Sur. *XVII Congreso Geológico Argentino*, Jujuy, Actas, **3**, 1171–1172.

YÁÑEZ, G., RANERO, G. R., VON HUENE, R. & DÍAZ, J. 2001. Magnetic anomaly interpretation across a segment of the Southern Central Andes (32–34°S): implications on the role of the Juan Fernández Ridge in the tectonic evolution of the margin during the upper Tertiary. *Journal of Geophysical Research*, **106**, 6325–6345.

YEATS, R. S., SIEH, K. & ALLEN, C. R. 1997. *The Geology of Earthquakes*. Oxford University Press, New York.

YRIGOYEN, M. R. 1994. Revisión estratigráfica del neógeno de las Huayquerías de Mendoza Septentrional, Argentina. *Ameghiniana, Revista de la Asociación Paleontológica Argentina*, **31**, 125–138.

ZAPATA, T. R. & ALLMENDINGER, R. W. 1996. Thrust-front zone of the Precordillera, Argentina: a thick-skinned triangle zone. *American Association of Petroleum Geologists Bulletin*, **80**, 359–381.

ZIEGLER, P. A., VAN WEES, J. D. & CLOETINGH, S. 1998. Mechanical controls on collision-related compressional intraplate deformation. *Tectonophysics*, **300**, 103–129.

ZIEGLER, P. A., BERTOTTI, G. & CLOETINGH, S. 2002. Dynamic processes controlling foreland development – the role of mechanical (de)coupling of orogenic wedges and forelands. *EGU Stephan Mueller Special Publication Series*, **1**, 17–56.

ZÖLLNER, W. 1950. Observaciones tectónicas en la Precordillera sanjuanina. Zona de Barreal. *Revista de la Asociación Geológica Argentina*, **5**, 111–126.

Quaternary evolution of the Cordillera Frontal piedmont between *c.* 33° and 34°S Mendoza, Argentina

M. A. ZÁRATE[1]*, A. MEHL[1] & L. PERUCCA[2]

[1]*Instituto de Ciencias de la Tierra y Ambientales de La Pampa (CONICET-Universidad Nacional de La Pampa), Uruguay 151, Santa Rosa, La Pampa, Argentina*

[2]*CONICET, Gabinete de Neotectónica INGEO-FCEFN Universidad Nacional de San Juan, Av. Ignacio de La Roza y Meglioli 5400, San Juan, Argentina*

*Corresponding author (e-mail: mzarate@exactas.unlpam.edu.ar)

Abstract: The piedmont of Cordillera Frontal between *c.* 33° and 34°S (Mendoza, Argentina) is a highly populated area deeply modified by human activities, known as Valle de Uco. It is situated within the borderland region of the geological provinces of Cordillera Frontal and Cuyo basin. The landscape is dominantly composed of both erosional and depositional landforms made of fluvio-aeolian deposits fractured and folded by tectonic processes together with some landforms of volcanic origin. Alluvial fans, related to several aggradational cycles of Quaternary age, are the most remarkable geomorphological units. Several tectonic features are present giving rise to conspicuous morphological features. Some of the streams are structurally controlled by faults while several drainage anomalies that indicate active tectonic processes have been identified. The Late Quaternary alluvial sequences, dominantly comprising sandy and silty deposits of volcaniclastic composition and secondarily metamorphic rocks, represent the fine-grained sedimentary facies of the fluvial systems accumulated in a distal fan environment. The alluvial deposits have been incised by several episodes of erosion since Pleistocene time.

The Andes Cordillera and the piedmont of Mendoza province (Argentina) are an active tectonic area characterized by a complex geological setting that determines a heterogeneous landscape. Of particular environmental and human significance is the piedmont of Cordillera Frontal between *c.* 33° and 34°S. Known as Valle de Uco, it is a highly populated area deeply modified by human activities and constitutes one of the three man-made agricultural oases of Mendoza province (Fig. 1). Consequently, the reconstruction of the piedmont evolution and the understanding of the processes involved are essential to evaluate possible environmental responses under the present climatic fluctuations in a densely populated area. Several contributions have demonstrated the environmental sensitivity of the region during Quaternary time as documented by the record of Pleistocene and Holocene glacial advances in the Andean headwaters of the fluvial systems (e.g. Espizúa 2004, 2005; Espizúa & Pitte 2009; Messager 2010) along with geomorphological, palaeoenvironmental and palaeoclimatic studies (Baker *et al.* 2009; Paez *et al.* 2010).

In order to understand the nature and characteristics of the present landscape, analysis of the Late Pleistocene and Holocene period is especially important. This key time interval covering the recent geological past includes a dramatic climatic change: the transition of the last glacial cycle to the present interglacial (Saltzman 2002). In addition, the Cordillera Frontal piedmont is considered as one of the source areas of the aeolian deposits of central Argentina (Zárate 2003; Mehl *et al.* 2012). Knowledge of the sedimentary record is therefore significant to validate the current models of aeolian sedimentation. These issues have renewed interest in the area and instigated multidisciplinary analysis by our research team. The studies, still under progress, include stratigraphical, sedimentological, geochronogical, morphostructural and palaeontological analysis, with the general aim of reconstructing the environmental and climatic conditions during Late Quaternary time across the Cordillera Frontal piedmont (e.g. Mehl & Zárate 2012; Rojo *et al.* 2012) and the eastern piedmont of San Rafael block, situated to the south (Tripaldi *et al.* 2011). Simultaneously, other authors (e.g. Pepin 2010; Pepin *et al.* 2013; Casa *et al.* 2011) have focused on tectonic, geomorphological and geochronological analyses in the study area (e.g. Cordón del Carrizalito piedmont, Las Tunas fluvial system).

The purpose of this paper is to provide a general overview of the Quaternary stratigraphy and evolution of the Cordillera Frontal piedmont at

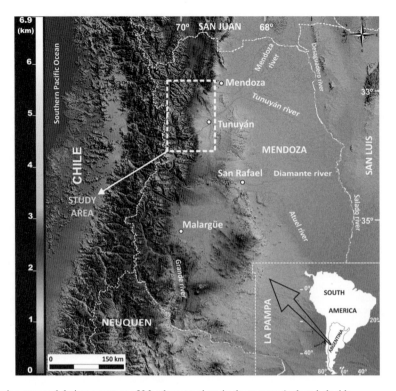

Fig. 1. Location map and drainage system of Mendoza province in the eastern Andean belt side.

Valle de Uco. Special emphasis is placed on Late Pleistocene and Holocene alluvial sequences that exhibit a suitable stratigraphic resolution for palaeoenvironmental reconstructions. The chronology and morphotectonic of the area are also analysed, along with the composition and provenance of the sediments. The final aim is to contribute to the understanding of the landscape dynamics in an active tectonic setting within the context of the Quaternary climate cycles.

Environmental setting

The piedmont sector of Valle de Uco, located c. 80 km south of Mendoza city, included the areas that Polanski (1963) called *Graben de Tunuyán* (Tunuyán Depression in this paper) and *Valle Extenso del Campo Bajo* (Figs 1, 2a, b). The current climate is dominated by arid–semiarid conditions resulting from the interaction of atmospheric circulation systems related to the South Pacific and the South Atlantic anticyclonic centres along with the low-pressure continental centre (Garreaud *et al.* 2009). The mean annual temperature is c. 12 °C and the average annual rainfall 350 mm at the locality of San Carlos (Fig. 2b). The vegetation cover is dominated by a xerophytic shrubland (Rojo *et al.* 2012) that belongs to the Chacoan phytogeographic domain (*Dominio Chaqueño*; Cabrera 1971). It is situated in an ecotone fringe called South American Arid Diagonal (Bruniard 1982), an environmentally sensitive area that has shown changes in its geographical extension in response to past climatic conditions (Abraham de Vazquez *et al.* 2000).

The piedmont at Valle de Uco is a relatively narrow plain, 30–50 km wide, that descends from c. 2000 m asl at the mountain front to nearly 700 m asl eastwards. The eastern limit is marked by a steep and dissected escarpment that determines the western margin of Meseta del Guadal (Fig. 2b). The piedmont is drained by the Tunuyán River and its tributaries that are streams of perennial discharge generated either by springs (Arroyo La Estacada and its affluents) or highly seasonal snowmelt in the mountain headwaters (e.g. Tunuyán River, Las Tunas River, Arroyo Papagayos). The Tunuyán River is a major tributary of the Desaguadero–Salado fluvial system located eastwards (Fig. 1).

Geological and tectonic setting

The study area is situated within the borderland region of the geological provinces of Cordillera Frontal and Cuyo basin (Ramos 1999).

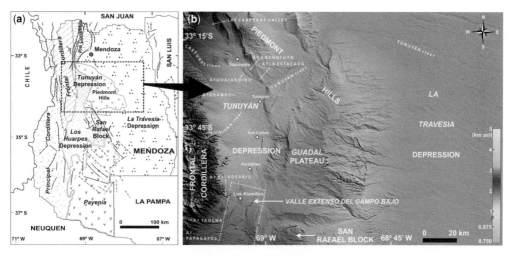

Fig. 2. (a) Main geomorphological units of Mendoza province (adapted from González Díaz & Fauqué 1993). (b) Digital elevation model of the Frontal Cordillera piedmont at 33°–34°S.

Cordillera Frontal is largely composed of Neopalaeozoic deposits that unconformably overlie a middle Proterozoic basement made up of gneissic rocks (Caminos 1993; Ramos 1999). The Neopalaeozoic sequence consists of marine and continental sediments (Ramos 1999). It is intruded by plutonic bodies (granodiorites, tonalites, gabbros) and interlayered with volcanic rocks that belong to the Choiyoi Group, a calcoalkaline orogenic complex including basalts, andesites, dacites and rhyolites (Llambías et al. 1993; Ramos 1999). Cordillera Frontal is a mountain block uplifted during the Late Miocene along the west-dipping fault system of La Carrera (Polanski 1963; Caminos 1979; Cortés 1993) (Fig. 2a, b).

A great part of the piedmont is developed across the Cuyo Basin, a Triassic rift generated by a general extension process that occurred at the southwestern margin of Gondwana; its location was structurally controlled by the suture of terranes accreted during Palaeozoic time (Kokogian et al. 1993). The Cuyo Basin was reactivated as a foreland basin during the Cenozoic (Ramos 1999), and underwent several successive episodes of deformation characterized by a dominant compressive regime during Late Neogene and Quaternary time that resulted in the formation of several morphostructural units (Figs 2, 3) (González Díaz & Fauqué 1993).

The Huarpes depression (Polanski 1963) is a 250-km-long tectonic basin along the eastern margin of Cordillera Frontal formed by the Miocene uplift (Fig. 2a). The basin was then filled with 1500–1800 m (Polanski 1963) of synorogenic deposits (Yrigoyen 1993; Irigoyen et al. 2002).

The Tunuyán depression (Tunuyán graben *sensu* Polanski 1963) is a tectonic basin formed in the northern part of the Huarpes depression during

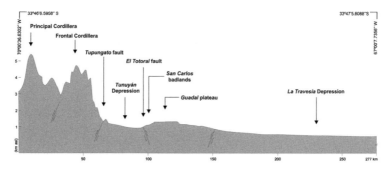

Fig. 3. Transverse Frontal Cordillera piedmont schematic profile showing the main morphostructural units. Vertical and horizontal distances are approximated; the vertical scale is exaggerated to highlight piedmont morphostructural features.

Pleistocene time (Fig. 2a, b). It was interpreted initially as a basin limited by normal faults (Polanski 1963). Later, other authors (Caminos 1979; Legarreta et al. 1992; Devizia 1993; Ploszkiewicz 1993) pointed out that the basin is bound by reverse faults. It evolved under east-vergent thrusts as suggested by its non-symmetrical shape and the eastern location of the depocentre (Perucca et al. 2011). The general basin architecture is due to deformation caused by thrusts that uplifted the Miocene–Pliocene deposits exposed at Meseta del Guadal (Cerrilladas Pedemontanas) where the active thrust front is located. It is hypothesized that the Tunuyán depression is a piggy-back basin formed on active thrust downstepping sheets (Perucca et al. 2011).

The Cerrilladas Pedemontanas (piedmont hills) is a positive morphostructural unit composed of Neogene synorogenic deposits intensely deformed by folding and thrusting (Yrigoyen 1993). The unit presently undergoes active fluvial dissection, which gives way to a badland topography.

La Travesía depression is a large sedimentary basin with its uppermost section composed of Neogene deposits and a blanket of exposed Quaternary deposits. At the latitudinal fringe of the study area, the western boundary is formed by the Cerrilladas Pedemontanas and the eastern boundary by the Desaguadero–Salado system.

To date, at a regional geomorphological scale, the Tunuyán depression and the Cerrilladas Pedemontanas are regarded as the proximal piedmont of Cordillera Frontal whereas La Travesía depression is the most distal piedmont environment (Figs 2b, 3).

Quaternary stratigraphy and general geomorphology

The piedmont landscape is dominantly composed of both erosional and depositional landforms consisting of fluvio-aeolian deposits that were fractured and folded by tectonic processes. In addition, piedmont volcanic activity generated minor cones at some sectors whereas major eruptions at Cordillera Frontal (see Pyroclastic deposits (APP) below) gave way to the accumulation of thick pyroclastic sequences in the southern part of the study area (Valle Extenso del Campo Bajo sensu Polanski 1963). Pediments are developed close to the mountain front at heights of c. 2000 m asl. These erosional surfaces cut through the Cordillera Frontal bedrock and form relatively wide interfluvial surfaces, partially covered by accumulations of coarse conglomerates at more distal locations (Fig. 4a; González Díaz & Fauqué 1993; Polanski 1963). Alluvial fans, laterally coalescing and corresponding to the Tunuyán River, Las Tunas River, the Arroyo Grande and the Arroyo Anchayuyo, are the most remarkable piedmont landforms (Fig. 5).

According to Polanski (1963), the evolution of the piedmont landscape was marked by the generation of four main Quaternary aggradational cycles that he grouped into lithostratigraphic units differentiated on the basis of their topographic positions, stratigraphic relationships and the general geological reconstruction of the region (Fig. 5). The resulting stratigraphic scheme (Table 1) was correlated with the global model of Quaternary glacial cycles (i.e. European glaciations, Mindel, Riss, Würm, North American Illinoian, Wisconsin) valid at the time. Except for some radiocarbon dates reported for Holocene deposits, no other numerical ages were then available. The stratigraphic scheme by Polanski (1963) has been widely accepted, and became an unavoidable reference for later geological studies. In the last years, geochronological analysis together with detailed field work has provided new evidence to partially reinterpret the Quaternary stratigraphy, and hence the evolution of the piedmont landscape during this interval. Note that detailed structural and geochronological analysis of Neogene deposits also provided new insights on the piedmont evolution during the Late Miocene and Pliocene (e.g. Yrigoyen 1993; Irigoyen et al. 2002). Notwithstanding, some important disagreements are evident in the field recognition of the units revealed by the interpretations and mapping carried out by later authors, as described below.

Aggradational cycles

The aggradational cycles make up major alluvial fan systems called fanglomerates by Polanski (1963) who interpreted that the alternating erosional cycles were the result of tectonic activity. The alluvial fan systems show a stepped geometry with successive terrace levels that yield a telescopic-like relationship (Bowman 1978; Janocko 2001), in which the younger fans lie at lower topographic position. This feature is distinctly illustrated at the basin of Arroyo Anchayuyo (Fig. 6).

The first aggradational cycle (Los Mesones Formation) consists of a series of isolated conglomerate outcrops in the mountain front area along with well-developed and preserved exposures in the piedmont (Fig. 4a, b; Polanski 1972). The highest remnants are found at an altitude of 2450–2400 m asl in the headwaters of La Estacada Basin (Arroyo Anchayuyo; Fig. 6). A Lower Pleistocene age was assigned to Los Mesones Formation considering that the unit unconformably overlies the deformed deposits of Los Mogotes Formation of likely Pliocene age (Polanski 1963). Yrigoyen (1993) later interpreted that the conglomerates exposed at various hills of the Tunuyán depression (Lomas

Fig. 4. (a) View of Los Mesones Formation at Estancia San Pablo, near its type area. The conglomerate deposits rest on a pediment on basement rocks at the foothills of Frontal Cordillera. (b) Las Carreras Valley: view of Los Mesones Formation lying on top of Neogene deposits, and of a narrow valley infilled with La Invernada Formation deposits. Frontal Cordillera is seen at the background. (c) Las Tunas Formation outcrop at the Las Tunas river banks. (d) El Zampal Formation and Regional Aggradational Plain at the Arroyo Anchayuyo. The present arroyo floodplain is seen in the foreground. (e) Deposits of the APP exposed in a quarry at the Valle Extenso del Campo Bajo area. (f) La Invernada Formation and the Asociación Piroclástica Pumícea (APP) at Los Alamitos locality in the Valle Extenso del Campo Bajo area defined by Polanski (1963).

de Gualtallary, Jaboncillo, del Peral), mapped as Los Mesones Formation by Polanski, belonged to Los Mogotes Formation. In turn, these same outcrops were mapped as La Invernada Formation by García (2004). Further north in the area of La Pilona anticline, Polanski (1963) mapped several exposures of Los Mesones Formation but none were identified there by Irigoyen *et al.* (2002).

The second aggradational cycle (La Invernada Formation) is a conglomerate deposit lithologically similar to that of Los Mesones Formation, but composed of generally smaller boulders and clasts (Polanski 1963, 1972). Field observations indicate a rather heterogeneous lithology that varies from conglomerates cemented and capped by calcium carbonate (Arroyo Yaucha in the surroundings of Los Alamitos; see location in Fig. 2) to conglomeratic fine-grained deposits (Las Carreras valley) (Figs 4b, 6). It was interpreted as a younger level than Los Mesones considering that the outcrops are located at lower topographic levels, accumulated in the Upper Pleistocene (Polanski 1963, 1972). The correlation of distant exposures of La Invernada Formation is rather problematic because its field recognition is mostly based on its relative altitudinal setting in relation to Los Mesones and Las Tunas formations. If one of these two units is not present, the identification of isolated exposures of La Invernada Formation becomes rather doubtful at some areas.

The third aggradational cycle (Las Tunas Formation) is an extensive unit across the piedmont composed of coarse and fine fluvial sediments (Fig. 4c). Its surface represents the Bajada Joven al Graben de Tunuyán according to Polanski (1963, 1972). In the type area of Las Tunas Formation at the mouth of Las Tunas River at the mountain front, Polanski identified three terraces interpreted as the result of cyclic erosion. Recently, the

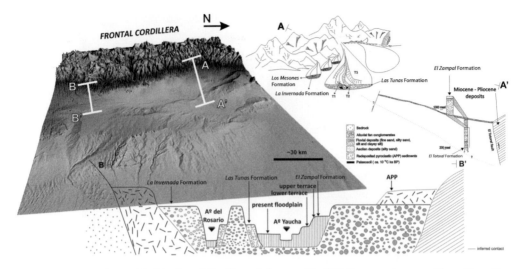

Fig. 5. Digital elevation model of the Frontal Cordillera piedmont at 33–34°S. Transversal topographic profiles with the main geomorphological and lithostratigraphic units distribution in two different tectonic locations of the piedmont: the Tunuyán Depression (A–A′) and the Valle Extenso del Campo Bajo (B–B′).

highest terrace level was mapped either as La Invernada Formation by García (2004) or Los Mesones Formation by Pepin (2010). Las Tunas Formation was chronologically assigned to the Upper Pleistocene, and placed at the end of the last interglacial just before the last maximum glaciation of Cordillera ('...postrimerías del último interglacial que precede la máxima glaciación de Cordillera...', Polanski 1972, p. 81). Recently, Pepin (2010) and Pepin et al. (2013) reported cosmogenic dates from the uppermost sections of the three terraces identified by Polanski (1963) at the piedmont apex of Las Tunas River, and an Ar/Ar date of pyroclastic deposits embedded in the unit. The highest

Table 1. *Aggradational cycles, lithostratigraphy and piedmont numerical ages according to Polanski (1963) and Zárate & Mehl (2008)*

Aggradational cycles, lithostratigraphy and ages (Polanski 1963)		Zárate & Mehl (2008)	Numerical ages	
Los Alamitos Formation El Zampal Formation La Estacada Formation Fourth aggradational cycle	Holocene (1400 ^{14}C yr BP) Holocene Early Holocene–Late Pleistocene	El Zampal Formation	↑	(Zárate & Mehl 2008; Mehl & Zárate 2012)
			~35 ka ~50 ka	
Las Tunas Formation/El Totoral Formation Third aggradational cycle	Late Pleistocene Wisconsin/Würm glaciation		>0.30 ka 0.60 Ma 0.70 Ma 1.2 Ma	(Pepin 2010)
Asociación Piroclástica Pumícea (APP)	Sangamon interglacial	—	c. 0.45 Ma	(Guerstein 1993)
La Invernada Formation Second aggradational cycle	Illinoian/Riss Beginning of the Late Pleistocene			None
Los Mesones Formation First aggradational cycle	Early Pleistocene			None

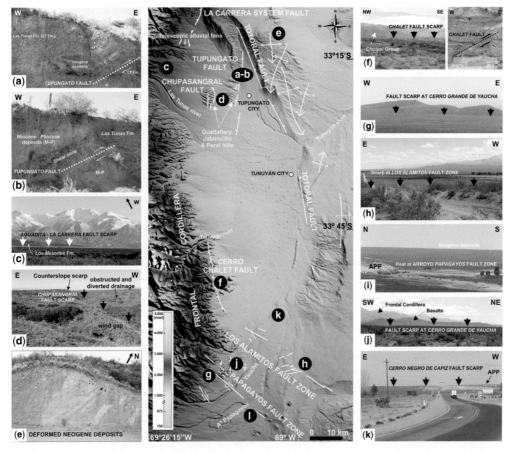

Fig. 6. Frontal Cordillera piedmont at 33–34°S, digital elevation model and (**a–k**) views of the faults, scarpt faults and thrusted-faulted deposits of the study area.

and then relatively oldest terrace of Las Tunas Formation (mapped as Los Mesones Formation by Pepin 2010) yielded an age suggesting that the unit is at least as old as at 1.2 Ma. In the uppermost sections of the lower two terraces of Las Tunas Formation, Pepin (2010) and Pepin *et al.* (2013) reported cosmogenic ages between *c.* 15 and 20 ka (15 500 ± 700 a T1 and 20 000 ± 800 a T2), while the pyroclastic deposits embedded in the unit yielded a date of 0.6 ± 0.2 Ma suggesting that the volcanic event occurred between 0.4 and 0.8 Ma. The numerical dates obtained by Pepin (2010) and Pepin *et al.* (2013) therefore widen considerably the time interval attributed to Las Tunas Formation by Polanski (1963).

The third aggradational cycle is completed by El Totoral Formation, defined on the basis of drilling information at the Tunuyán depression (Polanski 1963). The up-to-600-m-thick unit is composed of sandy, silty and clayey deposits with conglomeratic levels, and carbonate layers that include redeposited pyroclastic sediments of the APP at the lower part. Polanski (1963) attributed an Upper Pleistocene age to El Totoral Formation and interpreted that the unit, correlatable with Las Tunas Formation, recorded the finer sedimentary facies that filled up the Tunuyán depression.

The fourth aggradational cycle is represented by La Estacada and El Zampal formations which are considered to be accumulated at the very end of the Late Pleistocene and the Holocene (Polanski 1963). Zárate & Mehl (2008) redefined the stratigraphy of these deposits and grouped them into a single lithostratigraphic unit named El Zampal Formation that also includes the peaty deposits of Los Alamitos Formation (Polanski 1963) at Arroyo Yaucha.

El Zampal Formation is dominantly composed of fine fluvial facies, subordinated conglomerates and aeolian facies (Fig. 4d). The latter includes a surface loessial drape that also covers the surface of older units. Relatively thick (up to 20 m) and extensive outcrops are exposed along the riverbanks of the tributaries of the Tunuyán River, mainly the arroyos La Estacada, Anchayuyo and Las Torrecitas. As the lower stratigraphic contact is not exposed, it has been speculated that the sedimentation started likely during the last interglacial (Zárate & Mehl 2008). Alluvial aggradation dominated from c. 50 ka onwards according to the Optically Stimulated Luminescence (OSL) ages obtained at the alluvial sequence exposed at Arroyo Las Torrecitas (Toms et al. 2004). In turn, the nearby Arroyo La Estacada alluvial sequences record sedimentation since c. 35 ka to the present according to radiocarbon and OSL ages (Zárate 2002; Zárate & Páez 2002; De Francesco et al. 2007; Zárate & Mehl 2008; Mehl & Zárate 2012).

Polanski (1963) also mentioned the occurrence of an epicycle of erosion (epiciclo de erosión) that corresponds to the last fluvial incision and the formation of the present channels and floodplains.

Pyroclastic deposits (Asociación Piroclástica Pumícea or APP)

The deposits of the APP extend across an area of 23 000 km² in Chile and Argentina (Stern et al. 1984). They cover the southern sector of the study area (Valle Extenso del Campo Bajo *sensu* Polanski 1963), forming a wide pyroclastic plain dissected by the fluvial systems of the arroyos Papagayo and Yaucha (Fig. 4e, f). The APP has been genetically related to the Diamante stage, a catastrophic event related to the Diamante caldera formation at 0.45 Ma; this was followed by the Maipo stage that represents the evolution of the Diamante–Maipo volcanic complex during the last 100 ka (Sruoga et al. 2005). On the basis of drilling information, Polanski (1963) also reported the unit at the Tunuyán depression where the APP is interlayered and redeposited with fluvial sediments at the lower section of El Totoral Formation. In the southernmost part of the Tunuyán depression (Las Pareditas locality; see location in Fig. 2), drilling information (Abraham, J., pers. comm. 2010) indicates that the APP is at nearly 30 m of depth buried by fluvial conglomerates that are correlatable with Las Tunas Formation. Polanski (1963) considered that the APP was formed after the second aggradational cycle (La Invernada Formation) and prior to the fourth aggradational cycle (Las Tunas Formation), placing the unit during the Mindel–Riss interglacial (Table 1). However, the chronology of the APP remains controversial. Fission track analysis carried out on zircons yielded an age of 0.45 Ma (Stern et al. 1984). Ignimbritic flows (Pudahuel ignimbrite) exposed on the Chilean side and supposedly attributed to the APP yielded ages of 150 ka by U–Th–He on zircons from pumice pyroclasts (Lara et al. 2008), whereas ages as old as 2.3 Ma were previously obtained (Wall et al. 2001 in Lara et al. 2008).

Piedmont morphotectonic

Several tectonic features occur across the piedmont (Fig. 6), some of which give rise to remarkable morphological landforms. Polanski (1963) pointed out the occurrence of tectonic deformation features at several hills (Lomas del Gualtallary, Jaboncillo and Peral; Fig. 6), later interpreted as two folding structures (Yrigoyen 1993). More recently, these structures were considered anticline ridges formed by a very-low-angle thrust that deformed Cenozoic sedimentary rocks, active until Holocene time (García 2004; García et al. 2005).

Perucca et al. (2009, 2011) pointed out that the nearly north–south-trending faults of this area affected the middle and lower reaches of alluvial fans, most of them having the free face to the east (Tupungato fault; Fig. 6a–c) with some counter-slope scarps with the free face to the west (Chupasangral fault; Fig. 6d). A clear connection to the La Carrera fault system is not observed at the surface, although a mechanical linkage in the subsurface with the southern segments of the fault system might be possible (Casa et al. 2011), such as La Aguadita zone fault (Polanski 1963; Cortés et al. 1999).

García et al. (2005) analysed a digital elevation model and found that the drainage network located south of Lomas del Peral and Jaboncillo (Fig. 7) shows some anomalies, such as a subparallel stream pattern that incised the regional aggradation plain composed of El Zampal Formation (Zárate & Mehl 2008). Further south, the fluvial channels diverge slightly southwards towards the regional aggradational plain in response to sagging. This type of anomaly is frequently observed in association with growing anticlines or synclines (sagging areas; Audemard 1999). These observations allow the southern extension of the Tupungato thrust to be inferred, involved in the structuring of Lomas del Jaboncillo and Peral, and suggest a likely Holocene activity in the area (García & Cristallini 2011).

At the southern part of the Tunuyán depression, the Chalet reverse fault (Bastías et al. 1993; Tello 1994, 1998; Costa et al. 2000) uplifted Pleistocene fanglomerates and a group of aligned granitic hills (Cerro Chalet and others) (Fig. 6f). The downslope fault scarp dips 60°W and trends 320°. The highest scarps are located in the central section of the fault that exhibits a slightly sinuous pattern. No strike-slip displacement was observed during fieldwork since the free face along the scarp

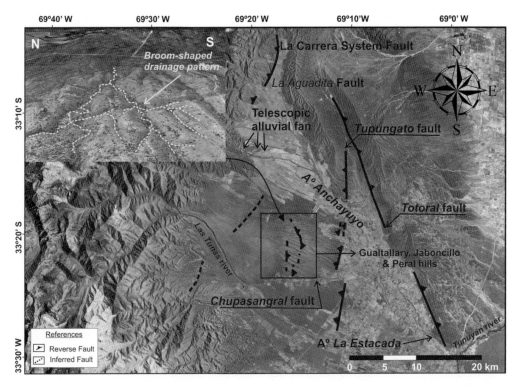

Fig. 7. Sketch of the main Quaternary faults developed on the northern tip of the Tunuyán depression (modified from Perucca *et al.* 2011) and (inset) Google Earth oblique view showing a broom-shaped river pattern in the counter-slope scarp of the Chupasangral.

looks to the east and the displacement is vertical (reverse). A lateral component related to the Papagayos fault zone (transtensive) (Fig. 6g, h) situated further south (Bastías *et al.* 1993; Tello 1994; Cortés & Sruoga 1998) might be possible. No evidence has been found to determine the kinematics of the Papagayos fault. However, the existence of scarps alternating in their direction of facing along-strike (scissoring), its rectilinear trace and the existence of associated volcanism point to a transtensive kinematic along pre-existent faults reactivated by Quaternary tectonics. New evidence of Quaternary deformation have been reported north of the mouth of Arroyo del Rosario related to the WNW-trending Los Alamitos fault zone (Fig. 6i), which is more than 25 km long (Casa *et al.* 2011).

Although thrust fault scarps are the most remarkable geomorphological evidence in compressive continental environments, in the study area a great deal of seismic events were generated by ruptures that did not reach the surface; they therefore do not show direct surface indicators. However, some geomorphological features are able to reflect very subtle topography modifications. In this regard, the analysis of the drainage characteristics is a valuable tool to study tectonic activity at fault and thrust belts since they are highly sensitive to vertical tectonic processes related to folds and thrusts (Audemard 1999). Ollarves *et al.* (2006) pointed out that second-order rivers are more efficient indicators of underlying tectonic activity. In this regard, several streams of the drainage system are structurally controlled by faults; this is illustrated by Arroyo La Estacada and its tributary the Arroyo Anchayuyo, close to a major fault trace (Totoral fault, Polanski 1963, figure 6) that bounds the Tunuyán depression on the NE. Further, drainage anomalies reported as indicators of active tectonic processes (Audemard 1999) have been identified in the piedmont area. They include river pattern inversion, river diversion, beheaded drainages and stream captures, changes in incision depth and channel gradient, abandoned river gaps, broom-shaped river patterns and the telescoping fan shapes mentioned in 'Agraddational cycles' (Fig. 7).

In general, the drainage pattern of the piedmont is dominantly dendritic-divergent, typical of alluvial fans. Where the rivers cross the counter-slope scarps at the tectonic landforms of Lomas del Jaboncillo and Peral, there is an increasing degree of channel incision together with a modification of the drainage pattern that includes palaeochannels abandoned by migration. A noticeable feature is illustrated by the broom-shaped drainage pattern developed by small tributaries that cross the counter-slope escarpment of the Chupasangral fault (Fig. 7). The minor streams are orthogonally arranged and gather into a single large river downstream to overcome the structure, opposed to the natural runoff. The broom-shaped pattern is also present at the mid-segment of the Arroyo Chupasangral alluvial fan (Fig. 7) where the small and deeply incised channels are grouped to cross the escarpment. At Lomas del Jaboncillo (Figs 6d & 7), drainage anomalies also occur including displaced, adapted and deflected channels caused by regressive erosion, capture or obturation. These anomalies were reported at active thrust systems of Venezuela and Colombia, emphasizing their importance as indicators of vertical motion produced by thrusts (Audemard 1999). Another example is found in the hanging block of the Chupasangral fault, where the channels are strongly incised and exhibit increased sinuosity downstream.

Schumm (1986) stated that stream channels are sensitive to variations in their longitudinal slope, probably in relation to lithological contrasts or as a result of tectonic activity. Hanging channels (hanging drainage) have been identified in the hanging block of the fault at Lomas del Jaboncillo. To the south, the Papagayos and Los Alamitos fault zones (Fig. 6h, i) may have controlled the location and formation of peat deposits as seen in arroyos Papagayos and Yaucha; these streams also exhibit increasing sinuosities and valley widening when crossing the faults. The existence of anomalous ponds and marshes and also the widening of the streams are likely indicators of active tectonics. Faults influencing the valley slope will affect the sinuosity of the channel, since the river tries to keep the channel slope constant in a self-organized manner; down-throwing faults result in increased meandering downstream (Ouchi 1985; Keller & Pinter 1996). Anomalous reaches that are not related to artificial controls or to tributary influences may be reasonably assumed to be the result of active tectonics (Schumm 1986; Schumm et al. 2002).

Late Quaternary sedimentary records and palaeoenvironmental conditions

From a morphostratigraphic perspective, the deposits of El Zampal Formation (Zárate & Mehl 2008) comprise several geomorphological units. The most extensive is the Regional Aggradational Plain (RAP), referred to as the Loessic Plain by Polanski (1963). It is a piedmont alluvial plain mostly comprising silty sands, silts and sandy silts resulting from the distal coalescence of alluvial fans (Fig. 8a, c, d). The aggradational process was mainly related to fluid sheet overflows that affected overbank areas and likely temporary inactive channels of sandy-like braided streams (Mehl & Zárate 2012). Secondary deposition by hyperconcentrated flows, channel lag deposits or longitudinal bar development has been inferred from the occurrence of gravel deposits included in the alluvial sequences (Mehl & Zárate 2012).

Alluvial sedimentation was punctuated by Andean volcanic eruptions that gave way to the presence of discrete tephra layers, along with soil forming intervals documented by palaeosols and lapses of significant limnic accumulation. An interval of more dominant aeolian deposition in overbank areas, represented by c. 1.5 m of sandy loess, is recorded between 17 110 $\pm$ 70 ^{14}C yr BP until c. 11 709–12 075 cal yr BP, suggests dominant arid conditions (Mehl & Zárate 2012). A soil-forming interval documented by a discrete palaeosol occurred between c. 11 709–12 075 cal yr BP and c. 10 685–11 144 cal yr BP (Mehl & Zárate 2012). It is traceable along nearly 12 km of the arroyos La Estacada and Anchayuyo and developed on top of the aeolian deposits. The resulting palaeosol is a remarkable stratigraphic layer that suggests a general stabilization of the fluvial basin during the Pleistocene–Holocene transition. Pollen spectra from the palaeosol and overlying sediments record a major vegetation change from halophytic plant communities towards Monte communities. The formation of the palaeosol therefore might indicate the prevailing influence of the Atlantic anticyclonic centre in central Argentina at the Early Holocene (Zárate & Páez 2002; Markgraf et al. 2009; Piovano et al. 2009). After the soil interval, alluvial sedimentation was characterized by numerous overbank episodes (flooding events) of different magnitude and the more frequent development of limnic layers, suggesting an increase in fluvial variability (Mehl & Zárate 2012). The occurrence of palaeosols at the top of the overbank deposits indicates sedimentary pauses between the overbank depositional episodes and hence stability intervals (Mehl & Zárate 2012). It is hypothesized that the flood events responsible for overbank depositional episodes might have been related to heavy summer rainfalls generated by storms from the Atlantic, as happens at present (Zárate & Páez 2002; Markgraf et al. 2009; Piovano et al. 2009). The higher vegetation productivity in the floodplains or overbank environments of the eastern Andean piedmont could have been related to these climatic conditions (Mehl & Zárate 2012). The RAP was incised up to c. 20–25 m by an erosional episode that occurred sometime during the interval bracketed between 8454–8968 cal yr BP and 5758–6186 cal yr BP.

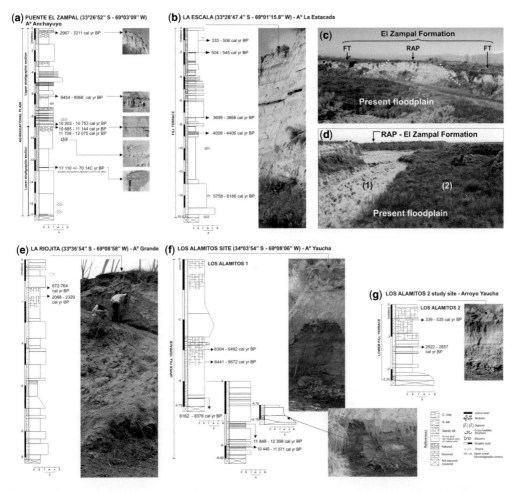

Fig. 8. (a) Regional Aggradational Plain (RAP) at Arroyo Anchayuyo: Puente El Zampal lithostratigraphic profile and calibrated radiocarbon ages (*). (b) Fill terrace (FT) at Arroyo La Estacada: La Escala lithostratigraphic profile and calibrated radiocarbon ages (*). (c) Arroyo La Estacada: RAP and FT deposits of El Zampal Formation. The stratigraphic contact between the RAP and the FT is indicated (dotted line). (d) Arroyo La Estacada: RAP deposits of El Zampal Formation and present floodplain dominated by (1) cortaderales and (2) xerophytics shrublands (adapted from Rojo *et al.* 2012). (e) Arroyo Grande deposits: La Riojita lithostratigraphic profile and calibrated radiocarbon ages (*). (f, g) Arroyo Yaucha Late Pleistocene and Holocene deposits: Los Alamitos 1 and 2 lithostratigraphic profiles and calibrated radiocarbon ages (*). (*Views of the deposits are shown).

As a result, the surface of the RAP became stable as recorded by a carbonate-gypsum crust of unknown origin. This was later blanketed by an aeolian mantle that is the parent material of the surface soil (Zárate & Mehl 2008; Mehl & Zárate 2012).

The incision event was followed by alluvial aggradation that makes a distinct filled terrace along Arroyo La Estacada composed of El Zampal Formation (Fig. 8b, c). Its surface is *c.* 18 m above the present channel and *c.* 6 m below the surface of the RAP. The terrace consists of a fining-upwards sedimentary sequence related to a sinuous stream. Gravel and coarse sand deposits at the lower part indicate the occurrence of fluvial processes restricted to the fluvial channel (e.g. channel fills or build-up of point bars) and likely the most proximal floodplain area to the channel belt (Mehl 2011; Mehl & Zárate 2012). The accumulation of these coarse facies occurred sometime before 5758–6186 cal yr BP as suggested by a radiocarbon date on vegetation remains 2 m above in the sequence. A palaeosol formed on top of the sandy fluvial deposits probably as a result of a cut-off event and the abandonment of a meandering belt. The end of the soil-forming interval occurred during

c. 4008–4406 cal yr BP. Dominant overbank depositional episodes followed including the frequent development of limnic levels. The occurrence of several palaeosols documents stabilization intervals of the floodplain surface during the aggradational process. The uppermost sedimentary layers were deposited during c. 333–506 cal yr BP at La Escala section. The pollen analysis of the alluvial record indicates the presence of hydrophytic and shrubby xerophytic communities which show variations in response to the fluvial dynamic of the arroyo (Rojo et al. 2012). The incision occurred sometime after c. 333–506 cal yr BP, resulting in the formation of the present floodplain and channel.

At Arroyo Grande (La Riojita site), El Zampal Formation makes up an alluvial landform deeply altered by human activities that have modified and masked its original morphology (Fig. 8e). Its surface is c. 10 m above the present channel. The deposits consist of a fining-upwards sequence with an unexposed basal contact, probably related to a sinuous fluvial system (Mehl 2011; Mehl & Zárate 2012). It mainly records overbank events, with organic matter inputs to the floodplain area reflected by limnic levels and two episodes of soil formation at 2068–2329 cal yr BP and 672–764 cal yr BP at the uppermost part followed by c. 2 m of sandy deposits. The chronology of the sequence suggests that is correlatable with the Holocene fill terrace of Arroyo La Estacada. The incision of the terrace and the development of the present channel and its floodplain took place very recently, sometime after the end of the soil-forming interval dated at 672–764 cal yr BP (Mehl 2011; Mehl & Zárate 2012).

In the northern part of the pyroclastic plain composed of the APP, two fill terraces composed of El Zampal Formation deposits occur at Arroyo Yaucha in the locality of Los Alamitos (Fig. 8f, g). These landforms are situated upstream from the intersection of Los Alamitos fault zone with Arroyo Yaucha. Both terraces consist of a general fining-upwards sedimentary sequence related to a sinuous fluvial system. The lower sections (basal contact not exposed) are dominated by gravel lenses of a stream channel overlain by fine sediment layers.

The relatively higher terrace with a surface 7 m above the present channel of Arroyo Yaucha includes numerous limnic levels formed during Early Holocene time that are dominant in the lower section as well as palaeosols, more frequently present at the middle and upper sections. This terrace was formed by an incision event that occurred sometime during a time interval bracketed between 6304–6492 cal yr BP and 2622–2857 cal yr BP (Mehl 2011; Mehl & Zárate 2012).

The lower terrace consists of Late Holocene deposits with abundant limnic levels and two conspicuous palaeosols. Its surface is c. 3 m above the present stream channel. The lower terrace is the result of a very recent incision event, placed sometime after the end (c. 339–535 cal yr BP) of the upper palaeosol formation (Mehl 2011; Mehl & Zárate 2012).

Compositional signature of Late Quaternary deposits

The compositional signature (rock fragments and mineral clasts) of the Late Quaternary piedmont deposits, expressed throughout the composition of the very fine-grained sand fraction, is dominated by the presence of abundant volcanic glass, mixed pyroclasts, rock fragments and quartz, all of them in association with heavy minerals such as amphiboles (green type hornblendes and basaltic hornblendes), pyroxenes, euhedral biotites (although scarce) and olivines (Fig. 9a, f). This mineral assemblage allows a volcaniclastic source to be inferred. The presence of mica grains, quartz grains with undulose extinction (tectonic fabric), hornblende and epidote grains also indicates a metamorphic source. The very fine-grained sand composition of the alluvial sediments therefore reflects two geologically distinct sources in the catchment areas at Cordillera Frontal volcaniclastic and metamorphic rocks. This compositional signature shows a positive correlation with both the tectonic regime of the region and also with the Q:F:Rf (being Rf rock fragments) composition of the Argentine Association of modern sands (Mehl et al. 2012), dominated by volcanic rock fragments (average Q:F:Rf ratio 26:18:56) and included in the Andean family of sands of South America (Potter 1994) (Fig. 9g).

The sediments come from the erosion of both types of source rocks at the headwaters of the drainage system in Cordillera Frontal and the subsequent transport towards the piedmont by fluvial and aeolian processes. However, some sediment could derive from the transport of primary and/or secondary volcanic ash (Mehl et al. 2012). Even when the three main volcanic zones of the Andes favour the production of volcaniclastic sands, the metamorphic outcrops of Cordillera Frontal are also a productive source of sediments, for example, the wide Proterozoic metamorphic exposures. Potter (1994) highlighted the importance of metamorphic rock fragments as a secondary but not insignificant component of the Andean family sands. Additionally, the extensive Neogene sedimentary deposits exposed to the east and north of the study area and the bedrock of the analysed piedmont fluvial systems constitute another plausible source for the Late Quaternary alluvial sequences. The main volcanic signature of the Late Pleistocene–Holocene alluvial lithic and mineral clasts supports

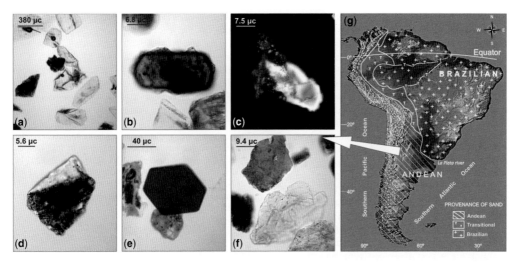

Fig. 9. View of sand clasts from the eastern Andean piedmont at 33–34°S (from Mehl *et al.* 2012). (**a**) Volcanic glass under plane-polarized light (PPL) from Puente El Zampal study site. (**b**) Mixed pyroclast under PPL from Los Alamitos 2 study site. (**c**) Polycrystalline quartz, under cross polarized light (XPL), from Los Alamitos 2 study site. (**d**) Lithic clast under PPL from Los Alamitos 2 study site. (**e**) Biotite under PPL from Los Alamitos 1 study site. (**f**) Muscovites under PPL from Puente El Zampal study site. (**g**) South American three great families of modern sands (according to Potter 1994).

the inferred Andean source of detritus currently proposed for the wide sedimentary Late Cenozoic cover of the central Argentina foreland (Zárate & Blasi 1993; Etchichury & Tófalo 2004; Iriondo & Kröhling 2007; Visconti 2007).

Final remarks

The Quaternary evolution of the Cordillera Frontal piedmont is the result of a complex history of tectonic and climatic processes, complemented by episodes of volcanic activity that generated a prominent depositional landform in the southern part (APP pyroclastic plain) and minor volcanic cones at a restricted sector. The stratigraphic relationship and timing of the aggradational cycles and erosional events is still debatable, primarily because the chronological calibration of the Quaternary piedmont record is a major hindrance, particularly for the pre-Late Quaternary deposits. The recognition and mapping of the aggradational units is also a major source of confusion with notorious different interpretations in the field recognition. As a result, the numerical ages and the new available information generate controversies and challenge the stratigraphic relationships and the chronology by Polanski (1963). As an example, the age attributed to the APP and those recently obtained for Las Tunas Formation are in disagreement with the stratigraphic relationships proposed by Polanski (1963). In Polanski's scheme the APP is older than Las Tunas Formation; new dates are evidently necessary to shed light on this discrepancy.

Tectonic processes are responsible for the generation of a major accommodation space represented by the formation of the Tunuyán depression. The resulting basin was filled by *c.* 600 m at least since the Diamante caldera event, if Polanski's interpretation (APP clasts redeposited at the lower section) is assumed correct. Further, tectonic activity seems to have caused several morphological changes in different landforms and the drainage system, including segmentation, head incision, development of erosional gullies, uplift of deposits and the formation of telescopic alluvial fans. Again, the chronology of tectonic events is currently somewhat uncertain. In general, the evidence suggests tectonic activity during the Quaternary *sensu lato* but a more detailed calibration is needed because of the still-poor chronological control of the stratigraphic record together with the difficulties in recognizing some lithostratigraphic units.

The uplift pulses of Cordillera Frontal seem to be the triggering mechanisms that initiated the main aggradational cycles (Polanski 1963). Los Mesones Formation, traditionally considered the first Quaternary cycle, is the oldest considering its topographic position, but its age is dubious (ranging from Pliocene to early Middle Pleistocene, according to different interpretations). Although likely accumulated during Quaternary time, the second aggradational cycle, La Invernada Formation, is interpreted as being younger than Los

Mesones Formation, with a stratigraphic position somewhat uncertain at several localities. Except for the descriptions by Polanski (1963), no detailed studies have been performed on these deposits at a regional scale. Las Tunas Formation, the third aggradational cycle, is the best exposed and most extensive. Glacial climatic conditions were apparently involved during the accumulation of Las Tunas (Polanski 1963; Pepin 2010; Pepin et al. 2013). The unit would record a much longer span than previously interpreted ranging from c. 1.2–0.6 Ma, comparing the mappings by Polanski (1963) and Pepin (2010), to the Late Pleistocene before c. 20–15 ka (Pepin 2010). Considering these ages, the accumulation of the uppermost section of Las Tunas Formation is synchronous with El Zampal Formation (Zárate & Mehl 2008). The dominant sandy and silty deposits of El Zampal Formation therefore represent the distal fine-grained sedimentary facies of the Late Quaternary fluvial systems; the proximal coarse-grained sedimentary facies dominantly composed of conglomerates would be documented by the uppermost part of the alluvial cones of Las Tunas Formation. In this context, El Zampal Formation is part of the aggradational cycle of Las Tunas Formation (Polanski 1963).

In relation to the fluvial terrace formations, the Pleistocene incisions of Las Tunas river terraces have been interpreted as the result of climatic changes, being formed at the end of glacial periods (Polanski 1963; Pepin 2010). Baker et al. (2009) also proposed that the fluvial terraces incision in the Diamante River further south corresponded to the beginning of interglacial periods. Further work is required to establish if these upstream episodes of erosion are reflected in some way, such as changes in sedimentation rate or grain size, in the downstream fine-grained alluvial sequences of El Zampal Formation.

Two other distinct Holocene episodes of incision are recorded in the fluvial system of the Tunuyán depression and the pyroclastic plain; the older occurred during Middle Holocene time and the younger at very recent times (less than 300–500 cal yr BP). Two Holocene terraces have also been reported at the Atuel River, south of the area under analysis (Zárate & Mehl 2011). The regional occurrence of the older fluvial terraces in different tectonic settings allows us to suggest that climate has been a possible controlling factor (Bull 1991). The process of incision was diachronic as revealed by the dates obtained, which likely reflects the variable responses of the different streams analysed. In this respect it is hypothesized that one of the degradation episodes might have been triggered by a climatic fluctuation occurring prior to c. 5700 cal yr BP (Mehl & Zárate 2012). The youngest degradational episode that gives way to the formation of the present channels and floodplain environments is regionally recorded at several streams of central Argentina. It is thought to be possibly related either to climatic conditions (Little Ice Age), human activities at the time of the Spanish arrival or a possible combination of both factors (Mehl & Zárate 2012).

The authors are grateful to Stella Marys Moreiras for her kind invitation to provide a contribution to this volume. Financial support from CONICET (PIP 5819 and PIP11-220100100123) and the University of La Pampa (EXA 2012-234) is kindly acknowledged. We would also like to thank two anonymous reviewers for their helpful suggestions.

References

ABRAHAM DE VAZQUEZ, E., GARLEFF, K. ET AL. 2000. Geomorphology and paleoecology of the Arid Diagonal in Southern South America. In: MILLER, H. & HERVÉ, F. (eds) Geoscientific Cooperation with Latin America Zeitschrift für Angewandte Geologie. Schweizerbart Science Publishers, Stuttgart, Sonderheft SH1, 55–62.

AUDEMARD, F. 1999. Morpho-structural expression of active thrust fault systems in the Humid Tropical Foothills of Colombia and Venezuela. Zeitschrift fur Geomorphologie, **118**, 1–18.

BAKER, S. E., GOSSE, J. C., MC DONALD, E. V., EVENSON, E. B. & MARTINEZ, O. 2009. Quaternary history of the piedmont reach of Río Diamante, Argentina. Journal of South American Earth Sciences, **28**, 54–73, http://dx.doi.org/10.1016/j.jsames.2009.01.001.

BASTÍAS, H., TELLO, G., PERUCCA, L. & PAREDES, J. 1993. Peligro Sísmico y Neotectónica. In: RAMOS, V. A. (ed.) Geología y Recursos Naturales de Mendoza. XII Congreso Geológico Argentino & II Congreso de Exploración de Hidrocarburos, Mendoza. Asociación Geológica Argentina, Buenos Aires, Relatorio, **VI.1**, 645–658.

BOWMAN, D. 1978. Determination of intersection points within a telescopic alluvial fan complex. Earth Surface Processes, **3**, 265–276.

BRUNIARD, E. D. 1982. La diagonal árida argentina: un límite climático real. Revista Geográfica. Instituto Panamericano de Geografía e Historia de México, **95**, 5–19.

BULL, W. B. 1991. Geomorphic Responses to Climate Change. Oxford University Press, New Cork.

CABRERA, A. L. 1971. Fitogeografía de la República Argentina. Boletín de la Sociedad Argentina de Botánica, **14**, 1–42.

CAMINOS, R. 1979. Cordillera Frontal. In :TURNER, J. C. M. (coord.) Geología Regional Argentina: II Simposio. Academia Nacional de Ciencias de Córdoba, Córdoba, **1**, 397–453.

CAMINOS, R. 1993. El basamento metamórfico Proterozoico-Paleozoico inferiorr. In: RAMOS, V. A. (ed.) Geología y Recursos Naturales de Mendoza. XII Congreso. Geológico Argentino & II Congreso

de Exploración de Hidrocarburos. Mendoza, **I.2**, 11–19.
CASA, A. L., CORTÉS, J. M. & RAPALINI, A. 2011. Fallamiento activo y modificación del drenaje en el piedemonte del Cordón del Carrizalito, Mendoza. *XVIII Congreso Geológico Argentino*, Neuquén, Argentina. Abstracts CD-ROM.
CORTÉS, J. M. 1993. El frente de corrimiento de la Cordillera Frontal y el extremo sur del valle de Uspallata, Mendoza. In: *Actas XII Congreso Geológico Argentino y II Congreso de Exploración de Hidrocarburos*, Mendoza. Buenos Aires. Asociación Geológica Argentina, **3**, 168–178.
CORTÉS, J. M. & SRUOGA, P. 1998. Zonas de fractura cuaternarias y volcanismo asociado en el piedemonte de la Cordillera Frontal (34°30′S), Argentina. *X Congreso Latinoamericano de Geología y VI Congreso Nacional de Geología Económica*, Buenos Aires, Actas, **2**, 116–121.
CORTÉS, J. M., VINCIGUERRA, P., YAMIN, M. & PASINI, M. M. 1999. Tectónica cuaternaria de la Región Andina del Nuevo Cuyo (28°–8° LS). *In:* CAMINOS, R. (ed.) *Geología Argentina*. Servicio Geológico Minero Argentino, Buenos Aires, Anales, **29**, 760–778.
COSTA, C., MACHETTE, M. N. ET AL. 2000. *Map and Database of Quaternary Faults and Folds in Argentina*. U.S. Geological Survey, Open File Report, Reston.
DE FRANCESCO, C. G., ZÁRATE, M. A. & MIQUEL, S. E. 2007. Late Pleistocene mollusc assemblages inferred from palaeoenvironments from the Andean Piedmont of Mendoza, Argentina. *Palaeogeography, Palaeoclimatology, Palaeoecology*, **257**, 461–469, http://dx.doi.org/10.1016/j.palaeo.2007.04.011
DEVIZIA, C. 1993. Yacimiento Piedras Coloradas. Estructura Intermedia. *In:* RAMOS, V. (ed.) *Geología y Recursos Naturales de Mendoza, Relatorio*. Congreso Geológico Argentino, XII, and Congreso de Exploración de Hidrocarburos, II, Mendoza. Asociación Geológica Argentina, Buenos Aires, 397–402.
ESPIZÚA, L. E. 2004. Pleistocene glaciations in the Mendoza Andes, Argentina. *In:* EHLERS, J. & GIBBARD, (eds) *Quaternary Glaciations -Extent and Chronology, Part III: South America, Asia, Africa, Australasia, Antarctica*. Elsevier, Cambridge, 69–73.
ESPIZÚA, L. E. 2005. Holocene glacier chronology of Valenzuela Valley, Mendoza Andes, Argentina. *The Holocene*, **15**, 1079–1085, http://dx.doi.org/10.1191/0959683605hl866rr
ESPIZÚA, L. E. & PITTE, P. 2009. The Little Ice Age glacier advance in the Central Andes (35°S), Argentina. *Palaeogeography, Palaeoclimatology, Palaeoecology*, **281**, 345–350, http://dx.doi.org/10.1016/j.palaeo.2008.10.032
ETCHICHURY, M. C. & TÓFALO, O. R. 2004. Mineralogía de arenas y limos en suelos, sedimentos fluviales y eólicos actuales del sector austral de la cuenca Chacoparanense. Regionalización y áreas de aporte. *Revista de la Asociación Geológica Argentina*, **59**, 317–329.
GARCÍA, V. H. 2004. *Análisis estructural y neotectónico de las lomas Jaboncillo y del Peral, departamento de Tupungato, provincia de Mendoza*. Tesis de licenciatura, Universidad de Buenos Aires, Buenos Aires.
GARCÍA, V. & CRISTALLINI, E. 2011. Evidencias estructurales y geomorfológicas de deformación cuaternaria en Los Sauces, Depresión de Tunuyán, Mendoza. *XVIII Congreso Geológico Argentino*. Neuquén, Actas CD-ROM.
GARCÍA, V., CRISTALLINI, E., CORTÉS, J. & RODRÍGUEZ, C. 2005. Structure and neotectonics of the Jaboncillo and Del Peral anticlines: New evidences of Pleistocene to ?Holocene deformation in the Andean piedmont, in Extended Abstracts. *International Symposium on Andean Geodynamics, 6th, Barcelona*. Paris, 301–304.
GARREAUD, R. D., VUILLE, M., COMPAGNUCCI, R. & MARENGO, J. 2009. Present-day South American climate. *Palaeogeography, Palaeoclimatology, Palaeoecology*, **281**, 180–195, http://dx.doi.org/10.1016/j.palaeo.2007.10.032
GONZÁLEZ DÍAZ, E. F. & FAUQUÉ, L. E. 1993. Geomorfología. *In:* RAMOS, V. A. (ed.) *Geología y Recursos Naturales de Mendoza. XII Congreso Geológico Argentino y II Congreso de Exploración de Hidrocarburos*. Buenos Aires, **I.17**, 217–234.
GUERSTEIN, P. G. 1993. *Origen y significado de la Asociación piroclástica Pumícea. Pleistoceno de la provincia de Mendoza entre los 33°30′ y 34°40′ L.S*. Tesis doctoral (inédita), Facultad de Ciencias Naturales y Museo. Universidad Nacional de La Plata, La Plata.
IRIGOYEN, M. V., BUCHAN, K. L., VILLENEUVE, M. E. & BROWN, R. L. 2002. Cronología y significado tectónico de los estratos sinorogénicos neógenos aflorantes en la región de Cacheuta-Tupungato, Provincia de Mendoza. *Revista de la Asociación Geológica Argentina*, **57**, 3–18.
IRIONDO, M. H. & KRÖHLING, D. M. 2007. Non-classical types of loess. *Sedimentary Geology*, **3**, 352–368, http://dx.doi.org/10.1016/j.sedgeo.2007.03.012
JANOCKO, J. 2001. Fluvial and alluvial fan deposits in the Hornád and Torysa river valleys; relationship and evolution. *Slovak Geological Magazine*, **7**, 221–230.
KELLER, E. A. & PINTER, N. 1996. *Active Tectonics. Earthquakes, Uplift, and Landscape*. Prentice-Hall, Upper Saddle River, New Jersey.
KOKOGIAN, D., FERNÁNDEZ SEVESO, F. & MOSQUERA, A. 1993. Las secuencias sedimentarias triásicas. *In:* RAMOS, V. A. (ed.) *Geología y Recursos Naturales de Mendoza. XII Congreso Geológico Argentino y II Congreso de Exploración de Hidrocarburos*. Buenos Aires, **I.7**, 65–78.
LARA, L. E., WALL, R. & STOCKLI, D. 2008. La ignimbrita Pudahuel (Asociación Piroclástica Pumícea) y la caldera Diamante (33° S): nuevas edades U–Th–He. *XVII Congreso Geológico Argentino*, Jujuy. Buenos Aires, Actas, **3**, 1365
LEGARRETA, L., KOKOGIAN, D. A. & DELLAPÉ, D. A. 1992. Estructuración terciaria de la Cuenca Cuyana: ¿Cuánto de inversión tectónica? *Revista de la Asociación Geológica Argentina*, **47**, 83–86.
LLAMBÍAS, E. J., KLEIMAN, L. E. & SALVARREDI, J. E. 1993. El magmatismo gondwánico. *In:* RAMOS, V. A. (ed.) *XII Congreso Geológico Argentino y II Congreso de Exploración de Hidrocarburos*. Geología y Recursos Naturales de Mendoza, Mendoza, **I.6**, 53–64.
MARKGRAF, V., WHITLOCK, C., ANDERSON, R. S. & GARCÍA, A. 2009. Late Quaternary vegetation and fire history in the northernmost Nothofagus forest region:

Mallín Vaca Lauquen, Neuquén Province, Argentina. *Journal of Quaternary Science*, **24**, 248–258, http://dx.doi.org/10.1002/jqs.1233

MEHL, A. E. 2011. *Sucesiones aluviales del Pleistoceno tardío – Holoceno, Valle de Uco (provincia de Mendoza): inferencias palaeoambientales y palaeoclimáticas*. PhD thesis, Universidad Nacional de La Plata (unpublished).

MEHL, A. E. & ZÁRATE, M. A. 2012. Late Quaternary alluvial records and environmental conditions in the eastern Andean piedmont of Mendoza (33°–34°S) Argentina. *Journal of South American Earth Sciences*, **37**, 41–59, http://dx.doi.org/10.1016/j.jsames.2012.01.003

MEHL, A., BLASI, A. & ZÁRATE, M. 2012. Composition and provenance of Late Pleistocene–Holocene alluvial sediments of the eastern Andean piedmont between 33 and 34° S (Mendoza Province, Argentina). *Sedimentary Geology*, **280**, 234–243, http://dx.doi.org/10.1016/j.sedgeo.2012.05.011

MESSAGER, G. 2010. *Signatures geomorphologiques de l'activite tectonique Plio-Quaternaire dans le sud des Andes Centrales, Argentine*. PhD thesis, Universite de Pau et des Pays de L'Adour.

OLLARVES, R., AUDEMARD, F. & LOPEZ, M. 2006. Morphotectonic criteria for the identification of active blind thrust faulting in alluvial environments: case studies from Venezuela and Colombia. *Zeitschrift für Geomorphologie*, **145**, 81–103.

OUCHI, S. 1985. Response of alluvial rivers to slow active tectonic movement. *Geological Society of America Bulletin*, **96**, 504–515.

PAEZ, M., NAVARRO, D., ROJO, L. & GUERCI, A. 2010. Vegetación y paleoambientes durante el Holoceno en Mendoza. *In*: ZÁRATE, M., NEME, G. & GIL, A. (eds) *Paleoambientes y ocupaciones humanas del centro-oeste de Argentina durante la transición Pleistoceno-Holoceno y Holoceno*. Sociedad Argentina de Antropología, Buenos Aires, 175–211.

PEPIN, E. 2010. *Interactions géomorphologiques et sédimentaires entre bassin versant et piedmont alluvial. Modélisation numérique et exemples naturels dans les Andes*. PhD thesis, Université de Toulouse, 282.

PEPIN, E., CARRETIER, S. *ET AL*. 2013. Pleistocene landscape entrenchment: a geomorphological mountain to foreland field case, the Las Tunas system, Argentina. *Basin Research*, **25**(5), 613–637, http://dx.doi.org/10.1111/bre.12019

PERUCCA, L., MEHL, A. & ZÁRATE, M. 2009. Neotectónica y sismicidad en el sector norte de la depresión de Tunuyán, provincia de Mendoza. *Revista de la Asociación Geológica Argentina*, **64**, 263–274.

PERUCCA, L., ZÁRATE, M. & MEHL, A. 2011. Quaternary tectonic activity in the piedmont of Cordillera Frontal (33°–34°S) Mendoza. *In*: SALFITY, J. & MARQUILLA, R. (eds) *Cenozoic Geology of the Central Andes of Argentina*. Instituto del Cenozoico, Salta, 317–328.

PIOVANO, E. L., ARIZTEGUI, D., CÓRDOBA, F., CIOCCALE, M. & SYLVESTRE, F. 2009. Hydrological variability in South America Below the Tropic of Capricorn (Pampas and Patagonia, Argentina) during the Last 13.0 Ka. *In*: VIMEUX, F., SYLVESTRE, F. & KHODRI, M. (eds) *Past Climate Variability in South America and Surrounding Regions (Developments in Palaeoenvironmental Research)*, Springer, Dordrecht, **4**, 323–351.

PLOSZKIEWICZ, J. V. 1993. Yacimiento tupungato. *In*: RAMOS, V. (ed.) *Geología y Recursos Naturales de Mendoza. Relatorio XII Congreso Geológico Argentino and II Congreso de Exploración de Hidrocarburos, Mendoza*. Asociación Geológica Argentina, Buenos Aires, 391–396.

POLANSKI, J. 1963. Estratigrafía, neotectónica y geomorfología del Pleistoceno pedemontano entre los ríos Diamante y Mendoza. *Revista de la Asociación Geológica Argentina*, **17**, 127–349.

POLANSKI, J. 1972. *Descripción Geológica de la Hoja 24ab, Cerro Tupungato, Provincia de Mendoza*. Boletín 128, Dirección Nacional de Geología y Minería, Buenos Aires.

POTTER, P. E. 1994. Modern sands of South America: composition, provenance and global significance. *International Journal of Earth Sciences*, **83**, 212–232.

RAMOS, V. A. 1999. Las provincias geológicas del territorio argentino. *In*: CAMINOS, R. (ed.) *Geología Argentina*. Instituto de Geología y Recursos Minerales, Buenos Aires, **29**, 41–96.

ROJO, L. D., MEHL, A., PAEZ, M. M. & ZÁRATE, M. A. 2012. Mid- to Late Holocene pollen and alluvial record of the arid Andean piedmont between 33°–34°S, Mendoza, Argentina: inferences about floodplain evolution. *Journal of Arid Environments*, **77**, 110–122, http://dx.doi.org/10.1016/j.jaridenv.2011.09.006

SALTZMAN, B. 2002. *Dynamical Paleoclimatology. Generalized Theory of Global Climate Change*. Academic Press, New York, London, International Geophysics Series **80**.

SCHUMM, S. A. 1986. Alluvial river response to active tectonics. *In*: WALLACE, R. E. (ed.) *Active Tectonics* Natational Academies Press, Washington, Studies in Geophysics 80–94.

SCHUMM, S. A., DUMONT, J. F. & HOLBROOK, J. M. 2002. *Active Tectonics and Alluvial Rivers*. Cambridge University Press, Cambridge.

SRUOGA, P., LLAMBÍAS, E. J., FAUQUÉ, L., SCHONWANDT, D. & REP, D. G. 2005. Volcanological and geochemical evolution of the Diamante Caldera–Maipo volcano complex in the southern Andes of Argentina (34°10′ S). *Journal of South American Earth Sciences*, **19**, 399–414, http://dx.doi.org/10.1016/j.jsames.2005.06.003

STERN, C. R., AMINI, H., CHARRIER, R., GODOY, E., HERVE, F. & VARELA, J. 1984. Petrochemistry and age of rhyolitic pyroclastic flows which occur along the drainage valleys of the Río Maipo and Río Cachapoal (Chile) and the Río Yaucha and Río Papagayos (Argentina). *Revista Geológica de Chile*, **23**, 39–52.

TELLO, G. 1994. Fallamiento Cuaternario y sismicidad en el piedemonte Cordillerano de la Provincia de Mendoza, Argentina. VII Congreso Geológico Chileno, Concepción, Santiago, Servicio Nacional de Geología y Minería, Actas, **1**, 380–384.

TELLO, G. 1998. *Actividad tectónica cuaternaria en el piedemonte cordillerano entre el Río Tunuyán y el Río Atuel y su vinculación con la sismicidad histórica del sur mendocino, Provincia de Mendoza, República Argentina*. PhD thesis, Universidad Nacional de San Juan.

TOMS, P. S., KING, M., ZÁRATE, M. A., KEMP, R. A. & FOIT, F. F., JR. 2004. Geochemical characterization, correlation and optical dating of tephra in alluvial sequences of central western Argentina. *Quaternary Research*, **62**, 60–75, http://dx.doi.org/10.1016/j.yqres.2004.05.005

TRIPALDI, A., ZÁRATE, M. A. & BROOK, G. A. 2011. Late Quaternary paleoenvironments and paleoclimatic conditions in the distal Andean piedmont, southern Mendoza, Argentina. *Quaternary Research*, **76**, 181–294, http://dx.doi.org/10.1016/j.yqres.2011.06.008

VISCONTI, G. 2007. *Sedimentología de la Formación Cerro Azul (Mioceno superior) de la provincia de La Pampa, Argentina*. PhD thesis, Universidad de Buenos Aires.

WALL, R. M., LARA, L. E. & PÉREZ DE ARCE, C. 2001. Upper Pliocene-Lower Pleistocene 40Ar/39Ar ages of Pudahuel Ignimbrite (Diamante-Maipo Volcanic Complex), Central Chile (33.5°S). *In: Simposio Sudamericano de Geología Isotópica*, Pucón, Chile, 3, Actas (CD-ROM).

YRIGOYEN, M. 1993. Los depósitos sinorogénicos terciarios. *Geología y Recursos Naturales de Mendoza. XII Congreso Geológico Argentino y II Congreso de Exploración de Hidrocarburos*. Mendoza, **1.11**, 217–234.

ZÁRATE, M. A. 2002. Geología y Estratigrafía del Pleistoceno tardío-Holoceno en el piedemonte de Tunuyán-Tupungato, Mendoza, Argentina. *In*: CABALERI, N., CINGOLANI, C., LINARES, E., LÓPEZ DE LUCHI, M. G., OSTERA, H. A. & PANARELLO, H. O. (eds) *XV Congreso Geológico Argentino*. El Calafate, Santa Cruz. Actas, **II**, 615–620.

ZÁRATE, M. A. 2003. Loess of southern South America. *Quaternary Science Reviews*, **22**, 1987–2006.

ZÁRATE, M. & BLASI, A. 1993. Late Pleistocene-Holocene eolian deposits of the southern Buenos Aires province, Argentina: a preliminary model. *Quaternary International*, **17**, 15–20.

ZÁRATE, M. A. & MEHL, A. 2008. Estratigrafía y geocronología de los depósitos del Pleistoceno tardío/Holoceno de la cuenca del Arroyo La Estacada, departamentos de Tunuyán y Tupungato (Valle de Uco), Mendoza. *Revista de la Asociación Geológica Argentina*, **63**, 407–416.

ZÁRATE, M. A. & MEHL, A. 2011. Evolución geomorfológica holocena de la cuenca media del río Atuel, Mendoza, Argentina. *XVIII Congreso Geológico Argentino*, Neuquen, 1372–1373.

ZÁRATE, M. A. & PÁEZ, M. M. 2002. Los paleoambientes del Pleistoceno tardío-Holoceno en la cuenca del Arroyo La Estacada, Mendoza. *In:* TROMBOTTO, D. & VILLALBA, R. (eds) *IANIGLA, 30 años de investigación básica y aplicada en ciencias ambientales*. Instituto Argentino de Nivología/Glaciología y Ciencias Ambientales, Mendoza, Argentina, 117–121.

Quaternary tectonics and seismic potential of the Andean retrowedge at 33–34°S

VÍCTOR H. GARCÍA[1]* & ANALÍA L. CASA[2]

[1]*Instituto de Investigación en Paleobiología y Geología, Universidad Nacional de Río Negro, Isidro Lobo y Belgrano (8332) General Roca, Río Negro, Argentina*

[2]*Servicio Geológico Minero Argentino, Av. General Paz 5445 (1650) San Martín, Buenos Aires, Argentina*

*Corresponding author (e-mail: vgarcia@unrn.edu.ar)

Abstract: The Andean retrowedge, located between 33°S and 34°S, lies in the transition region of the Pampean flat-slab subduction zone to the north and a normal subduction zone to the south. Neotectonic structures and shallow seismicity are very common north of this segment and become progressively less frequent southwards. The Frontal Cordillera and the Cerrilladas Pedemontanas are the main morphostructures involved in the Quaternary deformation of this region. The Frontal Cordillera is a thick-skinned fold-and-thrust belt uplifted since Late Miocene time. The Cerrilladas Pedemontanas are low-relief hills that represent the mild inversion of the Cuyo Triassic rift depocentre since Pliocene time. Middle Miocene–Holocene synorogenic strata cover the Cuyo basin and surrounding foreland areas. The Quaternary tectonic evolution of this area has been established through integration of new data from fieldwork in the Frontal Cordillera piedmont with subsurface information and previously published data. Mean Late Pleistocene uplift rates ranging between 0.21 and 0.92 mm a^{-1} and earthquake minimum moment magnitudes (M_w) of c. 6.4–6.7 have been estimated for the morphostructural units analysed in this manuscript.

Polanski (1963) was the first to attempt to characterize the neotectonics of the Frontal Cordillera piedmont between the Mendoza and Diamante rivers (33–34°45′S), through which the currently used Quaternary stratigraphy framework and the geomorphic description of the area were established. In addition, he interpreted the Tunuyán depression as a graben developed during Quaternary time and flanked by normal faults bounding the Frontal Cordillera and the Cerrilladas Pedemontanas (Fig. 1; Polanski 1963). On the basis of his work, neotectonic features such as the Jaboncillo hills and La Aguadita fault have been interpreted to be controlled by normal faulting (Fig. 2).

Subsequent studies based on structural field mapping and interpretation of 2D seismic lines demonstrated that the Tunuyán depression is an inter-montane basin developed since Middle Pliocene time and bounded by reverse faults that uplift the Frontal Cordillera and mildly inverted normal faults of Triassic Cuyo basin (Legarreta *et al.* 1992; Devizia 1993; Kozlowski *et al.* 1993; Ploszkiewicz 1993; Cortés *et al.* 1999). The Quaternary activity of the Frontal Cordillera thrust and the growth anticlines of its piedmont have been verified by recent field studies, tectonic geomorphology analysis and interpretation of 2D seismic lines (Borgnia 2004; García 2004; Casa 2005; García *et al.* 2005; Casa *et al.* 2010).

Previous main neotectonic studies in the Cerrilladas Pedemontanas were conducted by Instituto Nacional de Prevención Sísmica (INPRES 1995), who focused on the seismic microzonation of the Gran Mendoza, and Chiaramonte *et al.* (2000) and Brooks *et al.* (2003), who focused on the geometry and seismic potential of the Barrancas anticline in the extreme NE of this region (Fig. 2). The main evidence for Quaternary deformation reported for this morphostructure includes gently folded Middle Pliocene–Early Pleistocene alluvial conglomerates and small reverse faults affecting Late Pleistocene–Holocene deposits at the eastern flank of the fold (Chiaramonte *et al.* 2000). The epicentre of the M_b 6.0 Mendoza earthquake of 1985, with a depth of 12 km, was located in close proximity to the axis of the Barrancas anticline. The calculated focal mechanism indicates reverse faulting (INPRES 1985; Triep 1987) and its depth indicates a new structure growing towards the east (Zambrano 1979; Chiaramonte *et al.* 2000).

The main objective of the present study is to establish the Quaternary tectonic framework of the region through integration of previously published data on neotectonics with original data collected in

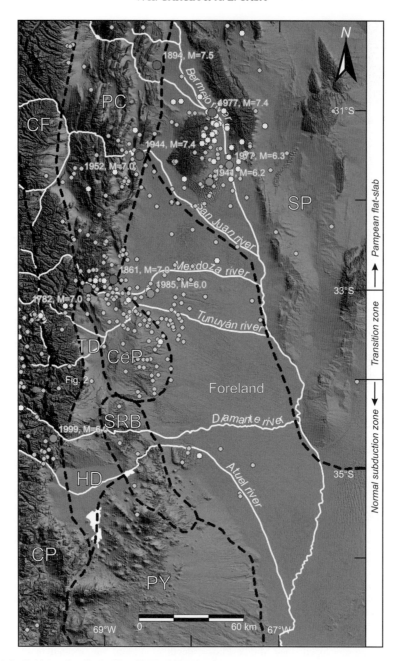

Fig 1. Shaded relief image based on a 90 m SRTM DEM showing the location of the studied area in the context of the main geological provinces or morphostructural units with respect to the subduction segments. The locations of historical and instrumental shallow (<30 km) earthquakes are indicated by dots. Green represents $4.0 \leq M_w < 5.0$; yellow $5.0 \leq M_w < 6.0$; orange $6.0 \leq M_w < 7.0$; and red $M_w \geq 7.0$ (USGS 2013). CP: Cordillera Principal; CF: Frontal Cordillera; PC: Precordillera; HD: Huarpes depression; TD: Tunuyán depression; CeP: Cerrilladas Pedemontanas; SRB: San Rafael block; PY: Payunia; SP: Sierras Pampeanas.

the field and interpretations of seismic lines and aerial photographs. Moreover, mean uplift rates and potential related earthquakes have been estimated for each neotectonic structure by using the available roughly calibrated ages for the deformed Quaternary units.

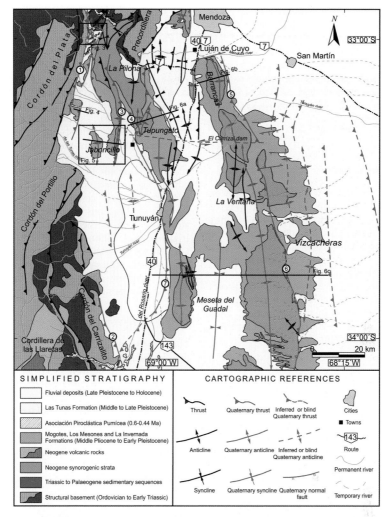

Fig. 2. Geological and neotectonic map of the Andean retrowedge between 33°S and 34°S. The main structures and stratigraphic units are indicated. The structures with evidence of Quaternary activity are mapped in red; dashed lines indicate inferred or blind neotectonic structures. (1) El Salto-La Aguadita fault; (2) Chalet fault; (3) Del Peral–Del Totoral fault; (4) Anchayuyo fault; (5) Barrancas–Lunlunta fault; (6) La Ventana fault; (7) Cerro Negro de Capiz fault. Compiled from Yrigoyen (1994), Milana & Zambrano (1996), Chiaramonte et al. (2000), Cristallini et al. (2000), Irigoyen et al. (2000), Giambiagi et al. (2003), Folguera et al. (2004), García (2004), Casa (2005), García et al. (2005), Casa et al. (2010) and Heredia et al. (2012).

Stratigraphic framework

The stratigraphy of the studied zone was divided into the following four sequences: (1) Palaeozoic and Permo-Triassic rocks that comprise the structural basement; (2) Triassic–Jurassic and Palaeogene continental sedimentary sequences; (3) Neogene synorogenic strata that fill the foreland basin; and (4) Plio-Quaternary alluvial sediments.

Structural basement

The structural basement is cropping out at the Frontal Cordillera and the Precordillera (Fig. 2). The eastern flanks of the Cordón del Plata and the Cordón del Portillo are composed of Neopalaeozoic marine sedimentary rocks (Polanski 1958) paraconformably covered by a thick succession of andesitic breccias, lava flows and rhyolitic ignimbrites from the Choiyoi Group (Polanski 1958; Llambías et al.

2003; Folguera et al. 2004; Martínez et al. 2006). This volcanic suite is also associated with subvolcanic and plutonic facies (Polanski 1958; Llambías et al. 2003). To the east, the basement of the foreland basin and the inverted Cuyo basin, which has been well documented from petroleum exploration, is composed of more than 2000 m of Ordovician–Devonian marine sedimentary rocks (Cuerda et al. 1989, 1993).

Triassic–Jurassic and Palaeogene sedimentary sequences

The Triassic sediments were deposited unconformably on the Palaeozoic substratum (Fig. 2) and are genetically related to an episode of rifting that induced fault-controlled subsidence. During the early depositional phase, the sediments were restricted to a series of partially isolated depressions filled by the Rio Mendoza and Las Cabras formations. The upper part of the Mesozoic–Palaeogene succession represents a rift-sag transition and a regional post-rift sag period. These sediments consist of fluvial, deltaic and lacustrine deposits (Kokogian & Mansilla 1989).

Neogene synorogenic strata

The Neogene synorogenic strata unconformably overlie the previous units and are composed of more than 3000 m of continental sedimentary rocks interbedded with volcanic ash levels (Fig. 2). This sequence begins with reddish sandstones and mudstones and minor conglomerates grouped into the Mariño Formation (Yrigoyen 1993; Irigoyen et al. 2000). The thickness of this unit reaches 1900 m SW of the Cacheuta range (Yrigoyen 1993) and 2200 m subsurface (García et al. 2005). Through magnetostratigraphic research, the depositional time span was established to be between 15.7 and 12.2 Ma (Irigoyen et al. 2000).

Whitish tuffs and biotitic ash levels alternating with brown-greyish sandstones and violaceous mudstones paraconformably overlie the Mariño Formation. These rocks have been grouped by Yrigoyen (1993) into the Tobas La Higuerita informal unit. The sequence reaches 200 m of thickness in some places and was deposited between 12.2 Ma and 11.5 Ma (Irigoyen et al. 2000).

Grey-greenish sandstones and multicoloured mudstones with some gypsum levels near the top have been grouped into the La Pilona Formation (Yrigoyen 1993). These rocks paraconformably overlie the Tobas La Higuerita unit and are more than 1000 m thick (García et al. 2005), showing some thinning in the axis of inverted anticlines in the Cuyo basin (Yrigoyen 1993). The age of this unit has been established to be between 11.5 Ma and 9.0 Ma (Irigoyen et al. 2000).

Paraconformably overlying the La Pilona Formation is a succession of more than 100 m of white tuffs and ash knows as Tobas Angostura (Yrigoyen 1993). These volcanic deposits have been dated by many researchers as between 9.7 Ma and 8.3 Ma (Yrigoyen 1993; Irigoyen et al. 2000; Pepin et al. 2013). The striking characteristics of this unit include contrasts in colour and seismic response with the surrounding deposits, which is useful for structural mapping.

The Río de los Pozos Formation, of age 8.3–3.8 Ma, is in conformable contact with Tobas Angostura and is composed of yellowish mudstones and tuffaceous clays with some conglomerates at the top of the sequence (Yrigoyen 1993; Irigoyen et al. 2000). Variability in thickness due to erosion is remarkable in the Cerrilladas Pedemontanas sector, with sections of 65 m to more than 500 m occurring along several kilometres of distance (Chiaramonte et al. 2000).

Plio-Quaternary alluvial sediments

The Plio-Quaternary deposits unconformably cover the Neogene synorogenic strata in the Cerrilladas Pedemontanas (Yrigoyen 1993) and the structural basement in the mountain front of the Frontal Cordillera (Fig. 2; Casa et al. 2010). The Mogotes Formation consists mainly of boulder conglomerates interbedded with red mudstones, sandstones and some ash levels (Yrigoyen 1994). The thickness of this unit varies considerably, reaching 2000 m in some depocentres and disappearing by erosion or no deposition at the axes of some growth anticlines (Yrigoyen 1993; García et al. 2005). These thick sediments were deposited between c. 3.8 Ma and 1.0 Ma (Yrigoyen 1993; Irigoyen et al. 1995).

To describe the Pleistocene stratigraphy of the Frontal Cordillera piedmont, Polanski (1963) proposed three alluvial units including Los Mesones, La Invernada and Las Tunas formations. The division of these boulder conglomerate deposits was established on the basis of their morphological expressions in the landscape (Polanski 1963) rather than lithological differences, due to their similarities in the sedimentary facies. Here, we simplify the Early Pleistocene stratigraphy by correlating the Los Mesones and La Invernada formations with the uppermost part of Mogotes Formation to estimate an age of c. 1.0 Ma for the top of this sequence (Pepin et al. 2013).

The Las Tunas Formation can be described as a fanglomerate of basement boulders and sandy matrix (Polanski 1963). The thickness of this unit reaches a maximum of 100 m. These sediments comprise the current bajada of the Frontal Cordillera and the alluvial filling of cut-and-fill terraces (Fig. 2). The pumiceous Asociación Piroclástica Pumícea (Polanski 1963) is interbedded several metres above the base of Las Tunas Formation (Pepin et al. 2013) near its namesake river. This volcanic ash deposit has been dated between 0.60 Ma and 0.44 Ma (Stern et al. 1984; Pepin et al. 2013), whereas the top of the youngest terrace of the Las Tunas river has been dated through terrestrial cosmogenic nuclide ^{10}Be analysis to be 20.0 ± 0.7 ka (Pepin 2010; Pepin et al. 2013). These ages constrain the depositional time span of the Las Tunas Formation to between 0.60 Ma and 0.02 Ma.

The youngest unit in the region, the El Zampal Formation, is composed of aeolian and fluvial silts that cover the current floodplains east of the Frontal Cordillera piedmont. These sediments are partially interfingered with the Las Tunas Formation and are younger than 120 ka (Toms et al. 2004; Zárate & Mehl 2008).

Tectonic setting

The compressive deformation of the Andean retrowedge at 33–34°S began in late Early Miocene time through the inversion of a Jurassic rift system at the Cordillera Principal, according to the deposition of the first synorogenic strata in the Alto Tunuyán basin (Giambiagi & Ramos 2002; Giambiagi et al. 2003, 2014). From Middle to Late Miocene time, the thin-skinned Aconcagua fold-and-thrust belt developed to begin the synorogenic deposition in the studied region (Mariño and La Pilona formations; Giambiagi & Ramos 2002; Irigoyen et al. 2002; Giambiagi et al. 2003, 2014).

Rapid uplift of the Frontal Cordillera began in late Late Miocene–Pliocene time through east-verging basement-cored thrust sheets (Caminos 1965; Kozlowski et al. 1993; Baldauf 1997; Giambiagi & Ramos 2002; Irigoyen et al. 2002; Giambiagi et al. 2003, 2014; Hoke et al. 2014). The inversion of the Cuyo basin began in Middle Pliocene time at a slow rate as evidenced by the angular unconformity between the Río de los Pozos and Mogotes formations (Yrigoyen 1993).

The progression of the deformation front towards the Frontal Cordillera piedmont and Cerrilladas Pedemontanas during Late Pliocene–Early Pleistocene time has been registered by the lack of deposition of the Mogotes/Los Mesones/La Invernada coarse conglomerates at the top of the growing anticlines and by the presence of progressive unconformities in the same sequence at their limbs (Yrigoyen 1993; García et al. 2005). Out-of-sequence thrusting has been documented in the Cordillera Principal during this time span in response to the eastwards migration of the tectonic front (Giambiagi & Ramos 2002; Giambiagi et al. 2003). The present morphostructural configuration of the region would have developed during the last 2 Ma (Giambiagi & Ramos 2002; Giambiagi et al. 2003). Major seismicity such as the M_w 6.0 Mendoza earthquake of 1985 (INPRES 1985; Triep 1987) occurs east of the Cerrilladas Pedemontanas in the apparently undeformed foreland, which indicates that, although the loci of deformation is migrating east, no geomorphic expression has been detected.

Quaternary tectonics of the region

This section describes the Quaternary tectonic framework and mean uplift rates assuming a 1.0 Ma age for the top of the combined Mogotes/Los Mesones/La Invernada conglomerates.

La Carrera fault system

Originally identified by Stappenbeck (1917), Polanski (1958) defined the Espolón de La Carrera in the eastern border of the Frontal Cordillera as an imbricated reverse structure with a general north–south strike (Fig. 2). These structures, which control the Cordón del Plata and the southern part of the Cordillera del Tigre, were redefined as the La Carrera fault system (Caminos 1965, 1979; Cortés 1993).

Fauqué et al. (2000, 2008) described and dated several Quaternary rock avalanche deposits on the northern edge of Cordón del Plata, whose rupture surfaces lie over north–south striking reverse faults included in the La Carrera fault system. The northern edge of the system is defined as an imbricated left-stepping arrangement of north–south-striking reverse faults connected through NW–NNW structures, which is a typical fabric orientation of the Late Palaeozoic San Rafael orogeny. At least two earthquakes recorded in this zone during the last 21 years are located above these NW faults oblique to the mountain front and have been interpreted as the reactivation of pre-Cenozoic mechanical anisotropies ((Fig. 1; Cortés et al. 2006; Cortés et al. 2014)

Several field studies provide additional evidence of Quaternary reverse activity of the La Carrera fault system (Borgnia 2004; Casa 2005; Casa et al. 2010). All of these tectonic geomorphology studies describe fault scarps that affect piedmont deposits and several geomorphic indices that reflect the continuity of tectonic activity during the Quaternary. The youngest piedmont deposits affected by Quaternary tectonic activity are correlated locally with 0.1–0.2 Ma Vallecitos till (Wayne & Corte 1983) and regionally with the Las Tunas Formation (Polanski 1963).

Casa *et al.* (2010) named the eastern principal fault of the La Carrera system as the Río Blanco fault and described a diverging splay that juxtaposes Miocene sedimentary rocks of the Mariño Formation over Plio-Pleistocene rocks of the Mogotes Formation. This splay clearly deforms the Late Pleistocene sediments (Fig. 3). Casa (2005) described a 6-m-high fault scarp and associated drainage anomalies. To the south, other similarly striking fault scarps with 4–5 m of relief affect the Quaternary alluvial cover and are interpreted as multiple fault scarps (Stewart & Hancock 1990).

The El Salto fault (Fig. 3) is defined as the easternmost splay of the principal fault, which records several pulses of Quaternary activity resulting in the development of three pediment levels over several small hills (Casa 2005; Casa *et al.* 2010). Each pediment surface records the unroofing of the mountain; the highest and oldest surfaces have high percentages of volcanic clasts typical of the Choiyoi Group, whereas the lower surfaces contain abundant blocks of Neopalaeozoic rocks. Casa (2005) interpreted this evolution on the basis of the Hamblin's (1976) model of episodic uplift. The continuity of spurs and triangular facets defines at least three periods of intense tectonic activity alternating with quiescence stages and pediment development. The minimum throw between blocks displaced by the fault is 230 m.

The activity of the Río Blanco fault was assessed through calculation of geomorphic indices of active tectonics between the La Manga and Blanco rivers (Fig. 3; Casa 2005; Casa *et al.* 2010). The mountain-front sinuosity index shows values of *c.* 1 between both valleys; however, the values were significantly higher at 1.14–1.18 north of the La Manga creek and south of the Blanco river; these valleys coincide with NW structures. Stream length-gradient index calculations (Hack 1973; Keller 1986) were conducted over all drainage basins of this mountain front segment. The results, evaluated over the same bedrock lithology, showed higher values between the La Manga creek and the Blanco river relative to those north and south of them respectively. The relationship between the width and height of the valley floor (valley floor width-to-height ratio; Bull & McFadden 1977) was determined over the valleys perpendicular to the

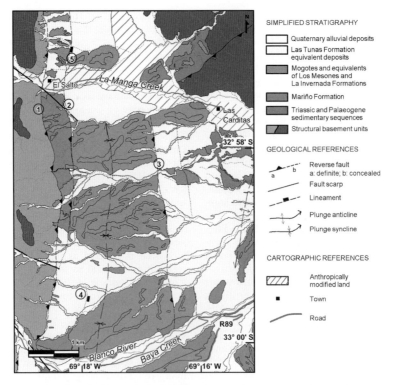

Fig. 3. Geological and neotectonic detailed map of the Cordón del Plata piedmont at the El Salto fault area. (1) Río Blanco fault; (2) splay of Río Blanco fault; (3) El Salto fault; (4) multiple fault scarps; (5) Late Pleistocene reactivation of the Río Blanco splay. Modified after Casa (2005) and Casa *et al.* (2010).

mountain front between the La Manga creek and the Blanco river, where the drainage basins are wine-glass shaped. Casa (2005) obtained low values, ranging between 0.09 and 0.26, which characterize tectonically active regions (Bull & McFadden 1977).

Northwards of the La Manga creek, the only recognized fault scarp is related to the splay of the Río Blanco fault. South of the Blanco river, geomorphic evidence of tectonic activity in the piedmont is less clear. The El Salto fault, situated between these valleys, shows a rocky fault scarp and the development of pediment levels that are absent south and north of the NW lineaments. The variations in the mountain-front geomorphic features, including the geomorphic indices considered and the attributes of drainage basins, correspond to the distribution of Quaternary tectonic activity in the piedmont, which consistently suggests the presence of segments with varying degrees of neotectonic activity along the La Carrera fault system related to NW- and NNW-striking structures (Cortés et al. 2006; Casa et al. 2010).

Chronological data of the deformed units and of those which are the seal of the activity based on regional correlations, together with constraints of the time span of faulting activity, allowed for estimation of the long-term uplift rates of the El Salto fault. Considering the evidence of activity of this fault between the deposition of the Asociación Piroclástica Pumicea and undeformed morraines dated at 0.1–0.2 Ma (Wayne & Corte 1983) and the height of the El Salto fault scarp near the Blanco river, Casa (2005) calculated a minimum mean uplift rate of 0.51 mm a^{-1}. Recent age data of a tuff in the Las Tunas river, correlated with the pyroclastic association at 0.6 ± 0.2 Ma (Pepin 2010; Pepin et al. 2013), indicate a minimum uplift rate of 0.34 mm a^{-1}.

In the piedmont of the southern part of Cordón del Plata, both Polanski (1963) and Caminos (1965) described La Aguadita as a normal fault with Pleistocene activity related to the development of the Tunuyán depression. This Quaternary fault was then considered a reverse fault scarp and was included as the easternmost expression of the La Carrera fault system in this area (Cortés et al. 1999). The La Aguadita fault marks the eastern limit of a rocky block parallel to the mountain front covered by the Mogotes/Los Mesones/La Invernada deposits and uplifted over the Las Tunas Formation (Fig. 4). This fault is believed to be connected northwards to the El Salto fault, the activity of which is pre- to syn-Las Tunas Formation (Casa et al. 2010). Considering a thickness of c. 70 m for this unit, c. 212 m of minimum throw can be measured on the basis of the top of the Mogotes/Los Mesones/La Invernada conglomerates (Fig. 4). Assuming an estimated age of 1.0 Ma for this level, a mean uplift rate of c. 0.21 mm a^{-1} was obtained.

Jaboncillo and Del Peral anticlines and Anchayuyo fault

The Jaboncillo and Del Peral hills are located in the piedmont of the Frontal Cordillera just west of Tupungato (Fig. 2). The Jaboncillo hills are composed mainly of alluvial conglomerates that can be correlated with the Los Mesones and La Invernada formations (Polanski 1963; García 2004; Perucca et al. 2009; Pepin et al. 2013). At the western flank of the hills, García (2004) measured a maximum dip of 20°/W (Fig. 5). In the extreme north-eastern region of the these hills, Bastias et al. (1993) recognized and named the Chupasangral fault, a NNW–SSE fault scarp that dislocates the conglomerates with western lower block (Fig. 5). In the Agua de los Chilenos creek this fault affects an alluvial terrace correlated with the Las Tunas Formation to generate a scarp of c. 2 m (García 2004). Moreover, García (2004) identified two additional subparallel scarps to the SW 1 km apart (Fig. 5). The vertical throw of these scarps measured with respect of the top of the older alluvial conglomerates is c. 10–20 m.

At the eastern side of the Del Peral hills, Neogene units dipping between 15°/W and 45°/W are unconformably covered by subhorizontal coarse conglomerates correlated with the Mogotes/Los

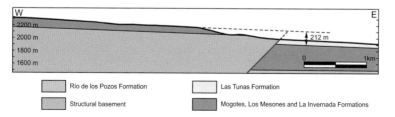

Fig. 4. Schematic cross-section of the La Aguadita fault at the southern extreme region of the La Carrera fault system. The location is shown in Figure 2.

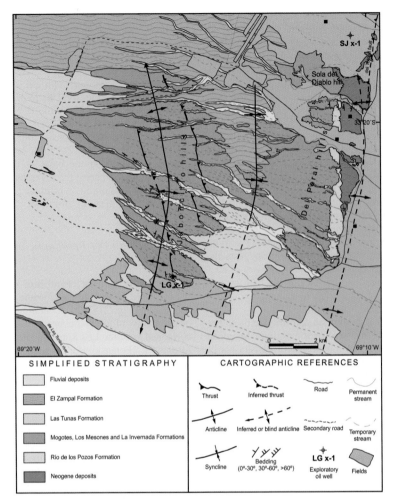

Fig. 5. Geological and neotectonic detailed map of the Jaboncillo and Del Peral hills at the piedmont of the Cordón del Plata. Modified after García (2004) and García et al. (2005).

Mesones/La Invernada sequence (Polanski 1963; García 2004; Perucca et al. 2009; Pepin et al. 2013). At the northern extreme of the Del Peral hills, Perucca et al. (2009) documented undifferentiated Neogene strata in the Sola del Diablo hill dipping 70°/W and thrusted over alluvial conglomerates older than Las Tunas Formation (Fig. 5).

Seismic 2D lines and petroleum exploration data were used by Garcia et al. (2005) to interpret the deep structure of these hills (Fig. 6a). The borehole LG x-1, located in the southern-most region of the Jaboncillo hills (Fig. 5), was used to establish the subsurface stratigraphic column of the area and to define the contacts between the Neogene units.

The Jaboncillo and Del Peral hills are thin-skinned anticlines formed above ramps located at various stratigraphic levels with common structural links. The Jaboncillo anticline is a fault-bend fold developed over a ramp dipping 30°/W and rooted at or near the contact between the Mariño Formation and the structural basement (Fig. 6a). By using the top of the Rio de los Pozos Formation as a reference level, the structural relief of this fold at its axis is determined to be c. 935 m. This unit does not exhibit important lateral variations in this region and can therefore be used as a local structural marker (García 2004). Considering the tops of the Mogotes/Los Mesones/La Invernada formations, the uplift oscillates c. 300 m; this indicates the syntectonic sedimentation of this succession (Fig. 6a) and an uplift rate of c. 0.3 mm a^{-1}, considering the age of 1.0 Ma for this horizon. The subparallel fault scarps are interpreted in this context as interstratal fold-accommodation backthrusts that merge

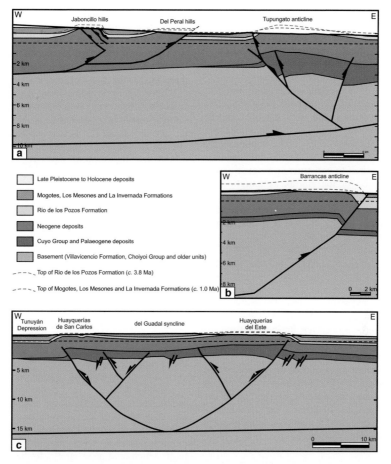

Fig. 6. Balanced structural cross-sections. (**a**) Jaboncillo and Del Peral anticlines and Anchayuyo fault (modified after García *et al.* 2005; Giambiagi & Moreiras 2012); (**b**) Barrancas anticline (modified after Chiaramonte *et al.* 2000; Brooks *et al.* 2003); and (**c**) Meseta del Guadal (modified after Cristallini *et al.* 2000). The location is shown in Figure 2.

with the main west-dipping ramp to form an incipient structural wedge (Fig. 6a; Mitra 2002). The uplift rate of these faults is *c.* 0.01–0.02 mm a^{-1} which indicates their secondary importance in relation to neotectonic folding.

The Del Peral anticline is modelled as a fault-propagation fold in its southern extreme and as a transported fault-propagation fold in the Sola del Diablo region (Figs 5 & 6a). The frontal limb has been completely eroded and only remnants of its backlimb are cropping out. The ramp dips 30°/W and is rooted in the upper detachment level of the Jaboncillo anticline at the middle section of Mariño Formation (Fig. 6a). The structural relief measured with respect to the top of Río de los Pozos Formation is *c.* 785 m in the middle section of the Del Peral hills (Fig. 6a). The top of the Mogotes/Los Mesones/La Invernada conglomerates is uplifted *c.* 225 m with respect to the undeformed basin (Fig. 6a), which indicates an uplift rate of approximately 0.23 mm a^{-1}. Although there is no evidence associated with the Late Pleistocene deformation of the Del Peral anticline, the southwards propagation of the fold could be active (García & Cristallini 2011).

The Anchayuyo fault is a partially inverted normal fault responsible for the uplifting of the Cuyo basin towards the west at this latitude (Figs 2, 6a). Giambiagi & Moreiras (2012) used 2D and 3D seismic records and borehole data to construct a balanced cross-section of this area and determined a structural relief of *c.* 1375 m on the basis of the top of the Río de los Pozos Formation (Fig. 6a). Considering the syntectonic sedimentation of the Mogotes/Los Mesones/La Invernada succession, the minimum uplift for the top of these deposits is *c.* 450 m indicating an uplift rate of *c.* 0.45 mm a^{-1} for the last 1.0 Ma.

Barrancas anticline

The Barrancas anticline has been previously studied to establish the provenance of the Neogene–Quaternary deposits outcropping in this foreland area (Chiaramonte 1996; Chiaramonte et al. 2000). On the basis of 2D seismic lines and borehole data, the geometry of the fold has been modelled as a fault-propagation fold with a ramp dipping between 32°/W and 45°/W (Chiaramonte 1996; Brooks 1999; Brooks et al. 2000; Chiaramonte et al. 2000). In the present study, we used the data of previous research to construct an original structural cross-section of the Barrancas anticline. In our model, the fault dips 50°/W in the sedimentary cover section and 35°/W in the basement to reach a basal décollement level c. 8.5 km deep that gently dips towards the west (Fig. 6b).

The geological and structural maps published by Chiaramonte et al. (2000) indicate that the thickness of the Río de los Pozos Formation is c. 1000 m at the forelimb and a maximum of 300 m at the backlimb (Fig. 6b). These measurements imply that substantial erosion of the Río de los Pozos Formation occurred during uplifting, as evidenced by the angular unconformity with the Mogotes Formation. The structural relief of the Barrancas anticline measured at the projected top of the Río de los Pozos Formation is c. 1660 m (Fig. 6b). Furthermore, folding of the top of the Mogotes/Mesones/La Invernada formations indicates an uplift of c. 480 m, which implies a mean rate of c. 0.48 mm a^{-1} during the last 1.0 Ma, as can be deduced from the structural section (Fig. 6b).

Meseta del Guadal

The Meseta del Guadal is a huge positive morphostructure located at the southern half of the Cerrilladas Pedemontanas (Fig. 2). Folded Neogene–Early Pleistocene deposits crop out at their eastern and western flanks (Figs 2, 6c). Yrigoyen (1994) described the structure of the Meseta del Guadal as having syncline geometry with very gently dipping flanks outlined by Plio-Pleistocene conglomerates. Cristallini et al. (2000) used subsurface information to construct a structural cross-section of this region to show bi-vergent basement-rooted middle-angle thrusts that uplift the flanks of the structure where anticlines expose the Neogene strata (Figs 2, 6c). These faults are more than 50 km long and strike roughly north–south (Fig. 2).

The only previous mention of neotectonic activity related to this morphostructure is the Manantianles fault, an east–west strike-slip fault that accommodates differential displacements between folds at the northern extreme of the Meseta del Guadal. Bastias et al. (1993) used geomorphic indicators such as displaced rivers to suggest the recent activity of this structure (Fig. 2).

The top of the Plio-Pleistocene units, including the Mogotes, Los Mesones and La Invernada formations, at the axis of the eastern anticline is c. 920 m above the same level at the undeformed foreland, whereas the western anticline produces a vertical uplift of c. 560 m (Fig. 6c). Considering the age of 1.0 Ma for the top of these units, uplift rates of c. 0.92 and 0.56 mm a^{-1} have been determined for the eastern and western anticlines, respectively.

Chalet fault

The NNW-striking Chalet fault was named by Bastias et al. (1993) and is defined as a reverse structure with east vergence and Quaternary activity that controls the mountain front of the Cordón del Carrizalito south of the Tunuyán river (Fig. 2). Tello (1994, 1998) reported deflected streams likely related to Holocene reactivation of the Chalet fault in addition to east-facing fault scarps between 2 and 23 m high. Perucca et al. (2008) detected expositions of this fault dipping 60°/230° near Cerro Chalet that clearly show reverse separation.

A clearer geomorphic expression of this fault appears south of the Del Cepillo creek, where the mountain front is straight and is characterized by the alignment of small outcrops of basement rocks. Approximately 3 km south of this creek, the mountain front is separated 2 km west from this fault, which continues parallel to the same strike.

Cortés & Sruoga (1998) linked the development of the Loma Grande de Yaucha syncline, located south of the Del Rosario creek, with activity of the Chalet fault and the termination of this structure against the NW-striking Casa de Piedra fault, which is part of the Papagayos fault zone (Cortés & Sruoga 1998; Cortés et al. 2014). The folded sediments correlate with the Los Mesones Formation. Due to the uncertainties of the age of the deformed sediments and diverse structural complexities, the mean uplift rates and seismogenic potential of this fault were not estimated.

Seismic potential

Despite the moderate magnitude (mean M_w 4.2) of the seisms that have characterized the studied area during the last 21 years and the absence of $M_w > 6$ earthquakes during the last c. 100 years (Fig. 1), the mean uplift rates of 0.21–092 mm a^{-1} for the last 1.0 Ma calculated for the Quaternary morphostructures are quite high to underestimate their seismic potential.

Two possible scenarios of seismic behaviour are considered in the present study to establish the seismic potential of the region: the first is reactivation of the entire length of the faults for estimating the maximum possible earthquakes and the second is reactivation by segments controlled by a minimum regional recurrence interval of c. 100 years, considering the aforementioned lack of $M > 6$ seismicity. This section details the results of potential earthquakes related to the capable faults (Azzaro 1998; Machette 2000; IAEA 2002) previously discussed, with the application of the well-known regressions of Wells & Coppersmith (1994).

El Salto–La Aguadita fault

Reactivation of the entire fault length. An eventual reactivation of the 45-km-long El Salto–La Aguadita fault could produce an earthquake of M_w 6.8 and maximum displacements of c. 1.6 m (Wells & Coppersmith 1994). Considering the 0.21 mm a^{-1} mean uplift rate calculated for the La Aguadita fault for the last 1.0 Ma (Fig. 4), a minimum recurrence interval of c. 4760 years is obtained for such seism sizes.

Reactivation by segments. The M_b 5.6 earthquake that occurred on 14 January 1980 with an estimated hypocentral depth of 14 km (USGS 2013) could be related to reactivation of a central-southern segment of the deep-rooted El Salto–La Aguadita fault (Fig. 2). According to Wells & Coppersmith (1994), the estimated mean uplift contributed by an earthquake of this size is c. 24 mm. Considering a recurrence interval of 100 years for such an earthquake type, the obtained mean uplift rate would be 0.24 mm a^{-1}, which surpassed the previously estimated mean uplift rate of 0.21 mm a^{-1} for its corresponding segment of the fault. Additional registered instrumental seismicity likely related to this structure includes two earthquakes of M_w 4.1 (USGS 2013). The contributed uplift of such isolated and moderate seisms is considered negligible for seismic potential analysis.

If the earthquake mechanism of the El Salto–La Aguadita fault is controlled by the rupture of discrete segments of c. 7 km long that produce seisms of M_w 5.6 (Wells & Coppersmith 1994), such events should occur with a recurrence of c. 114 years at each segment to reach the 0.21 mm a^{-1} estimated mean uplift rate. Considering that the fault is compartmentalized into seven segments, an earthquake of M_w 5.6 should be registered every 16 years. However, only one occurred during the last c. 100 years (USGS 2013).

One alternative hypothesis is such that the 10-km-long southern segment and 28-km-long central-northern segments would be reactivated by means of M_w 5.9 and M_w 6.5 earthquakes, respectively (Wells & Coppersmith 1994). Such events should have occurred at intervals of c. 265 and c. 1425 years to balance the estimated mean uplift rate. However, geological evidence in the northernmost El Salto fault segment suggests a rupture length of only 5 km (Casa et al. 2010).

Del Totoral–Del Peral fault

Reactivation of the entire fault length. The northward linking of the Del Peral and Del Totoral faults (Fig. 2) should be considered to estimate their seismic potential. In the same way, the maximum mean uplift rate of 0.3 mm a^{-1} obtained for the Jaboncillo anticline must be applied. According to the regression described by Wells & Coppersmith (1994), this 36-km-long fault can produce a maximum potential earthquake of M_w 6.7. The maximum vertical displacement related to such a seism size is c. 1 m (Wells & Coppersmith 1994), indicating a recurrence interval of c. 3330 years.

Reactivation by segments. Instrumental seismicity for the last 21 years includes only one earthquake of M_w 4.2 that can be related to this structure. The lack of $M_w > 6$ events during the last c. 100 years (USGS 2013) in the entire region suggests that if this capable structure were active, it would break in a maximum of two segments through M_w 6.4 earthquakes, with a recurrence interval of c. 750 years (Wells & Coppersmith 1994).

Anchayuyo fault

Reactivation of the entire fault length. The length of the Anchayuyo fault is 45 km (Fig. 2); considering the subsurface rupture length v. magnitude relationship reported by Wells & Coppersmith (1994), an earthquake of M_w 6.8 can be estimated. Such an event could develop a maximum displacement of c. 1.4 m with a recurrence interval of c. 3100 years.

Reactivation by segments. The instrumental seismicity related to the Anchayuyo fault includes five earthquakes occurring in the last 21 years and distributed as three M_w 4.1, one M_w 4.3 and one M_w 4.5. These events contribute 0.13 mm a^{-1} of the uplift rate of the morphostructure through creeping mechanisms. The remaining 0.32 mm a^{-1} should be attributed to more energetic earthquakes with recurrence intervals of more than 100 years given the lack of $M_w > 6$ events during that time span (USGS 2013).

Earthquakes of M_w 6.5 are capable of breaking fault segments of length c. 26 km every c. 940 years (Wells & Coppersmith 1994). Considering that the Anchayuyo fault is composed of two segments of 26 km long, an earthquake of M_w 6.5 should strike the zone every c. 540 years.

Barrancas–Lunlunta fault

Reactivation of the entire fault length. The Barrancas–Lunlunta fault is up to 35 km long (Fig. 2). According to Wells & Coppersmith (1994), this fault size could trigger an earthquake of M_w 6.7 and produce maximum surface displacements of c. 1 m. Considering the previously estimated mean uplift rate of c. 0.48 mm a^{-1}, a recurrence interval of c. 2080 years is obtained for events of this size.

Reactivation by segments. The seismological record associated with this structure includes 17 earthquakes of magnitudes between M_w 4 and M_w 4.6 during the last 21 years, in addition to one earthquake of M_w 5.6 that struck the region in August 2006 (USGS 2013). The sum of the first group of earthquakes gives a mean uplift rate of 0.48 mm a^{-1} (Wells & Coppersmith 1994), which compensates for the mean uplift rate calculated from the balanced cross-section (Fig. 6b). This fact could suggest that the rupture in the Barrancas–Lunlunta fault is more likely dominated by creeping mechanisms rather than by the occurrence of very energetic earthquakes.

Del Cerro Negro de Capiz fault

Reactivation of the entire fault length. This partially inverted normal fault of 50 km in length is responsible for the westward uplifting of the Meseta del Guadal (Yrigoyen 1994; Cristallini et al. 2000) and could produce M_w 6.9 earthquakes with maximum displacements of c. 2 m (Wells & Coppersmith 1994). The mean uplift rate of 0.56 mm a^{-1} obtained for the last 1.0 Ma established a minimum recurrence interval of c. 3570 years for such earthquakes.

Reactivation by segments. No record of instrumental nor historical seismicity related to this structure is available (Fig. 1). Considering the worldwide seismological catalogue time span of c. 100 years as the minimum recurrence interval of $M_w > 6$ quakes at this potentially active structure, events of at least M_w 6.7 that break segments of c. 36 km should occur every c. 935 years to attain the calculated mean uplift rate of 0.56 mm a^{-1}.

La Ventana fault

Reactivation of the entire fault length. This structure is at least 75 km long and uplifts the eastern border of the Meseta del Guadal towards the foreland (Fig. 2). Considering the obtained uplift rate, if the entire eastern thrust were to move, it could produce an earthquake of M_w 7.25 with maximum displacements of c. 5.7 m (Wells & Coppersmith 1994) and a recurrence interval of more than c. 6200 years.

Reactivation by segments. Up to 20 earthquakes between M_w 4 and M_w 4.7 have been registered as likely being related to the movement of this fault during the last 21 years (USGS 2013). The sum of the uplift generated by these creeping events gives a mean uplift rate of c. 0.51 mm a^{-1} (Wells & Coppersmith 1994). The remaining 0.41 mm a^{-1} should be attained by more energetic seismic events.

As used in previous analysis, a minimum recurrence interval of c. 100 years is employed to establish the minimum earthquake magnitude that could characterize such deformation in the area. Events of M_w 6.7 could break segments of c. 36 km and develop average vertical displacements of c. 0.52 m (Wells & Coppersmith 1994). Such events must occur every c. 1280 years to accommodate the remaining mean uplift rate of 0.41 mm a^{-1}.

Discussion

The uplift rates of the analysed morphostructures for the last 1.0 Ma have been obtained from balanced structural cross-sections based on the geometry of the top of the Mogotes/Los Mesones/La Invernada alluvial sequence (Fig. 6). Although the structural sections and the age of the top of this alluvial sequence can show slight differences locally, the results would not differ significantly with respect to those determined in the present study.

The potentially active Quaternary structures located in the eastern border of the Cerrilladas Pedemontanas show moderate to high mean uplift rates for the last 1.0 Ma (Barrancas-Lunlunta 0.48 mm a^{-1}; La Ventana 0.92 mm a^{-1}) in accordance with the closer location to the Andean orogenic front at these latitudes. The other two structures with moderate to high mean uplift rates are the Anchayuyo (0.45 mm a^{-1}) and Del Cerro Negro de Capiz (0.56 mm a^{-1}) faults, which limit the partially inverted Cuyo basin to the west. The neotectonic structures situated at the piedmont and at the mountain front of the Frontal Cordillera have moderate to low mean uplift rates (El Salto-La Aguadita 0.21 mm a^{-1}; Del Peral-Del Totoral 0.30 mm a^{-1}) in agreement with the more internal position to the Andean front.

This distribution of mean uplift rates for the last 1.0 Ma can be attributed to localization of the deformation in the borders of the Cuyo basin that acts as previous heterogeneity and concentrates orogenic wedge growth. This process has been verified by several analogue modelling experiments (e.g. Del Ventisette et al. 2006; Yagupsky et al. 2008). On the contrary, the Quaternary structures of the Tunuyán depression and the mountain front of the Frontal Cordillera are more likely to be active to compensate for the orogenic wedge growth to the east, tending to the restoration of the critical taper angle. The migration of the orogenic front towards the east likely produced the reactivation of the more internal neotectonic structures to balance the critical taper angle. These structural interpretations are supported by the instrumental seismicity of the region, which is more prominent at the eastern border of the Cerrilladas Pedemontanas and at the foreland and decreases in frequency to the west (Fig. 1).

The minimum magnitudes of earthquakes estimated for the capable structures studied are in the range of M_w 6.4–6.7, whereas the maximum potential earthquakes calculated are between M_w 6.7 and M_w 7.25. These proxies are based on the length of the potentially active faults and the obtained mean uplift rates. According to the diagram of Slemmons & de Polo (1986), the studied capable faults fall in the range B, which is related to moderate to well-developed geomorphic evidence of activity (Fig. 7). Moreover, the La Cal fault located c. 20 km north and studied recently by Salomon et al. (2013) by using palaeoseismological approaches is strongly related to the excellent geomorphic evidence of activity (range A; Fig. 7). Both the distribution of the active deformation along additional structures and the progressive decrease of the shortening rate southwards are plausible explanations for these observations.

The mean uplift rates for the last 1.0 Ma calculated for the neotectonic structures described in the present study are in the range of other neotectonic structures located 30–50 km northwards of the studied region (Schmidt et al. 2011; Salomon et al. 2013). Moreover, the latitudinal sum of

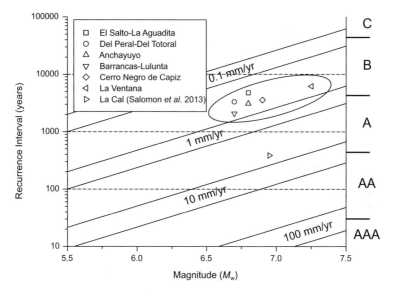

AAA: Extreme rate of activity, seldom developed, even on major plate boundaries; examples include subduction zones.

AA: Very high activity rate with excellent geomorphic evidence as at major boundary plate.

A: High activity rate with excellent geomorphic evidence as at major plate boundaries.

B: Moderate activity rate with moderate to well-developed evidence of activity.

C: Low activity rate with sparse geomorphic evidence of activity.

Fig. 7. Diagram of earthquake recurrence interval v. magnitude in areas in which tectonic rates exceed climatic rates of landscape modelling. The maximum earthquakes calculated for the studied faults are plotted in addition to values obtained by Salomon et al. (2013) for the La Cal fault. Modified from Slemmons & de Polo (1986).

1.44–1.69 mm a^{-1} for the mean uplift rates obtained are comparable to the rate of 1.5 mm a^{-1} derived from modelled GPS geodetical data for this latitude (Kendrick *et al.* 2006).

Concluding remarks

The Quaternary tectonics of the Tunuyán depression and Cerrilladas Pedemontanas are manifested mainly as fault scarps and broad folds involving Early Pleistocene alluvial conglomerates of the Mogotes, Los Mesones and La Invernada formations. Both thick- and thin-skinned styles of deformation coexist. However, basement structures are dominant and are displayed as either partially inverted normal faults of the Cuyo basin or deep-rooted thrust faults at the mountain front of the Frontal Cordillera.

The few reports of geomorphic and geological evidences of Late Pleistocene–Holocene deformation could be related to the long recurrence interval for large earthquakes estimated at 2000–6200 years in addition to the lack of detailed palaeoseismological studies in this region. Recent investigations demonstrate the importance of linking instrumental seismological data with the geological record (Salomon *et al.* 2013).

For a better understanding of the behaviour of each fault or fold, detailed palaeoseismological research is required to constrain the deformation periods, and to estimate local source parameters based both on geomorphic analysis and more accurate structural and geochronological data.

More importantly, the minimum earthquake magnitudes of M_w 6.4–6.7 calculated to compensate the mean uplift rates of the studied capable structures should be used in seismic risk assessment of the region on a short-term timescale (c. 100 years) rather than the present seismicity due to required time constrictions.

We especially thank Ernesto O. Cristallini and José M. Cortés for their valuable help during field and office work. This contribution was partially funded by UBACyT X221, X049 and X184 and PIP-CONICET 5422. Valuable comments and suggestions from Gregory Hoke and an anonymous reviewer significantly improved this paper. García participated in IGCP-586Y Geodynamic Processes in the Andes.

References

Azzaro, R. F. 1998. Environmental hazard of capable faults: the case of Pernicana Fault (Mt. Etna-Sicily). *Natural Hazards*, **17**, 147–162.

Baldauf, P. 1997. *Timing of the uplift of the Cordillera Principal, Mendoza Province, Argentina*. MSc thesis, George Washington University.

Bastias, H. E., Tello, G. E., Perucca, L. P. & Paredes, J. D. 1993. Peligro sísmico y neotectónica. *In*: Ramos, V. A. (ed.) *Geología y Recursos Naturales de Mendoza*. XII Congreso Geológico Argentino y II Congreso de Exploración de Hidrocarburos, Buenos Aires, Relatorio, **6**, 645–658.

Borgnia, M. 2004. *Neotectónica del piedemonte oriental del Cordón del Plata al norte del río Blanco, provincia de Mendoza*. Trabajo Final de Licenciatura, Facultad de Ciencias Exactas y Naturales, Universidad Nacional de Buenos Aires.

Brooks, B. 1999. *An inverse method to determine fault-related fold kinematic style from seismic reflection data: implications for cross-section construction and seismic hazard analysis*. PhD thesis, Cornell University, Ithaca.

Brooks, B. A., Sandvol, E. & Ross, A. 2000. Fold style inversion: placing probabilistic constraints on the predicted shape of blind thrust faults. *Journal of Geophysical Research*, **105**, 13281–13301.

Brooks, B. A., Bevis, M. *et al.* 2003. Crustal motion in the Southern Andes (26°–36°S): do the Andes behave like a microplate?. *Geochemistry, Geophysics, Geosystems*, **4**, http://dx.doi.org/10.1029/2003GC000505.

Bull, W. B. & McFadden, L. D. 1977. Tectonic geomorphology north and south of the Garlock Fault, California. *In*: Doehring, D. O. (ed.) *Geomorphology in Arid Regions*. Binghampton Symposia in Geomorphology International Series, **8**, 115–138.

Caminos, R. 1965. Geología de la vertiente oriental del Cordón del Plata, Cordillera Frontal de Mendoza. *Revista de la Asociación Geológica Argentina*, **20**, 351–392.

Caminos, R. 1979. Cordillera Frontal. *In*: Turner, J. C. M. (ed.) *Segundo Simposio de Geología Regional Argentina*. Academia Nacional de Ciencias, Córdoba, **1**, 397–453.

Casa, A. L. 2005. *Geología y neotectónica del piedemonte oriental del cordón del Plata en los alrededores de El Salto*. Trabajo Final de Licenciatura, Facultad de Ciencias Exactas y Naturales, Universidad de Buenos Aires.

Casa, A. L., Borgnia, M. M. & Cortés, J. M. 2010. Evidencias de deformación pleistocena en el sistema de falla de La Carrera (32°40′-33°15′ LS), Cordillera Frontal de Mendoza. *Revista de la Asociación Geológica Argentina*, **67**, 91–104.

Chiaramonte, L. 1996. *Estructura y sismotectónica del anticlinal Barrancas (Provincia de Mendoza)*. Trabajo Final de Licenciatura, Facultad de Ciencias Exactas y Naturales, Universidad de Buenos Aires.

Chiaramonte, L., Ramos, V. A. & Araujo, M. 2000. Estructura y sismotectónica del anticlinal Barrancas, cuenca Cuyana, provincia de Mendoza. *Revista de la Asociación Geológica Argentina*, **55**, 309–336.

Cortés, J. M. 1993. El frente de corrimiento de la Cordillera Frontal y el extremo sur del valle de Uspallata, Mendoza. *XII Congreso Geológico Argentino y II Congreso de Exploración de Hidrocarburos*, Buenos Aires, Actas, **3**, 168–178.

Cortés, J. M. & Sruoga, P. 1998. Zonas de fractura cuaternarias y volcanismo asociado en el piedemonte de la Cordillera Frontal (34°30′S), Argentina. *X Congreso Latinoamericano de Geología y VI Congreso Nacional de Geología Económica*, Buenos Aires, Actas, **2**, 116–121.

CORTÉS, J. M., VINCIGUERRA, P., YAMIN, M. & PASINI, M. M. 1999. Tectónica cuaternaria de la Región Andina del Nuevo Cuyo (28°–38° LS). *In*: CAMINOS, R. (ed.) *Geología Argentina*. Servicio Geológico Minero Argentino, Subsecretaría de Minería, Buenos Aires, Anales, **29**, 760–778.

CORTÉS, J. M., CASA, A., PASINI, M., YAMIN, M. & TERRIZANO, C. 2006. Fajas oblicuas de deformación neotectónica en Precordillera y Cordillera Frontal (31°30′–33°30′ LS). Controles paleotectónicos. *Revista de la Asociación Geológica Argentina*, **61**, 639–646.

CORTÉS, J. M., TERRIZZANO, C. M., PASINI, M. M., YAMIN, M. G. & CASA, A. L. 2014. Quaternary tectonics along oblique deformation zones in the Central Andean retro-wedge between 31°30′S and 35°S. *In*: SEPÚLVEDA, S. A., GIAMBIAGI, L. B., MOREIRAS, S. M., PINTO, L., TUNIK, M., HOKE, G. D. & FARÍAS, M. (eds) *Geodynamic Processes in the Andes of Central Chile and Argentina*. Geological Society, London, Special Publications, **399**. First published online February 5, 2014, http://dx.doi.org/10.1144/SP399.10

CRISTALLINI, E. O., BOGGETI, D. *ET AL*. 2000. *Cuenca Cuyana. Interpretación estructural regional*. Repsol-YPF, Internal report.

CUERDA, A. J., LAVANDAIO, E., ARRONDO, O. & MOREL, E. 1989. Investigaciones estratigráficas en la 'Formación Villavicencio', Canota, provincia de Mendoza. *Revista de la Asociación Geológica Argentina*, **43**, 356–365.

CUERDA, A. J., CINGOLANI, C. & BORDONARO, O. 1993. Las secuencias sedimentarias eopaleozoicas. *In*: RAMOS, V. A. (ed.) *Geología y recursos naturales de la provincia de Mendoza*. XII Congreso Geológico Argentino y II Congreso de Exploración de Hidrocarburos, Buenos Aires, Relatorio, **1**, 21–30.

DEL VENTISETTE, CH., MONTANARI, D., SANI, F. & BONINI, M. 2006. Basin Inversion and fault reactivation in laboratory experiments. *Journal of Structural Geology*, **28**, 2067–2083.

DEVIZIA, C. 1993. Yacimiento Piedras Coloradas – Estructura intermedia. *In*: RAMOS, V. A. (ed.) *Geología y recursos naturales de la provincia de Mendoza*. XII Congreso Geológico Argentino y II Congreso de Exploración de Hidrocarburos, Buenos Aires, Relatorio, **3**, 397–402.

FAUQUÉ, L. E., CORTÉS, J. M., FOLGUERA, A. & ETCHEVERRÍA, M. 2000. Avalanchas de roca asociadas a neotectónica en el valle del río Mendoza, al sur de Uspallata. *Revista de la Asociación Geológica Argentina*, **55** 419–423.

FAUQUÉ, L., CORTÉS, J. M. *ET AL*. 2008. Edades de las avalanchas de rocas ubicadas en el valle del río Mendoza aguas abajo de Uspallata. *XVII Congreso Geológico Argentino*, Buenos Aires, Actas, **1**, 282–283.

FOLGUERA, A., ETCHEVERRÍA, M. *ET AL*. 2004. *Hoja Geológica 3369–15 Potrerillos, provincia de Mendoza*. Servicio Geológico Minero Argentino, Boletín, Buenos Aires, **301**.

GARCÍA, V. H. 2004. *Análisis estructural y neotectónico de las lomas Jaboncillo y del Peral, departamento Tupungato, provincia de Mendoza*. Trabajo Final de Licenciatura, Facultad de Ciencias Exactas y Naturales, Universidad Nacional de Buenos Aires.

GARCÍA, V. H. & CRISTALLINI, E. O. 2011. Evidencias estructurales y geomorfológicas de deformación cuaternaria en Los Sauces, depresión de Tunuyán, Mendoza. *XVIII Congreso Geológico Argentino*, Buenos Aires, Actas en CD.

GARCÍA, V. H., CRISTALLINI, E. O., CORTÉS, J. M. & RODRÍGUEZ, C. 2005. Structure and neotectonics of the Jaboncillo and Del Peral anticlines: New evidences of Pleistocene to ? Holocene deformation in the Andean Piedmont. *VI International Symposium on Andean Geodynamics*, Extended Abstracts, 301–304.

GIAMBIAGI, L. & MOREIRAS, S. 2012. *Caracterización estructural del área Tupuntago: prospección por shale oil de la Formación Cacheuta*. Petrolera El Trebol S.A., Internal report, Mendoza.

GIAMBIAGI, L. B. & RAMOS, V. A. 2002. Structural evolution of the Andes in a transitional zone between flat and normal subduction (33°30′–33°45′S), Argentina and Chile. *Journal of South American Earth Sciences*, **15**, 101–116.

GIAMBIAGI, L. B., RAMOS, V. A., GODOY, E., ALVAREZ, P. P. & ORTS, S. 2003. Cenozoic deformation and tectonic style of the Andes, between 33° and 34° south latitude. *Tectonics*, **22**, http://dx.doi.org/10.1029/2001TC001354

GIAMBIAGI, L., TASSARA, A. *ET AL*. 2014. Evolution of shallow and deep structures along the Maipo–Tunuyán transect (33°40′S): from the Pacific coast to the Andean foreland. *In*: SEPÚLVEDA, S. A., GIAMBIAGI, L. B., MOREIRAS, S. M., PINTO, L., TUNIK, M., HOKE, G. D. & FARÍAS, M. (eds) *Geodynamic Processes in the Andes of Central Chile and Argentina*. Geological Society, London, Special Publications, **399**. First published online February 27, 2014, http://dx.doi.org/10.1144/SP399.14

HACK, J. T. 1973. Stream-profile analysis and stream-gradient index. *Journal of Research United States Geological Survey*, **1**, 421–429.

HAMBLIN, W. K. 1976. Patterns of displacement along the Wasatch fault. *Geology*, **4**, 619–622.

HEREDIA, N., FARIAS, P., GARCÍA-SANSEGUNDO, J. & GIAMBIAGI, L. 2012. The basement of the Andean Frontal Cordillera in the Cordón del Plata (Mendoza, Argentina): Geodynamic evolution. *Andean Geology*, **39**, 242–257.

HOKE, G. D., GRABER, N. R., MESCUA, J. F., GIAMBIAGI, L. B., FITZGERALD, P. G. & METCALF, J. R. 2014. Near-pure surface uplift of the Argentine Frontal Cordillera: insights from (U/Th)/He thermochronometry and geomorphic analysis. *In*: SEPÚLVEDA, S. A., GIAMBIAGI, L. B., MOREIRAS, S. M., PINTO, L., TUNIK, M., HOKE, G. D. & FARÍAS, M. (eds) *Geodynamic Processes in the Andes of Central Chile and Argentina*. Geological Society, London, Special Publications, **399**. First published online February 5, 2014, http://dx.doi.org/10.1144/SP399.10

IAEA 2002. Evaluation of seismic hazards for Nuclear Power Plants. IAEA Safety Standard Series, NS-G-3.3, Vienna.

INPRES 1985. El terremoto de Mendoza, Argentina, del 26 de enero de 1985. Informe General, Instituto Nacional de Prevención Sísmica, San Juan.

INPRES 1995. Microzonificación sísmica del Gran Mendoza. Instituto Nacional de Prevención Sísmica. Resumen Ejecutivo. Publicación Técnica 19, 269p.

IRIGOYEN, M. V., BROWN, R. L. & RAMOS, V. A. 1995. Magnetic polarity stratigraphy and sequence of thrusting: 33°S latitude, Mendoza province, Central Andes of Argentina. *COMTEC – ICL* Andean Thrust Tectonics Symposium, Program with abstracts, San Juan, 16–17.

IRIGOYEN, M. V., BUCHAN, K. L. & BROWN, R. L. 2000. Magnetostratigraphy of Neogene Andean foreland basin strata, lat. 33°S. Mendoza Province, Argentina. *Geological Society of America Bulletin*, **112**, 803–816.

IRIGOYEN, M. V., BUCHAN, K. L., VILLENEUVE, M. E. & BROWN, R. L. 2002. Cronología y significado tectónico de los estratos sinorogénicos neógenos aflorantes en la región de Cacheuta-Tupungato, Provincia de Mendoza. *Revista de la Asociación Geológica Argentina*, **57**, 3–18.

KELLER, E. A. 1986. Investigation of active tectonics: use of surficial earth processes. *In*: WALLACE, R. E. (ed.) *Active Tectonics*. National Academy Press, Washington, 136–147.

KENDRICK, E., BROOKS, B. A., BEVIS, M., SMALLEY, R., LAURÍA, E., ARAUJO, M. & PARRA, H. 2006. Active orogeny of the South-Central Andes studied with GPS geodesy. *Revista de la Asociación Geológica Argentina*, **61**, 555–566.

KOKOGIAN, D. & MANSILLA, O. 1989. Análisis estratigráfico secuencial de la Cuenca Cuyana. *In*: CHEBLI, G. & SPALLETTI, L. (eds) *Cuencas Sedimentarias Argentinas*. Universidad Nacional de Tucumán, serie Correlación Geológica, **6**, 169–210.

KOZLOWSKI, E., MANCEDA, R. & RAMOS, V. A. 1993. Estructura. *In*: RAMOS, V. A. (ed.) *Geología y Recursos Naturales de la Provincia de Mendoza*. XII Congreso Geológico Argentino y II Congreso de Exploración de Hidrocarburos, Relatorio, **1**, 235–256.

LEGARRETA, L., KOKOGIAN, D. A. & DELLAPÉ, D. A. 1992. Estructuración terciaria de la Cuenca Cuyana: ¿Cuánto de inversión tectónica? *Revista de la Asociación Geológica Argentina*, **47**, 83–86.

LLAMBÍAS, E. J., QUENARDELLE, S. & MONTENEGRO, T. 2003. The Choiyoi Group from central Argentina: a subalkaline transitional to alkaline association in the craton adjacent to the active margin of the Gondwana continent. *Journal of South American Earth Sciences*, **16**, 243–257.

MACHETTE, M. N. 2000. Active, capable, and potentially active faults – a paleoseismic perspective. *Journal of Geodynamics*, **29**, 387–392.

MARTÍNEZ, A. N., RODRÍGUEZ BLANCO, L. & RAMOS, V. A. 2006. Permo-Triassic magmatism of the Choiyoi Group in the Cordillera Frontal of Mendoza, Argentina: geological variations associated to changes in paleo-Benioff zone. Backbone of the Americas-Patagonia to Alaska, Abstracts, Mendoza, **18**.

MILANA, J. P. & ZAMBRANO, J. J. 1996. La Cerrillada Pedemontana Mendocina: un sistema geológico retrocorrido en vías de desarrollo. *Revista de la Asociación Geológica Argentina*, **51**, 289–303.

MITRA, S. 2002. Fold-accomodation faults. *Bulletin of the American Association of Petroleum Geologists*, **86**, 671–693.

PEPIN, E. 2010. *Interactions géomorphologiques et sédimentaires entre bassin versant et piémont alluvial. Modelisation numérique et examples naturels dans les Andes*. PhD thesis, Université de Toulouse.

PEPIN, E., CARRETIER, S. *ET AL.* 2013. Pleistocene landscape entrenchment: a geomorphological mountain to foreland field case (the Las Tunas system, Argentina). *Basin Research*, **25**, 613–637, http://dx.doi.org/10.1111/bre.12019.

PERUCCA, L., MEHL, A. & ZÁRATE, M. A. 2008. Fallamiento cuaternario en el piedemonte mendocino entre los 33°20′ y 34°14′S, provincia de Mendoza. XVII Congreso Geológico Argentino, Buenos Aires, Actas, **1**, 137–138.

PERUCCA, L., MEHL, A. E. & ZÁRATE, M. A. 2009. Neotectónica y sismicidad en el sector norte de la depresión de Tunuyán, provincia de Mendoza. *Revista de la Asociación Geológica Argentina*, **64**, 263–274.

PLOSZKIEWICZ, J. V. 1993. Yacimiento Tupungato. *In*: RAMOS, V. A. (ed.) *Geología y recursos naturales de la provincia de Mendoza*. XII Congreso Geológico Argentino y II Congreso de Exploración de Hidrocarburos, Relatorio, **3**, 391–396.

POLANSKI, J. 1958. El bloque varíscico de la Cordillera Frontal de Mendoza. *Revista de la Asociación Geológica Argentina*, **12**, 165–193.

POLANSKI, J. 1963. Estratigrafía, Neotectónica y Geomorfología del Pleistoceno Pedemontano entre los ríos Diamante y Mendoza. *Revista de la Asociación Geológica Argentina*, **17**, 127–349.

SALOMON, E., SCHMIDT, S., HETZEL, R., MINGORANCE, F. & HAMPEL, A. 2013. Repeated folding during Late Holocene earthquakes on the La Cal thrust fault near Mendoza city (Argentina). *Bulletin of the Seismological Society of America*, **103**, 936–949, http://dx.doi.org/10.1785/0120110335.

SCHMIDT, S., HETZEL, R., MINGORANCE, F. & RAMOS, V. A. 2011. Coseismic displacements and Holocene slip rates for two active thrust faults at the mountain front of the Andean Precordillera (~33°S). *Tectonics*, **30**, TC5011, http://dx.doi.org/10.1029/2011TC002932.

SLEMMONS, D. B. & DE POLO, C. M. 1986. Evaluation of active faulting and associated hazard. *In*: WALLACE, R. E. (ed.) *Active Tectonics*. National Academy Press, Washington, 45–62.

STAPPENBECK, R. 1917. Geologia de la falda oriental de la Cordillera del Plata. Ministerio de Agricultura. *Sección Geología, Mineralogía y Minería, Anales*, **12**, 9–49.

STERN, C. R., AMINI, H., CHARRIER, R., GODOY, E., HERVÉ, F. & VARELA, J. 1984. Petrochemistry and age of rhyolitic pyroclastic flows which occur along the drainage valleys of the Río Maipo and Río Cachapoal (Chile) and the Río Yaucha and Río Papagayos (Argentina). *Revista Geológica de Chile*, **23**, 39–52.

STEWART, I. S. & HANCOCK, P. L. 1990. What is a fault scarp? *Episodes*, **13**, 256–263.

TELLO, G. 1994. Fallamiento cuaternario y sismicidad en el piedemonte cordillerano de la provincia de Mendoza, Argentina. VII Congreso Geológico Chileno, Concepción (Chile), Actas, **1**, 380–384.

TELLO, G. 1998. *Actividad tectónica cuaternaria en el piedemonte cordillerano entre el río Tunuyán y el río Atuel y su vinculación con la sismicidad histórica del sur mendocino, provincia de Mendoza, República*

Argentina. PhD thesis, Facultad de Ciencias Exactas y Naturales, Universidad Nacional de San Juan.

TOMS, P. S., KING, M., ZÁRATE, M., KEMP, R. & FOIT, N., JR. 2004. Geochemical characterisation, correlation and dating of tephra in alluvial sequences of western Argentina. *Quaternary Research*, **62**, 60–75.

TRIEP, E. 1987. La falla activada durante el sismo principal de Mendoza de 1985 e implicaciones tectónicas. X Congreso Geológico Argentino, Actas, **1**, 199–202.

USGS 2013. World earthquake catalog. http://earthquake.usgs.gov/earthquakes.

WAYNE, W. J. & CORTE, A. E. 1983. Multiple Glaciations of the Cordón del Plata, Mendoza, Argentina. *Palaeogeography, Palaeoclimatology, Palaeoecology*, **42**, 185–209.

WELLS, D. L. & COPPERSMITH, K. V. 1994. New empirical relationships among magnitude, rupture lenght, rupture width, rupture area, and surface displacement. *Bulletin of the Seismological Society of America*, **84**, 974–1002.

YAGUPSKY, D., CRISTALLINI, E. O., FANTÍN, J., ZAMORA VALCARCE, G., BOTTESI, G. & VARADÉ, R. 2008. Oblique half-graben inversion of the Mesozoic Neuquén Rift in the Malargüe Fold and Thrust Belt, Mendoza, Argentina: new insights from analogue models. *Journal of Structural Geology*, **30**, 839–853.

YRIGOYEN, M. R. 1993. Los depósitos sinorogénicos terciarios. *In*: RAMOS, V. A. (ed.) *Geología y recursos naturales de la provincia de Mendoza*. XII Congreso Geológico Argentino y II Congreso de Exploración de Hidrocarburos, Relatorio, **1**, 123–148.

YRIGOYEN, M. R. 1994. Revisión estratigráfica del Neógeno de las Huayquerías de Mendoza septentrional, Argentina. *Ameghiniana*, **31**, 125–138.

ZAMBRANO, J. 1979. *Carta Geotectónica de Cuyo*. Instituto de Investigaciones Geológicas, Universidad Nacional de San Juan.

ZÁRATE, M. & MEHL, A. 2008. Estratigrafía y geocronología de los depósitos del Pleistoceno tardío/Holoceno de la cuenca del arroyo La Estacada, departamentos de Tunuyán y Tupungato (valle de Uco), Mendoza. *Revista de la Asociación Geológica Argentina*, **63**, 407–416.

Megalandslides in the Andes of central Chile and Argentina (32°–34°S) and potential hazards

STELLA M. MOREIRAS[1]* & SERGIO A. SEPÚLVEDA[2]

[1]*CONICET – IANIGLA (CCT), Avenue Dr Ruiz Leal s/n Parque, (5500), Mendoza, Argentina*
[2]*Department of Geology, University of Chile, Plaza Ercilla 803, Santiago, Chile*
*Corresponding author (e-mail: moreiras@mendoza-conicet.gob.ar)

Abstract: This review deals with an integration and update of the knowledge about large-volume landslides in the Central Andes at 32–34°S. An integrated landslide inventory for megalandslides in central Chilean and Argentinean Andean basins was developed, and dispersed chronological data on palaeolandslides were compiled, showing a dominance of Late Pleistocene and Holocene ages. Traditional hypotheses adopted for explaining landslide occurrence in the Central Andes are contrasted. Whereas seismic tremors have been widely suggested as the main triggering mechanism in Chilean collapses, palaeoclimatic conditions are considered as the main cause of Argentinean giant landslides. These different approaches denote the lack of multidisciplinary studies focused on the controversy about seismic or climate trigger mechanisms in the Central Andes. These studies are also essential to understand failure mechanisms and assessment of the related hazard and risk, which are essential to reduce social and economic impacts on vulnerable communities from future landslide events.

Megalandslides are large volume ($>10^6$ m^3) mass movements that are a common geomorphological feature of the Central Andes in Chile and Argentina at 32–34°S latitude. Landslides of different type and volume are usually present in the uplands, mainly located in glacial valleys. These include large rockslides and rock avalanches of volumes of the order of millions of cubic metres up to a few cubic kilometres. World global glaciations assumed by the research community in the early twentieth century misjudged the role of gravitational processes on landscape evolution. Modelling of erosion in the Central Andes was mainly attributed to glacial genesis. As a consequence, landslides began to be identified only recently (e.g. Abele 1984).

Geomorphological studies commonly missed landslides processes. They began to be described in the late 1960s and early 1970s (Salomón 1969). So far, recent analyses of these large failures have been mainly restricted to regional-scale studies (e.g. Chiu 1991; Moreno et al. 1991; Espizúa et al. 1993; Pereira 1995; Antinao & Gosse 2009; Moreiras 2009) or general descriptions (Hauser 2000), with a few exceptions for specific landslides (e.g. Casassa & Marangunic 1993; Hauser 2002; Espizúa 2005; Fauqué et al. 2005; Moreiras 2006a; Perucca & Moreiras 2006, 2008; González 2010; Moreiras 2010a; Welkner et al. 2010).

Later research has advanced significantly on the chronology of palaeolandslides, mainly due to the MAP-GAC project (Multi Andean Project: Geosciences for Andean Communities) that brought together the Geological Surveys of the Andean countries to work in collaboration on landslides, volcanic hazard and earthquake hazard (PMA:GCA 2007; Fauqué et al. 2009a, b). However, these data have been mainly communicated in specific local symposia without a broad audience or reports with limited circulation. Hence, in the framework of the IGCP586-Y project we have summarized and updated the present knowledge of megalandslides in this segment of the Central Andes, establishing the known chronologies and discussing the supposed triggering mechanisms, to identify new challenges for their further study. Furthermore, we have developed an inventory map of megalandslides at the latitude of the study area, revealing the importance of these processes in this portion of the Central Andes.

Study area

At the latitude of the study area (32–34°S), the Central Andes are composed of different geological provinces. The western Coastal Cordillera and the Main Cordillera comprise the Chilean Andes, while the Main Cordillera, Frontal Cordillera and Precordillera make up the eastern Argentinean side of Central Andes (Fig. 1).

The Coastal Cordillera (CC) is mainly formed by Palaeozoic, Mesozoic and Cenozoic intrusive bodies and volcano-sedimentary Mesozoic and Cenozoic rocks covered by Pliocene–Quaternary

From: SEPÚLVEDA, S. A., GIAMBIAGI, L. B., MOREIRAS, S. M., PINTO, L., TUNIK, M., HOKE, G. D. & FARÍAS, M. (eds) 2015. *Geodynamic Processes in the Andes of Central Chile and Argentina.* Geological Society, London, Special Publications, **399**, 329–344. First published online May 13, 2014, http://dx.doi.org/10.1144/SP399.18
© 2015 The Geological Society of London. For permissions: http://www.geolsoc.org.uk/permissions.
Publishing disclaimer: www.geolsoc.org.uk/pub_ethics

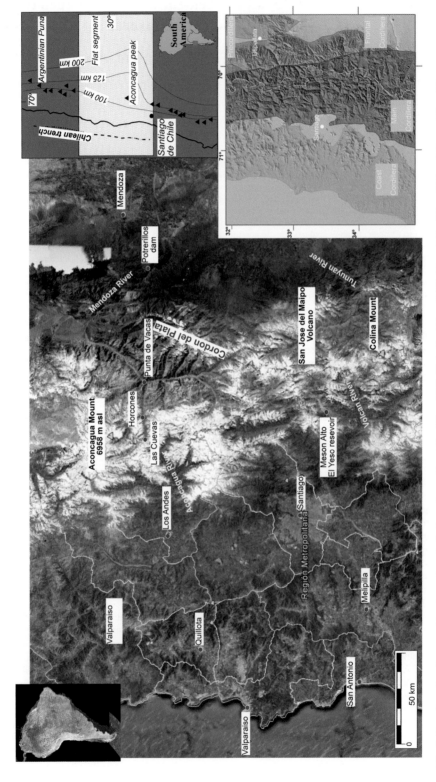

Fig. 1. Location of study area with enhanced geological provinces: Coast Cordillera (CC), Main Cordillera (MC), Frontal Cordillera (FC) and Precordillera (P).

alluvium (Gana et al. 1996; Wall et al. 1996, 1999; Sellés & Gana 2001). The Main Cordillera (MC) (Thiele 1980; Ramos 1996; Wall et al. 1999; Fock 2005) is characterized by Eocene to Miocene volcanic rocks of the Farellones and Abanico Formations in Chile and the Aconcagua volcanic complex in Argentina. These rocks overlie Jurassic and Cretaceous marine and volcaniclastic rocks. The marine sedimentary rocks in Chile correlate with those of the Mendoza Group on the Argentinean side formed by black shales, sandstones, limestones, and gypsum levels. Tertiary intrusive and synorogenic continental rocks known as the Santa Maria Formation, cropping out in the area of Mt Aconcagua (6958 m above sea-level (asl)) are also well developed in this geological unit. The border area between the two countries presents the current volcanic arc from about 33°S southwards, including some active volcanoes such as Tupungato-Tupungatito, Marmolejo-San José and Maipo. Structurally, this sector was developed by thin-skinned structures such as the Aconcagua fold-and-thrust belt (AFTB) (Ramos et al. 1996), some of which present seismic activity (Sepúlveda et al. 2008).

The Frontal Cordillera (FC) is composed of units formed during the Gondwana orogeny in the Late Palaeozoic to Early Mesozoic. These Permo-Triassic volcanic rocks, mainly of andesitic to silicic magmatic composition, are termed the Choiyoi Group. The subvolcanic rocks of this group are represented by granite bodies and rhyolitic porphyries and andesitic dikes. Structurally, the Frontal Cordillera behaved as a rigid block during the Andean deformation, as shown by the presence of thick-skinned thrusts (Ramos et al. 1996).

Older Palaeozoic rocks appear in the Precordillera range (P), an Andean thrust-and-fold belt sequence with a typical thin-skinned structure where Palaeozoic rocks crop out. Basically, the belt comprises an Early Cambrian to Middle Ordovician carbonate platform and a clastic marine Ordovician sequence unconformably covered by foreland basin deposits of Silurian to Ordovician/Carboniferous age.

Between 28–33°S, this segment of the Central Andes has a distinctive plate tectonic setting. The present convergence rate between the subducted Nazca plate and the South American plate averages about 6 to 7 centimetres per year (Khazaradze & Klotz 2003). Consequently, thrust tectonics predominate in this region since the primary shortening of the Andean orogen began about 22 Ma (Pilger 1984; Ramos 1988). This compressional force system is still active today, resulting in ongoing tectonic uplift of the Central Andes. Uplifting of 2 cm a^{-1} has been recently measured in the Mt Aconcagua peak (Mateo et al. 2009).

Moreover, as a consequence of progressive orogen-front migration towards to the east, the main active faulting is recorded in the eastern Andes foothills at present. This segment corresponds to the flat-slab subduction segment with an easterly dip of about 5° at c. 100 km depth where an intensive seismic activity exists. Approximately 90 per cent of Argentinean Quaternary deformation is concentrated in this segment which is characterized by intense neotectonic activity (Costa et al. 2000).

State-of-the-art investigations on megalandslides in the study area

As mentioned above, landslides began to be identified only recently in this sector of the Central Andes. The earliest description of megalandslides along the valley of the Las Cuevas and Mendoza rivers in Argentina, comes from a geomorphological study in the 1960s by Salomón (1969). Later, this kind of process was recognized in the Chilean Andes (Abele 1984) by reinterpretation of ancient moraines of the Aconcagua and Yeso valleys (Marangunic & Thiele 1971; Caviedes 1972). Following these ideas, several studies reinterpreting the glacial origin of Quaternary deposits were performed. At the same time revision of global glaciation epochs continues. The genesis of chaotic deposits corresponding to debris flows developed in alluvial fans of the Frontal Cordillera piedmont (Polanski 1958) was reinterpreted as moraines promoting the existence of four glaciations along the Rio Blanco river in Argentina (Wayne & Corte 1983).

Current investigations in the Argentinean Andes are reinterpreting several moraine deposits as collapses, especially in the Mendoza river basin. The Holocene deposit described as Horcones moraine (Espizúa 1993) is presently considered as the distal facies of a rock avalanche sourced in the south face of Mt Aconcagua. This debris material travelled 30 km before arriving at the Cuevas river where it generated a palaeolake (Fauqué et al. 2005, 2008a, c, 2009a; Fig. 2). Likewise, the existence of the Penitentes glaciation is being discussed. The outwash of this drift has been recently reinterpreted as the debris flow resulting from the El Abuelo–Mario Ardito rock avalanche (Fauqué et al. 2008b). Similarly, deposits originally attributed to glacial origin in the upper basin of the Las Cuevas river (Suárez 1983) are now described as five landslides (Goyete, Negro, Lagunita, Susanita and Matienzo) (Moreiras 2007; Moreiras et al. 2008a, 2012a; Lauro 2010).

Meanwhile, in the Chilean Andes glacial deposits seem to be restricted to the Late Pleistocene in the highest elevations. No Pleistocene drifts and few

Fig. 2. The Horcones rock avalanche generated in the south face of Mt Aconcagua (6958 m asl). This deposit is eroded by the Quebrada Blanca rock avalanche.

moraines have been identified along the valleys of the Maipo river basin, and geomorphological studies on outwash terraces identified along the Maipo river have only recently begun (Ormeño 2007). Several areas previously mapped as moraine deposits were reinterpreted in the Maipo basin as landslide deposits by Moreno et al. (1991) and Chiu (1991). For instance, at the confluence of the Colina and Morado rivers, the extensive area of La Engorda, interpreted as landslide deposits by these authors, was vaguely distinguished from chaotic material mapped as moraines by Thiele (1980) (Fig. 3a). Recently, González (2010) distinguished at least four different rock avalanches generated in the Morado valley covered by younger rockslides and rockfalls, all placed within older moraine deposits. On the other hand, some glacial deposits remapped as landslide material in the 1990s in Quebrada Morales were assigned back to moraine deposits by Moreiras et al. (2012a) after a field reconnaissance, while the largest deposit of the area, the Mesón Alto-El Yeso, was initially suggested by these authors and Deckart et al. (2014) to be a polygenetic deposit of rock avalanches covering moraine deposits (Fig. 3b).

In the Aconcagua river valley, the chaotic deposits of Portillo (Fig 3c), Potrero Alto and Salto del Soldado (Fig 3d) attributed to a glacial origin (Caviedes 1972) were reinterpreted as landslides (Abele 1984), except for the Guardia Vieja deposit that continues to be considered as a frontal moraine. Welkner et al. (2010) described and modelled in detail the Portillo deposits, concluding that at least two landslide events occurred in the area, damming a tributary of the Aconcagua river and forming the Laguna del Inca lake.

Inventory map

As the record of megalandslides was still incomplete for this portion of the Central Andes, an inventory map of this type of event was constructed for this review, through a compilation and check of previous publications (Moreiras 2005, 2009, 2010a; Moreiras et al. 2008a; Schachter 2008; Stumpf 2008; Antinao & Gosse 2009; González 2010) and unpublished mapping by the authors (Fig. 4). This compilation enabled the dimensions and areas affected by these phenomena to be determined. At a first approach at least 650 km^2 are affected by megalandslides in this area implying a spatial probability of occurrence of about 2 per cent.

From the inventory we can quickly deduce that megalandslides are concentrated in the Main and Frontal ranges where volcanic rocks of Tertiary and Jurassic–Cretaceous (Main Cordillera) and Permo-Triassic (Frontal Cordillera) ages, predominate. Some huge landslides are also related to main gypsum levels (Auquilco Formation) in the

Fig. 3. (a) La Engorda complex in the Volcán river, picture taken from Morales deposit, (b) view of the Mesón Alto–El Yeso rock avalanche from the Cortaderas slide, (c) deposit of the Portillo rock avalanche, and (d) the Salto del Soldado rock avalanche at both margins of the Aconcagua river.

Main Cordillera. However, hydrothermal alteration generated by Tertiary intrusive bodies seems to have greatly affected slope instability, probably more than any lithological aspects. In these high altitude areas, pronounced slopes of glacial valleys could also favour collapses. Furthermore, many huge landslides have been generated by unstable relict glacial material lying on valley hillslopes. This is the case of many complex slides in the Aconcagua Park (Moreiras et al. 2008a, b).

Meanwhile, large landslides are not recognized in the Coastal Cordillera and only a few in Precordillera geological provinces. The existence of narrow valleys in the latter seems to promote rapid debris mobilization disallowing preservation of these chaotic deposits at the latitude of study. Even though moderate landslides have been described in the Precordillera (Moreiras 2005, 2006a) as being able to generate impounded lakes (Fauqué et al. 2005; Perucca & Moreiras 2008), these landslides are mainly restricted to Silurian and Carboniferous rocks cropping out at mountain fronts generally affected by Quaternary faults. Meanwhile, the smoother relief and granitic composition of large parts of the Coastal Cordillera at this latitude apparently preclude the triggering of giant collapses, as was observed during the recent 2010 central Chile earthquake, which mainly induced shallow slides and falls despite the large size of the earthquake (M_w 8.8).

Chronology of palaeomegalandslides

Megalandslides older than the Late Pleistocene are rare in the Central Andes, as is clearly evident from Figure 4 and Table 1. These Pleistocene ages are commonly assigned by stratigraphy or relative dating techniques. The rock avalanches clustered in the Cordón del Plata range were assigned to the Middle Pleistocene according to tephrochronology and palaeontological content (Moreiras 2006a, 2010b, 2011). The Middle Pleistocene age of Coyanco, Cerro San Simón and El Ingenio rockslides is assigned by tephrochronology because they are overlaid by an ash layer dated 0.45 ± 0.07 Ma by fission-track age on zircon (Stern et al. 1984).

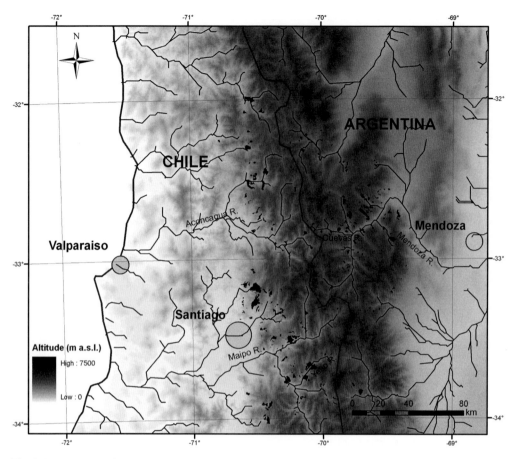

Fig. 4. Inventory map of megalandslides identified in the Central Andes (32°–34°S). (After Antinao 2008; Moreiras *et al.* 2008*a, b*).

However, the accuracy of this relative chronology is limited. This is the case of several huge landslides such as Cerro El Salto, Cabeza de Toro, Las Hualtatas, Cortaderas, Arboles, Quebrada Seca and Loma Huinganal for which certain ages are unknown (Table 1). Likewise, those dated by numerical methods, as the Macul rockslide, show a wide range in their cosmogenic ages (Antinao & Gosse 2009), and in other cases the age is assigned by a single dating. For example, the Estero Marques was dated by a unique sample resulting in 112 ± 14 ka (Rivano *et al.* 1993). Figure 4 shows a minor quantity of Pleistocene palaeolandslides where only data derived from numerical dating are included.

The Pleistocene age assumed by relative techniques is also doubtful when regional Quaternary stratigraphy has not been studied in a regional context or when tephrochronology is not enough. As an example, even though ash levels with similar aspect and geochemistry crop out along the Mendoza river valley, they correspond to very different numerical ages. Therefore, temporal correlation based on tephro-chronology provide ambiguous results. However, a Pleistocene age was generally assumed for the large events (from 10^6 m^3 up to 5 km^3) in this portion of the Central Andes. Despite this, recent studies mainly based on radiocarbon and cosmogenic dating reveal younger ages in some cases (Table 1).

For instance, the Amarillo and Negro landslides in the Mendoza river (Salomón 1969) were studied stratigraphically establishing that the former was earlier than the latter and it may have occurred during an interglacial stage from 125 ka to 75 ka probable correlated to MIS (Marine Isotope Scale) 5 (Espizúa 2005). Nonetheless, organic matter recently found in finer sequences of related impounded lakes reveals ages of 8640 ± 200 yr BP and $18\,660 \pm 330$ yr BP, respectively, for the Amarillo and Negro slides (Fauqué *et al.* 2008*c*; Fig. 5a).

Table 1. Rock avalanche and rockslide chronology in the Central Andes (32°–34°S)

Collapse	Vol. km³	Age		Dating technique	Authors
Cerro El Salto, Ch	1.34		Early Pleistocene	Relative dating	1
Cabeza de Toro, Ch	5.9		Early–Middle Pleist	Relative dating	1
Las Hualtatas, Ch			Early–Middle Pleist	Stratigraphy	2
Placetas Amarillas, Ar	1.6	$163.9 \pm 21.8/155 \pm 13.5/112.8 \pm 14.6\ ka$	Middle Pleistocene	**Cosmogenic**	3
Piedras Blancas, Ar	0.89	$>350 \pm 80 \pm ka$	Middle Pleistocene	**Cosmogenic**	3
Tigre Dormido, Ar	1.7	$46.7 \pm 2.7/46.4 \pm 3.7\ ka$	Middle Pleistocene	**Cosmogenic**	3
Cortaderas, Ar			Middle Pleistocene	Tefro-stratig	
Quebrada Seca, Ch			Middle Pleistocene	Relative dating	4
Loma Huinganal, Ch	2.6		Pleistocene	Stratigraphy	1
Macul, Ch		$8 \pm 0.3/19.1 \pm 0.6/80.4 \pm 2.6\ ka$	Late Pleist–Holoc	**Cosmogenic**	1
Potrerillos, Ch		$12.2 \pm 0.5/23.8 \pm 0.9/80.5 \pm 5\ ka$	Late Pleist–Holoc	Stratigraphy	6
Cristo Redentor, Ar			Late Pleistocene	**Cosmogenic**	5, 1
Salto del Soldado, Ch	0.2	$14.1 \pm 1.3/9.8 \pm 0.7\ ka$	Late Pleistocene	Stratigraphy	5, 7
Las Arpas, Ch	12.5		Late Pleistocene	Stratigraphy	3
Cerro Vizcachas, Ch	4.8		Late Pleistocene	Relative dating	1
Batuco – Tranquilla, Ch	2.5		Late Pleistocene	Relative dating	1
Peuco, Ch	3.2		Late Pleistocene	Relative dating	1
Llaretas del Rocin, Ch	0.5	$112 \pm 14\ ka$	Late Pleistocene	**Cosmogenic**	5, 1
Estero Maquis, Ch	3.9		Late Pleistocene	Relative dating	1
Cerro Guanaco, Ch	3.0		Late Pleistocene	Stratigraphy	8
El Ingenio, Ch	4.4		Late Pleistocene	Stratigraphy	9
Laurel, Ch	3.6		Late Pleistocene	Stratigraphy	
La Engorda, Ch	>0.1	$24.5 \pm 0.4\ ka$	Holoceno	**Radiocarbon**	10, 18
Acequion I, Ar		$7,497 \pm 157\ BP$	Holoceno	**Radiocarbon**	11
Acequion II, Ar		$>7,497 \pm 157\ BP$	Holoceno	**Radiocarbon**	11
Negro, Ar		$18,660 \pm 330\ BP$	Holoceno	**Radiocarbon**	12
Amarillo, Ar		$8,640 \pm 200\ BP$	Holoceno	**Radiocarbon**	12
Mario Ardito – Penitentes, Ar	0.18	$10.6 \pm 1.1/11.8 \pm 0.1\ ka$	Holoceno	**Cosmogenic**	13
Horcones, Ar		$11.1 \pm 0.7/8.1 \pm 1.2\ ka$	Holoceno	**Cosmogenic**	14
Tolosa I, Ar		$14.65 \pm 1.9\ ka$	Holoceno	**Cosmogenic**	15
Tolosa II, Ar		$1.38 \pm 1.5\ ka$	Holoceno	**Cosmogenic**	15
Portillo, Ch	1.1	$12.9 \pm 0.5, 13.2 \pm 0.9, 14.4 \pm 01.6,$ $11.5 \pm 0.5\ ka$	Holoceno	Stratigraphy	8, 17
San Nicolas-Cortaderas, Ch	2.4		Holoceno	Relative dating	16
Yeso–Meson Alto, Ch	4.5	$4.7 \pm 0.6\ ka$	Holoceno	**Cosmogenic**	1
Quebrada Blanca, Ar		8260–$8254\ BP$	Holoceno	Stratigraphy	12
El Abuelo, Ar			Holoceno	Stratigraphy	
Cortaderas (reactivación), Ch	0.02	1958	Holoceno	Historic	19
El Manzanito, Ch	0.004	1958	Holoceno	Historic	19
Parraguirre, Ch	0.006	1987	Holoceno	Historic	20

Authors: 1. Antinao & Gosse (2009). 2. Wall et al. (1999). 3. Fauqué et al. (2008d). 4. Antinao (2008). 5. Rivano et al. (1993). 6. Pereira (1995). 7. Charrier (1983). 8. Godoy et al. (1994b). 9. Moreno et al. (1991). 10. Moreno et al. (1991). 11. Perucca & Moreiras (2008). 12. Fauqué et al. (2008a). 13. Fauqué et al. (2008b). 14. Fauqué et al. (2008c). 15. Rosas et al. (2008). 16. Abele (1984). 17. Welkner et al. (2010). 18. González (2010). 19. Sepúlveda et al. (2008). 20. Hauser (2002).

Fig 5. Palaeolandslides along the valley of Las Cuevas and Mendoza rivers: (**a**) Amarillo and Negro slides, (**b**) El Abuelo–Mario Ardito rock avalanche and Penitentes moraine, (**c**) Quebrada Blanca rock avalanche sourced in gypsum levels of Jurassic Auquilco Formation being eroded by the Horcones rock avalanche (both deposits are showed in Figure 3), (**d**) Tolosa rock avalanches close to Las Cuevas village, (**e**) Cristo Redentor rotational slide, and (**f**) Puente del Inca sackung.

Holocene cosmogenic ages (Cosmogenic Nucleids (CN)) have also been found for the remaining palaeolandslides that have occurred along the Mendoza river valley. The El Abuelo-Mario Ardito rock avalanche close to the Penitentes ski centre was dated using cosmogenic $36Cl^{16}$ isotope at $13\,890 \pm 1300$, $10\,620 \pm 1120$ and $11\,820 \pm 1580$ years (Fauqué *et al.* 2008*b*; Fig. 5b), and blocks measured on the surface of the Horcones deposit at $11\,110 \pm 760$ and 8170 ± 1220 years old (Fauqué *et al.* 2008*c*). The last phenomenon dammed the Cuevas river valley generating an impounded lake which radiocarbon dating revealed to be 14.8 ± 14 ka (Fauqué *et al.* 2008*c*) and a U-series dating obtained from a travertine layer capping this lake sequence resulted in an age of 9.7 ± 5 ka (Espizúa 1999; Fig. 2). Likewise, cosmogenic dating of rupture planes generated on the western hillslope of

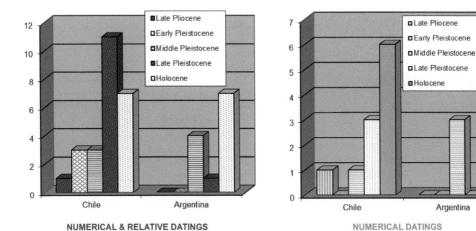

Fig. 6. Chronology of palaeolandslides. The first bar chart shows ages of megalandslides obtained by both relative and numerical dating techniques while the second chart shows results obtained using only numerical dating.

the Tolosa peak gave dates of 14 650 ± 1900 and 11 380 ± 1500 years, which match ages measured on block surfaces of 15.5 ± 2.1 ka and 13.9 ± 2.1 ka, respectively (Rosas et al. 2008). These findings support the occurrence of at least two rock avalanches during the last Quaternary period (Fig. 5d).

Furthermore, immense landslides recognized in the upper basin of Las Cuevas river (Moreiras 2007; Moreiras et al. 2008a, b), as well as those mapped in the Aconcagua Park region (Moreiras, et al. 2008b) are assigned to the Holocene. Even though no quantitative dating exists, these deposits are located above 3200 m asl, and they are without any evidence of being eroded by Pleistocene glacier masses.

The majority of Chilean palaeolandslides are also dated as Late Pleistocene–Holocene. The Portillo rock avalanche, damming at present the Inca lake, was assigned to Late Pleistocene–Holocene (Rivano et al. 1993); however, cosmogenic dating and geomorphological studies reveal that there are two rock avalanche events occurring at 4 and 13 ka, respectively (Welkner et al. 2010; Fig. 3c).

Furthermore, the Salto del Soldado rock avalanche was dated at 14.1 ± 1.3 ka (Rivano et al. 1993) and 9.8 ± 0.7 ka (Antinao & Gosse 2009; Fig. 3d). Similarly, the La Engorda composite deposit corresponds to the Late Pleistocene–Holocene, with a radiocarbon age 24.5 ± 0.4 ka in lacustrine deposits (Moreno et al. 1991). From geomorphological mapping, González (2010) proposed that this age corresponds to the oldest rock avalanche of the La Engorda landslide complex, the youngest being possibly Holocene, but further dating is needed to verify this hypothesis.

According to the last numerical dating, the Mesón Alto-El Yeso rock avalanche (Abele 1984) in the Yeso valley, with a volume of c. 4.5 km^3, is the youngest event in the region. Its estimated age is 4.7 ka (Antinao 2008; Antinao & Gosse 2009; Fig. 3b).

In this manner, direct numerical dating of palaeomegalandslides often result in younger ages than first suggested. The rock avalanches in the Cordón del Plata range that had been assigned to the Middle Pleistocene according to tephrochronology and palaeontological content (Moreiras 2006a, 2011), were reassigned to the Middle to Late Pleistocene when cosmogenic dating techniques were applied (Fauqué et al. 2008d; Moreiras 2010b; Table 1). On the other hand, most palaeolandslides thought to be of Holocene age (Fig. 6) could be a reflection of more recent collapses which might be masking features older than could be explained by erosion or shielding by subsequent hillslope movements.

Discussion

Megalandslide causes and possible triggering mechanisms

In the Argentinean Andes, the largest landslide events have been associated with warm, wet periods (Espizúa 2005), even though local evidence of such interstages is lacking. Moreiras (2006a) suggested that although the rock avalanches clustered in the Cordon del Plata range had a seismic origin, it is probable that they occurred in a wetter interstage prior to the Uspallata glaciation. But a geological record of this wetter period has not been found in the Mendoza river valley. Additionally, rock avalanches generated from the Tolosa mount were associated with a Holocene interstage, although this is not certainly proved by local climate proxies (Rosas et al. 2008).

At present, a climatic cause of palaeolandslides in the Central Andes is only weakly supported, as local records of glacial–interglacial periods seem to be vague. The Horcones and Penitentes glaciations established by Espizúa (1993) are presently being discussed due to recent reinterpretation of the genesis of these deposits (Fauqué et al. 2008b; Fauqué et al. 2009a, b). Thus, the palaeoclimate correlation established by Espizúa (1993) with glaciations of the Aconcagua valley established by Caviedes (1972) is obsolete as Portillo and Salto del Soldado are landslides (Moreiras 2010b).

Debuttressing combined with land isostatic rebounding after Pleistocene glacier mass retreat was proposed as the main conditioning factor in catastrophic geological processes in Las Cuevas valley, as has also been suggested for landslides in Argentinean Patagonia (González Díaz 2003, 2005). Abele (1974) has commented that greater permafrost degradation and thawing during the Holocene compared with the Pleistocene have had an influence on slope instability in the Central Andes. Notwithstanding this, climate influence has been mainly disregarded in the western slopes of the Andes; a seismic genesis is generally suggested for Chilean collapses, and possibly may apply to all those associated with neotectonic activity, particularly given the proximity of the failures to faults and statistical correlation of the landslide distribution with faults and shallow crustal seismic activity (Abele 1984; Antinao & Gosse 2009).

Bigger landslides in the Central Andes cluster along main active faults, for example, the Los Marquis and Arpa rock avalanches are located along the Pocuro fault. However, it is not clear if this correlation is due to earthquake activity rather than to tectonically disturbed and sheared rocks due to the presence of such faults. Nonetheless, Quaternary activity of regional faults has been rarely proved due to the lack of seismotectonic and palaeoseismological studies, and the low frequency of large magnitude crustal earthquakes in the region during the last century (Sepúlveda et al. 2008). However, geomorphological studies have demonstrated geological activity along the San Ramón fault, at the edge of the Santiago basin (Rauld 2002; Rauld et al. 2006; Armijo et al. 2010; Rauld 2011), and a M_w 6.3 earthquake struck the mountain region of the Maipo basin in 1958 triggering rockfalls and soil slides (Las Melosas earthquake: Sepúlveda et al. 2008, Alvarado et al. 2009). Further, low to medium magnitude seismicity close to several faults of the AFTB has been registered during the last decade (Pérez et al. 2010).

Neotectonic studies focused on this dilemma are beginning in Argentina. Seismicity is extended to the Holocene in the Precordillera (Perucca & Moreiras 2008) based on preservation of seismites in palaeolakes dammed by two rock avalanches in the Acequión river (San Juan province, 32°S). Rock avalanches clustered in the northern extreme of the Cordon del Plata (Cordillera Frontal) linked to evidence of Quaternary activity of the Carrera fault system prove the occurrence of $M > 6$ palaeoearthquakes at least until the Late Pleistocene (Moreiras 2006a, 2011). Uncertain Quaternary activity of north–south regional faults in the Aconcagua Park in the Mendoza valley do not allow the correlation of the occurrence of immense Holocene landslides with palaeoseismicity, yet these deposits are distributed along these Mesozoic structures (Moreiras et al. 2008b). Further south, Penna (2010) has recognized a close relationship between large landslides and faulting along the Antiñir–Copahue fault system (37°–38°S), and has proposed a possible seismic trigger.

The Mesón Alto–El Yeso, Cortaderas and San Nicolás rock avalanches in the Yeso river (Chilean Cordillera Principal) (Fig. 4d), are aligned with the Laguna Negra fault, an extension to which is the likely source of the 1958 Las Melosas earthquake. This shock triggered a relatively small reactivation of the Cortaderas landslide and other slope instabilities in the area (Sepúlveda et al. 2008). However, no seismic origin could be determined for the Parraguirre landslide that generated the Alfalfal debris flow along the Colorado river on 28 November 1987 (Casassa & Marangunic 1993). In this historical case, seismic shaking ($M_s \leq 4.5$) recorded during the event was associated with the collapse process itself. Then, thawing after intense snow precipitation during the warm phase of the 1987 El Niño phenomenon was suspected as the probable cause (Hauser 2002).

Without a more refined chronology, it is impossible to address the issue of the relative roles played by climate and earthquakes as conditioning or triggering factors causing the megalandslides in the study area. Even though it is well accepted that the distribution of failures is related to traces of regional faults, actual neotectonic activity of these faults has rarely been proved, particularly in the case of the AFTB where the largest Holocene rock avalanches occurred. Neither does any correlation exist between the progressive migrations of the orogenic front towards the east since the Miocene, and the chronology of palaeolandslides. At present, the orogenic front is located at the eastern foothills of the Andes. For this reason great landslides should also be identified in the Argentinean Precordillera. While not certain at the latitude of the study area, several megalandslides have been identified towards the north associated with Quaternary faults of the Maradona–Acequion fault system (Perucca & Moreiras 2006, 2008; Moreiras & Banchig 2008). A dozen slides are aligned to the inferred fault of the Saso river from 31°31′ to 31°41′S near Los Caracoles reservoir, San Juan province (Fauqué et al. 2008e). Nearby to the west, several slides generated at both margins of the Uruguay river have been mapped by these authors as well.

At present, several earthquake-induced landslides have been recorded with shallow earthquakes, but local seismic sources are not implicated in all landslides, especially smaller ones. Rockfalls close to faults of the AFTB were reported for the Maule earthquake (M_w 8.8), even though the epicenter was 270 km away (Wick et al. 2010). However, no large rockslide or rock avalanche such as those described here was recognized for this extremely large seismic event, thus the hypothesis of local shaking related to crustal earthquakes remains the most likely for the large-volume mass movements.

The tentative role of debuttressing and land isostatic rebounding after the retreat of Pleistocene glacier masses may be correlated with a Holocene age for a majority of palaeolandslides. But lack of local climate proxies, as was mentioned above, does not allow a proper analysis of the relationship between palaeoclimate and slope instability. Moreover, young ages below 5 ka such as for the Mesón Alto landslide would not be expected for these mechanisms, which should operate to a greater extent immediately after the glacier retreat, but are more likely to be related to long-term stress relaxation and weathering (Poschinger 2002). However, at present hillslope stability in the Central Andes is strongly affected by weather and climatic phenomena such as the warm phase of ENSO (El Niño Southern Oscilation) (Erickson & Högstedt 2004; Moreiras 2006b; Sepúlveda et al. 2006; Moreiras et al. 2012b).

Megalandslides and geological hazards

From the hazard point of view, megalandslides can be considered as extreme events; those of high magnitude–low frequency are not usually registered in the historic record of a community but are present in the natural, geological record of the area in which they have been emplaced (Nott 2006). The consequence of the occurrence of such extreme events can be catastrophic, causing disasters of very high social impact. The incorporation of prehistoric records is thus essential for a complete hazard and risk assessment of an area; short-term historical records do not reflect the long-term changes in the nature of a hazard, because the relationship between magnitude and frequency changes over time, a concept known as non-stationarity (Nott 2006). The occurrence of such large failures in the study area is rare, but at least two historic examples of landslides of volumes greater than one million cubic metres in Chile – the 1987 Parraguirre rock avalanche and debris flow (Hauser 2002) and two soil slides triggered by the 1958 Las Melosas earthquake (Sepúlveda et al. 2008) – indicate that although the probability of such large failures is low, they still must be considered as a significant hazard.

Historical records of catastrophic megalandslides are not available for the Argentinean side at the latitude of the study area. Reports of this kind of event are scarce as the population is mainly concentrated in cities located on the Andes foothills. High altitude mountain areas are remote; only a few villages exist along the main corridors located in mountain valleys. Nevertheless, a large-volume slide has been reported in Cañon del Atuel, southern Mendoza (35°S). According to reports of local people it occurred around 40 years ago and its triggering mechanism is unknown. A slow active slide is being monitored in the Panda mount, inside the Aconcagua Park region, by measurement of tensional crack dimension.

Apart from the direct hazard in slope areas and surroundings that can be reached by the sliding mass, large landslides can also induce indirect hazards far away from the landslide site (Sepúlveda & Moreiras 2013). For instance, large rock avalanches may travel several kilometres downstream, and if mixed with ice or water may generate debris flows that can travel for dozens of kilometres. The 1987 Parraguirre landslide and resulting debris flow in the Colorado river in Chile (Hauser 2002), that killed at least 37 people and caused serious damage to two hydroelectric plants, is a good example of such a process. Furthermore, megalandslides usually cause the formation of dams (Costa & Schuster 1988), but these dams often fail soon after lake formation, causing outburst floods and debris flows that can have catastrophic consequences, thereby posing a significant hazard (e.g. Harp & Crone 2006; Dunning et al. 2007). In these cases, dam stability can be assessed by geomorphological approaches (Ermini & Casagli 2003; Dong et al. 2009).

The Erizo lake dammed by a landslide generated in the Estrella peak, San Juan province (31°40′S), collapsed on 12 November 2005. The failure of this lake impounded by a 57 m high dam generated an outburst flood into the San Juan river. The resulting violent flow of 32.1 million cubic metres was discharged in 67 minutes and travelled 254 km in 12 hours (D'odorico Benites et al. 2009). This phenomenon caused severe damage downslope destroying bridges and roads, and isolating many people in the mountain areas. The just-built Caracoles dam was severely damaged. In the Ullum reservoir – located 180 km downstream – water become turbid in three minutes, affecting the availability of potable water for inhabitants of San Juan city.

The most drastic outburst recorded in Argentina happened in 1914 when the Carri Lauquen Lake collapsed generating an extraordinary outburst flood on the Barrancas river (36°30′S) (Moreiras & Coronato 2010). This lake had been formed by a rock avalanche (González Díaz et al. 2001) which has been dated cosmogenically at 2 ka (González Díaz et al. 2005) The original 21-km-long lake was reduced to 5.6 km, and its surface was lowered by c. 95 m. Two billion cubic metres of water were drained overnight (Groeber 1916). The resulting debris flow/flood dammed the Quili-Malalal river, forming a tiny lagoon 200 m long and 25 m wide, and flowed more than 300 km devastating downstream valleys. Several animal farms completely disappeared, and fields of wheat, corn and alfalfa were buried by debris. In addition, two small towns in this valley were devastated. This debris flow/flood also wiped out farms, railway stations, railway lines and roads located along the valley of the Colorado river, which is formed by the union of the Barrancas and Grande rivers. The overflow also triggered several landslides. In all, 175 fatalities were reported, and more than 100 people disappeared. Groeber (1933) noted that 20 years later the farmland in both valleys still had not recovered.

Similarly, large landslides occurring on the slopes of lakes or water reservoirs can produce catastrophic flooding downstream due to dam overtopping by the generated wave – like the famous Vajont case in Italy in 1963 (e.g. Kilburn & Petley 2003) – or dam failure. Smaller slides of up to half a million cubic metres have been identified in the shoreline of Potrerillos reservoir (32°59′S) possibly triggered by volume fluctuation due to its fill and discharge (Moreiras et al. 2010). This water supply is mainly used for irrigation north of Mendoza province.

Therefore, a good understanding of the generation mechanisms, recurrence and deposits of large-volume landslides in a mountainous, active tectonic area such as the Central Andes, is essential to correctly assess the hazard and risk for communities located close to, or downstream from, the potential landslide areas.

Concluding remarks

Traditional hypotheses on the triggering mechanisms of large megalandslides have been developed seperately for the Chilean and Argentinean Andes. Whereas a seismic cause is proposed for events on the Chilean border, palaeoclimate is assumed as the main conditioning factor for Argentinean landslides.

The uncertain genesis of megalandslides denotes the lack of multidisciplinary studies focused on the controversy over seismic or climate failure mechanisms in the Central Andes. In this framework, detailed geomorphological, neotectonic, and seismotectonic studies combined with dating of identified deposits, are fundamental to understanding hillslope behaviour in the Central Andes in the future. Many of the large identified palaeolandslides has not yet been dated so evaluation of causative mechanisms is extremely hard and uncertain.

There is no doubt that megalandslides, by their very volumes, have been a significant process in the geomorphological evolution of the Central Andes during the Holocene. These findings are essential to explain the relationship between the uplift associated with the evolution of the Andean orogeny and climate. Likewise, we note that recent studies on hillslope instability in the Central Andes have implications for a reinterpretation of glacial–interglacial stages of main valleys. These findings strongly impact on the regional palaeoclimatic understanding, as glaciations established for the valleys of Las Cuevas and Mendoza rivers have been widely used for regional climate correlations (Gosse 1994; Espizúa 1999; Baker et al. 2009). Till deposits that could be correlated with the Confluencia and Horcones drifts have not been determined in the Atuel river (35°S), or in the Diamante river terraces in the southwestern Mendoza province (Baker et al. 2009). From this approach we suggest an integrated and regional Quaternary study reviewing the past climate conditions at this latitude.

Therefore, despite considerable progress in our understanding of the impact of climate on many of the processes that contribute to slope instability, it is not yet possible to say whether climatic parameters alone will enhance or dampen landslide activity, or affect the frequency of landslides.

Understanding whether landslides in the Central Andes are of climatic or tectonic origin could also be enhanced by specific studies of related dammed palaeolakes: their fine-grain-size sequences are ideal environments for the preservation of seismites and climatic markers. A lack of such studies precludes the understanding of landscape evolution and estimation of erosion rates for this sector of the Central Andes. Furthermore, specific geotechnical analyses are essential for establishing a seismic-failure mechanism for these large instabilities.

In this context, detailed geomorphological analyses of large landslides, combined with geochronological studies to estimate their recurrence and to assess the vulnerability of potentially affected communities and strategic infrastructure, are fundamental to understanding failure mechanisms; such analyses are also vital for the development of predictive models on slope instability and for the assessment of the related hazard and risk, which in turn are essential to reduce social and economic impacts on vulnerable communities from future landslide events.

This work was done within the framework of several Chilean and Argentinean research projects, integrated under the IGCP586-Y project. Those headed by S. Moreiras are PIP 112-200801-00638 (2009–2011), SECYT 06/A441 (2009–2011), SECYT 06/G550 (2009–2011) and SECYT 06/A519 (2011–2013). Research by S.A. Sepulveda at the study area was mainly funded by Millenium Initiative P02-033.

References

ABELE, G. 1974. *Bergstürze in den Alpen: The Verbreitung, Morphologie und Folgeerscheinun-gen*. Wissenschaftliche Alpenvereinshefte, Heft 25, Munchen.

ABELE, G. 1984. Derrumbes de montaña y morenas en los Andes chilenos. *Revista de Geografía Norte Grande*, **11**, 17–30.

ALVARADO, P., BARRIENTOS, S., SAEZ, M., ASTROZA, M. & BECK, S. 2009. Source study and tectonic implications of the historic 1958 Las Melosas crustal earthquake, Chile, compared to earthquake damage. *Physics of the Earth and Planetary Interiors*, **175**, 26–36.

ANTINAO, J. L. 2008. *Quaternary landscape evolution of the Southern Central Andes of Chile quantified using landslide inventories, ^{10}Be and ^{36}Cl cosmogenic isotopes and (U–Th)/He thermochronology*. PhD thesis, Dalhousie University, Halifax.

ANTINAO, J. L. & GOSSE, J. 2009. Large rockslides in the Southern Central Andes of Chile (32–34.5°S): tectonic control and significance for Quaternary landscape evolution. *Geomorphology*, **104**, 117–133.

ARMIJO, R., RAULD, R. ET AL. 2010. The West Andean Thrust, the San Ramón Fault, and the seismic hazard for Santiago, Chile. *Tectonics*, **29**, doi:10.1029/2008TC002427.

BAKER, S. E., GOSSE, J. C., MCDONALD, E. V., EVENSON, E. V. & MARTINEZ, O. 2009. Quaternary history of the piedmont reach of Río Diamante, Argentina. *Journal of South American Earth Sciences*, **28**, 54–73.

CAVIEDES, C. 1972. Geomorfología del Cuaternario del valle de Aconcagua, Chile Central. *Freiburger Geographische Hefte*, **11**, 153.

CASASSA, G. & MARANGUNIC, C. 1993. The 1987 Río Colorado rockslide and debris flow, Central Andes, Chile. *Bulletin of the Association of Engineering Geologists*, **30**, 321–330.

CHARRIER, R. 1983. *Hoja El Teniente*. Carta geológica de Chile, Instituto de Investigaciones Geológicas, Santiago.

CHIU, D. 1991. *Geología del relleno Cuaternario de las hoyas de los ríos Yeso, Volcán y Maipo, este último entre las localidades de Guayacán y los Queltehues, Región Metropolitana, Chile*. BSc thesis, Departamento de Geología, Universidad de Chile, Santiago.

COSTA, C. H., GARDINI, C. E., DIEDERIX, H. & CORTÉS, J. M. 2000. The Andean orogenic front at Sierra de Las Peñas-Las Higueras, Mendoza, Argentina. *Journal of South American Earth Sciences*, **13**, 287–292.

COSTA, J E. & SCHUSTER, R L. 1988. The formation and failure of natural dams. *Geological Society of America Bulletin*, **100**, 1054–1068.

DECKART, K., PINOCHET, K., SEPÚLVEDA, S. A., PINTO, L. & MOREIRAS, S. M. 2014. New insights on the origin of the Mesón Alto deposit, Yeso Valley, central Chile: A composite deposit of glacial and landslide processes? *Andean Geology*, **41**, 248–258.

DONG, J. J., TUNG, Y. H., CHEN, C. C., LIAO, J. J. & PAN, Y. W. 2009. Discriminant analysis of the geomorphic characteristics and stability of landslide dams. *Geomorphology*, **110**, 162–171.

D'ODORICO BENITES, P. E., PÉREZ, D. J., SEQUEIRA, N. & FAUQUÉ, L. 2009. El Represamiento y aluvión del Río Santa Cruz, Andes Principales (31°40′s), Provincia de San Juan. *Revista de la Asociación Geológica Argentina*, **65**, 713–724.

DUNNING, S. A., MITCHELL, W. A., ROSSER, N. J. & & PETLEY, D. N. 2007. The Hattian Bala rock avalanche and associated landslides triggered by the Kashmir Earthquake of 8 October 2005. *Engineering Geology*, **93**, 130–144.

ERICKSON, I. & HÖGSTEDT, J. 2004. *Landslide hazard assessment and landslide precipitation relationship in Valparaiso, central Chile*. MSc thesis, Geography Department, Göteborg University, Göteborg.

ERMINI, L. & CASAGLI, N. 2003. Prediction of the behavour of landslide dams using a geomorphological dimensionless index. *Earth Surface Processes and Landforms*, **28**, 31–47.

ESPIZÚA, L. E. 1993. Quaternary glaciations in the Rio Mendoza Valley, Argentine Andes. *Quaternary Research*, **40**, 150–162.

ESPIZÚA, L. E. 1999. Chronology of Late Pleistocene glacier advances in the Río Mendoza Valley. *Argentina Global and Planetary Change*, **22**, 193–200.

ESPIZÚA, L. E. 2005. Megadeslizamientos pleistocénicos en el valle del río Mendoza, Argentina. *Proceedings of XVI Congreso geológico Argentino*, La Plata, Argentina, Tomo **3**, 477–482.

ESPIZÚA, L. E., BENGOCHEA, J. D. & AGUADO, C. 1993. Mapa de riesgo de remoción en masa en el valle del Río Mendoza. *Proceedings of XII Congreso Geológico Argentino y II Congreso de Exploración de Hidrocarburos, Mendoza, Argentina*, **VI**, 323–332.

FAUQUÉ, L., BAUMANN, V. ET AL. 2005. Evidencias de paleoendicamientos en la cuenca del río Mendoza. Provincia de Mendoza. Argentina. *Proceedings of XVI Congreso Geológico Argentino*, La Plata, Argentina, Tomo **3**, 507–514.

FAUQUÉ, L., HERMANNS, R. ET AL. 2008a. Megadeslizamientos de la pared sur del Cerro Aconcagua y su relación con la génesis del depósitos de Horcones, Mendoza, Argentina. *Proceedings of XVII Congreso Geológico Argentino*, Jujuy, 276–277.

FAUQUÉ, L., ROSAS, M., HERMANNS, R., BAUMANN, V., LAGORIO, S., WILSON, C. & HEWITT, K. 2008b. Origen y edad del depósito asignado al drift Penitentes,

Mendoza, Argentina. *Proceedings of XVII Congreso Geológico Argentino*, Jujuy, 278–279.

FAUQUÉ, L., HERMANNS, R., WILSON, C., CEGARRA, M., ROSAS, M. & BAUMMAN, V. 2008c. Paleorepresamientos del río Mendoza entre Polvaredas y Punta de Vacas, Mendoza, Argentina. *Proceedings of XVII Congreso Geológico Argentino*, Jujuy, 274–275.

FAUQUÉ, L., CORTES, J. M. *ET AL*. 2008d. Edades de las avalanchas de rocas ubicadas en el valle del Río Mendoza aguas debajo de Uspallata. *Proceedings of XVII Congreso Geológico Argentino*, Jujuy, 282–283.

FAUQUÉ, L., CEGARRA, M., WILSON, C., YAMIN, M., GAIDO, F. & ANSELMI, G. 2008e. Procesos geodinámicos en el área del proyecto hidroeléctrico Los Caracoles, San Juan, Argentina. *Proceedings of XVII Congreso Geológico Argentino, Jujuy*, 272–273.

FAUQUÉ, L., HERMANNS, R. *ET AL*. 2009a. Mega-landslide in the southern wall of Mt Aconcagua and its relationship with deposits assigned to pleistocene glaciations. *Revista de la Asociación Geológica Argentina*, **65**, 691–712.

FAUQUÉ, L., HERMANNS, R. L. & WILSON, C. G. J. 2009b. Mass removal in the Andean region. *Revista de la Asociación Geológica Argentina*, **65**, 687.

FOCK, A. 2005. *Cronología y tectónica de la exhumación en el Neógeno de los Andes en Chile central entre los 33°y 34°S. Tesis de Magíster (inédito)*, Departamento de Geología, Universidad de Chile, Santiago.

GANA, P., WALL, R. & GUTIÉRREZ, A. 1996. *Mapa Geológico del Area de Valparaíso-Curacaví*. Servicio Nacional de Geología y Minería, Mapas Geológicos 1.

GODOY, E., LARA, L. & BURMESTER, R. 1994. El 'lahar' cuaternario de Colon-Coya: Una avalancha de detritos pliocena. *Proceedings of the 7th Congreso Geológico Chileno*, Concepción, Chile, **1**, 305–309.

GONZÁLEZ DÍAZ, E. F. 2003. El englazamiento en la región de la caldera de Caviahue-Copahue (Provincia del Neuquén): Su reinterpretación. *Revista de la Asociación Geológica Argentina*, Buenos Aires, **58**, 356–366.

GONZÁLEZ DÍAZ, E. F. 2005. Geomorfología de la región del volcán Copahue y sus adyacencias (centro-oeste del Neuquén). *Revista de la Asociación Geológica Argentina*, Buenos Aires, **60**, 072–087.

GONZÁLEZ, P. 2010. *Geología y geomorfología del complejo de remoción en masa La Engorda, Chile Central*. Diploma thesis, Departamento de Geología, Universidad de Chile, Santiago.

GONZÁLEZ DÍAZ, E., GIACCARDI, A. & COSTA, C. 2001. La avalancha de rocas del río Barrancas (cerro Pelán), norte del Neuquén: su relación con la catástrofe del río Colorado (29/12/1914). *Revista de la Asociación Geológica Argentina*, **56**, 466–480.

GONZÁLEZ DÍAZ, E. F., FOLGUERA, A. & HERMANNS, R. 2005. La avalancha de rocas del cerro Los Cardos (37°10′S, 70°53′O) en la región norte de la provincia del Neuquén. *Revista de la Asociación Geológica Argentina*, Buenos Aires, **60**, 207–220.

GOSSE, J. C. 1994. *Alpine glacial reconstructions: late Quaternary geology of the Río Atuel valley, Argentine Andes, Mendoza*. PhD thesis, Lehigh University, Bethlehem, PA, USA.

GROEBER, P. 1916. Informe sobre las causas han producio las cescientes del Río Colorado (territorios del Neuquen y La Pampa) en 1914, Ministerio de Agricultura de la Nación (Argentina), Direccíon General de Minas, Buenos Aires, *Geologia e Hidrologia, Serie B (Geología), Bolletín No. 11*, 28 p. Buenos Aires.

GROEBER, P. 1933. Confluencia de los Río Grande y Barrancas (Mendoza y Neuquen), Ministerio de Agricultura de la Nacíon (Argentina), Dirección de Minas y Geología, Buenos Aires, Boletín No. 38, 72 p. Buenos Aires.

HARP, E. L. & CRONE, A. J. 2006. *Landslides triggered by the October 8, 2005, Pakistan earthquake and associated landslide-dammed reservoirs*. US Geological Survey Open-File Report **2006-1052**.

HAUSER, A. 2000. Remociones en Masa en Chile. Servicio Nacional de Geología y Minería, Boletín No. 45.

HAUSER, A. 2002. Rock avalanche and resulting debris flow in Estero Parraguirre and Río Colorado, Región Metropolitana, Chile. *In*: EVANS, S. G. & DEGRAFF, J. V. (eds) *Catastrophic Landslides: Effects, Occurrence, and Mechanisms*. Reviews in Engineering Geology, **15**, 135–148.

KHAZARADZE, G. & KLOTZ, J. 2003. Short and long-term effects of GPS measured crustal deformation rates along the South-central Andes. *Journal of Geophysical Research*, **108**, 1–13, http://dx.doi.org/10.1029/2002JB001879

KILBURN, C. R. J. & PETLEY, D. N. 2003. Forecasting giant, catastrophic slope collapse: lessons from Vajont, Northern Italy. *Geomorphology*, **54**, 21–32.

LAURO, C. 2010. *Estudio del origen y morfología de los paleo-represamientos de la quebrada Benjamín Matienzo, nacientes del río de las Cuevas, Mendoza*. Master Degree thesis. Universidad Nacional de Cuyo.

MARANGUNIC, C. & THIELE, R. 1971. Procedencia y determinaciones gravimétricas de espesor de la morrena de la Laguna Negra, Provincia de Santiago. *Comunicaciones*, **38**, 25pp.

MATEO, M. L., LENZANO, L. E. & MOREIRAS, S. M. 2009. Co. Aconcagua Geodynamics, Mendoza, Argentina: preliminary results. *Advances in Geosciences*, **22**, 167–172.

MOREIRAS, S. M. 2005. Landslide susceptibility zonation in the Rio Mendoza Valley, Argentina. *Geomorphology*, **66/1–4**, 345–357.

MOREIRAS, S. M. 2006a. Chronology of a probable neotectonic Pleistocene rock avalanche, Cordon del Plata (Central Andes), Mendoza, Argentina. *Quaternary International*, **148**, 138–148.

MOREIRAS, S. M. 2006b. Frequency of debris flows and rockfall along the Mendoza river valley (Central Andes), Argentina. *In*: PIOVANO, E. L., VILLALBA, R. & LEROY, S. A. G. (eds) Special Issue: Holocene environmental catastrophes in South America: from the lowlands to the Andes. *Quaternary International*, **158**, 110–121.

MOREIRAS, S. M. 2007. Grandes Colapsos de laderas en la Quebrada de Matienzo – Cordillera Principal, Provincia de Mendoza. *Proceedings of VI Jornadas Geológicas y Geofísicas Bonaerenses*, Mar del Plata, Argentina.

MOREIRAS, S. M. 2009. Análisis de las variables que condicionan la inestabilidad de las laderas en los valles de los ríos Las Cuevas y Mendoza. *Revista de la Asociación Geológica*, **65**, 780–790.

MOREIRAS, S. M. 2010a. Avances en el estudio geomorfológico de la quebrada de Matienzo, Mendoza. *Contribuciones Científicas GAEA*, **2**, 159–173.

MOREIRAS, S. M. 2010b. Geomorphologic evolution of the Mendoza River Valley. *In*: DEL PAPA, C. & ASTINI, R. (eds) *Field Excursion Guidebook, 18th International Sedimentological Congress*, 26 September–1 October 2010, Mendoza, Argentina, **FE-B2**, 20.

MOREIRAS, S. M. 2011. Clustering of Pleistocene rock avalanches in the Central Andes, Argentina. *En*: SALFITY, J. A. & MARQUILLAS, R. A. (eds) *Cenozoic Geology of Central Andes of Argentina*. SCS Publisher, Salta, 265–282. ISSN: 978-987-26890-0-1

MOREIRAS, S. M. & BANCHIG, A. L. 2008. Further evidences of Quaternary activity of Maradona Faulting (Precordillera Central), Argentina. *7th International Symposium on Andean Geodynamics*. Niza, Francia, 7–10 Setiembre 2008, 123.

MOREIRAS, S. M. & CORONATO, A. 2010. Landslide processes in Argentina. Natural hazards and human-exacerbated disasters. *In*: LATRUBESSE, E. (ed.), SHRODER, J. F. (series ed.) *Latin-America. Special Volume of Geomorphology: Developments in Earth Surface Processes* (1st ed). Elsevier, **13**, 301–331.

MOREIRAS, S. M., LENZANO, M. G. & RIVEROS, N. 2008a. Inventario de procesos de remoción en masa en el Parque provincial Aconcagua, provincia de Mendoza – Argentina. Multiequina. *Latin American Journal of Natural Resources*, **17**, 129–146.

MOREIRAS, S. M., OLMEDO, V. E. & DÍAZ, A. F. 2008b. Evidencias de grandes deslizamientos en la quebrada de Matienzo nacientes del Río de Las Cuevas (provincia de Mendoza). *Proceedings of XII Reunión Argentina de Sedimentología (XIIRAS)*, Buenos Aires, Argentina, 119.

MOREIRAS, S. M., GARDINI, C. & SALES, D. 2010. *Cálculo estimado de volumen del deslizamiento del sector 17, Presa potrerillos*. Informe elaborado para CEMPPSA.

MOREIRAS, S. M., LAURO, C. & MASTRANTONIO, L. 2012a. Stability analysis and morphometric characterization of palaeo-lakes of the Benjamin Matienzo Basin-Las Cuevas River, Argentina. *Natural Hazards*, **62**, 593–611, http://dx.doi.org/10.1007/s11069-012-0095-7

MOREIRAS, S. M., LISBOA, S. & MASTRANTONIO, L. 2012b. The role of snow melting upon landslides in the central Argentinean Andes. *In*: WOHL, E. & RATHBURN, S. (guest eds) *Earth Surface and Processes Landforms*. Special issue on Historical Range of Variability, Elsevier. http://dx.doi.org/10.1002/esp.3239

MORENO, H., THIELE, R. & VARELA, J. 1991. *Estudio geológico y de riesgo volcánico y de remoción en masa del proyecto hidroeléctrico Alfalfal II y Las Lajas*. Chilgener S.A., Technical Report, Santiago.

NOTT, J. 2006. *Extreme Events. A Physical Reconstruction and Risk Assessment*. Cambridge University Press.

ORMEÑO, A. 2007. *Geomorfologia dinámica del río Maipo en la zona cordillerana de Chile Central en la zona cordillerana de Chile Central e implicancias eotectónicas*. MSc thesis, Departamento de Geología, Universidad de Chile, Santiago.

PEREIRA, F. X. 1995. Esquema geomorfológico del sector norte del valle del río cuevas, entre Puente del Inca y Las Cuevas, provincia de Mednoza. *Revista de la Asociación Geológica Argentina*, **50**, 103–110.

PÉREZ, A., LEYTON, F., RAULD, R., CAMPOS, J., BARRIENTOS, S., VARGAS, G. & THIELE, R. 2010. Paper A15: Peligro sísmico en la Región Metropolitana: nuevas perspectivas en un contexto tectónico andino, caso Santiago de Chile. *X Congreso Chileno de Sismología e Ingeniería Antisísmica*, Santiago, 24–27 May 2010, paper A15.

PENNA, I. 2010. *Procesos de remoción en masa en el retroarco norneuquino (37°–38°S). Factores condicionantes y sus implicancias en el modelado del paisaje*. PhD thesis, Universidad de Buenos Aires.

PERUCCA, L. P. & MOREIRAS, S. M. 2006. Liquefaction phenomena associated with historic earthquakes in San Juan and Mendoza provinces, Argentina. Special issue: Holocene environmental catastrophes in South America: from the lowlands to the Andes. *Quaternary International*, **158**, 96–109.

PERUCCA, L. P. & MOREIRAS, S. M. 2008. Indicative structures of paleo-seismicity in the Acequión region, San Juan province, Argentina. *Geodinamica Acta*, **21**, 93–105, http://dx.doi.org/10.3166/ga

PILGER, R. H. 1984. Cenozoic plate kinematics, subduction and magmatism: South American Andes. *Journal of the Geological Society of London*, **141**, 793–802.

PMA:GCA 2007. *Movimientos en Masa en la Región Andina: Una Guía para la Evaluación de Amenazas*. Proyecto Multinacional Andino: Geociencias para las Comunidades Andinas. Servicio Nacional de Geología y Minería, Publicación Geológica Multinacional, No.4.

POLANSKI, J. 1958. El Bloque varíscico de la Cordillera Frontal de Mendoza. *Asociación geológica Argentina, Revista*, **12**, 165–193.

POSCHINGER, A. 2002. Large rockslides in the Alps: a commentary on the contribution of G. Abele (1937–1994) and review of some recent developments. *In*: EVANS, S. G. & DE GRAFF, J. V. (eds) *Catastrophic Landslides: Effects, Occurrence and Mechanisms*. Geological Society of America Reviews in Engineering Geology, Boulder, Colorado, **15**, 237–255.

RAMOS, V. A. 1996. Evolución tectónica de la alta cordillera de San Juan y Mendoza. *Subsecretaria de Minería de la Nación, Dirección Nacional del Servicio Geológico*, Anales, Buenos Aires, **24**, 447–460.

RAMOS, V. A., AGUIRRE URRETA, M. B. ET AL. 1996. Geología de la Región del Aconcagua. Provincias de San Juan y Mendoza. *Subsecretaria de Minería de la Nación. Dirección Nacional de Servicio Geológico*, Anales, Buenos Aires, **24**, 1–510.

RAMOS, V. A. 1988. The tectonics of the Central Andes; 30° to 33°S latitude. *In*: CLARK, S. P., JR, BURCHFIEL, C. & SUPPE, J. (eds) *Processes in Continental Lithospheric Deformation*. Geological Society of America, Special Paper, **218**, 31–54.

RAULD, R. 2002. *Analisis morfoestructural del frente cordillerano de Santiago Oriente, entre el río Mapocho y la Quebrada Macul*. BSc thesis, Departamento de Geología, Universidad de Chile, Santiago.

RAULD, 2011. *Deformación cortical y peligro sísmico asociado a la falla San Ramón en el frente cordillerano de Santiago, Chile Central (33°S)*. PhD thesis, Departamento de Geología, Universidad de Chile.

RAULD, R., VARGAS, G., ARMIJO, R., ORMEÑO, A., VALDERAS, C. & CAMPOS, J. 2006. Cuantificación de escarpes de falla y deformación reciente en el frente cordillerano de Santiago, *Actas XI Chilean Geological Congress*, 7–11 August 2006, Antofagasta.

RIVANO, S., SEPÚLVEDA, P., BORIC, R. & ESPIÑEIRA, P. 1993. *Mapa Geológico de la Hoja Quillota-Portillo (escala 1:250.000), V Región de Valparaíso*. Carta geológica de Chile, vol. **73**. Servicio Nacional de Geología y Minería, Santiago.

ROSAS, M., WILSON, C., HERMANNS, H., FAUQUÉ, L. & BAUMANN, V. 2008. Avalanchas de rocas de las Cuevas una evidencia de la desestabilización de las laderas como consecuencia del cambio climático del Pleistoceno superior. *Proceedings of XVII Congreso Geológico Argentino*, Jujuy, 313–314.

SALOMÓN, J. N. 1969. El alto valle del río Mendoza. Estudio de geomorfología. *Boletín de Estudios Geográficos*, **16**, 1–50.

SCHACHTER, P. 2008. *Análisis de susceptibilidad de remociones en masa con métodos estadísticos multivariados en el sector nororiente de la cuenca de Santiago*. Professional Degree Dissertation, Departamento de Geología, Universidad de Chile, Santiago.

SELLÉS, D. & GANA, P. 2001. *Geología del área Talagante-San Francisco de Mostazal*. Servicio Nacional de Geología y Minería. Carta Geológica de Chile, Serie Geología Básica, **74**.

SEPÚLVEDA, S. A. & MOREIRAS, S. M. 2013. Large volume landslides in the central Andes of Chile and Argentina (32ª–34ªS) and related hazards. *In*: GENEVOIS, R. & PRESTINZINI, A. (eds) *International Conference on Vajont 1963–2013: Thoughts and Analyses After 50 Years Since the Catastrophic Landslide. Italian Journal of Engineering Geology and Environment*, Book Series, **6**, 287–294.

SEPÚLVEDA, S. A., REBOLLEDO, S. & VARGAS, G. 2006. Recent catastrophic debris flows in Chile: geological hazard, climatic relationships and human response. *Quaternary International*, **158**, 83–95.

SEPÚLVEDA, S. A., ASTROZA, M., KAUSEL, E., CAMPOS, J., CASAS, E., REBOLLEDO, S. & VERDUGO, R. 2008. New findings on the 1958 Las Melosas Earthquake Sequence, Central Chile: implications for seismic hazard related to shallow crustal earthquakes in subduction zones. *Journal of Earthquake Engineering*, **12**, 435–455.

STERN, C. R., AMINI, H., CHARRIER, R., GODOY, E., HERVE, F. & VARELA, J. 1984. Petrochemistry and age of rhyolitic pyroclastic flows which occur along the drainage valleys of the Río Maipo and Río Cachapoal (Chile) and the Río Yaucha and Río Papagayos (Argentina). *Revista Geológica de Chile*, **23**, 39–52.

STUMPF, A. 2008. *Landslide susceptibility assessment in Central Chile. Application of probabilistic GSI-based method at eastern Santiago de Chile and the bordering Andes*. Diploma Thesis, Institut Für Geographie, Technische Universitat Dresden.

SUÁREZ, J. 1983. *Rasgos del modelado glaciario en la quebrada de Benjamín Matienzo*. Contribución al proyecto de Palinología 4.2.II.d. Editorial Inca.

THIELE, R. 1980. *Hoja Santiago*. Instituto de Investigaciones Geológicas, Carta Geológica de Chile No. 39.

WALL, R., GANA, P. & GUTIÉRREZ, A. 1996. *Mapa Geológico del Area de San Antonio-Melipilla*. Servicio Nacional de Geología y Minería, Mapas Geológicos, **2**.

WALL, R., SELLÉS, D. & GANA, P. 1999. *Area Tiltil-Santiago (1:100,000 scale map), Región Metropolitana*. Servicio Nacional de Geología y Minería, Mapas Geológicos, **11**.

WAYNE, W. J. & CORTE, A. E. 1983. Multiple glaciations of the Cordon del Plata, Mendoza, Argentina. *Palaeogeography, Palaeoclimatology, Palaeoecology*, **42**, 185–209.

WELKNER, D., EBERHART, E. & HERMANNS, R. L. 2010. Hazard investigation of the Portillo Rock Avalanche Site, central Andes, Chile, using an integrated field mapping and numerical modelling approach. *Engineering Geology*, **114**, 278–297.

WICK, E., BAUMANN, V. & JABOYEDOFF, M. 2010. Report on the impact of the 27 February 2010 earthquake (Chile, M_w 8.8) on rockfalls in the Las Cuevas valley, Argentina. *Natural Hazards and Earth System Sciences*, **10**, 1989–1993, http://dx.doi.org/10.5194/nhess-10-1989-2010

^{36}Cl terrestrial cosmogenic nuclide dating suggests Late Pleistocene to Early Holocene mass movements on the south face of Aconcagua mountain and in the Las Cuevas–Horcones valleys, Central Andes, Argentina

REGINALD L. HERMANNS[1]*, LUIS FAUQUÉ[2] & CARLOS G. J. WILSON[2]

[1]*Geological Survey of Norway, Leiv Eirikksons vei 39, 7491 Trondheim, Norway*

[2]*Servicio Geológico Minero de Argentina, 1322 Buenos Aires, Argentina*

*Corresponding author (e-mail: Reginald.Hermanns@ngu.no)

Abstract: The morphology, sedimentology and mineralogy of deposits that previously had been associated with glacial advances (the Penitentes, Horcones and Almacenes drifts) were reinvestigated and dated using the terrestrial cosmogenic nuclide (TCN) ^{36}Cl. These results indicate that the deposits previously associated with the Horcones and Almacenes drifts are actually deposits of a rock slope failure from the southern face of Aconcagua mountain forming a debris–ice avalanche that were deposited 10 490 ± 1120 years ago, while the deposits previously associated with the Penitentes drift is a rock avalanche from the Mario Ardito valley that deposited in the Las Cuevas valley 11 220 ± 2020 years ago. Earlier in the Late Pleistocene a further rock–ice avalanche sourced from Aconcagua mountain and deposited in the Las Cuevas valley, predating related lake sediments with a calibrated ^{14}C age of 14 798–13 886 years and travertine deposits with a U-series age of 24 200 ± 2000 years. In addition, three further rock-avalanche deposits were dated that sourced from Tolosa mountain, having ^{36}Cl mean ages of 14 740 ± 1950 years, 12 090 ± 1550 years and 9030 ± 1410 years. No deposits of massive rock slope failures were found in those parts of the valleys that date younger, suggesting that climatic conditions at the transition from the Late Pleistocene to the Holocene, that were different from today's, caused the slopes to fail. Alternatively, the rock slope failures could have been seismically triggered. We suggest that the slope failures at the southern face of Aconcagua mountain have caused or contributed to a reorganization of glacial ice flow from Aconcagua mountain that might ultimately be the cause of the surging behaviour of the Horcones Inferior glacier today. Our results indicate that the glacial stratigraphy of this part of the Central Andes is still poorly understood and requires detailed mapping and dating.

Supplementary material: Sample coordinates, sample porosity and density, Cl nuclide composition and geochemical composition are available at http://www.geolsoc.org.uk/SUP18753.

Gold Open Access: This article is published under the terms of the CC-BY 3.0 license.

The impact of climate warming on slope stability, mainly through thaw of alpine permafrost and debuttressing of glacially oversteepened, unstable rock slopes due to glacier retreat, has been the subject of much discussion (Abele 1974; Evans & Clague 1994; Noetzli *et al.* 2007; Huggel *et al.* 2010; Clague *et al.* 2012; Fischer *et al.* 2012). Except for the historic period, however, this causative relationship can only be demonstrated by dating large numbers of rock slope failures, although care has to be taken as both climate-related conditioning and seismic activity can result in multiple rock slope failures in a given region. In general multiple rock slope failures that occurred at the same time in the past are interpreted as indicators for palaeoseismic events, especially if additional independent indicators for such events exist (e.g. Adams 1981; Schuster *et al.* 1992; Moreiras 2006; Hermanns & Niedermann 2011). On the other hand, temporal clustering of rock slope failures coinciding or following climatic changes are commonly interpreted as an indication that stability conditions of rock slopes are primarily linked to climate change. Statistically representative examples have been published from Norway, the European Alps and Scotland (e.g. Prager *et al.* 2008; Soldati *et al.* 2004; Hermanns & Longva 2012; Ostermann & Sanders 2012; Ballantyne & Stone 2013, respectively).

In the Andes the first systematic regional inventory of ages of rock slope failures was presented by Hermanns *et al.* (2000) for NW Argentina. Previously 25 of 55 mapped deposits had been dated; this dataset was extended by Hermanns & Schellenberger (2008) to 33 dated deposits. None of the mapped rock slope failures had a source on glaciated slopes, even though alpine glaciers reached down to

4300 m above sea-level (asl) in the easternmost ranges during the Late Pleistocene (Haselton *et al.* 2002). Landslides occurred in the incised valleys mainly during periods of wetter climate; it is likely that enhanced runoff and lateral erosion of valley floors were the main causes of rock slope failures (Trauth *et al.* 2000; Hermanns & Schellenberger 2008). However, some large landslides apparently occurred during what are thought to have been dry phases of the Holocene. These exceptions occur near active faults and are interpreted to have been seismically triggered (Hermanns & Schellenberger 2008; Hermanns & Niedermann 2011).

At the transition between the Central and Patagonian Andes rock slope failures have occurred on a Plio-Pleistocene volcanic plateau incised by fluvial and glacial erosion by 200–1200 m deep valleys (Penna *et al.* 2011). About 80 percent of the landslide deposits occur in sections of the valleys that were eroded by glaciers and have higher local relief, indicating that glacial-toe erosion and debuttressing were important factors in conditioning the slopes for failure (Penna *et al.* 2011). However, because all of the rock slope failures predate the time of maximal glaciation in the area by more than 10 000 years, glacial erosion and debuttressing were conditioning, and not triggering, factors.

Due to similar landforms, large volumes, large aerial distribution, strong erosion and hence discontinuous deposits, deposits of rock slope failure are often misinterpreted as glacial landforms (e.g. Hewitt 1999; González Díaz 2003; Hewitt 2006). To recognize rock slope failures that fall onto glaciers is even more difficult, as the rock mass interacts with the glacial ice resulting in both run-out distances that are far beyond those of rock slope failures onto solid ground, and secondary processes once the ice melts within the sliding mass (Hauser 2002; Huggel *et al.* 2005; Evans *et al.* 2009). This has led to the misinterpretation of glacial advances and mountain valley evolution in the Himalaya and the Andes (e.g. Hewitt *et al.* 2011; González Díaz 2003, respectively). In the Central Andes the glacial advances were predominantly defined based on deposits in the Horcones, Las Cuevas and Mendoza valleys below Aconcagua mountain. With an altitude of 6959 m Aconcagua is the highest mountain in the Andes; due to their relatively easy access, the glacial deposits in Horcones, Las Cuevas and Mendoza valleys (Fig. 1) have been well studied (Espizua 1993; Espizua & Bigazzi 1998; Espizua 1999). However, the origin of the supposed glacial deposits has been under dispute for a long time (Pereyra & González Díaz

Fig. 1. The study area lies in Argentina at the border with Chile and includes the Horcones and Las Cuevas river catchments down to the confluence with the Tupungato river where both rivers form the Mendoza river. The box indicates the area shown in Figure 2.

1993). New sedimentological and geochronological data on those deposits suggest that a glacial origin is indeed an incorrect interpretation (Fauqué et al. 2009).

The glacial deposits of the Horcones, Las Cuevas and Mendoza valleys, although poorly constrained in age, are the reference section for glacial chronology in the Andes at c. 35°S (e.g. Heusser 2003; Zech et al. 2008). Zech et al. (2008) show that glacial advances in this area are asynchronous with glacial advances towards the north and south and suggest that this indicates that glaciations in the Aconcagua region are sensitive to precipitation while those further north and south are sensitive to temperature. We combine here data published in Spanish (Fauqué et al. 2009) with unpublished data to show that multiple deposits previously interpreted as glacial end moraines are massive landslide deposits and that therefore interpretation of the sensitivity of these deposits to various driving forces for glaciation must be made with care. Our data rather suggest that the timing of glacial advances in the area is not understood today and that an extensive effort needs to be made to establish a well-constrained glacial history of the Central Andes at 35°S. In turn this would also allow the driving forces of rock slope failures in this part of the Andes to be better constrained.

Geological setting

The study area lies along the Las Cuevas and Horcones valleys that lie at an elevation between 2500 m asl at the confluence of the Las Cuevas and the Tupungato rivers and the summit of Aconcagua mountain at 6959 m asl (Figs 1 & 2a). The study area does not include other valleys draining the Aconcagua massif that also contain deposits of large rock slope failures (e.g. Moreiras et al. 2008). The climate zones stretch from Tundra between 2700 and 4100 m altitude to Polar at altitudes above 4100 m, where mean monthly temperatures are below 0 °C (Minetti 1985); precipitation falls between May and October, mostly in the form of snow (Minetti 1986). The Las Cuevas and Horcones rivers are fed mainly by snowmelt and meltdown of glacial ice in the summer months. Glaciers exist in the study area on the south face of Aconcagua mountain and the lower Horcones valley.

This area lies at the transition from the Main Cordillera, characterized by Jurassic–Tertiary rocks, to the Frontal Cordillera, characterized by Permo-Triassic rocks (Giambiagi et al. 2003; Fig. 2b). The Main Cordillera is a thin-skinned thrust belt that exposes Jurassic–Tertiary clastic and volcaniclastic sediments and carbonates as well as Tertiary volcanic rocks (Ramos et al. 1996). The Frontal Cordillera is composed of Permo-Triassic volcanic rocks and Jurassic intrusive rocks (Ramos et al. 1996). Within the Main Cordillera is the Aconcagua massif, built up of dacitic to andesitic volcanic rocks of Tertiary age. The remaining units are volcanic, volcaniclastic, clastic and carbonate rocks of Jurassic–Cretaceous age. These lithological differences are an important tool for determining the source area of landslides in the area that have travelled a long distance. Quaternary deposits are fluvial and glacial valley-fill deposits, moraines of previous glaciations, rock glaciers and deposits of various types of mass movements.

In this study, most of the glacial and mass-movement deposits were mapped out in the Horcones and the upper and lower Las Cuevas valleys (Fig. 3). All large landslide deposits have been sampled for age determination. Our earlier detailed mapping and dating (Fauqué et al. 2009) have shown that various deposits that had previously been interpreted as being of glacial origin based on relatively scarce chronological data with huge uncertainties (Espizua 1993), are instead deposits of large mass movements while others indeed have a typical glacial morphology and are therefore of glacial age. This is suggested at least at two localities where lateral moraines were dated with single samples (Fauqué et al. 2009). Although this is not a statistically representative number of samples, these results give a first indication that the ages of these lateral moraines are c. 14 000 and 16 500 years old.

Methods

In this section, we first describe the morphology in connection with the lithology/mineralogy of the deposits. We then present multiple terrestrial cosmogenic nuclide (TCN) ages obtained with ^{36}Cl and accelerating mass spectrometry (AMS) ^{14}C ages of stratigraphically related deposits (Fauqué et al. 2009), as well as ^{36}Cl ages of multiple other large landslide deposits originated from massive rock slope failures in these valleys.

Mineralogical analyses

Mineral compositions (excepting clays) were determined by the analysis of 2–3 g (taken from 80 g of sample material, sieved to <35 μm) by X-ray powder diffraction using a D5000 diffractometer (Bruker AXS) at the Geological Survey of Canada Laboratory in Ottawa. Cu radiation and a secondary graphite monochromator were used. The diffraction data were collected from 4 to 70° 2θ with a step width of 0.02° and a counting time of 2 s per step. The generator settings were 40 kV and 30 mA. Quantitative phase analysis was determined

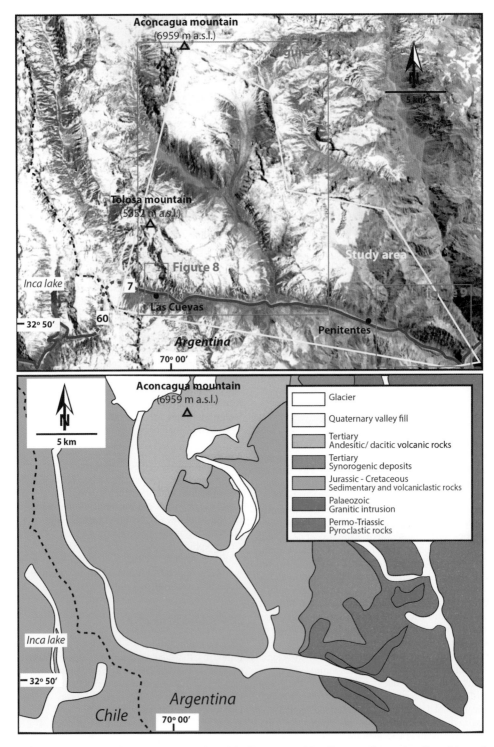

Fig. 2. (**a**) Satellite image of Aconcagua mountain and the Horcones and Las Cuevas valleys and outline of the study area. The boxes indicate the areas shown in Figures 3, 8 and 9. (**b**) Greatly simplified geological map of the study area showing main lithological differences between Aconcagua mountain and the Horcones and Las Cuevas valleys.

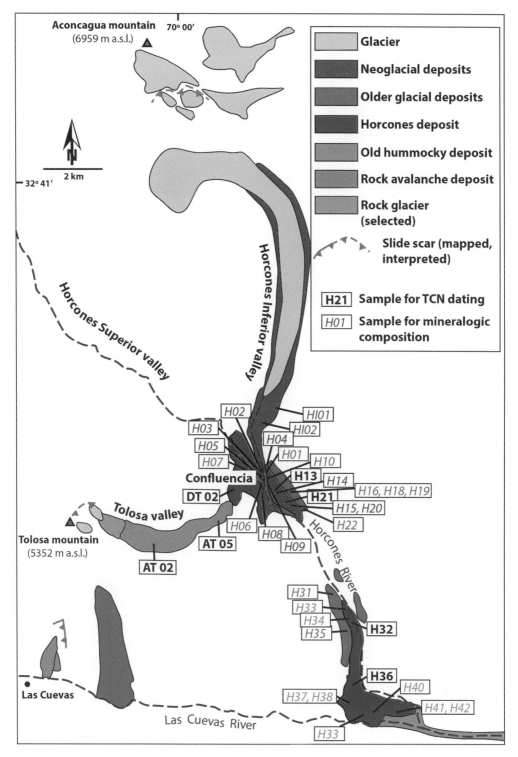

Fig. 3. Simplified and extended quaternary geological map after Fauqué *et al.* (2009) showing the distribution of selected quaternary deposits and sample sites for TCN dating and determination of mineralogical composition.

	HI 01	HI 02	H 01	H 02	H 03	H 04	H 05	H 06	H 07	H 08	H 09	H 10	H 14	H 15	H 16	H 18	H 19	H 20	H 22
Calcite	8.6	5.8	8.1					8.4	8.3	2.9		7.6			4.4	7.7	6.5		33.3
Dolomite																	3.3		5.2
Quartz	12.3	10.4	14.8	6.3	10.2	8.5	11.8	21.5	14.7	8.5	15.6	15.7	13.8	16.9	11.5	18.2	22.9	8.4	17.5
Plagioclase	58.7	59.4	43.2	70.6	79.4	70.9	77.2	58.6	49.9	77.4	72.6	46.9	61.2	68.8	50.9	44.3	40.0	78.3	32.7
Orthoclase			5.8																
Tremolite	6.6	4.6		8.2	9.2	6.5	4.4			6.9	6.7	1.4			1.6	1.0		9.1	
Clinochlore	12.0	2.9	6.7	13.7		12.6	4.1		17.8	2.8	1.7	6.1	11.8	9.0	7.2	6.0	7.0	3.1	7.0
Muscovite		4.2	7.1					6.7				17.8			15.4	17.9	16.3		2.9
Hastingsite			0.5																
Gypsum		1.2							3.8										
Hemetite	1.8	2.7	3.8	1.2	1.3	1.5	2.6	4.7	5.5	1.5		3.3	2.5	3.2	2.1	3.4	3.9	1.1	2.4
Ilmenite																			
Alunite											1.6		2.1	2.1	1.2	1.4			
Laumontite		9.0	10.0									1.2			5.5				

(a)

Fig. 4. Mineralogic composition in percent of sedimentological samples of supraglacial deposits (orange), basal and marginal moraine deposits (red), of the Horcones rockslide deposit in Confluencia (blue, see Fig. 4b), as well as of grey domains of the Horcones deposit in the confluence of the Horcones and Las Cuevas valleys (light blue, see Fig. 4b) and red domains in the same deposit (light red).

(b)

	HI 31	HI 33	H 34	H 35	H 37	H 38	H 39	H 40	H 41	H 42
Calcite	10.5	4.1		14.6		14.2			3.5	5.1
Dolomite	4.3			2.2						
Quartz	37.9	7.1	14.6	20.0	15.1	14.2	12.5	9.5	9.1	5.1
Plagioclase	36.7	60.7	72.1	49.9	68.1	58.8	72.9	74.6	76.5	55.9
Orthoclase										
Tremolite		6.0	7.0	1.9	9.1		8.2	11.3	8.4	5.5
Clinochlore	3.5	11.0	2.0		3.6		5.7	3.9	1.5	5.8
Muscovite	4.2	3.2		4.1		4.5				9.0
Hastingsite						4.4				
Gypsum						1.4				
Hemetite	2.9	2.2	2.0	2.2	1.7	2.1	0.7	0.7	1.1	3.8
Ilmenite						0.4				
Alunite		1.6								0.7
Laumontite		4.3	2.4		2.3					9.0

Fig. 4. *Continued.*

using the Rietveld analysis technique contained in the BGMN/AUTOQUAN software package (Bergmann *et al.* 1998) and results are given in percent (Fig. 4).

Method of TCN dating using ^{36}Cl

As part of the Multinational Andean Project: Geosciences for Andean Communities feasibility study on reopening the trans Andean railroad connecting Mendoza with Santiago, we collected samples for ^{36}Cl dating in order to help estimate the rock-avalanche hazard in the area. Given good preservation of the deposits, surface exposure dating with cosmogenic nuclides is an excellent method for obtaining the age of a rock-avalanche deposit, and eliminates the need for stratigraphic interpretation. Excellent summaries of this method were published by Lal (1991), Cerling & Craig (1994), Kurz & Brooke (1994) and Gosse & Phillips (2001) and the method has been repeatedly used for dating mass movements (e.g. Ballantyne *et al.* 1998; Hermanns *et al.* 2001, 2004; Bigot Cormier *et al.* 2005; Dortch *et al.* 2009; Antinao & Gosse 2009; Welkner *et al.* 2010; Blais-Stevens *et al.* 2011, and references therein). Here, we briefly review only the methods of sampling, principles of the method and correction factors, as they require some special considerations.

In selecting the boulders to be sampled, the geomorphological environment and surface characteristics were examined with great care. We selected boulders ranging from 1 to 30 m in diameter located in the central part of the deposit, away from any steps in the valley relief as well as from the failure surfaces of the Las Cuevas rock avalanches. We sampled when possible the uppermost 3–5 cm from central, horizontal surfaces of the boulders with a hammer and chisel to minimize any effects of boulder morphology on the resulting age (Masarik *et al.* 2000). We measured the shielding of the topography in steps of 30°. The local elevation was taken with a GPS and altimeter calibration in the morning. We therefore believe that the accuracy is within a margin of error of 20 m.

Sample density was measured in the Geological Survey of Canada Laboratory. Samples were prepared for isotopic concentration measurements at the PRIME Laboratory (Purdue University). Geochemical composition of rocks was analysed in

Table 1. *Results of ^{36}Cl cosmogenic nuclide ages of deposits in the Horcones and lower Las Cuevas valleys with different erosion rates*

	Erosion rate (mm thousand years^{-1})								
	0.00	0.56	1.11	1.67	2.22*	2.78	3.33	3.89	4.44	5.00
AT02										
Statistical mean	8890	8760	8650	8540	**8450**	8360	8270	8190	8120	8050
±σ_2	550	540	520	510	**500**	490	480	480	470	460
±Snow cover	1400	1400	1300	1300	**1300**	1200	1200	1200	1200	1200
AT05										
Statistical mean	10 050	9920	9810	9700	**9600**	9510	9430	9350	9280	9210
±σ_2	560	540	530	520	**510**	500	500	490	480	480
±Snow cover	600	600	500	500	**500**	500	500	500	500	500
DT02										
Statistical mean	9960	9930	9900	9870	**9850**	9820	9800	9780	9760	9740
±σ_2	830	830	820	820	**820**	810	810	810	800	800
±Snow cover	600	600	600	600	**600**	600	600	600	600	600
H13[†]										
Statistical mean	8300	8260	8230	8200	**8170**	8140	8120	8100	8070	8050
±σ_2	750	740	730	730	**720**	720	720	710	710	710
±Snow cover	500	500	500	500	**500**	500	500	500	500	500
H21										
Statistical mean	10 710	10 660	10 600	10 550	**10 510**	10 470	10 430	10 390	10 360	10 320
±σ_2	660	650	640	640	**630**	630	630	620	620	620
±Snow cover	600	600	600	600	**600**	600	600	600	600	600
H32										
Statistical mean	11 390	11 310	11 240	11 170	**11 110**	11 050	11 000	10 950	10 910	10 870
±σ_2	160	160	160	160	**160**	150	150	150	150	150
±Snow cover	700	700	600	600	**600**	600	600	600	600	600
H36[†]										
Statistical mean	8930	8850	8780	8710	**8640**	8580	8520	8470	8420	8370
±σ_2	710	700	690	680	**670**	660	650	650	640	630
±Snow cover	500	500	500	500	**500**	500	500	500	500	500
PE01										
Statistical mean	15 200	14 800	14 450	14 150	**13 890**	13 650	13 430	13 240	13 070	12 910
±σ_2	1540	1460	1400	1350	**1300**	1270	1240	1210	1190	1170
±Snow cover	1100	1000	1000	1000	**900**	900	900	900	800	800
PE02										
Statistical mean	11 020	10 900	10 800	10 710	**10 620**	10 540	10 470	10 400	10 330	10 270
±σ_2	1200	1180	1160	1140	**1120**	1110	1090	1080	1070	1060
±Snow cover	800	800	800	700	**700**	700	700	700	700	700
PE03										
Statistical mean	12 720	12 460	12 220	12 010	**11 820**	11 650	11 500	11 360	11 230	11 110
±σ_2	1610	1540	1490	1450	**1410**	1370	1340	1320	1300	1280
±Snow cover	900	900	800	800	**800**	800	800	700	700	700

*We assume that an erosion rate of 2.2 mm thousand years^{-1} is most representative. Ages are given with 2 sigma analytical uncertainty and an additional uncertainty based on snow cover of boulders.
[†]Not considered in calculation of the ages as interpreted too young due to boulder rotation (see text).

XRAL Laboratories (SGS Canada Inc.). Following these results, ^{36}Cl ages were calculated using a Microsoft Excel spreadsheet, CHLOE31, published by Phillips & Plummer (1996). One of the major uncertainties on the ages in this region is the estimation of snow cover. We used oral reports from park rangers of the Aconcagua Provincial Park, army employees from the Puente del Inca post, and the Argentine road authorities who also frequently transit the area in wintertime. Besides, this is an area of snow drift so that we can consider that the snow cover on large boulders is greater than that on top of smaller boulders. Taking into account these differences, we estimate an additional

Table 2. Results of ^{36}Cl cosmogenic nuclide ages of deposits in the upper Las Cuevas valley with different erosion rates

	Erosion rate (mm thousand years^{-1})									
	0.00	0.56	1.11	1.67	2.22*	2.78	3.33	3.89	4.44	5.00
CU0101										
Statistical mean	13 160	12 970	12 810	12 660	**12 530**	12 400	12 290	12 190	12 100	12 020
$\pm \sigma_2$	730	710	690	680	**670**	660	650	640	630	630
$\pm$ Snow cover	800	700	700	700	**700**	700	700	700	600	600
CU0102										
Statistical mean	12 630	12 420	12 240	12 070	**11 920**	11 780	11 650	11 540	11 430	11 330
$\pm \sigma_2$	670	650	630	610	**600**	590	580	570	560	560
$\pm$ Snow cover	600	600	500	500	**500**	500	500	500	500	500
CU0201										
Statistical mean	14 880	14 660	14 470	14 300	**14 140**	14 000	13 880	13 770	13 660	13 570
$\pm \sigma_2$	930	900	880	860	**850**	830	820	810	810	800
$\pm$ Snow cover	400	300	300	300	**300**	300	300	300	300	300
CU0202†										
Statistical mean	12 130	11 870	11 650	11 450	**11 270**	11 110	10 960	10 820	10 700	10 580
$\pm \sigma_2$	750	720	690	670	**650**	640	620	610	600	590
$\pm$ Snow cover	500	500	500	400	**400**	400	400	400	400	400
CUA01										
Statistical mean	12 400	12 240	12 090	11 960	**11 840**	11 730	11 630	11 540	11 460	11 380
$\pm \sigma_2$	1220	1190	1160	1140	**1120**	1100	1090	1080	1070	1060
$\pm$ Snow cover	1200	1100	1100	1100	**1100**	1000	1000	1000	1000	1000
CUA02										
Statistical mean	16 260	15 980	15 730	15 520	**15 330**	15 160	15 010	14 870	14 750	14 650
$\pm \sigma_2$	1500	1450	1410	1370	**1350**	1320	1300	1290	1280	1260
$\pm$ Snow cover	1500	1500	1400	1400	**1400**	1300	1300	1300	1300	1300
CU3										
Statistical mean	17 180	16 750	16 390	16 080	**15 810**	15 570	15 360	15 170	15 000	14 850
$\pm \sigma_2$	1730	1640	1580	1530	**1490**	1450	1420	1400	1380	1360
$\pm$ Snow cover	1700	1600	1600	1500	**1500**	1400	1400	1400	1400	1400
CU401										
Statistical mean	15 340	14 990	14 680	14 420	**14 190**	13 980	13 790	13 620	13 470	13 330
$\pm \sigma_2$	2160	2070	1990	1930	**1880**	1840	1800	1770	1740	1720
$\pm$ Snow cover	1400	1300	1300	1200	**1200**	1200	1200	1100	1100	1100
CU402										
Statistical mean	9700	9530	9380	9240	**9120**	9000	8890	8790	8700	8610
$\pm \sigma_2$	1460	1410	1370	1330	**1300**	1270	1240	1220	1200	1180
$\pm$ Snow cover	1200	1100	1100	1100	**1000**	1000	1000	1000	900	900

*We assume that an erosion rate of 2.2 mm thousand years^{-1} is most representative. Ages are given with 2 sigma analytical uncertainty and an additional uncertainty based on snow cover of boulders.
†Not considered in calculation of the ages as interpreted too young due to boulder rotation (see text).

uncertainty factor to the statistical uncertainty of analytical results. This uncertainty reflects the variation of snow cover as an estimate of the uncertainty resulting from the size of boulders in relation to the average snow cover in the area and the effect of wind on large boulders. This is necessary because we cannot know if a boulder of 5–10 m length and 10–15 m width is indeed snow-free if it overtops the surrounding area by the snow depth. Thus, we estimate an uncertainty due to snowdrifts by calculating the age without snow cover and a maximum estimated snow cover (Tables 1 & 2) following the principle outlined in Blais-Stevens et al. (2011). The resulting difference is significant and amounts to as much as 15 percent of the age. However, if snow cover had been more pronounced in the past than today or vice versa this would apply to all samples, hence the relative age difference is mainly expressed by the analytical uncertainty.

In the summary of the results in Tables 1 and 2 we present different ages corresponding to an assumption that there was no erosion of the boulders, and that there were different erosion rates up to 5 mm ka^{-1}. We interpret the ages marked in bold (erosion rate of 2.22 mm ka^{-1}) to be those closest to the true age of exposure. These erosion rates are similar to erosion rates used by Kaplan et al. (2004) and Costa & González Díaz (2007) in TCN studies, further south in the Argentinean Andes.

Results

Deposits at Confluencia, the confluence of the Horcones Superior and Inferior valleys

The Horcones Inferior glacier is a glacier which has had multiple surges in the past decades (e.g. Milana 2007; Fauqué et al. 2009; Pitte et al. 2009). At present the glacial terminus is c. 1.5 km NE of the Confluencia area (Fig. 3). However, several well-preserved interleaved lateral and frontal moraines up to 15 m in thickness indicate that the Horcones Inferior glacier has reached down to Confluencia in the recent past (Espizua 1993, Fig. 5). The mineralogical composition of supraglacial deposits of the Horcones glacier and the neoglacial deposits (samples HI 01, HI 02, and H 06 respectively, Fig. 4) were taken as representative samples for glacial deposits. These have a composition of 58 percent plagioclase, 10–20 percent quartz, 6–8.5 percent calcite as well as various amounts of tremolite, clinochlore, muscovite, gypsum, laumontite and 1.8–4.7 percent hematite, the latter giving the reddish colour to those deposits.

On the east side of the Horcones valley occurs a lateral moraine that is several tens of metres high and stretches over 2 km (Fig. 3). Espizua (1993) and Fauqué et al. (2009) agree upon the glacial origin of the deposit and Espizua (1999) maps this deposit as the Almacenes lateral moraine. She assigns a maximum age of 15 000 ± 2100 years of underlying fluvial sediments to this unit that is in agreement with a single ^{36}Cl age obtained from a boulder, (which suggests that this moraine is 13 900 ± 2200 years old; see Fauqué et al. (2009)). The mineralogical composition and the colour of the lateral moraine (samples H 16, H 18, and H 19 in Fig. 4) are similar to the neoglacial deposits but on average 10 percent lower in plagioclase, 5 percent higher in quartz and higher in muscovite.

In vertical cuts it is visible that the valley floor of Confluencia is furthermore covered by deposits of the same reddish colour (Figs 5 & 6). These deposits are a polymict, matrix-supported conglomerate, with grain sizes rarely exceeding a few tens of centimetres in diameter (lower part of photo in Fig. 5). The mineralogical composition of this basal deposit (samples H 01, H 07 and H10 in Fig. 4) is identical with the lateral moraine deposit. The deposit is interpreted by Espizua (1993) and Fauqué et al. (2009) as basal moraine. However, Espizua (1999) dates the deposit by thermoluminescence (TL) of quartz grains extracted from overlying fluvial deposits, up to 30 cm thick (photo in Fig. 5). These grains are older than 31 000 ± 3100 years.

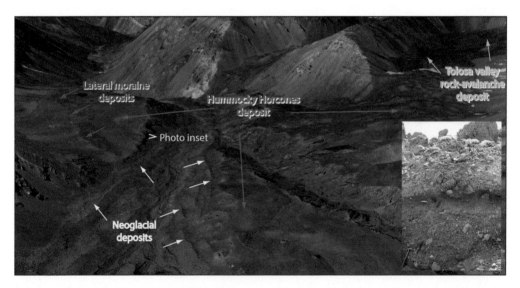

Fig. 5. Oblique satellite view towards SSW showing the Confluencia area (see Fig. 3 for location) with distribution and character of various deposits. Note the lower Aconcagua base camp east of the lateral moraine for scale. The location of the photo in the lower right corner is given by the open triangle and shows a lower reddish matrix-supported conglomerate covered by 30 cm fine-grained fluvial deposits in turn covered by a grey breccia that is matrix-supported in the lower part and clast-supported in the upper part (the hammer for scale is 29 cm high).

Fig. 6. Photo of site for sample AT02 over Quebrada del Tolosa rock avalanche towards hummocky deposit. Note that in the gorge eroded by Horcones Superior river it is visible that the greyish deposit is underlain by a reddish basal deposit.

Fauqué *et al.* (2009) AMS ^{14}C dated the same fluvial deposits with a whole organic content sample to a calibrated age of 13 543–12 098 years BP. It is unlikely that this fluvial deposit spans an interval of c. 18 kyr. However, based on missing additional ages for the underlying basal moraine it is not possible to determine whether (1) the TL age is too old and the quartz was not entirely bleached during transport or (2) the organic content of the fluvial deposit got contaminated by modern root penetration of the several-metre-thick overlying deposit. Alternatively, the deposit might have been formed at 31 000 ± 3100 years and has been on the surface until 13 543–12 098 years BP when covered by the overlying deposits.

Between the lateral moraine and the opposite valley side, the valley is entirely filled by a hummocky deposit. Hummocks are up to 20 m high with a high concentration of large boulders on their surfaces. The hummocky deposit also intrudes the Quebrada del Tolosa valley for a length of 1000 m stretching along the valley up to an altitude of 3570 m, which is 150 m above the elevation of the Confluencia area. In vertical cuts it is visible that this deposit is composed of grey breccias of several metres to several tens of metres thickness that are matrix-supported in its lower part and clast-supported in its upper part. The mineralogical composition (H 02, H 03, H 04, H05, H 08, H 09, H 14, H 15 in Fig. 4) is characterized by 60–80 percent plagioclase, 6–17 percent quartz, as well as tremolite, clinochlore and hematite. Calcite and muscovite are nearly absent from the deposit.

Espizua (1993, 1999) maps this deposit as the Almacenes moraine, and describes it as indistinguishable from the lateral moraine in Confluencia. We took three samples for TCN dating of the deposit using ^{36}Cl (samples H 13, H 21, DT 02 in Table 1). Ages obtained vary between 8170 ± 1220 and 10 510 ± 1230 years.

Further inside the Quebrada del Tolosa there is a massive deposit with a lobate form composed entirely of a volcanic conglomerate rock. The entire deposit is covered by a carapace of large boulders that vary in size between a few metres to tens of metres (Figs 3 & 6). The deposit spans two-thirds up the Tolosa valley, has lateral and frontal rims, and is thus mapped as a rock-avalanche deposit. The upper third of the Tolosa valley is covered by a rock glacier with typical concentric rings covering most of the deposit. It has ice close to the surface and is therefore interpreted as being active. The rock glacier connects to a niche in Tolosa mountain situated NE from the top that is today occupied by a glacier. This niche is interpreted as representing the scar area of the rock avalanche (Fig. 3).

These deposits have not been described previously. We dated two samples of selected boulders AT 02 and AT 05 that resulted in ages of 8450 ± 1850 and 9600 ± 1010 years, respectively.

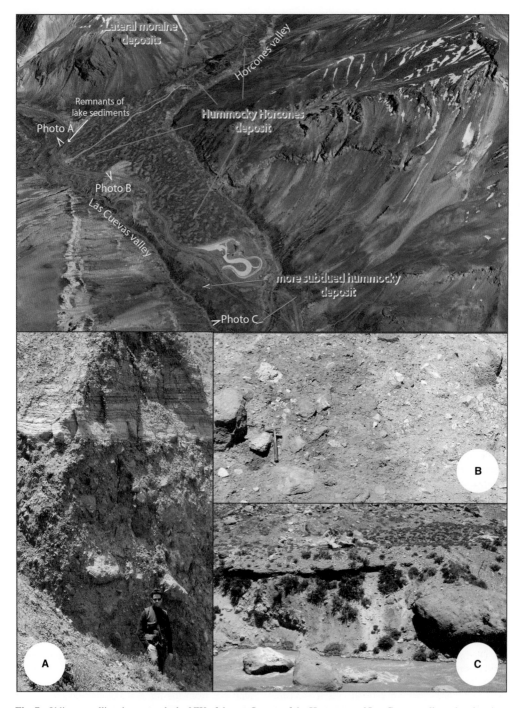

Fig. 7. Oblique satellite view towards the NW of the confluence of the Horcones and Las Cuevas valleys showing the hummocky deposit dominating the area. On the west side of the Horcones valley a lateral moraine deposit has an elevation 100 m higher than the Horcones valley. In the foreground there is a more subdued hummocky deposit, mainly covered by scree deposits. Note for scale that the lake at the foot of the moraine is 200 m long. Photo A shows breccias with greyish and reddish domains overlain by lacustrine deposits (person for scale is 175 cm tall). Photo B shows breccias of the hummocky deposit that have reddish domains with polymict clasts that are more rounded and a greyish

Deposits at the confluence of the Horcones and Las Cuevas valleys

In the area of the confluence of the Horcones and Las Cuevas valleys, a prominent lateral moraine exists on the NW slope of the Horcones valley (Fig. 3). The top of the moraine is c. 100 m higher than the valley infill. On the opposite valley side only patches of this moraine are preserved (Figs 3 & 7). This lateral moraine in the Horcones valley had not been dated but was associated by Espizua (1999) with a glacial advance called the Punta de Vacas moraine that has its frontal moraine further downstream in the Mendoza valley. There the deposits overlie alluvial fan sediments that contain a tephra layer dated by fission-track on glass shards to 134 000 ± 32 000 years. Fauqué et al. (2009) obtained a ^{36}Cl age on one boulder from that deposit of 16 510 ± 2110 years. The deposit has a reddish colour and is composed of 36–50 percent plagioclase, 20–38 percent quartz, 15–17 percent carbonates, 4 percent muscovite, 2–3 percent hematite as well as tremolite and clinochlore (samples H31 and H35 in Fig. 4). Although otherwise indicated on an overview map (Espizua 1993) no lateral moraines exist at this confluence inside the Las Cuevas valley (Rosas et al. 2007).

Similar to the valley fill at Confluencia, the valley floor here is also filled by a hummocky deposit (Figs 3 & 7). The deposit spans the entire Las Cuevas valley and thus caused the damming of the valley as indicated by lacustrine sediments directly overlying the rim of the hummocky deposit (Espizua 1993; Fauqué et al. 2009). Furthermore the deposit forms a lobe downriver within the Las Cuevas valley that is 2 km long and ends in an abrupt 20 m high front. Downriver more subdued hummocks continue that are strongly covered by rock-fall cones from the side of the valley. This surface is cemented by a travertine layer.

This hummocky deposit was previously interpreted as a frontal moraine (Espizua 1993, 1999) of a separate glacial advance (the Horcones moraine) but was reinterpreted as the deposit of a mass movement (Fauqué et al. 2009). The deposit has large boulders up to several metres in diameter on the surface. In erosional cuts along the Horcones and Las Cuevas rivers as well as along the road it is visible that the deposit is patchy with greyish and reddish domains. While the reddish domains are in general polymict, the greyish domains are rather monomict. The deposit is a matrix-supported breccia. The transition between the domains is often sharp (Fig. 7b) and the deposit was sampled for its mineralogy in both the greyish and reddish domains (light blue and light red respectively in Fig. 4, H 37–H 42). While the greyish domains have a higher concentration of plagioclase, tremolite and clinochlore, with carbonates and muscovite being nearly absent, the reddish domains in contrast contain both carbonates and muscovite as well as a higher concentration of hematite. Also, the deposit below the travertine layer is a breccia containing boulders more than 10 m in diameter and has the same greyish and reddish domains. Separated outcrops of the deposit can be found along the gorge of the Las Cuevas river down to its confluence with the Tupungato river, where the Mendoza river forms. Similar deposits can also be found locally in patches several hundred metres long along the Mendoza river (Fig. 2); however, the end of this older deposit is difficult to determine as side valleys have not been investigated and the deposits within Mendoza river could easily also have sourced from any other side valleys. To map out the lower limit of this lower deposit, detailed mapping of valley-fill deposits in all side valleys of the Mendoza river valley has to be carried out.

Espizua (1999) dated the travertine layer underlying the hummocky deposit in the Las Cuevas valley with U-series ages to 24 200 ± 2000 years and 22 800 ± 3100 years. A further travertine layer overlying the hummocky deposit was dated to 9700 ± 5000 years (Bengochea et al. 1987). This is congruent with AMS ^{14}C ages on whole organic carbon of the lake sediments that were deposited in the Las Cuevas valley that were dated to whole organic carbon to calibrated ages of 14 798–13 886 years BP obtained from sediments close to the bottom of the lacustrine sequence and of 8620–8254 years BP close to the top of this unit (Fauqué et al. 2009). Two samples were taken for ^{36}Cl dating of boulders from the top of the younger pristine hummocky deposit (H 32 and H 36 in Table 1). Ages obtained are 11 110 ± 760 years and 8640 ± 1170 years and therefore do not coincide within uncertainties.

Rock-avalanche deposits within the Las Cuevas valley

Rock-avalanche deposits at Las Cuevas. Ten kilometres upriver from the confluence of the Horcones and the Las Cuevas valleys massive deposits of mass movements exist at the Las Cuevas locality

Fig. 7. (*Continued*) domain that is mainly composed of the same lithology (hammer for scale is 29 cm high). Photo C shows the lower more subdued hummocky deposit. The top of that deposit is sealed by travertine. Also this breccias is patchy with greyish and reddish domains. The height of the travertine layer above the river is 18 m.

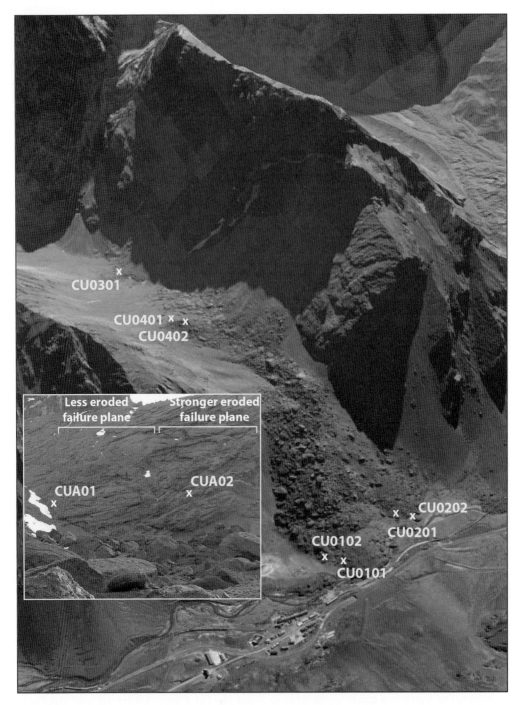

Fig. 8. Oblique satellite view towards NE showing the locality of Las Cuevas, lobate deposits of mass movements, rock-fall deposits within hanging side valley, west-facing failure surface, and sample locations (note for scale that the elongated building with a red roof in the foreground is 57 m long). The photo highlights the difference between the northern and southern parts of the failure surface.

Fig. 9. Oblique satellite view towards NE showing a massive deposit filling the Las Cuevas valley over a distance of 3 km and Mario Ardito side valley with landslide scar and filled with rock-avalanche deposits. The photo shows landslide deposits several tens of metres thick filling the Mario Ardito valley.

(Rosas et al. 2008) that were sourced from the southern side of Tolosa mountain (Fig. 3). Two lobate rock-avalanche deposits covered with boulders several metres to tens of metres in diameter partially overlie each other (Fig. 8). While the eastern deposit has a more subdued morphology, the western deposit is more pristine. The deposits lie directly below an important niche in the mountain that has a pronounced failure surface dipping c. 40° west towards a hanging side valley. Along the failure surface two distinctive domains can be discriminated based upon the development of erosional features (Rosas et al. 2008; Fig. 8). The northern part of the sliding plane is smooth and continuous while the southern part is partially eroded by channels and slabs that are missing. We sampled both domains of the sliding plane, and both lobate deposits within the valley as well as rock-fall deposits within the hanging side valley for ^{36}Cl TCN dating. While sampling the more eroded failure surface we took care over the most outstanding spurs to avoid sampling surfaces that were eroded post-failure. The sample from the more eroded and hence older failure surface resulted in an age of 15 330 ± 1350 years; this is slightly older than the age of the partially covered, more subdued and therefore older deposit that resulted in ages of 11 910 ± 1100 and 14 100 ± 1150 years (CUA02,

CU0201, and CU0202 in Table 2 and Fig. 7, respectively). The sample from the less eroded and therefore younger failure plane resulted in an age of 11 810 ± 1120 years that coincides well with the ages obtained for the samples taken from the overlying more pristine and therefore younger lobe that yield ages of 12 530 ± 1330 years and 11 920 ± 1100 years (CUA01, CU0101, and CU0102 in Table 2 and Fig. 7, respectively).

The samples CU3, CU401 and CU402 that were taken from rock-fall deposits from two distinct deposits (Table 2; Fig. 8) resulted in ages of 15 810 ± 2990 years, 14 190 ± 3080 years and 9120 ± 2300 years, respectively; these ages overlap within uncertainties. Hence, two rock avalanches sourced from that side of Tolosa mountain in the Late Pleistocene, and rock-fall activity has also been more active than in the Holocene.

Rock-avalanche deposit east of Penitentes. Eleven kilometres east of the confluence of the Horcones and Las Cuevas valleys and 1.5 km east of the locality of Penitentes, a massive deposit fills the Las Cuevas valley floor for a distance of 3 km (Figs 2 & 9). This deposit was previously interpreted as the end moraine of a glacial advance sourcing from the Horcones valleys and mapped as the Penitentes moraine (Espizua 1993). Travertine that lies between two deposits interpreted as tills was dated with a ^{230}Th/^{232}Th age to 38 300 ± 5300 years (Espizua 1999). The deposit is a breccia composed of material that is exposed within the Frontal Cordillera. No material sourcing from the Main Cordillera was recognized (Fauqué et al. 2009). In addition the texture and composition of the deposit is identical to a rock-avalanche deposit that fills the Mario Ardito side valley that enters the Las Cuevas valley at the upper end of the deposit and was therefore reinterpreted as belonging to the same rock-avalanche event. Although the boulder carapace of this deposit is much less developed than at the rock-avalanche deposits originating from Tolosa mountain, this landslide is interpreted, due to the run-out distance, to be a rock avalanche and is defined here as the Penitentes rock-avalanche deposit. As would be expected, boulder size is smaller as the source area coincides with a important thrust fault (Ramos et al. 1996) that caused a significant break-down of the source rock prior to its failure.

Three samples (PE01, PE02 and PE03) were taken for TCN dating (Table 1; Fig. 9) using ^{36}Cl, resulting in ages of 13 890 ± 2200, 10 620 ± 1820 and 11 820 ± 2.210 years, respectively. Ages coincide within uncertainty limits and the mean age of the three ages is 12 110 ± 2100 years.

Discussion

The source and genesis of the hummocky Horcones deposit

Neoglacial deposits in the lower Horcones valley have a mineralogical composition corresponding to a mixture of the lithologies cropping out in the Horcones valley that include both the andesitic south face of Aconcagua mountain and the sedimentological rocks that crop out in the valley. Besides the relative quartz/plagioclase/tremolite content showing the andesitic composition of the Aconcagua source, the carbonates are also ideal for determining the source of the sediment as they do not occur within the south face of Aconcagua mountain. The neoglacial deposits are closer to an andesitic composition than the lateral moraines preserved from earlier glaciations (Fig. 4). This is likely to be because the lower Horcones glacier does not erode the valley walls along the part where it overlies the sedimentary sequence (Fig. 3), thus fresh bedrock originates from the south face of Aconcagua mountain only; all other materials transported by the glacier are remobilized Quaternary sediments. The older glacial deposits are more enriched in carbonates and more distinct due to their andesitic composition and are interpreted to represent a more balanced average composition of the Horcones valley. These deposits have polymict clasts and are well homogenized as no spatial variation is visible. The lateral moraine deposits are also identical in colour and composition to the lower unit in Confluencia that fills the valley. Hence these deposits are interpreted as being of the same glacial origin and that they are basal moraines.

Beside the lateral moraines and on top of the basal moraine, hummocky deposit fills the entire valley over a distance of 2 km at Confluencia (Figs 3 & 5). This deposit has only andesitic boulders and its mineral composition is predominantly andesitic. Carbonate minerals as tracers for sedimentary rocks are nearly absent. Such hummocky deposits do not occur within the Horcones Superior and Horcones Inferior valleys; the only possible source for these andesitic rocks is Aconcagua mountain. The Horcones Superior valley is today ice-free and if this deposit had sourced out of that valley, any deposits should be preserved also inside the Horcones Superior valley. In contrast, the Horcones Inferior valley today is covered by the Horcones glacier and neoglacial deposits; a connection of the Horcones deposit to the southern face of Aconcagua mountain would therefore be masked. Indeed the southern face of Aconcagua mountain is represented by bedrock, glaciers and snow fields only and several large niches within the face exist from which large volumes could source (Fig. 10). We tentatively marked a niche that is today filled by two hanging glaciers as the potential source area (Figs 3 & 10). Most striking of the morphology of the south face of Aconcagua mountain is the missing catchment of the valley hosting the Ventisquero de Relinchos glacier (Fig. 10). This valley is c. 1.2 km wide in its lower parts but only a tiny glacier c. 10 m wide fills its upper part. While the northern slope of that valley is represented by the south face of Aconcagua mountain, a southern limiting slope is missing. Furthermore this hanging valley connects directly to a niche filled by the Superior glacier of the south face of Aconcagua mountain. This is a hanging glacier that is disconnected from the lower Horcones glacier but feeds that glacier by calving. The missing southern slope of the Ventisquero de Relinchos valley thus represents a missing andesitic rock mass on the south face of Aconcagua mountain that was eroded in the past from the south face. We postulate here that the andesitic deposit in Confluencia is at least part of that mass that had failed in one or more massive landslides into the Horcones Inferior valley.

A rockslide origin of the hummocky Horcones deposit is suggested not only by its morphology and purely andesitic origin but also by the position of this deposit within the Quebrada del Tolosa, being 150 m higher than its position in Confluencia

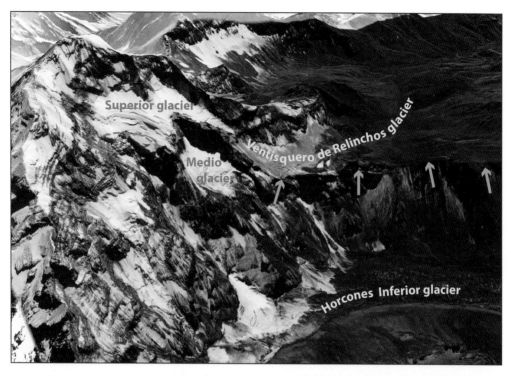

Fig. 10. Oblique satellite view of the south face of Aconcagua mountain showing the distribution of today's glaciers. Blue arrows mark abrupt termination of the Ventisquero de Relinchos valley floor that has no upper catchment matching the lower part of the valley. Note that the altitude difference between the Horcones Inferior glacier and the top of Aconcagua mountain is 2500 m.

and thus representing a run-up (Figs 3 & 5). In order for a glacier that sedimented in the Tolosa valley up to that altitude to have deposits sourced from the southern face of Aconcagua mountain, the glacier would need to have been at least 150 m thick at Confluencia; however, the lateral moraine at Confluencia is only a few tens of metres above the base of the valley. This further supports the landslide origin.

The hummocky deposit has the same morphology and texture at the confluence of the Horcones and Las Cuevas valleys (Figs 3 & 7). However, the composition has changed slightly, as expressed by a patchy distribution of reddish breccias with polymict clasts and greyish breccias with monomict clasts (Fig. 7b). The greyish domains have an andesitic composition nearly identical to the deposit at Confluencia while the reddish domains have a composition that is a mixture of the andesitic breccias and the glacial moraine deposits. Thus those reddish domains are interpreted to be glacial deposits entrained into the landslide.

The terminus of the pristine hummocky deposit lies within the Las Cuevas valley 21 km away from the foot of the southern face of Aconcagua mountain. However, downriver from that and cemented by travertine there is an identical deposit with more subdued hummocky morphology. The terminus of this is not mapped out yet but deposits have been found at least further 10 km downriver. This deposit is interpreted to be of the same source and type as the stratigraphically higher and more pristine hummocky deposit. It is not possible to locate a second niche in the south face of Aconcagua mountain as hanging glaciers occupy the massif and are likely to have reshaped its face. The travel distances of both hummocky deposits are very long. This is especially true as the velocity of the landslides was 54 m s^{-1} at Confluencia (calculated on the basis of the run-up height at Confluencia of 150 m following the principles of Crandell & Fahnestock (1965)) but not significant at the confluence of the Horcones and Las Cuevas valleys (no deposits showing run-up could be found). Therefore these deposits do not represent typical rock-avalanche deposits. This is not surprising as a failure on the south face of Aconcagua mountain would have fallen on the Horcones Inferior glacier. Such rock-slope failures onto glaciers in other parts of the Andes and the world have entrained large amounts of glacial ice (Hauser 2002; Huggel et al. 2005; Evans et al. 2009) that have always resulted in an excessive travel distance of the landslides and a change of landslide behaviour as ice started melting out. It can also be expected that a failure at the southern face of Aconcagua mountain would have entrained glacial ice into the flow. This ice would have melted not only during sliding but also out of the deposit after deposition (thermocast). Thermocast also explains the strong hummocky morphology of the Horcones deposit. Such a deposit is not as compact as a typical rock avalanche and erosion of hummocks and deposition into depressions left by melted glacial ice can be expected. Such mass redeposition on the surface would also affect the position of large boulders, causing rotation and toppling. This strongly influences cosmogenic nuclide production in boulders, as boulder surfaces that are horizontal today have not necessarily been horizontal in the geological past, and this would have resulted in lower irradiation and a lower age determination. For this reason, a large spread of ages of boulders on the deposit is to be expected (see below). We argue therefore, following Fauqué et al. (2009), that the pristine hummocky deposit at Confluencia and at the confluence of the Horcones and Las Cuevas valleys represents the deposit of the same landslide deposit (hereafter called the Horcones deposit, Fig. 3) that originated from the south face of Aconcagua mountain. Hence the deposit represents that of a debris-ice avalanche. The underlying deposit in the Las Cuevas valley that has a surface with more subdued hummocks, is interpreted similarly to be the deposit of a similar but much older debris-ice avalanche originating from the same source. The origin is therefore distinct from previous interpretations (Espizua 1993, 1999) in which these deposits were defined as frontal moraines of various glacial advances (Almacenes, Horcones). In the same way the Penitentes rock-avalanche deposit does not represent a glacial end moraine but rather the deposits of a rock avalanche from a side valley (Fauqué et al. 2009). Therefore the situation in the high Andes of Argentina is similar to the Himalaya where deposits of rock slope failures have been misinterpreted as glacial deposits (Hewitt 1999).

It is not possible to establish the failed volume that caused the younger and older debris-ice avalanche based on the deposits, as deposits are strongly eroded along several kilometre-long stretches of the valley, for example, the younger deposit is entirely eroded for more than 3 km in the narrowest part of the Horcones valley (Fig. 3). In addition we showed that significant entrainment of moraine material had occurred before the debris-ice avalanche arrived in the Las Cuevas valley. The volume also cannot be established based on the source area as we do not have information on the prefailure topography, and the niche in the south face of Aconcagua mountain is today almost entirely filled with a hanging glacier making a reconstruction of the failure surface very difficult. Furthermore, the travel distance of 21 km between the foot of Aconcagua mountain and the lower limit of the Horcones deposit is considerable. This is in line with run-out behaviour of rock slope failures on to ice in other parts of the Andes and the world (Hauser 2002; Huggel et al. 2005; Evans et al. 2009; Delaney & Evans 2014).

Quality of TCN ages

TCN is an ideal tool to date the age of rock-avalanche deposits as no further material than the deposit itself is needed and no stratigraphic interpretations with under- or overlying deposits have to be taken into account (e.g. Ballantyne et al. 1998; Hermanns et al. 2001; Hermanns et al. 2004; Dortch et al. 2009). However, Ivy-Ochs et al. (2009) were able to show a very wide range of ages by dating a large number of boulders on the same surface of a rock-avalanche deposit that was deposited in a single event. This is interpreted as being due to pre-exposure on the rock slope prior to failure in the case of older ages, or block rotation of the boulders following rock-avalanche deposition in the case of younger ages. Therefore, TCN ages that do not coincide with the average of multiple ages are not considered further when calculating the mean age of all samples for a given surface (Ballantyne & Stone 2013; Martin et al. 2014). Our multiple TCN ages from the various sampled deposits presented in this paper coincide in general within analytical uncertainty limits (Fig. 11). However, there are a few exceptions. Two boulders sampled from the Horcones deposit date as too young in comparison to three other samples. As discussed above, rotation of blocks is typical for surfaces that undergo thermocast and the younger ages are therefore excluded from data for calculating the age of the event. One of the boulders sampled from the Penitentes rock-avalanche deposit is dated as older than the other two. This sample is of a different lithololgy and is therefore interpreted as having been exposed before becoming entrained into the landslide. Also, one of the samples from the older lobe at Las Cuevas has a younger date and is again interpreted as representing a boulder that rotated after rock-avalanche deposition.

The Las Cuevas river upstream from the confluence of the Horcones and Las Cuevas valleys had been temporarily blocked in the Late Pleistocene into the Holocene (Figs 3 & 7); sedimentation rate was probably low in this lake basin in the Late Pleistocene as two additional blockages of

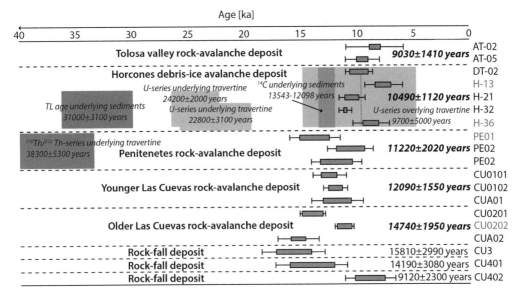

Fig. 11. Overview of [36]Cl ages of samples given in this study and average age for the deposit as well as previously published ages for each deposit: U-series, ^{230}Th/^{232}Th and TL ages from Bengochea *et al.* (1987) and Espizua (1999); ^{14}C age from Fauqué *et al.* (2009). The boxes represent the ages including analytical 2 sigma uncertainties, while the error bars represent the uncertainty due to snow cover. The age of the deposit is based on the samples represented with a black label. Sample ages given with sample number in grey are not considered as these ages are interpreted as being too old or too young, due to pre-exposure of the sample prior to landsliding or due to boulder rotation after deposition, respectively.

the Las Cuevas river existed, one due to rock-avalanche deposits at Las Cuevas, the other due to glacial deposits. The related lake sediments were dated by AMS ^{14}C dating of organic material extracted from the lake deposit (Figs 3 & 7; Fauqué *et al.* 2009), resulting in calibrated ages of 14 798–13 886 and 8620–8254 years BP. The older ages predate the pristine Horcones deposit, also suggesting an older event of similar type (see discussion above). The younger ages postdate the Horcones deposit. Furthermore, ^{14}C ages of organic materials within fluvial deposits directly underlying the Horcones deposit in Confluencia fit both with a calibrated age of 13 543–12 098 years BP and the stratigaphic position (Fig. 11). In addition, all our ages fit within the stratigraphic position and with previously published ages obtained by TL dating, U-series dating and ^{230}Th/^{232}Th dating. However, often there are considerable time gaps between the ages of the underlying strata and the deposit itself.

Ages of large rock slope failures in the Horcones and Las Cuevas valleys

The age of the Horcones and underlying older hummocky deposits. The age of the deposit underlying the hummocky Horcones deposit has not yet been directly determined. However multiple stratigraphic relationships give a minimum age. At Confluencia only one deposit exists that originated from the south face of Aconcagua mountain. A single TCN age of the lateral moraine at that locality suggests that the area was still covered by a glacier at 13 880 ± 830 years (Fauqué *et al.* 2009). At the same time the confluence of the Horcones and Las Cuevas valleys was dammed by a breccia that has reddish and greyish domains as suggested by overlying lacustrine sediments that yielded a calibrated ^{14}C age of 14 798–13 886 years BP. That this area was ice-free at that time is suggested by a single boulder from the lateral moraine in the Horcones valley dated to 16 510 ± 2110 years (Fauqué *et al.* 2009). The age of this deposit is further contrained by U-series dating of travertine that cements its surface, yielding two ages of 24 200 ± 2000 and 22 800 ± 3100 years (Espizua 1999).

The Horcones deposit, in contrast, postdates a fluvial phase in Confluencia dated with a calibrated ^{14}C age to 13 543–12 098 years BP (Fauqué *et al.* 2009). This coincides with the multiple [36]Cl TCN ages presented here that assign a mean age of the deposit to 10 490 ± 1120 years (Fig. 11). These ages are further supported by the top of the lacustrine sediments in the Las Cuevas valley having a calibrated ^{14}C age of 8620–8254 years BP (Fauqué et al. 2009), indicating that by that time this landslide barrier was still not eroded. Furthermore a U-series age of 9700 ± 5000 years of a travertine layer overlying the Horcones deposit is congruent with our age (Bengochea et al. 1987).

Ages of other rock avalanches from Tolosa mountain and from the Quebrada Mario Ardito. Similarly to the landslides sourced from the south face of Aconcagua mountain, all rock avalanches in the Las Cuevas and adjacent valleys that have been dated with ^{36}Cl TCN, date to the end of the Pleistocene. From Tolosa mountain two rock avalanches occurred on the south side; the failure planes could be differentiated by the degree of erosion and the deposits partially overlie each other. The older of the two events has an age of 14 740 ± 1950 years while the younger has an age of 12 090 ± 1550 years (Fig. 11). Hence the largest time interval of the exposure history of both sliding planes was common. The visible strong variation of erosional features of both planes might be explained by the three ages of rock-fall deposits sourcing from the same slope and dating between 15 810 ± 2990 years and 9120 ± 2300 years suggesting that geomorphic processes were much more intensive at the transition from the Pleistocene to the Holocene than throughout the Holocene (Fig. 11). This is also suggested by the age of the rock avalanche that occurred on the east side of Tolosa mountain that dates to 9030 ± 1410 years. Also the rock avalanche that sourced from the Mario Ardito valley and entered the lower part of the Las Cuevas valley has, with an age of 11 220 ± 2020 years, a Late Pleistocene age.

Possible causes for temporal distribution of slope failures in the Horcones and Las Cuevas valleys

It is not within the scope of this paper to redefine the glacial chronology of the Horcones and Las Cuevas valleys. However, we have shown that multiple deposits previously interpreted as representing frontal moraines are actually deposits of massive rock slope failures. Therefore the glacial chronology should be established on lateral moraines that exist in those valleys. So far only two of these moraines have been dated with single ^{36}Cl ages (Fauqué et al. 2009). One is the lateral moraine in the Horcones valley close to the confluence with the Las Cuevas valley (Figs 3 & 7) that has an age of 16 510 ± 830 years and the other the lateral moraine at Confluencia (Figs 3 & 5) that has an age of 13 880 ± 830 years (Fauqué et al. 2009).

The large rock slope failures dated in our paper have ages that follow the last glacial advance, while the Holocene was free of such events in the study area. This suggests that climatic changes towards warmer conditions might have conditioned the slopes to fail. Evans & Clague (1994) propose that (1) slope debuttressing by glacial erosion and glacial retreat, (2) permafrost melting, and (3) increased pore-water pressure within the slope might have conditioned slopes for failure in the Canadian Rockies. Special emphasis has been given in recent years to the effects of permafrost melting (Noetzli et al. 2007; Fischer et al. 2012; Huggel et al. 2010) that can condition large rock slopes to fail. The permafrost in the Central Andes lies today at an elevation of 4000 ± 200 m (Schrott 1991), and was not likely to have been lower at the end of the Pleistocene due to larger glacial extents. With elevations of the source areas of the rock avalanches on the south side of Tolosa mountain and in the the Mario Ardito valley lying at 3900 and 3500 m asl, respectively, this effect might have contributed to those failures. In contrast the interpreted source areas of the rock avalanche lie on the east side of Tolosa mountain and on the south face of Aconcagua mountain at elevations of 4300–5200 and 5200–5800 m asl, respectively. Hence neither permafrost melting nor enhanced pore-water pressure could have contributed to destabilizing rock slopes. However, both the uppermost Tolosa valley and the south face of Aconcagua mountain are occupied by glaciers today. Hence, increased glacial erosion at the foot of those mountains during the Late Pleistocene might have contributed to failure.

Rock-avalanche deposits overlying each other and failure surfaces with significant variation of erosional features clearly indicate two generations of events at Las Cuevas, although uncertainty margins are overlapping (Fig. 11). The younger event coincides in age within uncertainty margins with the age of the Horcones deposit and the age of the rock avalanche in the Tolosa valley, as well as the age of the Penitentes rock-avalanche deposit, and therefore they could have occurred simultaneously. The source area of all rock slope failures lies within a periphery of only 20 km. The older deposit at Las Cuevas coincides with the age of a rock-avalanche deposit damming the Inca lake in Chile (Welkner et al. 2010) in a distance of less than 10 km (Fig. 2). Applying the relationship of distance v. magnitude of triggering earthquakes given by Keefer (1984) all rock slope failures, except the older debris-ice avalanche deposit in the Las Cuevas valley, might have been triggered by two earthquakes within the study area that had a magnitude 6 or higher. Seismic triggering is also proposed as a possible cause for rock avalanches at Laguna del Inca (Welkner et al. 2010) and elsewhere in Chile close to our study area (Sepúlveda et al. 2012).

This temporal distribution of large rock slope failures is different outside the Aconcagua region in the Central Andes of both Chile and Argentina. In Chile, Antinao & Gosse (2009) indicated that rock slope failures distribute more or less evenly over the Late Pleistocene and Holocene, yet the number of events in the Middle Pleistocene was reduced. In contrast, in the Cordon de Plata south of Uspallata in Argentina, all rock slopes failures are several tens to hundreds of thousands of years old (Moreiras 2006; Fauqué et al. 2008). The only exceptions in this part of Argentina are three rock avalanches along the Mendoza river that date into the Late Pleistocene and Early Holocene (Hermanns & Longva 2012). This river also drains the glaciated part of the Andes, and enhanced erosional undercutting of valley slopes due to glacial meltdown might also act here as a conditioning factor.

Therefore the situation in this part of the Andes is different from the transition from the Central Andes to the Patagonian Andes 600 km to the south or in the northern Central Andes 800 km to the NNW. In the south most rock slope failures occurred on glacial oversteepened slopes in the Middle Holocene but postdate deglaciation by c. 10 kyr (Penna et al. 2011). In the north most rock slope failure in valleys correlate with phases of higher precipitation and runoff in the Late Pleistocene and Holocene (Trauth et al. 2000; Hermanns & Schellenberger 2008).

Surges of the Lower Horcones glacier

The lower Horcones glacier had multiple historical surges (e.g. Habel 1897; Milana 2007; Fauqué et al. 2009; Pitte et al. 2009; Wilson 2010). The last surge occurred in 2004 and followed a summer characterized by enhanced snow and ice avalanches on the south face of Aconcagua mountain (personal communication of park rangers of Aconcagua provincial park). This glacier is fed only by snow and ice avalanches from Aconcagua mountain as the lower catchment is too small to form a glacier of that size and lies just above the modern permafrost line. If the Ventisquero de los Relinchos valley had not been captured on the south face of Aconcagua mountain, this ice – which today drops onto the Horcones Inferior glacier – would flow off to the east into the Ventisquero de los Relinchos valley (Fig. 10). Hence ice distribution on Aconcagua mountain might have significantly changed due to slope failures. Because of this, the glacial history of Aconcagua mountain cannot be assessed by studying moraine deposits in the Horcones valley alone. Also, valleys draining to the east have to be analysed to understand past glacial dynamics on Aconcagua mountain.

Conclusions

We have studied the morphology, sedimentology and mineralogy of deposits previously interpreted as glacial deposits in the Horcones and Las Cuevas valleys as well as TCN-dated them using ^{36}Cl. Although some of these deposits contain moraine material, they are related to large rock slope failures that entrained moraine deposits and probably ice on their path. Our data suggest that two rock slope failures sourced from the south face of Aconcagua mountain. These failures might have rearranged glacial flow significantly. This is supported by a captured glacial valley that is missing its upper drainage on that side of Aconcagua mountain. These slope failures dropped into the Horcones Inferior valley and travelled over the glacier. They passed the glacial front and deposited a mixture of rock materials from the south face of Aconcagua mountain and entrained glacial deposits on their way further down-valley in the Horcones and Las Cuevas valleys. The first event had a higher mobility and predates $24\,200 \pm 2000$ years. The second event occurred at the transition from the Late Pleistocene to the Holocene. Four further rock slopes failures date to the time between 15 ka and the onset of the Holocene. No deposits of large younger rock slopes failures have been recognized in this section of the valleys suggesting that processes related to climatic changes in the Late Pleistocene caused the slopes to fail. Alternatively, all slope failures except the older debris-ice avalanche from the south face of Aconcagua mountain could have been triggered by two earthquakes with magnitudes higher than 6.

This project was financed through the Multinational Andean Project: Geosciences for Andean Communities through contributions of the Canadian International Development Agency and the Servicio Geólogic Minero Argentino. We would like to acknowledge M. Rosas and V. Baumann for support during field work as well as C. Jermyn for sample preparation and age calculations. The park rangers of the Aconcagua provincial park as well as the army post at Puente del Inca supported our work actively by providing information and helping with logistics in the field. Publication of this paper was financed by the Geological Survey of Norway.

References

ABELE, G. 1974. *Bergstürze in den Alpen*. Ausschüsse des Deutschen und Österreichischen München, Alpenvereins.

ADAMS, J. 1981. Earthquake-dammed lakes in New Zealand. *Geology*, **9**, 215–219.

ANTINAO, J. L. & GOSSE, J. 2009. Large rockslides in the Southern Central Andes of Chile (32–34.5°S): tectonic control and significance for Quaternary landscape evolution. *Geomorphology*, **104**, 117–133.

BALLANTYNE, C. K. & STONE, J. O. 2013. Timing and periodicity of paraglacial rock-slope failures in the Scottish Highlands. *Geomorphology*, **186**, 150–161.

BALLANTYNE, C. K., STONE, J. O. & FIFIELD, L. K. 1998. Cosmogenic Cl-36 dating of postglacial landsliding at The Storr, Isle of Skye, Scotland. *The Holocene*, **8**, 347–351.

BENGOCHEA, L. E., PORTER, S. C. & SCHWARZ, H. P. 1987. Pleistocene glaciation across the high Andes of Chile and Argentina. *XII International Congress of the International Union of Quaternary Research INQUA*, Ottawa, Canada. July 31–August 9, 1987, Abstracts, 105.

BERGMANN, J., FRIEDEL, P. & KLEEBERG, R. 1998. BGMN – a new fundamental parameters based Rietveld program for laboratory X-sources, its use in quantitative analysis and structure investigation. *PCPD Newsletter, International Union of Crystallography*, **20**, 5–8.

BIGOT CORMIER, F., BRAUCHER, R., BOURLES, D., GUGLIELMI, Y., DUBAR, M. & STEPHAN, J. F. 2005. Chronological constraints on processes leading to large active landslides. *Earth and Planetary Science Letters*, **235**, 141–150.

BLAIS-STEVENS, A., HERMANNS, R. L. & JERMYN, C. 2011. A 36Cl age determination for Mystery Creek rock avalanche and its implications in the context of hazard assessment, British Columbia, Canada. *Landslides*, **8**, 407–416, http://dx.doi.org/10.1007/s10346-011-0261-0

CERLING, T. E. & CRAIG, H. 1994. Geomorphology and in-situ cosmogenic isotopes. *Annual Review of Earth and Planetary Sciences*, **22**, 273–317.

CLAGUE, J. J., HUGGEL, C., KORUP, O. & MCGUIRE, B. 2012. Climate change and hazardous processes in high mountains. *Revista de la Asociación Geológica Argentina*, **69**, 328–338.

COSTA, C. H. & GONZÁLEZ DÍAZ, E. F. 2007. Age constraints and paleoseismic implication of rockavalanches in the Northern Patagonian Andes, Argentina. *Journal of South American Earth Sciences*, **24**, 48–57, http://dx.doi.org/10.1016/j.jsames.2007.03.001

CRANDELL, D. R. & FAHNESTOCK, R. K. 1965. Rockfalls and avalanches from Little Tahoma Peak on Mount Rainier, Washington. *US Geological Survey Bulletin*, **1221-A**, A1–A30.

DELANEY, K. B. & EVANS, S. G. 2014. The 1997 Mount Munday landslide (British Columbia) and the behavior of rock avalanches on glacial surfaces. *Landslides*, http://dx.doi.org/10.1007/s10346-013-0456-7

DORTCH, J. M., OWEN, L. A., HANEBERG, W. C., CAFFEE, M. W., DIETSCH, C. & KAMP, U. 2009. Nature and timing of large landslides in the Himalaya and Transhimalaya of northern India. *Quaternary Science Reviews*, **28**, 1037–1054, http://dx.doi.org/10.1016/j.quascirev.2008.05.002

ESPIZUA, L. E. 1993. Quaternary glaciations in the Río Mendoza valley, Argentine Andes. *Quaternary Research*, **40**, 150–162.

ESPIZUA, L. E. 1999. Chronology of late Pleistocene glacier advances in the Río Mendoza valley, Argentina. *Global and Planetary Change*, **22**, 193–200.

ESPIZUA, L. E. & BIGAZZI, G. 1998. Fission-track dating of the Punta de Vacas glaciation in the Río Mendoza valley, Argentina. *Quaternary Science Reviews*, **17**, 755–760.

EVANS, S. G. & CLAGUE, J. J. 1994. Recent climatic change and catastrophic geomorphic processes in mountain environments. *Geomorphology*, **10**, 107–128.

EVANS, S. G., BISHOP, N. F., FIDEL SMOLL, L., VALDERRAMA MURILLO, P., DELANEY, K. B. & OLIVER-SMITH, A. 2009. A re-examination of the mechanism and human impact of catastrophic mass flows originating on Nevado Huascarán, Cordillera Blanca, Peru in 1962 and 1970. *Engineering Geology*, **108**, 96–118.

FAUQUÉ, L., CORTÉS, J. M. ET AL. 2008. Edades de las avalanchas de rocas ubicados en el valle del río Mendoza aguas debajo de Uspallata. *XVII Congreso Geologico Argentino*, October 7–10, 2008, Jujuy. Actas I, 282–283.

FAUQUÉ, L., HERMANNS, R. L. ET AL. 2009. Megadeslizamientos de la pared sur del Cerro Aconcagua y su relación con depósitos asignados a la glaciación Pleistocena. *Revista de la Asociación Geológica Argentina*, **65**, 691–712.

FISCHER, L., HUGGEL, C., KÄÄB, A. & HAERBERLI, W. 2012. Slope failures and erosion rates on a glacierized high-mountain face under climatic changes. *Earth Surface Processes and Landforms*, **38**, 836–846, http://dx.doi.org/10.1002/esp.3355

GIAMBIAGI, L. B., ÁLVAREZ, P. P., GODOY, E. & RAMOS, V. 2003. The control of pre-existing extensional structures on the evolution of the southern sector of the Aconcagua fold and thrust belt, southern Andes. *Tectonophysics*, **369**, 1–19.

GONZÁLEZ DÍAZ, E. F. 2003. El englazamiento en la región de la caldera de Cavihuave-Copahue (Provincia de Neuquén): Su reinterpreatación. *Revista de la Asociación Geológica Argentina*, **58**, 356–366.

GOSSE, J. C. & PHILLIPS, F. M. 2001. Terrestrial in situ cosmogenic nuclids; theory and application. *Quaternary Science Reviews*, **20**, 1475–1560.

HABEL, J. 1897. *Ansichten aus Südamerika: Schilderung einer Reise am La Plata, in den argentinischen Anden und an der Westküste.* Verlag von D. Reimer (E. Vohsen).

HASELTON, K., HILLEY, G. & STRECKER, M. 2002. Average Pleistocene climatic patterns in the southern Central Andes: controls on mountain glaciations and paleoclimate implications. *Journal of Geology*, **110**, 211–226.

HAUSER, A. 2002. Rock avalanche and resulting debris flow in Estero Parraguirre and Río Colorado, region Metropolitana, Chile. *In*: EVANS, S. G. & DEGRAFF, J. V. (eds) *Catastrophic Landslides*. Geological Society of America Reviews in Engineering Geology, **15**, 135–148.

HERMANNS, R. L. & LONGVA, O. 2012. Rapid rock-slope failures. *In*: CLAGUE, J. J. & STEAD, D. (eds)

Landslides: Types, Mechanisms and Modeling. Cambridge University Press, Cambridge, UK, 59–70.

HERMANNS, R. L. & NIEDERMANN, S. 2011. Late Pleistocene–Early Holocene paleoseismicity deduced from lake sediment deformation and coeval landsliding in the Calchaquíes valleys, NW Argentina. In: AUDEMARD, F. A., MICHETTI, A. & MCCALPIN, J. P. (eds) Geological Criteria for Evaluating Seismicity Revisited: Forty Years of Paleoseismic Investigations and the Natural Record of Past Earthquakes. Geological Society of America, Boulder, Special Paper, **479**, 181–194.

HERMANNS, R. L. & SCHELLENBERGER, A. 2008. Quaternary tephrochronology helps define conditioning factors and triggering mechanisms of rock avalanches in NW Argentina. Quaternary International, **178**, 261–275.

HERMANNS, R. L., TRAUTH, M. H., NIEDERMANN, S., MCWILLIAMS, M. & STRECKER, M. R. 2000. Tephrochronologic constraints on temporal distribution of large landslides in northwest Argentina. Journal of Geology, **108**, 35–52.

HERMANNS, R. L., NIEDERMANN, S., VILLANUEVA GARCIA, A., SOSA GOMEZ, J. & STRECKER, M. R. 2001. Neotectonics and catastrophic failure of mountain fronts in the southern intra-Andean Puna Plateau, Argentina. Geology, **29**, 619–623.

HERMANNS, R. L., NIEDERMANN, S., IVY-OCHS, S. & KUBIK, P. W. 2004. Rock avalanching into a landslide-dammed lake causing multiple dam failure in Las Conchas valley (NW Argentina) – evidence from surface exposure dating and stratigraphic analyses. Landslides, **1**, 113–122.

HEUSSER, C. J. 2003. Ice Age Southern Andes: A Chronicle of Paleoecological Events. Elsevier, Amsterdam.

HEWITT, K. 1999. Quaternary moraines vs catastrophic rock avalanches in the Karakoram Himalaya, Northern Pakistan. Quaternary Research, **51**, 220–237.

HEWITT, K. 2006. Disturbance regime landscapes: Mountain drainage systems interrupted by large rockslides. Progress in Physical Geography, **30**, 365–393.

HEWITT, K., GOSSE, J. & CLAGUE, J. J. 2011. Rock avalanches and the pace of late Quaternary development of river valleys in the Karakoram Himalaya. Geological Society of America Bulletin, **123**, 1836–1850.

HUGGEL, C., ZGRAGGEN-OSWALD, S. ET AL. 2005. The 2002 rock/ice avalanche at Kolka/Karmadon, Russian Caucasus; assessment of extraordinary avalanche formation and mobility, and application of QuickBird satellite imagery. Natural Hazards and Earth System Sciences, **5**, 173–187.

HUGGEL, C., FISCHER, L., SCHNEIDER, D. & HAEBERLI, W. 2010. Research advances on climate-induced slope instability in glacier and permafrost high-mountain environments. Geographica Helvetica, **65**, 146–156.

IVY-OCHS, S., POSCHINGER, A. V., SYNAL, H.-A. & MAISCH, M. 2009. Surface exposure dating of the Flims landslide, Graubünden, Switzerland. Geomorphology, **103**, 104–112.

KAPLAN, M. R., ACKERT, R. P., JR., SINGER, B. S., DOUGLAS, D. C. & KURZ, M. D. 2004. Cosmogenic nuclide chronology of millenial-scale glacial advances during O-isotope stage 2 in Patagonia. Geological Society of America Bulletin, **116**, 308–321.

KEEFER, D. K. 1984. Landslides caused by earthquakes. Geological Society of America Bulletin, **95**, 406–421.

KURZ, M. D. & BROOKE, E. 1994. Surface Exposure Dating with Cosmogenic Nuclides, Dating in Exposed and Surface Contexts. University of New Mexico Press, Albuquerque, New Mexico, USA.

LAL, D. 1991. Cosmic ray labeling of erosion surfaces; in situ nuclide production rates and erosion models. Earth and Planetary Science Letters, **104**, 424–439.

MARTIN, S., CAMPEDEL, P. ET AL. 2014. Lavini di Marco (Trentino, Italy): ^{36}Cl exposure dating of a polyphase rock avalanche. Quaternary Geochronology, **19**, 106–116.

MASARIK, J., KOLLAR, D. & VANYA, S. 2000. Numerical simulation of in situ production of cosmogenic nuclides: effects of irradiation geometry. Nuclear Instruments and Methods in Physics Research B, **172**, 786–789.

MILANA, J. P. 2007. A model of the glacier Horcones inferior surge, Aconcagua region, Argentina. Journal of Glaciology, **53**, 565–572.

MINETTI, J. 1985. Régimen termométrico de San Juan. Centro de Investigaciones Regionales de San Juan (CIRSAJ)-(CONICET). Informe Técnico No. 9.

MINETTI, J. 1986. El régimen de precipitaciones en San Juan y suentorno. Centro de Investigaciones Regionales de San Juan (CIRSAJ)-(CONICET). Informe Técnico No. 8.

MOREIRAS, S. 2006. Chronology of a probable neotectonic Pleistocene rock avalanche, Cordon de Plata (Central Andes), Mendoza, Argentina. Quaternary International, **148**, 138–148.

MOREIRAS, S. M., LENZANO, M. G. & RIVEROS, N. 2008. Inventario de procesos de remoción en masa en el Parque provincial Aconcagua, Provincia de Mendoza, Argentina. Multequina. Latin American Journal of Natural Resources, **17**, 129–146.

NOETZLI, J., GRUBER, S., KOHL, T., SALZMAN, N. & HAEBERLI, W. 2007. Three-dimensional distribution and evolution of permafrost temperatures in idealized high-mountain topography. Journal of Geophysical Research, **112**, F02S13, http://dx.doi.org/10.1029/2006JF000545

OSTERMANN, M. & SANDERS, D. 2012. Post glacial rockslides in a 200 × 130 km area of the Alps: Characteristics, ages, and uncertainties. In: EBERHARDT, E., FROESE, C., TURNER, A. K. & LEROUEIL, S. (eds) Landslide and Engineered Slopes: Protecting Society through Improved Understanding. Taylor & Francis Group, London, 659–663.

PENNA, I., HERMANNS, R. L., FOLGUERA, A. & NIEDERMANN, S. 2011. Multiple slope failures associated with neotectonic activity in the southern central Andes (37°–37°30′S). Patagonia, Argentina. Geological Society of America Bulletin, **123**, 1880–1895.

PEREYRA, E. X. & GONZÁLEZ DÍAZ, E. F. 1993. Reinterpretación geomórfica de la llamada Morena de Los Horcones, Puente del Inca, Provincia de Mendoza. XII Congreso Geológico Argentino, October 10–15, Mendoza. Actas **4**, 73–79.

PHILLIPS, F. M. & PLUMMER, M. A. 1996. CHLOE: a program for interpreting in-situ cosmogenic nuclide data for surface exposure dating and erosion studies. *Abstracts of the 7th International Conference on Accelerator Mass Spectrometry.* Radiocarbon, **38**, 98–99.

PITTE, P., FERRI HIDALGO, L. & ESPIZUA, L. E. 2009. Aplicación de sensores remotos al estudio de glaciares en el Cerro Aconcagua. *Anais XIV Simposio Brasileiro de Sensoriamento Remoto, Natal,* Brasil, April 25–30, 2009, extended abstract 1473–1480.

PRAGER, C., ZANGERL, C., PATZELT, G. & BRANDNER, R. 2008. Age distribution of fossil landslides in the Tyrol (Austria) and its surrounding areas. *Natural Hazard and Earth System Sciences,* **8**, 377–407.

RAMOS, V. A., AGUIRRE_URRETA, M. B. *ET AL.* 1996. Geología de la región del Aconcagua. *Subsecretaria de Minería de la Nación, Dirección Nacional del Servicio Geológico,* Anales, **24**.

ROSAS, M., BAUMANN, V. *ET AL.* 2007. Estudio geoscientífico aplicado al ordenamiento territorial, Puente del Inca, Provincia de Mendoza, Publicación del Servicio Geologico Minero de Argentina, Buenos Aires, 73.

ROSAS, M., WILSON, C., HERMANNS, R. L., FAUQUÉ, L. & BAUMANN, V. 2008. Avalanchas de rocas de Las Cuevas una evidencia de la desestabilisacón de las laderas como consequencia del cambio climatic del Pleistoceno superior. *XVII Congreso Geologico Argentino,* October 7–10, 2008, Jujuy. Actas I, 313–315.

SCHROTT, L. 1991. Global solar radiation, soil temperature and permafrost in the Central Andes, Argentina: a progress report. *Permafrost Periglacial Processes,* **2**, 59–66. http://dx.doi.org/10.1002/ppp.3430020110

SCHUSTER, R. L., LOGAN, R. L. & PRINGLE, P. T. 1992. Prehistoric rock avalanches in the Olympic Mountains, Washington. *Science,* **258**, 1620–1621.

SEPÚLVEDA, S. A., FUENTES, J. P., OPPIKOFER, T., HERMANNS, R. L. & MOREIRAS, S. M. 2012. Analysis of a large-scale, stepped planar failure in the Central Andes uplands, Chile, using roughness profiles from terrestrial laser scanning. *In:* EBERHARDT, E., FROESE, C., TURNER, A. K. & LEROUEIL, S. (eds) *Landslides and Engineered Slopes: Protecting Society through Improved Understanding.* Taylor & Francis Group, London, 1243–1247.

SOLDATI, M., CORSINI, A. & PASUTO, A. 2004. Landslides and climate change in the Italian Dolomites since the Late Glacial. *Catena,* **55**, 141–161.

TRAUTH, M. H., ALONSO, R. A., HASELTON, K. R., HERMANNS, R. L. & STRECKER, M. R. 2000. Climate change and mass movements in the NW Argentine Andes. *Earth and Planetary Science Letters,* **179**, 243–256.

WELKNER, D., EBERHARDT, E. & HERMANNS, R. L. 2010. Hazard investigation of the Portillo Rock Avalanche site, central Andes, Chile, using an integrated field mapping and numerical modelling approach. *Engineering Geology,* **114**, 278–297.

WILSON, C. G. J. 2010. *Evolución geomorfológica del valle del Río Cuevas, Provincia de Mendoza.* República Argentina. Trabajo de licenciatura, Departamento de Ciencias Geológicas, Universidad de Buenos Aires.

ZECH, R., MAY, J.-H., KULL, C., ILGNER, J., KUBIK, P. W. & VEIT, H. 2008. Timing of the late Quaternary glaciation in the Andes from *c.* 15 to 40°S. *Journal of Quaternary Science,* **23**, 635–647.

Historical damage and earthquake environmental effects related to shallow intraplate seismicity of central western Argentina

STELLA MARIS MOREIRAS* & MARÍA SOLANGE PÁEZ

CONICET–IANIGLA (CCT), Avenida Dr Ruiz Leal s/n Parque, 5500, Mendoza, Argentina

*Corresponding author (e-mail: moreiras@mendoza-conicet.gob.ar)

Abstract: Central western Argentina is identified as the most hazardous seismic zone in the country. Historical earthquakes with magnitudes greater than $M_s > 6.0$ frequently occur in this territory and are associated with the subduction of the Nazca plate. However, seismic hazards have not been fully assessed in this region. No secondary seismic effects of a potential earthquake with destructive consequences have been considered, nor has the existence of shallow Quaternary blind faults been identified by seismic surveys. Neotectonic studies performed up to the present describe only those Quaternary faults with some surficial expression. Lacking proper hazard assessment limits strategies to reduce the economic impact of drastic seismic events. This chapter is focused on the impact of major destructive earthquakes that have occurred in central western Argentina in order to understand the incidence of these phenomena in the past and to consider the vulnerability of the region in light of its increased urbanization and changing agricultural practices.

Central western Argentina (CWA), between the southerly latitudes of 28° and 34°, is the most hazardous seismic zone in the country (Fig. 1), as evidenced by numerous high magnitude intraplate earthquakes reported since 1575 (INPRES 1985). This high seismic hazard is related to the subduction of a slab segment at these latitudes which includes the eastern extension of the Juan Fernandez Ridge and is characterized by shallow earthquakes above the Wadati–Benioff zone (Ramos 1996). Seismic movements usually occur in the mountainous area or the eastern piedmont with magnitudes just exceeding 7.0 (Santibañez Rodriguez 2006). The most violent earthquake recorded in Argentina's territory, known as the Great Argentinean earthquake (surface-wave magnitude (M_s) c. 8), occurred within this segment in 1894. Further destructive earthquakes were later reported in 1927, 1929, 1944 and 1977 (Fig. 1). All of these strong events led to secondary effects such as liquefaction features, terrain cracks or landslides; however, they were rarely reported. Usually, economic losses in the populated cities were estimated by taking into account damage to main buildings: churches, schools, homes of famous people, or government facilities. Specific studies made by geophysicists or researchers are less common than the economic reports, and past historic observations in remote mountain areas are extremely rare (Burmeister 1861; Forbes 1861; Bodenbender 1894; Loos 1907, 1926, 1928; Lünkenheimer 1930; Harrington 1944). Occasionally, foreign travellers, shocked by the earthquakes' impacts, mentioned secondary effects in their books (Rickard 1863; Morey 1938).

Lack of specific studies about secondary earthquake effects, together with failures in our collective memory of damage caused by past earthquakes, prevent an accurate seismic risk assessment for this region. The population increases that have occurred over the last centuries can be expected to translate into more people dying from earthquakes in the future. The aim of this chapter is to evaluate the probable catastrophic consequences of an extreme seismic event in this highly seismic zone of Argentina.

Geographical features

CWA encompasses part of the arid central Andes with peaks reaching 7 km in altitude. This high mountain landscape is an international tourist attraction. At least 7000 tourists visit the Aconcagua Park annually (Aconcagua mountain reaches 6958 m above sea level (a.s.l.)). The topography decreases towards the eastern piedmont (700 m a.s.l.) where the main capital cities are located: Mendoza with a population of 800 000 inhabitants and San Juan with 120 000 inhabitants. Both cities are irrigated oases where agriculture is the primary economic activity. The production of high quality wine destined for a global market has undergone a boom in the last two decades; this has increased the pressure on terrains for growing varietal grapes, which has resulted in the development of new vineyards just along the foot of the Andes. Intensive artificial irrigation is required due to the arid climate, and this causes a modification of the original water-table

From: SEPÚLVEDA, S. A., GIAMBIAGI, L. B., MOREIRAS, S. M., PINTO, L., TUNIK, M., HOKE, G. D. & FARÍAS, M. (eds) 2015. *Geodynamic Processes in the Andes of Central Chile and Argentina.* Geological Society, London, Special Publications, **399**, 369–382. First published online February 19, 2014, http://dx.doi.org/10.1144/SP399.6
© 2015 The Geological Society of London. For permissions: http://www.geolsoc.org.uk/permissions.
Publishing disclaimer: www.geolsoc.org.uk/pub_ethics

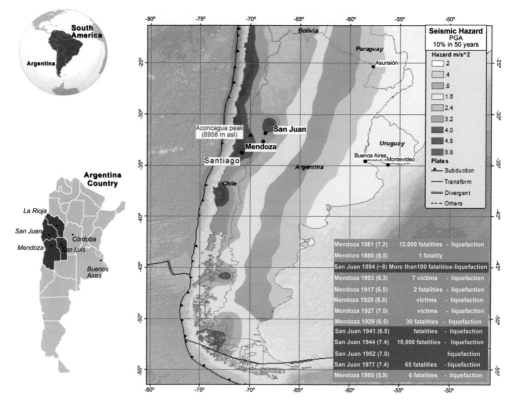

Fig. 1. Seismic hazard map of Argentina showing hazard as peak ground acceleration (PGA) with 10% probability of exceedence in 50 years (modified from the USGS website). A table with fatalities and liquefaction records that were documented following main historical earthquakes is presented, as well as the location of central western Argentina, encompassing Mendoza and San Juan provinces.

level, which is critical for liquefaction phenomena. Less populated villages exist in the vast remote mountain areas; however, only a few of these communities contain more than 1000 inhabitants (e.g. Uspallata village, Mendoza, with c. 10 000, or Barreal, San Juan, with c. 2600 (INDEC 2010)). These villages and their connecting roads have been widely affected by rock falls triggered by local and distant earthquakes. The Maule earthquake (movement magnitude (M_w) 8.8) in 2010, for example, triggered several rock falls in the Aconcagua region (Wick et al. 2011) despite the fact that the epicentre was 250 km away.

Tectonic setting

CWA experiences intense seismic activity as a consequence of the subducting slab fragment of the Nazca plate beneath the South American plate at these latitudes (28°–32°S), with a subduction angle lower than 14°. The flattening of the Nazca Plate started 8–10 Ma (Jordan & Gardeweg 1987; Kay et al. 1991), a process that is linked to the subduction of the eastern extension of the Juan Fernandez Ridge, the shutting off of the volcanic arc and the uplift of the Sierras Pampeanas basement.

The compressive stress generated by the westward displacement of the Nazca boundary at an absolute rate of 2.2 cm a^{-1} (Uyeda & Kanamori 1979) results in dislocations in local and regional faults (Strauder 1973; Barazanghi & Isacks 1976; Smalley & Isacks 1987, 1990; Smalley et al. 1993). This displacement rate could have been different in magnitude and direction during the geological past, thus giving rise to a complex mosaic of movements. In fact, the convergence of the Nazca and South American plates began about 200 Ma with the orogenic front migrating progressively towards the east.

The presently active orogenic front between 32° and 33°S is located in the piedmont of CWA (Fig. 2) (Bastías et al. 1984, 1990, 1993; Bastías 1985; Cortés et al. 1999; Costa et al. 2000a). This

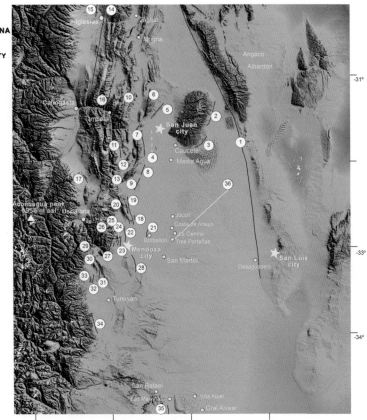

Fig. 2. Active orogenic front of central western Argentina between 32° and 33°S evidenced by faults with Quaternary activity: those with surficial expression in red (some are documented historical ruptures) and blind faults are in white.

area shows an intense neotectonic activity evidenced by Quaternary faulting and seismic activity (Costa *et al.* 2000a, b; Siame *et al.* 2006) that corresponds to shallow intraplate earthquakes with depths <30 km. A regional deformation rate of c. 3 mm a^{-1} has been established by GPS for this sector of the Precordillera. However, we do not know whether this deformation is concentrated within the 50 km width of the orogenic front or in some particular structure (Brooks *et al.* 2003; Kendrick *et al.* 2006; Schmidt *et al.* 2011; Salomon *et al.* 2013).

Structurally, the active orogenic front is composed of a series of parallel inverse faults with both east and west vergence that present geomorphological evidence of Quaternary displacements (Moreiras *et al.* 2013). Towards the south, these structures correspond to anticlines sourced from the tectonic inversion of Triassic basins and the low-angle imbrication affecting Quaternary deposits (Brooks *et al.* 2000; Chiaramonte *et al.* 2000; Vergés *et al.* 2007; Ahumada & Costa 2009). Quaternary faults and folds with surficial expression have been described in Costa *et al.* (2000a) but potential hazard assessment is ignored. Moreover, activity of blind faults identified by seismic surveys in the north of Mendoza province is commonly disregarded (Fig. 2).

Seismicity

Seismicity varies from western towards eastern Argentina, as well as from the northern to the southern portion of the Andes, being markedly higher in CWA (Fig. 3). In this region, there are two types of seismicity: (a) **interplate seismicity** associated with deeper earthquakes sourced just from the contact area of the Nazca and South American plates, about 125 km in depth at the latitude of Mendoza city (Ramos 1988) (Fig. 3c); and (b) **intraplate seismicity** associated with shallow earthquakes

Fig. 3. Seismicity of central western Argentina showing earthquakes with epicentre 0–70 km in depth in red, 7–299 km in depth in green, and 300–700 km in depth in blue: (**a**) location of $M_s > 4.0$ earthquakes during the last decades (since 1965: NEIC); (**b**) regional map showing existence of intraplate and interplate seismicity for the study area (USGS website); (**c**) depth of Nazca plate along the flat subduction segment (28°–32°S).

originating from active structures on the eastern piedmont where main cities are located (Fig. 3a, b).

The seismic source proposed for the most catastrophic earthquake of Mendoza city, which occurred on 22 March 1861, was a shallow intraplate quake located in this segment. It is linked to the reactivation of the northern sector of Cerro La Cal fault (Mignorance 2006; Schmidt *et al.* 2011).

At present, distribution of shallow earthquakes (<30 km in depth) shows a certain alignment with active sub-parallel faults with a north–south trend in the Andean piedmont. This seismicity is rarely greater than M_s 4.5, showing thrust mechanism earthquakes linked to regional compression stress. However, strike-slip and normal faulting mechanisms have been found for low-magnitude earthquakes in the Mendoza piedmont, probably responding to local effects like variations in cortical densities, post-seismic relaxation or accommodation of basal blocks (Moreiras *et al.* 2013).

Historical records and damage

Seismic records for CWA are very restricted due to the short documented history of South America, preserved only since the Spanish conquest in the seventeenth century. However, the studied region has suffered at least six destructive earthquakes of $M_s > 7.0$ over the last 150 years. The three most destructive earthquakes associated with surface ruptures occurred in this region in 1894 (M_s c. 8), 1944 (M_s 7.4) and 1977 (M_s 7.4), and had devastating effects on local settlements and capital cities nearby (Moreiras 2004) (Table 1).

The 1861 earthquake (M_s 7.2) resulted in 6000–12 000 fatalities, the numbers depending on the historical source, which represent at least one-third of the population of Mendoza at that time (Loos 1907). Moreover, thousands of people were injured. This shallow earthquake was estimated to have had an intensity of between IX–X (on the Modified Mercalli (MM) intensity scale), and destroyed all homes (*c.* 2000) and main churches in Mendoza city (Fig. 4). August Bravard, a French geologist who had predicted this catastrophic earthquake a year beforehand, died in this event.

The largest historic Argentinean earthquake occurred in 1894 and caused more than 100 fatalities and severe damage in San Juan, La Rioja, Córdoba and Mendoza provinces. Several old religious buildings were seriously damaged (La Merced, San Agustín, the chapel of Dolores and the Church of St Pantaleón, and Santo Domingo, the latter being under construction at the time). Many houses were in ruins and numerous public buildings destroyed, such as the Government facility, the legislature, the public market, the military barracks, Los Andes theatre and the Franklin library. A bridge was damaged in Alto de Sierra, and the breakdown of irrigation canals produced severe damage to cultivated fields in Angaco and Albardón departments (Bodenbender 1894).

Only two fatalities were reported during the 1917 earthquake (Panquehua, Mendoza) and some damage occurred in Mendoza city, and in Santa Rosa, Corralitos, and San Rafael departments. Las Heras department was the most severely affected. This earthquake was noticed in San Juan (Pocito and Caucete), Buenos Aires and Chile.

The epicentre of the 1920 earthquake (M_s 6.3) was Costa de Araujo, which was the most affected city, where the earthquake reached an intensity of X (MM), that is, total destruction. Also Lavalle, La Central, Tres Porteñas, Borbollón, Jocolí and Colonia André localities were severely damaged (VI MM). A total of 250 injuries were recorded, streets were flooded by groundwater coming to the surface at Jocolí, Corralitos, San Martín and Alto Verde localities. At least 1000 ha of cultivated land was flooded and marshy invaded Alto Verde locality.

The next dangerous quake occurred in 1927, impacting on Mendoza city. Government buildings and several schools were completely destroyed and the quake was catastrophic for Resbalón locality (Las Heras). Three died and more than seven people were injured. The movement was felt in La Rioja, Córdoba, and Buenos Aires provinces, and damage was reported in San Juan province. The Villa Atuel–Las Malvinas earthquake in 1929 also affected several provinces (Mendoza, San Juan, San Luis, Cordoba, Santa Fe, Buenos Aires, La Pampa and Neuquén).

The Caucete earthquake (M_s 6.8) in 1941 affected Caucete, 25 de Mayo, Albardón, Angaco and Sarmiento departments. Communication systems and the Gral Belgrano railway were severely damaged.

The hypocentre of the earthquake of 15 January 1944 (La Laja, San Juan; M_s 7.4) was located at 11 km depth (Alvarado *et al.* 2005; Alvarado & Beck 2006) and associated with La Laja fault in the eastern Precordillera system. It reached an intensity of IX (MM) and resulted in 6000–10 000 fatalities and the destruction of 80 per cent of homes. During this event, a vertical displacement of 22 cm and a horizontal displacement of 25 cm were measured (Harrington 1944). In the 1977 earthquake (Caucete, San Juan, M_s 7.4), surficial ruptures were documented along the Ampacama–Niquizanga fault (Volponi *et al.* 1978; Bastías 1985) with a vertical displacement of 30 cm. Fatalities were estimated at 125 and more than 40 000 were left homeless after the total destruction of Caucete city.

San Juan was hit again in 1952 (La Rinconada earthquake), with the most affected areas being Pocito, Zonda, and Ullum departments with an intensity of VIII (MM). Then, the shallow 1977 earthquake (Caucete–San Juan) (M_s 7.4, 17 km depth) resulted in 70 fatalities, up to 200 people injured and considerable structural damage (INPRES 1993). The most affected areas were the

Table 1. Damages and environment effects caused by historical earthquakes of Central western Argentina

Date	Epicentre	Seismic source *	S	W	D	Ms	I	Mw	Io	Damages and environment effects
Previous XX century										
02-03-1561	Mendoza-San Juan					7?	VI			NO SPECIFIC DATA
22-05-1782	Santa Rita (Mendoza)	Melocotón?	32°42'	69°12'	30	6.5/7	VIII			No victims. Damages in buildings.
10-10-1804	Mendoza-San Juan		31°40'	67°59'			VI			NO SPECIFIC DATA
23-11-1857	Mendoza									NO SPECIFIC DATA
20-03-1861	Mendoza	Cerro La Cal -Salagasta	32°54'	68°54'	30	7.2	IX-X	c. $7^{(1)}$	XI	12 000 victims and 747 people were reported as severely injured (Total population 18 000). Mendoza completely destroyed. Santo Domingo and San Francisco temples collapsed. Government facility was severely damaged. Main square disappeared. Fires and floods. Lacking of food and potable water. Liquefaction and landslides.$^{(1)}$
1876	Mendoza-San Juan									Several fatalities and injured people.
19-08-1880	Cacheuta Mendoza		33°	69°		5.5	VI			One fatality in the Melocotón, Tunuyán. Minor damages on buildings.
27-10-1894	La Punilla North San Juan	Punilla -Majaditas	29°45'	69°00'	30	c. 8(7.5)	IX	$7.8^{(2)}$	XII	Over 100 fatalities and severe damage in San Juan, La Rioja, Córdoba and Mendoza. Felt in Buenos Aires, Santiago, Chile, South Brazil, and Perú. Several religious buildings were seriously damaged: La Merced, San Agustín, and Santo Domingo, the chapel of Dolores and the Church of St. Pantaleón. Lot of houses were in ruins. Many public buildings destroyed: Government facility, legislature, public market, barrack, Los Andes theatre and Franklin library. A bridge was damaged in Alto de Sierra. Liquefaction and landslides were widely reported in Iglesia department. At least 50 holes 0.5 m-in with and 3–4 m deep appeared in Angaco. Several rockfall and cracks along Copiapó route. Landslides in the Cura valley and Colangüil. Temporal streams took into a new path generating lagoons. Generation of swamps in Blanco river.$^{(2)}$
12-04-1899	Jagüé (La Rioja)	Jagüé	28°39'	68°25'	30	6.4	VIII		VIII	Destroyed Jagüé village and caused important damages in several La Rioja towns.
XX Century										
12-08-1903	Uspallata (Mendoza)		32°06'	69°06'	70	6.3	VIII		IX	Damages in Mendoza city, Guaymallén and Godoy Cruz departments. Seven fatalities. Landslides in Uspallata, Puente del Inca and Punta de Vacas localities. Water ejection, fissures and cracks appeared in several places.

Date	Location	Fault/Source	S	W	D	Ms	Mw	I	Io	Description
26-07-1917	Panquehua (Mendoza)	Salinas	32°20'	68°54'	50	6.5		VII	VIII	2 fatalities. Several damages in buildings and churches In Mendoza and San Juan. Felt in Córdoba and Tucumán. Liquefaction phenomena in Mendoza.
17-12-1920	Costa de Araujo (Mendoza)	San Martín blind fault?	32°42'	68°24'	40	6.3 / 6.8		X - VI	X	250 injured people and building damages around 50 km of epicentre. Liquefaction and landslides. Swamps were generated.
19-03-1924	La Cantera (San Juan)	La Cantera	32°24'	69°18'	> 60	6.2		VIII	IX	Government buildings and several schools were completed destroyed. Damages in Chile. Liquefaction, terrain fissures and landslides.
14-04-1927	Uspallata (Mendoza)					7.1 / 7.5				
30-05-1929	Las Malvinas – Villa Atuel (Mendoza)	Malvinas	34°54'	68°	40	6.5		VII	VIII	30 fatalities. Building damages and several victims in Las Malvinas-Villa Atuel, San Rafael. Liquefaction features.
23-10-1936	Barrancas (Mendoza)	Barrancas				6		VI		Homes damages in Maipú and Rivadavia departments. Damages, liquefaction and landslides
3-07-1941	Caucete (San Juan)	Ampacama				6.8			VIII	Cañada Seca, Las Malvinas and Salto de Las Rosas villages were severely damaged.
5-06-1942	San Rafael (Mendoza)	Malvinas?						VI		
15-01-1944	La Laja (San Juan)	La Laja - Precordillera	31°24'	68°24'	30	7.4	7[3]	IX	IX	San Juan city was destroyed, 75% of houses collapsed. 10 000 fatalities (total population of 90 000). Liquefaction and landslides.[3]
11-06-1952	Rinconada (San Juan)	Rinconada-Precordillera	31°36'	68°35'	30	7.0	6.8[3]	VIII	IX	Liquefaction, landslides and building damages.[3]
24-10-1957	Villa Castelli (La Rioja)		28°54'	68°	37	6.0		VII		Building damages
25-04-1967	Uspallata (Mendoza)					5.4 / 5.7		VI		Several damages in Las Heras and Mendoza city.
23-11-1977	Caucete (San Juan)	Ampacama				7.4		IX	IX	Fatalities 65 and 300 injured persons. Liquefaction and landslides
26-01-1985	Lunlunta (Mendoza)	East Barrancas	33°06'	68°30'	12	5.9		VIII	VII	Six fatalities, 238 injured persons and 12 500 destroyed houses.
05-08-2006	Lunlunta (Mendoza)	East Barrancas	33°06'	68°30'		5.7		VII		Serious damages on buildings of Maipú and Luján departments.

S: South Latitude, W: West longitude, D: depth in km, Ms: Surface-wave magnitude, I: MM intensity, Mw: Movement magnitude, and Io: epicentral intensity (ESI, 2007).

*seismic sources taken from Costa et al. (2000a)
[1] Salomon et al. (2013)
[2] Pérez & Costa (2012)
[3] Alvarado & Beck (2006)

Fig. 4. Damage caused by the 1861 earthquake (M_s 7.2): (**a**) picture showing ruins of Mendoza city; (**b**) total destruction of the Main Square – today known as Pedro del Castillo – where social, economic and political activities took place; (**c**) Santo Domingo Church in 1880 after the earthquake; (**d**) fires (in the most modern stores that had a gas supply) caused by the earthquake and that lasted for four days; likewise floods were caused by the obstruction of canals.

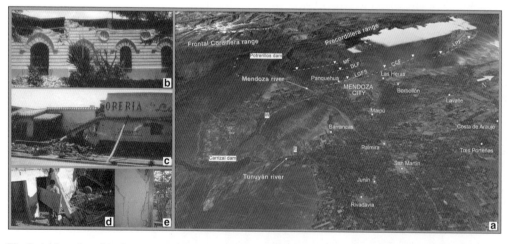

Fig. 5. (**a**) Location of the Barrancas anticline with Quaternary activity faults in both flanks; the eastern fault was reactivated in 1985, 2001, 2006 and 2012, affecting Mendoza city. Nearby active structures are indicated in dashed white lines: MF, Melocotón fault; DLF, Divisadero Largo fault; LGFS, La Gloria fault system; CCF, Cerro La Cal fault; LPF, Las Peñas fault; (**b**) Hospital del Carmen, Godoy Cruz department, severely damaged during the 1985 earthquake (M_s 5.9, 18.7 km depth); (**c**) store partially collapsed as a consequence of the same event; (**d**) a home affected in Maipú department during the earthquake of 2006 (M_s 5.7, 25 km depth); (**e**) damage caused to a building in Maipú department in the 2012 earthquake (M_s 5.3, 22.8 km depth).

departments of Caucete, Angaco, Valle Fértil, Albardón, 25 de Mayo, Pocito, San Martín, 9 de Julio, Sarmiento, and Santa Lucía. During this earthquake, a permanent ground deformation of 1.20 m was measured by geodetic levelling, and displacement of scarps was 0.30 m (Bastías 1985).

The 1985 earthquake (M_s 5.9) affecting Mendoza city was associated with a fault located on the eastern flank of the Barrancas anticline (Chiaramonte et al. 2000). Recent reactivation of this fault was recorded in 2001 (M_s 5.6), 2006 (M_s 5.7; 25 km depth) and 2012 (M_s 5.3, 22.8 km depth) (Fig. 5).

Ground seismic effects

Ground seismic phenomena have had a key role in the destructive earthquakes of CWA. Historic books widely described landslides as fire balls and terrain cracks, and liquefaction features as mud volcanoes. The 1861 earthquake (M_s 7.2) generated surficial ruptures, fissures and soil liquefaction in Buena Nueva and Las Ciénagas. The biggest hole reached 2 m in width. Sand volcanoes and mud flows were reported near Capilla del Rosario, Lavalle (Verdaguer 1929). After this strong quake, the Obispo baths were formed in the Borbollón area into which stinking gases and water flowed (Ponte 1987). Landslides were reported near Uspallata (Morey, in RJEHM 1936), and David Forbes (in RJEHM 1939), an English geologist who travelled from Uspallata to Puente del Inca during the evening of the 1861 earthquake, described huge landslides after a violent explosion, as well as fissures forming in the Precordillera.

Liquefaction phenomena spreading out to a distance of more than 200 km from the epicentre were described after the 1894 earthquake (Bodenbender 1894) (Fig. 6b, c & d). Water rich in sulphur and sulphates flowed through fissures up to 0.5 m wide and land collapsed. These features were identified in Angaco, Albardón, San Martín, and Caucete departments. Landslides, rock falls along cliffs and many fissures were reported along the Copiapo route (San Juan–Chile) (El Debate 1894; La Unión 1894). Terrain cracks were observed along 50 km across the Matagusanos Valley (Morey 1938). Temporal streams taking new paths generated dammed lagoons and several swamps appeared in the Blanco River. Landslides were reported in the Cura Valley and Colangüil. These types of processes, being observed up to 400 km from the epicentre zone of the 1894 earthquake, affected an area of 110 000 km^2 (Pérez & Costa 2012). Historically, a magnitude 8.2 was assigned to this event, being later reduced according to the length of the fault rupture to M_s 7.5 (Bastías 1985; Castano 1993), M_s 8 (Alvarado et al. 2005) and M_s 7.6 (Alvarado & Beck 2006). However, documented ground seismic effects led to an assignment of M_s c. 8 for this extraordinary event. Pérez & Costa (2012) calculated M_w 7.8 for this earthquake based on the relationship between epicentral distance and distribution of liquefaction phenomena and gravitational processes.

Terrain fissures and cracks from which warm water sprang appeared in Canota during the 1903 earthquake (Uspallata, Mendoza, at M_s 6.3), as well as in Panquehua during the 1917 earthquake (M_s 6.5) (Los Andes 1903, 1917). Sand volcanoes were also reported in Las Heras department (La Palabra 1917).

Numerous fissures and wells with abundant emerging sand and gushing water were generated at La Central and Tres Porteñas localities (San Martín, Mendoza) by the 1920 earthquake. The largest sand volcano was 2.5 m wide and 3 m deep. Many craters were spaced at 40–50 m from each other, and fissures with a north–south trend and up to 0.2–0.3 m wide opened up near the Mendoza River in Costa de Araujo (Fig. 6e). At Lavalle locality, many wells opened up, but in some places where water pressure was probably lower, the soil rose up forming a convex circle 4–5 m wide and 1 m high (Morey 1938; Los Andes 1920). A similar process took place at Resguardo locality (Las Heras) during the 1927 earthquake. An increase in the volume of water springing from hot spots in the Villavicencio thermal baths was reported as well.

A shallow water-table (0.95 m) and loess soils favoured the formation of craters with surging hot water (18 °C) in Villa Atuel during the 1929 earthquake (M_s 6.5) (Lünkenheimer 1930) (Fig. 6f). The largest, c. 2 m in length, were observed in the southern part of Villa Atuel. Embankment subsidence of up to 75 cm was observed in the Pacific Railway. Another important earthquake took place in 1941 (M_s 6.7); among the recorded effects were numerous liquefaction phenomena and landslides (Moreiras 2004; Perucca & Moreiras 2006).

During the 1944 earthquake (Fig. 6g, h), there were reports of liquefaction features (springs, sand volcanoes, and sandy craters) and fissures (Fig. 6i, j). These types of phenomena were widely observed in Albardón, and the neighbouring localities of Ampacama and Bermejo. A bridge crossing the San Juan River collapsed due to soil liquefaction. These features were also reported in the previous 1941 Caucete (M_s 6.8) earthquake. Cracks and fissures in cultivated fields and liquefaction phenomena were likewise observed at Zonda locality during the 1952 Rinconada earthquake.

Liquefaction features played a major role in the 1977 earthquake (M_s 7.4), causing great damage

Fig. 6. Pictures of damage and secondary effects reported during destructive earthquakes of central western Argentina: (a) San Juan city after the 1894 earthquake; (b) terrain crack in Mogna documented by Bodenbender (1894); (c–d) sand volcanoes in Angaco (Bodenbender 1894); (e) sand volcanoes from which water emerged at Tres Porteñas locality, caused by the Central Costa de Araujo earthquake (1920); a pencil is used to indicate scale (Los Andes 1920); (f) craters and volcanoes flowing with warm water (18 °C) at Villa Atuel locality in the 1929 earthquake, San Rafael (Lünkenheimer 1929); (g–h) view of San Juan city after the 1944 earthquake; (i) fissure on a road in 1944; (j) liquefaction feature caused by 1940 earthquake; (k–l) fissures on roads caused by 1977 earthquake; (m) liquefaction processes along the San Juan River; (n) a bridge that had collapsed during this event (INPRES 1977).

and economic loss (Fig. 6k–n); these features extended over 2000 km², being reported up to 70 km away from the epicentre. Lateral spreading, craters, sand volcanoes and spring waters were widely observed. The largest cracks were 100 m long and several metres wide (INPRES 1977).

Earthquakes have been closely related to changes in groundwater tables. A general rise of the groundwater table during 1919–1920 in Corralitos, Buena Nueva, Lavalle and San Martin was associated with the 1920 earthquake. Subsequent to this event, the water-table level went up markedly at Valle Hermoso and Tres Porteñas localities. Some artesian wells dried up and several marshes also desiccated (Loos 1928). During the 1927 earthquake, groundwater emerging in dry areas formed marshes that obstructed traffic in western and eastern Mendoza. Springs of La Laja locality showed a rapid increase in their water volume after the 1944 seismic event (Harrington 1944), and the volume doubled after the 1894 earthquake (El Debate 1894).

Regional seismic hazard

As described above, a large number of earthquakes have been reported in CWA during historical times; however, no clear relationship exists between the main Quaternary structures and seismic epicentres. Rarely have historical rupture surfaces been documented; records exist for La Laja fault in the 1944 earthquake and Niquizanga fault in the 1977 earthquake, and recent (1985) seismic activity has been linked to the fault located on the eastern flank of the Barrancas anticline. Absence of a surficial expression of Quaternary faults precludes characterization of their potential seismic risk using basic parameters. Besides, empirical relationships that have been observed worldwide and that are used in neotectonic studies for establishing earthquake magnitude (displacement, long rupture, etc.) could underestimate cortical earthquake magnitude in the study area (Costa *et al.* 1999; Costa 2005). Also, the existence of blind faults is commonly disregarded in this region. This geological dilemma, combined with uncertain earthquake prediction and a weak collective memory of earthquake impacts, makes Andean communities more vulnerable.

In fact, the growing population pressure during the last decades has led to expansion on to unstable terrains. Urbanization of Mendoza and San Juan cities advances on to the western piedmont, an area affected by potential sources of shallow seismicity (Frau & Saragoni 2005). Still, moderate seismic activity ($M_s > 4.0$) predominates in the region, and its morphological features allow determination of at least a maximum probable earthquake of M_s 6.4 in the case of the Mendoza piedmont (Frau *et al.* 2010) and M_s 6.4–6.9 in the eastern Precordillera fault system (Martos 1995; Perucca & Paredes 2003, 2004). Low recurrence has usually characterized this type of cortical fault but coseismic displacement of >1 m has been recorded in the region, showing a greater seismic capacity than suggested by international catalogues (Costa *et al.* 2006). Nonetheless, numerous neighbourhoods, schools and dams are located just above fault traces that are lacking in land planning that considers seismic hazard.

Likewise, the importance of ground seismic effects is ignored, so seismic-resistant buildings are not strong enough to mitigate shallow seismicity in CWA. As far as we know, liquefaction could be catastrophic near the epicentre zone when the terrain is prone to this phenomenon (Table 1). An earthquake's magnitude restricts the maximum distance of liquefaction features from the epicentre; however, historical liquefaction processes in CWA have occurred at a greater distance than expected for a determined magnitude (Perucca & Moreiras 2006). Sand volcanoes appeared in Desaguadero village (Bermejo River) in the 1977 earthquake (M_s 7.4) located 260 km away from the epicentre, greatly exceeding the distance of 125 km and 221 km estimated by Kuribayashi & Tatsuoka (1975) and Papadopulos & Lefkopulos (1993), respectively.

Besides, terrain conditions in cultivated areas have been shifted due to human management. At present, those areas liquefied during historical earthquakes have a deeper water-table level, whereas piedmont areas where the water-table is recharged are saturated by irrigation water. This implies a new scenario for liquefaction occurring in a potential quake of $M_s > 5.0$. Historical liquefaction effects are often almost exclusively restricted to alluvial plain, palaeochannels and floodplain deposits; however, present conditions could produce liquefaction in saturated alluvial deposits located near a potential seismic source.

Conclusion

Over the last centuries, numerous earthquakes of $M_s > 7.0$ have occurred in CWA, inflicting catastrophic damage to cities located in the Andes foothills, with a large negative impact on the regional economy. In addition, these earthquake events have greatly altered the landscape, and terrains have suffered from subsequent environmental effects. Liquefaction features have been documented in areas distant from the epicentre, whereas landslides have been reported in remote mountain areas. Unfortunately, a growing population and urban expansion into these areas increase the vulnerability of Andean communities who have short collective memories of earthquake shock and lack preventive measures to cope with seismic hazard. This critical situation may bring about a natural disaster in the decades to come.

This work was done within the framework of several research projects headed by S. Moreiras: PIP 112-200801-00638 (2009–2011), SECYT 06/A441 (2009–2011), SECYT 06/G550 (2009–2011) and SECYT 06/A519 (2011–2013). Much historical material has been obtained from the INPRES web page; data and location of shallow earthquakes were recorded from the NEIC (National Earthquake Information Centre: USGS). We are grateful to both anonymous reviewers for useful comments that helped improve the original manuscript and Ms Horak and C. Stern for their review of the English.

References

AHUMADA, E. A. & COSTA, C. H. 2009. Deformación cuaternaria en la culminación norte del corrimiento las Peñas, frente orogénico andino, Precordillera Argentina. *XII Congreso Geológico Chileno*, Santiago, Chile.

ALVARADO, P. & BECK, S. 2006. Source characterization of the San Juan (Argentina) crustal earthquakes of 15 January 1944 (M_w 7.0) and 11 June 1952 (M_w 6.8). *Earth and Planetary Science Letters*, **243**, 615–631.

ALVARADO, P., BECK, S., ZANDT, G., ARAUJO, M. & TRIEP, E. 2005. Crustal deformation in the south central Andes backarc terranes as viewed from regional broadband seismic waveform modelling. *Geophysical Journal International*, **163**, 580–598.

BARAZANGHI, M. & ISACKS, B. 1976. Spatial distribution of earthquakes and subduction of the Nazca plate beneath South America. *Geology*, **4**, 686–692.

BASTÍAS, H. 1985. *Fallamiento Cuaternario en la Región Sismotectónica de Precordillera*. Degree thesis. Universidad Nacional de San Juan, 147p.

BASTÍAS, H., WEIDMANN, N. & PÉREZ, A. 1984. Dos zonas de fallamiento Plio-Cuaternario en la Precordillera de San Juan. *In*: *IX Congreso Geológico Argentino, Río Negro*. 329–341.

BASTÍAS, H., ULIARTE, E., PAREDES, J., SANCHEZ, A., BASTIAS, J., RUZYCKI, L. & PERUCCA, L. 1990. Neotectónica de la provincia de San Juan. *In*: *11th Congreso Geológico Argentino & Relatorio de Geología y Recursos Naturales de la provincia de San Juan*. SEGEMAR, Argentina, 228–245.

BASTÍAS, H., TELLO, G. E., PERUCCA, J. L. & PAREDES, J. D. 1993. Peligro sísmico y neotectónica. 12th Congreso Geológico Argentino y 2th Congreso de Exploración de Hidrocarburos. *In*: RAMOS, V. A. (ed.) *Geología y Recursos Naturales de Mendoza*. Relatorio, Mendoza, **6**, 645–658.

BODENBENDER, G. 1894. El terremoto argentino. Del 27 de octubre de 1894. *Boletín de la Academia Nacional de Ciencias en Córdoba*, **14**, 293–329.

BROOKS, B. A., SANDVOL, E. & ROSS, A. 2000. Fold style inversion: placing probabilistic constraints on the predicted shape of blind thrust faults. *Journal of Geophysical Research*, **105**, 13281–13302.

BROOKS, B. A., BEVIS, M. ET AL. 2003. Crustal motion in the Southern Andes (26°–36°S): Do the Andes behave like a microplate? *Geochemical Geophysical Geosystems*, **4**, 1085, http://dx.doi.org/10.1029/2003GC000505

BURMEISTER, C. 1861. *Reise durch die La Plata-Staaten*. H.W. Schmift, Halle, **I**, 38–76.

CASTANO, J. 1993. *La verdadera dimensión del problema sísmico en la provincia de San Juan*. Publicación Técnica INPRES, San Juan, N°18.

CHIARAMONTE, L., RAMOS, V. A. & ARAUJO, M. 2000. Estructura y sismotectónica del anticlinal de Barrancas, cuenca cuyana, provincia de Mendoza. *Revista de la Asociación Geológica Argentina*, **55**, 309–336.

CORTÉS, J. M., VINCIGUERRA, P., YAMÍN, M. & PASINI, M. M. 1999. Tectónica cuaternaria de la Región Andina del Nuevo Cuyo (28°–38° LS). *In*: *Geología Argentina, Servicio Geológico Minero Argentino*. Subsecretaría de Minería, Buenos Aires, **29**, 760–778.

COSTA, C. 2005. The seismogenic potential for large earthquakes at the southernmost Pampean flat-slab (Argentina) from a geologic perspective. *In*: *Proceedings 5th International Symposium on Andean Geodynamics*, Barcelona, **6**, 211–214.

COSTA, C., ROCKWELL, T., PAREDES, J. & GARDINI, C. 1999. Quaternary deformations and seismic hazard at the Andean orogenic front (31°–33°, Argentina): a paleoseismological perspective. *In*: *Proceedings 4th International Symposium on Andean Geodynamics*, Barcelona, **6**, 187–191.

COSTA, C. H., MACHETTE, M. N. ET AL. 2000a. *Map and database of quaternary faults and folds in Argentina*. USGS Open-File Report 00-0108.

COSTA, C. H., GARDINI, C. E., DIEDERIX, H. & CORTÉS, J. M. 2000b. The Andean orogenic front at Sierra de Las Peñas-Las Higueras, Mendoza, Argentina. *Journal of South American Earth Sciences*, **13**, 287–292.

COSTA, C. H., AUDEMARD, F. A., BEZERRA, F. H., LAVENU, R. A., MACHETTE, M. N. & PARÍS, G. 2006. An overview of the main quaternary deformation of South America. *Revista Asociación Geologica Argentina*, **61**(4), 461–479.

El Debate newspaper, 29th October, 1894.

FORBES, D. 1861. Informe sobre el terremoto de Mendoza. *In*: *Revista de la Junta de Estudios Históricos de Mendoza*, **10**, 111–120, Primera Epoca (1938), Mendoza.

FRAU, C. & SARAGONI, G. R. 2005. Demanda sísmica de fuente cercana. Situación del Oeste Argentino. Congreso Chileno de Sismología e Ingeniería Antisísmica. IX Jornadas. Concepción de Chile.

FRAU, C., GALLUCI, R. A., MOREIRAS, S. M. & GIAMBIAGI, L. B. 2010. Estudio de Peligrosidad Sísmica para el Proyecto Urbano Palmares Valley. INFORME TÉCNICO N°16/010. CEREDETEC elaborado para Palmares Valley S.A.

HARRINGTON, H. 1944. El sismo de San Juan del 15 de enero de 1944. Corporación para la Promoción del Intercambio S.A. 79p.

INDEC. *Censo Nacional de Población, Hogares y Viviendas 2010*. Argentina.

INPRES. 1977. El terremoto de San Juan del 23 de noviembre de 1977. Informe Preliminar, 102p.

INPRES. 1985. El terremoto de Mendoza Argentina del 26 de enero de 1985. Informe general.
INPRES. 1993. *La verdadera dimensión del problema sísmico en la provincia de San Juan.* Publicación Técnica 18, San Juan.
JORDAN, T. & GARDEWEG, M. 1987. Tectonic evolution of the late Cenozoic Central Andes. *In*: BEN AVRAHAM, Z. (ed.) *Mesozoic and Cenozoic Evolution of the Pacific Margins.* Oxford University Press, New York, 193–207.
KAY, S. M., MPODOZIS, C., RAMOS, V. A. & MUNIZAGA, F. 1991. Magma source variations for mid–late Tertiary magmatic rocks associated with shallowing zone and thickening crust in the central Andes (28° to 33°S). *In*: HARMON, R. S. & RAPELA, C. W. (eds) *Andean Magmatism and its Tectonic Setting.* Geological Society of America, Special Paper, **265**, 113–137.
KENDRICK, E., BROOKS, B. A., BEVIS, M., SMALLEY, R., LAURIA, E., ARAUJO, M. & PARRA, H. 2006. Active orogeny of the south-central Andes studied with GPS geodesy. *Revista Asociación Geológica Argentina*, **61**, 555–566.
KURIBAYASHI, E. & TATSUOKA, F. 1975. Brief review of liquefaction during earthquakes in Japan. *Soils and Foundations*, **15**, 81–92.
LA PALABRA NEWSPAPER, 1917.
LA UNIÓN NEWSPAPER, 1894. Diario de la tarde, San Juan. Año XVI, N°2–20.
LOS ANDES NEWSPAPER, 1903.
LOS ANDES NEWSPAPER, 1917.
LOS ANDES NEWSPAPER, 1920.
LOOS, P. 1907. Estudios de sismología. Los movimientos sísmicos de Mendoza. *Anales del Ministerio de Agricultura, Sección Geología, Mineralogía y Minas*, **3**, 1–38.
LOOS, P. 1926. Los terremotos del 17 de diciembre de 1920 en Costa de Araujo, Lavalle, La Central, Tres Porteñas. *Contribuciones Geofísicas del Observatorio Astronómico de la Universidad de La Plata*, **1**, 129–158.
LOOS, P. 1928. El terremoto argentino-chileno del 14 de abril de 1927. *Contribuciones Geofísicas del Observatorio Astronómico de la Universidad de La Plata*, **2**, 67–106.
LÜNKENHEIMER, F. 1930. El terremoto surmendocino del 30 de mayo de 1929. *In*: *Contribuciones Geofísicas del Observatorio Astronómico de la Universidad Nacional de La Plata*, **3**, 89–143.
MARTOS, L. 1995. *Análisis morfo-estructural de la faja pedemontana oriental de las Sierras de Marquesado, Chica de Zonda y Pedernal.* Facultad de Ciencias Exactas, Físicas y Naturales. Ph.D. thesis, Universidad Nacional de San Juan, 555p.
MIGNORANCE, F. 2006. Morfometría de la falla histórica de la zona de falla La Cal. *Revista Asociación Geológica Argentina*, **61**, 620–638.
MOREIRAS, S. M. 2004. Landslide incidence zonification along the Rio Mendoza Valley, Mendoza, Argentina. *Revista Earth Surface Processes and Landforms*, **29**, 255–266.
MOREIRAS, S. M., GIAMBIAGI, L. B., SPAGNOTTO, S., NACIF, S., MESCUA, J. F. & TOURAL DAPOZA, R. 2013. Caracterización de fuentes sismogénicas en el frente orogénico activo de los Andes Centrales a la latitud de la ciudad de Mendoza (32°50′–33°S). *Andean Geology* (in press).
MOREY, F. 1938. *Los temblores de tierra.* Mendoza Sísmica. Mendoza. D'Accurzio
NEIC (NATIONAL EARTHQUAKE INFORMATION CENTRE) USGS WEBSITE http://earthquake.usgs.gov/earthquakes/world/argentina. Last view on 2012.
PAPADOPULOS, G. A. & LEFKOPULOS, G. 1993. Magnitude-distance relations for liquefaction in soil from earthquakes. *Bulletin, Seismological Society of America*, **83**, 925–938.
PÉREZ, I. & COSTA, C. 2012. Fenómenos secundarios inducidos por el terremoto de 1894, su aplicación en la determinación de la magnitud del evento. Proceedings of Tectonic meeting, San Juan. CD.
PERUCCA, L. P. & MOREIRAS, S. M. 2006. Liquefaction phenomena associated with historical earthquakes in San Juan and Mendoza Provinces, Argentina. *Quaternary International*, **158**, 96–109.
PERUCCA, L. & PAREDES, J. 2003. Fallamiento cuaternario en la zona de La Laja y su relación con el terremoto de 1944, Departamento Albardón, San Juan, Argentina. *Revista Mexicana de Ciencias Geológicas*, **20**, 20–26.
PERUCCA, L. & PAREDES, J. 2004. Descripción del Fallamiento activo en la provincia de San Juan. Tópicos de Geociencias. *In*: MIRANDA, S. & SISTERNA, J. (eds) *Un volumen de Estudios Sismológicos, Geodésicos y Geológicos en Homenaje al Ing. Fernando Séptimo Volponi*, Editorial UNSJ, San Juan, 269–309.
PONTE, J. R. 1987. *Mendoza aquella ciudad de barro.* Imprenta Municipal de la Ciudad de Mendoza, Mendoza.
RAMOS, V. A. 1988. The tectonics of the Central Andes; 30° to 33°S latitude. *In*: CLARK, S. P., JR., BURCHFIEL, C. & SUPPE, J. (eds) *Processes in Continental Lithospheric Deformation.* Geological Society of America, Special Paper, **218**, 31–54.
RAMOS, V. A. 1996. Evolución tectónica de la Alta Cordillera de San Juan y Mendoza. *In*: RAMOS, V. A. (ed.) *Geología de la región del Aconcagua, provincias de San Juan y Mendoza.* Subsecretaría de Minería de la Nación. Dirección Nacional del Servicio Geológico, Anales, **24**, 447–460.
RICKARD, F. I. 1863. *A Mining Journey Across the Great Andes, with Explorations in the Silver Mining District of the Province of San Juan and Mendoza and a Journey across the Pampa to Buenos Aires.* Smith Elder & Co, Cornhill, UK.
RJEHM (Revista de la Junta de Estudios Históricos de Mendoza), 2, 1936.
RJEHM (Revista de la Junta de Estudios Históricos de Mendoza), 15, 1939.
SALOMON, E., SCHMIDT, S., HETZEL, R., MINGORANCE, F. & HAMPEL, A. 2013. Repeated folding during Late Holocene earthquakes on the La Cal thrust fault near Mendoza city (Argentina). *Bulletin of the Seismological Society of America*, **103**, 936–949, April 2013, http://dx.doi.org/10.1785/0120110335
SANTIBAÑEZ RODRIGUEZ, D. H. 2006. Determinación del potencial de licuefacción de suelos no cohesivos saturados bajo cargas sísmicas usando el ensayo de penetración estándar. Degree thesis,

Escuela de Ingeniería en Construcción. Facultad de Ciencias de la Ingeniería. Universidad Austral de Chile, 125p.

SIAME, L., BELLIER, O. & SEBRIER, M. 2006. Active tectonics in the Argentine Precordillera and western Sierras Pampeanas. *Revista de la Asociación Geológica Argentina*, **61**, 604–619.

SCHMIDT, S., HETZEL, R., MINGORANCE, F. & RAMOS, V. A. 2011. Coseismic displacements and Holocene slip rates for two active thrust faults at the mountain front of the Andean Precordillera (33°LS). *Tectonics*, **30**, TC5011, http://dx.doi.org/10.1029/2011TC002932

SMALLEY, R. & ISACKS, B. 1987. A high resolution local network study of the Nazca Plate Wadati-Benioff Zone under Western Argentina. *Journal of Geophysical Research*, **92**, 13093–13912.

SMALLEY, R. F. & ISACKS, B. 1990. Seismotectonics of thin and thick skinned deformation in the Andean foreland from local network data: evidence for a seismogenic lower crust. *Journal of Geophysical Research*, **95**, 12487–12498.

SMALLEY, R., JR., PUJOL, J. ET AL. 1993. Basement seismicity beneath the Andean Precordillera thin-skinned thrust belt and implications for crustal and lithospheric behavior. *Tectonics*, **12**, 63–76.

STRAUDER, W. 1973. Mechanism and spatial distribution if Chilean earthquakes with relation to subduction of the oceanic plate. *Journal of Geophysical Research*, **78**, 5033–5061.

UYEDA, S. & KANAMORI, H. 1979. Back-arc opening and mode of subduction. *Journal of Geophysical Research*, **84**, 1049–1061.

VERDAGUER, J. A. 1929. Historia eclesiástica de Cuyo. *Premiata Scuola Tipográfica Salesiana, Milán*, **2**, 415–425.

VERGÉS, J., RAMOS, V. A., MEIGS, A., CRISTALLINI, E., BETTINI, F. H. & CORTÉS, J. M. 2007. Crustal wedging triggering recent deformation in the Andean thrust front between 31°S and 33°S Sierras Pampeanas-Precordillera interaction. *Journal of Geophysical Research*, **112**, B03S15, http://dx.doi.org/10.1029/2006JB004287.

VOLPONI, F., QUIROGA, M. & ROBLES, A. 1978. *El terremoto de Caucete del 23 de noviembre de 1977*. Instituto Sismológico Zonda, Universidad Nacional de San Juan.

WICK, E., BAUMANN, V. & JABOYEDOFF, M. 2011. Report on the impact of the 27 February 2010 earthquake (Chile, Mw 8.8) on rockfalls in the Las Cuevas Valley, Argentina. Brief communication. *Natural Hazards and Earth System Sciences*, **10**, 1989–1993.

Near pure surface uplift of the Argentine Frontal Cordillera: insights from (U–Th)/He thermochronometry and geomorphic analysis

GREGORY D. HOKE[1]*, NATHAN R. GRABER[1], JOSÉ F. MESCUA[2], LAURA B. GIAMBIAGI[2], PAUL G. FITZGERALD[1] & JAMES R. METCALF[1,3]

[1]*Department of Earth Sciences, Syracuse University, Syracuse, NY 13244, USA*

[2]*Instituto Argentino de Nivilogía, Glaciología y Ciencias Ambientales, CCT, Mendoza, Argentina*

[3]*Present Address: Department of Geosciences, University of Colorado Boulder, Boulder, CO 80309, USA*

*Corresponding author (e-mail: gdhoke@syr.edu)

Abstract: Apatite (U–Th)/He thermochronology from palaeosurface-bounded vertical transects collected in deeply incised river valleys with >2 km of relief, as well as geomorphic analysis, are used to examine the timing of uplift of the Frontal Cordillera and its relation to the evolution of the proximal portions of the Andean foreland between 32° and 34°S latitude. The results of apatite (U–Th)/He (AHe) analyses are complex. However, the data show positive age-elevation trends, with higher elevation samples yielding older AHe ages than samples at lower elevation. Slope breaks occur at c. 25 Ma in both profiles, separating very slow cooling and or residence within a partial retention zone (slope of c. 10 m/Myr) at the highest elevations from a slope of c. 60–100 m/Myr cooling rate at lower elevations. The older AHe ages suggest either (1) minimal burial of the Frontal Cordillera and/or (2) significant pre–middle Miocene local relief. Geomorphic analysis of the adjacent, east-draining Río Mendoza and Río Tunuyán catchments reveals a glacial imprint to the landscape at elevations above 3000 m, including greater channel steepness and lower profile concavities developed during glacial erosion. Detailed analysis of headwall heights provides evidence of ongoing rock uplift along the entire eastern flank of the Frontal Cordillera and in the eastern flank of the Principal Cordillera south of the slab dip transition.

Orogenic processes, manifest as rock uplift, result in the vertical displacement of rock material with respect to the geoid (England & Molnar 1990). The net effect of rock uplift on the Earth's topography depends on the relative efficiency of erosive processes, which in turn control the propagation of deformation (e.g. Davis & Dahlen 1983; Hilley et al. 2004; Whipple & Meade 2004). In the extreme end-member scenario, where the rate of exhumation is equal to rate of rock uplift, orogenic topography would be quasi-static (e.g. Taiwan or the Southern Alps of New Zealand; Willett 1999). At the other extreme, low erosion rates, coupled with long-term shortening or deep crustal processes, result in surface uplift elevating an entire region (e.g. Altiplano Plateau; Montgomery et al. 2001). Studies constraining the timing, magnitude and rate of rock uplift improve our ability to discriminate between the possible mechanisms resulting in the creation and maintenance of topography (e.g. Hoke & Garzione 2008), including mantle-driven effects (e.g. Moucha & Forte 2011), crustal shortening (e.g. McQuarrie 2002) or climate (e.g. Thomson et al. 2010).

The Frontal Cordillera of the Central Andes in the Mendoza Province of Argentina between 32° and 34°S latitude (Fig. 1) is located in a zone where several major tectonic transitions occur (Jordan et al. 1983; Giambiagi et al. 2011). These include a shift in the subduction geometry of the Nazca Plate (c. 5–10° plate dip at 100 km depth north of 33°S and a 30° dip to the south; Cahill & Isacks 1992; Anderson et al. 2007), the southern termination of the Precordillera terrain, and a two-fold decrease in Cenozoic crustal shortening, from c. 120 km at 32°S (Allmendinger et al. 1990; von Gosen 1992; Cristallini & Ramos 2000) to c. 60 km at 34°S (Giambiagi et al. 2012). The dramatic decrease in shortening does not produce the expected decrease in average range elevation (Fig. 1c). Hilley et al. (2004) compiled data on the timing and amount of shortening that bracket the age of uplift in the Frontal Cordillera between c. 9 and 6 Ma (Fig. 2). However, this compilation does not account for the variations in shortening magnitude and timing along the north–south strike of the mountain range (e.g. Jordan et al. 1993). Robust quantitative constraints on the timing and rates of

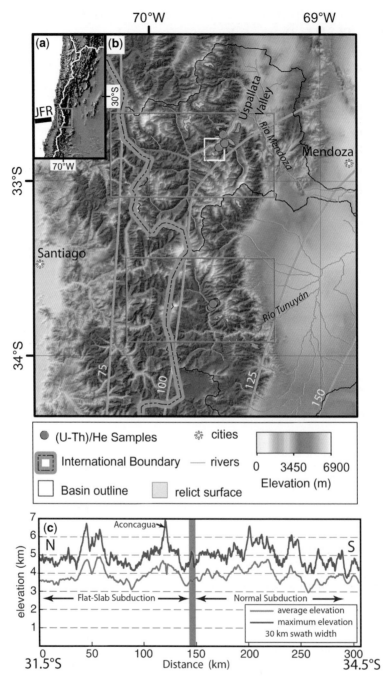

Fig. 1. Location maps of the study area. (a) Shaded relief map of the southern portion of the Central Andes and the political borders of Bolivia, Chile and Argentina. The red box outlines the area shown in (b). (b) Topographic map of the study areas with the Rio Menodza and Rio Tunuyán catchments outlined by black lines and major rivers shown in blue. The AHe transects in both catchments are denoted by red circles. Palaeosurface remnants are mapped as blue polygons and the white box marks the location of the 3D perspective view shown in Fig. 3. The red outlines are the extent of the swath profiles of the Tunuyán and Mendoza catchments illustrated in Fig. 7. Pink lines are the contours of the Wadati–Benioff Zone from Cahill & Isacks (1992) (c) North–south swath profile along the crest of the range showing the lack of decrease in topography from north to south.

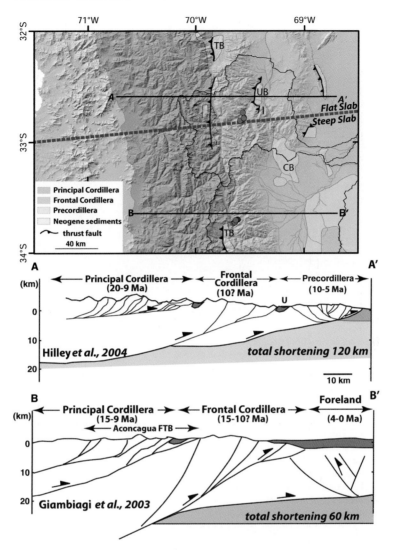

Fig. 2. Regional map of Frontal Cordillera (green), Principal Cordillera (purple) and Precordillera (pink), and exposed Miocene sediments (yellow). The Frontal Cordillera is nearly completely composed of the Permian to Triassic Choyoi Group extrusive and intrusive rocks. Lower panels are synthesized structural cross-sections adapted from Hilley et al. (2004) and Giambiagi et al. (2003) with additional timing constraints on the Precordillera from Walcek & Hoke (2012).

uplift initiation are few in this sector of the Frontal Cordillera (Irigoyen et al. 2000), yet they are essential for identifying and discriminating between processes that could maintain high topography south of 33°S in the absence high crustal shortening such as, for example, ridge subduction (Yáñez et al. 2002), recent steepening of the subducting slab (Kay et al. 2006) or significant pre-existing topography (Walcek & Hoke 2012). Improved constraints on the timing and rates of rock uplift and exhumation are necessary in order to determine which tectonic and geodynamics factors control the topographic growth of this complex part of the Andes. Our study focuses on the low-temperature thermochronology and geomorphology of two adjacent east-draining catchments: the Río Mendoza and Río Tunuyán.

The Frontal Cordillera is capped in some places by high-elevation pre-Miocene palaeosurfaces, representing a pre-uplift datum, which can be used to estimate the amount of surface uplift experienced by the Choiyoi Group (Fig. 3) since its development. These palaeosurfaces represent a regionally extensive pre-Miocene peneplain, developed

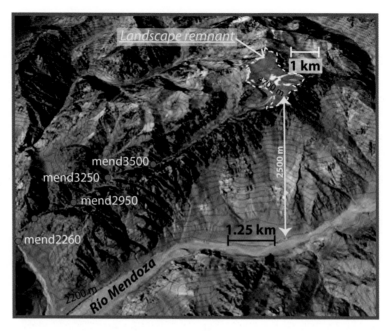

Fig. 3. A 3D oblique aerial view from Google Earth looking SW towards the palaeosurface capping the Río Mendoza vertical transect. Total vertical relief on the south wall of the Río Mendoza canyon is 2500 m.

on the surface of the Choiyoi Group granites and rhyolites prior to the initiation of Miocene sediment deposition. Uplift of the Frontal Cordillera and Precordillera (Walcek & Hoke 2012) has segmented the Miocene sedimentary depocentres into those seen today (Fig. 2a). In the standard model of thrust belt and foreland basin architecture (DeCelles & Giles 1996), foreland basin sediments would have been continuous and extend across what is today the Frontal Cordillera. Such a deposition scenario would elevate the temperature of the underlying Choiyoi Group via burial (Fig. 4a). However, if the foreland was characterized by discontinuous depozones, perhaps akin to a 'broken' foreland (Jordan 1995; Ramos et al. 2002) or the pre-existing 'old' topography remnant from different deformation episodes (e.g. the Precordillera, Walcek & Hoke 2012, Fig. 4b), the peneplains may have experienced little or no Miocene sedimentation. Our sampling strategy was one of vertical profiles from the high-relief Tunuyán and Mendoza river valleys. These adjacent valleys are bounded at high elevations by the palaeosurfaces at high elevation and incise deep canyons into the Choiyoi Group granites and rhyolites. As such, we are effectively sampling the thermal history of what we would infer to be uplift-related river incision of the Frontal Cordillera and thus test whether or not the foreland was continuous by exploring whether the thermal history is consistent with the removal of overlying sediments. If uplift and exhumation follows burial, as one end-member scenario suggests, an exhumed partial retention zone (PRZ) may be present in the upper section of the crustal profile (e.g. Fitzgerald et al. 2006). When sufficiently thick sediments unconformably overlie the basement, as shown schematically in Fig. 4, apatite in the basement may undergo partial or complete loss of He due to thermally activated volume diffusion (e.g. Farley 2000). For the (U–Th)/He (AHe) system to be reset completely, the rocks must be heated above c. 80 °C for c. 10^5 years (the higher the temperature, the faster the rate of helium loss). If the temperature is not elevated to this level, then it is possible only a partial loss will occur. It is possible to model whether the rocks reached the closure temperature by modelling the geothermal gradient for the case of burial and exhumation (Fig. 4c). Geomorphic analysis (Brocklehurst & Whipple 2007; Brocklehurst et al. 2008) of the three arc-second resolution Shuttle Radar Topography Mission (SRTM) digital elevation model (DEM) is employed to identify areas currently undergoing rock uplift.

Geological setting

The Andes Mountains formed since the Jurassic as the result of active-margin processes along the western margin of South America (Jordan et al.

1983; Mpodozis & Ramos 1989). This includes subduction of the Nazca Plate under the overriding South American continent. The angle of subduction changes abruptly at 33°S latitude. South of 33°S latitude, the downgoing slab has a dip of c. 30°. North of 33°S, the dip of the subducting slab is nearly horizontal. Between 28° and 33°S, this 'flat-slab' zone is characterized by a pronounced bench in the downgoing Nazca slab at 100 km depth and 260 km east of the trench (Fig. 1b; Cahill & Isacks 1992; Anderson et al. 2007). The location of the modern flat-slab segment is also spatially coincident with the subduction of the Juan Fernandez Ridge (JFR) (Gutscher 2002; Yáñez et al. 2002) and underlies the Principal and Frontal Cordilleras north of 33°S, as well as the southern termination of the Precordillera and the Sierras Pampeanas (Hilley et al. 2004).

Prior to break-up, South America was located in the southwestern part of Gondwana. Extension, related to the initial break-up of Gondwana, resulted in the formation of the regionally extensive Choiyoi Group volcanic province during the late Permian to early Triassic. The Choiyoi Group is a complicated package of rocks dominated by felsic shallow intrusive and extrusive igneous rocks. Rifting continued into the Triassic, producing the extensional Cuyo Basin (Kokogian et al. 1993). Later, back-arc extension in the Jurassic to Cretaceous gave way to the development of the Neuquén Basin, which covers 32–40°S of latitude and contains continental to deep marine rocks (Legarreta & Uliana 1999). Major plate reorganization took place in the mid Cretaceous and, by this time, the compressive Andean deformation had started in various parts of the Andes (Mpodozis & Ramos 1989). The largest pre-Andean thermal events that could have affected the Choiyoi Group are Mesozoic extensional events responsible for the Cuyo and Neuquén basins.

At the latitudes of this study, the Principal Cordillera (Fig. 2a) comprises the thin-skinned Aconcagua fold–thrust belt (AFTB), composed primarily of Mesozoic marine strata interlaced with Jurassic–Cretaceous volcanic rocks deposited in the northernmost extension of the Neuquén Basin (Legarreta & Uliana 1999). Geological constraints from the onset of foreland basin sedimentation indicate that the AFTB developed at c. 18 Ma (Jordan et al. 1996; Giambiagi et al. 2001; Ramos et al. 2002). Foreland basin sedimentation continued throughout the early to late Miocene (Fig. 2), with significant proximal foredeep deposits reaching 2–3 km in thickness in the Manantiales Basin (Jordan et al. 1996), 1.5 km in the longitudinal Uspallata Valley (Fig. 2; Cortés 1993) and 2 km in the Cacheuta Basin (Irigoyen et al. 2000). The distribution of Miocene sediments between the Frontal and Principal Cordilleras and the Frontal and Pre-Cordilleras (Fig. 2) would suggest some continuity to the early foreland (Irigoyen et al. 2000; Ramos et al. 2002). Alternatively, the foreland may have been compartmentalized by pre-existing highs, as observed in the Precordillera (Walcek & Hoke 2012).

As the deformation front propagated eastward towards the foreland during the middle to late Miocene, the Choiyoi Group and any overlying Miocene sedimentary deposits were uplifted, forming the thick-skinned Frontal Cordillera (Giambiagi & Ramos 2002; Ramos et al. 2002; Giambiagi et al. 2003, 2012). In addition to eroding any overlying Miocene sediments, rivers and glaciers (Espizua 2004) have incised an additional 1–2.5 km into the Frontal Cordillera since the onset of thrusting-related middle Miocene uplift.

Methods

AHe thermochronology and geothermal gradients

Low-temperature thermochronology is often used to constrain the cooling history of a rock as it is exhumed and ultimately exposed at the surface (e.g. Fitzgerald et al. 1995; Reiners et al. 2000; Ehlers & Farley 2003). With a nominal closure temperature of 70 °C, apatite AHe thermochronometry (e.g. Farley 2002) provides constraints on (1) the rate and timing of exhumation as rocks transit the upper c. 3km of the Earth's crust towards the surface, (2) resetting (or partial resetting) of the thermochronometer due to transient or burial heating, dependent on the evolution of the geothermal gradient (Ehlers 2005), or (3) the movement of isotherms (e.g. Braun 2002), including depression of isotherms in relation to river incision. Because of the low closure temperature, AHe can constrain the rates at which surface processes respond to tectonic or climatic changes in the landscape (e.g. Braun 2002; Berger et al. 2008; Thomson et al. 2010). We applied AHe in this study because of the extensive pre-Miocene palaeolandscape (palaeosurfaces) remnants (Fig. 3) that cap the Frontal and Precordillera, indicating that the overall amount of exhumation since the formation of these palaeosurfaces is low.

Constraining the thermal structure of the upper part of the crust facilitates the interpretation of thermochronometry data. Local crustal thermal structure is a product of the local rock properties, topography, radiogenic heat production and heat flow from the mantle. Although there are continental-scale heat flow data for South America (Hamza & Muñoz 1996), there are few constraints on the modern regional geothermal gradient in this part of the Andes. Data from the Chilean slope of the

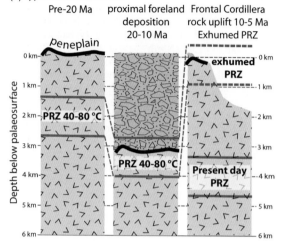
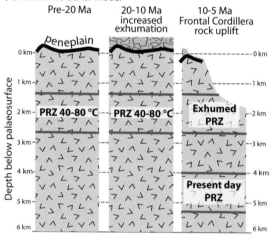
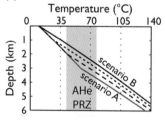
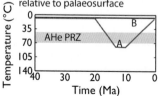

Table 1. *Geothermal gradient modelling parameters (from Turcotte & Schubert 2002)*

Material	Heat flux Q (mW m^{-2})	Density ρ (kg m^{-3})	Thermal conductivity k (W mK^{-1})	Heat generation H (W m^{-3})
Sandstone		2700	3.5	2×10^{-6}
Granite		2300	3	4×10^{-6}
Mantle	40			

Andes suggest a geothermal gradient of 28–30 °C/km south of 33°S (Maksaev *et al.* 2009). However, Andean foreland basin sediments appear to be cold with a geothermal gradient on the order of 15 °C/km (Collo *et al.* 2011).

In order to facilitate differentiation between high and low Miocene sedimentation scenarios for the Frontal Cordillera we model the geothermal gradient for two end-member models. The first model accounts for c. 3km of sediment deposited on top of the Choiyoi Group, similar to that preserved in the Manantales Basin (Jordan *et al.* 1996), while the other model assumes the palaeolandscape remnants were always exposed or only thinly covered with sediment (<1 km thickness; Fig. 4). In both cases, we assume the mean annual surface air temperature to be c. 10 °C. In the first scenario, the palaeolandscape is buried by sediment, which in turn increases the temperature of the sediment–palaeolandscape interface and the underlying rock (Fig. 4). Geotherms were constructed using a one- or two-layer models depending on the scenario as described in Turcotte and Schubert (2002) using reasonable values for thermal conductivity k and radiogenic heat production $-\rho H$ for both the sediments and felsic igneous rocks (see Table 1). The depth below the palaeosurfaces at which ages would be completely reset would be shallower than that of constantly exposed palaeosurface. The results for 0, 1, 2 and 3 km of sediment deposited on top of the palaeosurfaces are shown in Fig. 4c. The model results indicates that, for the 3 km burial scenario, temperatures due to burial heating sufficient to completely reset AHe ages in the Choiyoi Group granites occur c. 1 km below the palaeosurface (Fig. 4c). If no sediments were deposited on top of the palaeosurfaces, the 70 °C isotherm would occur at a depth of c. 2.7 km below the palaeosurface (Fig. 4c). This range of solutions has important implications for how we interpret our thermochronology data.

The AHe closure temperature varies according to rate of cooling, grain size and accumulation of radiation damage in the crystal lattice of the apatite grain (e.g. Wolf *et al.* 1996; Reiners & Farley 2001; Ehlers & Farley 2003; Flowers *et al.* 2009). The PRZ refers to a window of temperatures where the partial diffusive loss of He occurs. The faster a rock transits the PRZ, the less pronounced its effect on the age of a sample (Ehlers & Farley 2003). Studies using Durango apatite predict a 70 °C closure temperature for crystals with a radius of c. 70 μm assuming a cooling rate of 10 °C/Myr (e.g. Flowers *et al.* 2009). However, the Durango model is limited to interpreting cooling ages with a homogeneous distribution of uranium and thorium within the crystal. Recent studies have shown that the concentration and distribution of U and Th within a crystal is proportional to radiation damage accumulation and is referred to as 'effective uranium' (eU) (e.g. Shuster *et al.* 2006; Flowers *et al.* 2009). Apatites with high eU accumulate more radiation damage, which creates traps for He, changing the retentivity of He in the crystal and in effect altering the closure temperature in the AHe system. Zoning of the parent elements within an apatite crystal also results in problems with the alpha-particle correction (F_T correction, Farley 2002), a problem that is accentuated by residence in or slow cooling through a PRZ, which often results in a considerable spread of ages (Meesters & Dunai 2002; Fitzgerald *et al.* 2006). Alpha ejection correction is estimated for each grain based on grain dimensions (Farley 2002), but zoning is not taken

---

Fig. 4. End-member scenarios for the thermal history of the Frontal Cordillera starting with the same initial condition of a pre-Miocene lower relief palaeolandscape. Note the differences in the position of the PRZ in the eroded valley wall on the right-hand side of each figure. (**a**) A typical foreland with c. 3 km of burial by deposits of the proximal foredeep between 20 and 10 Ma. Post-10 Ma, uplift-related exhumation of the sediments and exposure of the Frontal Cordillera to river incision. Here the PRZ would be crossed c. 1 km below the palaeosurfaces. (**b**) The no burial scenario, where the Frontal Cordillera palaeolandscape undergoes rock uplift and river incision. (**c**) Models of geothermal gradients for different scenarios of 0, 1, 2 and 3 km of sediment overlying the palaeosurfaces. The thermal properties of the sediments reflect the geothermal gradients reported by Collo *et al.* (2011) and those of the granitoids are from Turcotte & Schubert (2002) and Maksaev *et al.* (2009). (**d**) Hypothetical time–temperature pathways relative to the palaeolandscape for the scenarios illustrated in (a) and (b).

into account. Another factor that may cause age variations is the 'bad neighbours' scenario, where daughter alpha-particles in the apatite grain are injected into the neighboring [U,Th]-rich grains (e.g. Spiegel *et al.* 2009). In addition, mineral inclusions can also create complications in the apatite AHe system. Although these inclusions may consist of a variety of minerals, zircon and monazite in particular (e.g. Farley 2000; Vermeesch 2008) can contain significant quantities of uranium and thorium. These inclusions may therefore produce erroneously old ages by contributing 'parentless daughter' to the host mineral, because when apatite is dissolved in HNO_3 to determine parent isotope concentrations, zircon inclusions are not dissolved, so the parent is not measured and an anomalously old age results.

Sampling. Fifteen samples were collected from two vertical profiles (Figs 1 & 2) in the Choiyoi Group. The northern profile was collected in the Río Mendoza valley between 2200 and 3500 m elevation. The southern profile was sampled in the Río Tunuyán valley between 2500 and 4600 m (Figs 1 & 2). Profile locations were chosen to span the maximum relief in the Choiyoi Group below the capping palaeosurface remnants. Samples of 7–5 kg in size were collected at *c.* 250 m elevation intervals.

Sample preparation and data acquisition. Rocks were crushed and milled with a 2 mm gap to maximize intact apatite yield and then sieved to between 250 and 65 μm. Non-magnetic minerals were isolated using a Frantz magnetic separator before apatite was separated from the non-magnetic fraction using lithium polytungstate heavy liquids at a density of 2.85 g/ml. Clear, euhedral apatite grains were handpicked in ethanol using a binocular microscope under polarized light. Individual apatite grains were inspected for mineral inclusions, photographed, measured (length and width) and described. He extraction was performed in the Syracuse University SUNGIRL facility. Grains were divided into single and multi-grain aliquots based on relative size and packaged into Pt tubes. Both single and multi-grains were run. Multi-grain aliquots have the advantage of a larger He signal, avoiding the problems with near blank measurements that may be encountered for small single crystals. However, multi-grain aliquots may be chemically heterogeneous crystals or contain U–Th-rich mineral inclusions, thus having a higher probability of a mixed age, making interpretation more difficult. Pt-tubules were heated to *c.* 1500 °C with a CO_2 laser to outgas He, which was measured in a quadrupole mass spectrometer. U, Th and Sm concentrations were determined by inductively coupled plasma–mass spectrometry (ICP-MS) at the University of Arizona Radiogenic He Dating Lab (ARHDL).

Topographic analysis

Analysis of the landscape can yield significant information about how geomorphic processes respond to tectonic, surface and climatic influences. Swath elevation profiles over widths of 50–100 km provide insight into the distribution of elevations in the landscape. All swath profiles were created using (Consultative Group on International Agriculture Research–Consortium for Spatial Information) CGIAR-CSI's 250 m DEM created from resampling 90 m SRTM data.

Rivers are sensitive to even small changes in topography and respond quickly to changes in relief (Whipple & Tucker 1999; Kirby & Whipple 2012). Under non-glacial conditions, river networks are the first elements in the landscape to respond to changes in surface elevation, generating areas of anomalously steep gradient referred to as knickpoints. Knickpoints propagate upstream like a wave, adjusting the entire river network to the new equilibrium conditions related to increased relief (Whipple & Tucker 1999; Whipple 2004). This commonly leads to the development of incised river canyons, which in many cases may provide constraints on the timing and magnitude of rock uplift (e.g. Schoenbohm *et al.* 2004; Hoke *et al.* 2007; Miller *et al.* 2012; for a complete discussion see Kirby & Whipple 2012). In high latitudes and/or high elevations, glaciers leave indelible marks on the landscape. The glacial buzzsaw hypothesis suggests glacial erosion is so efficient that it sets the upper limit for range height (e.g. Brozovic *et al.* 1997; Mitchell & Montgomery 2006; Brocklehurst & Whipple 2007). A recent study in the heavily glaciated Patagonian Andes shows a clear correlation between the equilibrium line altitudes (ELAs) with mean and maximum elevation (Thomson *et al.* 2010). The ELA is the place on a glacier with the maximum ice flux and where there is no net loss or gain of ice. Brocklehurst & Whipple (2007) examined the response of glaciated landscapes to tectonic uplift where independent constraints on rock uplift rates were available. Their results suggest that total headwall relief provides the most robust metric for detecting rock uplift (Brocklehurst & Whipple 2007; Brocklehurst 2010).

The stream profile analysis presented here used the CGIAR-CSI version 4 of NASA's 90 m SRTM digital elevation model. The 90 m DEM was conditioned for stream profile extraction in ArcGIS. We used the stream profiling software available at http://geomorphtools.org to analyse river profiles in MATLAB. Following profile extraction, knickpoints or knickzones were identified manually through visual inspection of the long profiles and slope–area plots, which are generated from the extracted stream data. Given the 90 m resolution of the SRTM data, mapped knickpoints had a minimum of 30 m of relief over length scales much shorter than a typical segment. Concavity and steepness indices (Whipple & Tucker 1999; Snyder et al. 2000) were determined for individual knickpoint-bounded longitudinal profile segments, which are also manually selected using the MATLAB interface (Wobus et al. 2006). Almost all of the tributary streams in both the Río Mendoza and Río Tunuyán have been glaciated and the effects of this glaciation are abundant in the landscape. However, streams with glaciers currently covering the upper reaches were avoided when selecting streams for this longitudinal profile analysis, because the glacier surface does not reflect a river. The concavities seen in stream segments depend on the surficial processes sculpting the terrain (Fig. 5). For rivers draining glaciated mountains, non-glaciated reaches have concavity values near c. 0.7, while in reaches with a glacially controlled morphology slope concavity is significantly lower at c. 0.4. The uppermost reaches of the stream, the headwalls, typically show a concavity at or near 0.0 (Fig. 5; Crosby & Whipple 2006; Brocklehurst & Whipple 2007).

Following the selection of the tributary streams feeding the Río Mendoza and Río Tunuyán, we identified the transition between the headwall and stream channel by drainage area and a constant slope in slope–area plots (Fig. 5). Typically, drainage areas of 10^6–10^7 m^2 mark the end of the headwall and the beginning of the stream channel in both streams. Once headwalls are isolated, the amount of headwall relief is measured (Fig. 5). Brocklehurst & Whipple (2007) suggest that areas of headwall relief <500 m indicate regions of active rock uplift.

Results

AHe data

The AHe data collected in this study comprised 25 single grains and 23 multi-grain aliquots, in addition to 9 Durango apatite standards. The Durango ages averaged 33.67 ± 1.77 Ma, within 2σ error of the published AHe age of Durango apatite (Farley 2000). From the 48 packets analysed, 18 were discarded for a number of reasons: (1) the grains failed the re-extraction test (more than two heating phases to extract a majority of the helium), (2) the grains were too small and hence the F_T correction uncertainties were too large (Ehlers & Farley 2003), and/or (3) the signal size of the ^{4}He or U–Th–Sm measurements were too small (very close to blank). Data were not included from grains that had F_T values of <0.6 alpha ejection correction, if

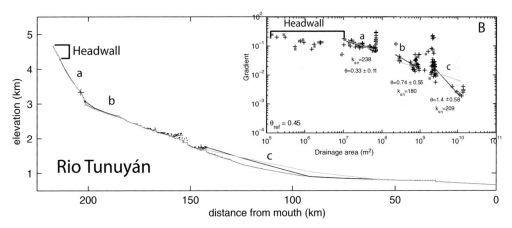

Fig. 5. Representative longitudinal profile of the steams draining the upper reaches of the Río Mendoza and the Río Tunuyán. The profile form is typical for rivers that have experienced significant glaciations and exhibit a large degree of headwall relief. The cross symbols represent knickpoints. The cyan and blue lines are stream power model best fits using a reference concavity of 0.45 and the actual profile data, respectively. The grey line is the smoothed, spike filtered longitudinal profiles and the green, dashed line is the raw data. A section of noisy data is visible where the Río Tunuyán traverses the Frontal Cordillera in a narrow canyon. Inset: plot of drainage area v. slope along the river profile. Knick points (open circles) correspond to abrupt jumps in slope along the overall negative in slope v. downstream cumulative drainage area. Individual stream segments were selected based on long profile form and coherent trends in slope–area plots. Crosses are individual gradient and slope values for 30 m vertical intervals and red squares represent log binned values. The upstream segment has a concavity of 0.33, typical of a glacially influenced reach.

⁴He was <2 times blank (< 0.002 ppm following blank correction) and U quantities <0.001 ppm. Poor coupling between the laser and platinum packet containing the sample can cause low quantities of He to be extracted during heating. However, coupling of the laser, that is, if the Pt tubule glowed during heating, is routinely monitored during He-extraction.

Sixteen aliquots both multi-grain and single-grain packets were analysed from the Río Mendoza sampling profile (Table 2). Of the sixteen aliquots, two were discarded for failing the re-extraction test and three were discarded for low He concentrations. Ages in the eleven remaining sample aliquots ranged from 105 Ma to 12 Ma (Table 2, Fig. 6a).

Thirty-two aliquots were analysed from the Río Tunuyán samples. Thirteen of the aliquots were discarded for either failing the reextraction test, yielding low levels of uranium. After inconsistent data were discarded, 19 aliquots remained viable for interpretation (Table 2). Ages from these samples range between 172 and 8 Ma (Fig. 6b). According to Fitzgerald et al. (2006), the youngest ages are typically more reliable as exhumation ages, as most complications with the AHe system result in erroneously old ages.

Topographic analysis

Relief varies from 1000–3500 m throughout the Frontal Cordillera, Principal Cordillera and Precordillera (Fig. 7). In the Precordillera, high relief (>1000 m) is confined to the easternmost extent of the range. The Principal Cordillera shows much broader high-relief extents, which are dispersed along deeply incised river valleys or appear as anomalous features centred on volcanic edifices. In contrast to the Precordillera and Principal Cordillera, the Frontal Cordillera exhibits much higher relief, between 2000 and 3500 m. This relief in the Frontal Cordillera is concentrated along the eastern front of the range and increases rapidly as the mountain front is approached from the east (Fig. 7).

Stream profile results

A majority of the streams analysed show typical glaciated stream morphologies with upper reaches dominated by low concavities, 0.3–0.4 (Fig. 5). The headwall relief in most streams in the study area is relatively low at c. 500 m (Fig. 7). Headwall relief in tributary streams drastically increases to as much as 2500 m along the eastern front of the Frontal Cordillera in both the Río Tunuyán and Río Mendoza sectors (Fig. 7). The Río Tunuyán profile (Fig. 5), despite having some artefacts, is steep with very low concavity where it traverses the Frontal Cordillera before exiting to the foreland. In contrast to the Río Tunuyán, the Río Mendoza trunk stream has a remarkably graded profile over a similar range of drainage area.

Discussion

AHe data interpretation

All ages from the Río Mendoza and the Río Tunuyán yielded ages younger than the Permo-Triassic rock crystallization ages of the Choiyoi Group (Table 2, Fig. 6). However, data from both the Río Mendoza and Río Tunuyán samples yielded considerable within-sample age variation, as commonly observed for apatite AHe ages, often due to many of the factors discussed above. If there is variable [eU] between grains, there is likely an age variation between those grains. This effect, elucidated in the RDDAM model (Shuster et al. 2006; Flowers et al. 2009) attributes the correlation between increasing [eU] and increasing AHe age as due to radiation damage within the crystal. The radiation damage acts as a trap for the daughter ⁴He, which does not then diffuse out of the crystal. In our data there is no correlation between AHe age and [eU], indicating that other factors, as described above, are creating the age dispersion (Fig. 8).

Slow cooling through the PRZ creates complex AHe data (e.g. Fitzgerald et al. 2006; Flowers & Kelley 2011) and presents the best explanation for the complexity of our data. First, age dispersion is to be expected in settings where samples have resided within (or slowly cooled through) a PRZ for any period of time because grain size, zonation and small effects in helium diffusivity amplify the age variation (Reiners & Farley 2001; Meesters & Dunai 2002; Fitzgerald et al. 2006; Flowers & Kelley 2011). The last potentially large regional thermal perturbation, Jurassic to early Cretaceous extension leading to the Neuquén Basin, occurred nearly 100 Myr before Andean mountain building (Meigs et al. 2006). There is more scatter in ages for higher elevations from both profiles, which suggests that samples collected from the upper part of the profile through the Choiyoi Group spent a substantial amount of time within the PRZ prior to later, rapid exhumation. This implies that either (1) the early exhumation of the Frontal Cordillera was very slow, followed by more rapid exhumation, (2) the Miocene foreland deposits, which are thought to have buried the Choiyoi, were of insufficient thicknesses (depending on the geothermal gradient) to reset the AHe system in apatites, and thus these samples resided within a PRZ before later more rapid exhumation and/or (3) the geothermal gradient of the Choiyoi Group was low, less than 20–25 °C/km and these samples

Table 2. *Results of apatite (U–Th)/He analysis*

Sample	Latitude	Longitude	Elevation (m)	Packet	Length (mm)	Width (mm)	He (nmol g^{-1})	U (ppm)	Th (ppm)	Sm (ppm)	Ft	Age (Ma)	±	eU
Mendo2260E	−69.5589	−32.7773	2260	S	104.95	66.08	1.18	10.75	48.16	514.23	0.64	12.99	2.05	22.06
Mendo2260D			2260	S	152.38	93.63	0.58	3.96	13.06	400.29	0.74	14.15	1.01	7.03
Mendo2260B			2500	M	206.39	160.53	0.01	0.02	0.08	1.07	0.61	34	2.41	0.04
Mendo2950B	−69.5616	−32.7855	2500	M	82.32	64.94	0.003	0.02	0.05	1.4	0.59	25.11	4.12	0.03
Mendo2950C			3100	S	284.88	193.75	0.1	0.08	0.29	11.85	0.86	84.97	6.52	0.15
Mendo3250A	−69.5820	−32.8037	3100	S	246.8	149.08	1.09	3.97	10.8	288.74	0.72	27.31	1.93	6.51
Mendo3250B			3400	M	205.08	94.73	0.01	0.01	0.04	1.47	0.81	45.92	3.26	0.02
Mendo3500A	−69.5927	−32.8072	3400	S	251.83	132.68	2.26	4.72	12.22	324.04	0.79	49.98	3.54	7.59
Mendo3500B			3400	M	220.45	140.98	0.03	0.04	0.16	2.97	0.69	70.26	4.97	0.08
Mendo3500D			3400	S	191.5	72.95	5.62	6.2	15.02	516.38	0.78	108.1	13.52	9.73
Mendo3500E			3400	S	180.88	113.28	2.22	4.86	13.26	304.94	0.83	50.18	3.58	7.97
Tunu 2600A	−69.6537	−33.7293	2600	S	180.38	164.28	2.68	3.86	24.57	61.29	0.73	58.3	4.45	9.63
Tunu 2600E			2600	S	198.83	88.95	10.76	22.48	116.18	806.83	0.76	47.94	3.53	49.78
Tunu 3280A	−69.6308	−33.7026	3280	S	238.98	103.78	5.51	17.48	97.04	726.18	0.65	28.97	2.05	40.29
Tunu 3280B			3280	M	210.02	72.87	0.02	0.05	0.26	1.83	0.8	46.94	5.69	0.11
Tunu 3280E			3280	S	186.55	127.43	0.11	1.32	5.61	58.55	0.81	7.84	0.56	2.64
Tunu 3565A	−69.6260	−33.6997	3565	S	174.53	140.48	13.76	26.44	47.74	215.06	0.79	79.3	5.65	37.66
Tunu 3770A	−69.6214	−33.6976	3770	S	183.3	118.48	7.5	10.14	32.19	133.32	0.74	93.41	6.63	17.7
Tunu 3770B			3770	S	203.9	93.35	25.25	25.84	85.75	631.1	0.73	121.88	83.67	46
Tunu 3770C			3770	S	278.68	88.4	8.23	20.19	107.22	1204.36	0.74	37.62	2.79	45.39
Tunu 3770D			3770	S	215.93	96.75	41.76	47.62	269.46	1768.97	0.72	82.76	5.86	110.95
Tunu 3770E			3770	M	222.1	94	0.04	0.06	0.3	2.98	0.8	72.93	5.16	0.13
Tunu 4010B	−69.6160	−33.6959	4100	S	187.85	125.1	11.25	64.07	7.3	426.11	0.84	37.79	2.72	65.78
Tunu 4200A	−69.6133	−33.6948	4200	M	257.86	155.56	0.03	0.03	0.003	0.45	0.83	175.37	12.58	0.03
Tunu 4200B			4200	M	211.94	160.5	0.05	0.15	0.01	1.6	0.79	61.18	4.39	0.16
Tunu 4200C			4200	M	249	126.88	0.06	0.09	0.01	0.84	0.82	127.07	9.13	0.09
Tunu 4200D			4200	M	326.56	138	0.00097	0.09	0.01	0.77	0.86	2.34	0.17	0.09
Tunu 4200E			4200	S	308.5	187.28	12.62	17.86	8.82	186.59	0.81	125.29	8.96	19.93
Tunu 4200F			4200	S	160.93	140.85	5.47	10.7	0.77	133.82	0.66	104.07	7.47	10.88
Tunu 4580E	−69.6154	−33.6956	4580	S	131.63	69.68	133.53	263.52	10.39	499.25	0.87	137.83	18.95	265.96
Tunu 4580D			4580	M	280.08	232.33	0.31	0.29	0.01	1.55	0.88	216.83	15.6	0.29
Tunu 4580C			4580	M	289.74	250.57	0.22	0.18	0.01	0.96	0.82	237.03	17.05	0.18
Tunu 4580B			4580	M	231.81	166.85	0.1	0.12	0.005	0.57		176.71	12.71	0.12

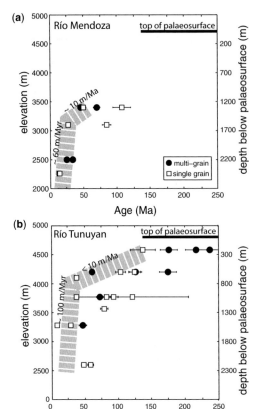

Fig. 6. Graphs of AHe age v. elevation for the (**a**) Río Mendoza and (**b**) Río Tunuyán areas. The dashed lines represent the best-fit-by-eye lines to the data. Error bars are 1-sigma uncertainties.

were therefore not (partially) reset even if the Miocene foreland deposits were very thick.

As discussed above, younger AHe ages from a single sample are typically more reliable. Therefore, in the age–elevation plots for the Tunuyán and Mendoza vertical profiles we use the trend of the youngest ages plus the age dispersion to interpret the data. We acknowledge that following rejection of some sample aliquot ages there is not as much data as would normally be desirable and it is likely that we have not captured the full variation of intra-sample AHe ages. Nevertheless, the AHe data contain significant trends that can be used to test regional tectonic models.

In both the Río Mendoza and Río Tunuyán, fit-by-eye interpreted age–elevation plots (Fig. 6) indicate a significant change in the rate of cooling at 30–25 Ma. We interpret this change in slope at c. 3000 m elevation at the Río Mendoza profile, and at c. 4000 m in the Tunuyán profile, as the base of the exhumed PRZs. The gentle slope above the break in slope indicates these samples resided for long periods of time within the PRZ and the gentle slope is not representative of an apparent exhumation rate. Meanwhile, the steeper slope below the break indicates faster cooling and is representative of an apparent exhumation rate. The c. 1000 m difference in the elevation of the slope breaks observed between profiles suggests that the Río Mendoza profile may have occupied a lower part of the landscape (e.g. a river valley) than the Río Tunuyán profile or had a thicker, but still <1 km, sedimentary veneer.

At the Río Mendoza profile, the slope of the profile above the break is c. 10 m/Myr (similar to many exhumed PRZs), whereas below the c. 25 Ma break, the apparent slope is 60–70 m/Myr. This steeper part of the profile can be interpreted as an exhumation rate; however, this exhumation rate is below the lowest rates reported in other arid parts of the Andes (Barnes et al. 2008) and far below rates reported in Chile (Farías et al. 2008; Spikings et al. 2008; Maksaev et al. 2009). Sedimentation rates in the Cacheuta Basin, based on the magnetostratigraphy of Irigoyen et al. (2000), are two to four times higher than the apparent exhumation rate in the Río Mendoza vertical profile. However, these sediments are derived from both the Frontal and Principal Cordilleras (Irigoyen 1998). Extrapolating the lower steeper slope of the Río Mendoza profile 'zero' age, the y-intercept is at an elevation of c. 3000 m below the palaeosurface or c. 500m below the present river level. The youngest age of the lowermost sample at 2400 m below the palaeosurfaces is 12 Ma. Our interpretation of samples in the Río Mendoza profile is therefore very slow cooling (long-term residence in the PRZ) from at least c. 50 Ma until c. 25 Ma, then more rapid cooling, at exhumation rates of 60–70m/Myr, until c. 12 Ma, followed by more rapid cooling and exhumation until the present.

The Tunuyán profile can be interpreted in a similar fashion to the Río Mendoza. A break in the slope in the age elevation profile is evident at c. 25 Ma. The slope of the profile above this break at c. 3000 m is <10m/Myr, extending back to c. 150 Ma. As with the Río Mendoza profile, this gentle slope is similar to many exhumed PRZs. Below the c. 25 Ma break, the slope of the age elevation profile is on the order of 100m/Myr. Parallel to the interpretation of the Río Mendoza profile, if we extrapolate the apparent slope, it intercepts the zero AHe age c. 1000 m below river level, thus also suggesting that this average exhumation rate has increased since the age of the youngest AHe age, which is c. 7 Ma. Combining the interpretation from the Río Tunuyán profile with the Rio

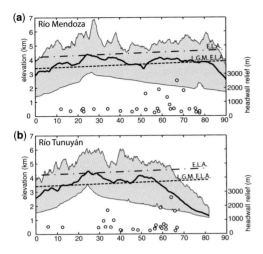

Fig. 7. East–west topographic swath profiles constructed from 90 m SRTM data of areas within the Río Mendoza (**a**) and the Río Tunuyán (**b**) catchments shown in Fig. 1. The grey envelope represents the range in elevations with distance along the profile and the black line is the average range elevation. The dashed line marks the approximate position of the modern and last glacial maximum ELAs in each area based on Clapperton 1994. Headwall relief is plotted along the east–west swath profile distance. Areas with high headwall relief indicate regions of increased uplift. Both the Río Tunuyán and Río Mendoza areas exhibit increased headwall relief along the eastern margin of the Frontal Cordillera. The second distinct increase in headwall relief visible in (b) suggests that rock uplift is ongoing in the Principal Cordillera south of the slab–dip transition.

Mendoza profile our conclusion is that these profiles record very slow cooling (long-term residence within a PRZ) from at least c. 50 Ma and perhaps since c. 150 Ma until c. 30–20 Ma, followed by increased cooling (exhumation at rates rate of c. 60–100m/Myr) until c. 10Ma, followed by more rapid cooling and exhumation until the present. Due to the complicated nature of the data, the timing and rates of exhumation are approximate. Nevertheless, they do provide quite robust trends to test and also to compare to the known geological history.

The apparent increase in inferred exhumation rates at c. 25 Ma in both profiles suggests that the initiation of uplift and exhumation in the Frontal Cordillera began at that time. This timing for initial uplift and exhumation in the Frontal Cordillera is earlier than previously proposed (9–6 Ma in Ramos et al. 2002), and appears to be synchronous with the development of the AFTB in the Principal Cordillera in the early or middle Miocene (Jordan et al. 1983; Ramos et al. 1996; Giambiagi et al. 2003). While we suggest the c. 25 Ma break in slope reflects the onset of uplift and exhumation in the Frontal Cordillera, associated with the initiation of activity in the AFTB, the oldest record of Neogene sedimentation in the region is c. 18Ma (Jordan et al. 1996). Oligocene extension and basin formation has been documented previously in northern Patagonia (Jordan et al. 2001), but not at the latitudes of our study area. In addition, the AHe data also suggest the initiation of more rapid exhumation at c. 10 Ma. This interpretation coincides with the accepted 9–6 Ma onset of deformation in the Frontal Cordillera (Ramos et al. 2002; Hilley et al. 2004). Simple thermal models (Ter Voorde et al. 2004) demonstrate that topographic cooling, the cooling caused solely by block uplift without accompanying erosion, will cause the 'uplift' of the PRZ, but will not record the steeper part of the profile below the base of an exhumed PRZ until cooling associated with erosion (i.e. exhumation) actually starts. If we assume the palaeosurface preserved on top of the Choiyoi Group was a steady-state landscape, any rock located between c. 1.6 and 3.0 km depth below the surface would be residing in an AHe PRZ, assuming an average geothermal gradient of 25 °C/km, and little to no sediment loading. Therefore, to be below the PRZ and yield a completely reset AHe age, the rock of the Choiyoi Group would need to be at a depth of c. 3.1 km beneath the palaeosurfaces (Fig. 6). This explains the results we see in the data for both the Río Mendoza and Río Tunuyán age elevation profiles; in essence, there has not been sufficient river incision to expose the post Late Miocene exhumation ages associated with most recent uplift of the Frontal Cordillera.

There are two conclusions to be gleaned from the results presented here. The first is that early uplift and exhumation along the entire Frontal Cordillera likely began at c. 25 Ma based on the timing of the break in slope observed in our AHe age–palaeodepth data, reaching its current level of exhumation between 9 and 7Ma (Fig. 6). The latter half of this age window coincides with the southward migration of the JFR, which could drive an increase in crustal shortening and rock uplift in the Frontal Cordillera. Spikings et al. (2008) present data constraining the onset of rapid exhumation at c. 7.5 Ma along the western flank of the Principal Cordillera between 35° and 38°S. Their data document a northward increase in exhumation in the principal Cordillera of c. 1 km to 5.5 km between 38°S and 35°S, which they relate to the subduction of the JFR (Spikings et al. 2008). Spikings et al.'s (2008) 7.5 Ma age is similar to the youngest ages in our Tunuyán samples, but our data are not strong enough to draw a conclusive link that exhumation in the Principal Cordillera affected the Frontal Cordillera as well. Second, we see little evidence for Miocene burial

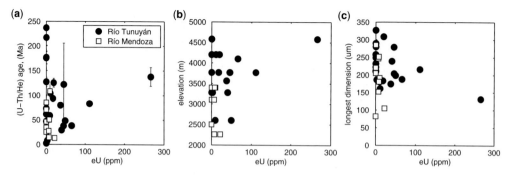

Fig. 8. Effective uranium concentration of apatite grains from the Tunuyán and Mendoza vertical profiles v. (**a**) AHe age, (**b**) elevation and (**c**) grain size.

heating to temperatures within the AHe PRZ in the the Choiyoi Group of the Frontal Cordillera, Together, this suggests that this area did not have the archetypal foreland basin architecture where sedimentation is high in the foredeep and proximal foredeep (DeCelles & Giles 1996), as has been suggested by prior studies (Irigoyen et al. 2000; Giambiagi et al. 2003). Rather, sedimentation was restricted to basins between the local highs of the proto Frontal and Precordillera, and the presence of Choiyoi Group clasts found in early parts of the Argentine foreland sequence become less difficult to explain (Irigoyen 1998).

A previous study in the eastern Frontal Cordillera using apatite fission track (AFT) analysis and a horizontal sampling strategy in the Choiyoi Group (Ávila & Chemale 2005) yielded pre-Neogene ages. This supports our notion that the sediments of the proximal foredeep adjacent to the Principal Cordillera did not attain sufficient thicknesses necessary to fully reset the AHe (our data) or AFT (Ávila & Chemale 2005) thermochronometers (Fig. 4b). Our data set and that of Ávila & Chemale (2005) also underscore the relatively low amount of exhumation on the eastern flanks of the Andes at these latitudes.

Geomorphic evidence for uplift in the Frontal Cordillera

The erosional efficiency of alpine glaciers can potentially set an upper limit on topographic development (Brozovic et al. 1997; Mitchell & Montgomery 2006). Thomson et al. (2010) show that global Cenozoic cooling may have a profound impact of the development of the Andes correlating 5–7 Ma glaciations to the retreat of the eastern-most thrust front between 38° and 49°S. In our study area, however, glaciation does not appear to exert a strong control on the regional topography. Neither the last glacial maximum (LGM) equilibrium line altitude (ELA) or the present-day ELA (Clapperton 1994; Condom et al. 2007) appear to correlate with maximum or mean elevations in the Río Mendoza and Río Tunuyán sectors of the study area (Fig. 7). The preservation of palaeosurfaces across the landscape in the Frontal Cordillera (Figs 1 & 3) also suggests that although alpine glaciers may have had a profound impact on stream morphology, they are not a dominant force sculpting the region (Bissig et al. 2002; Aguilar et al. 2011).

We interpret the pronounced increase in the average headwall relief to reflect active rock uplift occurring along the entire eastern front of the Frontal Cordillera (Fig. 7). In addition to elevated headwall relief along the Frontal Cordillera in both the Mendoza and Tunuyán catchments, a second, less pronounced increase in headwall relief corresponding to ongoing rock uplift in the Principal Cordillera is observed within the Tunuyán valley. This suggests that this region of the Principal Cordillera is actively uplifting. While the thermochronology data presented suggest rapid uplift of the Frontal Cordillera since c. 10 Ma, our geomorphic analysis of headwall relief in the Frontal and Principal Cordilleras confirms ongoing active uplift.

Conclusion

Apatite (U-Th)/He thermochronology and geomorphic DEM analysis of the Frontal Cordillera between 32°S and 34°S latitude provide insight into the evolution of the Andean orogen and challenge some of the basic assumptions of how the Andes evolved over the Neogene. Despite complicated AHe results, a positive trend in age elevation plots in the northern and southern sectors of the study is clear (Fig. 6). The absence of Miocene foreland deposition (totally absent or <1 km thick) sufficient to cause burial reheating and age resetting of the Choiyoi granites and rhyolites is perhaps the most significant result as it suggests that the Miocene sediments found adjacent to the Principal and Frontal Cordilleras were not part of a continuous foreland basin, but rather deposited in isolated sub-basins between topographic highs. The data from the Río Mendoza and Río Tunuyán vertical profiles are interpreted to indicate relatively slow rock uplift with relatively slow rates of 60–100 m/Myr. However, rapid rock uplift and river incision must have occurred since c. 10 Ma, consistent with previous constraints on the uplift of the Frontal Cordillera. Stream profile analysis shows specific regions of the Frontal Cordillera where headwall relief is very pronounced, which we interpret to reflect a zone of active rock uplift, similar to that observed by Brocklehurst et al. (2008). This zone of active uplift also corresponds to where the highest non-volcanic relief occurs in both catchments.

This research was supported by an NSF International Research Fellowship (OISE-0601957) to GDH. NG was supported by a teaching assistantship at Syracuse University. We thank two anonymous reviewers for constructive and thoughtful comments, which served to significantly improve the presentation of our work. LBG, GDH and JFM are part of IGCP-586Y, 'Geodynamic Processes in the Andes'.

References

AGUILAR, G., RIQUELME, R., MARTINOD, J., DARROZES, J. & MAIRE, E. 2011. Variability in erosion rates related to the state of landscape transience in the semi-arid Chilean Andes. *Earth Surface Processes and Landforms*, **36**, 1736–1748.

ALLMENDINGER, R. W., FIGUEROA, D., SNYDER, D., BEER, J., MPODOZIS, C. & ISACKS, B. L. 1990. Foreland shortening and crustal balancing in the Andes at 30°S latitude. *Tectonics*, **9**, 789–809.

ANDERSON, M., ALVARADO, P., ZANDT, G. & BECK, S. 2007. Geometry and brittle deformation of the subducting Nazca Plate, Central Chile and Argentina. *Geophysical Journal International*, **171**, 419–434.

ÁVILA, J. & CHEMALE, F. 2005. Thermal evolution of inverted basins: Constraints from apatite fission track thermochronology in the Cuyo Basin, Argentine Precordillera. *Radiation Measurements*, **39**, 603–611.

BARNES, J. B., EHLERS, T. A., MCQUARRIE, N., O'SULLIVAN, P. B. & TAWACKOLI, S. 2008. Thermochronometer record of central Andean Plateau growth, Bolivia (19.5°S). *Tectonics*, **27**, TC3003.

BERGER, A. L., GULICK, S. P. S. ET AL. 2008. Quaternary tectonic response to intensified glacial erosion in an orogenic wedge. *Nature Geoscience*, **1**, 793–799.

BISSIG, T., CLARK, A. H., LEE, J. K. W. & HODGSON, C. J. 2002. Miocene landscape evolution and geomorphologic controls on epithermal processes in the El Indio-Pascua Au–Ag–Cu belt, Chile and Argentina. *Economic Geology and the Bulletin of the Society of Economic Geologists*, **97**, 971–996.

BRAUN, J. 2002. Quantifying the effect of recent relief changes on age–elevation relationships. *Earth and Planetary Science Letters*, **200**, 331–343.

BROCKLEHURST, S. H. 2010. Tectonics and geomorphology. *Progress in Physical Geography*, **34**, 357–383.

BROCKLEHURST, S. H. & WHIPPLE, K. X. 2007. Response of glacial landscapes to spatial variations in rock uplift rate. *Journal of Geophysical Research–Earth Surface*, **112**, F02035.

BROCKLEHURST, S. H., WHIPPLE, K. X. & FOSTER, D. 2008. Ice thickness and topographic relief in glaciated landscapes of the western USA. *Geomorphology*, **97**, 35–51.

BROZOVIC, N., BURBANK, D. & MEIGS, A. 1997. Climatic limits on landscape development in the northwestern Himalaya. *Science*, **276**, 571–574.

CAHILL, T. & ISACKS, B. L. 1992. Seismicity and shape of the subducted Nazca Plate. *Journal of Geophysical Research–Solid Earth*, **97**, 17 503–17 529.

CLAPPERTON, C. 1994. The Quaternary glaciation of Chile: a review. *Revista Chilena De Historia Natural*, **67**, 369–383.

COLLO, G., DÁVILA, F. M., NÓBILE, J., ASTINI, R. A. & GEHRELS, G. 2011. Clay mineralogy and thermal history of the Neogene Vinchina Basin, central Andes of Argentina: analysis of factors controlling the heating conditions. *Tectonics*, **30**, TC4012.

CONDOM, T., COUDRAIN, A., SICART, J. & THERY, S. 2007. Computation of the space and time evolution of equilibrium-line altitudes on Andean glaciers (10°N–55°S). *Global and Planetary Change*, **59**, 189–202.

CORTÉS, J. M. 1993. El frente de corrimiento de la Cordillera Frontal y el extremo sur del valle de Uspallata, Mendoza. In: *XII° Congreso Geológico Argentino y II Congreso de Exploración de Hidrocarbouros*, Buenos Aires, pp. 168–178.

CRISTALLINI, E. O. & RAMOS, V. A. 2000. Thick-skinned and thin-skinned thrusting in the La Ramada fold and thrust belt: crustal evolution of the High Andes of San Juan, Argentina (32°SL). *Tectonophysics*, **317**, 205–235.

CROSBY, B. & WHIPPLE, K. 2006. Knickpoint initiation and distribution within fluvial networks: 236 waterfalls in the Waipaoa River, North Island, New Zealand. *Geomorphology*, **82**, 16–38.

DAVIS, D. & DAHLEN, F. A. 1983. Mechanics of fold-and-thrust belts and accretionary wedges. *Journal of Geophysical Research*, **88**, 1153–1127.

DECELLES, P. G. & GILES, K. A. 1996. Foreland basin systems. *Basin Research*, **8**, 105–123.

EHLERS, T. A. 2005. Crustal thermal processes and the interpretation of thermochronometer data. *Low-Temperature Thermochronology: Techniques, Interpretations, and Applications*, **58**, 315–350.

EHLERS, T. A. & FARLEY, K. A. 2003. Apatite (U–Th)/He thermochronometry: methods and applications to problems in tectonic and surface processes. *Earth and Planetary Science Letters*, **206**, 1–14.

ENGLAND, P. & MOLNAR, P. 1990. Surface uplift, uplift of rocks, and exhumation of rocks. *Geology*, **18**, 1173–1177.

Espizua, L. E. 2004. Pleistocene glaciations in the Mendoza Andes, Argentina. *Developments in Quaternary Science*, **2**, 69–73.

Farías, M., Charrier, R. *et al.* 2008. Late Miocene high and rapid surface uplift and its erosional response in the Andes of Central Chile (33°–35°S). *Tectonics*, **27**, TC1005.

Farley, K. A. 2000. Helium diffusion from apatite; general behavior as illustrated by Durango fluorapatite. *Journal of Geophysical Research*, **105**, 2903–2914.

Farley, K. A. 2002. (U–Th)/He dating: techniques, calibrations, and applications. *Reviews in Mineralogy and Geochemistry*, **47**, 819–844.

Fitzgerald, P., Sorkhabi, R. & Redfield, T. 1995. Uplift and denudation of the central Alaska Range: a case study in the use of apatite fission track thermochronology to determine absolute uplift parameters. *Journal of Geophysical Research*, **100**, 20 175–20 191.

Fitzgerald, P., Baldwin, S., Webb, L. & O'Sullivan, P. 2006. Interpretation of (U–Th)/He single grain ages from slowly cooled crustal terranes: a case study from the Transantarctic Mountains of southern Victoria Land. *Chemical Geology*, **225**, 91–120.

Flowers, R. M. & Kelley, S. A. 2011. Interpreting data dispersion and 'inverted' dates in apatite (U–Th)/He and fission-track datasets: an example from the US midcontinent. *Geochimica et Cosmochimica Acta*, **75**, 5169–5186.

Flowers, R. M., Ketcham, R. A., Shuster, D. L. & Farley, K. A. 2009. Apatite (U–Th)/He thermochronometry using a radiation damage accumulation and annealing model. *Geochimica et Cosmochimica Acta*, **73**, 2347–2365.

Giambiagi, L. B. & Ramos, V. A. 2002. Structural evolution of the Andes in a transitional zone between flat and normal subduction (33°30′–33°45′ S), Argentina and Chile. *Journal of South American Earth Sciences*, **15**, 101–116.

Giambiagi, L. B., Tunik, M. A. & Ghiglione, M. 2001. Cenozoic tectonic evolution of the Alto Tunuyán foreland basin above the transition zone between the flat and normal subduction segment (33°30′–34°S), western Argentina. *Journal of South American Earth Sciences*, **14**, 707–724.

Giambiagi, L. B., Ramos, V. A., Godoy, E., Alvarez, P. P. & Orts, S. 2003. Cenozoic deformation and tectonic style of the Andes, between 33° and 34° south latitude. *Tectonics*, **22**, 1041.

Giambiagi, L. B., Mescua, J., Bechis, F., Martinez, A. & Folguera, A. 2011. Pre-Andean deformation of the Precordillera southern sector, southern Central Andes. *Geosphere*, **7**, 219–239.

Giambiagi, L. B., Mescua, J., Bechis, F., Tassara, A. & Hoke, G. 2012. Thrust belts of the southern Central Andes: along-strike variations in shortening, topography, crustal geometry, and denudation. *Geological Society of America Bulletin*, **124**, 1339–1351.

Gutscher, M. 2002. Andean subduction styles and their effect on thermal structure and interplate coupling. *Journal of South American Earth Sciences*, **15**, 3–10.

Hamza, V. M. & Muñoz, M. 1996. Heat flow map of South America. *Geothermics*, **25**, 599–646.

Hilley, G. E., Strecker, M. R. & Ramos, V. A. 2004. Growth and erosion of fold-and-thrust belts with an application to the Aconcagua fold-and-thrust belt, Argentina. *Journal of Geophysical Research–Solid Earth*, **109**, B01410.

Hoke, G. D. & Garzione, C. N. 2008. Paleosurfaces, paleoelevation, and the mechanisms for the latest Miocene topographic development of the Altiplano Plateau. *Earth and Planetary Science Letters*, **271**, 192–201.

Hoke, G. D., Isacks, B. L., Jordan, T. E., Blanco, N., Tomlinson, A. J. & Ramezani, J. 2007. Geomorphic evidence for post-10 Ma uplift of the western flank of the Central Andes (18°30′–22°S). *Tectonics*, **26**, TC5021.

Irigoyen, M. V. 1998. *Magnetic polarity stratigraphy and geochronological constraints on the sequence of thrusting in the Principal and Frontal cordilleras and the Precordillera of the Argentine Central Andes (33°S latitude)*. PhD thesis, Carleton University.

Irigoyen, M. V., Buchan, K. L. & Brown, R. L. 2000. Magnetostratigraphy of Neogene Andean foreland-basin strata, lat 33°S, Mendoza Province, Argentina. *Geological Society of America Bulletin*, **112**, 803–816.

Jordan, T. 1995. Retroarc foreland and related basins. *In*: Busby, C. & Ingersoll, R. (eds) *Tectonics of Sedimentary Basins*. Blackwell, Boston, 331–362.

Jordan, T., Isacks, B. L., Allmendinger, R. W., Brewer, J. A., Ramos, V. A. & Ando, C. J. 1983. Andean tectonics related to geometry of subducted Nazca Plate. *Geological Society of America Bulletin*, **94**, 341–361.

Jordan, T., Allmendinger, R. W., Damanti, J. F. & Drake, R. E. 1993. Chronology of motion in a complete thrust belt – the Precordillera, 30–31°S, Andes mountains. *Journal of Geology*, **101**, 135–156.

Jordan, T., Tamm, V., Figueroa, G., Flemings, P. B., Richards, D., Tabbutt, K. & Cheatham, T. 1996. Development of the Miocene Manantiales foreland basin, Principal Cordillera, San Juan, Argentina. *Revista Geologica De Chile*, **23**, 43–79.

Jordan, T., Burns, W. M., Veiga, R., Pangaro, F., Copeland, P., Kelley, S. & Mpodozis, C. 2001. Extension and basin formation in the southern Andes caused by increased convergence rate: a mid-Cenozoic trigger for the Andes. *Tectonics*, **20**, 308–324.

Kay, S., Burns, M., Copeland, P. & Ramos, V. 2006. Upper Cretaceous to Holocene magmatism and evidence for transient Miocene shallowing of the Andean subduction zone under the northern Neuquén Basin. *In*: Ramos, V. & Kay, S. (eds) *Evolution of an Andean Margin: A Tectonic and Magmatic View from the Andes to the Neuquén Basin (35°–39°S lat)*. Geological Society of America, Boulder, **407**, 19–60.

Kirby, E. & Whipple, K. X. 2012. Expression of active tectonics in erosional landscapes. *Journal of Structural Geology*, **44**, 54–75.

Kokogián, D. A., Fernández Seveso, F. & Mosquera, A. 1993. Las secuencias sedimentarias triásicas. *In*: *XII Congreso Geológico Argentino y II Congreso de Exploración de Hidrocarburos*, Mendoza.

Legarreta, L. & Uliana, M. 1999. El Jurásico y Cretácico de la Cordillera Principal. *In*: Caminos, R. (ed.) *Geologia Argentina*. Servicio Geologico Minero Argentino, Buenos Aires, **19**, 399–432.

Maksaev, V., Munizaga, F., Zentilli, M. & Charrier, R. 2009. Fission track thermochronology of Neogene plutons in the Principal Andean Cordillera of central Chile (33–35°S): implications for tectonic evolution and porphyry Cu–Mo mineralization. *Andean Geology*, **36**, 153–171.

McQuarrie, N. 2002. Initial plate geometry, shortening variations, and evolution of the Bolivian orocline. *Geology*, **30**, 867–870.

Meesters, A. G. C. A. & Dunai, T. J. 2002. Solving the production–diffusion equation for finite diffusion domains of various shapes. *Chemical Geology*, **186**, 333–344.

Meigs, A., Krugh, W., Schiffman, C., Verges, J. & Ramos, V. 2006. Refolding of thin-skinned thrust sheets by active basement-involved thrust faults in the eastern Precordillera of western Argentina. *Revista de la Asociacion Geologica Argentina*, **61**, 589–603.

Miller, S. R., Baldwin, S. L. & Fitzgerald, P. G. 2012. Transient fluvial incision and active surface uplift in the Woodlark Rift of eastern Papua New Guinea. *Lithosphere*, **4**, 131–149.

Mitchell, S. G. & Montgomery, D. R. 2006. Influence of a glacial buzzsaw on the height and morphology of the Cascade Range in central Washington State, USA. *Quaternary Research*, **65**, 96–107.

Montgomery, D., Balco, G. & Willett, S. 2001. Climate, tectonics, and the morphology of the Andes. *Geology*, **29**, 579–582.

Moucha, R. & Forte, A. M. 2011. Changes in African topography driven by mantle convection. *Nature Geoscience*, **4**, 707.

Mpodozis, C. & Ramos, V. 1989. The Andes of Chile and Argentina. *In*: Ericksen, G. E., Cañas Pinochet, M. T. & Reinemund, J. A. (eds) *Geology of the Andes and its Relation to Hydrocarbon and Mineral Resources*. Circum-Pacific Council for Energy and Mineral Resources, Houston, **11**, 56–90.

Ramos, V. A., Cegarra, M. & Cristallini, E. 1996. Cenozoic tectonics of the High Andes of west-Central Argentina (30–36°S latitude). *Tectonophysics*, **259**, 185–200.

Ramos, V. A., Cristallini, E. & Pérez, D. J. 2002. The Pampean flat-slab of the Central Andes. *Journal of South American Earth Sciences*, **15**, 59–78.

Reiners, P. W. & Farley, K. A. 2001. Influence of crystal size on apatite (U–Th)/He thermochronology: an example from the Bighorn Mountains, Wyoming. *Earth and Planetary Science Letters*, **188**, 413–420.

Reiners, P. W., Brady, R., Farley, K. A., Fryxell, J. E., Wernicke, B. & Lux, D. 2000. Helium and argon thermochronometry of the Gold Butte block, south Virgin Mountains, Nevada. *Earth and Planetary Science Letters*, **178**, 315–326.

Schoenbohm, L. M., Whipple, K. X., Burchfiel, B. C. & Chen, L. 2004. Geomorphic constraints on surface uplift, exhumation, and plateau growth in the Red River region, Yunnan Province, China. *Geological Society of America Bulletin*, **116**, 895–909.

Shuster, D. L., Flowers, R. M. & Farley, K. A. 2006. The influence of natural radiation damage on helium diffusion kinetics in apatite. *Earth and Planetary Science Letters*, **249**, 148–161.

Snyder, N. P., Whipple, K. X., Tucker, G. E. & Merritts, D. J. 2000. Landscape response to tectonic forcing: digital elevation model analysis of stream profiles in the Mendocino triple junction region, northern California. *Geological Society of America Bulletin*, **112**, 1250–1263.

Spiegel, C., Kohn, B., Belton, D., Berner, Z. & Gleadow, A. 2009. ScienceDirect.com – Earth and Planetary Science Letters – Apatite (U–Th–Sm)/He thermochronology of rapidly cooled samples: the effect of He implantation. *Earth and Planetary Science Letters*, **285**, 105–114.

Spikings, R., Dungan, M., Foeken, J., Carter, A., Page, L. & Stuart, F. 2008. Tectonic response of the central Chilean margin (35–38°S) to the collision and subduction of heterogeneous oceanic crust: a thermochronological study. *Journal of the Geological Society*, **165**, 941.

Ter Voorde, M., De Bruijne, C. H., Cloetingh, S. & Andriessen, P. 2004. Thermal consequences of thrust faulting: simultaneous versus successive fault activation and exhumation. *Earth and Planetary Science Letters*, **223**, 395–413.

Thomson, S. N., Brandon, M. T., Tomkin, J. H., Reiners, P. W., Vásquez, C. & Wilson, N. J. 2010. Glaciation as a destructive and constructive control on mountain building. *Nature*, **467**, 313–317.

Turcotte, D. L. & Schubert, G. 2002. *Geodynamics*. Cambridge University Press, Cambridge.

Vermeesch, P. 2008. Three new ways to calculate average (U–Th)/He ages. *Chemical Geology*, **249**, 339–347.

von Gosen, W. 1992. Structural evolution of the Argentine Precordillera – the Rio-San-Juan Section. *Journal of Structural Geology*, **14**, 643–667.

Walcek, A. & Hoke, G. 2012. Surface uplift and erosion of the southernmost Argentine Precordillera. *Geomorphology*, **153–154**, 156–168.

Whipple, K. X. 2004. Bedrock rivers and the geomorphology of active orogens. *Annual Review of Earth and Planetary Sciences*, **32**, 151–185.

Whipple, K. X. & Meade, B. 2004. Controls on the strength of coupling among climate, erosion, and deformation in two-sided, frictional orogenic wedges at steady state. *Journal of Geophysical Research*, **109**, F01011.

Whipple, K. X. & Tucker, G. E. 1999. Dynamics of the stream-power river incision model: implications for height limits of mountain ranges, landscape response timescales, and research needs. *Journal of Geophysical Research–Solid Earth*, **104**, 17 661–17 674.

Willett, S. 1999. Orogeny and orography: the effects of erosion on the structure of mountain belts. *Journal of Geophysical Research*, **104**, 28 957–28 982.

Wobus, C., Whipple, K. et al. 2006. Tectonics from topography: procedures, promise and pitfalls. *In*: Willett, S. D., Hovius, N., Brandon, M. T. & Fisher, D. M. (eds) *Tectonics, Climate and Landscape Evolution*. Geological Society of America, Denver, **398**, 55–74.

Wolf, R. A., Farley, K. A. & Silver, L. T. 1996. Helium diffusion and low-temperature thermochronometry of apatite. *Geochimica et Cosmochimica Acta*, **60**, 4231–4230.

Yáñez, G., Cembrano, J., Pardo, M., Ranero, C. & Selles, D. 2002. The Challenger–Juan Fernández–Maipo major tectonic transition of the Nazca–Andean subduction system at 33–34°S: geodynamic evidence and implications. *Journal of South American Earth Sciences*, **15**, 23–38.

Erosion in the Chilean Andes between 27°S and 39°S: tectonic, climatic and geomorphic control

S. CARRETIER[1]*, V. TOLORZA[2], M. P. RODRÍGUEZ[2], E. PEPIN[1], G. AGUILAR[3],
V. REGARD[1], J. MARTINOD[1], R. RIQUELME[4], S. BONNET[1], S. BRICHAU[1], G. HÉRAIL[1],
L. PINTO[2], M. FARÍAS[2], R. CHARRIER[2,5] & J. L. GUYOT[1]

[1]*Geosciences Environnement Toulouse, OMP, UPS, CNRS, IRD, Université de Toulouse, France*
[2]*Departamento de Geología, Universidad de Chile, Santiago, Chile*
[3]*Advanced Mining Technology Center, Facultad de Ciencias Fsicas y Matemticas, Universidad de Chile, Santiago, Chile*
[4]*Departamento de Ciencias Geológicas, Facultad de Ingeniería y Ciencias Geológicas, Universidad Católica del Norte, Antofagasta, Chile*
[5]*Universidad Andres Bello, Santiago, Chile*
*Corresponding author (e-mail: sebastien.carretier@get-obs-mip.fr)

Abstract: The effect of mean precipitation rate on erosion is debated. Three hypotheses may explain why the current erosion rate and runoff may be spatially uncorrelated: (1) the topography has reached a steady state for which the erosion rate pattern is determined by the uplift rate pattern; (2) the erosion rate only depends weakly on runoff; or (3) the studied catchments are experiencing different transient adjustments to uplift or to climate variations. In the Chilean Andes, between 27°S and 39°S, the mean annual runoff rates increase southwards from 0.01 to 2.6 m a^{-1} but the catchment averaged rates of decadal erosion (suspended sediment) and millennial erosion (^{10}Be in river sand) peak at *c.* 0.25 mm a^{-1} for runoff *c.* 0.5 m a^{-1} and then decrease while runoff keeps increasing. Erosion rates increase non-linearly with the slope and weakly with the square root of the runoff. However, sediments trapped in the subduction trench suggest a correlation between the current runoff pattern and erosion over millions of years. The third hypothesis above may explain these different erosion rate patterns; the patterns seem consistent with, although not limited to, a model where the relief and erosion rate have first increased and then decreased in response to a period of uplift, at rates controlled by the mean precipitation rate.

To what extent does the mean precipitation rate or tectonic uplift rate control the erosion rate in mountain ranges? Recent models suggest that climate, through its effect on erosion, plays a determinant role in localizing deformation, and in controlling mountain elevation and uplift rate (Whipple 2009). In addition, variations in palaeoerosion rates (Charreau *et al.* 2011) and in palaeosedimentation rates (e.g. Metivier *et al.* 1999; Clift 2006; Uba *et al.* 2007) potentially record variations in the mean precipitation rate (Castelltort & van den Driessche 2003). The role of climate in driving mountain erosion has become a central question in tectonics, geomorphology and sedimentology (Allen 2008).

Because it is difficult to reconstruct the evolution of the erosion rate in mountains over 100 ka to Ma, the evolution of the sediment outflux from mountain ranges has been studied using numerical and physical modeling (e.g. Kooi & Beaumont 1994; Tucker & Slingerland 1996; Bonnet & Crave 2003; Whipple & Meade 2006; Stolar *et al.* 2007). A tectonic uplift is predicted to generate erosion, the amplitude of which varies according to a timescale called the response time (Kooi & Beaumont 1996; Whipple 2009). The response time is thought to be modulated by climatic conditions (Bonnet & Crave 2003; Stolar *et al.* 2006; Whipple & Meade 2006; Tucker & vanderBeek 2013). Consequently, the relationship between erosion and precipitation rates is predicted to depend on the timescale over which the erosion rate is analysed. In the simplest ideal case of non-glaciated mountain ranges where the uplift is held constant, the cumulative erosion at a given time (the time integral of the erosion rate since the onset of the uplift) is greater where the climate is wetter simply because the response time is less and the slopes are smaller in this case (Bonnet & Crave 2003). In some circumstances, decadal or millennial erosion rates can be greater where the climate is drier. This is predicted when the drainage network grows slowly, leading to steep hillslopes, deep valleys (high fluvial relief or mean incision), and high catchment mean erosion rates exceeding the uplift rate (Carretier et al. 2009). In the case of a topography responding to a pulse of uplift, the erosion response is predicted to first increase and then to decrease, as illustrated by Figure 1 (e.g. Kooi & Beaumont 1996). While the mean erosion rate of a dry catchment reaches its maximum, the erosion rate of a wetter catchment may already be decreasing because its response time is shorter. This is another situation where the current erosion rate of a dry catchment may be greater than a wet catchment (e.g. Barnes & Pelletier 2006). Recent studies document that catchment mean erosion rates increase non-linearly with slope, so that high transient slopes in dry catchments can lead to high erosion rates (e.g. Binnie et al. 2007; Roering et al. 2007; Carretier et al. 2013). A decadal or millennial catchment mean erosion rate may not depend significantly on the precipitation rate once the topography has reached a steady state, either because the erosion rate balances the rock uplift rate (Burbank et al. 2003), or because the slopes and erosion decrease to near zero in declining reliefs (Frankel & Pazzaglia 2006). In both cases, differences in precipitation rates have been compensated by differences in hillslope angle or river width (e.g. Riebe et al. 2001; Burbank et al. 2003; von Blanckenburg et al. 2004; Stolar et al. 2007). Finally, the absence of a correlation between the catchment mean erosion rate and the mean precipitation rate may also be explained by a weak relationship between both parameters. This has been suggested by a correlation analysis between the suspended sediment yields in rivers and the mean precipitation rates or runoff (e.g. Dadson et al. 2003; Aalto et al. 2006; Syvitsky & Milliman 2007; Pepin et al. 2010). The influence of other erosion controls such as the mean slope, lithology or vegetation may be stronger (e.g. Summerfield & Hulton 1994).

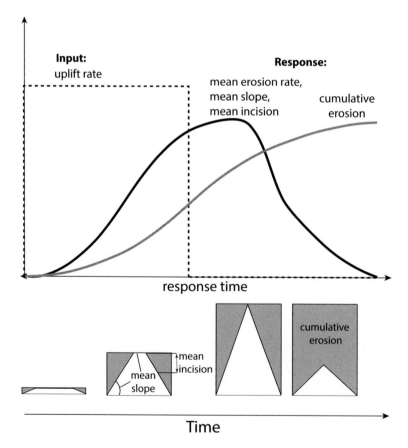

Fig. 1. Conceptual model of the co-evolution of the hillslope gradient (mean slope), mean incision, cumulative erosion and current erosion rate of an initial surface in response to an uplift. The time needed by erosion and relief to reach a maximum defines a response time. We hypothesize that the studied Andean catchments of Central Chile correspond to different steps of this evolution, for which the response time depends on the mean annual precipitation rate.

In this paper, we analyse the catchment mean decadal and millennial erosion rates for central Chile (but include two sites in Argentina) between 27° and 39°S with respect to the long-term evolution of the topography over Ma. These data include previously published catchment mean decadal and millennial erosion rates (Pepin et al. 2010, 2013; Aguilar et al. 2014; Carretier et al. 2013) and new data in the Biobío catchment (37–39°S). Along this range, the precipitation rate increases from north to south, with a catchment mean annual runoff ranging between 0.01 and 2.6 m a^{-1} (Pepin et al. 2010). The topography also shows significant variations (Rehak et al. 2010). Decadal and millennial erosion rates are anticorrelated to mean runoff south of 32°S (Pepin et al. 2010; Carretier et al. 2013). We discuss the three explanations cited above for this anticorrelation, namely, that the topography and erosion have reached a steady state controlled by a spatially variable uplift rate, that the functional relationship between the mean erosion rate and the mean runoff is weak, and that the studied catchments are experiencing different evolutions after an uplift period that occurred Ma ago. In the latter case, we hypothesize that the mean precipitation rate may have controlled the erosion response time of these catchments, and thus the pattern of the decadal and millennial erosion rates. This region may provide a space-to-time conversion framework that permits the analysis of erosion response to uplift under different climates.

In order to analyse the functional relationship between the erosion rate and other parameters, we explore the correlation between the decadal and millennial erosion rates, vegetation, runoff, lithology and topographic parameters. Then we compare the north–south pattern of these erosion rates with cumulative erosion estimates over Ma with structural and thermochronological data in order to explore the possibility that the erosion pattern mimics the uplift rate pattern or that it corresponds to different stages of the erosion response to ancient uplift.

Methodology

In order to document the north–south pattern of decadal erosion, we used the average annual flux of suspended sediment leaving the mountain front at the gauging stations of the Chilean Direccion General de Aguas (DGA, www.dga.cl) located at the foot of the main Cordillera, and previously published in Pepin et al. (2010). These fluxes represent a low estimate of the total sediment flux because the bedload has not been quantified. The station records span periods of 3–42 years (Table 1). The mean annual water discharge (m^3 a^{-1}) at each gauging station was calculated by Pepin et al. (2010) by averaging the daily water discharge over the measurement period. The mean annual runoff (m a^{-1}) was calculated by dividing the mean annual water discharge by the catchment area (Pepin et al. 2010).

The mean annual sediment discharge at each station was calculated by Pepin et al. (2010), by averaging the daily suspended sediment discharge over the measurement period. The mean annual catchment erosion rates were obtained by dividing the mean annual sediment discharge by a rock density of 2700 kg m^{-3} and by the catchment area determined from the Shuttle Radar Topography Mission (SRTM) digital elevation model. For the Rucalhue station in the Biobío River (37–39°S), affected by the building of the Pangue Dam in 1996, two mean annual erosion rates and corresponding geomorphic parameters were recalculated. One set of values corresponds to the pre-dam period and the other to the post-dam period (excluding the area above the dam: see Table 1). For the Puente Perales station in the Lara River, new data between December 2006 and August 2010 were added, which decrease the mean erosion rate calculated by Pepin et al. (2010) by 8%. For this catchment, geomorphic parameters were calculated without considering the area that is above the natural lakes. The uncertainties affecting the mean annual erosion rates include hourly variations in the sediment concentration, variable and unknown proportions of bedload, periods with missing data, and the potential differences between the survey practices. These uncertainties are difficult to estimate and may vary from one station to another. A global 1σ of $\pm 5\%$ for mean runoff and of $\pm 30\%$ for sediment discharge was assumed by Pepin et al. (2010).

In order to estimate the millennial erosion rates, we used the ^{10}Be concentrations published by Carretier et al. (2013), two samples published by Pepin et al. (2013) from the Las Tunas catchment in Argentina and two new samples in the Biobío catchment in the south of the zone studied by Carretier et al. (2013) (Table 2). For the new Biobío data, the area used for the following calculation excludes the area above natural lakes of glacial origin. These data correspond to the ^{10}Be concentration analyses of river sand quartz sampled at the outlet of 15 catchments (Fig. 1). From these concentrations, the mean catchment erosion rates were calculated following the same procedure, assuming that the ^{10}Be concentration has reached a steady state on the hillslopes, and allowing the use of a model linking the mean ^{10}Be concentration with the mean catchment erosion rate, modified from the initial models given by Brown et al. (1995) and Granger et al. (1996) (Table 3). In this model, a catchment mean ^{10}Be concentration production rate P (atoms g^{-1} a^{-1}) is calculated by averaging the surface ^{10}Be production rate of each catchment

Table 1. Data corresponding to catchments with suspended sediment measurements (from Pepin et al. (2010))

River	Station	Lat.(°)	Long. (°)	Elev. asl (m)	Lat. centroid catch. (°)	Long. centroid catch. (°)	Period	Nb. of records	Catch. area (km²)	Granitoids area (km²)	Mean elev. (m)	Hillslope gradient (m m⁻¹)	Catch. incision (m)	Green veget. cover (% area)	Runoff (m a⁻¹)	Eros. rate (mm a⁻¹)
Jorquera	Vertedero	−28.04	−69.96	1250	−27.75	−69.47	1967–2006	11017	4169	664	3790	0.30	559	1.20	6.30E-03	1.70E-03
Pulido	Vertedero	−28.09	−69.94	1310	−28.20	−69.68	1967–2006	11374	2018	1041	3567	0.37	598	1.50	2.50E-02	4.60E-03
Huasco	Algodones	−28.73	−70.50	600	−29.10	−70.12	1994–2006	4519	7189	3585	3396	0.41	711	1.40	3.50E-02	4.00E-03
Turbio	Huanta	−29.84	−70.39	1195	−29.99	−70.12	1972–1986	4462	2787	989	3619	0.43	766	1.00	6.90E-02	1.10E-02
Claro	Montegr^{de}	−30.09	−70.49	1120	−30.28	−70.36	1972–1986	4439	1249	1010	3332	0.45	854	1.70	8.80E-02	8.50E-04
Hurtado	Angostura de Pangue	−30.44	−71.00	485	−30.45	−70.58	1967–2006	10419	1876	712	2501	0.36	621	2.40	5.40E-02	5.20E-03
Grande	Puntilla San Juan	−30.71	−70.92	420	−30.88	−70.58	1964–2006	10471	3541	1538	2476	0.40	607	3.40	9.50E-02	1.20E-02
Cogoti	Embalse Cogoti	−31.03	−71.04	670	−31.17	−70.80	1967–2006	7804	741	206	2094	0.34	707	4.00	1.00E-01	1.10E-02
Illapel	Las Burras	−31.51	−70.81	1079	−31.43	−70.67	1965–2006	12301	608	607	3130	0.39	658	1.70	1.40E-01	2.90E-03
Choapa	Salamanca	−31.81	−70.93	500	−31.85	−70.67	1974–1986	2973	2228	604	2622	0.38	682	3.70	2.20E-01	2.80E-02
Putaendo	Resguardo Los Patos	−32.50	−70.58	1218	−32.43	−70.43	1966–2006	12803	964	108	2868	0.40	686	2.50	2.70E-01	4.30E-02
Colorado	Colorado	−32.86	−70.41	1062	−32.66	−70.27	1965–1994	8045	834	93	3253	0.44	739	2.10	4.00E-01	8.70E-02
Aconcagua	Rio Blanco	−32.91	−70.30	1420	−32.96	−70.18	1966–1998	5204	890	73	3428	0.53	894	1.60	8.00E-01	1.10E-01
Maipo	El Manzano	−33.59	−70.38	850	−33.81	−70.13	1965–2006	11645	4863	239	3185	0.48	898	2.20	7.40E-01	2.40E-01
Cachapoal	Puente Termas de Cauquene	−34.25	−70.57	700	−34.33	−70.31	2003–2006	1299	2472	175	2627	0.45	746	4.10	5.40E-01	1.70E-01
Tinguiriri	Bajo Los Briones	−34.71	−70.82	518	−34.77	−70.49	1989–2006	4042	1449	231	2537	0.43	735	5.80	1.20E+00	1.80E-01
Teno	D.J. Claro	−34.99	−70.82	900	−35.04	−70.59	1976–2006	10383	1208	49	2191	0.44	645	7.70	1.60E+00	1.20E-01
Claro	Los Quenes	−35.00	−70.81	900	−35.15	−70.72	1977–2006	9754	354	15	1856	0.42	832	10.70	2.10E+00	1.00E-01
Nuble	San Fabien 2	−36.58	−71.52	420	−36.68	−71.23	1985–2006	5353	1630	286	1620	0.38	541	15.40	2.10E+00	2.80E-02
Laja*	Puente Perales	−37.23	−72.54	65	−37.28	−71.66	1988–2010	6389	2622	519	1021	0.24	397	24.70	1.70E+00	2.50E-02
Bio Bio*	Rucalhue	−37.71	−71.90	245	−37.93	−71.47	1985–1995	2460	6795	505	1383	0.26	435	16.90	1.80E+00	2.60E-02
Bio Bio*	Rucalhue	−37.71	−71.90	245	−37.93	−71.47	1996–2006	5271	1818	429	1210	0.32	439	23.90	1.80E+00	9.50E-02
Cautin	Rari Ruca	−38.43	−72.00	400	−38.49	−71.41	1985–2006	6666	1291	43	1153	0.23	391	24.90	2.20E+00	4.10E-02
Trancura	A.J. Rio Llafenco	−39.33	−71.82	386	−39.32	−71.55	1985–2006	6544	1365	513	1141	0.31	477	25.80	2.60E+00	1.80E-02

*Recalculated.

Table 2. *Data corresponding to ^{10}Be samples*

Sample	Lat. (°)	Long. (°)	Mass of quartz (g)	^{10}Be/^{9}Be	±1σ	^{9}Be (atoms g^{-1})	^{10}Be (atoms g^{-1})	±1σ
SAN1*	−27.20	−69.92	44.83	1.91E-12	2.80E-13	2.41E+19	1.03E+06	1.54E+05
HUA12*	−28.60	−70.73	49.52	1.24E-12	3.60E-14	2.39E+19	5.99E+05	2.50E+04
HUA10*	−28.70	−70.55	37.35	1.08E-12	3.03E-14	2.03E+19	5.89E+05	1.66E+04
HUA7*	−28.80	−70.46	16.71	6.91E-13	4.43E-14	2.01E+19	8.33E+05	5.35E+04
HUA1*	−28.99	−70.28	34.25	8.08E-13	2.28E-14	2.04E+19	4.80E+05	1.36E+04
ELK1*	−29.85	−70.49	41.98	3.09E-13	3.97E-14	2.40E+19	1.77E+05	2.33E+04
HUR1*	−30.31	−70.73	23.22	6.85E-13	4.46E-14	2.01E+19	5.93E+05	3.86E+04
CHO0823S*	−31.60	−71.40	15.45	1.67E-13	6.98E-15	2.03E+19	2.18E+05	9.45E+03
ILL1*	−31.60	−71.11	18.87	4.36E-13	1.24E-14	2.04E+19	4.69E+05	1.35E+04
CHO0820*	−31.66	−71.22	21.78	2.54E-13	1.16E-14	2.03E+19	2.35E+05	1.08E+04
CHO0822S*	−31.66	−71.30	35.38	3.47E-13	9.91E-15	2.03E+19	1.98E+05	5.80E+03
CHO1*	−31.69	−71.27	23.02	2.22E-13	7.52E-15	2.04E+19	1.96E+05	6.71E+03
ACO1*	−32.83	−70.54	40.11	1.72E-13	2.13E-14	2.36E+19	1.01E+05	2.92E+03
MAI1*	−33.58	−70.44	38.32	1.66E-13	9.46E-15	2.02E+19	8.70E+04	5.01E+03
CAC1*	−34.21	−70.53	46.83	1.79E-13	2.03E-14	2.39E+19	9.14E+04	1.07E+04
TIN1*	−34.68	−70.87	32.58	1.60E-13	8.42E-15	2.03E+19	9.94E+04	5.28E+03
TEN1*	−34.99	−70.86	0.47	2.82E-15	8.91E-16	2.05E+19	7.33E+04	4.81E+04
LON1*	−35.18	−71.12	1.02	4.37E-15	1.26E-15	2.03E+19	6.44E+04	2.91E+04
MAU1*	−35.73	−71.02	44.54	2.40E-13	2.65E-14	2.40E+19	1.29E+05	1.48E+04
D1-1†	−37.59	−72.15	22.68	9.52E-14	1.55E-14	2.06E+19	8.63E+04	1.40E+04
Bbm1-2†	−37.67	−72.01	20.33	1.16E-13	1.62E-14	2.05E+19	1.17E+05	9.46E+03
CLA2‡	−33.28	−69.55	26.51	1.47E-13	1.33E-14	2.06E+19	1.14E+05	1.00E+04
CLA4‡	−33.29	−69.48	19.56	1.16E-13	1.35E-14	2.05E+19	1.22E+05	1.40E+04

*From Carretier et al. (2013).
†This study.
‡From Pepin et al. (2013).

pixel using the production model of Stone (2000), and a sea-level high-elevation production rate of 4.5 atoms g^{-1} (Balco et al. 2008). The production rate at each pixel was multiplied by an estimate of the topographic shielding factor ranging between 0 and 1 using the method of Codilean (2006), and by the relative proportion of quartz in the underlying lithology in order to limit the bias due to lithological variations (Safran et al. 2005). This relative proportion is given by $\chi_i / \sum_1^n \chi_i$ where χ_i is the percentage of quartz in the lithology of pixel i and n is the number of pixels (Safran et al. 2005). Catchment lithologies were obtained from the 1:1 000 000 geological maps of Chile, from which an estimate of the proportion of quartz minerals was determined, as follows: granitoid rocks, 25%; rhyolitic volcanic rocks, 5%; undifferentiated detritic rocks, 5%; ignimbrites, 2%; other lithologies, 0%. We also calculated the mean surface ^{10}Be production rate without correcting for the quartz content ('Ptotal' in Table 3), and the relative difference is 15% on average. Then, the mean catchment erosion rates ε [L/T] were calculated using the following equation:

$$\varepsilon = \frac{\Lambda_n f_n P}{\rho [^{10}Be]} + \frac{\Lambda_{\mu s} f_{\mu s} P}{\rho [^{10}Be]} + \frac{\Lambda_{\mu f} f_{\mu f} P}{\rho [^{10}Be]} \quad (1)$$

where $\rho = 2700$ kg m^{-3}; $\Lambda_n = 160$ g cm^2, $\Lambda_{\mu s} = 1500$ g cm^2, and $\Lambda_{\mu f} = 5300$ g cm^2 are the effective apparent attenuation lengths for neutrons, negative muons, and fast muons, respectively; and $f_n = 0.9785$, $f_{\mu f} = 0.0015$ and $f_{\mu f} = 0.0065$ (Braucher et al. 2003). The 1σ uncertainty for the erosion rates were calculated by propagating the analytical uncertainty of the ^{10}Be concentration and a 15% uncertainty assumed for the production rate (Table 2). Note that erosion rates of Carretier et al. (2013) were recalculated by adding the effect of fast muons, which leads to erosion rates c. 10% lower.

The calculated mean erosion rate applies for a time period, or integration time. The integration time depends inversely on the calculated erosion rate and reaches several thousands to tens of thousands of years in this study. It is defined as the time necessary to erode 0.6 m, a value close to the neutron mean free path in rocks.

In order to calculate the topographic parameters, we used the SRTM digital elevation model. The mean hillslope gradient corresponds to the average of the slopes calculated in the steepest direction, excluding pixels from the drainage network. The network is defined for the drainage area above a critical drainage area of 8.1 km^2. This area was chosen via a visual inspection to exclude the pixels of deep valleys. Other slope calculations were investigated by Carretier et al. (2013) (Data Repository) and show a difference of less than 10%. The catchment mean incision is obtained by calculating a tensile surface draped over the divide of a catchment (s.surf.tps function of GRASS with a tension parameter of 60). The elevation differences between this surface and the catchment elevations are averaged and divided by the catchment area. The resulting value has the dimension of a length that we call the mean incision. In the simplest case of a block uplift with relicts of the uplifted peneplain (e.g. Walcek & Hoke 2012), the mean incision is an estimate of the volume of sediment eroded below the envelope surface normalized by the catchment area. The studied catchments probably do not correspond to this simple situation. The mean incision is the same as the R_{va} index used by Frankel & Pazzaglia (2006), except for the details regarding the calculation of the envelope surface (by selecting divide pixels around the local maximum in a circular moving window with a 5 km radius in their case). These authors proposed R_{va} as a key metric to quantify the morphological differences between growing and decaying reliefs (see fig. 9 in Frankel & Pazzaglia (2006)). A plateau incised by canyons corresponds to a small mean incision value (a large part of the surface is not incised), a topography with a dentritic river network and deep valleys has a large mean incision value, and a topography with a dentritic network of shallow valleys has a small mean incision value.

The percentage of green vegetation cover FCOVER is derived from the VEGETATION sensor aboard SPOT4 and SPOT5, providing a spatial resolution of approximately 1.15 km (Baret et al. 2007). The CYCLOPES products and associated detailed documentation are available at http://postel.medias france.org. We used data from December 2003 in order to minimize the snow cover (summer).

The geology underlying the studied catchment is mainly composed of granitoid rocks and volcano-detritic rocks. We hypothesize that these two categories erode differently. Carretier et al. (2013) found an inverse relationship between the erosion rates and the proportion of granitoid rocks. We reanalysed this relationship, adding new data from Pepin et al. (2013) and from the Biobío basin. The proportion of area occupied by granitoid rocks was quantified in each catchment using the 1:1 000 000 geological map of Chile (Carretier et al. 2013).

Patterns of erosion rates and control factors

In this section we combine the erosion rates obtained from data published in previous studies (Pepin et al. 2010; Carretier et al. 2013; Pepin et al. 2013), with two new ^{10}Be concentrations and recalculated decadal erosion rates in the Biobío catchment. Then we evaluate the scaling relationship between the catchment mean erosion rate and runoff, and other parameters.

Figure 2 displays maps of decadal and millennial erosion rates. Millennial erosion rates calculated by Walcek & Hoke (2012) for three catchments of the Argentine Precordillera are also shown for comparison, bearing in mind that these catchments are two to three orders of magnitude smaller than the others. The erosion rates vary between $0.17 \pm 0.05 \; 10^{-2}$ mm a^{-1} and 0.32 ± 0.06 mm a^{-1}. The ^{10}Be-derived erosion rate is at its maximum near 33–34°S (c. 0.25 mm a^{-1} in Chile and c. 0.3 mm a^{-1} in Argentina), in a region where the catchments are steep, the precipitation rates are moderate (at the transition between the arid north and wet south in Chile and between the arid south and wet north in Argentina), and the vegetation cover is low. Antinao & Gosse (2009) obtained similar values (c. 0.3 mm a^{-1}) in the Chilean Maipo valley at these latitudes from the analysis of rockslides produced by short-term (20-year) seismicity. In Chile, this peak corresponds to catchments where the proportion of granitic rocks is the lowest. The other rocks are mainly volcano-detritic rocks, which suggests that these rocks may be eroded more easily. Although this lithological control on erosion is possible, it is difficult to prove here, as already pointed out by Carretier et al. (2013): the sector with small granitoid areas corresponds to a steep zone where the precipitation rates increase sharply. Another argument seems to moderate the lithological effect (Fig. 3): the maximum millennial erosion-rate value occurs for two catchments (33°S) sharing the same high mean hillslope gradient (c. 0.5 m m^{-1}) but with different lithologies that are mostly volcanic on the western side and mainly granitic or gneissic on the eastern side (Polanski 1963). The hillslope gradient seems to exert a primary control: the erosion rates increase non-linearly when the mean hillslope gradient is close to a critical slope of 0.53 m m^{-1} (Fig. 4), as noted by Carretier et al. (2013) and Walcek & Hoke (2012) in this region or nearby, and confirming previous observations in other mountain ranges (Montgomery & Brandon 2002; Binnie et al. 2007; Roering et al. 2007; Ouimet et al. 2009; DiBiase et al. 2010; DiBiase & Whipple 2011).

The new millennial erosion rate values in the south are similar to the decadal erosion rates (Fig. 2). This similarity extends to a wetter region as per the observations made by Carretier et al. (2013), and further suggests that the contribution of rare and unrecorded extreme hydrological events to erosion (Kirchner et al. 2001) decreases

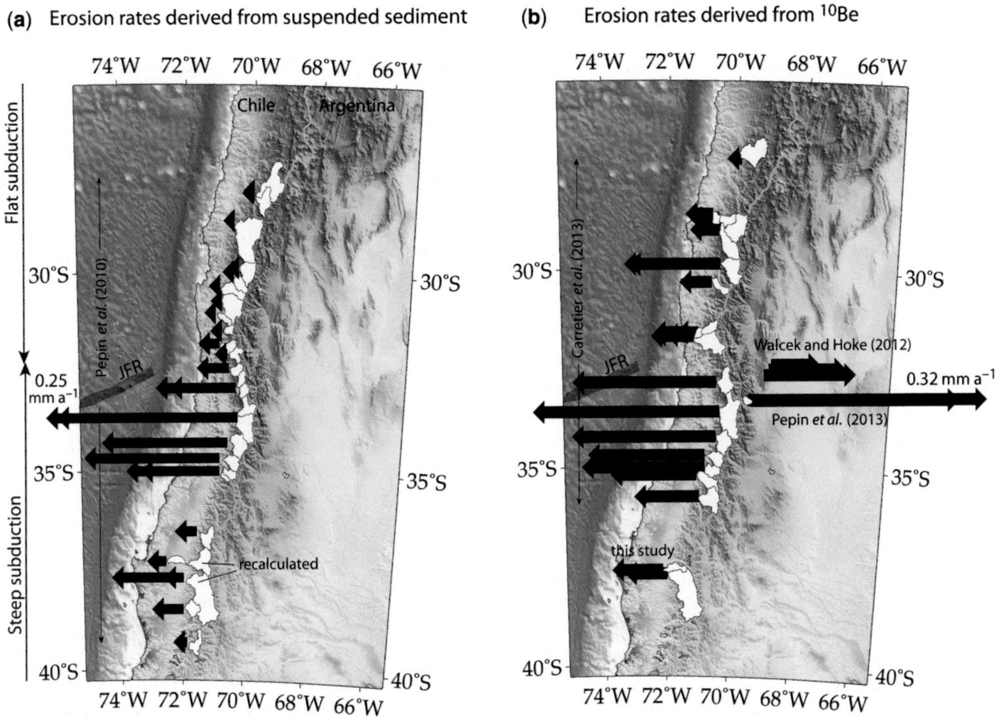

Fig. 2. (**a**) Decadal catchment erosion rates derived from suspended sediment at gauging stations (from Pepin *et al.* (2010) except for the two rates recalculated here). Note the same scale in A and B indicated by the maximum value. Studied catchments are in white. JFR is Juan Fernandez Ridge. (**b**) Millennial catchment mean erosion rates derived from ^{10}Be concentrations of river sand (data source indicated on the graph). The data for three erosion rates given by Walcek & Hoke (2012), corresponding to actively incising catchments (mean upstream slopes between 0.27 and 0.38 m m^{-1}), are shown for comparison. Note that these catchments are two to three orders of magnitude smaller than the other catchments, and hence are not visible.

toward wetter climates (see discussion in Carretier *et al.* (2013)).

The vegetation is significantly correlated with runoff, so that the effect of the former is difficult to establish. The increase of vegetation cover may contribute to the erosion rate decrease to the south of 34°S, by stabilizing the hillslopes. As suggested by Langbein & Schumm (1958), erosion may increase with precipitation rate until the vegetation cover begins to protect the hillslopes from erosion, implying a decrease in erosion rate while precipitation rates continue to increase.

In order to analyse the scaling relationship between the erosion rates and some possible controls including the catchment mean annual runoff, we carried out a multivariate analysis of variance, assuming a model for erosion rates E and selecting the control parameters: the mean hillslope gradient (HSlope), the runoff, and a lithological parameter represented by the ratio between the granitoids area and the catchment area Area$_g$/Area. The assumed model combines a non-linear function of the mean hillslope gradient, and power laws of the other parameters:

$$E = C \frac{\text{HSlope}}{1 - (\text{HSlope}/S_c)^2} \text{Runoff}^\alpha \left(\frac{\text{Area}_g}{\text{Area}}\right)^\beta \quad (2)$$

where C is a constant. As an alternative to the power law function for the slope or relief used in other analyses (e.g. Aalto *et al.* 2006; Syvitsky & Milliman 2007), we introduced a mean hillslope gradient function suggested by the hillslope erosion model presented by Roering *et al.* (1999), where S_c is a critical slope equal here to 0.53 m m^{-1} (Fig. 4). The power law function for runoff is suggested by the analysis of a large suspended sediment yield database (e.g. Syvitsky & Milliman 2007). In order to carry out the analysis of variance, we moved the slope term to the left-hand side and

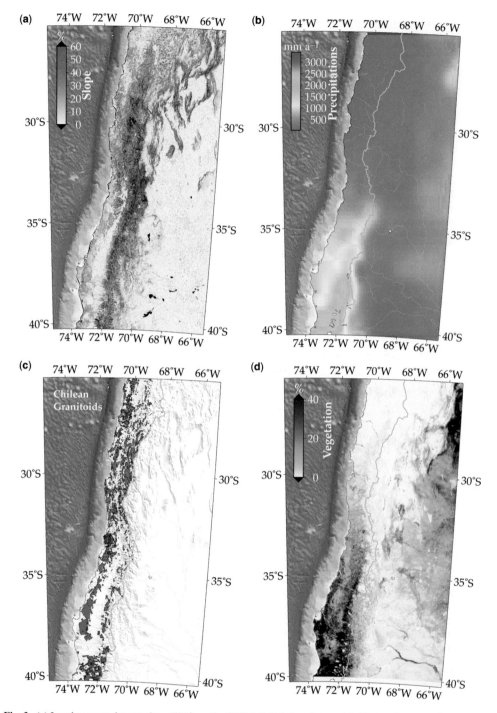

Fig. 3. (**a**) Local steepest-descent slope (%) from the SRTM digital elevation model. The maximum catchment mean slopes are located near 33°–34°S. (**b**) Mean annual precipitation rates from Matsuura & Willmott (2011) world database with a resolution of 0.5°, interpolated with a resolution of 0.1° here. (**c**) Granitoids from the 1:1 000 000 geological map of Chile. (**d**) Percentage of green vegetation cover from the FCOVER index (Baret *et al.* 2007). Data from December 2003 are shown here.

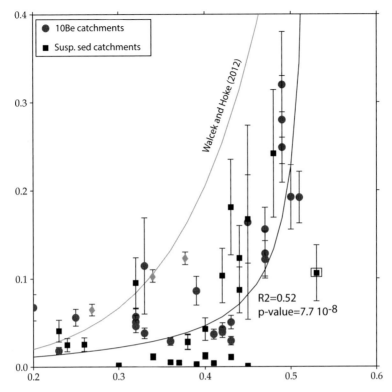

Fig. 4. Erosion rate v. mean catchment slope. The solid line is a model on the form erosion rate α (Slope/1 − (Slope/S_c)^2) with $S_c = 0.53$. R^2 and p-value apply to squares and circles but do not consider the point with the slope equal to S_c (surrounded by a square). Millennial erosion rates of three small catchments studied by Walcek & Hoke (2012) (grey diamonds) and their erosion-slope model are plotted for comparison. These catchments are two to three orders of magnitude smaller than the other ones.

take the logarithm of this modified equation:

$$\ln[E/\frac{\text{HSlope}}{1-(\text{HSlope}/S_c)^2}]$$
$$= \ln(C) + \alpha \ln(\text{Runoff}) + \beta \ln\left(\frac{\text{Area}_g}{\text{Area}}\right). \quad (3)$$

A regression analysis was then carried out using the decadal erosion rates on the one hand, and the millennial erosion rates on the other hand, for E. The data of Walcek & Hoke (2012), which correspond to much smaller catchments, are not included in this analysis. In both cases, the results indicate cases in which only $\ln(C)$ ($\ln(C) = -4.14 \pm 0.58$, p-value = 2.9E-7 for the decadal erosion rates and $\ln(C) = 0.9 \pm 0.4$, p-value = 0.04 for the millennial erosion rates) and $\ln(\text{Runoff})$ ($\alpha = 0.57 \pm 0.14$, p-value = 6.0E-4 for the decadal erosion rates and $\alpha = 0.50 \pm 0.08$, p-value = 7.6E-6 for the millennial erosion rates) were significant predictors of the left-hand side of the modified equation. The logarithm Area$_g$/Area was not a significant predictor in either case ($\beta = -0.52 \pm 0.26$, p-value = 0.06 for the decadal erosion rates and $\beta = 0.03 \pm 0.16$, p-value = 0.82 for the millennial erosion rates). The overall model fit was $R^2 = 0.65$ for the decadal erosion rates and $R^2 = 0.77$ for the millennial erosion rates. This model suggests that the control of runoff on the erosion rates is weaker (less-than-linear dependence with Runoff$^{c.0.5}$) than the control of slope, which is consistent with other databases (e.g. Aalto et al. 2006; Syvitsky & Milliman 2007).

Comparison with geological features

Figure 5 displays the latitudinal variations of some parameters: the thickness of sediment in the trench (drawn schematically after Bangs & Cande (1997)), runoff, the catchment mean incision, the erosion rates (a square for the decadal erosion rates; a circle for the millennial erosion rates; the numbers are shown as an integration time in ka), and a crustal profile based on the mean elevation of the

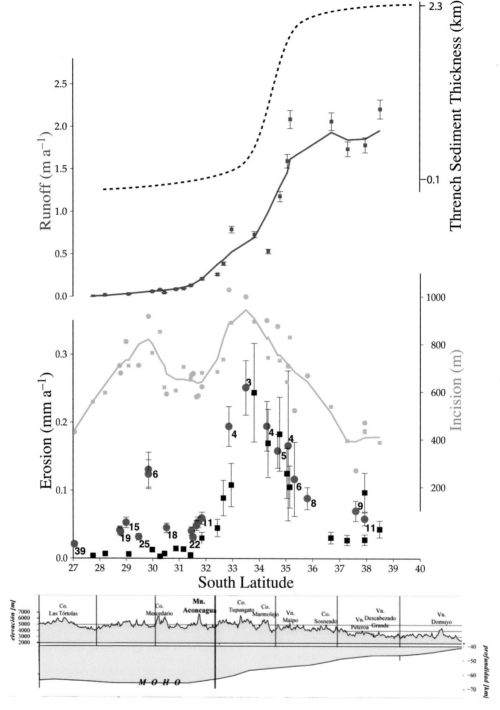

Fig. 5. From top to bottom: schematic sediment thickness in the trench after Bangs & Cande (1997), runoff, mean catchment incision, decadal (squares) and millennial (circles) erosion rates, crustal section after Tassara et al. (2006) (from Farías (2007)). Between 37° and 38°S, the millennial erosion rates are new data and the decadal erosion rates were recalculated from Pepin et al. (2010) to account for new measurements between 2006 and 2010 (Puente Perales station, Table 1) and for the periods before and after the dams were built (Rucalhue station).

Cordillera and on the Moho depth after Tassara *et al.* (2006). A striking observation is that the erosion rates peak at the inflection point for the runoff and crustal thickness near 34°S, and then decrease southwards, whereas the thickness of the sediment accumulated in the trench, representing a cumulative erosion over a timescale much longer than thousands of years, follows the southward runoff increase. The erosion rates are strongly correlated with the mean catchment incision and with mean catchment slope.

Discussion

Previous results have shown a weak relationship between the decadal and millennial erosion rates with the mean annual runoff. Is this observation sufficient to explain that the pattern of mean erosion rates and mean annual runoff are uncorrelated to the south of 33°S? Does this imply that the mean precipitation rate has a minor effect on the topographic evolution over Ma? In the following, we investigate three hypotheses that could explain the difference between the decadal, millennial and longer timescale erosion rates and the mean annual runoff, replacing these data in a temporal frame of several Ma (Fig. 6). We define three sectors which may correspond to different evolutionary stages of erosion in response to a Mio-Pliocene surface uplift (Fig. 6): the northern sector corresponds to latitudes between 27° and 33°S, the central sector between 33° and 34°S and the southern sector between 35° and 39°S.

Hypothesis 1: The topography has reached a dynamic equilibrium between the erosion rate and the rock uplift rate, so that the decadal and millennial erosion rates reflect the spatial differences in the rock uplift rates (Fig. 6a). The crustal thickness and the mean elevation decrease southwards (Fig. 5). If the tenfold increase in the erosion rates between the north and central sectors represents differences in the uplift rates, we would expect a southward increase in both the topography of the range and in the crustal thickness. The mean annual runoff increase between the north and central sectors seems too small to explain the observed decrease in crustal thickness. Alternatively, the decrease in the crustal thickness and in the erosion rate to the south of 34°S may be consistent with a decrease in the uplift rate. At these latitudes, a southward decrease of Neogene shortening and shortening rate of the upper crust is observed in Argentina (Giambiagi *et al.* 2012). According to Arriagada *et al.* (2013), this decrease in the Neogene shortening is consistent with the change in the palaeomagnetic rotation angle that they observe between the north and south of 33°S. Both observations

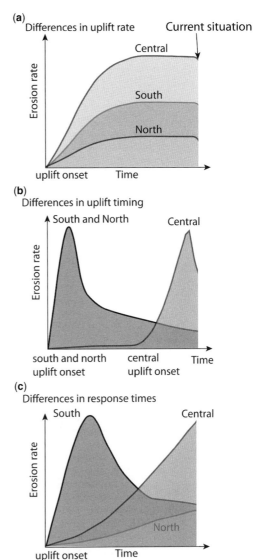

Fig. 6. Three hypotheses were investigated to explain the patterns in Figure 5. The north refers to latitudes between 27° and 33°S, the centre to latitudes between 33° and 35°S and the south to latitudes between 35° and 39°S. (**a**) In hypothesis 1, the Andes reached a dynamic equilibrium with different uplift rates, yielding different erosion rates. (**b**) In hypothesis 2, uplift occurred more recently in the centre than to the north and south, so that erosion rates are higher in the centre. (**c**) In hypothesis 3, uplift occurred at the same period but the difference in the precipitation rates brought the topography to different evolution stages, determining the differences in the erosion rate.

may be consistent with a southward decrease in the rock uplift rate on the Chilean side. This possibility requires additional thermochronological data to be

Table 3. Data corresponding to catchments with ^{10}Be samples

Sample	Lat. centroid catch. (°)	Long. centroid catch. (°)	Catch. area (km^2)	Granitoid area (km^2)	Hillslope gradient ($m\,m^{-1}$)	Mean catch. incision (m)	Green veget. cover (% of area)	Runoff ($m\,a^{-1}$)	P (atoms $g^{-1}\,a^{-1}$)	Ptotal (atoms $g^{-1}\,a^{-1}$)	(^{10}Be) Eros. rate ($mm\,a^{-1}$)	±1σ	Integr. time (ka)
SAN1*	−27.05	−69.60	3788	335	0.23	433	2.1	1.58E-02	32.5	30.8	1.9E-02	3.9E-03	39
HUA12*	−28.80	−70.24	7834	3735	0.41	680	3.8	3.47E-02	37.4	37.7	3.7E-02	5.7E-03	20
HUA10*	−28.76	−70.13	7245	3567	0.42	711	3.8	3.47E-02	39	40	3.9E-02	6.0E-03	19
HUA7*	−29.48	−70.20	2914	1193	0.43	713	3.5	3.47E-02	41.5	42.6	2.9E-02	4.8E-03	25
HUA1*	−29.00	−70.09	3176	1886	0.43	813	3.6	3.47E-02	41	43.9	5.0E-02	7.7E-03	15
ELK2*	−29.85	−70.19	2921	1172	0.47	919	2.7	6.94E-02	38.6	45.8	1.2E-01	2.1E-02	6
ELK1*	−29.85	−70.19	2921	1172	0.47	919	2.7	6.94E-02	38.6	45.8	1.3E-01	2.6E-02	6
HUR1*	−30.53	−70.37	1119	521	0.42	593	4.1	5.36E-02	42.9	37.2	4.3E-02	7.0E-03	18
CHO0823S*	−31.65	−70.95	5998	2129	0.32	581	12.8	7.57E-02	17	20.2	4.6E-02	7.2E-03	13
ILL1*	−31.53	−70.79	1231	360	0.36	679	10.8	7.57E-02	22.9	24.7	2.9E-02	4.4E-03	22
CHO0820*	−31.47	−70.88	1985	725	0.33	673	11	7.57E-02	15.2	18.7	3.8E-02	6.0E-03	13
CHO0822S*	−31.71	−70.90	5903	2119	0.32	586	12.8	7.57E-02	17.2	20.4	5.1E-02	7.8E-03	12
CHO1*	−31.85	−70.83	3757	1310	0.32	623	13.4	6.62E-02	18.9	21.9	5.7E-02	8.8E-03	11
ACO1*	−32.87	−70.28	2123	218	0.51	1031	7.3	5.14E-01	33	37	1.9E-01	2.9E-02	4
MAI1*	−33.50	−70.14	4935	247	0.49	1002	6.9	7.38E-01	36.8	39.9	2.5E-01	4.0E-02	3
CAC1*	−34.29	−70.26	2163	168	0.50	903	7.7	5.42E-01	29.9	30.6	1.9E-01	3.7E-02	4
TIN1*	−34.70	−70.51	1465	240	0.47	878	16.2	1.19E+00	26.3	27.5	1.6E-01	2.5E-02	5
TEN1*	−35.09	−70.61	1225	66	0.45	711	20.7	1.61E+00	20.4	19.9	1.6E-01	1.1E-01	4
LON1*	−35.33	−70.73	1789	77	0.33	525	32.5	1.90E+00	12.5	17.3	1.1E-01	5.5E-02	6
MAU1*	−35.81	−70.68	2693	508	0.39	671	16.6	1.90E+00	18.9	21.8	8.6E-02	1.6E-02	8
D1-1†	−37.60	−71.60	1099	285	0.20	273	69	1.79E+00	9.9	8.4	6.8E-02	1.5E-02	9
Bbm1-2†	−37.93	−71.47	7616	942	0.25	473	43	1.79E+00	11.1	10.7	5.6E-02	9.5E-03	11
CLA2‡	−33.25	−69.60	258	258	0.49	678	1.8	Not avail.	58.1	58.1	3.2E-01	6.0E-02	2
CLA4‡	−33.27	−69.50	311	311	0.49	672	2.8	Not avail.	53.7	53.7	2.8E-01	5.0E-02	2

*From Carretier et al. (2013).
†This study.
‡From Pepin et al. (2013).
Runoff refers to the nearest gauging station. P is the catchment mean ^{10}Be production rate corrected for differences in lithologies. Erosion rates are calculated using P (recalculated from Carretier et al. (2013), adding the effect of fast muons which leads to erosion rates c. 10% lower). Ptotal is the catchment mean ^{10}Be production rate without correcting for lithology, given for comparison.

tested. Nevertheless, the following arguments contradict a near-dynamic equilibrium between uplift and erosion.

Between 27° and 33°S on the western side of the Andes, there are remnants of palaeosurfaces (Mortimer 1973; Farías et al. 2008; Riquelme et al. 2008; Bissig et al. 2002; Bissig & Riquelme 2009; Rehak et al. 2010; Aguilar et al. 2011) showing that the topography is transient and thus has not reached a dynamic equilibrium. In particular, this is the case where the erosion rate is the highest, near 33°S (Farías et al. 2008). This precludes interpretation of erosion rates in terms of uplift rates in the north and central sectors. North of 24°S, outside the studied area, Barnes & Ehlers (2009) and Jordan et al. (2010) concluded that the Neogene surface uplift of the western central Andes has been rather progressive. In contrast, several authors have argued that uplift has been discontinuous between 27° and 33°S. In the extreme north sector, Bissig et al. (2002) and Bissig & Riquelme (2009) proposed a renewed uplift in the late Miocene based on supergene copper mineralization ages. Between 29° and 32°S, apatite fission-track (AFT) and (U–Th)/He-apatite thermochronological data indicate a progressive exhumation, but with a main cooling event between 20 Ma and 15 Ma (Cembrano et al. 2003; Rodríguez et al. 2012a, Rodríguez et al. in review), which may be associated with the tectonic inversion of the Abanico basin (Charrier et al. 2007) during the same period further to the south. Another cooling event occurred at c. 8 Ma in the north sector (Rodríguez et al. 2012a). In the central sector, Farías et al. (2008) proposed that rock and surface uplifts have not been continuous and have mainly occurred between c. 10 and 4.6 Ma near 33°S in the main Cordillera (see also Maksaev et al. (2009)), although some uplift may have occurred earlier during the tectonic inversion of the Abanico basin (Charrier et al. 2007) and during the Pleistocene (Lavenu & Cembrano 2008; Armijo et al. 2010).

Hypothesis 2: Uplift did not occur at the same time along the studied area, implying that the decadal and millennial erosion rates reflect different stages of the topographic evolution, with a minor influence from the mean annual precipitation rate (Fig. 6b). As mentioned above, surface uplift between 29° and 32°S seems to pre-date the main surface uplift that affected the Chilean Andes near Santiago (33–34°S) (Farías et al. 2008). Nevertheless, the similar, post c. 10 Ma uplift event identified near 29°S (Bissig et al. 2002), around 31°S (Rodríguez et al. 2012a; Rodríguez et al. in review) and near 33–34°S (Farías et al. 2008) shows some synchronism in the north and central sectors. This event might be consistent with the southward migration of the flat-slab segment associated with the Juan Fernandez Ridge (Fig. 2) that reached its current position 10 Ma ago (Yañez et al. 2002, and references therein). The proximity of the southern edge of the flat slab and the decadal and millennial erosion-rate peak suggest an influence of the southward migration of the flat slab. This would also be consistent with the idea that the subduction of oceanic plateaus drives some rock uplift in the Cordillera (Martinod et al. 2010), but which has not occurred in the southern sectors. However, a post-10 Ma uplift is also proposed for the north of Chile and south of Peru, not influenced by the flat slab, which suggests another cause (e.g. Garcia & Hérail 2005; Thouret et al. 2007; Schildgen et al. 2009; Rodríguez et al. in review). Because the topographic response time to uplift is of the order of Ma in this sector, as shown (for Chile) by Farías et al. (2008) and (for the Argentine Precordillera) by Walcek & Hoke (2012), then the mean erosion rate may have just reached the peak illustrated in Figure 1. Between 35° and 38°S, the AFT data provided by Spikings et al. (2008) suggest that a first tectonic exhumation event occurred between 18 and 15 Ma. As in the northern sector, this event may be associated with the tectonic inversion of the Abanico basin (Charrier et al. 2007). The large erosion rate observed near 33° and 34°S may thus reflect a younger uplift pulse and a rapid erosion response (due to higher slopes) compared to an older uplift to the north of 32°S and to the south of 35°S. However, an older uplift in the southern sector is not clearly established. Although exhumation or provenance data between 32° and 39°S are sparse, they suggest that the main exhumation associated with the tectonics and erosion occurred between c. 10 and 3 Ma in the Santiago region (Farías et al. 2008; Maksaev et al. 2009; Rodríguez et al. 2012a, b), and around 8 to 5 Ma between 35° and 38°S (Spikings et al. 2008), with a large range of AFT and (U-Th)/He ages between 20 and 6 Ma. This is synchronous with, or slightly after, the peak of the upper-crust deformation in the Argentina slope (15–7 Ma, Giambiagi et al. 2012). Consequently, differences in the timing of the uplift do not seem to explain the erosion rate pattern between 32° and 39°S.

Hypothesis 3: A period of surface uplift occurred in the Mio-Pliocene and the mean precipitation rate controlled the erosion response time. Modern erosion rates correspond to different stages of the topographic evolution (Fig. 6c). In this hypothesis, the short-term erosion pattern (decadal or millennial) reflects different stages of humped erosion curves such as those displayed in Figure 1. It is assumed that the erosion response time is shorter under a wetter climate: the catchments in the northern sector correspond to the growing stage of topography and the erosion rate (moderate-to-high incision and slope but low erosion rate); those in the central part have reached the erosion rate peak (high incision, slope and erosion rate); while those in the southern sector are in the decreasing stage (low incision, slope and erosion rate). Latitudinal variations in the incision and erosion rates are consistent with this model (Fig. 5). Large slope variations are expected during the topographic adjustment shown in Figure 1. The strong correlation of the erosion rate with the slope and the weaker correlation with the runoff, as suggested by the correlation analysis, implies that the erosion rate should vary significantly during this slope adjustment (Fig. 5). Across the entire dataset, there is also a good correlation between the mean catchment slope and the mean catchment incision, which is consistent with the evolution model shown in Figure 1. Although the uplift of the northern sector may have occurred earlier than in the central and southern sectors, the topography seems less evolved. The northern catchments present large remnants of perched Miocene palaeosurfaces (Bissig et al. 2002; Aguilar et al. 2011; Rodríguez et al. 2014). The averaged erosion rate calculated from the incision of these surfaces is similar to the millennial and decadal erosion rates, which is consistent with a slow increase in the erosion rate and a long response time of erosion to surface uplift (Aguilar et al. 2011, 2014). In the central sector, Farías et al. (2008) showed that the retreating river incision, driven by the 10 to 4.6 Ma uplift, reached the catchment head in the Pleistocene. This situation corresponds to the peak in Figure 1 and, as expected, the largest erosion rates are located in these catchments. Finally, this explanation requires that the total erosion (exhumation) over millions of years is higher in the southern sector than in the central and northern sectors (illustrated by the grey surface below the curves in Fig. 6c). Thermochronological data are lacking to verify this point, in particular between 33° and 39°S (see the data presented by Spikings et al. (2008)). Nevertheless, the volume of sediment trapped in the trench clearly increases southwards (Bangs & Cande 1997), with a sharp gradient in the central sector (Fig. 5). Although the age of the sediment is not well constrained, the larger volume of sediment deposited in the south strongly suggests that there has been more cumulative erosion in the south than in the north. The increase in the sedimentation rate in the trench of the southern sector from the Pliocene (Melnick & Echtler 2006), leading to c. 1.5 km of sediment thickness, is consistent with a rapid erosion response after the Mio-Pliocene rock uplift of the southern sector. The increase in glacial erosion from 6 Ma, as suggested by Melnick & Echtler 2006, may be responsible for this rapid erosion response. In addition, the thermochronological ages given by Thomson et al. (2010) at 38° to 39°S (the southernmost extremity of the region covered by our data) are mainly distributed between 1 and 5 Ma for U–Th/He on apatite and between 1 and 15 Ma for AFTs. These Mio-Pliocene ages suggest an exhumation associated with an uplift pulse. The young ages between 1 and 5 Ma suggest much higher erosion rates than those determined with ^{10}Be, the former being probably associated with glacial erosion (Thomson et al. 2010) and consistent with Pliocene trench sediment. It is possible that the succession of glacial and interglacial periods rendered the erosion rate highly variable during the Plio-Pleistocene, so that the decadal and millennial erosion rate represents a 'pause' in this evolution at these latitudes. Alternatively, this apparent inconsistency between the Plio-Pleistocene thermochronological ages and the low ^{10}Be-derived erosion rates may be consistent with a rapid response to uplift in the Plio-Pleistocene enhanced by glacial erosion, and then a rapid decrease in the erosion rate leading to low millennial and modern erosion rates.

Finally, these three hypotheses correspond to end-member models and are not mutually exclusive. It remains possible that differences in the response time (hypothesis 3) is the main explanation for the difference between the northern and central sectors, whereas a difference in the magnitude (hypothesis 1) and the timing of the uplift (hypothesis 2) accentuates the differences between the central and southern sectors. Yet, only hypothesis 3 seems to be consistent with the available information. Future exhumation and structural data as well as dating of palaeosurfaces should help to clarify the relative weight of the other hypotheses.

Conclusion

The decadal and millennial erosion patterns between 27° and 39°S span two orders of magnitude and do not fit the current precipitation rate pattern, which increases monotonically southwards. The erosion pattern is well correlated and increases non-linearly with the mean slope or the mean incision pattern and scales with the square root of the mean annual runoff. In contrast, sediment trapped in the subduction trench suggests a correlation between the current runoff pattern and erosion over Ma. The differences between these erosion patterns and the runoff pattern may be explained by a simple model of growing and relaxing topography after a period of Mio-Pliocene surface uplift, whose response time is controlled by the mean annual precipitation rates. According to this model, catchments in the north may be in the growing stage (low erosion rate), the catchment in the central sector may have reached its peak (high erosion rate), and the southern catchments may be in the relaxing phase (low erosion rate). Differences in the timing and magnitude of the uplift may also play a role but more exhumation and structural data are needed to evaluate their influence, in particular between 33° and 39°S. Overall, this contribution suggests that even if the modern erosion rate is decoupled from the precipitation rate, climate may still play a fundamental role in the cumulative erosion of a mountain range by accelerating the erosion response to uplift.

This study was funded by the Agence Nationale pour la Recherche (#ANR-06-JCJC0100) and Institut de Recherche pour le Développement. It is also a contribution to the IGCP-UNESCO 586-Y Project awarded to L. Pinto (Universidad de Chile) and L. Giambiagi (CONICET, Argentina), to the FONDECYT projects #11085022 and #1120272 to the Laboratoire Mixte International COPEDIM. We thank G. Hoke and O. Korup for detailed reviews, although we alone are responsible for any errors or misconceptions.

References

AALTO, R., DUNNE, T. & GUYOT, J. 2006. Geomorphic controls on Andean denudation rates. *The Journal of Geology*, **114**, 85–99.

AGUILAR, G., RIQUELME, R., MARTINOD, J., DARROZES, J. & MAIRE, E. 2011. Variability in erosion rates related to the state of landscape transience in the semi-arid Chilean Andes. *Earth Surface Processes and Landforms*, **36**, 1736–1748, http://dx.doi.org/10.1002/esp.2194

AGUILAR, G., CARRETIER, S., REGARD, V., VASSALLO, R., RIQUELME, R. & MARTINOD, J. 2014. Grain size-dependent ^{10}Be concentrations in alluvial stream sediment of the Huasco Valley, a semi-arid Andes region. *Quaternary Geochronology*, **19**, 163–172. http://dx.doi.org/10.1016/j.quageo.2013.01.011

ALLEN, P. 2008. Time scales of tectonic landscapes and their sediment routing systems. *In*: GALLAGHER, K., JONES, S. J. & WAINWRIGHT, J. (eds) *Landscape Evolution: Denudation, Climate and Tectonics Over Different Time and Space Scales*. Geological Society, London, Special Publications, **296**, 7–28.

ANTINAO, J. & GOSSE, J. 2009. Large rockslides in the Southern Central Andes of Chile (32–34.5°S): tectonic control and significance for Quaternary landscape evolution. *Geomorphology*, **104**, 117–133.

ARMIJO, R., RAULD, R. ET AL. 2010. The West Andean Thrust, the San Ramon Fault, and the seismic hazard for Santiago, Chile. *Tectonics*, **29**, 10.1029/2008TC002427

ARRIAGADA, C., FERRANDO, R., CORDOVA, L., MORATA, D. & ROPERCH, P. 2013. The Maipo orocline: a first scale structural feature in the Miocene to Recent geodynamics evolution in the central Chilean Andes. *Andean Geology*, **40**, 419–437.

BALCO, G., STONE, J., LIFTON, N. & DUNAI, T. 2008. A complete and easily accessible means of calculating surface exposure ages or erosion rates from Be and Al measurements. *Quaternary Geochronology*, **3**, 174–195.

BANGS, J. & CANDE, S. 1997. Episodic development of a convergent margin inferred from structures and processes along the southern Chile margin. *Tectonics*, **16**, 489–503.

BARET, F., HAGOLLE, O. ET AL. 2007. LAI, fAPAR and fCover CYCLOPES global products derived from VEGETATION. *Remote Sensing of Environment*, **110**, 305–315, http://dx.doi.org/10.1016/j.rse.2007.02.018

BARNES, J. & EHLERS, T. 2009. End member models for Andean Plateau uplift. *Earth and Planetary Science Letters*, **97**, 105–132, http://dx.doi.org/10.1016/j.earscirev.2009.08.003

BARNES, J. & PELLETIER, J. 2006. Latitudinal variation of denudation in the evolution of the Bolivian Andes. *American Journal of Science*, **506**, 1–31.

BINNIE, S. A., PHILLIPS, W. M., SUMMERFIELD, M. A. & FIFIELD, L. K. 2007. Tectonic uplift, threshold hillslopes, and denudation rates in a developing mountain range. *Geology*, **35**, 743–746.

BISSIG, T. & RIQUELME, R. 2009. Contrasting landscape evolution and development of supergene enrichment in the El Salvador porphyry Cu and Potrerillos-El Hueso Cu-Au districts, northern Chile. *In*: TITLEY, S. (ed.) *Supergene Environments, Processes and Products*. Society of Economic Geologists, Special Publication, **14**, 59–68.

BISSIG, T., CLARK, A., LEE, J. & HODGSON, C. 2002. Miocene landscape evolution and geomorphologic controls on epithermal processes in the El Indio-Pascua Au–Ag–Cu belt, Chile and Argentina. *Economic Geology and the Bulletin of the Society of Economic Geologists*, **97**, 971–996.

BONNET, S. & CRAVE, A. 2003. Landscape response to climate change: insights from experimental modeling and implications for tectonic versus climatic uplift of topography. *Geology*, **31**, 123–126, http://dx.doi.org/10.1130/0091–7613(2003)031<0123:LRTCCI>2.0.CO;2

BRAUCHER, R., BROWN, E., BOURLÈS, D. & COLIN, F. 2003. In situ produced Be measurements at great depths: implications for production rates by fast muons. *Earth and Planetary Science Letters*, **211**, 251–258, http://dx.doi.org/10.1016/S0012–821X(03)00205-X

BROWN, E. T., STALLARD, R. F., LARSEN, M. C., RAISEBECK, G. M. & YIOU, F. 1995. Denudation rates determined from the accumulation of in situ-produced Be in the Luquillo Experimental Forest, Puerto Rico. *Earth and Planetary Science Letters*, **129**, 193–202.

BURBANK, D., BLYTHE, A. ET AL. 2003. Decoupling of erosion and precipitation in the Himalayas. *Nature*, **426**, 652–655.

CARRETIER, S., POISSON, B., VASSALLO, R., PEPIN, E. & FARÍAS, M. 2009. Tectonic interpretation of erosion rates at different spatial scales in an uplifting block. *Journal of Geophysical Research*, **114**, F02003, http://dx.doi.org/10.1029/2008JF001080

Carretier, S., Regard, V. et al. 2013. Slope and climate variability control of erosion in the Andes of central Chile. Geology, 41, 195–198, http://dx.doi.org/10.1130/G33735.1

Castelltort, S. & van den Driessche, J. 2003. How plausible are high-frequency sediment supply-driven cycles in the stratigraphic record? Sedimentary Geology, 157, 3–13, http://dx.doi.org/10.1016/S0037-0738(03)00066-6

Cembrano, J., Zentilli, M., Grist, A. & Nez, G. Y. 2003. Nuevas edades de trazas de fisión para Chile Central (30°-40°): implicancias en el Alzamiento y exhumación de Los Andes desde el Cretácico, In: Abstracts of the 10th Chilean Geological Congress. Universidad de Concepción 6–10 de Octubre 2003.

Charreau, J., Blard, P. et al. 2011. Paleo-erosion rates in Central Asia since 9 Ma: a transient increase at the onset of Quaternary glaciations? Earth and Planetary Science Letters, 304, 85–92, http://dx.doi.org/10.1016/j.epsl.2011.01.018

Charrier, R., Pinto, L. & Rodríguez, M. 2007. Tectonostratigraphic evolution of the Andean orogen in Chile. In: Moreno, T. & Gibbons, W. (eds) Geology of Chile. Geological Society, London, 21–116.

Clift, P. 2006. Controls on the erosion of Cenozoic Asia and the flux of clastic sediment to the ocean. Geology, 241, 571–580, http://dx.doi.org/10.1016/j.epsl.2005.11.02

Codilean, A. 2006. Calculation of the cosmogenic nuclide production topographic shielding scaling factor for large areas using DEMs. Earth Surface Processes and Landforms, 31, 785–794, http://dx.doi.org/10.1002/esp.1336

Dadson, S. J., Hovius, N. et al. 2003. Links between erosion, runoff, variability and seismicity in the Taiwan orogeny. Nature, 426, 648–651.

DiBiase, R. & Whipple, K. 2011. The influence of erosion thresholds and runoff variability on the relationships among topography, climate, and erosion rate. Journal of Geophysical Research, 116, F04036, http://dx.doi.org/10.1029/2011JF002095

DiBiase, R., Whipple, K., Heimsath, A. & Ouimet, W. 2010. Landscape form and millennial erosion rates in the San Gabriel Mountains, CA. Earth and Planetary Science Letters, 289, 134–144, http://dx.doi.org/10.1016/j.epsl.2009.10.036

Farías, M. 2007. Tectónica y erosión en la evolución del relieve de los Andes de Chile Central durante el Neógeno. PhD thesis, Universidad de Chile/Universitée de Toulouse.

Farías, M., Charrier, et al. 2008. Late Miocene high and rapid surface uplift and its erosional response in the Andes of Central Chile (33–35°S). Tectonics, 27, TC1005, http://dx.doi.org/10.1029/2006TC002046

Frankel, K. L. & Pazzaglia, F. J. 2006. Mountain fronts, base-level fall, and landscape evolution: insights from the southern Rocky Mountains, In: Willet, D., Hovius, N., Brandon, M. & Fisher, D. (eds) Tectonics, Climate, and Landscape Evolution. Geological Society of America, Special Paper, 398, 419–434, http://dx.doi.org/10.1130/2006.2398(26)

Garcia, M. & Hérail, G. 2005. Fault-related folding, drainage network evolution and valley incision during the Neogene in the Andean Precordillera of Northern Chile. Geomorphology, 65, 279–300, http://dx.doi.org/10.1016/j.geomorph.2004.09.007

Giambiagi, L., Mescua, J., Bechis, F., Tassara, A. & Hoke, G. 2012. Thrust belts of the southern Central Andes: along-strike variations in shortening, topography, crustal geometry, and denudation. Geological Society of American Bulletin, 124, 1339–1351, http://dx.doi.org/10.1130/B30609.1

Granger, D., Kircher, J. & Finkel, R. 1996. Spatially averaged long-term erosion rates measured from in situ-produced cosmogenic nuclides in alluvial sediment. The Journal of Geology, 104, 249–257.

Jordan, T., Nester, P., Blanco, N., Hoke, G., Dávila, F. & Tomlinson, A. 2010. Uplift of the Altiplano Puna plateau: a view from the west. Tectonics, 29, TC5007, http://dx.doi.org/10.1029/2010TC002661

Kirchner, J. W., Finkel, R., Riebe, C., Granger, D., Clayton, J., King, J. & Megahan, W. 2001. Mountain erosion over 10-year, 10000-year, and 10000 000-year timescales. Geology, 29, 591–594.

Kooi, H. & Beaumont, C. 1994. Escarpment evolution on high-elevation rifted margins: insights derived from a surface processes model that combines diffusion, advection, and reaction. Journal of Geophysical Research, 99, 12 191–12 209.

Kooi, H. & Beaumont, C. 1996. Large-scale geomorphology: classical concepts reconciled and integrated with contemporary ideas via surface processes model. Journal of Geophysical Research, 101, 3361–3386.

Langbein, W. & Schumm, S. 1958. Yield of sediment in relation to mean annual precipitation. Transactions of the American Geophysical Union, 39, 1076–1084.

Lavenu, A. & Cembrano, J. 2008. Deformación compresiva cuaternaria en la Cordillera Principal de Chile central (Cajón del Maipo, este de Santiago). Revista Geologica de Chile, 35, 233–252.

Maksaev, V., Munizaga, F., Zentilli, M. & Charrier, R. 2009. Fission track thermochronology of Neogene plutons in the Principal Andean Cordillera of central Chile (33–35°S): implications for tectonic evolution and porphyry Cu–Mo mineralization. Andean Geology, 36, 153–171.

Martinod, J., Husson, L., Roperch, P., Guillaume, B. & Espurt, N. 2010. Horizontal subduction zones, convergence velocity and the building of the Andes. Earth and Planetary Science Letters, 299, 299–309, http://dx.doi.org/10.1016/j.epsl.2010.09.010

Matsuura, K. & Willmott, C. J. 2011. http://climate.geog.udel.edu/climate/html_pages/download.html

Melnick, D. & Echtler, M. 2006. Inversion of forearc basins in south-central Chile caused by rapid glacial age trench fill. Geology, 34, 709–712.

Metivier, F., Gaudemer, Y., Tapponnier, P. & Klein, M. 1999. Mass accumulation rates in Asia during the Cenozoic. Geophysical Journal International, 137, 280–318, http://dx.doi.org/10.1046/j.1365-246X.1999.00802.x

Montgomery, D. R. & Brandon, M. T. 2002. Topographic controls on erosion rates in tectonically active mountain ranges. Earth and Planetary Science Letters, 201, 481–489.

Mortimer, C. 1973. The Cenozoic history of the southern Atacama Desert. Journal of the Geological Society, London, 129, 505–526.

OUIMET, W. B., WHIPPLE, K. X. & GRANGER, D. E. 2009. Beyond threshold hillslopes: channel adjustment to base-level fall in tectonically active mountain ranges. *Geology*, **37**, 579–582, http://dx.doi.org/10.1130/G30013A.1

PEPIN, E., CARRETIER, S., GUYOT, J. & ESCOBAR, F. 2010. Specific suspended sediment yields of the Andean rivers of Chile and their relationship to climate, slope and vegetation. *Hydrological Sciences Journal*, **55**, 1190–1205, http://dx.doi.org/10.1080/02626667.2010.512868

PEPIN, E., CARRETIER, S. ET AL. 2013. Pleistocene landscape entrenchment: a geomorphological mountain to foreland field case, the Las Tunas system, Argentina. *Basin Research*, **25**, 1–25, http://dx.doi.org/10.1111/bre.12019

POLANSKI, J. 1963. Estatigrafía neotectónica y geomorphología del pleistoceno pedemontano entre los rios diamante y mendoza, provincia de Mendoza. *Revista de la Asociacion Geologica Argentina*, **17**, 127–349.

REHAK, K., BOOKHAGEN, B., STRECKER, M. & ECHTLER, H. P. 2010. The topographic imprint of a transient climate episode: the western Andean flank between 15.5 and 41.5°S. *Earth Surface Processes and Landforms*, **35**, 1516–1534, http://dx.doi.org/10.1002/esp.1992

RIEBE, C., KIRCHNER, J., GRANGER, D. & FINKEL, R. 2001. Minimal climatic control on erosion rates in the Sierra Nevada, California. *Geology*, **29**, 447–450, ISI:000168445700018

RIQUELME, R., DARROZES, J., MAIRE, E., HÉRAIL, G. & SOULA, J. C. 2008. Long-term denudation rates from the Central Andes (Chile) estimated from a digital elevation model using the black top hat function and inverse distance weighting: implications for the Neogene climate of the Atacama Desert. *Revista Geologica de Chile*, **35**, 105–121.

RODRÍGUEZ, M., CHARRIER, R., BRICHAU, S. & CARRETIER, S. 2012a. Thermochronometric constraints for Cenozoic exhumation in north central Chile (29°S-32°S), In: *Abstracts of the XII Chilean Geological Congres, Antofagasta*, 5–9 August 2012.

RODRÍGUEZ, M., PINTO, L. & ENCINAS, A. 2012b. Cenozoic erosion in the Andean forearc in Central Chile (33-34°S): sediment provenance inferred by heavy mineral studies. *Geological Society of America, Special Papers*, **487**, 141–162, http://dx.doi.org/10.1130/2012.2487(09)

RODRÍGUEZ, M. ET AL. in review. Thermochronometric contraints on the development of the Andean topographic front in north central Chile (28.5–32S). *Tectonics*.

RODRÍGUEZ, M. P., AGUILAR, G., URRESTY, C. & CHARRIER, R. 2014. Neogene landscape evolution in the Andes of north-central Chile between 28 and 32°S: interplay between tectonic and erosional processes. In: SEPÚLVEDA, S. A., GIAMBIAGI, L. B., MOREIRAS, S. M., PINTO, L., TUNIK, M., HOKE, G. D. & FARÍAS, M. (eds) *Geodynamic Processes in the Andes of Central Chile and Argentina*. Geological Society, London, Special Publications, **399**. First published online April 4, 2014, http://dx.doi.org/10.1144/SP399.15

ROERING, J. J., KIRCHNER, J. W. & DIETRICH, W. E. 1999. Evidence for nonlinear, diffusive sediment transport on hillslopes and implications for landscape morphology. *Water Resources Research*, **35**, 853–870.

ROERING, J. J., PERRON, J. T. & KIRCHNER, J. W. 2007. Functional relationships between denudation and hillslope form and relief. *Earth and Planetary Science Letters*, **264**, 245–258, http://dx.doi.org/10.1016/j.epsl.2007.09.035

SAFRAN, E., BIERMAN, P., AALTO, R., DUNNE, T., WHIPPLE, K. & CAFFEE, M. 2005. Erosion rates driven by channel network incision in the Bolivian Andes. *Earth Surface Processes and Landforms*, **30**, 1007–1024.

SCHILDGEN, T., HODGES, K., WHIPPLE, K., PRINGLE, M., VAN SOEST, M. & CORNELL, K. 2009. Late Cenozoic structural and tectonic development of the western margin of the central Andean Plateau in southwest Peru. *Tectonics*, **28**, TC4007.

SPIKINGS, R., DUNGAN, M., FOEKEN, J., CARTER, A., PAGE, L. & STUART, F. 2008. Tectonic response of the central Chilean margin (35–38°S) to the collision and subduction of heterogeneous oceanic crust: a thermochronological study. *Journal of the Geological Society, London*, **165**, 941–953.

STOLAR, D. B., WILLETT, S. D. & ROE, G. H. 2006. Climatic and tectonic forcing of a critical orogen. In: WILLETT, S. D., HOVIUS, N., BRANDON, M. T. & FISHER, D. M. (eds) *Tectonics, Climate, and Landscape Evolution*. Geological Society of America Special Paper, **398**, 241–250, http://dx.doi.org/10.1130/2006.2398(14)

STOLAR, D., ROE, G. & WILLETT, S. 2007. Controls on the patterns of topography and erosion rate in a critical orogeny. *Journal of Geophysical Research*, **112**, F04002.

STONE, J. 2000. Air pressure and cosmogenic isotope production. *Journal of Geophysical Research*, **105**, 23753–23759, ISI: 000089895700027

SUMMERFIELD, M. A. & HULTON, N. J. 1994. Natural controls of fluvial denudation rates in major world drainage basins. *Journal of Geophysical Research*, **99**, 13 871–13 883.

SYVITSKY, J. & MILLIMAN, J. 2007. Geology, geography, and humans battle for dominance over the delivery of fluvial sediment to the coastal ocean. *The Journal of Geology*, 115.

TASSARA, A., GOTZE, H., SCHMIDT, S. & HACKNEY, R. 2006. Three-dimensional density model of the Nazca plate and the Andean continental margin. *Journal of Geophysical Research*, **111**, B09404, http://dx.doi.org/10.1029/2005JB003976

THOMSON, S., BRANDON, M., TOMKIN, J., REINERS, P., VÁSQUEZ, C. & WILSON, N. 2010. Glaciation as a destructive and constructive control on mountain building. *Nature*, **467**, 313–317, http://dx.doi.org/10.1038/nature09365

THOURET, J. C., WORNER, G., GUNNELL, Y., SINGER, B., ZHANG, X. & SOURIOT, T. 2007. Geochronologic and stratigraphic constraints on canyon incision and Miocene uplift of the Central Andes in Peru. *Earth and Planetary Science Letters*, **263**, 151–166.

TUCKER, G. & SLINGERLAND, R. 1996. Predicting sediment flux from fold and thrust belts. *Basin Research*, **8**, 329–349.

TUCKER, G. & VANDERBEEK, P. 2013. A model for post-orogenic development of a mountain range and its foreland. *Basin Research*, **24**, 241–259, http://dx.doi.org/10.1111/j.1365-2117.2012.00559.x

UBA, C., STRECKER, M. & SCHMITT, A. 2007. Increased sediment accumulation rates and climatic forcing in the central Andes during the late Miocene. *Geology*, **35**, 979–982.

VON BLANCKENBURG, F., HEWAWASAM, T. & KUBIK, P. 2004. Cosmogenic nuclide evidence for low weathering and denudation in the wet, tropical highlands of Sri Lanka. *Journal of Geophysical Research*, **109**, F03008, ISI: 000224573800001

WALCEK, A. & HOKE, G. 2012. Surface uplift and erosion of the southernmost Argentine Precordillera. *Geomorphology*, **153–154**, 156–168.

WHIPPLE, K. X. 2009. The influence of climate on the tectonic evolution of mountain belts. *Nature Geoscience*, **2**, 97–104, http://dx.doi.org/10.1038/ngeo413

WHIPPLE, K. & MEADE, B. 2006. Orogen response to changes in climatic and tectonic forcing. *Earth and Planetary Science Letters*, **243**, 218–228, http://dx.doi.org/10.1016/j.epsl.2005.12.022

YAÑEZ, G., CEMBRANO, J., PARDO, M., RANERO, C. & SELLES, D. 2002. The Challenger-Juan Fernandez-Maipo major tectonic transition of the Nazca-Andean subduction system at 33-34°S: geodynamic evidence and implications. *Journal of South American Earth Sciences*, **15**, 23–38, http://dx.doi.org/10.1016/S0895-9811(02)00004-4

Neogene landscape evolution in the Andes of north-central Chile between 28 and 32°S: interplay between tectonic and erosional processes

MARÍA PÍA RODRÍGUEZ[1,2]*, GERMÁN AGUILAR[2], CONSTANZA URRESTY[1] & REYNALDO CHARRIER[1,3]

[1]*Departamento de Geología, Facultad de Ciencias Físicas y Matemáticas, Universidad de Chile, Santiago, Chile*

[2]*Advanced Mining Technology Center (AMTC), Facultad de Ciencias Físicas y Matemáticas, Universidad de Chile, Santiago, Chile*

[3]*Escuela de Ciencias de la Tierra, Universidad Andrés Bello, Campus República, Santiago, Chile*

*Corresponding author (e-mail: maria.p.rodriguez@gmail.com)

Abstract: We combine geomorphological analysis of palaeosurfaces and U–Pb zircon geochronology of overlying tuffs to reconstruct the Neogene landscape evolution in north-central Chile (28–32°S). Prior to the Early Miocene, a pediplain dominated the landscape of the present-day Coastal Cordillera. The pediplain was offset during the Early (Middle?) Miocene, leading to uplift of the present-day eastern Coastal Cordillera and to the formation of a secondary topographic front. During the Late Miocene, the entire Coastal Cordillera was uplifted, with resulting deposition taking place within river valleys similar to those of the present day. A new pediplain developed on top of these deposits between the Early to Middle Pleistocene and was finally uplifted post-500 ka. These three major uplift stages correlate with episodes of increased deformation widely recognized throughout the Central Andes, starting after a Late Oligocene–Early Miocene episode of increased plate convergence. North of 30°S, the previous palaeotopography along the western Coastal Cordillera probably influenced Neogene landscape evolution. The presence of an inherited palaeotopography together with a strong decrease of precipitation to the north of 30°S would have determined differences in landscape development between this area and the area to the south of 30°S since the Early Miocene.

The morphology of active mountain belts results from the interplay between tectonic processes, which deform the lithosphere and result in uplifted regions of the Earth's surface, and erosional processes, which are mainly controlled by climate and rock type (Strecker *et al.* 2007). In the Central Andes (15–34°S, Fig. 1), along-strike variations of topography and the amount of shortening have been mostly related to north–south-changing tectonic features. The most widely mentioned correspond to subduction geometry (Jordan *et al.* 1983; Isacks 1988), the age of the Nazca Plate (Ramos *et al.* 2004) and interplate coupling (Lamb & Davis 2003). It is thought that these tectonic factors together with the pre-Neogene geological history (Ramos *et al.* 1996; Lamb *et al.* 1997; Tassara & Yañez 2003; Giambiagi *et al.* 2012) may have played a dominant role in the first-order topography and structure of the Andes (Hilley & Coutand 2010). However, it is also likely that erosional processes may influence the kinematics of deformation (Sobel & Strecker 2003; Hilley *et al.* 2004) and control the response time to uplift (Aguilar *et al.* 2011; Carretier *et al.* 2013) at the scale of morphostructural segments (Fig. 1; Hilley & Coutand 2010). Here, we present the case of the Central Andes of north-central Chile between 28° and 32°S, whose Neogene landscape evolution may have been influenced by several factors. Firstly, this region is located on the Pampean or Chilean flat-slab segment (27–33°S), within which the subduction angle between the Nazca and South American Plates is *c.* 10°, contrary to the adjacent regions of the Central Andes where this angle is *c.* 30° (Cahill & Isacks 1992). Slab flattening is thought to be related to the subduction of the buoyant Juan Fernández Ridge (Fig. 1), which migrated from the northern to the southern part of the study area between *c.* 16 and 12 Ma (Yañez *et al.* 2001). However, it is unclear if Neogene crustal thickening and uplift are a consequence of the subduction of the Juan Fernández Ridge (Cembrano *et al.* 2003) or are better related to changes in the convergence parameters between the Nazca

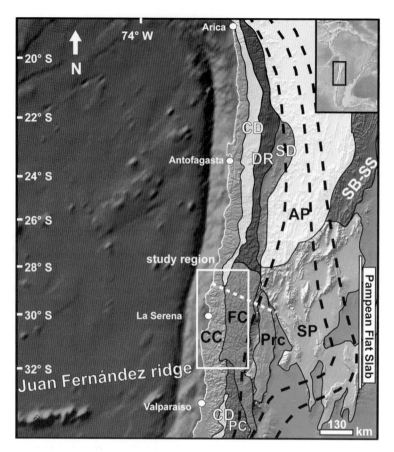

Fig. 1. Morphostructural units and tectonic setting of the Central Andes (15–34°S). Dashed black lines mark depth contour lines of the Nazca Plate underneath the South America Plate at 100, 150 and 200 km (Cahill & Isacks 1992). Dashed white line marks the symmetry axis of the Vallenar Orocline (Arriagada et al. 2009). CC, Coastal Cordillera; CD, Central Depression; DR, Domeyko Cordillera; SD, Subandean Depression; AP, Altiplano-Puna (including the Western and Eastern Cordilleras); SB, Subandean Ranges; SS, Santa Bárbara System; FC, Frontal Cordillera; Prc, Precordillera; SP, Sierras Pampeanas; PC, Principal Cordillera.

and South American Plates (Kay & Mpodozis 2002; Charrier et al. 2013).

Secondly, the study area also records a transition in the pre-Neogene topography that is reflected in the presence of the Vallenar Orocline at this latitude of the Central Andes (Fig. 1; Arriagada et al. 2009). The Vallenar Orocline is thought to mark the southernmost extent of Eocene–Oligocene deformation (Arriagada et al. 2009), associated with the so-called Incaic Range (Steinmann 1929; Charrier & Vicente 1972; Cornejo et al. 2003). During the Eocene and Oligocene the Incaic Range was the main palaeogeographic feature to the north of 31°S (Fig. 1; Maksaev & Zentilli 1999; Charrier et al. 2007; Arriagada et al. 2009, 2013; Bissig & Riquelme 2010), while an extensional volcano-sedimentary basin, known as the Abanico Basin (Charrier et al. 2002), developed south of 32°S. However, it is unclear how the presence of the Incaic relief may have influenced the subsequent landscape evolution throughout the studied area.

Thirdly, the region of north-central Chile has a semi-arid climate, which is transitional between the hyperarid conditions of the southern Atacama Desert north of 27°S and the more humid conditions of central Chile south of 33°S (Fig. 2a). The Southeast Pacific anticyclone (SEP) is the main factor responsible for the hyperaridity north of 27°S, whereas the penetration of the southern hemisphere westerlies results in the more humid conditions of central Chile (Veit 1996). In particular, along the studied region a north-to-south rise in precipitation occurs at 30°S related to the influence of the SEP (Fig. 2a). During the Palaeogene, the climate in north-central Chile was warmer and more humid than at present as indicated by the woody components of palaeoflora from fossiliferous localities just south of La Serena (Fig. 2a; Villagrán et al. 2004). Since c. 21 to 15 Ma subtropical vegetation has been replaced by sclerophytic shrubs indicating a warm, seasonal climate receiving scarce rainfall from both the east and the west (Villagrán et al. 2004). The transition between a hyperarid climate to the north of 27°S and a humid climate south of 33°S occurred after the Middle Miocene (Le Roux 2012). During this period the combination of a series of events, including glaciations in West Antarctica, formation of the Humboldt Current and uplift of the Andes, are thought to have been responsible for the development of the latitudinal precipitation gradient throughout the study area (Le Roux 2012). The role, if any, that the present-day along-strike increase in precipitation and/or climatic changes throughout the Miocene could have played in shaping the landscape in north-central Chile is largely unknown.

In this study we combine geomorphological analysis of sub-planar palaeosurfaces in the Coastal Cordillera (Fig. 1) with the U–Pb zircon geochronology of overlying tuffs to reconstruct the Neogene history of uplift and incision of these palaeosurfaces in north-central Chile (28–32°S). Our results are discussed considering previous data on sub-planar palaeosurfaces in the Coastal Cordillera (Rodríguez et al. 2013) and the higher Frontal Cordillera (e.g. Bissig et al. 2002; Nalpas et al. 2009). Finally, we discuss the roles that tectonic and erosional processes have played in the development of the present-day topography in north-central Chile.

Regional framework

The large-scale geomorphology of the study area is characterized by a marked rise in mean elevation along west–east transects (Fig. 2b, d; Aguilar et al. 2011, 2013). This first-order geomorphological feature represents a topographic front separating two north–south-elongated morphostructural units corresponding to the Coastal Cordillera and Frontal Cordillera from west to east (Fig. 2b, c). Contrary to what is observed in the Andean segments to the north of 27°S and to the south of 33°S, no continuous Central Depression is observed to the east of the Coastal Cordillera in the study region (Figs 1 & 2c). Only north of 30°S is there an area which corresponds to that of relatively lower topography within the Coastal Cordillera (the Domeyko Depression; Fig. 2c).

The Coastal Cordillera is characterized to the west by a series of shore platforms displaying low slope values (<20°) (Fig. 2c; Paskoff 1970; Ota et al. 1995; Benado 2000; Saillard et al. 2009; Rodríguez et al. 2013). To the east, the Coastal Cordillera reaches a maximum elevation of c. 3200 m above mean sea-level (a.s.l.) (Fig. 2b). The Coastal Cordillera mainly corresponds to an east-dipping homoclinal block of Triassic–Lower Cretaceous volcano-sedimentary rocks that unconformably cover a Devonian–Carboniferous (Permian?) metamorphic and sedimentary basement (Fig. 3; Rivano & Sepúlveda 1991). Both features are intruded by Triassic–Early Cretaceous north–south-elongated plutonic belts with increasing ages to the east (Fig. 3; Rivano & Sepúlveda 1991; Emparán & Pineda 2006; Arévalo et al. 2009). Towards the border with the Frontal Cordillera, the Lower Cretaceous rocks at the top of the east-dipping homoclinal block are unconformably covered by subhorizontal Upper Cretaceous–Paleocene volcano-sedimentary rocks and intruded by a plutonic belt of similar age (Fig. 3; Pineda & Emparán 2006; Pineda & Calderón 2008). Along the coast and within the main valleys, Neogene marine and continental sedimentary rocks are exposed (Fig. 4; Rivano & Sepúlveda 1991; Le Roux et al. 2004, 2005, 2006; Emparán & Pineda 2006; Arévalo et al. 2009). As will be explained later, these deposits are closely related to the development of the pediplains studied here. The Coquimbo Formation corresponds to a shallow marine-to-transitional sedimentary succession exposed along the coast near the localities of Punta Choros and Tongoy (Figs 3 & 4). It records continuous marine deposition from the Early Miocene (c. 23 Ma) to the Early Pleistocene (c. 1 Ma) (Le Roux et al. 2004, 2005, 2006). South of 30°S, the Coquimbo Formation grades laterally towards the east into the continental Confluencia Formation (Figs 3 & 4). The Confluencia Formation is composed of fluvial and alluvial facies exposed along the lower and middle courses of the main valleys (Figs 3 & 4). The fluvial deposits change laterally towards the valley walls into the alluvial deposits (Fig. 4). In some areas the latter overlie the fluvial deposits (Fig. 4). No geochronological constraints exist for the Confluencia Formation, but based on its relationship with the Coquimbo Formation, a general Miocene–Pleistocene age can be assumed (Emparán & Pineda 2006). The alluvial facies within the Confluencia Formation present an interbedded ash bed south of Tongoy (Figs 3 & 4), which has been correlated with a similar level within the marine Coquimbo Formation exposed just to the west (Figs 3 & 4; Emparán & Pineda 2006) and dated at c. 6 Ma (Emparán & Pineda 2000). North of 30°S, the Domeyko Gravels are exposed within the Domeyko Depression (Figs 3 & 4). The Domeyko Gravels are alluvial deposits interpreted to have accumulated in a closed basin with a local sediment source (Arévalo et al. 2009).

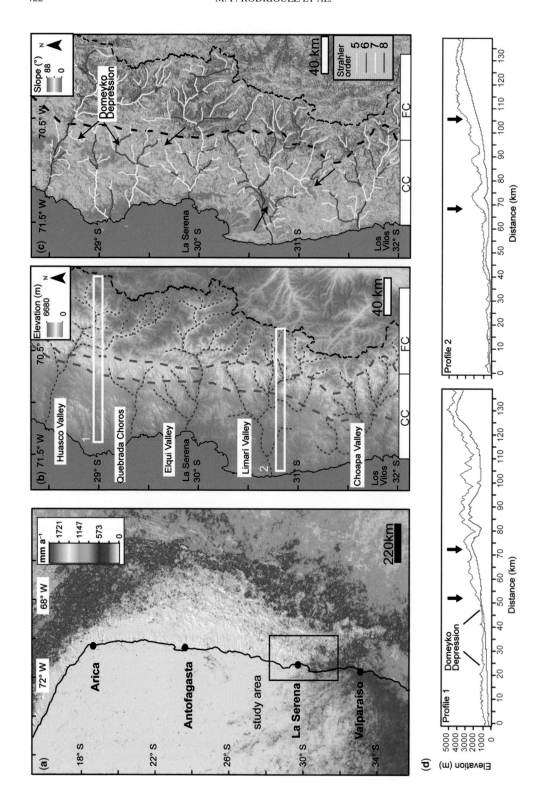

There are no chronostratigraphic or geochronological constraints available for the Domeyko Gravels. However, they are thought to be of Middle Miocene age (Arévalo et al. 2009) according to regional correlations with the Atacama Gravels at c. 27°S (Mortimer 1973). Deposition of the Atacama Gravels started c. 17 Ma and ended by c. 10 Ma (Cornejo et al. 1993), finally leading to regional pedimentation and development of the Atacama Pediplain on top. Also exposed within the Domeyko Depression are alluvial and colluvial deposits that crop out attached to relatively higher topographic areas and that overlie the Domeyko Gravels (Fig. 4; Arévalo et al. 2009). No direct geochronological constraints are available for these deposits, but they have been correlated with similar deposits at 27°S (Arévalo et al. 2009) presenting intercalated ash units with ages between c. 7 and 3 Ma (Fig. 4; Arévalo et al. 2009). The alluvial and colluvial deposits exposed north of 30°S are correlated with the alluvial facies of the Confluencia Formation exposed south of 30°S.

The Frontal Cordillera reaches elevations as high as c. 6700 m a.s.l. It is formed by a core of Carboniferous to Permian magmatic units (Fig. 3; Nasi et al. 1990; Pineda & Calderón 2008), which here is referred to as the central Frontal Cordillera (Fig. 5). The core is covered to the west by a dominantly west-dipping block of Triassic–Upper Cretaceous folded volcano-sedimentary rocks, intruded by a Late Cretaceous–Early Paleocene magmatic belt (Mpodozis & Cornejo 1988; Nasi et al. 1990; Pineda & Emparán 2006; Pineda & Calderón 2008). This area will be referred to below as the western Frontal Cordillera (Fig. 5). To the east, the basement core is intruded by, or in faulted contact with, a block composed mostly of Permo-Triassic magmatic and volcanic rocks unconformably overlain by Oligocene–Miocene folded volcano-sedimentary rocks (Fig. 3; Maksaev et al. 1984; Nasi et al. 1990; Martin et al. 1999; Bissig et al. 2001; Winocur et al. 2014). These rocks are unconformably covered by Miocene subhorizontal volcanic rocks and intruded by a north–south-trending Oligocene magmatic belt (Fig. 3; Maksaev et al. 1984; Nasi et al. 1990; Martin et al. 1999; Bissig et al. 2001; Winocur et al. 2014). The area of the Frontal Cordillera to the east of the basement core is referred to here as the eastern Frontal Cordillera (Fig. 5). Finally, a NNE–SSW-trending magmatic belt of Eocene age intrudes the areas of the central and western Frontal Cordillera (Figs 2 & 5). South of 31.5°S, the area to the east of the main topographic front corresponds to the Principal Cordillera (Fig. 5), which is defined by a core of Oligocene–Miocene folded volcano-sedimentary rocks (Charrier et al. 2002; Mpodozis et al. 2009; Jara & Charrier 2014) flanked to the east by a fold-and-thrust belt of Mesozoic sedimentary and volcanic rocks (Fig. 5). These rocks are unconformably covered by Miocene subhorizontal volcanic rocks and intruded by a north–south-trending Miocene magmatic belt (Fig. 3; Mpodozis et al. 2009; Jara & Charrier 2014).

Crustal thickening processes in the study area began with the Late Cretaceous tectonic inversion of volcano-sedimentary extensional basins of a Lower Jurassic–Lower Cretaceous arc–back-arc system (Emparán & Pineda 2000; Arancibia 2004; Emparán & Pineda 2006; Charrier et al. 2007; Salazar 2012). Late Cretaceous inversion reactivated pre-existing normal faults along the Coastal and Frontal Cordilleras (Fig. 3; Emparán & Pineda 2000; Arancibia 2004; Emparán & Pineda 2006; Pineda & Emparán 2006; Arévalo et al. 2009). Eocene–Oligocene compression throughout the study area is associated with the Incaic Orogenic Phase (Steinmann 1929; Charrier & Vicente 1972; Cornejo et al. 2003). The Incaic Orogenic Phase corresponds to an important episode of shortening, uplift and exhumation widely recognized throughout the Domeyko Cordillera in northern Chile during the Eocene and Oligocene. Palaeomagnetic data indicate that Eocene–Oligocene clockwise palaeomagnetic rotations become mostly zero south of 31°S (Arriagada et al. 2009, 2013). Therefore, it has been interpreted that the study area includes the southern limit of Incaic deformation (Arriagada et al. 2009, 2013). According to structural and geochronological data, Eocene compression in the Huasco Valley was associated with inversion of previous Lower Cretaceous extensional basins by a series of low-angle faults located between the San Félix and La Totora Faults in the western and central Frontal Cordillera (Figs 3 & 5, Salazar 2012). At the latitude of the Elqui River

Fig. 2. (a) Shaded relief image map colour-coded for mean annual precipitation downloaded from http://climate.geog.udel.edu/~climate/html_pages/download.html. (b) Elevation map throughout study area based in the SRTM DEM. Dashed red lines mark the position of topographic fronts. Dashed blue lines mark the main rivers and tributaries. Dashed black line marks the international border. (c) Slope map throughout study area derived from the SRTM DEM. Thick dashed black line marks the position of the main topographic front. Thin dashed line marks the position of the international border. Arrows mark depressed areas within the Algarrobillo pediplain south of 30°S. (d) Maximum (red) and minimum (blue) elevation profiles in a 5 km diameter swath. Arrows mark the position of topographic fronts. Trace of profiles in Figure 2b.

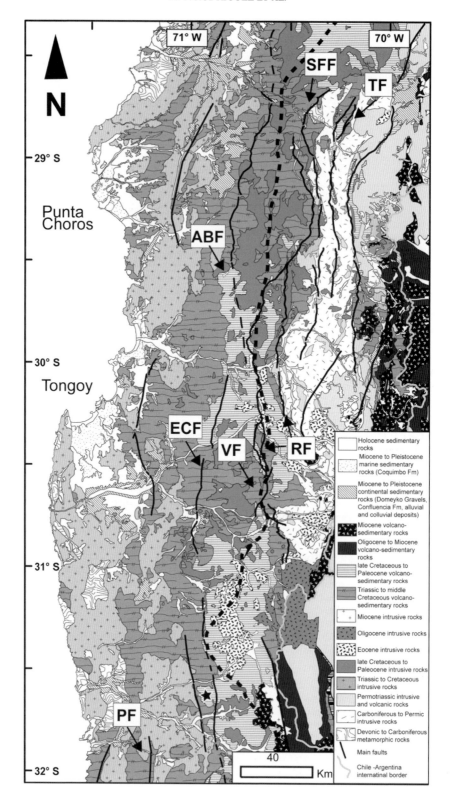

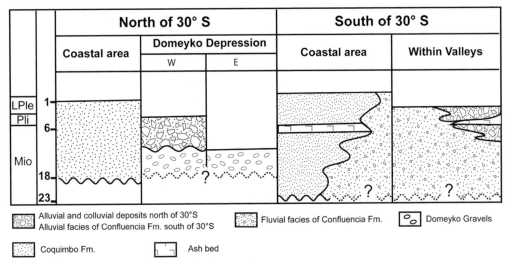

Fig. 4. Chronostratigraphic chart for Neogene and Quaternary sedimentary units exposed in the Coastal Cordillera north and south of 30°S. Mio, Miocene; Pli, Pliocene; LPle, Lower Pleistocene.

valley, the Eocene compression was related to a pop-up system formed by the closely spaced west-vergent Vicuña Reverse Fault and the east-vergent Rivadavia Reverse Fault in the western Frontal Cordillera (Figs 3 & 5). Finally, contractional tectonics affected the eastern Frontal Cordillera and the Principal Cordillera from the Early Miocene to at least the Late Miocene (Nasi et al. 1990; Rivano & Sepúlveda 1991; Bissig et al. 2001; Mpodozis et al. 2009; Winocur 2010; Jara & Charrier 2014; Winocur et al. 2014).

Large- to medium-scale geomorphological features

The topographic front that separates the Coastal Cordillera from the Frontal Cordillera defines two areas differing in their slope and hypsometry (Aguilar et al. 2013). The Frontal Cordillera presents contrasting slope values, with very high values (>45°) associated with canyons and low values (<20°) mostly observed at the high elevations of watersheds in the eastern Frontal Cordillera along the international border between Chile and Argentina (Fig. 2c). The Coastal Cordillera has homogeneous and lower slope values compared with the Frontal Cordillera, although high slope values are observed locally within river valleys and along the edges of low-slope areas (Fig. 2c). Another abrupt rise in the mean elevation present throughout west–east transects within the Coastal Cordillera defines a secondary topographic front (Fig. 2b). It is characterized by c. 600 to 1000 m of relief (difference in elevations) and separates the Coastal Cordillera into two areas referred to here as the western Coastal Cordillera and the eastern Coastal Cordillera (Fig. 5). Hypsometric integral values show a progressive increase from the Coastal Cordillera to the Frontal Cordillera, revealing that the zone between the secondary and main topographic fronts is an ancient mountain front, which probably evolved as a degradational feature carved during the Neogene (Aguilar et al. 2013). The low-slope areas throughout the Coastal and eastern Frontal Cordilleras are also generally characterized by low relief, forming sub-planar inter-river areas (i.e. the interfluves). These sub-planar surfaces resemble the morphology of palaeo-surfaces widely described in the Central Andes forearc to the north and south of the study area (Mortimer 1973; Tosdal et al. 1984; Clark et al. 1990; Farías et al. 2005; García & Hérail 2005; Quang et al. 2005; Hoke et al. 2007; Riquelme et al.

Fig. 3. Geological map of the study area, based on Sernageomin (2003). The black dashed line marks the main topographic front line that separates the Coastal Cordillera to the west, from the Frontal Cordillera to the east. Black stars show location of tuffs dated by U–Pb zircon geochronology on top of the La Silla Pediplain and the Algarrobillo Pediplain in this study. VF, Vicuña Fault; RF, Rivadavia Fault; SFF, San Félix Fault; TF, La Totora Fault; ABF, Agua de los Burros Fault; ECF, El Chape Fault; PF, Pupio Fault.

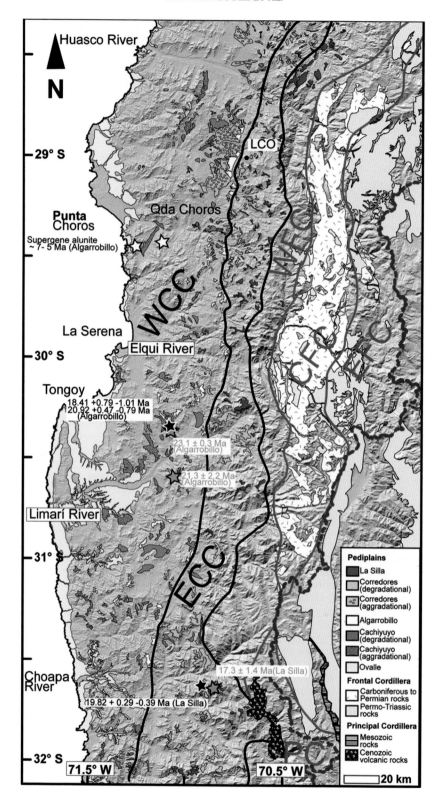

2007; Farías et al. 2008a, b; Hall et al. 2008). Their low relief and slope indicate that incision was mostly inhibited during landform formation. However, their present-day location at hundreds of metres above the river thalwegs implies that they were initially graded to an ancient base level. Therefore, they are generally interpreted as palaeosurfaces displaced from their original location due to regional forearc uplift or tilting (Mortimer 1973; Tosdal et al. 1984; Farías et al. 2005; Hoke et al. 2007; Riquelme et al. 2007; Farías et al. 2008a, b). These types of sub-planar palaeosurfaces have been mostly classified as pediplains (Mortimer 1973; Tosdal et al. 1984), that is, extensive surfaces formed due to the coalescence of multiple pediments. Pediments correspond to abraded bedrock surfaces covered by a thin veneer of alluvial debris or weathered material (Cooke et al. 1993). It has been recognized that pediplains may contain degradational and aggradational counterparts, with degradational parts corresponding to bedrock surfaces and aggradational parts corresponding generally to the top surface of fluvial and/or alluvial deposits that represent the erosional material formed due to bedrock surface degradation (Mortimer 1973; Tosdal et al. 1984; Riquelme et al. 2003; Riquelme et al. 2007).

In the Coastal Cordillera of the study area, four to six pediplains have already been mapped in the area of the Domeyko Depression (Garrido 2009; Urresty 2009). South of 30°S, a geomorphological marker formed by marine and continental landforms that have been uplifted c. 150 m was dated using cosmogenic ^{10}Be (Rodríguez et al. 2013). The marine landforms correspond to shore platforms partly developed on top of the older Coquimbo Formation (Figs 3 & 4; Le Roux et al. 2006). The continental landforms correspond to a high strath terrace and a pediment that form a single continental planation surface mostly carved into older fluvial gravels from the Miocene to Pleistocene Confluencia Formation (Fig. 4; Emparán & Pineda 2006).

Pediplains have been identified throughout the western Frontal Cordillera (Aguilar et al. 2013) and were mapped and dated in the eastern Frontal Cordillera along the international border between Chile and Argentina (Bissig et al. 2002; Nalpas et al. 2009). No pediplains have been identified within the central Frontal Cordillera.

Here we have mapped pediplains mostly in the Coastal Cordillera (Figs 5 & 6a–c). Importantly, the present study is the first attempt to regionally map and correlate the pediplains of the Coastal and Frontal Cordilleras in north-central Chile in order to characterize the processes involved in their formation, uplift and incision.

Methods

Geomorphological mapping

Satellite images and elevation, slope and geological maps together with field observations were used to map pediplains. The satellite images used included the panchromatic band of the Landsat 7 ETM+ presenting a resolution of 15 m per pixel. Elevation and slope maps were extracted from the Shuttle Radar Topography Mission digital elevation model (SRTM DEM, 90 m resolution per pixel) using ArcGis 9.3 and Envi 4.2 software packages. The geological maps used range in scale between 1:250 000 and 1:100 000. Flow grids were extracted from the SRTM DEM using the software RiverTools to visualize the drainage network and the thalweg profiles of the main channels that incise the pediplains. In order to standardize the geomorphological mapping, criteria for surface recognition were defined, which are similar to a protocol used by Clark et al. (2006) to recognize remnant surfaces of an ancient landscape throughout the eastern Tibetan Plateau. As previously mentioned, the low relief and slope of the studied surfaces allow us to deduce that they formed graded to their respective original base-level surfaces. Therefore, in order to map these surfaces it was necessary to establish maximum values for relief and slope. The maximum relief for surface recognition was established as c. 600 m (Clark et al. 2006) whereas only surfaces presenting moderately low slopes (<20°) were mapped. It was also necessary to put some constraints on other geomorphological or sedimentological features of these surfaces that indicate they were actually displaced from their original base levels, as they lack significant active

Fig. 5. Remnants of pediplains throughout the study area in shaded relief image of the SRTM DEM. Black stars show location of tuffs dated by U–Pb zircon geochronology overlying the La Silla Pediplain in Cerro Carrizo and the Algarrobillo Pediplain in Quebrada Higuerillas. Grey stars show location of volcanic deposits overlying the La Silla and Algarrobillo Pediplains dated in previous studies (Rivano & Sepúlveda 1991; Bissig 2001; Emparán & Calderón 2008). White stars show location of supergene alunite samples dated by ^{40}Ar/^{39}Ar geochronology by Creixell et al. (2012). Black lines mark the position of topographic fronts. Red lines mark the boundaries between the different blocks composing the Frontal Cordillera. The blue dashed line marks the international border between Chile and Argentina. WCC, western Coastal Cordillera; ECC, eastern Coastal Cordillera; WFC, western Frontal Cordillera; CFC, central Frontal Cordillera; EFC, eastern Frontal Cordillera; PC, Principal Cordillera; LCO, Las Campanas Astronomical Observatory.

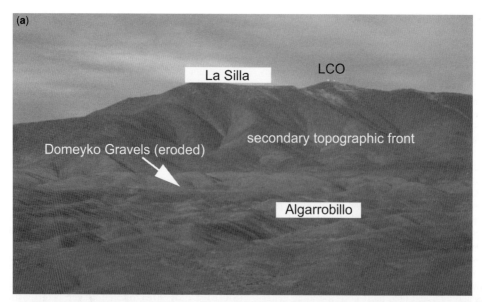

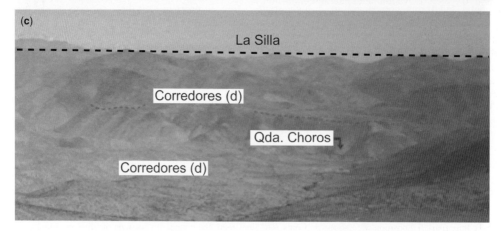

Fig. 6. (**a**) View to the NE of the secondary topographic front in the area of the Domeyko Depression. The Las Campanas Astronomical Observatory (LCO) (*c.* 2300 m a.s.l.) is observed on top of remnants of the La Silla Pediplain. LCO location in Figure 5. Remnants of the Algarrobillo Pediplain are exposed at the foot of the secondary topographic front underlying eroded exposures of the Domeyko Gravels. (**b**) View to the east of the secondary topographic front in

sedimentation, and that they are related to knick points downstream (Clark *et al.* 2006).

Geochronology

Three samples from tuff layers covering two different bedrock pediplains presented in the following section, the La Silla and the Algarobillo Pediplains, were collected from the localities of Cerro Carrizo and Quebrada Higuerillas (Fig. 7). The samples were crushed and sieved to obtain the 250–1000 μm fraction. Mineral separation was obtained according to standard laboratory techniques in the Mineral Separation Laboratory of the Geology Department of the University of Chile. At least 50 zircons from each sample were mounted in epoxy and polished for laser ablation analyses at the University of Arizona. The U–Pb geochronology of zircons was carried out using laser ablation-multicollector-inductively coupled plasma-mass spectrometry (LA-MC-ICP-MS) at the Arizona LaserChron Center (University of Arizona). Ages were calculated using the subroutine 'Zircon Age extractor' of Isoplot (Ludwig 2009), which implements an algorithm ('TuffZirc') for extracting reliable ages and age errors from suites of ^{238}U–^{206}Pb dates on complex single-zircon populations to finally provide a best estimate for the magmatic age of the tuff (Fig. 7).

Results

The pediplains studied here correspond to gently undulating bedrock and aggradational surfaces, which are exposed as patches that can be correlated based on their elevation and lateral connection. Furthermore, the pediplains had to meet the criteria defined in the previous section. A total of five levels of pediplains are recognized within the study area (Figs 5 & 8a, b). These five levels are systematically observed throughout the entire study area, but they display some differences between the regions located to the north and to the south of 30°S (Fig. 5). The two highest degradational pediplains, here named as La Silla and Corredores Pediplains, are located in the eastern Coastal Cordillera just to the east of the secondary topographic front (Figs 5 & 8a, b). The Corredores Pediplain is also composed of an aggradational part exposed within the western Coastal Cordillera just to the

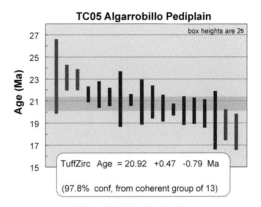

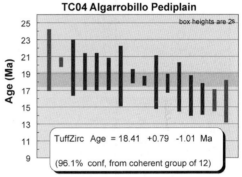

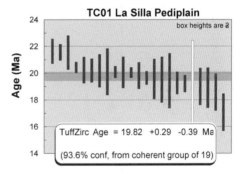

Fig. 7. LA-ICPMS U–Pb zircon ages obtained for tuffs covering the Algarobillo and La Silla Pediplains. Errors are $\pm 2\sigma$.

west of the secondary topographic front north of 30°S (Figs 5 & 8a). The Algarobillo Pediplain is exposed within the western Coastal Cordillera just to the west of the secondary topographic front

---

Fig. 6. (*Continued*) the area of the Domeyko Depression. The LCO (*c.* 2300 m a.s.l.) is observed on top of remnants of the La Silla Pediplain. Remnants of the aggradational part of the Corredores Pediplain (Corredores (a) on top of the Domeyko Gravels) are exposed at the foot of the secondary topographic front. Remnants of the Algarobillo Pediplain underlying the Domeyko Gravels are also observed. (**c**) View to the SE of remnants of the degradational part of the Corredores (Corredores (d)) Pediplain at both flanks Quebrada Choros.

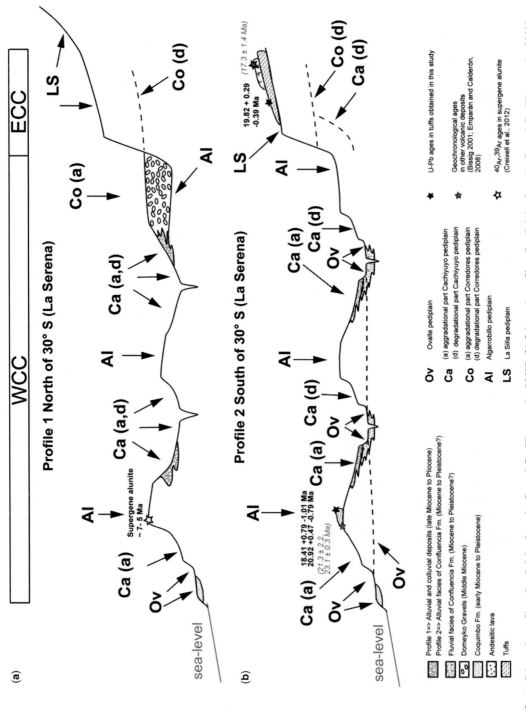

Fig. 8. (a) Schematic profiles of pediplains from the Coastal Cordillera north of 30°S. (b) Schematic profiles of pediplains from the Coastal Cordillera south of 30°S.

(Figs 5 & 8a, b). The Algarrobillo Pediplain underlies the aggradational deposits related to the development of the Corredores Pediplain north of 30°S (Figs 5, 6a, b & 8a). The Cachiyuyo Pediplain, which has a lower elevation in relation to the Algarrobillo Pediplain, is exposed in both the western and eastern Coastal Cordillera (Figs 5 & 8a, b). It presents both degradational and aggradational counterparts (Figs 5 & 8a, b). Finally, the lowest pediplain observed within the study area, the Ovalle Pediplain, occurs within the western Coastal Cordillera (Figs 5 & 8a, b).

La Silla Pediplain

The La Silla Pediplain corresponds to a degradational bedrock surface always exposed just to the east of the secondary topographic front (Figs 5, 6a–c & 8a, b). North of 30°S the pediplain is present in the eastern Coastal Cordillera and in some areas of the western Frontal Cordillera (Fig. 5). South of 30°S the pediplain forms the highest summits of the eastern Coastal Cordillera (Figs 5 & 8b). The range of elevations of the La Silla Pediplain is constant throughout the study area, lying between 3200 and 1800 m a.s.l. It is carved independent of the lithology into Upper Cretaceous and Paleocene volcano-sedimentary rocks and Paleocene–Eocene granitoids. Importantly, remnants of the La Silla Pediplain are exposed both west and east of the Vicuña Fault near the town of Hurtado (Fig. 9). The youngest rocks cross-cut by the La Silla Pediplain correspond to a granitoid with a U–Pb zircon age of 48.1 ± 0.4 Ma (Table 1; Emparán & Pineda 2006). One tuff sample interpreted as an ash fall overlying the La Silla Pediplain within the Choapa Valley was collected (Figs 5, 7 & 8b). The U–Pb zircon age obtained for this tuff sample is 19.82 + 0.29–0.39 Ma (Table 1; Figs 5, 7 & 8b). Additionally, an andesitic lava with a K–Ar age of 17.3 ± 1.4 Ma covers the dated tuff c. 2 km south of the sampled site and within the same surface remnant (Table 1; Figs 5 & 8b; Rivano & Sepúlveda 1991).

Corredores Pediplain

The Corredores Pediplain is composed of both degradational and aggradational counterparts (Figs 5, 6b, c & 8a, b). Its degradational part is a bedrock surface only exposed in the eastern Coastal Cordillera and always incised into the La Silla Pediplain (Figs 5 & 6c). The elevation of this bedrock surface ranges between 2000 and 1200 m a.s.l. throughout the entire study area. Similar to the La Silla Pediplain, it is mainly carved into the Upper Cretaceous and Paleocene volcano-sedimentary rocks, but it is also well developed on top of Paleocene granitoids. The aggradational part of the Corredores Pediplain is exposed to the west of the secondary topographic front and only in the area located to the north of 30°S, within the Domeyko Depression (Figs 5, 6b & 8a). It has an elevation range between 1400 and 800 m a.s.l., corresponding to the surface on top of the alluvial deposits of the Domeyko Gravels of probable Middle Miocene age (Table 1; Figs 5, 6b & 8a). Importantly, the aggradational part of the Corredores Pediplain is not observed south of 30°S (Figs 5 & 8b).

Algarrobillo Pediplain

The Algarrobillo Pediplain is a degradational bedrock surface that is separated by the secondary topographic front from the La Silla Pediplain and the degradational part of the Corredores Pediplain throughout the entire study area (Figs 5, 6a & 8a, b). North of 30°S, remnants of the Algarrobillo Pediplain are exposed on top of the NNE–NNW-trending ranges of the Coastal Cordillera and within the Domeyko Depression (Fig. 5). Remnants exposed at the summits of the NNE–NNW-trending ranges have elevations as high as 1800 m a.s.l. that diminish seawards to 1200 m a.s.l. Within the Domeyko Depression the elevation of the Algarrobillo Pediplain is c. 1700 m a.s.l. at the foothills of the secondary topographic front just north of 30°S. Further north it diminishes to 1200 m a.s.l. and plunges underneath the Domeyko Gravels of probable Middle Miocene age (Figs 5, 6a, b & 8a). Remnants of the Algarrobillo Pediplain at the western border of the Domeyko Depression are at elevations that are slightly lower (c. 1500–1200 m a.s.l.), but within the same elevation range as those on top of the NNE–NNW-trending ranges (c. 1800–1200 m a.s.l.). As no dislocation or encasement is observed between low relief/slope remnant surfaces of both areas, they are correlated here as part of the same original pediplain. South of 30°S, the Algarrobillo Pediplain's remnants form the summits of the western Coastal Cordillera in a range of elevations between 1600 and 1100 m a.s.l. that diminish progressively seawards. Here, exposures of the Algarrobillo Pediplain are present as close as c. 3 km to the present-day coastline (Fig. 5). The Algarrobillo Pediplain is carved mainly into Jurassic and Lower Cretaceous volcano-sedimentary and intrusive rocks. Two samples were collected from an ignimbritic deposit at the top of the Algarrobillo Pediplain (Figs 5, 7 & 8b). The U–Pb zircon geochronological determinations for these samples give two similar ages of 20.92 + 0.47–0.79 Ma and 18.41 + 0.79–1.01 Ma (Figs 5, 7 & 8b). In other studies two ages of 23.07 ± 0.33 ($^{40}Ar/^{39}Ar$ biotite) and 21.3 ± 2.2 (K–Ar biotite) were obtained for tuffs overlying

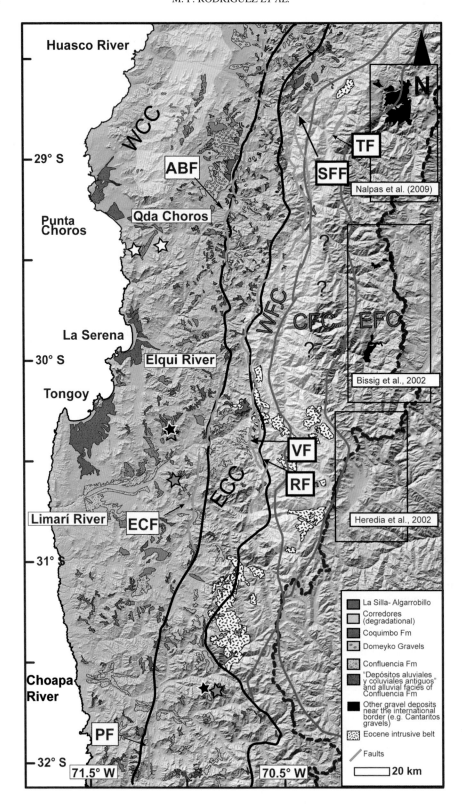

the Algarrobillo Pediplain just to the south of La Serena (Table 1; Figs 5 & 8b; Bissig 2001; Emparán & Calderón 2008).

Cachiyuyo Pediplain

The Cachiyuyo Pediplain is composed of both aggradational and degradational bedrock counterparts (Figs 5 & 8a, b). They are incised within north–south-trending tributaries draining the Algarrobillo and Corredores Pediplains (Figs 5 & 8a, b). The elevation of the Cachiyuyo Pediplain mostly ranges between 1000 and 700 m a.s.l. north of 30°S and between 1100 and 500 a.s.l. south of 30°S. The degradational part of the Cachiyuyo Pediplain is carved mainly into Jurassic– Lower Cretaceous volcano-sedimentary and intrusive rocks and to a lesser degree into the Triassic succession of volcanic rocks and the Palaeozoic metamorphic basement. Within the Domeyko Depression the aggradational part of the Cachiyuyo Pediplain corresponds to the surface on top of alluvial and colluvial sediments of probable Late Miocene–Pliocene age (Table 1; Figs 8a & 9; Arévalo et al. 2009). South of 30°S, the aggradational part of the Cachiyuyo Pediplain corresponds to the surface on top of the alluvial facies within the Confluencia Formation (Table 1; Figs 8b & 9; Emparán & Pineda 2006), correlated with the alluvial and colluvial deposits exposed to the north of 30°S. In both areas the alluvial and colluvial deposits are adjacent to topographically higher areas corresponding mostly to remnants of the Algarrobillo and the aggradational part of the Corredores Pediplains (Fig. 8a, b).

Ovalle Pediplain

The Ovalle Pediplain is exposed south of 30°S as a single planation surface formed by morphologically continuous marine and continental landforms already described and dated using cosmogenic ^{10}Be (Rodríguez et al. 2013). These ^{10}Be cosmogenic age determinations indicate that the Ovalle Pediplain formed between c. (1200?) 800 and 500 ka (Early–Middle Pleistocene, Table 1). Its elevation varies from c. 100 m a.s.l. near the coast to c. 400 m a.s.l. near the secondary topographic front. The Ovalle Pediplain is incised into the Algarrobillo and Cachiyuyo Pediplains (Figs 5 & 8a, b) and has been uplifted c. 150 m above the present-day thalwegs (Rodríguez et al. 2013). Whereas the marine landforms mainly correspond to a wide shore platform, the continental landforms correspond to a high fluvial terrace and a pediment morphologically connected and systematically exposed throughout the lower and middle courses of present-day river valleys in the area south of 30°S (Figs 5 & 8b, Rodríguez et al. 2013). In this area, the Ovalle Pediplain cross-cuts Jurassic granitoids and the Palaeozoic metamorphic basement, and the older alluvial and fluvial facies of the Confluencia Formation within the valleys. According to the interpretation of the concentration of cosmogenic ^{10}Be in samples from the high fluvial terrace, this landform corresponds to an older aggradational terrace related to fluvial deposition of the Confluencia Formation later modified during the pedimentation event leading to the development of the Ovalle Pediplain (Rodríguez et al. 2013). North of 30°S the Ovalle Pediplain is restricted to the coastal region (Figs 5& 8a) where it is exposed as a wide shore platform or *rasa* (Regard et al. 2010). Near the coast in both areas, the shore platform is carved into Jurassic and Triassic granitoids, the Miocene–Pleistocene marine deposits of the Coquimbo Formation (Fig. 8a, b) and the Palaeozoic metamorphic basement.

Discussion

Age of formation and incision of pediplains

The age of pediplains is generally constrained by the youngest geological unit overlain by the pediment and the geological units covering the surface

---

Fig. 9. Shaded relief image of the study area showing the trace of main faults and showing remnants of the Corredores and Algarrobillo Pediplains, as well as outcrops of Lower Miocene–Pleistocene continental and marine deposits. Black stars show location of tuffs dated by U–Pb zircon geochronology overlying the La Silla Pediplain in Cerro Carrizo and the Algarrobillo Pediplain in Quebrada Higuerillas. Grey stars show location of volcanic deposits overlying the La Silla and Algarrobillo Pediplains dated in previous studies (Rivano & Sepúlveda 1991; Bissig 2001; Emparán & Calderón 2008). White stars show location of supergene alunite samples dated by ^{40}Ar/^{39}Ar geochronology by Creixell et al. (2012). Black lines mark the position of topographic fronts. Red lines mark the boundaries between the different blocks composing the Frontal Cordillera. Blue lines mark the trace of main faults. The areas in transparent white indicate the probable extension of an Eocene positive relief in the western Coastal Cordillera and of the Main Incaic Range along the western and central Frontal Cordillera. WCC, western Coastal Cordillera; ECC, eastern Coastal Cordillera; WFC, western Frontal Cordillera; CFC, central Frontal Cordillera; EFC, eastern Frontal Cordillera; ABF, Agua de los Burros Fault; ECF, El Chape Fault; PF, Pupio Fault; VF, Vicuña Fault; RF, Rivadavia Fault; SFF, San Félix Fault; TF, La Totora Fault.

Table 1. Geochronological and relative ages used to constrain the development of pediplains from the Coastal Cordillera in north-central Chile (28–32°S)

Name	Location*	Description	Geochronological constraints
La Silla	ECC (some areas of WFC)	Degradational	Andesitic lava (other studies) => >17.3 ± 1.4 Ma (K–Ar whole rock) overlying tuffs (this study) => 19.82 +0.29 −0.39 Ma (U–Pb zircon) youngest rocks cross-cut (other studies) => 48.1 ± 0.4 Ma U–Pb zircon
Corredores	ECC	Degradational and aggradational (Domeyko Gravels)	Probable Middle Miocene age for Domeyko Gravels => probable Late Miocene age for Corredores Pediplain
Algarrobillo	WCC	Degradational	Supergene alunite (incision timing) => c. 7–5 Ma $^{40}Ar/^{39}Ar$ overlain by Domeyko Gravels overlying tuffs (this study) => >20.92 +0.47 −0.79 Ma and 18.41 +0.79 −1.01 Ma (U–Pb zircon), overlying tuffs (other studies) => 23.07 ± 0.33 ($^{40}Ar/^{39}Ar$ biotite) and 21.3 ± 2.2 Ma (K–Ar biotite)
Cachiyuyo	WCC (some areas of the WFC)	Degradational and aggradational (alluvial facies of Confluencia Fm and Depósitos aluviales and coluviales antiguos)	Probable Late Miocene–Early Pliocene age for 'Depósitos aluviales and coluviales antiguos' => probable Late Pliocene age for Cachiyuyo Pediplain
Ovalle	WCC	Degradational	^{10}Be cosmogenic ages point to a formation period between c. (1200?) 800 and 500 ka (Rodríguez et al. 2013)

*ECC, eastern Coastal Cordillera; WCC, western Coastal Cordillera; WFC, western Frontal Cordillera.

Elevation: + (top) to − (bottom)

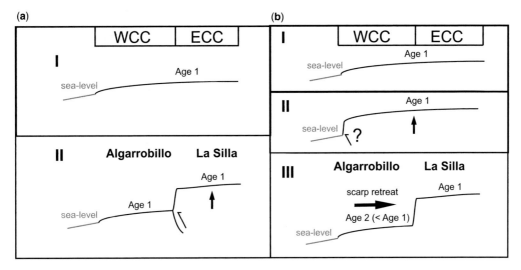

Fig. 10. Schematic profiles showing possible origins for the original La Silla–Algarrobillo pediplain. (**a**) Original La Silla–Algarrobillo pediplain forming near sea-level and offset by north–south-trending faults aligned with the secondary topographic front. (**b**) Original La Silla–Algarrobillo pediplain forming near sea-level with the secondary topographic front forming due to scarp retreat.

(e.g. Bissig *et al.* 2001). Whereas the age of the youngest geological unit eroded by the pediment constrains the maximum age of initiation of pediplain development, the ages of the oldest units covering this surface are used to constrain the minimum age of development of the pediplain. Finally, it is also important to consider the relative ages given by the relationship of incision between two pediplains (Table 1).

Regardless of the mechanism by which the tuffs were deposited on top of the Algarobillo and La Silla Pediplains, the ages obtained (Table 1) indicate that both surfaces were sub-planar components of the landscape by the Early Miocene. However, it is known that ignimbritic flows or ash falls are able to surge up valley flanks. This could imply that the Algarobillo and La Silla Pediplains were not necessarily graded to base level when the tuffs were deposited on top. The La Silla Pediplain is also covered by an andesitic lava of c. 17 Ma (Table 1; Figs 5 & 8b; Rivano & Sepúlveda 1991). As lava is not able to surge up valley flanks, the La Silla Pediplain was graded to its base level when both deposits covered this surface. The Early Miocene ages of several tuffs (Table 1; Figs 5 & 8b) covering the Algarrobillo Pediplain are in good agreement with the underlying position of this surface with respect to the Domeyko Gravels of probable Middle Miocene age within the Domeyko Depression (Fig. 8a). The fact that the Algarrobillo Pediplain served as a depocentre for the Domeyko Gravels also suggests that it was graded to its base level by the Early Miocene, just before deposition of this unit. Importantly, exposures of the La Silla Pediplain are systematically separated from the Algarrobillo Pediplain remnants by a secondary topographic front (Fig. 8a, b). This scarp could present two different origins. One implies that both pediplains formed a once-continuous surface that was displaced after c. 17 Ma by a series of north–south faults, namely, the Agua de los Burros, El Chape and Pupio Faults (Fig. 10a; Moscoso *et al.* 1982; Rivano & Sepúlveda 1991; Pineda & Emparán 2006; Arévalo *et al.* 2009), which are spatially correlated with the secondary topographic front (Fig. 9). The other possibility is that this feature results from scarp retreat after regional uplift of a single surface (Fig. 10b). In such a case, the La Silla Pediplain would be older than the Algarrobillo Pediplain (Fig. 10b). However, the ages between c. 23 and 18 Ma of tuffs overlying the Algarrobillo Pediplain are slightly older than the age of c. 19 to 17 Ma of volcanic deposits overlying the La Silla Pediplain. Therefore, the geomorphological and geochronological data described here strongly suggest that the La Silla and Algarrobillo Pediplains once formed a single low-relief/slope surface that was later offset by north–south faults that displaced the La Silla Pediplain to higher elevations (Fig. 10a). According to the estimated age of the Domeyko Gravels, offset from the original La Silla–Algarrobillo single surface would have occurred after c. 17 Ma and prior to the Middle Miocene. After the offset of this original surface, the degradational and aggradational parts of the Corredores Pediplain developed on top of the

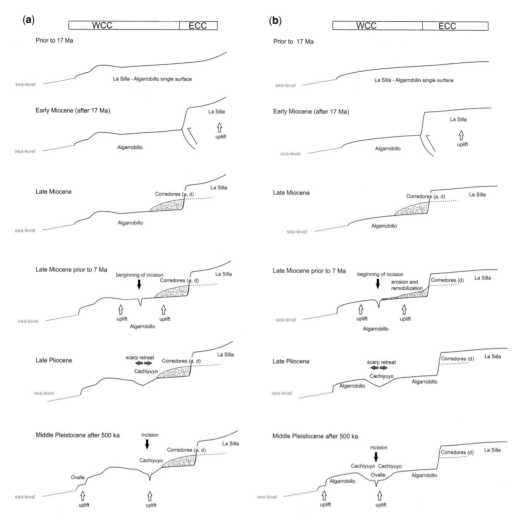

Fig. 11. Landscape evolution model for the Coastal Cordillera from the Early Miocene to the Middle Pleistocene: (**a**) north of 30°S; and (**b**) south of 30°S. WCC, western Coastal Cordillera: ECC, eastern Coastal Cordillera.

La Silla and Algarrobillo Pediplains, respectively (Fig. 11a, b). According to the probable Middle Miocene age of the Domeyko Gravels, the Corredores Pediplain was already formed by the Late Miocene (Table 1; Fig. 11a, b). Presently the Corredores Pediplain's remnants are located several hundreds of metres above the river thalwegs (Fig. 8a, b). Similarly, the Algarrobillo Pediplain's remnants, which underlie the aggradational part of the Corredores Pediplain (Fig. 8a, b), are also located several hundreds of metres above the present-day river's thalwegs (Fig. 8a, b). Thus, both surfaces were incised after the Late Miocene (Fig. 11a, b). A maximum age for incision of the Algarrobillo and Corredores Pediplains is given by the age of the Cachiyuyo Pediplain. The estimated age for the alluvial to colluvial deposits related to the aggradational part of the Cachiyuyo Pediplain is Late Miocene–Early Pliocene (Table 1). Thus, the age of the Cachiyuyo Pediplain is younger than Early Pliocene, probably Late Pliocene (Table 1). Supergene alunite $^{40}Ar/^{39}Ar$ ages in the range c. 7 to 5 Ma were obtained from samples collected from the Algarrobillo Pediplain and valleys incising this surface near Quebrada Choros (Table 1; Figs 5 & 8a; Creixell et al. 2012). Episodes of supergene copper enrichment are thought to occur under semi-arid conditions beneath pediplains due to abrupt descents in the water table and concomitant pediment incision (Sillitoe et al. 1968; Mortimer 1973; Tosdal et al. 1984; Quang et al. 2005). It is not clear if such abrupt descents are related to tectonic uplift (Mortimer 1973) or to changing climatic conditions (Arancibia *et al.* 2006). Regardless of the cause of water table descents, the ages of supergene alunite are consistent with incision of the Algarrobillo Pediplain taking place between *c.* 7 and 5 Ma (Late Miocene–Early Pliocene, Table 1), shortly before the development of the Late Pliocene Cachiyuyo Pediplain (Table 1). Finally, the ^{10}Be cosmogenic age determinations made by Rodríguez *et al.* (2013) indicate the Ovalle Pediplain formed between *c.* (1200?) 800 and 500 ka (Early–Middle Pleistocene, Table 1).

In the eastern Frontal Cordillera near the international border between 29 and 30°S at the head of the Huasco River, three once-continuous levels of pediplains are well preserved (Bissig *et al.* 2002). The higher pediplain, the Frontera–Deidad surface (4600–5300 m a.s.l.), intersects intrusive bodies with an ^{40}Ar/^{39}Ar age of *c.* 18 Ma (Bissig *et al.* 2002), which constrains its maximum age. The minimum age of this surface is inferred to be 15 Ma (Bissig *et al.* 2002). Farther north within the Huasco River's headwaters a smooth surface (4350–4500 m a.s.l.) is defined by the top of a package of gravels, informally named the Cantarito gravels, previously correlated with the Atacama Gravels (Fig. 9; Cancino 2007) exposed within the Central Depression at 27°S (Mortimer 1973). This surface probably corresponds to the aggradational part of the Frontera–Deidad surface as an ignimbrite at the base of the gravels was dated at 22 ± 0.6 Ma (K–Ar, Cancino 2007). Entrenched into the Frontera–Deidad surface, that of the Azufrera–Torta is observed (4300–4600 m a.s.l.). The minimum age of the Azufrera–Torta surface is constrained by an overlying dacitic tuff with an ^{40}Ar/^{39}Ar age *c.* 12.7 Ma (Bissig *et al.* 2002). Finally, the pedimentation process finished with the incision of the palaeovalley formed by the Los Rios surface, whose minimum age is defined by an ignimbrite on top presenting an ^{40}Ar/^{39}Ar age of *c.* 6 Ma (Bissig *et al.* 2002). The few peaks that arise on top of the Frontera–Deidad surface would correspond to vestiges of an older uplifted surface, the Cumbre surface, which represents a local back-scarp of the Frontera–Deidad surface (Bissig *et al.* 2002). According to the rocks that cross-cut this surface and its relationship with the Frontera–Deidad surface (Bissig *et al.* 2002), the probable age of the Cumbre surface is Late Oligocene–Early Miocene. Between 30 and 31°S, Heredia *et al.* (2002) recognized an extensive planation surface on top of rocks of the Permo-Triassic basement and covered by Early Miocene lavas (*c.* 18 Ma), which may also be correlated with the Cumbre surface surface (Fig. 9).

Independently of the mechanisms involved in pediplain development at both locations, the similarity between the ages presented here and the ages obtained by Bissig *et al.* (2002) indicates that tectonic and climatic conditions were favourable for pediplain formation in the Coastal and in the eastern Frontal Cordillera since the Early Miocene. A priori, some chronological correlations among the pediplains from both areas can be suggested based on the age information and regional correlations given here. The Early Miocene minimum age of the original La Silla–Algarrobillo pediplain indicates that this surface formed coevally with the Late Oligocene–Early Miocene Cumbre surface. Deposition of the Domeyko Gravels and the development of the probable Late Miocene Corredores Pediplain would have outlasted the formation of both, the Frontera-Deidad and the Azufrera-Torta Pediplains, which formed after the Early Miocene (*c.* 22–18 Ma) and during the Late Miocene (*c.* 12 Ma), respectively. Importantly, the timing of incision of the Algarrobillo Pediplain (*c.* 7–5 Ma) coincides with the incision of the palleovalley of the Los Rios surface (*c.* 6 Ma). Finally, the age correlations between pediplains from the western Coastal Cordillera and the eastern Frontal Cordillera made here are only preliminary as further geomorphological and geochronological data are needed to support these correlations.

Location of pediplains with respect to the Incaic Range

Taking the constraints on Eocene–Oligocene upper plate deformation as indicators of the Incaic Range (see Regional Framework), we suggest that within the Huasco Valley the Incaic Range formed a NNE–NNW-trending belt between the San Félix and La Totora Faults in the western and central Frontal Cordillera (Salazar 2012) (Fig. 9). Within the Elqui and the northern Limarí Valleys, constraints on Eocene deformation and exhumation (Cembrano *et al.* 2003) indicate that the Incaic Range was shifted to the west with respect to further north. Here, mountain building during the Eocene would have been mostly focused in the area between the Vicuña and Rivadavia Faults (Fig. 9; Cembrano *et al.* 2003; Emparán & Pineda 2006; Pineda & Calderón 2008). However, exposures of Late Eocene–Early Oligocene arc-like stocks (Bocatoma unit, Mpodozis & Cornejo 1988) are observed further to the east (Fig. 9). The geochemical signature of these rocks indicates that they developed in a compressional-arc tectonic regime (Bissig *et al.* 2003). Thus, the Incaic Range probably extended further to the east in this area (Fig. 9), occupying the present-day western and central Frontal Cordilleras. To the south of 31°S, no evidence of Eocene–Oligocene contractional deformation is reported along Frontal/Principal Cordilleras of the Choapa Valley (Fig. 9). Importantly, apatite fission-track and U–Th/He thermochronology indicate that rocks from the Coastal Cordillera to the west of the Domeyko Depression to the north of 30°S were exhumed in response to uplift by the Middle to Late Eocene (Maksaev et al. 2009). In contrast, to the south of 30°S, apatite fission-track ages between c. 120 and 80 Ma in the western Coastal Cordillera (Cembrano et al. 2003) indicate that this area was exhumed in response to uplift by the Early to Late Cretaceous, prior to the Incaic Orogenic Phase. Thus, the Eocene–Oligocene palaeotopography along the study region was characterized by the presence of a Main Incaic Range along the western and central Frontal Cordilleras to the north of 31°S, but also by the presence of a positive relief in the western Coastal Cordillera to the west of the Domeyko Depression to the north of 30°S (Fig. 9). No evidence exists of the presence of an Eocene–Oligocene mountainous range along the Frontal/Principal Cordilleras south of 31°S nor the western Coastal Cordillera to the south of 30°S (Fig. 9).

The mentioned constraints on Eocene to Oligocene contractional deformation and exhumation are in good agreement with the spatial distribution of Neogene pediplains described here for the Coastal Cordillera and described in previous works for the eastern Frontal Cordillera (Fig. 9). North of 31°S, the La Silla Pediplain seems to follow the western border of the Main Incaic Range as their remnants are exposed along the western Frontal Cordillera in the area of the Huasco Valley; whereas within the Elqui and Limarí Valleys the La Silla Pediplain is exposed farther to the west forming the highest summits of the eastern Coastal Cordillera (Fig. 5). In the high Frontal Cordillera of the same area (north of 31°S), previous works indicate that Neogene pediplains developed along the eastern Frontal Cordillera, to the east of the position of the Main Incaic Range inferred here (Fig. 9). Thus, Neogene pediplains developed to the west and east of the Main Incaic Relief (Fig. 9). No evidence of Incaic deformation has been described south of 31°S along the Frontal/Principal Cordilleras. However, remnants of the La Silla Pediplain are anyway located immediately to the west of the exposures of Eocene to Oligocene magmatic rocks that mark the position of the Eocene-Oligocene volcanic arc (Fig. 9). Finally, with respect to the pediplains of the western Coastal Cordillera, the differences on exhumation timing to the north and to the south of 30°S are in good agreement with the slight relief display by the Algarrobillo Pediplain north of 30°S and the absence of such relief south of 30°S (Figs 8a, b & 9).

The spatial relationship between the La Silla Pediplain and the areas affected by Eocene–Oligocene uplift and deformation in the Frontal Cordillera suggests a strong control of previous palaeotopography on the Neogene landscape evolution of both the Coastal and Frontal Cordilleras in the study area. Moreover, Neogene pediplains developed to the west and to the east of the position proposed here of the Main Incaic Range (Fig. 9). This is consistent with the previous proposition of Charrier et al. (2007), which suggested that the Main Incaic Range may have acted as the Eocene–Oligocene watershed.

Constraints on the original base level and the timing of uplift in the Coastal Cordillera

Low relief/slope surfaces can develop at high elevations (above sea-level) if downstream aggradation occurs, allowing the establishment of a new and higher base level and the concomitant reduction of the erosive efficiency of the drainage system, all of which finally induce the progressive smoothing of the relief upstream (Babault et al. 2005).

The geomorphological and geochronological data described above strongly suggest that the La Silla and Algarrobillo Pediplains once formed a single low relief/slope surface. By the Early Miocene this single surface dominated the landscape of the present-day Coastal Cordillera throughout the entire study area (Fig. 11a, b). Importantly, the La Silla–Algarrobillo surface displayed a slight relief north of 30°S, with NNE–NNW-oriented ranges to the west of the Domeyko Depression at relatively higher elevations (Fig. 11a). The presence of a higher topography in this area prior to the Early Miocene is consistent with apatite fission-track and U–Th/He thermochronology (Maksaev et al. 2009) indicating that rocks from the western Coastal Cordillera to the west of the Domeyko Depression were exhumed in response to uplift by the Middle–Late Eocene. Moreover, the sedimentology of the Domeyko Gravels also indicates that they accumulated in a closed basin with a local sediment source (Arévalo et al. 2009). According to Le Roux et al. (2004, 2005, 2006), shallow marine sedimentation related to the Coquimbo Formation was already taking place in the Tongoy Bay and at Punta Choros, immediately to the west of the La Silla–Algarrobillo pediplain by the Early Miocene around c. 23 and 18 Ma, respectively (Figs 4 & 9; Coquimbo Formation). This indicates that the original base level for this surface would probably correspond to sea-level at that time. Disruption of the original La Silla–Algarrobillo surface and relative uplift of the eastern Coastal Cordillera with respect to the western Coastal Cordillera occurred after c. 17 Ma. It is unclear whether or not the Domeyko Gravels correspond to syntectonic deposits (Garrido 2009). Therefore, more sedimentological and geochronological work is needed to establish if disruption of the original La Silla–Algarrobillo surface occurred between the Early to Middle Miocene or extended into the Middle Miocene. A similar period of uplift in the Early–Middle Miocene is interpreted from apatite fission-track and U–Th/He data that indicate accelerated cooling affecting the central Frontal Cordillera between c. 20 and 15 Ma (Cembrano et al. 2003; Rodríguez et al. 2012). Moreover, Early–Middle Miocene contractional deformation and related uplift would have extended into the eastern Frontal Cordillera according to structural (Winocur 2010; Winocur et al. 2014) and geomorphological data (Bissig et al. 2002).

After disruption of the original Algarrobillo–La Silla surface, the development at high elevation of the degradational part of the Corredores Pediplain is explained by the geomorphological connection with the top of the Domeyko Gravels. However, the Corredores Pediplain is not geomorphologically continuous with any aggradational surface south of 30°S. One possibility is that in this area the Corredores Pediplain was tectonically uplifted to its present-day elevation by the same north–south faults which previously displaced the La Silla Pediplain (Fig. 9). Nevertheless, south of 30°S the Corredores Pediplain usually presents the same range of elevations (2000–1200 m a.s.l.) as in the Domeyko Depression and it is also always entrenched within the La Silla Pediplain. Therefore, the most probable explanation is that the Corredores Pediplain was actually formed at high elevations due to aggradation to the west of the secondary topographic front throughout the entire study area, but these deposits were later removed south of 30°S. The presence of a topographic barrier to the west of the Domeyko Depression north of 30°S (Figs 8a & 11a) would allow preservation of the Domeyko Gravels after incision of the Corredores Pediplain. In contrast, the absence of such a barrier south of 30°S (Figs 8b & 11b) probably allowed erosion and remobilization of deposits associated with the Corredores Pediplain overlying the Algarrobillo Pediplain. South of 30°S the western border of the Corredores Pediplain coincides with the maximum extension to the east of the Miocene–Pleistocene fluvial facies of the Confluencia Formation (Figs 8b & 9; Emparán & Pineda 2000). These deposits may correspond to the aggradational deposits related to the Corredores Pediplain, later remobilized from the top of the Algarrobillo Pediplain due to incision, and redeposited within the river valleys that incised this surface (Figs 8b & 11b). With respect to the Algarrobillo Pediplain, it is known that marine deposition of the Coquimbo Formation was still taking place to the west by the Late Miocene when incision on top of this pediplain started, as suggested by supergene alunite ages (Figs 5 & 8a; Creixell et al. 2012). Since sea-level was the base level for the Algarrobillo Pediplain during most of the Miocene, this surface was necessarily uplifted to its present-day high elevations. Therefore, the incision ages between 7 and 5 Ma indicate that uplift of the Algarrobillo Pediplain started before 7 Ma. Thus, uplift of the Algarrobillo Pediplain probably occurred in the Late Miocene before 7 Ma (Fig. 11a, b). Importantly, the transition between a hyperarid climate to the north of 27°S and a humid climate south of 33°S occurred c. 15 Ma (Le Roux 2012). Thus, aggradation of the Middle Miocene Domeyko Gravels and development at high elevations of the degradational part of the Corredores Pediplain may be related, at least in part, to a climatically driven decrease of the transport capacity of rivers (Fig. 11a, b). Most probably, later incision on top of the aggradational/degradational Corredores Pediplain is a consequence of the Late Miocene uplift of the underlying Algarrobillo Pediplain (Fig. 11a, b). Late Miocene uplift of these pediplains is consistent with chronostratigraphic analysis performed in the marine Coquimbo Formation, which has provided evidence of a period of generalized uplift affecting the coastal areas next to the Algarrobillo Pediplain by the Late Miocene (Le Roux et al. 2005). Finally, uplift of the Algarrobillo and Corredores Pediplains indicates the present-day western and eastern Coastal Cordilleras were co-evally uplifted by the Late Miocene (Fig. 11a, b).

By the Late Pliocene, the Cachiyuyo Pediplain had developed at the foot of the Algarrobillo and Corredores Pediplains (Fig. 11a, b). The local base level for development of the Cachiyuyo Pediplain is given by the aggradation related to the alluvial and colluvial deposits north of 30°S (Arévalo et al. 2009) and the alluvial facies within the Confluencia Formation south of 30°S (Figs 8a, b & 9; Emparán & Pineda 2006). South of 30°S these deposits interfinger towards the centre of the present-day valleys with fluvial facies within the Confluencia Formation (Fig. 8b). Therefore, the base level for the Cachiyuyo Pediplain probably corresponded to the original surface on top of this facies of the Confluencia Formation that was later modified by the pedimentation event leading to formation of the Ovalle Pediplain (Fig. 8b; Rodríguez et al. 2013). The Cachiyuyo Pediplain was formed during the Late Pliocene (Table 1). There is no evidence to indicate whether incision of the Cachiyuyo Pediplain has a tectonic or climatic origin. According to geohistorical analysis of the Coquimbo Formation, strong uplift of the coastal area from c. 2 Ma has led to the emergence of the Tongoy Bay sediments (Fig. 9; Le Roux et al. 2006). Thus, one possibility is that uplift by c. 2 Ma would have also affected the Coastal Cordillera to the west of the Tongoy Bay. However, there is no direct evidence pointing to tectonic-related incision of the Cachiyuyo Pediplain (Fig. 11a, b). Finally, according to Rodríguez et al. (2013) the Ovalle Pediplain was uplifted c. 150 m, after c. 500 ka (Fig. 11a, b).

Tectonic versus erosional controls

The similarity in elevation and the latitudinal continuity of the different levels of pediplains described here show that the timing of Neogene surface uplift was similar north and south of the city of La Serena (30°S). Nevertheless, two important differences in pediplain development and preservation are observed between both areas:

(1) North of 30°S the Algarrobillo Pediplain is covered by the Domeyko Gravels (Fig. 8a) that were probably deposited in the Middle Miocene within a basin disconnected from the sea and flanked to the west by NNE ranges. In contrast, south of 30°S the same pediplain is uncovered and more incised according to hypsometric analysis (Aguilar et al. 2013). In this area, the Miocene–Pleistocene continental deposits from the Confluencia Formation are encased within the broad valleys that incise the Algarrobillo Pediplain (Fig. 8b). Part of these deposits probably corresponds to material remobilized after the Middle Miocene from an original position on top of the Algarrobillo Pediplain; otherwise the Corredores Pediplain could not have developed at high elevations. Near the coast, the Confluencia Formation changes laterally towards the west into the marine Coquimbo Formation (Fig. 4).

(2) South of 30°S the Ovalle Pediplain is a wide Early–Middle Pleistocene planation surface composed of marine and continental erosion landforms, with the continental erosional surfaces developed on top of the older fluvial gravels of the Confluencia Formation (Fig. 8b). In contrast, to the north of 30°S the Ovalle Pediplain forms a much narrower strip next to the coast that is mainly composed of shore platforms mostly disconnected from continental erosion surfaces inland (Fig. 8a).

In summary, a morphological and sedimentological connection between river and coastal systems is observed south of 30°S which has existed since at least the Early Miocene; however, it is not detected further north. This indicates that the drainage system south of 30°S has presented a larger capacity to incise and transport material towards the sea than the drainage system to the north. According to the sedimentological features of the Domeyko Gravels (Arévalo et al. 2009) and the palaeotopography of the Algarrobillo Pediplain, by the Early Miocene the ability of rivers to incise and transport was inhibited by the blocking of the drainage exerted by high NNE-trending ranges just to the west of the Domeyko Depression.

Presently, in spite of the lower elevation of north–south-oriented ranges in the western Coastal Cordillera south of 30°S relative to the NNE–NNW-oriented ranges to the north, low-slope, depressed areas aligned with the Domeyko Depression are also locally observed within the Algarrobillo Pediplain south of La Serena (Fig. 2c). In Figure 2c it is also shown how rivers draining these depressions are captured by higher-order channels within the Elqui and Limarí Valleys. Therefore, the differences in pediplain development after the Early Miocene in both areas seem to be related to the ability of the main channels to capture lower-order channels (Farías 2007). There are three possible explanations for this: (1) the rocks are easier to erode south of 30°S (Farías 2007), (2) the slope is higher south of 30°S (Carretier et al. 2013) or (3) the water flow is higher than further north (Whipple & Tucker 1999). The first possibility is ruled out because depressions of both areas are developed on top of the same lithological units, Early Cretaceous granitoids to the west and Lower to Upper Cretaceous volcano-sedimentary rocks to the east. The second possibility is rejected because the regional slope would depend on previous topography (before the Early Miocene) being dominated by the Main Incaic Range that, according to palaeomagnetic data, diminishes in importance south of 31°S. The last possibility is the favoured explanation because water flow depends on the drainage area and precipitation. Indeed, the areas drained by the Elqui, Limarí and Choapa Rivers are evidently higher than the area drained by the rivers in the Domeyko Depression (Fig. 2c). However, the high order of the main channel of the Huasco Valley indicates that its drainage area is also significant and similar to the areas drained by the Elqui, Limarí and Choapa Rivers. This suggests that drainage area could be an important, but not dominant, factor controlling landscape evolution in the study area. However, it is observed that precipitation rises from <100 mm a^{-1} to $>200-300$ mm a^{-1} south of 30°S (Fig. 2a). This latitude marks the northernmost penetration of the southern hemisphere westerlies, which bring moisture from southern latitudes, opposite to the effect of the Southeast Pacific Anticyclone, the main factor responsible for the hyperaridity of the Atacama Desert to the north. The latitudinal precipitation gradient was acquired after the Middle Miocene by a combination of a series of events including glaciations in West Antartica, formation of the Humboldt Current and uplift of the Andes (Le Roux 2012). Thus, it is proposed here that a rise in water flow due to higher precipitation south of 30°S would have played an important role in determining the differences in geomorphological evolution observed north and south of 30°S since the Middle Miocene. Such a precipitation gradient could have been superimposed on the previous palaeotopography that presented an inherited Incaic component along the Coastal Cordillera, north of 30°S, and along the Frontal Cordillera, north of 31°S. Thus, the palaeotopography inherited from the Eocene–Oligocene (Incaic) phase of uplift and deformation would correspond to a dominant factor controlling Neogene landscape development in the study area.

Uplift timing throughout the study area closely correlates with episodes of increased deformation recognized throughout the western flank of the Andes to the north of 27°S and to the south of 33°S. The Early (Middle?) Miocene uplift stage of the eastern Frontal Cordillera correlates with a period of intense deformation along the western border of the Altiplano of northern Chile (Pinto et al. 2004; Victor et al. 2004; Farías et al. 2005) that is also recognized in southern Perú (Megard 1984) and with the tectonic inversion of the extensional volcano-sedimentary Abanico Basin in central Chile south of 32°S (Charrier et al. 2002). The Late Miocene uplift stage of the western and eastern Coastal Cordilleras correlates with regional uplift of the forearc region recognized in southern Perú (Tosdal et al. 1984; Clark et al. 1990; Quang et al. 2005; Schildgen et al. 2007), northern Chile (Hoke et al. 2007), the Southern Atacama Desert (Riquelme et al. 2007) and central Chile south of 33°S (Farías et al. 2008a, b; Maksaev et al. 2009). The post-500 ka uplift of the Ovalle Pediplain correlates with a renewal of uplift of marine landforms along the Pacific coast post 400 ± 100 ka after an Early–Middle Pleistocene period of relatively slow uplift identified to the north of La Serena (30°S) by Regard et al. (2010). Uplift of the Ovalle Pediplain also correlates with Pleistocene–Holocene uplift of pediments and other continental landforms along the forearc of southern Peru and northern Chile (González et al. 2003, 2006; Kober et al. 2007; Hall et al. 2008; Saillard et al. 2009; Jordan et al. 2010).

Increased deformation by the Early (Middle?) Miocene would be explained by a more intense stress transmission and widespread strain due to an increased plate convergence rate (Charrier et al. 2009, 2013) after break-up of the Farallon Plate into the Nazca and Cocos Plates (Pardo-Casas & Molnar 1987; Somoza 1998). The driving forces for Late Miocene and Middle Pleistocene uplifts are still unclear and a matter of great debate in the case of the Late Miocene (Garzione et al. 2006; Barnes & Ehlers 2009), but determining these driving forces is beyond the scope of the study. However, the fact that Late Miocene and Middle Pleistocene uplifts are recognized for such a vast area from southern Perú to central Chile suggests that they were controlled by first-order tectonic features.

Conclusions

Prior to c. 17 Ma an extensive pediplain sloping down to sea-level dominated the landscape west of an inherited Main Incaic Range mainly exposed in the present-day Frontal Cordillera (Fig. 11a, b). North of 30°S, this pediplain extended across the entire present-day Coastal Cordillera and was not completely sub-planar as it already presented a slight inherited relief flanking the Domeyko Depression to the west (Fig. 11a). In contrast, south of 30°S the pediplain extended across the present-day Coastal Cordillera, progressively diminishing in elevation towards the sea (Fig. 11b).

In the Early (Middle?) Miocene the pediplain was offset by a series of north–south-trending faults (Fig. 11a, b). The eastern Coastal Cordillera was uplifted with respect to the western Coastal Cordillera throughout the entire study area (Fig. 11a, b). Importantly, offset of the original pediplain is co-eval with a significant period of uplift and exhumation in the Frontal Cordillera to the east. Uplift led to the formation of a secondary topographic front within the present-day Coastal Cordillera and concomitant deposition next to the scarp. Aggradation to the west of the scarp and development of degradational pediplains at high elevations by the Late Miocene (Fig. 11a, b) may have been favoured by the establishment of the observed latitudinal precipitation gradient throughout the study region after the Middle Miocene (Le Roux 2012). North of 30°S the aggradational deposits accumulated within the Domeyko Depression (Fig. 11a). In contrast, south of 30°S the absence of a topographic barrier to the west and probable higher precipitation rate compared to the north prevented preservation of the aggradational deposits, which were later remobilized (Fig. 11b).

By the Late Miocene, the entire Coastal Cordillera was uplifted (Fig. 11a, b). South of 30°S, uplift generated a rejuvenation of the drainage system, generating material that had previously accumulated to the west of the secondary topographic front and which was remobilized and redeposited by fluvial systems similar to those of the present-day (Fig. 11b). Between the Early and Middle Pleistocene a new pediplain, partly carved on top of these deposits and connected with shore platforms towards the coast, developed southwards from La Serena, whereas to the north only the shore platforms developed (Fig. 11b). Finally, this pediplain was uplifted post-500 ka.

The uplift stages recognized for the study area correlate with episodes of increased deformation widely recognized throughout the western flank of the Andes both to the north and to the south of the study area. Uplift by the Early (-Middle?) Miocene correlates with the Late Oligocene–Early Miocene episode of increase plate convergence between the Nazca and South American Plates, but the driving forces for Late Miocene and Middle Pleistocene uplift remain unclear. The differences in landscape development recognized between the regions located to the north and south of 30°S, respectively, are probably related to differences in the previous palaeotopography along the western Frontal Cordillera and the first-order climatic transition at 30°S that defines the southernmost reaches of the Atacama Desert.

This study was supported in part by the Chilean Government through the Comisión Nacional de Ciencia y Tecnología CONICYT (Anillo ACT-18 project, AMTC), the Advanced Mining Technology Center (AMTC) of the Facultad de Ciencias Físicas y Matemáticas, Universidad de Chile; FONDECYT Project 11085022 'Interacción Clima-Tectónica en el Alzamiento Andino de Chile Central durante el Neógeno'; FONDECYT N° 11121529, FONDECYT N° 1120272, the Ecos-Conicyt proyect C11U02 'Levantamiento de la costa pacífica de Sud América y acoplamiento interplaca en la zona de subducción'. Additional support was obtained by IRD Project 'Erosión en los Andes'.

This study is part of the PhD theses of M.P. Rodríguez and G. Aguilar, both of whom were supported by 4-year grants from CONICYT, the Universidad Católica del Norte and Égide-France (Bourse d'excellence Eiffel). Useful discussions with S. Carretier, M. Farías and V. Tolorza helped develop and clarify our ideas. The authors would like to express their thanks for the constructive comments by two anonymous reviewers which led to significant improvements in the manuscript.

References

Aguilar, G., Riquelme, R., Martinod, J., Darrozes, J. & Maire, E. 2011. Variability in erosion rates related to the state of landscape transience in the semi-arid Chilean Andes. *Earth Surface Processes and Landforms*, **36**, 1736–1748, http://dx.doi.org/10.1002/esp.2194

Aguilar, G., Riquelme, R., Martinod, J. & Darrozes, J. 2013. Role of climate and tectonics in the geomorphologic evolution of the Semiarid Chilean Andes between 27–32°S. *Andean Geology* **40**, 79–101, http://dx.doi.org/10.5027/andgeoV40n1-a04

Arancibia, G. 2004. Mid-Cretaceous crustal shortening: evidence from a regional-scale ductile shear zone in the Coastal Range of central Chile (328 S). *Journal of South American Earth Sciences*, **17**, 209–226.

Arancibia, G., Matthews, S. J. & Pérez de Arce, C. 2006. K-Ar and $^{40}Ar/^{39}Ar$ geochronology of supergene processes in the Atacama Desert, Northern Chile: tectonic and climatic relations. *Journal of the Geological Society*, **163**, 107–118, http://dx.doi.org/10.1144/0016-764904-161.

Arévalo, C., Mourgues, F. A. & Chávez, R. 2009. *Geología del Área Vallenar-Domeyko, Región de Atacama*. Servicio Nacional de Geología y Minería, Carta Geológica de Chile.

Arriagada, C., Mpodozis, C., Yañez, G., Roperch, P., Charrier, R. & Farías, M. 2009. Rotaciones Tectónicas en Chile Central: El Oroclino de Vallenar y el 'Megakink' del Maipo. *XII Congreso Geológico Chileno*, Santiago, Chile, 23–26 November, Departamento de Geología de la Universidad de Chile.

Arriagada, C., Ferrando, R., Córdova, L., Morata, D. & Roperch, P. 2013. The Maipo Orocline: a first scale structural feature in the Miocene to Recent geodynamic evolution in the central Chilean Andes. *Andean Geology*, **40**, 419–437, http://dx.doi.org/10.5027/andgeoV40n3-a02.

Babault, J., Van Den Driessche, J., Bonnet, S., Castelltort, S. & Crave, A. 2005. Origin of the highly elevated Pyrenean peneplain. *Tectonics*, **24**, TC2010, http://dx.doi.org/10.1029/2004TC001697

Barnes, J. B. & Ehlers, T. A. 2009. End member models for Andean Plateau uplift. *Earth Science Review* **97**, 105–132.

Benado, D. E. 2000. *Estructuras y estratigrafía básica de terrazas marinas en sector costero de Altos de Talinay y Bahía Tongoy, implicancias neotectónicas*. Msc Thesis, Departamento de Geología, Universidad de Chile, Santiago.

Bissig, T. 2001. *Metallogenesis of the Miocene El Indio-Pascua gold-silver-copper belt, Chile/Argentina: Geodynamic, Geomorphological and Petrochemical controls on epithermal mineralization*. PhD thesis, Queen's University.

Bissig, T. & Riquelme, R. 2010. Andean uplift and climate evolution in the southern Atacama Desert deduced from geomorphology and supergene alunite-group minerals. *Earth and Planetary Science Letters*, **299**, 447–457.

Bissig, T., Lee, J. K. W., Clark, A. H. & Heather, K. B. 2001. The Cenozoic History of volcanism and hydrothermal alteration in the Central Andes Flat-Slab Region: New ^{40}Ar-^{39}Ar constraints from the El Indio-Pascua Au (-Ag, Cu) Belt, 29°20′–30°30′S. *International Geology Review*, **43**, 1–29.

Bissig, T., Clark, A. H., Lee, J. K. W. & Hodgson, C. J. 2002. Miocene landscape evolution and geomorphologic controls on epithermal processes in the El Indio-Pascua Au–Ag–Cu belt, Chile and Argentina. *Economic Geology and the Bulletin of the Society of Economic Geologists*, **97**, 971–996, http://dx.doi.org/10.2113/97.5.971

Bissig, T., Clark, A. H., Lee, J. K. W. & von Quadt, A. 2003. Petrogenetic and Metallogenetic responses to Miocene slab flattening: new constraints from the El Indio-Pascua Au-Ag-Cu belt, Chile/Argentina. *Mineralium Deposita*, **38**, 844–862.

Cahill, T. & Isacks, B. L. 1992. Seismicity and shape of the subducted Nazca plate. *Journal of Geophysical Research*, **97**, 17503–17529.

Cancino, G. 2007. *Hoja El Transito mapa de compilación, 1:250 000*, SERNAGEOMIN.

Carretier, S., Regard, V. et al. 2013. Slope and climate variability control of erosion in the Andes of central Chile. *Geology*, **41**, 195–198, http://dx.doi.org/10.1130/G33735.1

Cembrano, J., Zentilli, M., Grist, A. & Yañez, G. 2003. Nuevas edades de trazas de fisión para Chile Central (30°–34°S): Implicancias en el alzamiento y exhumación de los Andes desde el Cretácico. *10° Congreso Geológico Chileno*, Concepción, Chile, 6–10 October, Departamento de Ciencias de la Tierra, Universidad de Concepción.

Charrier, R. & Vicente, J. C. 1972. Liminary and geosynclinal Andes: major orogenic phases and synchronical evolution of the central and Magellan sectors of the Argentine-Chillean Andes. *In: International Upper Mantle Project Conference on Solid Earth Problems, Proceedings, 26–31 October, Buenos Aires*, Comité Argentino del Manto Superior, **2**, 451–470.

Charrier, R., Baeza, O. et al. 2002. Evidence for Cenozoic extensional basin development and tectonic inversion south of the flat-slab segment, southern Central Andes, Chile (33°–36°S.L.). *Journal of South American Earth Sciences*, **15**, 117–139.

Charrier, R., Pinto, L. & Rodríguez, M. P. 2007. Tectonostratigraphic evolution of the Andean Orogen in Chile. *In*: Moreno, T. & Gibbons, W. (eds) *The Geology of Chile*. The Geological Society, London, 21–114.

Charrier, R., Farias, M. & Maksaev, V. 2009. Evolución tectónica, paleogeográfica y metalogénica durante el Cenozoico en los Andes de Chile norte y central e implicaciones para las regiones adyacentes de Bolivia y Argentina. *Revista de la Asociación Geológica Argentina*, **65**, 5–35.

Charrier, R., Herail, G., Pinto, L., Garcia, M., Riquelme, R., Farias, M. & Muñoz, N. 2013. Cenozoic tectonic evolution in the Central Andes in northern Chile and west central Bolivia: implications for paleogeographic, magmatic and mountain building evolution. *International Journal of Earth Sciences*, **102**, 235–264.

Clark, A. H., Tosdal, R. M., Farrar, E. & Plazolles, A. 1990. Geomorphologic environment and age of supergene enrichment of the Cuajone, Quellaveco, and Toquepala Porphyry Copper-Deposits, Southeastern Peru. *Economic Geology and the Bulletin of the Society of Economic Geologists*, **85**, 1604–1628.

Clark, M. K., Royden, L. H., Whipple, K. X., Burchfiel, B. C., Zhang, X. & Tang, W. 2006. Use of a regional, relict landscape to measure vertical deformation of the eastern Tibetan Plateau. *Journal of Geophysical Research*, **111**, F03002, http://dx.doi.org/10.1029/2005JF000294

Cooke, R., Warren, A. & Goudie, A. 1993. *Desert Geomorphology*. UCL Press, London.

Cornejo, P., Mpodozis, C., Ramírez, C. F. & Tomlinson, A. J. 1993. *Estudio geológico de la Región de Potrerillos y El Salvador (26°–27° Latitude S)*. Servicio Nacional de Geología y Minería - Corporación del Cobre, Informe Registrado, IR-93-01, Volume **1**, Santiago.

Cornejo, R., Matthews, S. & Pérez de Arce, C. 2003. The 'K-T' compressive deformation event in northern Chile (24–27°S). *In*: *10° Congreso Geológico Chileno*, Concepción, Chile, 6–10 October, Departamento de Ciencias de la Tierra, Universidad de Concepción.

Creixell, C., Ortiz, M. & Arévalo, C. 2012. *Geología del área Carrizalillo-El Tofo, Región de Atacama, Servicio Nacional de Geología y Minería, 1 mapa escala 1:100 000*. Carta Geológica de Chile, Serie Geología Básica.

Emparán, C. & Calderón, M. 2008. *Geología del area Ovalle-Peña Blanca, Región de Coquimbo*. Servicio Nacional de Geología y Minería, Santiago.

Emparán, C. & Pineda, G. 2000. *Geología del area La Serena-Higuerillas, Región de Coquimbo*. Servicio Nacional de Geología y Minería, Santiago.

Emparán, C. & Pineda, G. 2006. *Geología del Área Andacollo-Puerto Aldea, Región de Coquimbo*. Carta Geológica de Chile, Serie Geológica Básica.

Farías, M. 2007. *Tectónica y erosión en la evolución del relieve de los Andes de Chile Central durante el Neógeno*. PhD thesis, Universidad de Chile y Université de Toulouse III, (inédito), Santiago–Toulouse.

Farías, M., Charrier, R., Comte, D., Martinod, J. & Herail, G. 2005. Late Cenozoic deformation and uplift of the western flank of the Altiplano: evidence from the depositional, tectonic, and geomorphologic evolution and shallow seismic activity (northern Chile at 19 degrees 30°S). *Tectonics*, **24**, TC4001, http://dx.doi.org/10.1029/2004TC001667

Farías, M., Charrier, R. et al. 2008a. Late Miocene high and rapid surface uplift and its erosional response in the Andes of central Chile (33°–35°S). *Tectonics*, **27**, Tc1005, http://dx.doi.org/10.1029/2006tc002046

Farías, M., Charrier, R. et al. 2008b. No subsidence in the development of the Central Depression along the Chilean margin. *7th Symposium on Andean Geodynamics*, Nice, France, 2–4 September, 206–209, Institute de recherche pour le développement (IRD).

García, M. & Hérail, G. 2005. Fault-related folding, drainage network evolution and valley incision during the Neogene in the Andean Precordillera of Northern Chile. *Geomorphology*, **65**, 279–300.

Garrido, G. 2009. *Evolución geomorfológica de la Depresión de Domeyko entre los 28°45'–29°00'S durante el Neógeno*. Thesis, Departamento de Geología, Universidad de Chile.

Garzione, C. N., Molnar, P., Libarkin, J. C. & MacFadden, B. 2006. Rapid late Miocene rise of the Andean plateau: evidence for removal of mantle lithosphere. *Earth and Planetary Science Letters*, **241**, 543–556.

Giambiagi, L., Mescua, J., Bechis, F., Tassara, A. & Hoke, G. 2012. Thrust belts of the southern Central Andes: along-strike variations in shortening, topography, crustal geometry, and denudation. *Geological Society of America Bulletin*, **124**, 1339–1351, http://dx.doi.org/10.1130/B30609.1

GONZÁLEZ, G., CEMBRANO, J., CARRIZO, D., MACCI, A. & SCHNEIDER, H. 2003. Link between forearc tectonics and Pliocene-Quaternary deformation of the Coastal Cordillera, Northern Chile. *Journal of South American Earth Sciences*, **16**, 321–342.

GONZÁLEZ, G., DUNAI, T., CARRIZO, D. & ALLMENDINGER, R. 2006. Young displacements on the Atacama Fault System, northern Chile from field observations and cosmogenic ^{21}Ne concentrations. *Tectonics*, **25**, TC3006, http://dx.doi.org/10.1029/2005TC001846.

HALL, S. R., FARBER, D. L., AUDIN, L., FINKEL, R. C. & MÉRIAUX, A.-S. 2008. Geochronology of pediment surfaces in southern Peru: implications for Quaternary deformation of the Andean forearc. *Tectonophysics*, **459**, 186–205.

HILLEY, G. E. & COUTAND, I. 2010. Links between topography, erosion, rheological heterogeneity, and deformation in contractional settings: insights from the Central Andes. *Tectonophysics*, **95**, 78–92.

HILLEY, G. E., STRECKER, M. R. & RAMOS, V. A. 2004. Growth and erosion of fold-and-thrust belts, with an application to the Aconcagua Fold-and-Thrust Belt, Argentina. *Journal of Geophysical Research, Solid Earth*, **109**, B01410, http://dx.doi.org/10.1029/2002JB002282.

HOKE, G. D., ISACKS, B. L., JORDAN, T. E., BLANCO, N., TOMLINSON, A. J. & RAMEZANI, J. 2007. Geomorphic evidence for post-10 Ma uplift of the western flank of the central Andes 18°30′–22°S. *Tectonics*, **26**, TC5021, http://dx.doi.org/10.1029/2006TC002082.

ISACKS, B. 1988. Uplift of the central Andean plateau and bending of the Bolivian orocline. *Journal of Geophysical Research: Solid Earth*, **93**, 3211–3231.

JARA, P. & CHARRIER, R. 2014. Nuevos antecedentes estratigráficos y geocronológicos para el Meso-Cenozoico de la Cordillera Principal de Chile entre 32° y 32°30'S: Implicancias estructurales y paleogeográficas. *Andean Geology*, **41**, 174–209, http://dx.doi.org/10.5027/andgeoV41n1-a07.

JORDAN, T. E., ISACKS, B. L., ALLMENDINGER, R. W., BREWER, J. A., RAMOS, V. A. & ANDO, C. J. 1983. Andean tectonics related to geometry of subducted Nazca plate. *Geological Society of America Bulletin*, **94**, 341–361.

JORDAN, T. E., NESTER, P. L., BLANCO, N., HOKE, G. D., DÁVILA, F. & TOMLINSON, A. J. 2010. Uplift of the Altiplano-Puna plateau: a view from the west. *Tectonics*, **29**, TC5007, http://dx.doi.org/10.1029/2010TC002661

KAY, S. & MPODOZIS, C. 2002. Magmatism as a probe to the Neogene shallowing of the Nazca plate beneath the modern Chilean flatslab. *Journal of South American Earth Science*, **15**, 39–59.

KOBER, F., IVY-OCHS, S., SCHLUNEGGER, F., BAUR, H., KUBIK, P. W. & WIELER, R. 2007. Denudation rates and a topography-driven rainfall threshold in northern Chile: Multiple cosmogenic nuclide data and sediment yield budgets. *Geomorphology*, **83**, 97–120.

LAMB, S. & DAVIS, P. 2003. Cenozoic climate change as a possible cause for the rise of the Andes. *Nature*, **425**, 792–797, http://dx.doi.org/10.1038/nature02049

LAMB, S., HOKE, L., KENNAN, L. & DEWEY, J. 1997. Cenozoic evolution of the Central Andes in Bolivia and northern Chile. *In*: BURG, J. -P. & FORD, M. (eds) *Orogeny Through Time*. Geological Society, London, Special Publications, **121**, 237–264.

LE ROUX, J. P. 2012. A review of Tertiary climate changes in southern South America and the Antarctic Peninsula. Part 2: continental conditions. *Sedimentary Geology*, **247**, 21–38.

LE ROUX, J. P., GÓMEZ, C., FENNER, J. & MIDDLETON, H. 2004. Sedimentological processes in a scarp-controlled rocky shoreline to upper continental slope environment, as revealed by unusual sedimentary features in the Neogene Coquimbo Formation, north-central Chile. *Sedimentary Geology*, **165**, 67–92.

LE ROUX, J. P., GÓMEZ, C. ET AL. 2005. Neogene–Quaternary coastal and offshore sedimentation in north-central Chile: record of sea level changes and implications for Andean tectonism. *Journal of South American Earth Sciences*, **19**, 83–98.

LE ROUX, J. P., OLIVARES, D. M., NIELSEN, S. N., SMITH, N. D., MIDDLETON, H., FENNER, J. & ISHMAN, S. E. 2006. Bay sedimentation as controlled by regional crustal behaviour, local tectonics and eustatic sea-level changes: Coquimbo Formation (Miocene–Pliocene), Bay of Tongoy, central Chile. *Sedimentary Geology*, **184**, 133–153.

LUDWIG, K. R. 2009. *SQUID 2 (rev. 2.50)*, A user's manual, Berkeley Geochronology Center. Special Publications, **5**.

MAKSAEV, V. & ZENTILLI, M. 1999. Fission track thermochronology of the Domeyko Cordillera, northern Chile: implications for Andean tectonics and porphyry copper metallogenesis. *Exploration and Mining Geology*, **8**, 65–89.

MAKSAEV, V., MOSCOSO, R., MPODOZIS, C. & NASI, C. 1984. Las unidades volcánicas y plutónicas del Cenozoico superior entre la Alta Cordillera del Norte Chico (29°–31°S), geología, alteración hidrotermal y mineralización. *Revista Geológica de Chile*, **21**, 11–51.

MAKSAEV, V., MUNIZAGA, F., ZENTILLI, M. & CHARRIER, R. 2009. Fission track thermochronology of Neogene plutons in the Principal Andean Cordillera of central Chile (33–35°S). Implications for Tectonic Evolution and Porphyry Cu-Mo Mineralization. *Andean Geology*, **36**, 153–171.

MARTIN, M. W., KATO, T. T., RODRIGUEZ, C., GODOY, E., DUHART, P., MCDONOUGH, M. & CAMPOS, A. 1999. Evolution of the late Paleozoic accretionary complex and overlying forearc-magmatic arc, south central Chile (38°–41°S): constraints for the tectonic setting along the southwestern margin of Gondwana. *Tectonics*, **18**, http://dx.doi.org/10.1029/1999TC900021

MEGARD, F. 1984. The andean orogenic period and its major structures in Central and Northern Peru. *Journal of the Geological Society*, **141**, 893–900.

MORTIMER, C. 1973. The Cenozoic history of the southern Atacama Desert, Chile. *Journal of the Geological Society, London*, **129**, 505–526, http://dx.doi.org/10.1144/gsjgs.129.5.0505

MOSCOSO, R., NASI, C. & SALINAS, P. 1982. *Hoja Vallenar y parte norte de La Serena, geological map, 1:250 000*. SERNAGEOMIN.

MPODOZIS, C. & CORNEJO, P. 1988. *Hoja Pisco Elqui. IV Región de Coquimbo.* Servicio Nacional de Geología y Minería, Carta Geológica de Chile, No. 68, Santiago.

MPODOZIS, C., BROCKWAY, H., MARQUARDT, C. & PERELLÓ, J. 2009. Geocronología U/Pb y tectónica de la región de Los Pelambres-Cerro Mercedario: implicancias para la evolución cenozoica de Los Andes del centro de Chile y Argentina. *In: XII Congreso Geológico Chileno*, **12**, Santiago, Chile, 23–26 November, Departamento de Geología de la Universidad de Chile.

NALPAS, T., DABARD, M.-P., PINTO, L. & LOI, A. 2009. Preservation of the Miocene Atacama Gravels in Vallenar area, northern Chilean Andes: climate, stratigraphic or tectonic control? *XII Congreso Geológico Chileno*, Santiago, 23–26 November, Departamento de Geología de la Universidad de Chile.

NASI, C., MOSCOSO, R. & MAKSAEV, V. 1990. *Hoja Guanta, Regiones de Atacama y Coquimbo*, Sernageomin, Santiago, Chile.

OTA, Y., MIYAUCHI, T., PASKOFF, R. & KOBA, M. 1995. Plio–Quaternary terraces and their deformation along the Altos de Talinay, North–Central Chile. *Revista Geologica de Chile*, **22**, 89–102.

PARDO-CASAS, F. & MOLNAR, P. 1987. Relative motion of the Nazca (Farallón) and South American plates since Late Cretaceous time. *Tectonics*, **6**, 233–248.

PASKOFF, R. 1970. *Recherches géomorphologiques dans le Chili semi-aride*. Biscaye Freres, Bordeaux.

PINEDA, G. & CALDERÓN, M. 2008. *Geología del área Monte Patria-El Maqui, región de Coquimbo, Escala 1:100 000.* Carta Geológica de Chile, Serie Geología Básica, n.116, SERNAGEOMIN: 44 h. Santiago.

PINEDA, G. & EMPARÁN, C. 2006. *Geología del area Vicuña-Pichasca, Región de Coquimbo.* Servicio Nacional de Geología y Minería, Santiago.

PINTO, L., HÉRAIL, G. & CHARRIER, R. 2004. Sedimentación sintectónica asociada a las estructuras neógenas en el borde occidental del plateau andino en la zona de Moquella (19°15′S, Norte de Chile). *Revista Geológica de Chile*, **31**, 19–44.

QUANG, C. X., CLARK, A. H., LEE, J. K. W. & HAWKES, N. 2005. Response of supergene processes to episodic Cenozoic uplift, pediment erosion, and ignimbrite eruption in the porphyry copper province of southern Peru. *Economic Geology*, **100**, 87–114, http://dx.doi.org/10.2113/100.1.0087

RAMOS, V., CEGARRA, M. & CRISTALLINI, E. 1996. Cenozoic tectonics of the High Andes of west-central Argentina (30–36°S latitude). *Tectonophysics*, **259**, 185–200.

RAMOS, V., ZAPATA, T., CRISTALLINI, E. & INTRACASO, A. 2004. The Andean thrust system-latitudinal variations in structural styles and orogenic shortening. *In:* MCCLAY, K. R. (ed.) *Thrust Tectonics and Hydrocarbon Systems.* AAPG Memoir, **82**, 30–50.

REGARD, V., SAILLARD, M. *ET AL.* 2010. Renewed uplift of the Central Andes Forearc revealed by coastal evolution during the Quaternary. *Earth and Planetary Science Letters*, **297**, 199–210.

RIQUELME, R., MARTINOD, J., HÉRAIL, G., DARROZES, J. & CHARRIER, R. 2003. A geomorphological approach to determining the Neogene to Recent tectonic deformation in the Coastal Cordillera of northern Chile (Atacama). *Tectonophysics*, **361**, 255–275.

RIQUELME, R., HÉRAIL, G., MARTINOD, J., CHARRIER, R. & DARROZES, J. 2007. Late Cenozoic geomorphologic signal of Andean forearc deformation and tilting associated with the uplift and climate changes of the Southern Atacama Desert (26°S–28°S). *Geomorphology*, **86**, 283–306, http://dx.doi.org/10.1016/j.geomorph.2006.09.004

RIVANO, S. & SEPÚLVEDA, P. 1991. *Hoja Illapel, Región de Coquimbo.* Carta Geológica de Chile, Sernageomin, Santiago, Chile.

RODRÍGUEZ, M. P., CHARRIER, R., CARRETIER, S., BRICHAU, S. & FARÍAS, M. 2012. Alzamiento y exhumación Cenozoicos en el Norte Chico de Chile (30 a 33°S). *XIII Congreso Geológico Chileno*, Antofagasta, Chile, 5–9 August.

RODRÍGUEZ, M. P., CARRETIER, S. *ET AL.* 2013. Geochronology of pediments and marine terraces in north-central Chile and their implications for Quaternary uplift in the Western Andes. *Geomorphology*, **180–181**, 33–46.

SAILLARD, M., HALL, S. R. *ET AL.* 2009. Non-steady long-term uplift rates and Pleistocene marine terrace development along the Andean margin of Chile (31°S) inferred from 10Be dating. *Earth and Planetary Science Letters*, **277**, 50–63.

SALAZAR, E. 2012. *Evolución tectonica-estratigrafica post-Paleozoica de la Cordillera de Vallenar.* Thesis, Departamento de Geología, Universidad de Chile.

SCHILDGEN, T. F., HODGES, K. V., WHIPPLE, K. X., REINERS, P. W. & PRINGLE, M. S. 2007. Uplift of the western margin of the Andean plateau revealed from canyon incision history, southern Peru. *Geology*, **35**, 523–526.

SERNAGEOMIN 2003. Carta Geológica de Chile (escala 1:1 000 000) Servicio Nacional de Geología y Minería.

SILLITOE, R. H., MORTIMER, C. & CLARK, A. H. 1968. A chronology of landform evolution and supergene mineral alteration, Southern Atacama Desert, Chile. *Institute of Mining and Metallurgy Transactions (Section B)*, **27**, 166–169.

SOBEL, E. R. & STRECKER, M. R. 2003. Uplift, exhumation, and precipitation: tectonics and climatic control of Late Cenozoic landscape evolution in the northern Sierras Pampeanas, Argentina. *Basin Research*, **15**, 431–451.

SOMOZA, R. 1998. Updated Nazca (Farallon)-South America relative motions during the last 40 My: implications for mountain building in the central Andean region. *Journal of South American Earth Sciences*, **11**, 211–215.

STEINMANN, G. 1929. *Geologie von Peru.* Kart Winter, Heidelberg.

STRECKER, M. R., ALONSO, R. N., BOOKHAGEN, B., CARRAPA, B., HILLEY, G. E., SOBEL, E. R. & TRAUTH, M. H. 2007. Tectonics and climate of the southern central Andes. *Annual Review of Earth and Planetary Sciences*, **35**, 747–787, http://dx.doi.org/10.1146/annurev.earth.35.031306.140158

TASSARA, A. & YAÑEZ, G. 2003. Relación entre el espesor elástico de la litosfera y la segmentación tectónica del margen andino (15–47°S). *Revista Geológica de Chile*, **30**, 159–186.

TOSDAL, R. M., CLARK, A. H. & FARRAR, E. 1984. Cenozoic polyphase landscape and tectonic evolution of the Cordillera Occidental, Southernmost Peru. *Geological Society of America Bulletin*, **95**, 1318–1332.

URRESTY, C. 2009. *Evolución geomorfológica de la parte sur de la Depresión de Domeyko (29°00′–29°40′S) durante el Neógeno*. Thesis, Departamento de Geología, Universidad de Chile.

VEIT, H. 1996. Southern westerlies during the Holocene deduced from geomorphological and pedological studies in the Norte Chico, northern Chile (27–33°S). *Palaeogeography, Palaeoclimatology, Palaeoecology*, **123**, 107–119.

VICTOR, P., ONCKEN, O. & GLODNY, J. 2004. Uplift of the western Altiplano plateau (Northern Chile). *Tectonics*, **23**, TC4004, http://dx.doi.org/101029/2003TC001519

VILLAGRÁN, C., LEON, A. & ROIG, F. A. 2004. Paleodistribution of the alerce and cypres of the Guaitecas during the interstadial stages of the Llanquihue glaciation: Llanquihue and Chiloe provinces, Los Lagos Region, Chile. *Revista Geológica de Chile*, **31**, 133–151.

WHIPPLE, K. X. & TUCKER, G. E. 1999. Dynamics of the stream-power river incision model: implications for height limits of mountain ranges, landscape response timescales and research needs. *Journal of Geophysical Research*, **104**, 17661–17674.

WINOCUR, D. 2010. *Geología y estructura del Valle del Cura y el sector central del Norte Chico, provincia de San Juan y IV Región de Coquimbo, Argentina y Chile*. PhD thesis. Universidad de Buenos Aires, (inédito), Buenos Aires.

WINOCUR, D. A., LITVAK, V. & RAMOS, V. 2014. Magmatic and tectonic evolution of the Oligocene Valle del Cura basin, Main Andes of Argentina and Chile: evidence for generalized extension. *In*: SEPÚLVEDA, S. A., GIAMBIAGI, L. B., MOREIRAS, S. M., PINTO, L., TUNIK, M., HOKE, G. D. & FARÍAS, M. (eds) *Geodynamic Processes in the Andes of Central Chile and Argentina*. Geological Society, London, Special Publications, **399**. First published online February 17, 2014, http://dx.doi.org/10.1144/SP399.2

YAÑEZ, G. A., RANERO, R. & HUENE, V. 2001. Magnetic anomaly interpretation across the southern central Andes (32°–34°S). The role of the Juan Fernández Ridge in the late Tertiary evolution of the margin, *Journal of Geophysical Research–Solid Earth*, **196**, 6325–6345.

Index

Page numbers in *italic* denote Figures. Page numbers in **bold** denote Tables.

Abanico extensional basin 2, 4, 68, *70*, 71, *72*, 420
  basin width analogue modelling 4, 84, 95, 99
Abanico Formation 39, 40, 71, 163
accommodation systems tracts 226, *227*, 228, *234*, *235*, 237
accretionary prism, Choapa Metamorphic Complex 20–21, 25
Aconcagua fold and thrust belt *18*, 41, 69, *70*, *72*, 96, 97–98
  deformation 74, 76
  out-of-sequence structures 99–100
Aconcagua mountain *3*, 40, *348*, *349*
  landslides 7, 331, *332*, 333, 346–365
    as source of hummocky deposits 360–362
  TCN ^{36}Cl dating 363
aeolian deposits, Frontal Cordillera piedmont 299, 302–303
Aetostreon 206, 207, *209*, *212*
aggradation 226, *227*, *234*, 236
  cycles, Frontal Cordillera piedmont 296–300
Agrio fold and thrust belt 215, *216*
Agrio Formation 133, *134*, 147–148, 203, 205–213, *206*
  ammonoids 205, 206–211
  stratigraphy *33*, 205–211
Agua de la Mula Member 133, *134*, 205, 211, 213
Agua de los Burros Fault *424*, 435
Agua Dulce Metaturbidites 21, 23–24
Agua Las Muñeras Creek *248*, 254, *255*, 261
Agua Salada Volcanic Complex 32
Airy- Keiskanen compensation model 170, *171*, 176, 177, 193
Ajial Formation 31
Alfalfal debris flow 338
Algarrobal Formation 27, 34
Algarrobillo Pediplain *426*, *428*, 429, *430*, 431–433
  geochronology **434**, 435, 436–437
  incision 440
  uplift 438–439
alluvial fans 5, 38, 249, 274, *278*, *279*
  Frontal Cordillera piedmont 296–300, 302–304
  see also Regional Aggradational Plain (RAP)
Almacenes lateral moraine 354, 355
Alojamiento Fault *273*, 274, 275, 277
Alto del Tigre High *28*, 32
Alto Tunuyán basin *72–73*, 74, 75
Altos de Hualmapu Formation 31, *33*
Altos de Talinay Plutonic Complex 29
Alvarado depocentre 32
Amarillo rockslide 334, **335**, *336*
ammonoids, Agrio Formation 205, 206–211
Amphidonte 206, 207, 208, *209*
analogue modelling
  basin width and tectonic inversion 84–87
  comparison with natural basins 95–102
  results 87–95
Anchayuyo fault 69, 75, 319, 321–322
  seismic potential 321–322

Andacollo Group *132*, 133, *134*
Andean margin
  kinematic model 67–68
  thermomechanical model 65, 67
Andean Orogen
  development 1, 3
    deformation 1, 3, 4
    tectonic and surface processes 1, 3
  elevation *3*
  geodynamics and evolution 3–5
    tectonic cycles 13–43
  uplift and erosion 7–8
Andean tectonic cycle *14*, 29–43
  Cretaceous 32–36
  early period 30–35
  Jurassic 29–32
  late period 35–43
andesite
  Agrio Formation 205, 206, 207, *209*, *210*
  Chachahuén Group 214
  Neuquén Basin *161*, 162
Angualasto Group 20, 22, 23
apatite
  fission track dating 40, 71, 396, 438
  (U–Th)/He thermochronology 40, 75, 387–397
Ar/Ar age
  Abanico Formation 71
  Barreal-Las Peñas Deformation Zone 277, 279, 286
  Doña Ana Group 110
  Neogene pediplains 431, *432*, **434**, 437
  Río La Sal Formation 111
  Valle del Cura Formation 111, *113*, 115–117
Arboles megalandslide 334
arc magmatism
  Andean tectonic cycle 30, 31, 40–43, *42*
  Gondwanian tectonic cycles 21–22
  retroarc 22
    basin deposits 22–23
  western Sierras Pampeanas 19, 20
Argentina, central western
  geographical features 369–370
  seismicity 369, 371–380, *372*
    ground effects 377, 379
    historical records 373, **374–375**, *376*, 377
    seismic hazard *370*, 379
  tectonic setting 370–371
Arqueros Formation 32, 33, 34
Arrayán Formation 21, 23, 24
Arroyo Anchayuyo, alluvial fans 296, *297*
Arroyo del Zancarrón 115, 116
Arroyo Guanaco Zonzo 117, *120*
Arroyo Malo Formation 32
Arroyo Yaucha, fill terraces 299, *303*, 304
Asociación Piroclástica Pumícea (APP) 285, 296, *297*, **298**, 300, 304, 305, 315
Atacama Gravels 423
Auquilco Formation 31, 32, *33*
Austral Basin, U–Pb zircon dating 149

avalanches *see* rock avalanche deposits
Avilé Member lowstand wedge 205
   U–Pb zircon dating 147–148
Azufrera-Torta surface 437

backarc basins
   Andean tectonic cycle 4, 30, 31, 34–35
   inversion 69, 71
Bajada del Agrio Group 37, 133, *134*
Baños Colorados creek thrust section *248*, **252**, 259, *273*, 274, 278
Baños del Flaco Formation 31, 35
Barrancas Anticline 320
   1985 earthquake **375**, *376*, 377
Barrancas-Lunlunta fault, seismic potential 322
Barreal Block 280, *281*
Barreal Fault 274, 279, 280
Barreal-Las Peñas Deformation Zone 5, *269*, 271, 273, 274–284
   earthquakes 282
   fault scarps 279
   Late Cenozoic kinematics 275–276
   Late Cenozoic tectonics ages 276–282
   palaeoseismological records 278–282
   Quaternary deformation and relief construction 282–284
   Quaternary faults *272*, 274–275
   structural highs 283–284
   uplift 283
Barreal-Uspallata Basin *268*, 270, *272*
basins
   pre-Andean tectonic cycle 3, 26–29
   see also backarc basins; forearc basin deposits; piggyback basins; retroarc basin deposits
^{10}Be age determination 40, 278, 279, 286, 315
   pediplains 433, **434**, 437
^{10}Be concentration, catchment erosion rates 403, 405, *407*, **412**
Beazley Basin, vertical gravity gradient 191
Bermejo Basin 26, *28*, 178, 239
   rigidity 196
   vertical gravity gradient 191
Boca Lebu Formation 37
Bolivian Orocline *16*
Bouguer anomaly 169, *194*, *195*, 196
   effect of Andean root 190–191, 196
   flat slab transition zone *170*, 171–172, *173*
   inverse flexure modelling 186, *187*, *188*, 189
Brownish-red Clastic Unit 37

^{14}C dating
   Barreal-Las Peñas Deformation Zone 281, **282**, 286
   Horcones valley deposits 363
Cabeceras Creek *273*, 279, *280*, 281
Cabeza de Toro megalandslide 334, **335**
Cacheuta Basin *72*–*73*, 74
Cacheuta Formation 28–29
Cachiyuyo Pediplain *426*, *428*, *430*, 431, 433
   geochronology **434**, 436
   uplift 439–440
Cajón de Troncoso beds 31
Cajón Las Leñas alluvial cone deposits *161*
Caleta Horcón Formation 38
Caleu pluton 36

Cambrian
   Famatinian tectonic cycle 17
   Pampean tectonic cycle *14*, 15
Campanario Formation 40
Cañon del Atuel landslide 339
Cantarito Gravels 437
Capdeville Anticline *273*
Caracol basin 228, 230, 237
Caracol Range, uplift 239, *241*
carbonates, Cuyo Precordillera 17, 19
Carboniferous, Gondwanian tectonic cycles 22–23, 24, 25
Carri Lauquen Lake failure 340
Carrizalito Tonalite 17
Casleo Formation 279–282
Caucete 1941 earthquake 373, **375**
Caucete 1977 earthquake 373, **375**, 377, 379
Caucete Group 19
Cavilolén unit 31
Cenozoic
   Andean tectonic cycle 36–43
   Barreal-Las Peñas Deformation Zone
     kinematics 275–276
     tectonic ages 276–282
Central Depression 15, *16*, 38, *70*, *73*, 76
   geological setting 65
Cerrilladas Pedemontanas thrust belt *269*, 270, *272*, 311, 314
   geological setting 65, 296
Cerro Calera Formation 31
Cerro Corrales 205, *208*, *209*
   comparison with Agrio Formation 211, 213
Cerro de la Totora Formation 17
Cerro de las Cabras Formation 28–29
Cerro El Salto megalandslide 334, **335**
Cerro Manantial Fault *273*, 274, 275, *278*, 279, 284
Cerro Mesón Alto pluton *73*, 75
Cerro Salinas Anticline 277
Cerro Salinas-Montecito thrust system *246*, 247
Cerros Colorados frontal range 255, 259, 260
Cesco Formation 279, 283
Chacay Melehue Formation 133, *134*
Chacayal thrust system *66*, 69, *70*
Chachahuén Groups 214
Chachahuén volcanic complex *206*, 213–214
   deformation and uplift 215–217
   Early Cretaceous deposits 203, 205–213
   structural setting 214–215
   magma components 214
   tectonic setting 204–205
Chalet fault *299*, 300, 320
Chanic Orogeny 4, *14*, 17, 25
   deformation 20, 23
Chanic unconformity 17
Chihuidos High 215, *216*
Chilenia terrane *14*, 20
Choapa Metamorphic Complex, accretionary prism 20–21, 25
Choiyoi Group 22, 24, 25, *132*, 133, *134*, 146, 331
   thermochronometry 389–397
   uplift 385–387
Choiyoi Magmatic Province 29
Chos-Malal fold and thrust belt *18*
Chupasangral fault 317

INDEX

^{36}Cl, TCN dating 336–337, 347, *349*, 351–354, 357, 359, 360, 363–364
climate
  control on erosion and uplift 401–415
  and landscape evolution 420–421
  and landslides 337–338, 345
Coastal Batholith 23, 24, 25
Coastal Cordillera 1, *2*, 15, *16*, *70*
  crustal thickening 423
  geological setting 64–65, 329, *330*, 331, *424*, *425*
  gravity anomaly 191
  influence of precipitation 440–441
  pediplains *426*, 427, *428*, 429–442
  structure 68, *70*
  topography and geomorphology 421, *422*, 425, 427
    mapping 427
  uplift 8, 74, 76–77, 439–440
Cobquecura pluton 29
Cogotí superunit 36
Colangüil batholith 22
Colimapu Formation 35, *161*, 163, 164
Colohuincul Complex 133, 141
Conceptión Group 37
Confluencia Formation 38, 421, *425*, 433, 439, 440
conoids 223, 232, *233*
continental breakup *14*
Contreras Formation 71, 74
Coquimbo Formation 38, 421, *425*, *430*, 438, 439
coquina 206–209, *210*, 211, 213
Cordillera de la Brea 111, 113, *118*
Cordillera de la Ortiga 110, 111, 113, 117, *121*, *122*
Cordillera del Viento *132*, 133, 141, 142
  depocentre 32
Cordillera del Viento Formation 133, *134*
Córdoba terrane 15, 17
Cordón de Barda Negra 259
Cordón del Plata piedmont 315, *316*, 317
Cordón del Plata rock avalanches 337, 338
Corredores Pediplain *426*, *428*, 429, *430*, 431, *432*
  elevation 439
  geochronology **434**, 435–436
Cortaderas megalandslide *6*, 334, **335**, 338
Costa de Araujo 1920 earthquake 373, **375**, 377, *378*, 379
Cretaceous
  Andean tectonic cycle 32–36
  Early
    Neuquén Basin north-eastern margin 203–217
    Neuquén Basin north-western margin 155–164
    Neuquén Basin sediments, dating 146–148, 149
  Late, Neuquén Basin north-western margin 163–164
Cretaceous-Paleocene, Maipo-Tunuyán transect, deformation 69
Crioceratites 206, 207, *208*, *209*
Crioceratites diamantensis 162, 208, *209*, *210*, *212*, 213
Cristo Redentor rotational slide **335**, *336*
crust
  elastic thickness 170, 183–184, 192–196
  flexural rigidity 183, 197
Cuartitos Unit, faulting *98*, 100, *101*
Cucullaea gabrielis 208, *209*
Cuesta del Tambolar 19

Cuesta del Viento Formation 228, *231*
  piggyback basin
    evolution 239–241
    sedimentation 237–239, *238*, 242
Cumbre surface 437
Cura-Mallín Formation 39–40
Curanilahue Formation 37
Curepto-Bio Bío-Temuco basin 26, 27
Cuyania terrane *14*, 20, 167–168, 175, 176
Cuyo Basin 26, 27, *28*, 69, *268*, 271, 273, 295
Cuyo Group 32, 133, *134*
Cuyo Precordillera 17, 19

dams, sedimentary 230–232, *233*, *234*, *236*, 237, 279
debuttressing, and landslides 338, 339
deformation
  deep crustal 64
    conceptual models 64
    Maipo-Tunuyán transect 69, 71–77
      kinematic model 67–68
      thermomechanical modelling 65, 67
  Famatinian tectonic cycle 19–20
  Quaternary retro-wedges 267–287
  Quaternary thrust fronts 245–263
Del Cerro Negro de Capiz fault, seismic potential 322
Del Peral Anticline 317–319
Del Peral fault, seismic potential 321
Del Totoral fault, seismic potential 321
density models 168–179
  2D models 174–176
depocentres
  Andean tectonic cycle 29, 32, 168
  pre-Andean tectonic cycle 27, *28*, 29
Devonian, Famatinian tectonic cycle 17, 20
Diaguita Phase deformation 277–278, 284
Diamante deformation zone *see* Los Alamitos-Papagayos-Diamante Deformation Zone
Diamante stage 300
'Dioritas Gnéisicas de Cartagena' 29
Dolores Creek 279–281, **282**
Domeyko Depression 421, *422*, 423, *425*
  pediplains 427, 431, 433
Domeyko Gravels 421, 423, *428*, *430*, 431, 435, 437, 438–439, 440
Domeyko Range *16*, 423
Doña Ana Formation 109, 110
Doña Ana Group 100
  structural setting 117, 127–128
  volcanism 109, 110, 125–127
    geochemistry 119, 122–125, **124**, *126*

Early Chachahuén Group 214
earthquakes 5, 64, 76
  Caucete (1941) 373, **375**, 377, *378*
  central western Argentina 369, 371–380, *372*
    ground effects 377, 379
    historical records 373, **374–375**, *376*, 377
  Costa de Araujo (1920) 373, **375**, 377, *378*, 379
  Frontal Cordillera retro-wedge 320–322, 323
  Great Argentinean Earthquake (1894) 369
  La Laja (1944) 373, **375**, *378*, 379
  La Rinconada (1952) 373, **375**
  and landslides 338–339

earthquakes (*Continued*)
  Mendoza (1861) 279, 282, 372, 373, **374**, *376*
  Panquehua (1917) 282, 373, **375**
  San Juan (1894) 373, **374**, 377, *378*
  Uspallata (1927) 373, **375**, 377, 379
  Villa Atuel (1929) 373, **375**, 377, *378*
Eastern Coastal Cordillera 425, *426*, *432*
  pediplains 431, *436*, 438–439
  uplift 441
Eastern Cordillera *16*
Eastern Frontal Cordillera 423, 425, *426*, *432*
  pediplains 427, 437
Eastern Precordillera thrust system 271, 278–279
EGM2008 gravity field model 184, 185, 189–190, 196, 197, *198*
El Abuelo-Mario Ardito rock avalanche 331, **335**, 336, 360
  see also Quebrada Mario Ardito rock avalanche
El Chape Fault *424*, 435
El Cóndor-Las Peñas section *248*, **252**, 255–257
El Culenar Beds 33
El Diablo fault 39, 40, *96*, 99–100
El Freno Formation 32
El Infiernillo Creek *248*, 254
El Jarillal Creek *248*, 252–253, *254*
El Nino Southern Oscillation, and landslides 338, 339
El Quereo-Los Molles Basin 26, 27
El Ratón Formation 23
El Salto Fault 285, 316, 317
  seismic potential 321
El Sauce unit 31
El Teniente porphyry deposit 39, 40, 75–76
El Teniente Volcanic Complex 71, *73*, 75
El Tigre Fault 279
El Totoral Formation **298**, 299, 300
El Tránsito Metamorphic Complex 22
El Volcán plutonic unit 25
El Zampal Formation 297, *297*, **298**, 299–300, 302, 303–304, 306, 315
elastic thickness 170, 183–184, 192–197
Elqui plutonic complex 21, 25, 29
Elqui valley 21–22, 32
Eocene, Andean tectonic cycle 36–37
Eocene-Early Miocene, Maipo-Tunuyán transect, extension 69, 71
Eriphyla 208, *209*
Erizo lake failure 339–340
erosion 7–8
  decadal/millennial rate 406–409, 411, 413–415
    geological features 409–410
    vegetation 407
  response time 401–402
  role of climate 401–415
Escabroso Formation 100, 109, 110, 125–126
  geochemistry 119, 122–125, **124**, *126*
Escondida Creek *248*, 259, *273*, 274, *276*, 278
Estero Cristales Beds 37
Esteros Marques landslide 334, **335**
extension, Maipo-Tunuyán transect 4, 69, 71

Famatinian tectonic cycle *14*, 17, 19–20, 270
fanglomerates
  Frontal Cordillera piedmont 296–300
  Las Peñas-Las Higueras range 247, 249, 260
Farellones Formation 38, 39, 71, 74

Farellones volcanic arc 71, *72*
faulting
  Cuartitos Unit *98*, 100, *101*
  and landslides 338–339
  Norte Chico Region *98*, 100, *101*
  Quaternary 5, 371
    Barreal-Las Peñas Deformation Zone 274–275
faults, blind 5, 371
fissures, seismic 377, *378*
flat-slab subduction segment 2, 15, *16*, *268*
  Andean tectonic cycle 40–43
  collision with Juan Fernández Ridge 183, *184*, 185
  lithospheric characteristics 193
  normal subduction transition zone
    crustal discontinuities 167–179
    density model 168–179
      2D models 174–176
    isostatic gravity anomaly 170–172, 176–178
      decompensative 172, *174*, 177, 178
      Tilt method 173–174, *175*, 178
flexural rigidity 183
  inverse modelling 186–188
flexural strength 185–186
floodplains 234–235
fluvial terrace incision 302–304, 306
fluvio-aeolian deposits 5, 38, 42, 296, 299
fold-thrust belts 4, 39, 41
forearc basin deposits
  Andean 32–34
  Gondwanan 21
Forearc Precordillera *16*
Foreland 2, 40–43
Frontal Cordillera 1, *2*, *16*
  Andean tectonic cycle 40
  crustal thickening 423
  elevation *3*
  Famatinian tectonic cycle 17, 19
  geological setting 65, *330*, 331, *424*
  granitoid batholith 24
  influence of precipitation 440–441
  megalandslides 332–333
  piedmont 293–306
    drainage pattern 300–302
    Quaternary sediments
      aggradational cycles 296–300, *299*
      compositional signature 304–305
      and palaeoenvironment 302–305
      piedmont morphotectonics 300–302
      pyroclastic deposits 300
      stratigraphy and geomorphology 296–302
  retro-wedge 311–324
  structure 68–69, *70*, 109
  topography 423, *424*, 425
  topography and geomorphology 425, 427
    mapping 427
  uplift 383–397
    and deformation 441
    and erosion 7–8, 75–76, 305–306
    geomorphic evidence 396
    (U–Th)/He thermochronometry *384*, 386–397
  see also Eastern Frontal Cordillera; Western Frontal Cordillera
Frontera-Deidad surface 437

geohazards *see* earthquakes; landslides; rock avalanche
     deposits; seismic hazard
Gervilella 209
Gervillaria alatior 207, *209*
GOCE gravimetric satellite data 184, 185, 189–190, 196,
     197, *198*
Gondwana, assembly *14*
Gondwanan Orogenic Cycle 4
Gondwanan tectonic evolution 24–25
Gondwanian tectonic cycle *14*, 20–25
     northern segment 20–23
     southern segment 23–25
granitoids
     Andean tectonic cycle 31, 34, 36
     Famatinian 19
     Frontal Cordillera 24
     Gondwanian 22, 24
     pre-Andean tectonic cycle 29
gravimetric satellite data, GOCE 184
gravity
     inversion 186
     measurement 168–170
     vertical field gradient 183, 188, *194*, *195*, 196
          models 185, 189–191
               EGM2008 184, 185, 189
               GOCE data 184, 185, 189
gravity anomaly 183, 197–198
     correction 169–170
     decompensative 172, *174*, 177, 178
     isostatic 170–171, 176–178
          decompensative 172, *174*, 177, 178
          Tilt method 173–174, *175*, 178
     see also Bouguer anomaly
Great Argentinean Earthquake (1894) 369
groundwater, effect of earthquakes 379
Guanaco Sonso Formation 21–22
Guaraco Norte Formation 133, *134*, 142
Guarguaraz Complex 17
gypsum
     Andean tectonic cycle 35
     landslides 332

Hamulinites? 208, *209*, *212*
Hauterivian, Neuquén Basin
     lowstand wedge 146–148
     palaeogeography 162
Hoplitocrioceras gentilii 206, *209*, *210*, 211
Horcones Inferior glacier 354, *361*
     surges 365
Horcones River valley *348*, *349*
     hummocky deposits *354*, 355, *356*, 357, 360–362
     mineralogy *349*, *350–351*, 354–355, 357
     moraines 355, *356*, 357
     rock avalanche deposits 331, *332*, **335**, *336*, 346–365
          confluence with Las Cuevas valley *356*, 357
          Confluencia 354–355
          TCN dating *349*, 351–354, 362–364
          temporal distribution 364–365
Horqueta Formation 31, 32, 156
Huaco River, megafan *232*, 236
Huarpes depression 295
Huentelauquén Formation 21, 22, 25
Huincul High *132*, 141, 145, 146
Hurtado Formation 21

Ibáñez Formation, U–Pb zircon dating 149
Illapel Plutonic Complex 36
Incaic Orogeny *14*, 36, 37, 423
Incaic Range 420, *432*
     and location of pediplains 437–438
incised valleys 226, *227*, 229–230, *234*, *236*, 237
Infiernillo fault *66*, 68, 74
inversion *see* backarc basins, inversion; gravity,
     inversion; tectonic inversion

Jaboncillo Anticline 317–319
Jáchal River area
     megafan *232*, 236
     Quaternary piggyback basins, sediments
          228–237, 242
Jarillal Fault *273*, 274, *275*
Jocoli Basin 168
     density modelling 174–176, 178
     rigidity 196
     vertical gravity gradient 191
Juan Fernández Ridge 15
     collision with Chilean margin 183, *184*, 185, 277
     Moho 191–193, *195*
     and Nazca Plate subduction 193–196, 419
     subduction 184–185
     vertical gavity gradient 189–196
Juncal Formation 35
Jurassic
     Andean tectonic cycle 29–32, *33*
     pre-Andean tectonic cycle 27
     Upper, Neuquén Basin sediments 131–150, *132*
          U–Pb zircon dating 132, 135, 137, 139–141
Jurassic-Cretaceous boundary, Neuquén Basin
     sediments 149–150

K–Ar age
     Barreal-Las Peñas Deformation Zone 277
     Valle del Cura Formation 110, 111, *112*, 116
K–T Orogeny *14*, 36
Kimmeridgian
     lowstand wedge 146–148
     Neuquén Basin, north-western margin 156, 158
knickpoints 390, 391

La Aguadita fault 317
     seismic potential 321
La Cal fault 246, *248*, *273*, 279
La Carrera Fault system 284, 285, 315–317
La Cueva Formation 38
La Engorda landslide deposits 332, *333*, **335**, 337
La Estacada Formation **298**, 299
La Gloria pluton 75
La Invernada Formation 297, **298**, 314
La Laja 1944 earthquake 373, **375**, *378*, 379
La Lajuela Formation 32, 33
La Manga Creek 316–317
La Manga Formation 32, *33*, 133, *134*
La Obra pluton 39, 71, *72*, 74
La Ollita Formation 111
La Pilona Formation 43, 314
La Ramada basin 26, 27, 32
La Ramada fold and thrust belt *18*, 95–97
     out-of-sequence structures 99
La Rinconada 1952 earthquake 373, **375**

La Silla Pediplain *426*, *428*, 429, *430*, 431, *432*
 geochronology **434**, 435, *436*, 437
 and Incaic Range 438
 uplift 438–439
La Totora Fault 423, *424*, 437
La Tranca Basin 228, *229*, 230, 237
 Miocene piggyback basin 237–241
La Travesía Depression 296
La Ventana fault, seismic potential 322
Laguna gabbro 29
Laguna Verde unit 31
Lagunillas Formation 31, *33*, 35
 U–Pb zircon dating 149
Laja lahar 38
Lanalhue lineament 23, 24, 25
landscape evolution
 Neogene 7–8, 419–442
  drainage area as control 440
landslides 5, *6*, 7, 329
 ^{36}Cl TCN dating 351–354
 climate or seismic triggering 7
 dating 345–347
 misinterpretation as glacial landforms 329, 331–332, 346–347
 seismic 377
 see also megalandslides; rock avalanche deposits
Lapa Formation 133, *134*
Las Arpas rock avalanche **335**, 338
Las Breas Formation 27
Las Chacras-Los Guanacos thrust section *248*, 250–252, **252**, *253*
Las Chilcas Formation 34, 36
Las Cuevas River valley *348*, *349*
 rock avalanche *6*, *330*, 331, *336*, 337, 338, 346–365
 ^{36}Cl TCN dating 351–354, **352–353**, 362–364
 temporal distribution 364–365
Las Higueras thrust system *246*, 247, *248*, 249, *258*, 279
Las Hualtatas megalandslide 334, **335**
Las Lagunitas Formation 17
Las Llaretas, faulting *98*
Las Melosas 1958 earthquake *6*, 76, 338, 339
Las Peñas Fault 275
Las Peñas Thrust System 5, *246*, 247, *248*, 249–263
 bedrock control 260–262
 Quaternary activity 249–259
 scarp evolution 260
 shortening rates 259–260
 thrust activity 261–262
 thrust splays 260
Las Peñas-Baños Colorados creeks thrust section *248*, **252**, 257–259
Las Peñas-Las Higueras range 245, *246*, 247, *248*
 geology and tectonics 247–249
 Las Peñas thrust system 249–263
Las Tunas Formation 297–299, 306, 314–315
Late Chachahuén Group 214
Lautaro Formation, faulting 100
Licancheo Formation 38
Limache unit 31
liquefaction 5, 7, 377, *378*
Liquiñe-Ofqui Fault Zone *16*
lithosphere, thermomechanical model 65, 67
lithosphere-asthenosphere boundary 67

Lliu Lliu unit 31
Lo Prado forearc basin 30, 32–34
Lo Prado Formation 32, 33, 34
Lo Prado magmatic arc 32, 34
Lo Valdés Formation 35, *158*, 160, *161*, 162
Lo Valle Formation 36, 37
Loma Huinganal megalandslide 334, **335**
Lomas Bayas High 276, *281*, 283
Lomas de Jocolí 259, 274, *276*
Lomas del Inca Anticline 273, *277*, 279, *280*, 281, **282**, 283
Lomas del Inca Formation 277
Lomas del Jaboncillo 302
Lomitas Negras High 276, *277*
Los Alamitos Formation **298**
Los Alamitos-Papagayos-Diamante Deformation Zone *269*, 271, 273, 285–286
Los Bañitos pyroclastic flow *115*
Los Guanacos-El Cóndor creeks thrust section *248*, 252–255
Los Loros Creek *248*, 250, *253*
Los Marquis rock avalanche 338
Los Mesones Formation 296, *297*, **298**, 314
Los Molles Formation 30, *33*
Los Patillos Formation 32
Los Pelambres Formation 35
Los Rios surface 437
Los Tilos sequence 27
Lotena Group 32, *33*, 133, *134*, 135
lowstand wedges, Neuquén Basin 5, 131, 146–148

Macul rockslide 334, **335**
Main Cordillera *see* Principal Cordillera
main décollement *66*, 67, *68*, 71, 74, 75, 76
Maipo Basin, landslide deposits 332
Maipo drainage 155, *156*
Maipo Orocline *16*, 38
Maipo stage 300
Maipo-Tunuyán transect 64–77
 Early Miocene 71, *72*, 74
 Eocene-Early Miocene extensional event 69, 71, *72*
 geological setting 64–65, *66*
 kinematic model 67–77
 Late Cretaceous-Paleocene deformation 69, *72*
 Late Early Miocene *72*, 74
 Late Miocene *73*, 75
 Middle Miocene *72–73*, 74–75
 Pliocene *73*, 75–76
 Quaternary *73*, 76
 structure 68–69
 thermomechanical model 65, 67
Malargüe fold and thrust belt *18*, 96, 215, *216*
Malargüe Formation 37
Malargüe rift deposits 27
MAP-GAC project 329
Mariño Formation 27, 42, *256*, *257*, *258*, 260, 314
Mario Ardito *see* El Abuelo-Mario Ardito rock avalanche
Marmolejo volcano *66*, 68
MASH zone 71, *72–73*
Matahuaico Formation 22
Maule 2010 earthquake 5, 339, 370
megafans *232*, 236, 242

megalandslides 3, *6*, 7, 329–341
  causes and triggering 337–339
  as geological hazards 339–340
  inventory map 332–333, *334*
  palaeomegalandslides chronology 333–337
  see also landslides; rock avalanche deposits
Melipilla fault 68
Mendoza 1861 earthquake 279, 282, 372, 373, **374**, *376*
Mendoza Group 133, *134*, 146
Mendoza Province 294, *295*
Mendoza River valley, landslides *6*, 7, *330*, 331, 334
Mendoza-Neuquén backarc basin 30, 32, 34–35
  see also Neuquén Basin
Mercedes Basin
  rigidity 196
  vertical gravity gradient 191
Meseta del Guadal 320
Méson Alto-El Yeso rock avalanche 332, *333*, **335**, 337, 338, 339
metamorphism
  Famatinian tectonic cycle 17, 19–20
  Gondwanian tectonic cycle 21, 23, 25
  pre-Andean tectonic cycle 29
Millahue unit 29
Millongue Formation 37
Mimachlamys 207, *209*, *210*
Miocene
  Andean tectonic cycle 38–40
  Early
    Maipo-Tunuyán transect 71, *72*, 74
    Río La Sal Formation 111
  Late, Maipo-Tunuyán transect *73*, 75
  Middle, Maipo-Tunuyán transect *72–73*, 74–75
  piggyback basin sedimentation 237–241
Mogotes Formation 249, 255, 259, 277–278, 314, 316
Moho
  depth 193, *194*, 195, 196
  flat slab transition zone 170–171, 176–177, 186–188
  inverse flexure modelling 186–188, *189*
  Juan Fernández Ridge path 191–193
  Nazca Plate subduction 193
Montecito anticline 255, 257, 260
Morado valley, landslide deposits 332
Mostazal Formation 34
mud flows 377
Mulichinco Formation *33*, 133, *134*
Myoconcha transatlantica 208, *209*

Nacientes del Bio Bío Formation 31
Nacientes del Teno Formation 31, *33*
Naranjo Block 274
Navidad Basin *72–73*, 74
Navidad Formation 38, 71, 75
Nazca Plate subduction 1, 3, *68*, 183, 246, 267, 370, 419, *420*
  age of plate 193
  flat slab transition zone, density model 168–179
  and Juan Fernández Ridge 193, 419
  vertical gravity gradient 189–190
Negro rockslide 334, **335**, *336*
Neocomiceramus curacoensis 206, 208, *209*, *210*, 212

Neogene
  deformation 276–277
  landscape evolution 419–442
    drainage area as control 440
    pediplains 421–442
Neuquén Basin 2, 4–5, *28*, 30, 32
  geological setting and stratigraphy 132–137
    basement 133, *134*
  lowstand wedges 146–148
  north-eastern margin, Early Cretaceous, Chachahuén 203–217
  north-western margin 155–164
    Kimmeridgian 156, 158
    Late Cretaceous 163–164
    Tithonian-Hauterivian *157*, 158, 160–163
  tectonic evolution 133, 146
  Upper Jurassic sediments 131–150, *132*
    provenance studies 135, 139, 141–148
    U–Pb zircon dating 132, 135, 137, 139–150
  see also Mendoza-Neuquén backarc basin
Neuquén Group *33*, 37, 133, *134*, 205
Nieves Negras depocentre 32, 155
Nieves Negras Formation 31
Ñireco Formation 133, *134*
Norte Chico Region
  faulting *98*, 100, *101*
  Tertiary stratigraphy *114*

ocean-island basalts 17, 205, 214
Ocloyic Orogeny *14*, 17
  deformation 20
O'Higgins seamount *184*, 189
Ojo de Agua thrust *208*, 211, 214
Olcostephanus 162
Olenellus trilobites 17
Oligocene, Andean tectonic cycles 38, 39
Oligocene-Early Miocene
  Río La Sal Formation sedimentation 111
  Valle del Cura Formation
    structural setting 117, 127–128
    volcanism 111, 113, 115, 125–126
      geochemistry 118–119, 121–125, *126*, *127*
ophiolites, Famatinian 17, 19
Ordovician, Famatinian tectonic cycle 17, 19–20
oroclines 84
orogens, curved, structural change 83–84
Ouachita embayment 20
Ovalle Pediplain *430*, 433, 440
  geochronology **434**, *436*, 437
  uplift 440, 441

Pachimoco Basin 228, *231*, 235
palaeomegalandslides 329
  chronology 333–337
palaeosol, Frontal Cordillera piedmont 302, 303
Palauco depocentre 32
Palauco Formation 43
Palomares fault system *70*
Palomares ramp *72–73*, 74, 75
Pampa de Los Avestruces Granite 17
Pampa de los Burros Anticline 276, *277*, *281*, 283
Pampa de los Burros Shear Zone 274, 275, *277*
Pampa Seca Basin *273*, 274, *275*
Pampa Tril anticline *214*, 215

Pampean flat-slab *see* flat-slab subduction segment
Pampean Orogeny *14*
Pampean Ranges *see* Sierras Pampeanas
Pampean tectonic cycle *14*, 15, 17
Pampia terrane *14*, 17, 167, 175, 176
Pan de Azúcar Formation 30
Pangea, breakup *14*
Panquehua 1917 earthquake 282, 373, **375**
Papagayos deformation zone *see* Los Alamitos-Papagayos-Diamante Deformation Zone
parched basins 223
Parraguirre landslide **335**, 338, 339
partial retention zone, exhumation 386, *388*, 389, 392, 394–396
Patagual-El Venado unit 23, 29
Payenia volcanic province *18*, *42*, 43, 204, 205, *207*
   mantle source 205
   palaeoflat-slab subduction 43
Payún Matrú caldera 43, 205, *207*
Pedernal thrust system *246*, 247
pediments 228, *230*, 235, 316, 427
pediplains *426*, 427, *428*, *430*
   age of formation and incision 433–437
   location with respect to Incaic Range 437–438
   tectonic v. erosional control 440–441
Pehuenche Orogeny *14*, 36, 39
Peñasco Block *273*, 274, 275
Penitentes moraine/rock avalanche deposit 331, *336*, 359–360, 362
Peñuelas unit 31
Permian, Gondwanian tectonic cycle 21–22
Peruvian Orogeny *14*, 30, 35, 36, 37
Pichidangui Formation 27
Pie de Palo Complex 19
   vertical gravity gradient 191
piedmont systems 223, *224*, 226, 227, 228, 249, 255, *283*
   Frontal Cordillera, Quaternary 293–306
   see also Frontal Cordillera, piedmont
Piedra Santa Formation 133, *134*, 141
piggyback basins 4, 221, 223, 241–242
   accommodation systems tracts 226, 227, 228, *234*, *235*, 237
   aggradation 226, *227*, *234*, 236
   Miocene, La Tranca 237–241, 242
   modelling 226, *227*, 228
   Quaternary Precordillera 228–237, 242
      accommodation stages 235–237
      incised valleys 229–230, *236*, 237
      Jáchal River Valley *231*, 232–235
      sedimentary dams 230–232, *233*, *234*, *236*, 237
   sedimentary dynamics 223–226
   A/S ratio 225, 235–237
Pilmatué member 205, 211, 213
Pilpilco Formation 37
Pinna 207, *209*
Pliocene, Maipo-Tunuyán transect *73*, 75–76
Pocuro fault *96*, 100
porphyry Cu–Mo ore deposits *14*, 39, 75–76
Portillo fault system 69, *70*, 76
Portillo landslide deposits 332, *333*, **335**, 337

Potrerillos depocentre 27
Potrerillos Formation 28–29
Potrero Alto Beds 38
Potrero Alto landslide deposits 332
pre-Andean tectonic cycle *14*, 25–29
pre-Cuyo cycle *132*, 133, *134*, 141, 146
precipitation
   and erosion 401, 402–403
   and landscape evolution 420–421, *422*, 440–441
   Southeast Pacific anticyclone (SEP) 420
Precordillera 1, *2*, *16*, 168, *222*, *330*, 331
   elevation *3*
   gravity anomaly 191
   Moho depth 193, 196
   piedmont associations *224*
   Quaternary deformation 284
   Quaternary piggyback basins 228–242
   Quaternary thrust systems *246*, 247, *269*, 270, 271, 272
Principal Cordillera 1, *2*, 15, *16*, 38–40
   geological setting 65, *330*, 331
   gravity anomaly 190–191
   megalandslides 332–333
   Moho depth 193, 196
   seismicity 76
   structure 68, *70*
   uplift 74
Profeta Formation 30
Pterotrigonia coihuicoensis 208, *209*
Ptychomya koeneni 208, *209*
Pucalume Formation 32, 34
Pudahuel-Machalí Ignimbrite 38, 300
Puente del Inca *336*
Puerto Oscuro unit 31
Puesto Araya Formation 32
Puesto Uno Formation 277
Punta de Vacas moraine 357
Pupio Fault *424*, 435
pyroclastic deposits *see* Asociación Piroclástica Pumícea

Quaternary
   deformation, Central Andean retro-wedge 267–287
   Frontal Cordillera piedmont 293–306
   Maipo-Tunuyán transect *73*, 76
   piggyback basins 228–237, 242
   tectonics 5
   thrust fronts 245–263
Quebrada Blanca rock avalanche *6*, *332*, **335**, *336*
Quebrada Choros Pediplain *428*, *436*
Quebrada del Sapo Formation *134*, 135, *136*, 145, 149
Quebrada Mal Paso Beds 21
Quebrada Mario Ardito rock avalanche 364
   see also El Abuelo-Mario Ardito rock avalanche
Quebrada Marquesa Formation 32, 33
Quebrada Seca megalandslide 334, **335**
Quiriquina Formation 37

ramps 66, 67, 69, 74
   Palomares *72*–*73*, 74, 75
   Western Cordillera *66*, 71, 74, 75
   Yeguas Muertas *72*–*73*, 74, 75

INDEX

Rancho de Lata Formation 32
Rapel Formation 38
Rayoso Formation *33*, 205, 209
red beds, Neuquén Basin 162, 163, 164
Regional Aggradational Plain (RAP) 302–303
Remoredo Formation 133, *134*
retro-wedges
  Central Andes
    Quaternary deformation 267–287, *268*, *269*
      distribution patterns *269*, 271–273
      mechanical anisotropies 270, 271, 272–273
      morphotectonic units *269*, 271
      tectonic framework 270–271
  Frontal Cordillera 311–324
    Neogene synorogenic strata 314
    Neopalaeozoic sediments 313–314
    Plio-Quaternary alluvial sediments *313*, 314–315
    Quaternary tectonics 315–320
    seismic potential 320–322, 323
    tectonic setting 315
    Triassic-Palaeogene sediments *313*, 314
    uplift 322–324
retroarc basin deposits 22–23, 30
rhyolites 17
ridge push 63
Riedl shears 274, *276*
rifting, pre-Andean tectonic cycle 27, *28*
Rincón Blanco 29
Rinconada 1952 earthquake 373, **375**, 377
Río Atuel-La Valenciana depocentre 32
Río Blanco fault 316, 317
Río Blanco Formation 28
Río Colina Formation 31
Río Damas Formation 31, *33*, 35, 135, 156, 158
Río de La Plata Craton, collision 15, 17
Río de Los Pozos Formation 255, 314, 320
Río del Cobre depocentre 32
Río La Sal Formation, Oligocene-early Miocene sedimentation 111, *114*, 117, *121*
Río Mendoza
  apatite (U–Th)/He analysis 392, **393**, 394–397
  topographic analysis 390–391, 392
Río Mendoza Formation 28–29
Río Mendoza-Tupungato Deformation Zone *269*, 271, 273
  Quaternary deformation 284–285
Río Tascadero Formation 32, 35
Río Tunuyán
  apatite (U–Th)/He analysis 392, **393**, 394–397
  topographic analysis 390–391, 392
Riquiliponche-Las Chacras thrust section *248*, 250, **252**
Rivadavia Fault *424*, 425
Robinson Crusoe Island *184*, 189, 191
rock avalanche deposits 329
  ^{36}Cl TCN dating 351–354
  Horcones River valley 331, *332*, *336*, 346–365
Rodeo-Iglesia Basin *222*, 228
Rodinia breakup 14

Salar Depressions *16*
Saldeño Formation 37

Salinas Basin
  density modelling 175, 178
  rigidity 196
  vertical gravity gradient 191
Salto del Soldado landslide 332, *333*, **335**, 337
San Bartolo Fault 284
San Bartolo half graben *278*, 284
San Félix Fault 423, *424*, 437
San Gabriel pluton 39, *73*, 75
San José Formation 35
San Juan 1894 earthquake 373, **374**, 377, *378*
San Juan 1944 earthquake 373, **375**, 377, *378*
San Juan 1952 earthquake 373, **375**, 377
San Nicolás rock avalanche **335**, 338
San Pedro Ridge 4, 174, 178
San Rafael Block *3*, *18*, *268*, *269*, 270, 272
San Rafael Orogeny 4, *14*, 22, 25
San Ramón Fault 5, 39, *66*, 68–69, 74, *96*, 100
sand volcanoes 377, *378*, 379
Santa Barbara System *16*
Santa Clara Block *273*, 274, 275, 279
Santa Elena Member 31
Santa Juana Formation 23, 29
satellite basins 223
sediment discharge 403, **404**
seismic hazard, central western Argentina 5, *370*, 379
seismicity
  central western Argentina 369, 371–380, *372*
    ground effects 377, 379
    historical records 373, **374–375**, *376*, 377
    interplate 371
    intraplate 5, 371–372
    seismic hazard 5, *370*, 379
  and landslides 338–339
  Principal Cordillera 76
Sierra Azul depocentre 32
Sierra de Ansilta Fault 274, 275
Sierra de Chachahuén *see* Chachahuén volcanic complex
Sierra de las Peñas *272*, 274
Sierra de Pie de Palo *18*, 19, 168
Sierra de Reyes depocentre 32
Sierras Pampeanas 1, *2*, 16, *16*, *268*, 270
  elastic thickness 196
  elevation *3*
  Moho depth 193–196
  Pampean tectonic cycle 15
  see also Western Sierras Pampeanas
slab pull 63
Somún Curá Massif *132*, 143, 145
South American Plate 3, *68*, 183, *184*
Southeast Pacific anticyclone (SEP) 420
Southern Precordillera morphotectonic subunit *268*, *269*, 271, *272*, 273
  palaeoseismological records 278–282
  Quaternary deformation 284
  uplift 277
Southern Volcanic Zone 40, 205
spilites, Volcán valley 160
Spitidiscus shale 206, *208*, *209*, 211
spring waters 377, *378*, 379
stream profile analysis 390–391, 392
Subandean Systems *16*

subduction
　Andean tectonic cycle 30, 43
　driving forces 63–64
　Famatinian tectonic cycle 19, 20
　flat-slab subduction segment transition zone, crustal discontinuities 167–179
　Juan Fernández Ridge 184–185
　pre-Andean 25–26

Tarapacá Basin, U–Pb zircon dating 149
tectonic cycles 13–43
　Andean *14*, 29–43
　Famatinian *14*, 17, 19–20
　Gondwanian *14*, 20–25
　Pampean *14*, 15, 17
　pre-Andean *14*, 25–29
tectonic inversion 83, 117, 204–205
　and basin width, analogue modelling 84–87
　　comparison with natural basins 95–102
　　results 87–95
Teno lahar 38
terrain cracks, seismic 377, *378*, 379
terrestrial cosmogenic nuclide (TCN) dating *349*, 351–354, 362–364
　see also ^{36}Cl, TCN dating
Thalassinoides 209, *209*, *210*, 211
thermochronometry *see* U–Th/He thermochronometry
thrust fault scarps 300–301
thrust systems, Quaternary 245–263
thrust-top basins 223
Tilito Formation 100, 109, 110, 117, *120*, 125–126
　geochemistry 119, 122–125, **124**, *126*, *127*
Tilt method, gravity anomaly enhancement 173–174, *175*, 178
Tinguiririca lahar 38
Tithonian, Neuquén Basin
　palaeogeography 155, *157*
　sediments 149–150
Tithonian-Hauterivian, Neuquén Basin, north-western margin *157*, 158, 160–163
Tobas Angostura Formation 277, 314
Tobas La Higuerita unit 314
Tobas Multicolores Formation 111
Tolosa rock avalanches **335**, *336*, 337–338, *349*, 355
　dating 364
Tordillo Formation 5, 31, 34, 131, *140*, 158
　depocentres and depositional systems 135, *136*
　geology 133, 135, *138*
　provenance studies 135, 139, 141–148
　stratigraphy *134*, *136*, *159*
　U–Pb zircon dating 132, 135, 137, 139–150
Tranquilla unit 29
transference systems 223, 224–225, 226, *227*, *234*, 236–237
transgressive-regressive deposits 27, 30, 31, 37
Trapa-Trapa Formation 40
travertine, Horcones River valley 357
Tres Cruces Formation 27
Tres Esquina Formation 32
Triassic
　Andean tectonic cycle 32
　Gondwanian tectonic cycles 21, 23, 24
　pre-Andean tectonic cycle 25–29
Trihueco Formation 37

Tubul Formation 38
Tunuyán Basin *268*, 270
Tunuyán Depression 294, 295–296, *298*, 300, *301*
Tunuyuán lineament 4, 174, 178

U–Pb zircon dating
　Farellones Formation 71
　Neogene pediplains 429, 431, *432*, **434**, 438
　Neuquén Basin sediments 132, *134*, 135, 137, 139–150
U–Th/He thermochronometry 7, *384*, 386–397
uplift 37, 383
　Barreal-Las Peñas Deformation Zone 283
　and erosion 401–415
　　Miocene 7–8, 74, 75, 441–442
　Frontal Cordillera, U–Th/He thermochronometry 383–397
　Southern Precordillera morphotectonic subunit 277
Uspallata 1927 earthquake 373, **375**, 377, 379
Uspallata Group 27, 28

Vaca Muerta Formation *33*, 133, *134*, 135, *136*, 149
Valle de Uco piedmont 293, 294–296
Valle del Cura *110*
　geology 110, *112*–*113*
　Oligocene-Early Miocene
　　structural setting 117, *120*, *121*, *122*, 127–128
　　volcanism 111, 113, 115
　stratigraphy 109–110, *114*
　volcanism 111, 113, 115, 125–126
　geochemistry 118–119, 121–125, *126*, *127*
Valle del Cura Formation 111, 113, 115, *116*, 117, *118*, 119, 122–128
Valle Extenso del Campo Bajo 294, *295*, *298*, 300
Valle Grande Formation 31
Vallecitos beds 17, 20
Vallenar Orocline 420
vegetation
　effect on erosion rate 407
　as indicator of palaeoclimate 420–421
Veta Negra Formation 33, 34
Vicuña Fault *424*, 425, 431
Villa Atuel (1929), earthquakes 373, **375**, 377, *378*
Villa Atuel-Las Malvinas 1929 earthquake 373, **375**, 377, *378*
Villavicencio Formation 27
Viñita Formation 36
Virgatosphinctes andensensis 160
Vizcachas Group 214
Vizcacheras Basin *278*, 284
Volcán valley 155, *156*, 160, *163*
volcanism
　Doña Ana Group 109, 110
　subduction-related 29, 31–35, 39–40, *41*, 43
　Valle del Cura Formation 111, 113, 115

Weavericeras vacaense 206, *208*, *209*, 211
wedge-top depozone 223, 237
wells, seismic 377
Western Coastal Cordillera 38, 425, *426*, *432*
　pediplains *436*
　uplift 441

Western Cordillera ramp *16*, *66*, 71, 74, 75
Western Frontal Cordillera 423, *426*, *432*
  pediplains 427
Western Sierras Pampeanas
  calc-alkaline magmatic arc 19, 20
  Cretaceous rifting 168
  Cuyana terrane 168
  elastic thickness 196
  Famatinian tectonic cycle 19

Yalguaraz Deformation Zone *283*, 284
Yeguas Muertas depocentre 32
Yeguas Muertas ramp *72–73*, 74, 75
'Yeso Principal' 31, *33*
'Yeso Secundario' 35
Yeso valley 155, *156*, *161*

Zanja Honda basin 228, 230, 237